Principles of

INFORMATION SYSTEMS,

A Managerial Approach

Sixth Edition

Ralph M. Stair
Florida State University

George W. Reynolds
The University of Cincinnati

THOMSON

COURSE TECHNOLOGY

Australia • Canada • Mexico • Singapore • Spain • United Kingdom • United States

THOMSON
™
COURSE TECHNOLOGY

Principles of Information Systems, A Managerial Approach,
by Ralph Stair, Florida State University, and George
Reynolds, The University of Cincinnati

Senior Vice President, Publisher:
Kristen Duerr

Executive Editor:
Jennifer Locke

Product Manager:
Barrie Tysko

**Project Management and
Development:**
Elm Street Publishing Services, Inc.

Associate Product Manager:
Janet Aras

Editorial Assistant:
Christy Urban

Marketing Manager:
Jason Sakos

Text Design:
Ann Small-Wills

Cover Designer:
Betsy Young

Manufacturing Coordinator:
Denise Powers

Composition House:
GEX Publishing Services

Photo Researcher:
Abby Reip

Copyright © 2003 Course Technology, a
division of Thomson Learning, Inc.
Thomson Learning™ is a trademark
used herein under license.

Printed in Canada

1 2 3 4 5 6 7 8 9 TC 06 05 04 03 02

Library of Congress Cataloging-in-
Publication Data

Stair, Ralph M.

Principles of information systems: a
managerial approach / Ralph M. Stair,
George W. Reynolds—6th ed.
p. cm.

Includes bibliographical references and
index.
ISBN 0-619-06489-7
1. Management information systems. I.
   Reynolds, George Walter, 1944-II.
   Title.

T58.6.S72 2003
658.4'038—dc21

For more information, contact Course
Technology, 25 Thomson Place, Boston,
Massachusetts, 02210.
Or find us on the World Wide Web at:
www.course.com

For Lila and Leslie
—RMS

To Ginnie, Tammy, Kim, Kelly, and Kristy
—GWR

Education in information systems is critical for workers in every discipline. Today, information systems are used for business processes from communications to order processing to customer support and in business functions ranging from marketing to human resources to accounting and finance. Chances are, regardless of your future occupation, you need to understand what information systems can and cannot do and be able to use them to help you accomplish your work. You may even be asked to suggest new uses for information systems and participate in the design of solutions to business problems employing information systems. You will be challenged to identify and evaluate information systems options. To be successful, you must be able to view information systems from the perspective of business and organizational needs. For your solutions to be accepted, you must identify and address their impact on fellow workers, customers, suppliers, and other key business partners. For these reasons, a course in information systems is essential for students in today's high-tech world.

Principles of Information Systems: A Managerial Approach, Sixth Edition, continues the tradition and approach of the previous editions. Our primary objective is to provide the best information systems text and accompanying materials for the first information technology course required of all business students. Through surveys, questionnaires, focus groups, and feedback that we have received from current and past adopters, as well as others who teach in the field, we have been able to develop the highest-quality set of teaching materials available.

Principles of Information Systems: A Managerial Approach, Sixth Edition, stands proudly at the beginning of the IS curriculum and remains unchallenged in its position as the only IS principles text offering the basic IS concepts that every business student must learn to be successful. In the past, instructors of the introductory course faced a dilemma. On one hand, experience in business organizations allows students to grasp the complexities underlying important IS concepts. For this reason, many schools delayed presenting these concepts until students completed a large portion of the core business requirements. On the other hand, delaying the presentation of IS concepts until students have matured within the business curriculum often forces the one or two required introductory IS courses to focus only on personal computing software tools and, at best, merely to introduce computer concepts.

This text has been written specifically for the principles course in the IS curriculum. *Principles of Information Systems: A Managerial Approach,* Sixth Edition, treats the appropriate computer and IS concepts together with a strong managerial emphasis.

 APPROACH OF THE TEXT

Principles of Information Systems: A Managerial Approach, Sixth Edition, offers the traditional coverage of computer concepts, but it places the material within the context of business and information systems. Placing IS concepts in a business context has always set the text apart from general computer books and makes it appealing not only to MIS majors but also to students from other courses of study. It approaches MIS from a general management perspective. The text isn't overly technical but rather deals with the role that information systems play in an organization and the general concepts a manager needs to be aware of to be successful. The text stresses principles of IS, which are brought together and presented in a way that is both understandable and relevant. In addition, this book offers an overview of the entire IS discipline, as well as solid preparation for further study in advanced IS courses. It serves both general business students and those who will become IS professionals. In particular, this book provides a solid groundwork from which to build advanced courses in such areas as programming, project management, database management, data communications, Web site and systems development, electronic commerce applications, and decision support.

The overall vision, framework, and pedagogy that made the previous editions so popular have been retained in the sixth edition, offering a number of benefits to students. We continue to present IS concepts with a managerial emphasis. While much of the fundamental vision of this market-leading text remains unchanged, the sixth edition more clearly highlights established principles and draws out new ones that have emerged as a result of corporate and technological change.

IS PRINCIPLES FIRST, WHERE THEY BELONG

Exposing students to fundamental IS principles provides a service to those who do not later return to the discipline for advanced courses. Since most functional areas in business rely on information systems, an understanding of IS principles helps students in other course work. In addition, introducing students to the principles of information systems helps future functional area managers avoid mishaps that often result in unfortunate consequences. Furthermore, presenting IS principles at the introductory level creates interest among general business students who will later choose information systems as a field of concentration.

AUTHOR TEAM

Ralph Stair and George Reynolds have teamed up again for the sixth edition. Together, they have more than fifty years of academic and industrial experience. Ralph Stair brings years of writing, teaching, and academic experience to this text. He has written more than twenty books and a large number of articles while at Florida State University. George Reynolds brings a wealth of computer and industrial experience to the project, with more than thirty years of experience working in government, institutional, and commercial IS organizations. He has also authored fourteen texts and is an adjunct professor at the University of Cincinnati, teaching the introductory IS course. The Stair and Reynolds team brings a solid conceptual foundation and practical IS experience to students.

⊙ GOALS OF THIS TEXT

Because *Principles of Information Systems: A Managerial Approach,* Sixth Edition, is written for all business majors, we believe it is important not only to present a realistic perspective on IS in business but also to provide students with the skills they can use to be effective business leaders in their organization. To that end, *Principles of Information Systems: A Managerial Approach,* Sixth Edition, has four main goals:

1. To provide a core of IS principles with which every business student should be familiar
2. To offer a survey of the IS discipline that will enable all business students to understand the relationship of IS courses to their curriculum as a whole

3. To present the changing role of the IS professional
4. To show the value of the discipline as an attractive field of specialization

Achieving these goals will enable students, regardless of their major, to understand and use fundamental information systems principles so that they can function efficiently and effectively as business employees and managers. *Principles of Information Systems, A Managerial Approach*, Sixth Edition, is written for all business majors and presents a realistic perspective of IS in business that can provide students with the knowledge and understanding that they can use to be effective leaders in their companies.

IS PRINCIPLES

Principles of Information Systems: A Managerial Approach, Sixth Edition, although comprehensive, cannot cover every aspect of the rapidly changing IS discipline. The authors, recognizing this, provide students with an essential core of guiding IS principles to use as they face the career challenges ahead. Think of principles as basic truths, rules, or assumptions that remain constant regardless of the situation. As such, they provide strong guidance in the face of tough decisions. A set of IS principles is highlighted in the chapter opener of each chapter. The application of these principles to solve real-world problems is driven home from the opening vignettes to the end-of-chapter material. The ultimate goal of *Principles of Information Systems* is to develop effective, thinking employees by instilling them with principles to help guide their decision making and actions.

SURVEY OF THE IS DISCIPLINE

This text not only offers the traditional coverage of computer concepts but also stresses the broad framework to provide students with solid grounding in business uses of technology. In addition to serving general business students, this book offers an overview of the entire IS discipline and solidly prepares future IS professionals for advanced IS courses and their careers in the rapidly changing IS discipline.

CHANGING ROLE OF THE IS PROFESSIONAL

As business and the IS discipline have changed, so too has the role of the IS professional. Once considered a dedicated specialist, the IS professional now operates as an internal consultant to all functional areas, being knowledgeable about various needs and competent in bringing the power of information systems to bear throughout the business. The IS professional views issues through a global perspective that encompasses the entire organization and the broader industry and business environment in which it operates.

The scope of responsibilities of an IS professional today ranges not only throughout the organization but also throughout the entire interconnected network of suppliers, customers, competitors, and other entities, no matter where they are located. This broad scope offers IS professionals a new challenge: how to help an organization survive in a highly interconnected, highly competitive global environment. In accepting that challenge, the IS professional plays a pivotal role in shaping the business itself and ensuring its success. To survive, businesses must now strive for ultimate customer satisfaction and loyalty through competitive prices and ever-improving product and service quality. The IS professional assumes the critical responsibility of determining the organization's approach to both overall cost and quality performance and therefore plays an important role in the ongoing survival of the organization. This new duality in the role of the IS employee—a professional who exercises a specialist's skills with a generalist's perspective—is reflected throughout the book.

IS AS A FIELD FOR FURTHER STUDY

A career in IS can be exciting, challenging, and rewarding! It is important to show the value of the discipline as an appealing field of study and that the IS graduate is no longer a technical recluse. Today, perhaps more than ever before, the IS professional must be able to align IS and corporate goals and to ensure that IS investments are justified from a business perspective. The need to draw bright and interested students into the IS discipline is part of our ongoing responsibility. Upon graduation, IS graduates at many schools are among the highest paid of all business graduates. Throughout this text, the many challenges and opportunities available to IS professionals are highlighted and emphasized.

CHANGES IN THE SIXTH EDITION

We have implemented a number of exciting changes to the text based on user feedback on ways the text can be aligned even more closely with how the IS principles and concepts course is now being taught. A summary of these changes follows:

- *International Emphasis.* In this edition, we stress the global aspects of information systems as a major theme. As organizations increasingly find themselves competing in a global marketplace, they must recognize the resulting implications for their information systems. Globalization is profoundly changing businesses, markets, and society. With its years of service to the information systems discipline, this text retains the traditions and strengths of past successes while helping future managers and decision makers face tomorrow's global challenges.

- *New World Views Cases.* While the text has always stressed the global factors affecting information systems, these factors are emphasized even more in this edition through the introduction of a new feature, *World Views Cases.* These cases, written by instructors outside the United States, provide the reader with real insight into the IS issues facing foreign-based or multinational companies.

- *All New Vignettes Emphasize International Aspects.* In addition to the World Views Cases, all of the chapter-opening vignettes raise actual issues from foreign-based or multinational companies.

- *All New "IS Principles in Action" Boxes.* Closely tied to each chapter's principles, these supplemental Special Interest Boxes show how organizations have followed information systems principles to improve decision making and achieve organizational goals.

- *New Self-Assessment Tests.* End-of-chapter self-assessment tests help students review and test their understanding of key chapter concepts.

- *All New Cases.* Three new end-of-chapter cases provide a wealth of practical information for students and instructors. Each case explores a chapter concept or problem that a real-world company or organization has faced. The cases can be assigned as individual homework exercises or serve as a basis for class discussion.

- *Thoroughly Revised End-of-Chapter Material.* The material at the end of each chapter has been thoroughly updated. Summaries linked to the principles, key terms, review questions, discussion questions, problem-solving exercises, team activities, and Web exercises have been replaced and revised to reflect the theme of the sixth edition and to give students the opportunity to explore the latest technology in a business setting.

- *Database Normalization Supplement.* A brief supplement on data normalization has been developed for those instructors who want to discuss this topic in the database chapter. The supplement is available for download, for instructors and students, at www.course.com, via the "Students Download" link, on the web page for this book.

WHAT WE HAVE RETAINED FROM THE FIFTH EDITION

The sixth edition builds on what has worked well in the past; it retains the focus on IS principles and strives to be the most current text on the market.

- *Overarching Principle.* This book continues to stress a single all-encompassing theme: The right information, if it is delivered to the right person, in the right fashion, and at the right time, can improve and ensure organizational effectiveness and efficiency.
- *Information System Principles.* Information System Principles summarize key concepts that every student should know. Presented at the start of each chapter, this important feature is showcased in a convenient summary of key ideas.
- *Learning Objectives Linked to Principles.* Carefully crafted learning objectives are included with every chapter. The learning objectives are linked to the Information System Principles and reflect what a student should be able to accomplish after completing a chapter.
- *Summary Linked to Principles.* Each chapter includes a detailed summary, and each section of the summary is tied to an Information System Principle.
- *"Ethical and Societal Issues" Special Interest Boxes.* Each chapter includes an "Ethical and Societal Issues" box, which presents a timely look at ethical challenges and the societal impact of information systems. The boxes are related to the issues discussed in the chapters.
- *Current Examples, Boxes, Cases, and References.* As in each edition, we take great pride in presenting the most recent examples, boxes, cases, and references throughout the text. Some of these examples were developed at the last possible moment, literally weeks before the book went into publication. Information on new hardware and software, the latest operating systems, application service providers, the Internet, electronic commerce, ethical and societal issues, terrorism's effects on business and information systems, and many other current developments can be found throughout the text. Our adopters have come to expect the best and most recent material. We have done everything we can to meet or exceed these expectations.

INSTRUCTOR RESOURCES

The teaching tools that accompany this text offer many options for enhancing a course. In the sixth edition, we emphasize the importance of distance learning. And, as always, we are committed to providing one of the best teaching resource packages available in this market. Here are the options.

ELECTRONIC INSTRUCTOR'S MANUAL WITH SOLUTIONS

This all-new updated *Instructor's Manual* provides valuable chapter overviews; highlights key principles and critical concepts; offers sample syllabi, learning objectives, and discussion topics; and features possible essay topics, further readings or cases, and solutions to all of the end-of-chapter questions and problems, as well as suggestions for conducting the team activities. Additional end-of-chapter questions are also included.

EXAMVIEW®

ExamView® is a powerful objective-based test generator that enables instructors to create paper-, LAN-, or Web-based tests from test banks designed specifically for their Course Technology text. Instructors can utilize the ultra-efficient QuickTest Wizard to create tests in less than five minutes by taking advantage of Course Technology's question banks or customizing their own exams from scratch.

POWERPOINT PRESENTATIONS

This book comes with impressive Microsoft PowerPoint slides for each chapter. These slides are included to serve as a teaching aid for classroom presentation, to make available to students on the network for chapter review, or to be printed for classroom distribution. Instructors can add their own slides for additional topics they introduce to the class.

FIGURE FILES

Figure Files allow instructors to create their own presentations using figures taken directly from the text.

CLASSIC CASES

A frequent request from adopters is that they wish to have a broader selection of cases from which to choose. To meet this need, a set of seventy cases from the fourth and fifth editions of the text are included in the Instructor Resources. These classics are the authors' choices of the "best cases" from these editions and span a broad range of companies and industries.

DISTANCE LEARNING

Course Technology, the premiere innovator in management information systems publishing, is proud to present online courses in WebCT and Blackboard, as well as at MyCourse 2.0 to provide the most complete and dynamic learning experience possible.

- *MyCourse 2.0.* MyCourse 2.0 is a flexible, easy-to-use management tool that gives instructors true customization over the online components of their course. It allows them to personalize their course home page, schedule course activities and assignments, post messages, administer tests, and much more. MyCourse 2.0 is hosted by Thomson Learning, allowing for hassle-free maintenance and student access at all times.
- *Blackboard and WebCT Level 1 Online Content.* If you use Blackboard or WebCT, the test bank for this textbook is available at no cost in a simple, ready-to-use format. Go to www.course.com and search for this textbook to download the test bank.
- *Blackboard and WebCT Level 2 Online Content.* Blackboard Level 2 and WebCT Level 2 are also available for *Principles of Information Systems,* Sixth Edition. Level 2 offers course management and access to a Web site that is fully populated with content for this book. Students purchase the *Blackboard User Guide* (ISBN 0-7895-6165-4) or the *WebCT User Guide* (0-7895-6163-8). The *User Guides* include a password that allows student access to Level 2.

For more information on how to bring distance learning to your course, instructors should contact their Course Technology sales representative.

ONLINE DATABASE NORMALIZATION SUPPLEMENT

By reviewer request, a database normalization supplement is available for download, for instructors and students, from www.course.com on the Web page for this book. This supplement takes the reader through the three initial steps of data normalization and makes this process easy to understand through a useful set of figures demonstrating the process for a hypothetical database.

CNN FOR MIS

A video package, developed with CNN, includes 12 video clips on a range of MIS topics from all over the world. The videos are free to instructors or may be bundled with the text for a small additional cost.

ACKNOWLEDGMENTS

A book of this size and undertaking requires a strong team effort. We would like to thank all of our fellow teammates at Course Technology and Elm Street Publishing Services for their dedication and hard work. Special thanks to Barrie Tysko, our Product Manager. Our appreciation goes out to all the many people who worked behind the scenes to bring this effort to fruition, including Janet Aras, our Associate Product Manager, and Christine Spillett, Associate Production Manager. We would like to acknowledge and thank the folks at Elm Street Publishing Services for their hard work on the manuscript. Karen Hill, our development editor, deserves special recognition for her tireless effort and help in all stages of this project. Heather Johnson, our project editor, shepherded the book through the production process. Melissa Morgan, Angel Chavez, Leah Strauss, and Jan Huskisson helped with the illustrations, production, text permissions, and the final stages of the book.

We are grateful to the sales force at Course Technology and Thomson Learning in the U.S. and around the globe, whose efforts make this all possible. You helped to get valuable feedback from current and future adopters. As Course Technology product users, we know how important you are.

While we had input from many reviewers, we would like especially to recognize Professor Gordon Everest of the University of Minnesota for his many valuable suggestions for the sixth edition and our new database normalization supplement. We would also like to thank Ken Baldauf for his excellent help in writing many of the boxes and cases for this edition. Ken also provided invaluable feedback for many topics discussed in the book.

Ralph Stair would like to thank the Department of Information and Management Sciences, College of Business Administration, at Florida State University for their support and encouragement. He would also like to thank his family, Lila and Leslie, for their support.

George Reynolds thanks his family, Ginnie, Tammy, Kim, Kelly, and Kristy, for their patience and support in this major project. He would also like to thank Kristen Duerr and Ralph Stair for asking him to join the writing team back in 1997.

TO OUR PREVIOUS ADOPTERS AND POTENTIAL NEW USERS

We sincerely appreciate our loyal adopters of the previous editions and welcome new users of *Principles of Information Systems: A Managerial Approach*, Sixth Edition. As in the past, we truly value your needs and feedback. We can only hope the sixth edition continues to meet your high expectations.

We would especially like to thank reviewers of the sixth edition, focus group members, and reviewers of previous editions.

In addition, we would like to thank the faculty at the University of Wollongong, the University of Melbourne, and La Trobe University in Australia. We are grateful for the time and feedback that a number of faculty at these schools gave Ralph Stair on a recent trip to Australia. We appreciate your support and encouragement. The new World Views Cases were developed because of your insights into the international nature of information systems and need for IS textbooks to take a global perspective. We would also like to thank Melissa Traverso, E. P. Wee, and Jonathan Fredman at Thomson Learning for setting up the meetings.

REVIEWERS FOR THE SIXTH EDITION

We are indebted to the following individuals for their perceptive feedback on early drafts of this text:

Jill Adams, *Navarro College*

Cynthia C. Barnes, *Lamar University*

John Melrose, *University of Wisconsin—Eau Claire*
Bertrad P. Mouqin, *University of Mary Hardin—Baylor*
Pamela Neely, *Marist College*
Mahesh S. Raisinghani, *University of Dallas*
Anne Marie Smith, *LaSalle University*
Patricia A. Smith, *Temple College*
Herb Snyder, *Fort Lewis College*

REVIEWERS FOR THE FIRST, SECOND, THIRD, FOURTH, AND FIFTH EDITIONS

The following people shaped the book you hold in your hands by contributing to previous editions:

Robert Aden, *Middle Tennessee State University*
A. K. Aggarwal, *University of Baltimore*
Sarah Alexander, *Western Illinois University*
Beverly Amer, *University of Florida*
Noushin Asharfi, *University of Massachusetts*
Yair Babad, *University of Illinois—Chicago*
Charles Bilbrey, *James Madison University*
Thomas Blaskovics, *West Virginia University*
John Bloom, *Miami University of Ohio*
Warren Boe, *University of Iowa*
Glen Boyer, *Brigham Young University*
Mary Brabston, *University of Tennessee*
Jerry Braun, *Xavier University*
Thomas A. Browdy, *Washington University*
Lisa Campbell, *Gulf Coast Community College*
Andy Chen, *Northeastern Illinois University*
David Cheslow, *University of Michigan—Flint*
Robert Chi, *California State University—Long Beach*
Carol Chrisman, *Illinois State University*
Miro Costa, *California State University—Chico*
Caroline Curtis, *Lorain County Community College*
Roy Dejoie, *USWeb Corporation*
Sasa Dekleva, *DePaul University*
Pi-Sheng Deng, *California State University—Stanislaus*
Roger Deveau, *University of Massachusetts—Dartmouth*
John Eatman, *University of North Carolina*
Gordon Everest, *University of Minnesota*
Juan Esteva, *Eastern Michigan University*
Badie Farah, *Eastern Michigan University*
Karen Forcht, *James Madison University*
Carroll Frenzel, *University of Colorado—Boulder*
John Gessford, *California State University—Long Beach*
Terry Beth Gordon, *University of Toledo*
Kevin Gorman, *University of North Carolina—Charlotte*
Costanza Hagmann, *Kansas State University*
Bill C. Hardgrave, *University of Arkansas*
Al Harris, *Appalachian State University*
William L. Harrison, *Oregon State University*
Dwight Haworth, *University of Nebraska—Omaha*
Jeff Hedrington, *University of Wisconsin—Eau Claire*
Donna Hilgenbrink, *Illinois State University*
Jack Hogue, *University of North Carolina*
Joan Hoopes, *Marist College*

Donald Huffman, *Lorain County Community College*
Patrick Jaska, *University of Texas at Arlington*
G. Vaughn Johnson, *University of Nebraska—Omaha*
Grover S. Kearns, *Morehead State University*
Robert Keim, *Arizona State University*
Karen Ketler, *Eastern Illinois University*
Mo Khan, *California State University—Long Beach*
Michael Lahey, *Kent State University*
Jan de Lassen, *Brigham Young University*
Robert E. Lee, *New Mexico State University—Carlstadt*
Joyce Little, *Towson State University*
Herbert Ludwig, *North Dakota State University*
Jane Mackay, *Texas Christian University*
Al Maimon, *University of Washington*
James R. Marsden, *University of Connecticut*
Roger W. McHaney, *Kansas State University*
Lynn J. McKell, *Brigham Young University*
John Melrose, *University of Wisconsin—Eau Claire*
Michael Michaelson, *Palomar College*
Ellen Monk, *University of Delaware*
Bijayananda Naik, *University of South Dakota*
Leah R. Pietron, *University of Nebraska—Omaha*
John Powell, *University of South Dakota*
Maryann Pringle, *University of Houston*
John Quigley, *East Tennessee State University*
Mary Rasley, *Lehigh-Carbon Community College*
Earl Robinson, *St. Joseph's University*
Scott Rupple, *Marquette University*
Dave Scanlon, *California State University-Sacramento*
Werner Schenk, *University of Rochester*
Larry Scheuermann, *University of Southwest Louisiana*
James Scott, *Central Michigan University*
Vikram Sethi, *Southwest Missouri State University*
Laurette Simmons, *Loyola College*
Janice Sipior, *Villanova University*
Harold Smith, *Brigham Young University*
Herb Snyder, *Fort Lewis College*
Alan Spira, *University of Arizona*
Tony Stylianou, *University of North Carolina*
Bruce Sun, *California State University—Long Beach*
Hung-Lian Tang, *Bowling Green State University*
William Tastle, *Ithaca College*
Gerald Tillman, *Appalachian State University*
Duane Truex, *Georgia State University*
Jean Upson, *Lorain County Community College*
Misty Vermaat, *Purdue University—Calumet*
David Wallace, *Illinois State University*
Michael E. Whitman, *University of Nevada—Las Vegas*
David C. Whitney, *San Francisco State University*
Goodwin Wong, *University of California-Berkeley*
Amy Woszczynski, *Kennesaw State University*
Judy Wynekoop, *Florida Gulf Coast University*
Myung Yoon, *Northeastern Illinois University*

FOCUS GROUP CONTRIBUTORS FOR THE THIRD EDITION
Mary Brabston, *University of Tennessee*
Russell Ching, *California State University—Sacramento*
Virginia Gibson, *University of Maine*
Bill C. Hardgrave, *University of Arkansas*
Al Harris, *Appalachian State University*
Stephen Lunce, *Texas A & M International*
Merle Martin, *California State University—Sacramento*
Mark Serva, *Baylor University*
Paul van Vliet, *University of Nebraska—Omaha*

OUR COMMITMENT

We are committed to listening to our adopters and readers and to developing creative solutions to meet their needs. The field of IS continually evolves, and we strongly encourage your participation in helping us provide the freshest, most relevant information possible.

We welcome your input and feedback. If you have any questions or comments regarding *Principles of Information Systems: A Managerial Approach*, Sixth Edition, please contact us through Course Technology or your local sales representative, via e-mail at mis@course.com, via the Internet at www.course.com.

BRIEF CONTENTS

CONTENTS

4

7

The Internet, Intranets, and Extranets 272

PART 3   Business Information Systems 317

11

Specialized Business Information Systems: Artificial Intelligence, Expert Systems, Virtual Reality, and Other Specialized Systems 460

PART 4 **Systems Development 503**

PART 5   Information Systems in Business and Society 613

AN OVERVIEW

PART 1

An Introduction to Information Systems

PRINCIPLES	LEARNING OBJECTIVES
The value of information is directly linked to how it helps decision makers achieve the organization's goals.	• Distinguish data from information and describe the characteristics used to evaluate the quality of data.
Models, computers, and information systems are constantly making it possible for organizations to improve the way they conduct business.	• Name the components of an information system and describe several system characteristics. • Identify four basic types of models and explain how they are used.
Knowing the potential impact of information systems and having the ability to put this knowledge to work can result in a successful personal career, organizations that reach their goals, and a society with a higher quality of life.	• Identify the basic types of business information systems and discuss who uses them, how they are used, and what kinds of benefits they deliver.
System users, business managers, and information systems professionals must work together to build a successful information system.	• Identify the major steps of the systems development process and state the goal of each. • Discuss why it is important to study and understand information systems.

[Merck-Medco]

A Pharmacy for the Future

In the not-too-distant future, the corner pharmacy and the pleasant pharmacist who fills your prescriptions each month may become a fond memory from bygone days; they will, that is, if Merck-Medco has anything to say about it. Merck-Medco is one of the country's largest pharmacy-benefits managers (PBMs) and a pioneer in a brand-new method for distributing prescription drugs.

The pharmaceutical industry is big business today and will continue growing over the next several years. "Some of the growth stems from new drugs being introduced in the market, but the larger factor is that the population keeps growing older," says Eric Veiel, an analyst with Deutsche Bank. "The older we get, the more drugs we tend to take." Merck-Medco believes that by restructuring the traditional prescription medicine distribution system and automating the process of filling prescriptions, it can satisfy the increasing demand more efficiently.

Merck-Medco contracts with large employers and unions and collects a fee for processing patients' prescriptions and billing the patients' health-plan providers. Merck-Medco customers include United Airlines and its 80,000 U.S. employees, General Motors' 300,000 employees, and Oxford Health Plans' 1.5 million members, to name a few. Prescriptions are placed by physicians via e-mail, phone, or fax to any of Merck-Medco's 13 pharmacies and may be conveniently refilled by the patient at Merck-Medco's Web site or by phone. Once a prescription request is submitted, it starts a chain of events that involves several employees and processes at multiple locations.

A prescription request begins its journey at one of Merck-Medco's processing pharmacies, such as the Liberty Lake plant in Washington state. Here prescriptions are entered into a proprietary electronic system, which first checks the patient's records to see if it is time to refill the medication and then evaluates the prescription from a clinical standpoint. Are there harmful side effects if the drug is combined with other medications the patient is taking? Is there a generic equivalent or a more effective medication? In about one-third of the cases, a Merck-Medco pharmacist will phone the physician to discuss other options for the patient or to ask for clarification. Bob Blyskal, senior vice president of operations, says this initial screening is all-important. "We use technology to dramatically improve quality of care through accuracy, and it allows us to remove pharmacists from counting pills and devote them to real value-added types of activities—drug utilization, generic substitution." Once the prescription is approved and in the system, it is transmitted to a dispensary such as the one at Willingboro, New Jersey.

"Willingboro is the world's largest pharmacy and the most advanced," company president Richard T. Clark said. In its 280,000 square feet, as big as six football fields, it dispenses about 200,000 mail-order prescriptions a week. This number is expected to quadruple by 2003. The dispensing process starts in the command center, a central monitoring system that manages the entire dispensing process and allows employees to track a bottle through various machines. The first step is to print the medication documentation, up to eight pages of patient-specific directions, drug warnings, and billing information. Then the command center gives instructions to the automated pill-counting mechanism and instructs a printer to generate a label and apply it to an empty bottle. The bottle, held in a rack with 23 others, moves to the dispensing lane on a two-mile conveyor belt. Once the bottle gets to the proper dispensing channel, the command center releases the pills. A bar code reader ensures that the right bottle is under the right dispensing channel. Once filled, the bottle, along with the documentation, is put in a bag, heat-sealed, and bar-code labeled. Scanners do a final accuracy check, and the package goes to the mail-sorting area. The whole trip down the conveyor belt takes about 15 minutes.

Merck-Medco provides its customers with a Web site that gives useful information concerning their prescriptions and other health-care issues. The site, located at www.merck-medco.com, provides easy and secure methods for members to fill prescriptions, review

Information systems are everywhere. An advanced information system can be used to obtain prescriptions from automated kiosks: Customers respond to questions on a touch screen, pay with a credit card, and the dispenser delivers the medication.

(Source: AP/Wide World Photos.)

information system (IS)
a set of interrelated components that collect, manipulate, and disseminate data and information and provide a feedback mechanism to meet an objective

prescription prices and copayments based on their health plans, check prescription records and benefits, receive e-mail reminders for refills, order nonprescription general health items through its relationship with on-line drugstore CVS.com, and research a wide variety of health-care information.

Through seamless interaction among its Web site, pharmacist screening, and automated dispensing systems, Merck-Medco is often able to mail prescription refills on the same day they are requested. Are we wise to trust machines to fill our prescriptions? Merck-Medco claims that its system has never put a wrong pill in a wrong bottle.

As you read this chapter, consider the following:
◉ In what ways has Merck-Medco's new process for filling prescriptions improved the business of the organizations (such as United Airlines) that make use of its services, the lives of employees/members of those organizations, the practice of physicians, and the work of pharmacists and other employees of Merck-Medco? Are there any drawbacks to this new system?
◉ Besides pharmacies, what other traditional businesses might profit from this type of automation?

An **information system (IS)** is a set of interrelated components that collect, manipulate, and disseminate data and information and provide a feedback mechanism to meet an objective. We all interact daily with information systems, both personally and professionally. We use automatic teller machines at banks, checkout clerks scan our purchases using bar codes and scanners, we access information over the Internet, and we get information from kiosks with touchscreens. Major *Fortune* 500 companies are spending in excess of $1 billion per year on information technology. In the future, we will depend on information systems even more. General Motors, for example, has teamed up with Fidelity Investments to allow people to get information on investments and trade stocks using voice commands from a car or truck.[1] Knowing the potential of information systems and having the ability to put this knowledge to work can result in a successful personal career, organizations that reach their goals, and a society with a higher quality of life.

Computers and information systems are constantly changing the way organizations conduct business. We saw in the opening vignette, for example, how Merck-Medco was able to speed the delivery of prescriptions while maintaining a high degree of accuracy. Today we live in an information economy. Information itself has value, and commerce often involves the exchange of information rather than tangible goods. Systems based on computers are increasingly being used to create, store, and transfer information. Investors are using information systems to make multimillion-dollar decisions, financial institutions are employing them to transfer billions of dollars around the world electronically, and manufacturers are using them to order supplies and distribute goods faster than ever before. Computers and information systems will continue to change our society, our businesses, and our lives. In this chapter, we present a framework for understanding computers and information systems and discuss why it is important to study information systems. This understanding will help you unlock the potential of properly applied information systems concepts.

INFORMATION CONCEPTS

Information is a central concept throughout this book. The term is used in the title of the book, in this section, and in almost every chapter. To be an effective manager in any area of business, you need to understand that information is one of an organization's most valuable and important resources. This term, however, is often confused with the term *data*.

DATA VERSUS INFORMATION

Data consists of raw facts, such as an employee's name and number of hours worked in a week, inventory part numbers, or sales orders. As shown in Table 1.1, several types of data can be used to represent these facts. When these facts are organized or arranged in a meaningful manner, they become information. **Information** is a collection of facts organized in such a way that they have additional value beyond the value of the facts themselves. For example, a particular manager might find the knowledge of total monthly sales to be more suited to his or her purpose (i.e., more valuable) than the number of sales for individual sales representatives. Providing information to customers can also help companies increase revenues and profits. Uniglobe.com, Inc. provides fast and accurate information to people considering a cruise.[2] According to Mike Dauberman, senior vice president of business operations, "If you can grab a customer while they're on their peak of interest in a cruise, they're considerably more likely to buy it."

Data represents real-world things. As we have stated, data—simply raw facts—has little value beyond its existence. For example, consider data as pieces of railroad track in a model railroad kit. In this state, each piece of track has little value beyond its inherent value as a single object. However, if some relationship is defined among the pieces of the track, they will gain value. By arranging the pieces of track in a certain way, a railroad layout begins to emerge (Figure 1.1, top). Information is much the same. Rules and relationships can be set up to organize data into useful, valuable information.

The type of information created depends on the relationships defined among existing data. For example, the pieces of track could be rearranged to form different layouts (Figure 1.1, middle). Adding new or different data means relationships can be redefined and new information can be created. For instance, adding new pieces to the track can greatly increase the value—in this case, variety and fun—of the final product. We can now create a more elaborate railroad layout (Figure 1.1, bottom). Likewise, our manager could add specific product data to his sales data to create monthly sales information broken down by product line. This information could be used by the manager to determine which product lines are the most popular and profitable.

Turning data into information is a **process**, or a set of logically related tasks performed to achieve a defined outcome. The process of defining relationships

data
raw facts, such as an employee's name and number of hours worked in a week, inventory part numbers, or sales orders

information
a collection of facts organized in such a way that they have additional value beyond the value of the facts themselves

process
a set of logically related tasks performed to achieve a defined outcome

TABLE 1.1

Types of Data

Data	Represented By
Alphanumeric data	Numbers, letters, and other characters
Image data	Graphic images and pictures
Audio data	Sound, noise, or tones
Video data	Moving images or pictures

FIGURE 1.1

*Defining and Organizing
Relationships among Data Creates
Information*

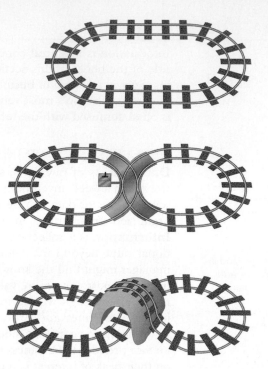

knowledge
an awareness and understanding
of a set of information and ways
that information can be made
useful to support a specific task
or reach a decision

among data to create useful information requires knowledge. **Knowledge** is an awareness and understanding of a set of information and ways that information can be made useful to support a specific task or reach a decision. Part of the knowledge needed for building a railroad layout, for instance, is understanding how large an area is available for the layout, how many trains will run on the track, and how fast they will travel. The act of selecting or rejecting facts based on their relevancy to particular tasks is also based on a type of knowledge used in the process of converting data into information. Therefore, information can be considered data made more useful through the application of knowledge. Trimac, a Canadian bulk hauling company, for example, computerized its data to produce useful information to help it analyze profit potential.[3] According to a Trimac manager, "This technology facilitates the implementation of data ... used for trip analysis, haul analysis, and profitability, either by customer or equipment."

In some cases, data is organized or processed mentally or manually. In other cases, a computer is used. In the earlier example, the manager could have manually calculated the sum of the sales of each representative, or a computer could calculate this sum. What is important is not so much where the data comes from or how it is processed but whether the results are useful and valuable. This transformation process is shown in Figure 1.2.

THE CHARACTERISTICS OF VALUABLE INFORMATION

To be valuable to managers and decision makers, information should have the characteristics described in Table 1.2. These characteristics also make the information more valuable to an organization. United Parcel Service (UPS) is able to determine the exact location of every package in its system.[4] This increased accuracy saves the company both time and money. According to one UPS executive,

FIGURE 1.2

*The Process of Transforming Data
into Information*

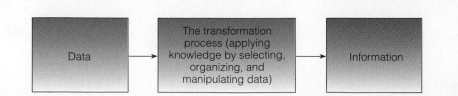

| Data | → | The transformation process (applying knowledge by selecting, organizing, and manipulating data) | → | Information |

"That saves us from having to unload 20 trucks to find one little package, the way we had to do it five years ago." In addition, if an organization's information is not accurate or complete, people can make poor decisions, costing organizations and individuals thousands, or even millions, of dollars. Many believe, for example, that the collapse of energy-trading firm Enron in 2001 was a result of inaccurate accounting and reporting information, which led investors and employees alike to misjudge the actual state of the company's finances and suffer huge personal losses. As another example, if an inaccurate forecast of future demand indicates that sales will be very high when the opposite is true, an organization can invest millions of dollars in a new plant that is not needed. Furthermore, if information is not pertinent to the situation, not delivered to decision makers in a timely fashion, or too complex to understand, it may be of little value to the organization.

Useful information can vary widely in the value of each of these quality attributes. For example, with market-intelligence data, some inaccuracy and incompleteness is acceptable, but timeliness is essential. Market intelligence may alert us that our competitors are about to make a major price cut. The exact details and timing of the price cut may not be as important as being warned far enough in advance to plan how to react. On the other hand, accuracy, verifiability, and completeness are critical for data used in accounting for company assets such as cash, inventory, and equipment.

TABLE 1.2

Characteristics of Valuable Data

Characteristics	Definitions
Accurate	Accurate information is error free. In some cases, inaccurate information is generated because inaccurate data is fed into the transformation process (this is commonly called garbage in, garbage out [GIGO]).
Complete	Complete information contains all the important facts. For example, an investment report that does not include all important costs is not complete.
Economical	Information should also be relatively economical to produce. Decision makers must always balance the value of information with the cost of producing it.
Flexible	Flexible information can be used for a variety of purposes. For example, information on how much inventory is on hand for a particular part can be used by a sales representative in closing a sale, by a production manager to determine whether more inventory is needed, and by a financial executive to determine the total value the company has invested in inventory.
Reliable	Reliable information can be depended on. In many cases, the reliability of the information depends on the reliability of the data collection method. In other instances, reliability depends on the source of the information. A rumor from an unknown source that oil prices might go up may not be reliable.
Relevant	Relevant information is important to the decision maker. Information that lumber prices might drop may not be relevant to a computer chip manufacturer.
Simple	Information should also be simple, not overly complex. Sophisticated and detailed information may not be needed. In fact, too much information can cause information overload, whereby a decision maker has too much information and is unable to determine what is really important.
Timely	Timely information is delivered when it is needed. Knowing last week's weather conditions will not help when trying to decide what coat to wear today.
Verifiable	Information should be verifiable. This means that you can check it to make sure it is correct, perhaps by checking many sources for the same information.
Accessible	Information should be easily accessible by authorized users to be obtained in the right format and at the right time to meet their needs.
Secure	Information should be secure from access by unauthorized users.

THE VALUE OF INFORMATION

The value of information is directly linked to how it helps decision makers achieve their organization's goals. For example, the value of information might be measured in the time required to make a decision or in increased profits to the company. Consider a market forecast that predicts a high demand for a new product. If market forecast information is used to develop the new product and the company is able to make an additional profit of $10,000, the value of this information to the company is $10,000 minus the cost of the information. National Semiconductor Corporation uses information to achieve its goal of speedy delivery of chips to customers.[5] A box of computer chips manufactured in Singapore can be shipped to computer manufacturers in the United States in less than 12 hours. Valuable information can also help managers decide whether to invest in additional information systems and technology. A new computerized ordering system may cost $30,000, but it may generate an additional $50,000 in sales. The *value added* by the new system is the additional revenue from the increased sales of $20,000. Most corporations have cost reduction as a primary goal. BASF, a large chemical company headquartered in Germany, spends nearly $100 million in distributing its inventory to North American customers.[6] Using a computerized inventory information system, BASF was able to reduce annual inventory distribution costs by 6 percent, or $6 million. Using the computerized information system also allowed BASF to identify a one-time cost savings of nearly $10 million.

⦿ SYSTEM AND MODELING CONCEPTS

system
a set of elements or components that interact to accomplish goals

Like information, another central concept of this book is that of a system. A **system** is a set of elements or components that interact to accomplish goals. The elements themselves and the relationships among them determine how the system works. Systems have inputs, processing mechanisms, outputs, and feedback. For example, consider an automatic car wash. Obviously, tangible *inputs* for the process are a dirty car, water, and the various cleaning ingredients used. Time, energy, skill, and knowledge are also needed as inputs to the system. Time and energy are needed to operate the system. Skill is the ability to successfully operate the liquid sprayer, foaming brush, and air dryer devices. Knowledge is used to define the steps in the car wash operation and the order in which those steps are executed (See Figure 1.3).

FIGURE 1 ⦿ 3

Components of a System

A system's four components consist of input, processing, output, and feedback.

The *processing mechanisms* consist of first selecting which of the cleaning options you want (wash only, wash with wax, wash with wax and hand dry, etc.) and communicating that to the operator of the car wash. Note that there is a

Input ⎯⎯⎯⎯⎯⎯⎯⎯→ Processing ⎯⎯⎯⎯⎯⎯⎯⎯→ Output

Feedback

| System | Elements | | | Goal |
	Inputs	Processing mechanisms	Outputs	
Coffee Shop 	Coffee beans, tea bags, water, sugar, cream, spices, pastries, other ingredients, labor, management	Brewing equipment	Coffee, tea, pastries, other beverages and food items	Quickly prepared delicious coffees, teas, and various food items
College 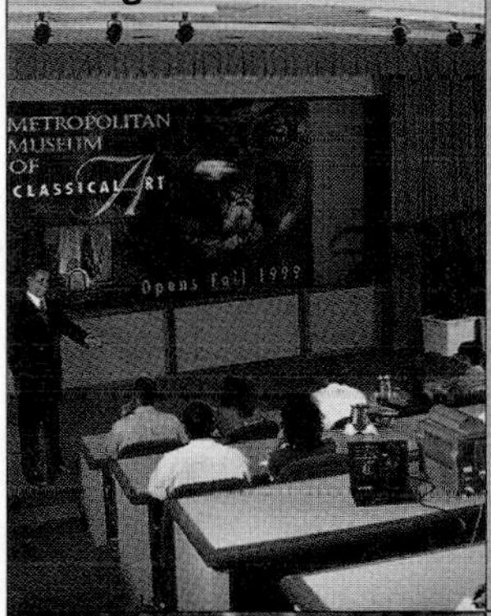	Students, professors, administrators, textbooks, equipment	Teaching, research, service	Educated students; meaningful research; service to community, state, and nation	Acquisition of knowledge
Movie 	Actors, director, staff, sets, equipment	Filming, editing, special effects, film distribution	Finished film delivered to movie theaters	Entertaining movie, film awards, profits

FIGURE 1.4

Examples of Systems and Their Goals and Elements

(Sources: © Steve Smith/Taxi; courtesy of 3M Visual Systems Division; image copyright © 1998 Photodisc.)

feedback mechanism (your assessment of how clean the car is). Liquid sprayers shoot clear water, liquid soap, or car wax depending on where your car is in the process and which options you selected. The *output* is a clean car. It is important to note that independent elements or components of a system (the liquid sprayer, foaming brush, and air dryer) interact to create a clean car. Figure 1.4 shows a few systems with their elements and goals.

SYSTEM COMPONENTS AND CONCEPTS

Figure 1.3 shows a typical system diagram—a simple automatic car wash. The primary purpose of the car wash is to clean your automobile. The **system boundary** defines the system and distinguishes it from everything else (the environment).

system boundary
the limits of the system; it defines the system and distinguishes it from everything else (the environment)

The way system elements are organized or arranged is called the *configuration*. Much like data, the relationships among elements in a system are defined through knowledge. In most cases, knowing the purpose or desired outcome of a system is the first step in defining the way system elements are configured. For example, the desired outcome of our system is a clean car. Based on past experience, we know that it would be illogical to have the liquid sprayer element precede the foaming brush element. The car would be rinsed and then soap would be applied, leaving your car a mess. As you can see from this example, knowledge is needed both to define relationships among the inputs to a system (your dirty car and instructions to the operator) and to organize the system elements used to process the inputs (the foaming brush must precede the liquid sprayer).

System Types

Systems can be classified along numerous dimensions. They can be simple or complex, open or closed, stable or dynamic, adaptive or nonadaptive, permanent or temporary. Table 1.3 defines these characteristics.

Classifying Organizations by System Type

Most companies can be described using the classification scheme in Table 1.3. For example, a janitorial company that cleans offices after business hours most likely represents a simple, stable system because there is a constant and fairly steady need for its services. A successful computer manufacturing company, however, is typically complex and dynamic because it operates in a changing environment. If a company is nonadaptive, it may not survive very long. Many of the early computer companies, including Osborne Computer, which manufactured one of the first portable computers, and VisiCorp, which developed the first spreadsheet program, did not adapt rapidly enough to the changing market for computers and software. As a result, these companies did not survive. On the other hand, IBM was able to reinvent itself from a manufacturer of large, mainframe computers to a manufacturer of all classes of computers and a software and services provider.

SYSTEM PERFORMANCE AND STANDARDS

System performance can be measured in various ways. **Efficiency** is a measure of what is produced divided by what is consumed. It can range from 0 to 100 percent. For example, the efficiency of a motor is the energy produced (in terms of work done) divided by the energy consumed (in terms of electricity or fuel). Some motors have an efficiency of 50 percent or less because of the energy lost to friction and heat generation.

Efficiency is a relative term used to compare systems. For example, a gasoline engine is more efficient than a steam engine because, for the equivalent amount of energy input (gas or coal), the gasoline engine produces more energy output.

efficiency
a measure of what is produced divided by what is consumed

TABLE 1.3

Systems Classifications and Their Primary Characteristics

Simple ⟷	Complex
Has few components, and the relationship or interaction between elements is uncomplicated and straightforward	Has many elements that are highly related and interconnected
Open ⟷	**Closed**
Interacts with its environment	Has no interaction with the environment
Stable ⟷	**Dynamic**
Undergoes very little change over time	Undergoes rapid and constant change over time
Adaptive ⟷	**Nonadaptive**
Is able to change in response to changes in the environment	Is not able to change in response to changes in the environment
Permanent ⟷	**Temporary**
Exists for a relatively long period of time	Exists for only a relatively short period of time

The energy efficiency ratio (energy input divided by energy output) is high for gasoline engines when compared with that of steam engines.

Effectiveness is a measure of the extent to which a system achieves its goals.[7] It can be computed by dividing the goals actually achieved by the total of the stated goals. For example, a company may have a goal to reduce damaged parts by 100 units. A new control system may be installed to help achieve this goal. Actual reduction in damaged parts, however, is only 85 units. The effectiveness of the control system is 85 percent (85/100 = 85%). Effectiveness, like efficiency, is a relative term used to compare systems.

Evaluating system performance also calls for the use of performance standards. A **system performance standard** is a specific objective of the system. For example, a system performance standard for a particular marketing campaign might be to have each sales representative sell $100,000 of a certain type of product each year (Figure 1.5a). A system performance standard for a certain manufacturing process might be to have no more than 1 percent defective parts (Figure 1.5b). Once standards are established, system performance is measured and compared with the standard. Variances from the standard are determinants of system performance.

SYSTEM VARIABLES AND PARAMETERS

Parts of a system are under direct management control, while others are not. A **system variable** is a quantity or item that can be controlled by the decision maker. The price a company charges for its product is a system variable because it can be controlled. A **system parameter** is a value or quantity that cannot be controlled, such as the cost of a raw material. The number of pounds of a chemical that must be added to produce a certain type of plastic is another example of a quantity or value that is not controlled by management; it is controlled by the laws of chemistry.

MODELING A SYSTEM

The real world is complex and dynamic. So when we want to test different relationships and their effects, we use models of systems, which are simplified, instead of real systems. A **model** is an abstraction or an approximation that is used to represent reality. Models enable us to explore and gain an improved understanding of real-world situations.

Since the beginning of recorded history, people have used models. A written description of a battle, a physical mock-up of an ancient building, and the use of symbols to represent money, numbers, and mathematical relationships are all examples of models. Today, managers and decision makers use models to help them understand what is happening in their organizations and make better decisions.

There are various types of models. The major ones are narrative, physical, schematic, and mathematical, as shown in Figure 1.6. A *narrative model,* as the name implies, is based on words; thus, it is a logical and not a physical model. Both verbal and written descriptions of reality are considered narrative models. In an organization, reports, documents, and conversations concerning a system are all important narratives. A *physical model* is a tangible representation of reality. Many physical models are computer designed or constructed. An engineer may develop a physical model of a chemical reactor to gain important information about how a large-scale reactor might perform, or a builder may develop a scale model of a new shopping center to give a potential investor information about the overall appearance and approach of the development. A *schematic model* is a graphic representation of reality. Graphs, charts, figures, diagrams, illustrations, and pictures are all types of schematic models. Schematic models are used extensively in developing computer programs and systems. A blueprint for a new

effectiveness
a measure of the extent to which a system achieves its goals; it can be computed by dividing the goals actually achieved by the total of the stated goals

system performance standard
a specific objective of the system

system variable
a quantity or item that can be controlled by the decision maker

system parameter
a value or quantity that cannot be controlled, such as the cost of a raw material

model
an abstraction or an approximation that is used to represent reality

FIGURE 1•5

System Performance Standards

building, a graph that shows budget and financial projections, electrical wiring diagrams, and graphs that show when certain tasks or activities must be completed to stay on schedule are examples of schematic models used in business. A *mathematical model* is an arithmetic representation of reality. Computers excel at solving mathematical models. Retail chains, for example, have developed mathematical models to identify all the activities, effort, and time associated with planning, building, and opening a new store so that they can forecast how long it will take to complete a store.

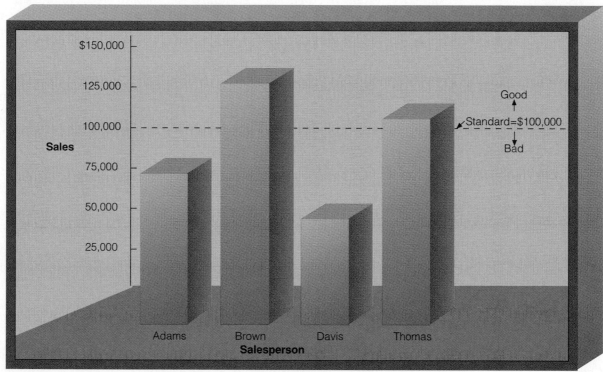

(a)

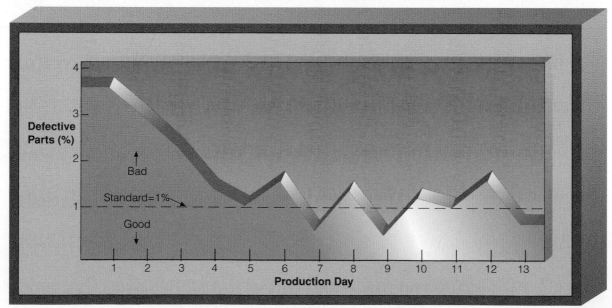

(b)

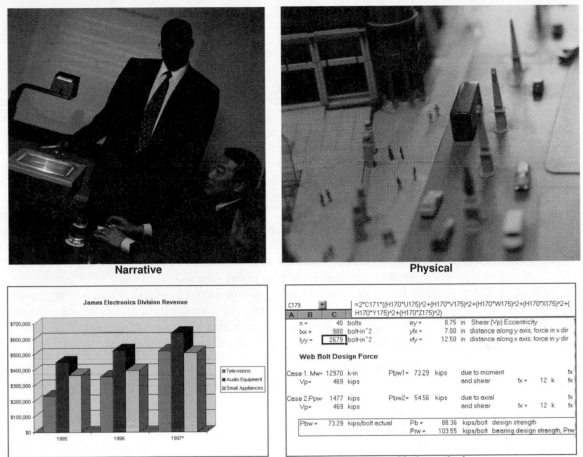

FIGURE 1.6

Four Types of Models

Narrative (words, spoken or written), physical (tangible), schematic (graphic), and mathematical (arithmetic) models.

(Source: Images copyright © 1998 PhotoDisc.)

In developing any model, accuracy is critical. An inaccurate model will usually lead to an inaccurate solution to a problem. Most models contain many assumptions, and it is important that they be as realistic as possible. Potential users of the model must be aware of the assumptions under which the model was developed.

WHAT IS AN INFORMATION SYSTEM?

An information system is a specialized type of system and can be defined in a number of different ways. As mentioned previously, an information system (IS) is a set of interrelated elements or components that collect (input), manipulate (process), and disseminate (output) data and information and provide a feedback mechanism to meet an objective (see Figure 1.7).

FIGURE 1.7

The Components of an Information System

Feedback is critical to the successful operation of a system.

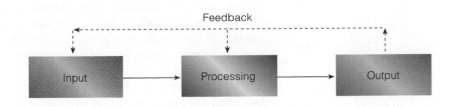

INPUT, PROCESSING, OUTPUT, FEEDBACK

Input

input
the activity of gathering and
capturing raw data

In information systems, **input** is the activity of gathering and capturing raw data. In producing paychecks, for example, the number of hours every employee works must be collected before paychecks can be calculated or printed. In a university grading system, individual instructors must submit student grades before a summary of grades for the semester or quarter can be compiled and sent to the students.

Input can take many forms. In an information system designed to produce paychecks, for example, employee time cards might be the initial input. In a 911 emergency telephone system, an incoming call would be considered an input. Input to a marketing system might include customer survey responses. Car manufacturers are experimenting with a fingerprint identification input device in their car security systems.[8] You may soon be able to gain entry to a car and start it with the touch of a finger. This unique input device will also adjust mirrors, the steering-wheel position, the temperature, and the radio for an individual's size and preferences. Regardless of the system involved, the type of input is determined by the desired output of the system.

Input can be a manual or automated process. A scanner at a grocery store that reads bar codes and enters the grocery item and price into a computerized cash register is a type of automated input process. Regardless of the input method, accurate input is critical to achieve the desired output.

Processing

processing
converting or transforming data
into useful outputs

In information systems, **processing** involves converting or transforming data into useful outputs. Processing can involve making calculations, making comparisons and taking alternative actions, and storing data for future use. Processing data into useful information is critical in business settings. Airline manufacturer Boeing, for example, streamlined its processing operations for needed parts.[9] According to Candace Ismael, Director of Supplier Management and Procurement, "The vision was to create a single process and supporting system for purchasing indirect parts."

Processing can also be done manually or with computer assistance. In the payroll application, each employee's number of hours worked must be converted into net, or take-home, pay. The required processing can first involve multiplying the number of hours worked by the employee's hourly pay rate to get gross pay. If weekly hours worked exceed 40 hours, overtime pay may also be included. Then deductions—for example, federal and state taxes, contributions to health and life insurance or savings plans—are subtracted from gross pay to get net pay.

Output

output
production of useful information,
usually in the form of documents
and reports

In information systems, **output** involves producing useful information, usually in the form of documents and reports. Outputs can include paychecks for employees, reports for managers, and information supplied to stockholders, banks, government agencies, and other groups. In some cases, output from one system can become input for another. For example, output from a system that processes sales orders can be used as input to a customer billing system. Often, output from one system can be used as input to control other systems or devices. For instance, the design and manufacture of office furniture is complicated with many variables. The salesperson, customer, and furniture designer can go through several design iterations to meet the customer's needs. Special computer programs and equipment create the original design and allow the designer to rapidly revise it. Once the last design mock-up is approved, the computer creates a bill of materials that goes to manufacturing to produce the order.

Output can be produced in a variety of ways. For a computer, printers and display screens are common output devices. Output can also be a manual process involving handwritten reports and documents.

feedback
output that is used to make changes to input or processing activities

Feedback

In information systems, **feedback** is output that is used to make changes to input or processing activities. For example, errors or problems might make it necessary to correct input data or change a process. Consider a payroll example. Perhaps the number of hours an employee worked was entered into a computer as 400 instead of 40 hours. Fortunately, most information systems check to make sure that data falls within certain ranges. For number of hours worked, the range might be from 0 to 100 hours because it is unlikely that an employee would work more than 100 hours for any given week. So, the information system would determine that 400 hours is out of range and provide feedback, such as an error report. The feedback is used to check and correct the input on the number of hours worked to 40. If undetected, this error would result in a very high net pay on the printed paycheck!

Feedback is also important for managers and decision makers. For example, bedding maker Sealy Corp. used a computerized feedback system to link its suppliers and plants.[10] According to Jim Packer, director of procurement, "We're building 90% of our products to order, and this communications link between us, our suppliers, and the transportation network and plants closes the loop." For Sealy, output from an information system might indicate that inventory levels for a few items are getting low—a potential problem. A manager could use this feedback to decide to order more inventory from a supplier. The new inventory orders then become input to the system. In addition to this reactive approach, a computer system can also be proactive—predicting future events to avoid problems. This concept, often called **forecasting**, can be used to estimate future sales and order more inventory before a shortage occurs.

forecasting
predicting future events to avoid problems

MANUAL AND COMPUTERIZED INFORMATION SYSTEMS

As discussed earlier, an information system can be manual or computerized. For example, some investment analysts manually draw charts and trend lines to assist them in making investment decisions. Tracking data on stock prices (input) over the last few months or years, these analysts develop patterns on graph paper (processing) that help them determine what stock prices are likely to do in the next few days or weeks (output). Some investors have made millions of dollars using manual stock analysis information systems. Of course, today many excellent computerized information systems have been developed to follow stock indexes and markets and to suggest when large blocks of stocks should be purchased or sold (called *program trading*) to take advantage of market discrepancies.

Program trading systems allow traders to keep up with swift changes in stock prices and make better decisions for their investors.

(Source: © Reuters NewMedia Inc./CORBIS)

Many information systems begin as manual systems and become computerized. For example, consider the way the U.S. Postal Service sorts mail. At one time most letters were visually scanned by postal employees to determine the ZIP code and were then manually placed in an appropriate bin. Today the bar-coded addresses on letters passing through the postal system are read electronically and automatically routed to the appropriate bin via conveyors. The computerized sorting system results in speedier processing time and provides management with information to help plan transportation needs. It is important to stress, however, that simply computerizing a manual information system does not guarantee improved system performance. If the underlying information system is flawed, the act of computerizing it might only magnify the impact of these flaws.

FIGURE 1.8

*The Components of a Computer-
Based Information System*

COMPUTER-BASED INFORMATION SYSTEMS

**computer-based information
system (CBIS)**
consists of hardware, software,
databases, telecommunications,
people, and procedures that are
configured to collect, manipulate,
store, and process data into
information

A **computer-based information system (CBIS)** consists of hardware, software, databases, telecommunications, people, and procedures that are configured to collect, manipulate, store, and process data into information. For example, a company's payroll systems, order entry system, or inventory control systems are examples of a CBIS. The components of a CBIS are illustrated in Figure 1.8. A business's **technology infrastructure** includes all the hardware, software, databases, telecommunications, people, and procedures that are configured to collect, manipulate, store, and process data into information. The technology infrastructure is a set of shared IS resources that form the foundation of each individual computer-based information system.

technology infrastructure
all the hardware, software, databases, telecommunications, people,
and procedures that are configured
to collect, manipulate, store, and
process data into information

Hardware

hardware
computer equipment used to perform input, processing, and output
activities

Hardware consists of computer equipment used to perform input, processing, and output activities. Input devices include keyboards, automatic scanning devices, equipment that can read magnetic ink characters, and many other devices. Investment firm T. Rowe Price, for example, uses voice response to allow customers to get their balances and other information using ordinary spoken sentences.[11] Processing devices include the central processing unit and main memory. Wal-Mart spent about $50 million to upgrade its central processing units and related equipment.[12] There are many storage and output devices, including secondary storage devices, printers, and computer screens. One company, for example, uses computer hardware in its stores to allow customers to order items that are not on store shelves.[13] The hardware helps the company "save the sale" and increase revenues.

Software

software
the computer programs that govern the operation of the computer

Software consists of computer programs that govern the operation of the computer. These programs allow a computer to process payroll, to send bills to customers, and to provide managers with information to increase profits, to reduce costs, and to provide better customer service. There are two basic types of software: system software, such as Windows XP, which controls basic computer operations such as start-up and printing, and applications software, such as Office XP, which allows specific tasks to be accomplished, such as word processing or tabulating numbers.[14]

Databases

database
an organized collection of facts and information

A **database** is an organized collection of facts and information. An organization's database can contain facts and information on customers, employees, inventory, competitors' sales information, online purchases, and much more. Most managers and executives believe a database is one of the most valuable and important parts of a computer-based information system.[15]

Telecommunications, Networks, and the Internet

telecommunications
the electronic transmission of signals for communications; enables organizations to carry out their processes and tasks through effective computer networks

networks
connected computers and computer equipment in a building, around the country, or around the world to enable electronic communications

Telecommunications is the electronic transmission of signals for communications, which enables organizations to carry out their processes and tasks through effective computer networks. Bob Evans Farms, for example, uses a telecommunications system and satellites to link its 459 restaurants to its plants and headquarters in Columbus, Ohio, to speed credit card authorization and report sales and payroll data.[16] **Networks** are used to connect computers and computer equipment in a building, around the country, or around the world to enable electronic communications. Merrill Lynch uses a wireless network that sends data through the air to connect 2,000 people between Manhattan and New Jersey.[17] Mike Brady, Merrill's first vice president for global network services, described the network equipment: "They look like searchlights on small refrigerators," he joked.

Internet
the world's largest computer network, actually consisting of thousands of interconnected networks, all freely exchanging information

Telecommunications and networks help people communicate using electronic mail (e-mail) and voice mail. These systems also help people work in groups. The **Internet** is the world's largest computer network, actually consisting of thousands of interconnected networks, all freely exchanging information. Research firms, colleges, universities, high schools, and businesses are just a few examples of organizations using the Internet. *PC Magazine* listed seven organizations that made excellent use of the Internet (see Table 1.4).[18] But anyone who can gain access to the Internet can communicate with anyone else on the Internet, including those who are in flight. American and Delta Air Lines announced plans to launch Internet service on 1,500 aircraft.[19]

The World Wide Web is a network of links on the Internet to documents containing text, graphics, video, and sound. Information about the documents and access to them are controlled and provided by tens of thousands of special computers called *Web servers*. The Web is one of many services available over the Internet and provides access to literally millions of documents.

intranet
an internal network based on Web technologies that allows people within an organization to exchange information and work on projects

extranet
a network based on Web technologies that allows selected outsiders, such as business partners and customers, to access authorized resources of the intranet of a company

The technology used to create the Internet is now also being applied within companies and organizations to create an **intranet**, which allows people within an organization to exchange information and work on projects.[20] An **extranet** is a network based on Web technologies that allows selected outsiders, such as business partners and customers, to access authorized resources of the intranet of a company. Lisa Boothe, the global e-business leader at Du Pont, reported that her company plans to move all of its fabric, chemical, and biotechnology businesses to its extranet site for corporate customers.[21] Many people use extranets every day without realizing it—to track shipped goods, order products from their suppliers, or access customer assistance from other companies. Log on to the FedEx site to check the status of a package, for example, and you are using an extranet.

People

People are the most important element in most computer-based information systems. Information systems personnel include all the people who manage, run, program, and maintain the system.[22] Bank One Corp., for example, recently hired 600 information systems personnel to speed up its computer-related projects.[23] Users are any people who use information systems to get results. Users include financial executives, marketing representatives, manufacturing operators, and many others. Certain computer users are also IS personnel.

Organization	Objective	Description of Internet Usage
Godiva Chocolatier	Increase sales and profits	The company developed a very profitable Internet site that allows customers to buy and ship chocolates. According to Kim Land, director of Godiva Direct, "This was set up from the beginning to make money." In two years, online sales have soared by more than 70 percent each year.
Environmental Defense	Alert the public to environmental concerns	The organization, formerly the Environmental Defense Fund, successfully used the Internet to alert people to the practice of catching sharks, removing their fins for soup, and returning them to the ocean to die. The Internet site also helped people fax almost 10,000 letters to members of Congress about the practice. According to Fred Krupp, the executive director of the Environmental Fund, "The Internet is the ultimate expression of 'think global, act local.'"
Buckman Laboratories	Better employee training	The company used the Internet to train employees to sell speciality chemicals to paper companies, instead of bringing them to Memphis for training. According to one executive, "Our retention rate is much higher, and we removed a week in Memphis, which meant big savings." Using the Internet lowered the hourly cost of training an employee from $1,000 to only $40.
Siemens	Reduce costs	Using the Internet, the company, which builds and services power plants, was able to reduce the cost of entering orders and serving customers. The Internet solution cost about $60,000 compared with a traditional solution that would have cost $600,000.
Goldman Industrial Group	Save time	The company makes machine tools and was able to slash the time it takes to fill an order from three or four months to about a week using the Internet to help coordinate parts and manufacturing with its suppliers and at its plants.
Partnership America	Make better decisions	The company developed an Internet site for wholesalers of computer equipment and supplies. The wholesalers use the Internet site to make better decisions about the features and prices of various pieces of computer equipment. The system allows wholesalers to connect to Partnership America's Internet site using cell phones. "When many of our customers need information, they're not at their desks," says one company representative.
Altra Energy Technologies	Get energy to companies that need it	The company developed an Internet site to help companies buy oil, gas, and wholesale power over the Internet.

TABLE 1.4

Uses of the Internet

procedures
the strategies, policies, methods, and rules for using a CBIS

Procedures

Procedures include the strategies, policies, methods, and rules for using the CBIS. For example, some procedures describe when each program is to be run or executed. Others describe who can have access to facts in the database. Still other procedures describe what is to be done in case a disaster, such as a fire, an earthquake, or a hurricane, renders the CBIS unusable.

Now that we have looked at computer-based information systems in general, we briefly examine the most common types used in business today. These IS types are covered in more detail in Part III.

BUSINESS INFORMATION SYSTEMS

The most common types of information systems used in business organizations are electronic commerce systems, transaction processing systems, management information systems, and decision support systems. In addition, some organizations employ special-purpose systems such as artificial intelligence systems, expert systems, and virtual reality systems. Together, these systems help employees in organizations accomplish both routine and special tasks—from recording sales, to processing payrolls, to supporting decisions in various departments, to providing alternatives for large-scale projects and opportunities.

ELECTRONIC COMMERCE

e-commerce

any business transaction executed electronically between parties such as companies (business-to-business), companies and consumers (business-to-consumer), business and the public sector, and consumers and the public sector

E-commerce involves any business transaction executed electronically between parties such as companies (business-to-business), companies and consumers (business-to-consumer), business and the public sector, and consumers and the public sector. People may assume that e-commerce is reserved mainly for consumers visiting Web sites for on-line shopping. But Web shopping is only a small part of the e-commerce picture; the major volume of e-commerce—and its fastest-growing segment—is business-to-business transactions that make purchasing easier for corporations. This growth is being stimulated by increased Internet access, user confidence, better payment systems, and rapidly improving Internet and Web security. E-commerce offers opportunities for small businesses, too, by enabling them to market and sell at a low cost worldwide, thus offering them an opportunity to enter the global market right from start-up.

Consumers who have tried on-line shopping appreciate the ease of e-commerce. They can avoid fighting the crowds in the malls, shop on-line at any time from the comfort of their home, and have goods delivered to them directly. In addition, under current laws governing on-line purchases, state sales taxes do not need to be paid. However, e-commerce is not without its downside. Consumers continue to have concerns about sending credit card information over the Internet to sites with varying security measures where high-tech criminals could obtain it. In addition, denial-of-service attacks that overwhelm the capacity of some of the Web's most established and popular sites have raised new concerns for the future growth of e-commerce. There are additional concerns about what data is gathered when a consumer visits a Web site and what companies do with the collected data; some have sold data to multiple sources, leading marketing companies to know more than we would like. The "Ethical and Societal Issues" box discusses privacy on the Internet.

Already a huge portion of the e-commerce market, business-to-business transactions—such as this paycheck service by Automated Data Processing—are projected to pass the $1 trillion mark at the beginning of 2003.

Yet, in spite of the concerns, e-commerce offers many advantages for streamlining work activities. Figure 1.9 provides a brief example of how e-commerce can simplify the process for purchasing new office furniture from an office supply company. Under the manual system, a corporate office worker must get approval for a purchase that costs more than a certain amount. That request goes to the purchasing department, which generates a formal purchase order to procure the goods from the approved vendor. Business-to-business e-commerce automates that entire process. Employees go directly to the supplier's Web site, find the item in its catalog, and order what they need at a price prenegotiated by the employee's company. If approval is required, the approver is notified automatically. As the use of e-commerce systems grows, companies are phasing out their more traditional systems. The resulting growth of e-commerce is creating many new business opportunities.

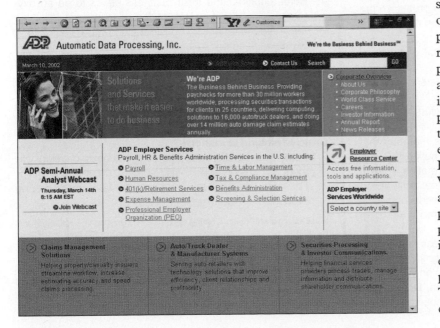

DoubleClick Tempers Marketing with Customer Privacy

As e-commerce becomes an integral part of virtually all companies, many are finding themselves faced with new ethical dilemmas. The hottest ethical issues in e-commerce have to do with privacy.

Society, in general, is conducting more of its daily activities electronically on the Internet. We use the Internet for communication, for access to information, and to purchase merchandise. The Internet is wonderfully convenient, putting the world at our fingertips. For marketing specialists, the Internet is also a wonderful tool for collecting telltale data that define user interests and trends.

Business-to-consumer (B2C) retailers are able to track each contact a customer makes with their organization on the Net. By storing information on the customer's computer, in data packages called *cookies,* businesses can track a customer's movement around the Web site. By recording the amount of time a customer views each page, a practice known as *collecting click-stream data,* a company can build a fairly accurate customer profile that defines the interests of that customer. By enticing the customer to supply his or her name and other personal information, for instance, by signing up for a "free" user account to access special services, the retailer is able to connect the user profile with a name and to store that information in its database. Through a technique called *data mining,* companies can sift through the combined information of any one customer or group of customers to recognize trends and tendencies—and ultimately pitch products and services specifically for that customer's interests.

This type of data collection takes place on a variety of levels in a variety of environments: retailers track the movements of customers around the Web, cell phone companies store information about the location of cell phones (and presumably their owners) around geographic regions, employers may track their employees' e-mail and Internet use on the corporate network. But all of this information collection raises some concerns, among them:

- How far can organizations go in the pursuit of personal information?
- Do customers need to be aware of what personal data is being collected?
- Once a company has gathered this information, how can it be used?

In general, consumer privacy advocates maintain that consumers have the right to know what information is being stored about them and to control how that information is used.

Take, for example, New York–based Internet advertising company DoubleClick. DoubleClick is a marketing company that, among other things, provides Internet-based advertising services through direct e-mail and banner ads on the Web. DoubleClick has taken some heat over allegations that it used technology in its banner ads placed on other companies' Web sites to collect personally identifiable information on Web users. By placing banner ads on numerous Web sites, DoubleClick can collect customer information from multiple sources to get a complete picture of a person's buying patterns and preferences. Consumer advocates were further enraged when it was discovered that DoubleClick intended to share customer information with an off-line marketing firm.

To ease public concern and regain customer confidence, DoubleClick changed course, joined the National Advertising Initiative (NAI), and adopted its principles. The NAI worked with the Federal Trade Commission and the U.S. Department of Commerce to develop regulations concerning consumer profiling, whereby member companies would police themselves. Under the NAI Principles, DoubleClick and other members must provide consumers with a clear explanation of the types of data they collect and the way they use them, as well as the ability to opt out of data-collecting efforts if users choose not to participate. Concurrent with joining NAI, DoubleClick decided to discontinue its Intelligent Targeting product, stating, "Given the focus of our business, we have decided that the Intelligent Targeting product is not something we plan to pursue in today's environment."

The balance between effective marketing and respect for customer privacy is difficult to reach. Like DoubleClick, most e-commerce retailers, or e-tailers, have published privacy policies by which they are legally bound. The Federal Trade Commission has ruled that, unless otherwise stated, these policies extend to a company's off-line data practices as well. To deal with these ethical dilemmas, some e-commerce firms are hiring ethics consultants such as Tom Shanks. Shanks says he hopes to find ways to help companies curb the abuse of consumer information by offering incentives to respect end users' privacy and by providing them with up-to-date information about privacy legislation.

Discussion Questions

1. Some consider the term *business ethics* to be an oxymoron. Does the new information economy provide more fertile ground for ethical considerations in business? If you were an ethics consultant, how would you sell your client on the benefits of applying ethical principles to business practices?
2. What methods of collecting consumer information were used prior to the Internet? How do you think the Internet has changed marketing approaches in general?

Critical Thinking Questions

3. Aside from public concerns over customer profiling, are there any customer benefits to profiling? List some.
4. How can the Web be used successfully for marketing products without upsetting consumers?

Sources: Zachary Tobias, "Putting the Ethics in E-Business," *ComputerWorld,* November 6, 2000, http://www.computerworld.com; "DoubleClick Drops 'Intelligent Targeting' Product," *Newsbytes,* January 9, 2002, http://www.washingtonpost.com/wp-dyn/technology; Brian Krebs, "Online Privacy Policies Apply to Offline Data Practices—FTC," *Newsbytes,* December 10, 2001; http://www.networkadvertising.org, follow links to "About NAI" and "Principles," accessed January 27, 2002; http://www.doubleclick.net, follow links to "Privacy Policy," accessed January 27, 2002.

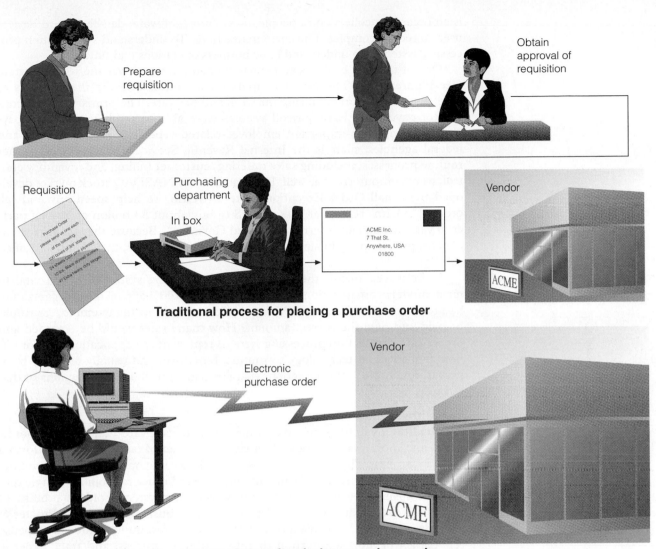

Traditional process for placing a purchase order

E-commerce process for placing a purchase order

FIGURE 1.9

E-commerce greatly simplifies the purchasing process.

One study reported that e-commerce could have a positive impact on stock prices and the market value of firms.[24] Today, several e-commerce firms have teamed up with more traditional brick-and-mortar firms to draw from each's strengths. Amazon.com, for example, is joining forces with Circuit City Stores, Inc.[25] With the new venture, customers will be able to order products through Amazon on the Internet and pick up products at one of the 600 local Circuit City stores or get them shipped to an address through Amazon.com's home or office delivery system.

TRANSACTION PROCESSING SYSTEMS, WORKFLOW SYSTEMS, AND ERP

Transaction Processing Systems

Since the 1950s, computers have been used to perform common business applications. The objective of many of these early systems was to reduce costs by automating many routine, labor-intensive business systems. A **transaction** is any business-related exchange such as payments to employees, sales to customers, or payments to suppliers. Thus, processing business transactions was the first application of computers for most organizations. A **transaction processing system (TPS)**

transaction
any business-related exchange such as payments to employees, sales to customers, or payments to suppliers

transaction processing system (TPS)
an organized collection of people, procedures, software, databases, and devices used to record completed business transactions

is an organized collection of people, procedures, software, databases, and devices used to record completed business transactions. To understand a transaction processing system is to understand basic business operations and functions.

One of the first business systems to be computerized was the payroll system (see Figure 1.10). The primary inputs for a payroll TPS are the numbers of employee hours worked during the week and pay rate. The primary output consists of paychecks. Early payroll systems were able to produce employee paychecks, along with important employee-related reports required by state and federal agencies, such as the Internal Revenue Service. Simultaneously, other routine processes, including sales ordering, customer billing, and inventory control, were computerized as well. For example, the NASDAQ stock market developed the Small Order Execution System (SOES) to help speed buy and sell orders.[26] DaimlerChrysler uses its TPS to buy about $3 billion of needed parts each year through an Internet site called Covisint.[27] Because these systems handle and process daily business exchanges, or transactions, they are all classified as TPSs.

In improved forms, these transaction processing systems are still vital to most modern organizations. Consider what would happen if an organization had to function without its TPS for even one day. How many employees would be paid and paid the correct amount? How many sales would be recorded and processed? Transaction processing systems represent the application of information concepts and technology to routine, repetitive, and usually ordinary business transactions, but transactions that are critical to the daily functions of that business.

Workflow Systems

workflow system
rule-based management software that directs, coordinates, and monitors execution of an interrelated set of tasks arranged to form a business process

A **workflow system** is rule-based management software that directs, coordinates, and monitors execution of an interrelated set of tasks arranged to form a business process. The primary purpose of workflow systems is to provide employees with tracking, routing, document imaging, and other capabilities designed to improve business processes. Transactional workflow systems hold the promise of improving the productivity and dependability of business processes. Procter & Gamble, a major U.S. consumer goods manufacturer, implemented an expense reporting workflow application to enter, submit, process, and track expense reports. The system streamlines the reimbursement process by simplifying expense entries and automating the approval process. The system cuts the time employees spend filling out expense reports and reduces the manual retyping and editing typically associated with expense report reconcilation.

Enterprise Resource Planning

enterprise resource planning (ERP) system
a set of integrated programs capable of managing a company's vital business operations for an entire multisite, global organization

An **enterprise resource planning (ERP) system** is a set of integrated programs that is capable of managing a company's vital business operations for an entire multisite, global organization. Although the scope of an ERP system may vary from company to company, most ERP systems provide integrated software to support the manufacturing and finance business functions of an organization.

FIGURE 1.10

A Payroll Transaction Processing System

The inputs (numbers of employee hours worked and pay rates) go through a transformation process to produce outputs (paychecks).

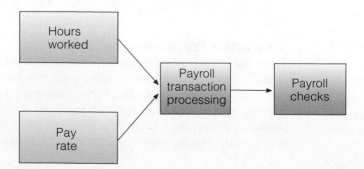

SAP AG, a German software company, is one of the leading suppliers of ERP software. The company controls 15 percent of an estimated $40 billion worldwide ERP market.

In such an environment, a demand forecast is prepared that estimates customer demand for several weeks. The ERP system checks what is already available in finished product inventory to meet the projected demand. Any shortcomings then need to be produced. In developing the production schedule, the ERP system checks the raw material and packing material inventory and determines what needs to be ordered to meet the planned production schedule. Most ERP systems also have a purchasing subsystem that orders the items required. In addition to these core business processes, some ERP systems may be capable of supporting additional business functions such as human resources, sales, and distribution. PeopleSoft, for example, recently launched an Internet-based ERP to manage customer relationships.[28] The primary benefits of implementing an ERP system include easing adoption of improved work processes and improving access to timely data for operational decision making.

MANAGEMENT INFORMATION AND DECISION SUPPORT SYSTEMS

The benefits provided by an effective transaction processing system are tangible and can be used to justify their cost in computing equipment, computer programs, and specialized personnel and supplies. They speed business activities and reduce clerical costs. Although early accounting and financial transaction processing systems were already valuable, companies soon realized that the data stored in these systems can be used to help managers make better decisions in their respective business areas, whether human resources, marketing, or administration. Satisfying the needs of managers and decision makers continues to be a major factor in developing management information and decision support systems.

Management Information Systems

management information system (MIS)

an organized collection of people, procedures, software, databases, and devices used to provide routine information to managers and decision makers

A **management information system (MIS)** is an organized collection of people, procedures, software, databases, and devices used to provide routine information to managers and decision makers. The focus of an MIS is primarily on operational efficiency. Marketing, production, finance, and other functional areas are supported by management information systems and linked through a common database. Management information systems typically provide standard reports generated with data and information from the transaction processing system (see Figure 1.11).

Management information systems were first developed in the 1960s and are characterized by the use of information systems to produce managerial reports. In most cases, these early reports were produced periodically—daily, weekly, monthly, or yearly. Today, Foxwoods Resort Casino generates daily reports that tell the company what specific customers like.[29] Foxwoods knows whether a customer likes flowers in her room or a drink in his hand and can accommodate individual needs or desires. Periodic reports such as Foxwoods' are printed regularly, so they are called *scheduled reports*. Scheduled reports help managers perform their duties. For example, a summary report of total payroll costs might help an accounting manager control future payroll costs. Because of their value to managers, MISs have proliferated throughout the management ranks. For instance, the total payroll summary report produced initially for an accounting

FIGURE 1.11

Functional management information systems draw data from the organization's transaction processing system.

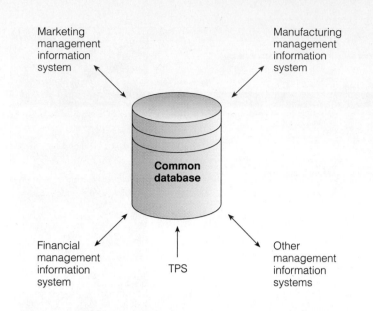

decision support system (DSS)

an organized collection of people, procedures, software, databases, and devices used to support problem-specific decision making

Decisioneering provides decision support software called Crystal Ball, which helps businesspeople of all types assess risks and make forecasts. Shown here is the Standard Edition being used for oil field development.

(Source: Crystal Ball screenshot courtesy of Decisioneering, Inc.)

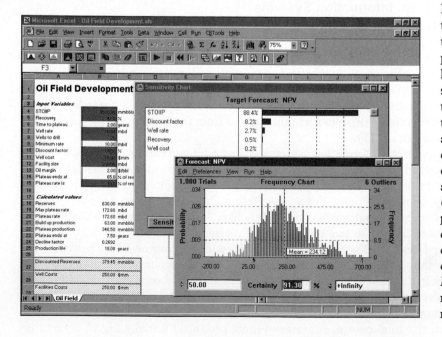

manager might also be useful to a production manager to help monitor and control labor and job costs. Other scheduled reports are used to help managers from a variety of departments control customer credit, payments to suppliers, the performance of sales representatives, inventory levels, and more.

Other types of reports were also developed during the early stages of management information systems. *Demand reports* were developed to give decision makers certain information upon request. For example, prior to closing a sale, a sales representative might seek a demand report on how much inventory exists for a particular item. This report would tell the representative whether enough inventory of the item is on hand to fill the customer order. *Exception reports* describe unusual or critical situations, like low inventory levels. The exception report is produced only if a certain condition exists—in this case, inventory falling below a specified level. For example, in a bicycle manufacturing company, an exception report might be produced by the MIS if the number of bicycle seats is too low and more should be ordered.

Decision Support Systems

By the 1980s, dramatic improvements in technology resulted in information systems that were less expensive but more powerful than earlier systems. People at all levels of organizations began using personal computers to do a variety of tasks; they were no longer solely dependent on the information systems department for all their information needs. So, people quickly recognized that computer systems could support additional decision-making activities. A **decision support system (DSS)** is an organized collection of people, procedures, software, databases, and devices used to support problem-specific decision making. The focus of a DSS is on decision-making effectiveness. Whereas an MIS helps an organization "do things right," a DSS helps a manager "do the right thing."

A DSS supports and assists all aspects of problem-specific decision making. As seen in the "IS Principles in Action" box, a DSS can also support customers by rapidly responding to their phone and e-mail inquiries. A DSS goes beyond a traditional management information system. A DSS can provide immediate assistance in solving complex problems that are not supported by a traditional MIS. Many of these problems are unique and not straightforward. For instance, an auto manufacturer might try to determine the best location to build a new manufacturing facility, or an oil company might want to discover the best place to drill for oil. Chevron, for example, uses a DSS to track and manage projects and employees in 40 countries.[30] Traditional MISs are seldom used to solve these types of problems; a DSS can help by suggesting alternatives and assisting final decision making.

Decision support systems are used when the problem is complex and the information needed to make the best decision is difficult to obtain and use. So, a DSS also involves managerial judgment. In addition, managers often play an active role in the development and implementation of the DSS. A DSS operates from a managerial perspective, and it recognizes that different managerial styles and decision types require different systems. For example, two production managers in the same position trying to solve the same problem might require different information and support. The overall emphasis is to support rather than replace managerial decision making.

The essential elements of a DSS include a collection of models used to support a decision maker or user (model base), a collection of facts and information to assist in decision making (database), and systems and procedures (user interface) that help decision makers and other users interact with the DSS (see Figure 1.12).

FIGURE 1.12

Essential DSS Elements

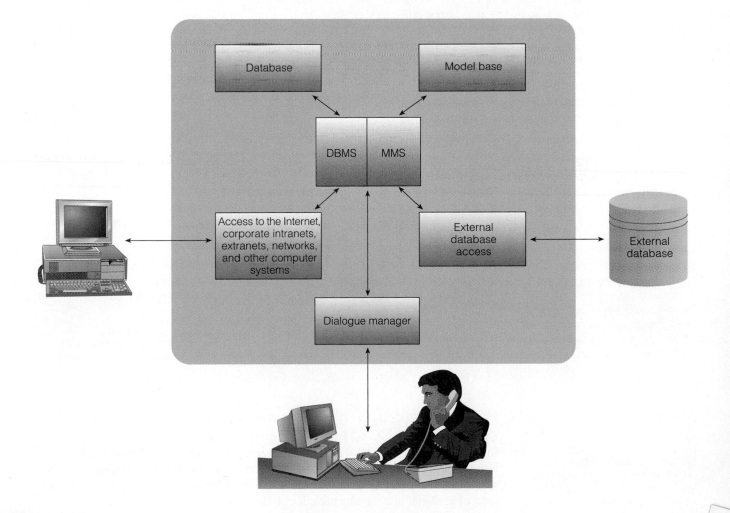

PRINCIPLE

● The value of information is directly linked to how it helps decision makers achieve the organization's goals.

Home Depot Invests Big in Information Systems

Although 2002 was a year of caution for most businesses when it came to information system spending, research shows that hard-goods retailers may increase their investment in information systems. Greg Buzek, president of IHL Consulting Group, which produced a report on the subject, stated that "leading-edge home improvement and electronics retailers like Home Depot, Lowes, Best Buy, and Circuit City will continue to invest heavily in new information technology. These retailers achieved their dominant market positions by deploying advanced technology during the slower economic times of the early 1990s. We are looking at another period where there will be favorable vendor and lender terms on capital IT spending. As a result, we expect to see these types of retailers maintain or even increase their level of capital spending to take advantage of these favorable terms."

Home Depot's recent commitments to major information system overhauls certainly bear this out. The Atlanta-based retailer of home improvement goods has embarked on a sweeping plan to tie together thousands of its software applications, stores, and systems in real time. The new unified system will provide relevant and up-to-date data and information for use in the company's MISs and DSSs to assist top management in making critical corporate decisions. Charlie Weston, director of information services at Home Depot, said the EAI implementation will probably run into millions of dollars but should pay for itself in the next several years.

Home Depot isn't stopping there—it is also changing the way it handles customer phone and e-mail inquiries. In the past, when customers phoned their local Home Depot with questions, the operator would connect them with a salesperson on the floor. Overburdened salespeople sometimes had to choose between either leaving the person waiting on the phone or making customers on the floor wait. Now Home Depot routes its calls to regional call centers. The Tampa call center employs 1,000 customer agents who handle customer calls in the company's southern division. These agents are able to access specific, detailed information for each store location using customer relationship management (CRM) solutions from Avaya Inc. Avaya's Interaction Management software links multiple databases in local stores and then delivers information to the regional agents, who seamlessly respond to customers by phone, e-mail, or other electronic media. By handling all contacts regionally, Home Depot frees store associates to spend more time with in-store customers.

Customers contacting the company by phone or e-mail also receive more efficient and accurate service, with little or no time spent on hold or waiting for a response.

"For us, contact center technology is more than getting a call to an agent," said Ed Buter, Senior Manager—Information Services, Home Depot. "We have approximately a million products in more than 1,200 stores, with numerous databases kept at the stores for things like product orders, delivery and installation schedules, tool rentals, or promotional events. Our challenge is to get this information to the agents while quickly and effectively integrating all channels of communication into a system that supports the business."

It's clear that Home Depot understands that the quality of information flow within an organization is critical to the organization's ability to gain a competitive advantage. By connecting all its stores in a real-time network, Home Depot can more effectively view its stores (expected to number more than 2,300 by 2005) as a single entity. Such collective information allows the chain's management to examine up-to-the-minute nationwide statistics and trends, which ultimately supports more effective and timely decisions. With the addition of regional call centers, Home Depot makes more effective use of its employees' time to serve their customers. Add to this the services offered at the Web site, homedepot.com, and it's hard to imagine what more this company could do to improve itself.

Discussion Questions

1. What types of information, statistics, and trends might Home Depot managers be interested in tracking with their new system in order to meet the organization's goals?
2. Before you visit homedepot.com, list some valuable services you think Home Depot can offer its customers on the Web.

Critical Thinking Questions

3. If you were a Home Depot competitor, what type of services might you offer customers above and beyond those mentioned here so that your organization could compete with Home Depot?
4. Consider the expense involved in opening regional call centers. How does this investment assist Home Depot in being a more efficient and effective company?

Sources: "Home Improvement and Electronics Retailers Lead IT Spending Growth in Retail," *Business Wire*, January 8, 2002; Marc Songini, "Home Depot Launches Major Integration," CRM projects, *Computerworld*, June 29, 2001, http://www.computerworld.com; Avaya Web site, http://www.avaya.com.

SPECIAL-PURPOSE BUSINESS INFORMATION SYSTEMS: ARTIFICIAL INTELLIGENCE, EXPERT SYSTEMS, AND VIRTUAL REALITY

artificial intelligence (AI)
a field in which the computer system takes on the characteristics of human intelligence

In addition to TPSs, MISs, and DSSs, organizations often use special-purpose systems. One of these systems is based on the notion of **artificial intelligence (AI)**, where the computer system takes on the characteristics of human intelligence. The field of artificial intelligence includes several subfields (see Figure 1.13).

FIGURE 1⦿13

The Major Elements of Artificial Intelligence

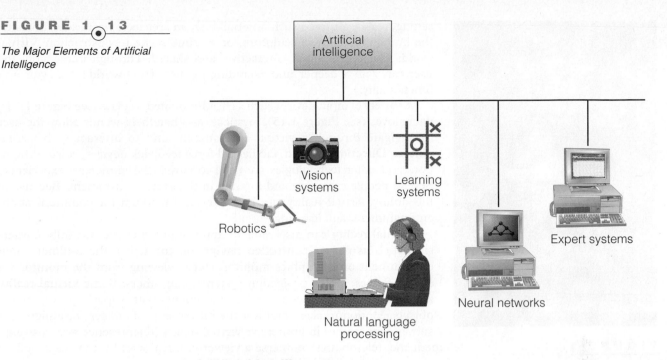

Artificial Intelligence

Robotics is an area of artificial intelligence in which machines take over complex, routine, or boring tasks, such as welding car frames or assembling computer systems and components. Vision systems allow robots and other devices to have "sight" and to store and process visual images. Natural language processing involves the ability of computers to understand and act on verbal or written commands in English, Spanish, or other human languages. Learning systems give computers the ability to learn from past mistakes or experiences, such as playing games or making business decisions, and neural networks is a branch of artificial intelligence that allows computers to recognize and act on patterns or trends. Some successful stock, options, and futures traders use neural networks to spot trends and make more profitable investments.

Expert Systems

expert system
a system that gives a computer the ability to make suggestions and act like an expert in a particular field

Expert systems give the computer the ability to make suggestions and act like an expert in a particular field.[31] The unique value of expert systems is that they allow organizations to capture and use the wisdom of experts and specialists. Therefore, years of experience and specific skills are not completely lost when a human expert dies, retires, or leaves for another job. Expert systems can be applied to almost any field or discipline. Expert systems have been used to monitor complex systems such as nuclear reactors, perform medical diagnoses, locate possible repair problems, design and configure information system components, perform credit evaluations, and develop marketing plans for a new product or new investment strategies. The collection of data, rules, procedures, and relationships that must be followed to achieve value or the proper outcome is contained in the expert system's **knowledge base**.

knowledge base
the collection of data, rules, procedures, and relationships that must be followed to achieve value or the proper outcome

The 1980s and 1990s brought advances in both artificial intelligence and expert systems. More and more organizations are using these systems to solve complex problems and support difficult decisions. However, many issues remain to be resolved and more work is needed to refine their meaningful uses.

Virtual Reality

virtual reality
originally, the term referred to immersive virtual reality, which means the user becomes fully immersed in an artificial, three-dimensional world that is completely generated by a computer

Originally, the term **virtual reality** referred to immersive virtual reality, which means the user becomes fully immersed in an artificial, three-dimensional world that is completely generated by a computer. The virtual world is presented in full scale and relates properly to the human size. It may represent any three-dimensional

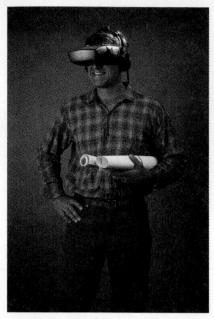

FIGURE 1.14

A Head-Mounted Display

The head-mounted display (HMD) was the first device of its kind providing the wearer with an immersive experience. A typical HMD houses two miniature display screens and an optical system that channels the images from the screens to the eyes, thereby presenting a stereo view of a virtual world. A motion tracker continuously measures the position and orientation of the user's head and allows the image-generating computer to adjust the scene representation to the current view. As a result, the viewer can look around and walk through the surrounding virtual environment.

(Source: Courtesy of Virtual Research Systems, Inc.)

setting, real or abstract, such as a building, an archaeological excavation site, the human anatomy, a sculpture, or a crime scene reconstruction. Virtual worlds can be animated, interactive, and shared. Through immersion, the user can gain a deeper understanding of the virtual world's behavior and functionality.

A variety of input devices such as head-mounted displays (see Figure 1.14), data gloves (see Figure 1.15), joysticks, and handheld wands allow the user to navigate through a virtual environment and to interact with virtual objects. Directional sound, tactile and force feedback devices, voice recognition, and other technologies are used to enrich the immersive experience. Several people can share and interact in the same environment. Because of this ability, virtual reality can be a powerful medium for communication, entertainment, and learning.

Virtual reality can also refer to applications that are not fully immersive, such as mouse-controlled navigation through a three-dimensional environment on a graphics monitor, stereo viewing from the monitor via stereo glasses, stereo projection systems, and others. Some virtual reality applications allow views of real environments with superimposed virtual objects. Motion trackers monitor the movements of dancers or athletes for subsequent studies in immersive virtual reality. Telepresence systems (e.g., telemedicine, telerobotics) immerse a viewer in a real world that is captured by video cameras at a distant location and allow for the remote manipulation of real objects via robot arms and manipulators. Many believe that virtual reality is reshaping the interface between people and information technology by offering new ways to communicate information, visualize processes, and express ideas creatively.

Useful applications of virtual reality include training in a variety of areas (military, medical, equipment operation, etc.), education, design evaluation (virtual prototyping), architectural walk-throughs, human factors and ergonomic studies, simulation of assembly sequences and maintenance tasks, assistance for the handicapped, study and treatment of phobias (fear of flying), entertainment, and, of course, virtual reality games.

It is difficult to predict where information systems and technology will be in 10 to 20 years. It seems, however, that we are just beginning to discover the full range of their usefulness. Technology has been improving and expanding at an increasing rate; dramatic growth and change are expected for years to come. Without question, a knowledge of the effective use of information systems will be critical for managers both now and in the long term.

SYSTEMS DEVELOPMENT

systems development
the activity of creating or modifying existing business systems

Systems development is the activity of creating or modifying existing business systems. Developing information systems to meet business needs is highly complex and difficult—so much so that it is common for information systems projects to overrun budgets and exceed scheduled completion dates. Business managers would like the development process to be more manageable, especially with predictable costs and timing. One strategy for improving the results of a systems development project is to divide it into several steps, each step with a well-defined goal and set of tasks to accomplish (see Figure 1.16). These steps are summarized next.

SYSTEMS INVESTIGATION AND ANALYSIS

systems investigation
a stage of systems development that has as its goal to gain a clear understanding of the problem to be solved or opportunity to be addressed

The first two steps of systems development are systems investigation and analysis. The goal of the **systems investigation** is to gain a clear understanding of the problem to be solved or opportunity to be addressed. Royal Caribbean, for

systems analysis
a stage of systems development during which the problems and opportunities of the existing system are defined

systems design
a stage of systems development that determines how the new system will work to meet the business needs defined during systems analysis

systems implementation
a stage of systems development during which the various system components (hardware, software, databases, etc.) defined in the design step are created or acquired and then assembled and the new system is put into operation

example, launched a systems investigation to determine whether a development project was feasible to automate purchasing at ports around the world.[32] Once an organization understands the problem, the next question to be answered is "Is the problem worth solving?" Given that organizations have limited resources—people and money—this question deserves careful consideration. If the decision is to continue with the solution, the next step, **systems analysis**, defines the problems and opportunities of the existing system.

SYSTEMS DESIGN, IMPLEMENTATION, AND MAINTENANCE AND REVIEW

Systems design determines how the new system will work to meet the business needs defined during systems analysis. **Systems implementation** involves creating or acquiring the various system components (hardware, software, databases, etc.) defined in the design step, assembling them, and putting the new system into operation. The purpose of **systems maintenance and review** is to check and modify the system so that it continues to meet changing business needs.

WHY LEARN ABOUT INFORMATION SYSTEMS?

systems maintenance and review
a stage of the systems development process that has as its goal to check and modify the system so that it continues to meet changing business needs

Studies have shown that the involvement of managers and decision makers in all aspects of information systems is a major factor for organizational success, including higher profits and lower costs. A knowledge of information systems will help you make a significant contribution on the job. It will also help you advance in your chosen career or field. Managers are expected to identify opportunities to implement information systems to improve the business. They are also expected to be able to lead information system projects in their area of the business.

FIGURE 1.15

A Data Glove

Realistic interactions with virtual objects via such devices as a data glove that senses hand position allow for manipulation, operation, and control of virtual worlds.

(Source: Courtesy of Virtual Technologies, Inc.)

FIGURE 1.16

An Overview of Systems Development

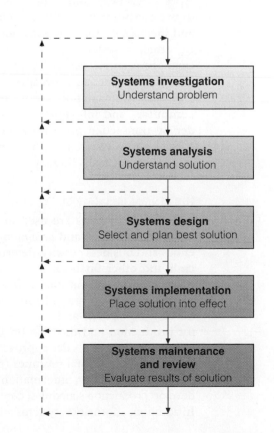

Information systems play a fundamental and ever-expanding role in all business organizations. If you are to have a solid understanding of how organizations operate, it is imperative that you understand the role of information systems within these organizations. Moreover, in this new century, business survival and prosperity continues to become more difficult. For example, increased mergers among former competitors to create global conglomerates, continued downsizing of corporations to focus on their core businesses and to improve efficiencies, efforts to reduce trade barriers, and the globalization of capital all point to the increased internationalization of business organizations and markets. In addition, business issues and decisions are becoming more complex and must be made faster. An understanding of information systems will help you cope, adapt, and prosper in this challenging environment.

Regardless of your chosen field or the organization for which you may work, it is likely that you will use information systems. Why study information systems? A knowledge of information systems will help you advance in your career, solve problems, realize opportunities, and meet your own personal goals.

COMPUTER AND INFORMATION SYSTEMS LITERACY

You must acquire both computer literacy and information systems literacy to be able to use information systems to meet personal and organizational goals. **Computer literacy** is a knowledge of computer systems and equipment and the ways they function. It stresses equipment and devices (hardware), programs and instructions (software), databases, and telecommunications.

Information systems literacy goes beyond a knowledge of the fundamentals of computer systems and equipment. **Information systems literacy** is a knowledge of how data and information are used by individuals, groups, and organizations. It includes knowledge of not only computer technology but also aspects of the broader range of information technology. Most important, however, it encompasses *how* and *why* this technology is applied in business. Knowing about various types of hardware and software is an example of computer literacy. Knowing how to use hardware and software to increase profits, cut costs, improve productivity, and increase customer satisfaction is an example of information systems literacy. Information systems literacy can involve a knowledge of how and why people (managers, employees, stockholders, and other individuals) use information technology; a knowledge of organizations, decision-making approaches, management levels, and information needs; and a knowledge of how organizations can use computers and information systems to achieve their goals. Knowing how to deploy transaction processing, management information, decision support, and special-purpose systems to help an organization achieve its goals is a key aspect of information systems literacy.

INFORMATION SYSTEMS IN THE FUNCTIONAL AREAS OF BUSINESS

Information systems are used in all functional areas and operating divisions of business. In *finance* and *accounting*, information systems are used to forecast revenues and business activity, determine the best sources and uses of funds, manage cash and other financial resources, analyze investments, and perform audits to make sure the organization is financially sound and that all financial reports and documents are accurate. In *sales* and *marketing*, information systems are used to develop new goods and services (product analysis), determine the best location for production and distribution facilities (place or site analysis), determine the best advertising and sales approaches (promotion analysis), and set product prices to get the highest total revenues (price analysis).

In *manufacturing*, information systems are used to process customer orders, develop production schedules, control inventory levels, and monitor product quality. In addition, information systems are used to design products (*computer-assisted design,*

---

computer literacy
knowledge of computer systems and equipment and the ways they function; it stresses equipment and devices (hardware), programs and instructions (software), databases, and telecommunications

information systems literacy
knowledge of how data and information are used by individuals, groups, and organizations

or *CAD*), manufacture items (*computer-assisted manufacturing*, or *CAM*), and integrate multiple machines or pieces of equipment (*computer-integrated manufacturing*, or *CIM*). Information systems are also used in *human resource management* to screen applicants, administer performance tests to employees, monitor employee productivity, and more. *Legal information systems* are used to analyze product liability and warranties and to develop important legal documents and reports.

INFORMATION SYSTEMS IN INDUSTRY

Information systems are used in almost every industry or field. The *airline industry* employs Internet auction sites to offer discount fares and increase revenue. *Investment firms* use information systems to analyze stocks, bonds, options, the futures market, and other financial instruments, as well as to provide improved services to their customers. *Banks* use information systems to help make sound loans and good investments. The *transportation industry* uses information systems to schedule trucks and trains to deliver goods and services at the least cost. *Publishing companies* use information systems to analyze markets and to develop and publish newspapers, magazines, and books. *Healthcare organizations* use information systems to diagnose illnesses, plan medical treatment, and bill patients. HMOs have begun to use Web technology to access patients' insurance eligibility and other information held in databases to cut patient costs. *Retail companies* are using the Web to take customer orders and provide customer service support. *Power management* and *utility companies* use information systems to monitor and control power generation and usage. *Professional services* firms employ information systems to improve the speed and quality of services they provide to customers. *Management consulting firms* use intranets and extranets to provide information on products, services, skill levels, and past engagements to its consultants. These industries will be discussed in more detail as we continue through the book.

SUMMARY

PRINCIPLE *The value of information is directly linked to how it helps decision makers achieve the organization's goals.*

Data consists of raw facts; information is data transformed into a meaningful form. The process of defining relationships between data requires knowledge. Knowledge is an awareness and understanding of a set of information and how that information can be made useful to support a specific task. To be valuable, information must have several characteristics: it should be accurate, complete, economical to produce, flexible, reliable, relevant, simple to understand, timely, verifiable, accessible, and secure. The value of information is directly linked to how it helps people achieve their organization's goals.

PRINCIPLE *Models, computers, and information systems are constantly making it possible for organizations to improve the way they conduct business.*

A system is a set of elements that interact to accomplish a goal or set of objectives. The components of a system include inputs, processing mechanisms, and outputs. Systems also contain boundaries that separate them from the environment and each other. Feedback is used by the system to monitor and control its operation to make sure it continues to meet its goals and objectives. Systems may be classified in many ways. They may be considered simple or complex. A stable, nonadaptive system does not change over time, while a dynamic, adaptive system does. Open systems interact with their environments; closed systems do not. Some systems exist temporarily; others are considered permanent.

System performance is measured by its efficiency and effectiveness. Efficiency is a measure of what is produced divided by what is consumed; effectiveness is a measure of the extent to which a system achieves its goals. A systems performance standard is a specific objective. A system variable is a quantity or item that can be controlled by the decision maker, such as how much of a product to produce, while a system parameter is a value or quantity that cannot be controlled, such as the cost of raw material.

• • •

There are four basic types of models: narrative, physical, schematic, and mathematical. These models serve as an abstraction or an approximation that is used to represent reality. Models enable us to explore and gain an improved understanding of real-world situations. The narrative model provides a verbal description of reality. A physical model is a tangible representation of reality, often computer designed or constructed. A schematic model is a graphic representation of reality such as a graph, chart, figure, diagram, illustration, or picture. A mathematical model is an arithmetic representation of reality.

PRINCIPLE *Knowing the potential impact of information systems and having the ability to put this knowledge to work can result in a successful personal career, organizations that reach their goals, and a society with a higher quality of life.*

Information systems are sets of interrelated elements that collect (input), manipulate and store (process), and disseminate (output) data and information. Input is the activity of capturing and gathering new data; processing involves converting or transforming data into useful outputs; and output involves producing useful information. Feedback is the output that is used to make adjustments or changes to input or processing activities.

The components of a computer-based information system include hardware, software, databases, telecommunications and the Internet, people, and procedures. CBISs play an important role in today's businesses and society. The key to understanding the existing variety of systems begins with learning their fundamentals. The types of business information systems used within organizations can be classified into four basic groups: (1) e-commerce, (2) TPS, (3) MIS and DSS, and (4) special-purpose business information systems.

E-commerce involves any business transaction executed electronically between parties such as companies (business-to-business), companies and consumers (business-to-consumer), business and the public sector, and consumers and the public sector. The major volume of e-commerce and its fastest-growing segment is business-to-business transactions that make purchasing easier for big corporations. E-commerce offers opportunities for small businesses by enabling them to market and sell at a low cost worldwide, thus enabling them to enter the global market right from start-up.

The most fundamental system is the transaction processing system (TPS). A transaction is any business-related exchange. The TPS handles the large volume of business transactions that occur daily within an organization. A workflow system is rule-based management software that directs, coordinates, and monitors execution of an interrelated set of tasks arranged to form a business process. The primary purpose of workflow systems is to provide end users with tracking, routing, document imaging, and other capabilities designed to improve business processes. An enterprise resource planning (ERP) system is a set of integrated programs that is capable of managing a company's vital business operations for an entire multisite, global organization.

The management information system (MIS) uses the information from a TPS to generate information useful for management decision making. Management information systems produce a variety of reports. Scheduled reports contain prespecified information and are generated regularly. Demand reports are generated only at the request of the user. Exception reports contain listings of items that do not meet a predetermined set of conditions.

A decision support system (DSS) is an organized collection of people, procedures, databases, and devices used to support problem-specific decision making. The DSS differs from an MIS in the support given to users, the decision emphasis, the development and approach, and system components, speed, and output.

The special-purpose business information systems include artificial intelligence systems, expert systems, and virtual reality systems. Artificial intelligence (AI) includes a wide range of systems, in which the computer system takes on the characteristics of human intelligence. Robotics is an area of artificial intelligence in which machines take over complex, routine, or boring tasks, such as welding car frames or assembling computer systems and components. Vision systems allow robots and other devices to have "sight" and to store and process visual images. Natural language processing involves the ability of computers to understand and act on verbal or written commands in English, Spanish, or other human languages. Learning systems give computers the ability to learn from past mistakes or experiences, such as playing games or making business decisions, while neural networks is a branch of artificial intelligence that allows computers to recognize and act on patterns or trends. The expert system (ES) is designed to act as an expert consultant to a user who is seeking advice about a specific situation. Originally, the term *virtual reality* referred to immersive virtual reality, in which the user becomes fully immersed in an artificial, three-dimensional world that is completely generated by a computer. Virtual reality can also refer to applications that are not fully immersive, such as mouse-controlled navigation through a three-dimensional environment on a graphics monitor, stereo viewing from the monitor via stereo glasses, stereo projection systems, and others.

PRINCIPLE *System users, business managers, and information systems professionals must work together to build a successful information system.*

Systems development involves creating or modifying existing business systems. The major steps of this process and their goals include systems investigation (gain a clear understanding of what the problem is); systems analysis (define what the system must do to solve the problem); systems design (determine exactly how the system will work to meet the business needs); systems implementation (create or acquire the various system components defined in the design step); and systems maintenance and review (maintain and then modify the system so that it continues to meet changing business needs).

• • •

Information systems play a fundamental and ever-expanding role in all business organizations. Computer and information systems literacy are prerequisites for numerous job opportunities. Computer literacy (a knowledge of computer systems and equipment) and information systems literacy (a knowledge of how data and information are used by individuals, groups, and organizations) are needed to get the most from any information system. Today, information systems are used in all the functional areas of business, including accounting, finance, sales, marketing, manufacturing, human resource management, and legal information systems. Information systems are also used in every industry, such as airlines, investment firms, banks, transportation companies, publishing companies, healthcare, retail, power management, professional services, and more.

Effective information systems can have a major impact on corporate strategy and organizational success. Businesses around the globe are enjoying better safety and service, greater efficiency and effectiveness, reduced expenses, and improved decision making and control because of information systems. Individuals who can help their businesses realize these benefits will be in demand well into the future.

CHAPTER 1 SELF-ASSESSMENT TEST

The value of information is directly linked to how it helps decision makers achieve the organization's goals.

1. A (An) _____ is a set of interrelated components that collect, manipulate, and disseminate data and information and provide a feedback mechanism to meet an objective.

2. Numbers, letters, and other characters are represented by
   A. image data.
   B. numeric data.
   C. alphanumeric data.
   D. symmetric data.

3. The value of data is measured by the increase in revenues. True False

Models, computers, and information systems are constantly making it possible for organizations to improve the way they conduct business.

4. A (An) _____ is a set of elements or components that interact to accomplish a goal.

5. Which of the following is a way to classify systems?
   A. Permanent—Temporary
   B. Simple—Dynamic
   C. Input—Output
   D. Open—Adaptive

6. Graphs, charts, and figures are examples of physical models. True False

Knowing the potential impact of information systems and having the ability to put this knowledge to work can result in a successful personal career, organizations that reach their goals, and a society with a higher quality of life.

7. A (An) _____ consists of hardware, software, databases, telecommunications, people, and procedures.

8. Computer programs that govern the operation of a computer system are called
   A. feedback.
   B. feedforward.
   C. software.
   D. transaction processing system.

9. Payroll and order processing are examples of a computerized management information system. True False

10. What type of system is used when the problem is complex and the information needed to make the best decision is difficult to obtain?
    A. TPS
    B. MIS
    C. DSS
    D. AI

11. Robotics and neural networks are examples of _____.

System users, business managers, and information systems professionals must work together to build a successful information system.

12. What determines how a new system will work to meet the business needs defined during systems investigation?
    A. Systems implementation
    B. Systems review
    C. Systems development
    D. Systems design

13. _____ literacy is a knowledge of how data and information are used by individuals, groups, and organizations.

Chapter 1 Self-Assessment Test Answers

1.information system, 2. C, 3. False, 4. system, 5. A, 6. False, 7. computer-based information system (CBIS), 8. C, 9. False, 10. C, 11. artificial intelligence, 12. D, 13. information systems

KEY TERMS

REVIEW QUESTIONS

1. What is an information system? What are some of the ways information systems are changing our lives?
2. How would you distinguish data and information? Information and knowledge?
3. Identify at least six characteristics of valuable information.
4. Define the term *system*. What is the difference between a stable system and a dynamic system?
5. What are the components of any information system?
6. How is system performance measured?
7. What is a model? What is the purpose of using a model?
8. What is a computer-based information system? What are its components?
9. Define efficiency and effectiveness as they relate to information systems.
10. Identify three functions of a transaction processing system.
11. What is the difference between an intranet and an extranet?
12. What is a workflow system? How is it different from a transaction processing system?
13. What are the most common types of computer-based information systems used in business organizations today? Give an example of each.
14. Identify three elements of artificial intelligence.
15. What are computer literacy and information systems literacy? Why are they important?
16. What are some of the benefits organizations seek to achieve through using information systems?
17. Identify the five steps in the systems development process and state the goal of each.

DISCUSSION QUESTIONS

1. Why is the study of information systems important to you? What do you hope to learn from this course to make it worthwhile?
2. How could a workflow system simplify the approval of travel expense reports? What are the benefits of such a system?
3. What is the difference between an MIS and a DSS?
4. Suppose you are a teacher assigned the task of describing the learning processes of preschool children. Why would you want to build a model of their learning processes? What kinds of models would you create? Why might you create more than one type of model?

5. Describe the "ideal" automated auto license plate renewal system for the drivers in your state. Describe the input, processing, output, and feedback associated with this system.

6. How is it that useful information can vary widely from the quality attributes of valuable information?

7. Discuss the potential use of virtual reality to enhance the learning experience for new automobile drivers.

How might such a system operate? What are the benefits and potential disadvantages of such a system?

8. Discuss how information systems are linked to the business objectives of an organization.

9. What are your career goals and how can a computer-based information system be used to achieve them?

PROBLEM-SOLVING EXERCISES

1. Prepare a data disk and a backup disk for the problem-solving exercises and other computer-based assignments you will complete in this class. Create one directory for each chapter in the textbook (you should have 14 directories). As you work through the problem-solving exercises and complete other work using the computer, save your assignments for each chapter in the appropriate directory. On the label of each disk, be sure to include your name, course, and section. On one disk write "Working Copy"; write "Backup" on the other.

2. Search through several business magazines (*Business Week, Computerworld, PC Week*, etc.) for a recent article that discusses the use of information technology to deliver significant business benefits to an organization. Now use other resources to find additional information about the same organization (Reader's Guide to Periodical Literature, on-line search capabilities available at your school's library, the company's public relations department, Web pages on the Internet, etc.). Use word processing software to prepare a one-page summary of

the different resources you tried and their ease of use and effectiveness.

3. Create a table that lists all the courses you are taking in the first column. The other columns of the table should be the weeks of the semester or quarter, such as Week 1, Week 2, etc. The body of the table should contain the actual assignments, quizzes, exams, the final exam, etc. for each course. Place the table into a database and print the results. Create a table in the database for the first three weeks of class and print the results. Create another table in the database for your two hardest classes for all weeks and print the results.

4. Do some research to obtain estimates of the rate of growth of the Internet (e.g., number of computers connected to the Internet, number of Internet Web sites, etc.). Use the plotting capabilities of your spreadsheet or graphics software to produce a bar chart of that growth over a number of years. Share your findings with the class.

TEAM ACTIVITIES

1. Before you can do a team activity, you need a team! The class members may self-select their teams, or the instructor may assign members to groups. Once your group has been formed, meet and introduce yourselves to each other. You will need to find out the first name, hometown, major,

and e-mail address and phone number of each member. Find out one interesting fact about each member of your team as well. Come up with a name for your team. Put the information on each team member into a database and print enough copies for each team member and your instructor.

2. With the other members of your group, use word processing software to write a one-page summary of what your team hopes to gain from this course and what you are willing to do to accomplish these goals. Send the report to your instructor via e-mail.

WEB EXERCISES

1. Throughout this book, you will see how the Internet provides a vast amount of information to individuals and organizations. We will stress the World Wide Web, or simply the Web, which is an important part of the Internet. Most large universities and organizations have an address on the Internet, called a Web site or home page. The address of the Web site for this publisher is http://www.course.com. You can gain access to the Internet through a browser, such as Internet Explorer or Netscape. Using an Internet browser, go to the Web site for this publisher. What did you find? Try to obtain information on this book. You may be asked to develop a report or send an e-mail message to your instructor about what you found.

2. Go to an Internet search engine, such as www.yahoo.com, and search for information about a company, including its Web site. Write a report that summarizes the size of the company, number of employees, its products, the location of its headquarters, and its profits (or losses) for last year. Would you want to work for this company?

3. Find a Web site whose subject is your favorite movie actor or actress, your hobby, or your profession. After checking out this Web site, find at least two other sites containing information on the same subject.

CASES

CASE 1

States Vie for #1 Technology Rating

In this chapter you have learned that information systems play a fundamental and ever-expanding role in all business organizations. City and state governments have also recognized the benefits of investing in modern information systems, and many are making it a number-one priority. Illinois is a good, if not extreme, example.

In 1998 Illinois was ranked forty-ninth in the Digital State Survey conducted by the Center for Digital Government. The Center for Digital Government is a national research and advisory institute providing government, industry, and education leaders with decision support, research, and educational resources to help them effectively incorporate new technologies in the twenty-first century. The survey measures the extent to which state governments utilize information systems to streamline government processes—a concept commonly known as e-government.

Under the leadership of Governor George H. Ryan, Illinois took the technological bull by the horns and worked to build a high-tech infrastructure that attracts and encourages cutting-edge research, new applications, and marketable products. In 1999 Illinois established the Illinois Technology Office, responsible for managing and coordinating technology initiatives in various areas such as education and research. An executive order was instated to establish VentureTECH, a five-year, $2 billion comprehensive strategy for investing state resources in education and advanced research and development, health sciences and biotechnology, and cutting-edge information technology programs. In just two years, Illinois moved from forty-ninth in the Digital State Survey to fourth, earning the distinction of being the "most improved" state. In 2001 Illinois moved to first place, tied with Kansas. Not surprisingly, in the same year, Chicago was ranked number two of all cities with populations over 250,000 in the Digital Cities survey conducted by the same organization. Honolulu was number one.

Illinois and Chicago are not alone in their efforts. At the close of the twentieth century, most cities and states were working frantically to build a more efficient government through technology. What specifically can cities and states improve on using technology? The Center for Digital Government examines eight areas in their survey of states:

1. Law Enforcement and the Courts: The utilization of digital technologies by the judicial system, including on-line access to court opinions, the use of digital communications by police agencies, and the availability of digital signature capability for contracts and filings.
2. Social Services: The availability of on-line information regarding program eligibility and application procedures and the application of digital technologies such as electronic benefit transfer (EBT) systems and smart cards for benefits delivery.
3. Electronic Commerce/Business Regulation: The availability of regulations, forms, and on-line assistance, and the ability to submit required paperwork using the Internet.
4. Taxation/Revenue: The ability of taxpayers to obtain information, submit returns, and correspond with revenue authorities on-line, and the ability of states to use digital technologies to store and retrieve taxpayer information.
5. Digital Democracy: The application of digital technologies to permit Internet access to laws, government officials, and other sources of information on the functions of various branches of government.
6. Management/Administration: The adoption of new information technologies with applicability across programs and agencies, and investment in long-term information technology infrastructure.
7. Education: The utilization of digital technologies for education purposes, including providing students and teachers with computers and access to the Internet and administrative functions like admissions, financial aid, and course registration.
8. Geographic Information Systems/Transportation: The utilization of digital technologies as a management tool for functions such as making decisions in economic development, law enforcement and firefighting, and a mapping tool for storing, analyzing, and printing data.

When ranking cities, the Center for Digital Government examined how city governments have progressed in adopting and utilizing digital technologies to improve the delivery of services to their citizens.

Historically, governments have been criticized as being slow and inefficient bureaucracies. What has brought on this recent interest in lean and efficient government operations? Perhaps it's smarter, more technically savvy politicians, or could it be that politicians are feeling the pressure from a more technically savvy populace?

Discussion Questions

1. What city or state policies or procedures have frustrated you—renewing a driver's license, paying for a traffic violation? Propose an e-government solution for these problems.
2. What types of information and services would you expect to see on your hometown's Web site?

Critical Thinking Questions

3. What effect does e-government have on state employees? Do you think they are pleased by development in these areas?
4. How do improvements in the eight areas listed assist us in our development as a society?

Sources: "Illinois Moves from 49th to 1st in 'Digital State' Rankings," *Illinois Government News Network*, January 3, 2002; "Honolulu Ranked as Nation's Top Digital City," *Honolulu Government News*, November 8, 2001; http://www.centerdigitalgov.com.

CASE 2

Technology in Abundance at the 2002 Winter Olympics

With 3,500 athletes and officials, 10,000 media personnel, and 26,000 Olympic staff and volunteers, the 2002 Winter Olympics in Salt Lake City was the biggest sporting and entertainment event of the year. In this day and age, information technology is critical to the success of any Olympic Games. So, behind the scenes technology partners devote extensive resources, time, and effort to deliver the technology solutions that make the games possible.

IKANO Communications, based in Salt Lake City, Utah, was the Official Data Networking Services Supplier for the Olympic Winter Games and Paralympic Winter Games of 2002. As discussed in the chapter, networks are used to connect computers and computer equipment in a building, around the country, or around the world to enable electronic communications. IKANO provided all data-networking services and accompanying support for the 2002 Olympic Winter Games. The high-performance network supplied data and information from Olympic events to staff, athletes, officials, the host broadcaster, and the Olympic Organizing Committee Web site.

Information distributed over the network included event schedules and real-time results, athlete profiles, weather, and performance data. "The data network services provided by IKANO are a critical component in our overall Games technology solution," said Mitt Romney, SLOC president and CEO. Among the several software systems that delivered information to each of the many Olympic venues were On Venue Results (OVR) and the Commentator Information System (CIS). On Venue Results provided real-time event scores and results, while the Commentator Information System provided commentators with a database of useful information about the athletes.

Installing the network to support fourteen venues scattered around Utah was in itself a challenge for IKANO. Once installed, IKANO turned its attention to keeping the system up and running and safeguarding it from security breaches. A secondary backup network was installed for use in case of a primary network failure. The backup network could be used for several minutes while technicians repaired the primary network. IKANO staged several network disasters prior to the games to train the staff for any conceivable emergency. IKANO was also understandably concerned about hackers breaking into the network and wreaking havoc. As a precaution against such attacks, IKANO installed several network security systems and armed its network with the vigilant eyes of an army of network specialists at 12 network management stations around the games.

In addition to the internal network IKANO provided to allow communication between the Olympic sites, a high-speed network was installed by Qwest to provide video, voice, and data communications locally and between the games and the outside world. Qwest provided 650 miles of fiber-optic cable to support a data transfer rate of 388 trillion bits of data per second. This high-speed network, supported by 158 Qwest trucks and 600 technicians, provided support for 483,840 phone and fax lines, 400 video circuits, and more than 1,000 high-speed connections to the Internet. It is through the Qwest network that the Olympics were able to be broadcast over stations such as NBC for audiences around the world.

The 2002 Winter Olympics offered unique challenges for information and telecommunication systems. A high-visibility event such as this created a high-pressure environment where network problems would not be tolerated. By seeing what goes on behind the scenes of the games, you can get a better understanding of how much our world depends on information systems and telecommunications and how vulnerable these systems can be.

Discussion Questions
1. What other types of events might be equally demanding to data network providers?
2. Who do you think got a better view of the 2002 Winter Olympics: those who were at the games or those who watched from home? Where would you rather be?

Critical Thinking Questions
3. What types of security issues do you think were major concerns for IKANO at the 2002 Winter Olympics? Prioritize your list.
4. How has information management changed at the Olympics of the past 20 years? List several areas from your observation of events.

Sources: "Olympic Technology: A Behind-the-Scenes Look," *PR Newswire*, January 9, 2002; "Final Olympic Technology Rehearsal Tests Games-Time Readiness of the Olympic Data Network," http://www.ikano.com, December 28, 2001; "Ride the Light to the 2002 Olympic Winter Games," http://www.qwest.com.

Expert Systems Simplify Life for Automotive Manufacturers

Infomedia is a successful Australian software developer that specializes in designing information systems for the automotive industry. The company's primary product, Microcat, has developed the reputation of being the best electronic parts catalogue (EPC) for the global automotive industry. It is shipped to more than 32,000 subscribers in over 100 countries and 22 languages. Infomedia produces versions of its electronic parts catalogue products for the majority of leading car manufacturers in Australia, including Daewoo, Daihatsu, Ford, General Motors Holden, Honda, Hyundai, Isuzu, Mitsubishi, Suzuki, and Toyota. International versions have been produced for Daihatsu, Ford, Hyundai, and Land Rover.

Microcat is popular in part because it adds "intelligence" and communications to the cataloguing methods, transforming them from simple table/code reference tools to interactive technical selling systems. By programming expert knowledge into the system, users can quickly and accurately identify repair jobs, and identify and request related parts and fluids for the required repairs. Infomedia works with the manufacturer to custom-design a Microcat system for cataloguing automotive parts. For example, in the case of a European automaker, details on more than 500,000 parts and 19.5 million vehicles

are stored and processed within Microcat. An examination of Infomedia's development process provides examples of many of the concepts discussed in this chapter.

Step 1—Preparation: Manufacturers who choose to implement the Microcat electronic parts sales and cataloguing system for their replacement parts operations begin with a thorough analysis of their situation and requirements with a Microcat data engineer. This engineer is assigned to work with the manufacturer's implementation team. This team generally includes a representative from cataloguing, replacement parts marketing, and information services. During this early stage, business rules, logic algorithms, interpretation rules, and data parsing gathered from the human experts are programmed into the system and tested on a reasonably sized data sample. After much tweaking, when the system provides intelligent and accurate responses to parts inquiries, it is time to load in the parts data.

Step 2—Source Material: To create the initial electronic catalogue, data is gathered from a number of original sources including electronic data and images, microfiche images, paper drawings, and photographs. The Microcat production team devises a "harmonizing" approach to represent all of the manufacturer's data in a consistent and intuitive fashion.

Step 3—Preprocessing: With the expert system and parts-related data in hand, Infomedia's production processing team goes to work preparing the first Microcat master catalogue implementation of the manufacturer's product lines. At this stage, Infomedia's preprocessing systems integrate the two—expert system and parts data—into one uniform digital continuum. Secondary preprocessing then optimizes all the data for speed and encrypts it for security. This step can take anywhere from a few weeks to several months.

Step 4—Preproduction: Before a broad release to parts centers, a "beta," or trial, version of the new Microcat system is prepared. Infomedia trainers work with the manufacturer and selected parts center staff to test and fine-tune the beta system. User recommendations and observations are given to the Microcat data engineers to consider and incorporate, where appropriate.

Step 5—Communications: Microcat engineers work to integrate the Microcat system with existing systems in the corporation. From Microcat, employees can access customer information, pricing and stock availability, and other business-related data. This approach eliminates much time-consuming duplication of effort.

Step 6—Periodic Update Process: Once the product is completed, it is delivered to each of the manufacturers' facilities on CD/DVD-ROM. The software is updated to reflect changes in prices, inventory, parts specifications, and other fluctuating variables at agreed-upon time intervals. For example, a manufacturer may receive an updated CD every month. The on-line and network versions of Microcat can update information in real time.

Microcat's development process is typical of the systems development process used in most information systems. The goals of the Microcat system are also typical of most information systems:

- Improve access to the data to improve customer service
- Use Microcat's interpretation capabilities to rely less on an individual's component knowledge

- Reduce chances of inaccurate or out-of-date data
- Increase sales of genuine parts by freeing up more time for sales staff to sell
- Reduce credit returns caused by inaccuracy of spare parts selection
- Interface with other computer systems
- Provide the opportunity to add specific or local knowledge to the data
- Simplify distribution of all the manufacturer's parts information
- Take advantage of technological advancements as they can be introduced.

Even if automotive parts are not your cup of tea, there is a lot to be learned about information systems from this very successful Australian company.

Discussion Questions

1. What advantages does an automotive manufacturer who uses Microcat have over a manufacturer who has developed and maintains its own electronic parts catalog?
2. What types of products might Microcat be enhanced to handle? If you owned Microcat, what industry would you pursue next?

Critical Thinking Questions

3. Microcat is an expert system because it includes the knowledge, logic, algorithms, and interpretation rules of human experts. How will the implementation of this tool affect the jobs of those human experts who helped design it?
4. How does Microcat allow automotive manufacturers to be more effective and efficient?

Sources: "Smart Cars Put Infomedia in Fast Lane," *australianIT.com*, January 4, 2002; "Infomedia and Toyota Motor Europe Sign Five-Year Data License Agreement," *Business Wire*, January 3, 2002; "Infomedia Releases the Next Generation in Automotive Dealership Management Systems," *Business Wire*, October 18, 2001; http://www.infomedia.com.au.

NOTES

Sources for the opening vignette on page 3: "Taking the Drugstore to the Customers: A New Automated Pharmacy; Merck-Medco's Opened N.J. Facility to Dispense Mail-Order Prescriptions," *Investor's Business Daily,* January 14, 2002; "'Pharmacy for the Future' Officially Opens in Willingboro, N.J.," *The Philadelphia Inquirer,* November 13, 2001; "Merck-Medco Discovers Prescription for Success While Others Struggle," *The Associated Press State & Local Wire,* November 28, 2001; Merck-Medco Web site, http://www.merck-medco.com.

1. Disabantino, Jennifer, "GM's OnStar Puts Stock in Service," *Computerworld,* February 26, 2001, p. 48.
2. Schwartz, Mathew, "The Care and Keeping of Customers," *Computerworld,* January 8, 2001, p. 58.
3. Rosencrance, Linda, "Data Warehouse Gives Trimac Information for the Long Haul," *Computerworld,* July 2, 2001, p. 47.
4. Haddad, Charles, "How UPS Delivered Through the Disaster," *Business Week,* October 1, 2001, p. 66.
5. Haddad, Charles, "Ground Wars," *Business Week,* May 21, 2001, p. 64.
6. Sery, Slava et al., "Optimization Models for Restructuring BASF," *Interfaces,* May–June 2001, p. 55.
7. Vranica, Suzanne, "Web Site Seeks to Turn Data into Dollars," *The Wall Street Journal,* July 27, 2001, p. B8.
8. Wright, Allison, "Custom Cars for Every Driver," *Computerworld,* March 12, 2001, p. 60.
9. Vijayan, J. "Procurement Network Harnesses Buying Power," *Computerworld,* June 4, 2001, p. 33.
10. Gladwin, Lee, "Users Extend Use of Web Portals to Supply Chain for Materials Procurement," *Computerworld,* June 11, 2001, p. 7.
11. Francis, Theo, "T. Rowe Price Rolls Out Voice Response," *The Wall Street Journal,* May 30, 2001, p. C21.
12. Mearian, Lucas, "Wal-Mart Deal Boosts IBM," *Computerworld,* October 22, 2001, p. 19.
13. Collett, Stacy, "Retailers, Travel Companies Deploy Thousands of Kiosks," *Computerworld,* August 6, 2001, p. 18.
14. Greene, Jay, "Microsoft: How It Became Stronger Than Ever," *Business Week,* June 4, 2001, p. 75.
15. Verton, Dan, "Oracle Launches Technology Offensive," *Computerworld,* June 25, 2001, p. 8.
16. Lais, Sami, "Satellites Link Bob Evans Farms," *Computerworld,* July 2, 2001, p. 51.
17. Berman, Dennis, "Disaster Gives New Life to Wireless Telecom Firms," *The Wall Street Journal,* October 3, 2001, p. B1.
18. Stevens, Larry, "The Best of the Web: Seven That Made It Work for Them," *PC Magazine,* September 4, 2001, p. 5.
19. Brewin, Bob, "Airlines Take Internet to the Skies," *Computerworld,* June 18, 2001, p. 10.
20. Randall, Neil, "Instant Intranet," PC Magazine, August 2001, p. 76.
21. Meehan, Michael, "Execs: Building Consensus Is Biggest B2B Challenge," *Computerworld,* June 18, 2001, p. 14.
22. Stevens, Laura, "Job Hunting," *Forbes,* June 25, 2001, p. 76.
23. Mearian, Lucas, "Bucking the Trend, Bank to Hire 600 IT Workers," *Computerworld,* December 3, 2001, p. 61.
24. Subramani, Mani and Walden, Eric, "The Impact of E-Commerce Announcements on the Market Value of Firms," *Information Systems Research,* June 2001, p. 135.
25. "Amazon.com Looks for Sales Boost in Circuit City," *Information Week Online,* August 21, 2001.
26. Mearian, Lucas, "Nasdaq Launches Revised Order System," *Computerworld,* July 9, 2001, p. 8.
27. Konicki, Steve, "DaimlerChrysler Spends $3B on Parts Through Covisint," *Information Week,* May 15, 2001, p. 1.
28. Songini, Marc, "PeopleSoft Kicks Off Web-Based CRM Suite," *Computerworld,* June 4, 2001, p. 14.
29. Nash, Kim, "Casinos Hit Jackpot with Customer Data," *Computerworld,* July 2, 2001, p. 16.
30. Verton, Dan, "Chevron Tightens Control on Net Access," *Computerworld,* July 30, 2001, p. 6.
31. Landro, Laura, "New Medical Software Gives Physicians Clues When They're Stumped," *The Wall Street Journal,* June 29, 2001, p. B1.
32. "Royal Caribbean Launches an IT Upgrade," *Information Week Online,* July 6, 2001.

Information Systems in Organizations

PRINCIPLES	LEARNING OBJECTIVES
The use of information systems to add value to the organization is strongly influenced by organizational structure, culture, and change.	• Identify the value-added processes in the supply chain and describe the role of information systems within them. • Provide a clear definition of the terms *organizational structure*, *culture*, and *change* and discuss how they affect the implementation of information systems.
Because information systems are so important, businesses need to be sure that improvements or completely new systems help lower costs, increase profits, improve service, or achieve a competitive advantage.	• Identify some of the strategies employed to lower costs or improve service. • Define the term *competitive advantage* and discuss how organizations are using information systems to gain such an advantage. • Discuss how organizations justify the need for information systems.
Information systems personnel are the key to unlocking the potential of any new or modified system.	• Define the types of roles, functions, and careers available in information systems.

Seeking Competitive Advantage with a Major Information System Upgrade

Businesses are learning that by improving the flow of information within their organization, they can become more efficient and effective. So to gain a competitive edge, businesses commit a considerable portion of their overall budget to maintain and improve their information systems. This reliance on information systems is particularly true in the banking industry, where the business depends entirely on the accuracy and efficient management of information.

Despite economic swings and continued mergers and acquisitions, banks continue to increase information system spending according to a poll of CIOs at 25 of the 100 largest U.S. banks. The report states that information system budgets account for 20 to 25 percent of their overall budgets. "In part, banks have little choice in this; a very large proportion of their spending is on maintaining existing infrastructure," said Octavio Marenzi, the report's author. "Banks are also increasingly viewing technology as a competitive differentiator, as a tool with which to wrest market share from their competition."

Maintaining and improving information systems in large organizations takes considerable time and effort. And that effort becomes increasingly complex as an organization expands into other countries or operates globally. Citibank has firsthand knowledge of the problems that can develop when working globally. The company now manages corporate offices in 100 countries.

In the 1970s Citibank was growing rapidly and expanding around the world. Citibank provided its new overseas branches with an information system that could be adapted to accommodate local currencies, regulatory rules, and business processes. Tailoring the system was convenient for quickly getting the overseas branches up and running, but the changes ultimately resulted in dozens of country-specific proprietary systems with little in common—an arrangement that Citibank could no longer support.

In what analysts estimate is a project worth between $100 and $500 million, Citibank is replacing its decades-old overseas corporate banking systems with FLEXCUBE, an enterprise banking software suite it purchased from India-based technology vendor, I-Flex Solutions Ltd. FLEXCUBE will provide a standard interface to all users. "As a global financial services company which is using technology to improve productivity and lower operating costs, the use of FLEXCUBE will allow us to gain efficiencies by standardizing our operating environment," said Bob Druskin, chief operations and technology officer at Citibank's New York–based parent, Citigroup.

FLEXCUBE allows Citibank to simply change parameters in the software to incorporate a particular country's language, government regulations, and currency conversions. It automates the general ledger, as well as customer accounting, deposits and withdrawals, and interest on accounts, among other services.

In spite of the benefits, making a transition to a new system generates a lot of tension within an organization. The 250 Citibank systems staff dedicated to the project will no doubt experience years of restless sleep. Started in 2001, the project is expected to be complete sometime in 2004. Citibank information specialists have designed a battery of tests to be continuously run on the new system, as it is gradually brought on-line in stages. The tests will insure the system is running as expected and correctly making the connections over the network. The new system will be run in parallel to the old system for a while so that if there are problems, the old system is available as a backup.

As the new system is implemented, Citibank is taking an opportunity to examine and improve its basic business processes. Once the new system is in place, these processes will be standardized across the entire organization. As a result, the transition to FLEXCUBE has created some rivalries within the organization. Local "operations people defend the way they do things. We keep saying, 'Yes, we know these countries are different, but they're more similar than different,' " stated Jeff Berg, executive director of program management at Citigroup. "In terms of adopting the Citibank culture, these banks aren't as different as we think."

Among the many advantages of FLEXCUBE, the new efficiencies will enable Citibank to downsize its European data centers from 18 to about 4. Citibank is also saving money by outsourcing this project to an Indian software company. The bank has charted an 18-month return on investment with this project.

As you read this chapter, consider the following:
- ◉ How will Citibank's new information system improve the decision-making ability of its upper management?
- ◉ What types of resistance might this system meet from the employees of Citibank? What can Citibank do to gain acceptance among employees and make the transition go more smoothly?

Technology's impact on business is growing steadily. Once used to automate manual processes, technology has now transformed the nature of work and the shape of organizations themselves. During the late 1960s and early 1970s, many computerized information systems were developed to provide reports for business decision makers. The information in these reports helped managers monitor and control business processes and operations. For example, reports that listed the quantity of each inventory item in stock could be used to monitor inventory levels. Unfortunately, many of these early computer systems did not take the overall goals of the organization and managerial problem-solving styles into consideration. Some decision makers wanted detailed inventory reports of all the items, and others wanted a list of inventory items only when the number on hand was very low. Even more important, these early systems were not developed as part of the business process itself. As a result, many of the early systems failed or were not utilized to their potential. As seen in the opening vignette, businesses such as Citibank now recognize that both important organizational concepts and processes must be considered and supported by effective information systems. Citibank's FLEXCUBE enterprise banking suite allows the company to make rapid changes to its systems to react to a changing environment.

◉ ORGANIZATIONS AND INFORMATION SYSTEMS

organization
a formal collection of people and other resources established to accomplish a set of goals

An **organization** is a formal collection of people and other resources established to accomplish a set of goals. The primary goal of a for-profit organization is to maximize shareholder value, often measured by the price of the company stock. Nonprofit organizations include social groups, religious groups, universities, and other organizations that do not have profit as the primary goal.

An organization is a system. Money, people, materials, machines and equipment, data, information, and decisions are constantly in use in any organization. As shown in Figure 2.1, resources such as materials, people, and money are input to the organizational system from the environment, go through a transformation mechanism, and are output to the environment. The outputs from the transformation mechanism are usually goods or services. The goods or services produced by the organization are of higher relative value than the inputs alone. Through adding value or worth, organizations attempt to achieve their goals.

How does this increase in value occur? Within the transformation mechanism, various subsystems contain processes that help turn specific inputs into goods or services of increasing value. These value-added processes increase the relative worth of the combined inputs on their way to becoming final outputs of the organization. Let us reconsider our simple car wash example from Chapter 1 (Figure 1.4). The first value-added process might be identified as washing the car. The output of this system—a clean, but wet, car—is worth more than the mere collection of ingredients (soap and water), as evidenced by the popularity

FIGURE 2.1

A General Model of an Organization

Information systems support and work within all parts of an organizational process. Although not shown in this simple model, input to the process subsystem can come from internal and external sources. Just prior to entering the subsystem, data is external. Once it enters the subsystem, it becomes internal. Likewise, goods and services can be output to either internal or external systems.

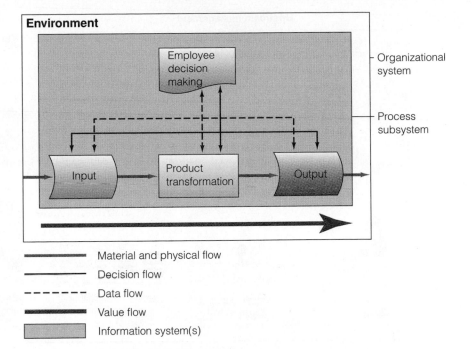

Environment

Employee decision making

Input → Product transformation → Output

Organizational system

Process subsystem

———— Material and physical flow
———— Decision flow
– – – – Data flow
━━━━ Value flow
▢ Information system(s)

---

value chain

a series (chain) of activities that includes inbound logistics, warehouse and storage, production, finished product storage, outbound logistics, marketing and sales, and customer service

---

Information systems are an integral part of the value-added process of American Blind and Wallpaper Factory. The Web site allows customers to purchase a wide variety of home decorating items—from window treatments to linens to wallpaper.

of automatic car washes. Consumers are willing to pay more for the skill, knowledge, time, and energy required to wash their car. The second value-added process can be identified as drying—the transformation of the wet car into a dry one with no water spotting. Again, consumers are willing to pay more for the additional skill, knowledge, time, and energy required to accomplish this transformation. In general, organizations establish these value-added processes to achieve their goals by exploiting opportunities and solving problems.

All business organizations contain a number of value-added processes. Providing value to a stakeholder—customer, supplier, manager, or employee—is the primary goal of any organization. The value chain, first described by Michael Porter in a 1985 *Harvard Business Review* article, is a concept that reveals how organizations can add value to their products and services. The **value chain** is a series (chain) of activities that includes inbound logistics, warehouse and storage, production, finished product storage, outbound logistics, marketing and sales, and customer service (Figure 2.2). Each of these activities is investigated to determine what can be done to increase the value perceived by a customer. Managing these activities is often called *supply chain management.*[1] Depending on the customer, value may mean lower price, better service, higher quality, or uniqueness of product.[2] The value comes from the skill, knowledge, time, and energy invested by the company. By adding a significant amount of value to their products and services, companies will ensure further organizational success.

What role does an information system play in these value-added processes?

Upstream management

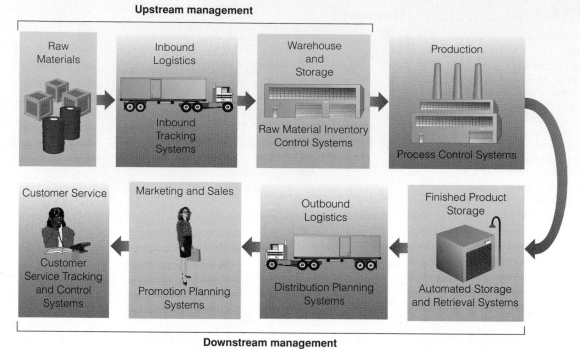

Downstream management

FIGURE 2 ⊙ 2

*The Value Chain of a
Manufacturing Company*

*The management of raw materials,
inbound logistics, and warehouse
and storage facilities is called*
upstream management, *and the
management of finished product
storage, outbound logistics, marketing and sales, and customer service
is called* downstream management.

A traditional view of information systems holds that they are used by organizations to control and monitor value-added processes to ensure effectiveness and
efficiency. An information system can turn feedback from the value-added
process subsystems into more meaningful information for employees' use
within an organization. This information might summarize the performance
of the systems and be used as the basis for changing the way the system operates. Such changes could involve using different raw materials (inputs),
designing new assembly-line procedures (product transformation), or developing new products and services (outputs). In this view, the information system is external to the process and serves to monitor or control it.

A more contemporary view, however, holds that information systems are
often so intimately intertwined with the underlying value-added process that
they are best considered *part of* the process itself. From this perspective, the
information system is internal to and plays an integral role in the process,
whether providing input, aiding product transformation, or producing output.
Consider a phone directory business that creates phone books for international
corporations. A corporate customer requests a phone directory listing all steel
suppliers in Western Europe. Using its information system, the directory business
can sort files to find the suppliers' names and phone numbers and organize them
into an alphabetical list. The information system itself is an integral part of this
process. It does not just monitor the process externally but works as part of the
process to transform raw data into a product. In this example, the information
system turns raw data input (names and phone numbers) into a salable output (a
phone directory). The same system might also provide the input (data files) and
output (printed pages for the directory).

The latter view brings with it a new perspective on how and why information
systems can be used in business. Rather than searching to understand the value-
added process independently of information systems, we consider the potential
role of information systems within the process itself, often leading to the discovery of new and better ways to accomplish the process. Thus, the way an organization views the role of information systems will influence the ways it
accomplishes its value-added processes.

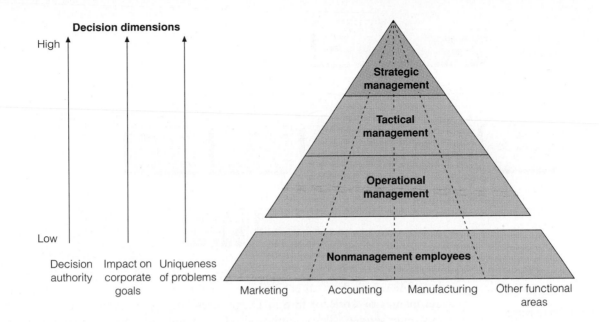

FIGURE 2.3

A simplified model of the organization, showing the managerial pyramid from top-level managers to nonmanagement employees.

organizational structure
organizational subunits and the way they are related to the overall organization

traditional organizational structure
organizational structure in which major department heads report to a president or top-level manager

flat organizational structure
organizational structure with a reduced number of management layers

empowerment
giving employees and their managers more responsibility and authority to make decisions, take certain actions, and have more control over their jobs

ORGANIZATIONAL STRUCTURE

Organizational structure refers to organizational subunits and the way they relate to the overall organization. Depending on the goals of the organization and its approach to management, a number of structures can be used. An organization's structure can affect how information systems are viewed and used. Although there are many possibilities, organizational structure typically falls into one of these categories: traditional, project, team, multidimensional, or virtual.

Traditional Organizational Structure

In the type of structure known as **traditional organizational structure**, also called a *hierarchical structure,* a managerial pyramid shows the hierarchy of decision making and authority from the strategic management to operational management and nonmanagement employees. The strategic level, including the president of the company and vice presidents, has a higher degree of decision authority, more impact on corporate goals, and more unique and one-of-a-kind problems to solve (see Figure 2.3). In most cases, major department heads report to a president or top-level manager. The major departments are usually divided according to function and can include marketing, production, information systems, finance and accounting, research and development, and so on (see Figure 2.4). The positions or departments that are directly associated with making, packing, or shipping goods are called *line positions.* A production supervisor who reports to a vice president of production, for example, works in a line position. Other positions may not be directly involved with the formal chain of command but may assist a department or area. These are *staff positions,* such as a legal counsel reporting to the president.

Today, the trend is to reduce the number of management levels, or layers, in the traditional organizational structure. A structure with a reduced number of management layers, often called a **flat organizational structure**, empowers employees at lower levels to make decisions and solve problems without needing permission from midlevel managers. **Empowerment** gives employees and their managers more responsibility and authority to make decisions, take certain actions, and in general have more control over their jobs. For example, an empowered salesclerk would be able to respond to certain customer requests or problems without needing permission from a supervisor. On a factory floor, empowerment can mean that an assembly-line worker has the ability to stop the production line to correct a problem or defect before the product is passed to the next station.

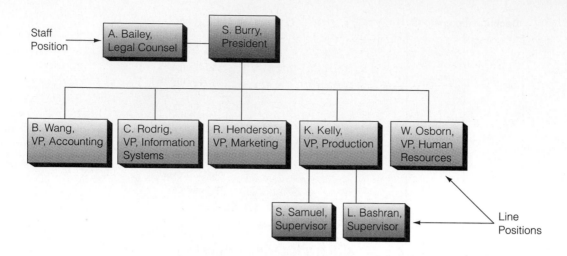

FIGURE 2.4

A Traditional Organizational Structure

U.S. Microbics, Inc., an environmental solutions provider named one of the best companies to work for in San Diego, uses employee empowerment to focus on customer service.[3] The company also has an employee profit sharing program and a belief that "none of us is as smart as all of us."

Information systems can be a key element in empowering employees. Often, information systems make empowerment possible by providing information directly to employees at lower levels of the hierarchy. The employees may also be empowered to develop or use their own personal information systems, such as a simple forecasting model or spreadsheet.

Project Organizational Structure

project organizational structure
structure centered on major products or services

A **project organizational structure** is centered on major products or services. For example, in a manufacturing firm that produces baby food and other baby products, each line is produced by a separate unit. Traditional functions like marketing, finance, and production are positioned within these major units (see Figure 2.5). Many project teams are temporary—when the project is complete, the members go on to new teams formed for another project.

Team Organizational Structure

team organizational structure
structure centered on work teams or groups

The **team organizational structure** is centered on work teams or groups. In some cases, these teams are small; in others, they are very large. Typically, each team has a leader who reports to an upper-level manager in the organization. Depending on the tasks being performed, the team can be either temporary or permanent.

FIGURE 2.5

A Project Organizational Structure

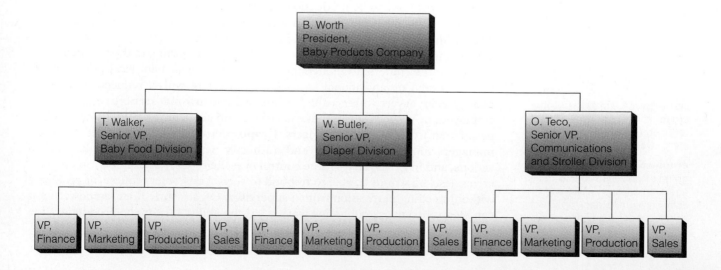

Multidimensional Organizational Structure

A **multidimensional organizational structure**, also called a *matrix organizational structure*, may incorporate several structures at the same time. For example, an organization might have both traditional functional areas and major project units. When diagrammed, this structure forms a matrix, or grid (see Figure 2.6).

One advantage of the multidimensional organizational structure is the ability to simultaneously stress both traditional corporate areas and important product lines. A potential disadvantage is multiple lines of authority. Employees have two bosses or supervisors: one functional boss and one project boss. As a result, conflicts may occur when one boss wants one thing and the other boss wants something else. For example, the functional boss might want the employee to work on a new product in the next two days, while the project boss might want the employee to fly to a two-day meeting. Obviously, the employee cannot do both. One way to resolve this problem is to give one boss priority if there are problems or conflicts.

Virtual Organizational Structure

A **virtual organizational structure** employs individuals, groups, or complete business units in geographically dispersed areas.[4] The individuals, groups, or complete business units can involve people in different countries operating in different time zones. The people may never meet face to face in the same room, which explains the use of the word *virtual*. Despite this separation, they can collaborate on any aspect of a project, such as supplying raw materials, producing goods and services, and delivering goods and services to the marketplace.[5] In some cases, a virtual organization is temporary, lasting only a few weeks or months. In others, it can last for years or decades.

A virtual organizational structure can be used within a firm. GTECH, a company that makes lottery equipment, used a virtual organizational structure with dispersed workers who had distinct skills and abilities.[6] GTECH saved $3 million by using its virtual structure—not having to relocate workers and being able to search anywhere for the best talent. Often, however, virtual structures are formed with individuals or groups outside a company. G5 Technologies, for example, formed an organization called the Virtual Corporate Management System to assemble several small manufacturing businesses to compete with larger businesses.[7] The idea for this virtual organization came from a research project at Lehigh University, and it benefited participants. One small business that manufactured military communications equipment saw a dramatic increase in revenues when it joined the Virtual Corporate Management System. In addition to reducing costs or increasing revenues, a virtual organizational structure can provide an extra level of security.[8] Many companies are now dispersing employees and using a virtual structure in case of a terrorist attack or a disaster. If a disaster strikes at the primary location, the company still has sufficient employees at other locations to keep the company running.

FIGURE 2.6

A Multidimensional Organizational Structure

Employees in each group may have two bosses—a project boss and a functional boss.

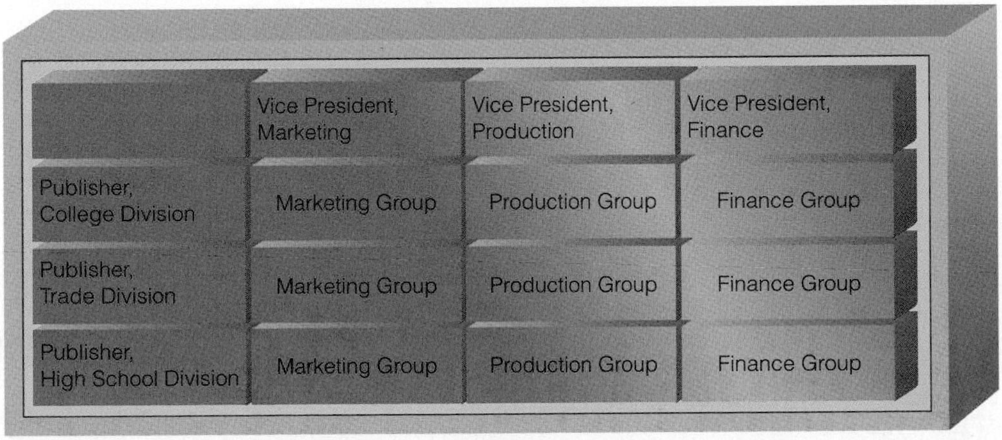

	Vice President, Marketing	Vice President, Production	Vice President, Finance
Publisher, College Division	Marketing Group	Production Group	Finance Group
Publisher, Trade Division	Marketing Group	Production Group	Finance Group
Publisher, High School Division	Marketing Group	Production Group	Finance Group

There are a number of keys to successful virtual organization structures. One strategy is to have in-house employees concentrate on the firm's core businesses and use virtual employees, groups, or businesses to do everything else. Using information systems to coordinate the activities of a virtual structure is essential. According to the manager of collaborative systems for Bechtel in Houston, Texas, "We use tools like e-mail, Outlook scheduling, and videoconferencing to keep projects going." Even with sophisticated information systems tools, face-to-face meetings are usually needed, especially at the beginning of new projects.[9]

ORGANIZATIONAL CULTURE AND CHANGE

culture
set of major understandings and assumptions shared by a group

organizational culture
the major understandings and assumptions for a business, a corporation, or an organization

Culture is a set of major understandings and assumptions shared by a group, for example, within an ethnic group or a country. **Organizational culture** consists of the major understandings and assumptions for a business, a corporation, or an organization. The understandings, which can include common beliefs, values, and approaches to decision making, are often not stated or documented as goals or formal policies. Employees, for example, might be expected to be clean-cut, wear conservative outfits, and be courteous in dealing with all customers. Sometimes organizational culture is formed over years. In other cases, it can be formed rapidly by top-level managers—for example, implementation of a "casual Friday" dress policy.

Like organizational structure, organizational culture can significantly affect the development and operation of information systems within an organization. A procedure associated with a newly designed information system, for example, might conflict with an informal procedural rule that is part of organizational culture. Organizational culture might also influence a decision maker's perception of the factors and priorities that must be considered in setting objectives. For example, there might be an unwritten understanding that all inventory reports must be prepared before ten o'clock Friday morning. Because of this understood time deadline, the decision maker may reject a cost-reduction option that required compiling the inventory report over the weekend.

organizational change
the responses that are necessary for profit and nonprofit organizations to plan for, implement, and handle change

Organizational change deals with how for-profit and nonprofit organizations plan for, implement, and handle change. Change can be caused by internal or external factors. Internal factors include activities initiated by employees at all levels. Top management at Kansas City Southern Railroad, for example, decided to launch a new $50 million management control system.[10] The objective of the new system was to guarantee on-time shipments for customers. External factors include activities wrought by competitors, stockholders, federal and state laws, community regulations, natural occurrences (such as hurricanes), and general economic conditions. Many European countries, for example, adopted the Euro, a single currency that changed how many financial companies do business and how they used their information systems. Introducing or modifying an information system also causes change because it affects the underlying activities and tasks related to the process. Often, this means changing the way individuals, groups, and the enterprise work.

Overcoming resistance to change can be the hardest part of bringing information systems into a business. According to an executive of Kansas City Southern Railroad concerning its new $50 million management control system, "User acceptance has greatly improved, but at various times we've had all the typical reactions of new users: scratched heads, anger, some very negative, and others very positive." But many potential improvements have failed because managers and employees were not prepared for change. Occasionally, employees even attempt to sabotage a new information system because they do not want to learn the new procedures and commands. In most of these instances, the employees were not involved in the decision to implement the change, nor were they fully informed about the reasons the change was occurring and the benefits that would accrue to the organization.

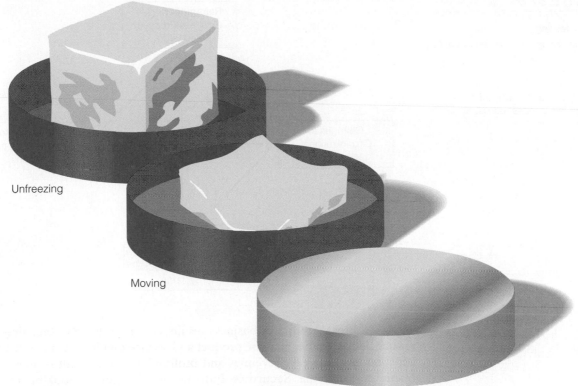

Unfreezing

Moving

Refreezing

FIGURE 2.7

A Change Model

change model
representation of change theories that identifies the phases of change and the best way to implement them

organizational learning
adaptations to new conditions or alterations of organizational practices over time

reengineering (process redesign)
the radical redesign of business processes, organizational structures, information systems, and values of the organization to achieve a breakthrough in business results

The dynamics of change can be viewed in terms of a change model. A **change model** is a representation of change theories that identifies the phases of change and the best way to implement them. Kurt Lewin and Edgar Schein propose a three-stage approach for change (see Figure 2.7). *Unfreezing* is the process of ceasing old habits and creating a climate receptive to change. *Moving* is the process of learning new work methods, behaviors, and systems. *Refreezing* involves reinforcing changes to make the new process second nature, accepted, and part of the job.[11] When a company introduces a new information system, a few members of the organization must become agents of change—champions of the new system and its benefits. Understanding the dynamics of change can help them confront and overcome resistance so that the new system can be used to maximum efficiency and effectiveness.

Organizational learning is closely related to organizational change. According to the concept of **organizational learning**, organizations adapt to new conditions or alter their practices over time. So, assembly-line workers, secretaries, clerks, managers, and executives learn better ways of doing business and incorporate them into their day-to-day activities. Collectively, these adjustments based on experience and ideas are called organizational learning. In some cases, the adjustments can be a radical redesign of business processes, often called *reengineering*. In other cases, these adjustments can be more incremental, a concept called *continuous improvement*.

REENGINEERING

To stay competitive, organizations must occasionally make fundamental changes in the way they do business. In other words, they must change the activities, tasks, or processes that they use to achieve their goals. **Reengineering**, also called **process redesign**, involves the radical redesign of business processes, organizational structures, information systems, and values of the organization to

FIGURE 2.8

Reengineering

Reengineering involves the radical redesign of business processes, organizational structure, information systems, and values of the organization to achieve a breakthrough in business results.

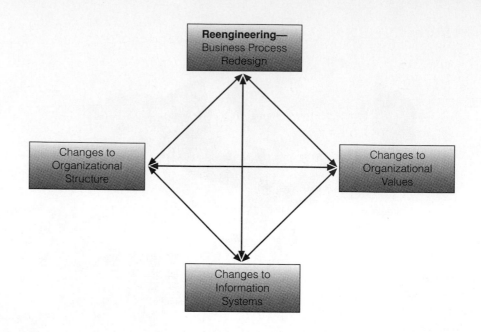

achieve a breakthrough in business results (see Figure 2.8). Reengineering can reduce delivery time, increase product and service quality, enhance customer satisfaction, and increase revenues and profitability. As a result of increased electronic trading, the Securities Industry Association is looking to radically reengineer the communications systems used for stock and securities trading.[12] The new communications systems will be fully implemented by 2005. The reengineering also calls for a new settlement system, called Straight-Through Processing. According to Arthur Thomas, chairman of Merrill Lynch Security Services, "Straight-Through Processing is actually the streamlining of processing to create better efficiencies and to reduce costs internally at individual firms."

A business process includes all activities, both internal (such as thinking) and external (such as taking action), that are performed to transform inputs into outputs. It defines the way work gets done. A few companies still process a customer order manually using several different people. The order moves from one step to the next, allowing people to make numerous errors and create misunderstandings. Today, most companies have computerized customer ordering, saving money and reducing possible errors. This simple example illustrates the fundamental changes reengineering creates in the way things are done, often across multiple departments. But asking people to work differently often meets with stiff resistance, and change is difficult to maintain—the values of the organization

With the increased volume of securities trading in the global marketplace, the Securities Industry Association has released a new model to reengineer the process for settling and clearing transactions that reduces costs and saves time.

(Source: AP/Wide World Photo.)

and its employees must be changed also. In the previous example of order processing, the original work process may have evaluated employees on how many orders were entered each day. Under the reengineered process, they may be evaluated on different factors associated with customer service—percentage of orders delivered on time or accuracy of customer bills. Helping employees understand the benefits of the new system is a major hurdle.

In contrast to simply automating the existing work process, reengineering challenges the fundamental assumptions governing their design. It requires finding and vigorously challenging old rules blocking major business process changes.

Rule	Original Rationale	Potential Problem
Small orders must be held until full-truckload shipments can be assembled.	Reduce delivery costs.	Customer delivery is slowed—lost sales.
No order can be accepted until customer credit is approved.	Reduce potential for bad debt.	Customer service is poor—lost sales.
All merchandising decisions are made at headquarters.	Reduce number of items carried in inventory.	Customers perceive organization has limited product selection—lost sales.

TABLE 2.1

Selected Business Rules That Affect Business Processes

These rules are like anchors weighing a firm down and keeping it from competing effectively. Examples of such rules are given in Table 2.1. Today, many companies use reengineering to increase their competitive position in the market.

Many more companies are now doing business over the Internet. Customers, the company, and suppliers can all be linked through an effective Internet site. Although some consumers still aren't yet ready to buy goods over the Internet, people are becoming more and more receptive to e-commerce. As a result, companies are reengineering their sales processes to take advantage of this opportunity. Along the way they are learning that the cost to process an order can be cut in half—and orders entered by customers themselves tend to have fewer errors than those dictated over the phone. This efficiency enables companies to offer price discounts to e-commerce customers. Amazon.com, which sells books, videotapes, and other consumer items over the Internet, typically sells these items for from 10 to more than 20 percent less than typical retail stores. In addition, some believe that Internet sales can reduce energy costs and pollution. These changes to value-added business processes occurred because organizations considered information system components as integral parts of their business process.

CONTINUOUS IMPROVEMENT

continuous improvement
constantly seeking ways to improve the business processes to add value to products and services

The idea of **continuous improvement** is to constantly seek ways to improve the business processes to add value to products and services. This continual change in turn will increase customer satisfaction and loyalty and ensure long-term profitability. Manufacturing companies make continual product changes and improvements. Service organizations regularly find ways to provide faster and more effective assistance to customers. By doing so, these companies increase customer loyalty, minimize the chance of customer dissatisfaction, and diminish the opportunity for competitive inroads.

To improve its operations, National Foot Care Program, Inc., launched a continuous improvement program, called the Quality Management & Improvement Program (QMIP).[13] The new program helps manage foot and ankle healthcare services for DaimlerChrysler's Medicare-eligible retirees. National Foot Care Program is the largest U.S. foot and ankle care provider and assists over 3,800 medical practitioners across the country.

Organizational commitment to goals such as continuous improvement can be supported by the strategic use of information systems. Continuous improvement involves constantly improving and modifying products and services to remain competitive and to keep a strong customer base. In doing so, companies can increase the quality of their products and services. Low-quality products can turn companies that once were the leaders in their industry into laggards that have lower profits and reduced market share. Without question, quality will continue to be an important factor for profitability and survival. Table 2.2 compares reengineering and continuous improvement.

Business Process Reengineering	Continuous Improvement
Strong action taken to solve serious problem	Routine action taken to make minor improvements
Top-down driven by senior executives	Worker driven
Broad in scope; cuts across departments	Narrow in scope; focus is on tasks in a given area
Goal is to achieve a major breakthrough	Goal is continuous, gradual improvements
Often led by outsiders	Usually led by workers close to the business
Information system integral to the solution	Information systems provide data to guide improvement team

TABLE 2.2

Comparing Business Process Reengineering and Continuous Improvement

technology diffusion
a measure of how widely technology is spread throughout the organization

technology infusion
the extent to which technology is deeply integrated into an area or department

technology acceptance model (TAM)
a model that describes the factors that lead to higher levels of acceptance and usage of technology

quality
the ability of a product (including services) to meet or exceed customer expectations

TECHNOLOGY DIFFUSION, INFUSION, AND ACCEPTANCE

To be effective, reengineering and continuous improvement efforts must be accepted and used throughout an organization. The extent to which an organization uses technology can be a function of technology diffusion, infusion, and acceptance. **Technology diffusion** is a measure of how widely technology is spread throughout an organization. An organization in which computers and information systems are located in most departments and areas has a high level of technology diffusion.[14] Some on-line merchants, such as Amazon.com, have a high level of diffusion and use computer systems to perform most of their business functions, including marketing, purchasing, and billing. **Technology infusion**, on the other hand, is the extent to which technology permeates an area or department. In other words, it is a measure of how deeply imbedded technology is in an area of the organization. Some architectural firms, for example, use computers in all aspects of designing a building or structure. This design area thus has a high level of infusion. Of course, it is possible for a firm to have a high level of infusion in one aspect of its operations and a low level of diffusion overall. The architectural firm may use computers in all aspects of design (high infusion in the design area) but may not use computers to perform other business functions, including billing, purchasing, and marketing (low diffusion).

Although an organization may have a high level of diffusion and infusion, with computers throughout the organization, it does not necessarily mean that information systems are being used to their full potential. In fact, the assimilation and use of expensive computer technology throughout organizations varies greatly.[15] One reason is a low degree of acceptance and use of the technology among some managers and employees. Research has attempted to explain the important factors that enhance or hinder the acceptance and use of information systems.[16] A number of possible explanations of technology acceptance and usage have been studied.[17] The **technology acceptance model (TAM)** specifies the factors that can lead to higher acceptance and usage of technology in an organization, including the perceived usefulness of the technology, the ease of its use, the quality of the information system, and the degree to which the organization supports the use of the information system.[18] Companies hope that a high level of diffusion, infusion, and acceptance will lead to greater performance and profitability.[19]

TOTAL QUALITY MANAGEMENT

The definition of the term *quality* has evolved over the years. In the early years of quality control, firms were concerned with meeting design specifications—that is, conformance to standards. If a product performed as designed, it was considered a high-quality product. A product can perform its intended function, however, and still not satisfy customer needs. Today, **quality** means the ability of a product (including services) to meet or exceed customer expectations.[20] For example, a computer that not only performs well but is easy to maintain and repair would be considered a high-quality product. This view of quality is completely customer oriented. A high-quality product will satisfy customers by functioning correctly and reliably, meeting needs and expectations, and being delivered on time with courtesy and respect.

To help them deliver high-quality goods and services, some companies have adopted continuous improvement strategies that require each major business process to follow a set of total quality management guidelines. **Total quality management (TQM)** consists of a collection of approaches, tools, and techniques that offers a commitment to quality throughout the organization. TQM involves developing a keen awareness of customer needs, adopting a strategic vision for quality, empowering employees, and rewarding employees and managers for producing high-quality products. As a result, processes may be redefined and restructured. The Plantronics Tijuana Manufacturing Facility, the world leader in communications headsets, was able to use TQM approaches to improve overall quality and win the Baja California Quality Award.[21] According to the president of Plantronics, "The Baja California Quality Award recognizes the effort and commitment that our employees have made to enrich our Total Quality Management (TQM) program. Our Total Quality Management program provided the environment for addressing the eight most important elements to the award—leadership, clients, planning, information, personnel development/human capital, process administration, social responsibility, and results."

Information systems are fully integrated into business processes in organizations that adhere to continuous improvement or TQM strategies. Capturing and analyzing customer feedback and expectations and designing, manufacturing, and delivering quality products and services to customers around the world are only a few ways computers and information systems are helping companies pursue their goals of quality and continuous improvement. Connecticut Hospice, for example, implemented a state-of-the art information system that has improved both quality and productivity.[22] "The No. 1 reason for the project was quality of patient care," says Marcel Blanchet, the chief information officer (CIO) of Connecticut Hospice. The system uses a sophisticated computer network and wireless palmtop computers to connect nurses and other healthcare personnel to critical patient information and back-office accounting systems.

total quality management (TQM)
a collection of approaches, tools, and techniques that offers a commitment to quality throughout the organization

OUTSOURCING AND DOWNSIZING

In an effort to control costs, organizations have looked at the number of people they have on the payroll. A significant portion of an organization's expenses are used to hire, train, and compensate talented staff. So organizations today are trying to determine the number of employees they need to maintain high-quality goods and services. With fierce competition in the marketplace, it is critical for organizations to use their resources wisely. Two strategies to contain costs are outsourcing and downsizing (sometimes called *rightsizing*).

Outsourcing involves contracting with outside professional services to meet specific business needs.[23] Often a specific business process is outsourced, such as employee recruiting and hiring, development of advertising materials, product sales promotion, or global telecommunications network support. One reason organizations outsource a business process is to enable them to focus more closely on their core business—and target limited resources to meet strategic goals.

Other reasons for outsourcing are to obtain cost savings or to benefit from the expertise of the service provider. IBM, for example, decided to outsource the manufacturing of its personal computers.[24] The company, which introduced some of the first personal computers twenty years ago, had losses of about $70 million on about $3 billion in sales of personal computers and printers to corporate customers in 2001. According to CEO Louis Gerstner, "The PC industry is maturing rapidly and moving to commoditization."

outsourcing
contracting with outside professional services to meet specific business needs

Ford Resorts to Downsizing and *In*-sourcing

In 2002 Ford Motor Company was mired in a financial crisis. The company posted a $5.1 billion loss for the fourth quarter of 2001, compared with earnings of $1.1 billion the same period a year earlier. Drastic measures were necessary to try to save the automotive giant.

In an effort to cut costs and improve financial results, Ford restructured its operations—cutting 35,000 jobs worldwide, closing five plants, and discontinuing four vehicle models. Ford then turned its attention to streamlining its business processes. "Ford understands the only way to improve their financial position is to streamline internal processes, because that has the highest potential for cutting costs," said Thilo Koslowski, an analyst at Gartner Inc. in Stamford, Connecticut. "IT is the only way Ford will get back on its feet. It's the only variable left where they can save money—they've already cut jobs, asked for cost reductions from suppliers, and closed plants."

So, Ford decided to move from outsourced information systems management to in-house employees to save the cost of expensive contracted services. At the time, Ford had roughly 5,000 information-technology workers, handling duties such as data storage and e-commerce projects and decisions on software and PC purchases. Approximately 60 to 70 percent of those IT workers were not Ford employees, even though they were assigned to work at or near Ford offices. They were instead employed by large technology companies such as IBM Corp., and Compuware Corp., or dozens of smaller technology firms. Through an evaluation system called Project Renaissance, Ford chose what it thought were the best and most valuable of the contract workers and converted them to full-time Ford employees. Ford said it had become less expensive to employ these workers as full-time Ford employees, rather than paying an

agency or another company those workers' salaries plus an administrative fee.

Ford isn't alone in looking to keep its information system talent intact while trimming costs by either eliminating consultants or converting those workers into regular staffers. Companies are doing whatever they can to streamline their information systems and services while maintaining and if possible improving the quality of the technology infrastructure to gain a competitive advantage.

Discussion Questions

1. What are the benefits of outsourcing the development of a computer-based information system? How do these benefits change when the system is in place?
2. Is Ford wise to be hiring new information system staff while closing plants and firing thousands of other employees? What is the logic behind this decision?

Critical Thinking Questions

3. What types of concerns and issues do you think Ford faces when hiring consulting staff as Ford employees? Now that these employees are on the Ford payroll, will the quality of information systems development degrade?
4. As corporations work to streamline information systems to get a higher return on their investment, will this negatively or positively influence innovation in the technology industry?

Sources: Lee Copeland, "Despite Fiscal Troubles, Ford Beefing Up IT Staff," *Computerworld*, January 18, 2002, http://www.computerworld.com; Linda Rosencrance, "Bulletin: Ford Cuts 35,000 Jobs Worldwide, Closes Five Plants," *Computerworld*, January 11, 2002, http://www.computerworld.com; Jeffrey McCracken, "Some Ford Contract Jobs Will Be Moved In-House," *Detroit Free Press*, November 8, 2001, p.

Companies that are considering outsourcing to cut the cost of their IS operations need to review this decision carefully. A growing number of organizations are finding that outsourcing does not necessarily lead to reduced costs. See the "IS Principles in Action" box, which discusses Ford Motor Company's decision to bring IS personnel in-house. One of the primary reasons for cost increases is poorly written contracts that result in additional charges from the outsourcing vendor for each additional task identified. There can be other disadvantages of outsourcing. According to Michael Porter, Harvard Business School professor and consultant, "The short-term cost savings of outsourcing were very apparent, very attractive, and very seductive. ... But when you outsource something, you tend to make it more generic. ... That creates strategic vulnerabilities and also tends to commoditize your product. You're sourcing to people who also supply your competitors."[25]

downsizing
reducing the number of employees to cut costs

Downsizing involves reducing the number of employees to cut costs. The term "rightsizing" is also used. Rather than pick a specific business process to be downsized, companies usually look to downsize across the entire company. Downsizing clearly drives down wages. However, there are also often many bad side effects. Employee morale hits rock bottom. Lines of communication within

In the restaurant industry, competition is fierce because entry costs are low. So a small restaurant that enters the market can be a threat to existing restaurants.

(Source: © Owen Franken/CORBIS.)

the company are weakened. Employee productivity drops. Often, high-priced consultants must be hired to help patch the business back together. The lost time, waning productivity, and devastated morale create hidden costs, which can far outweigh the usual cost savings predicted from a layoff.

Employers need to be open to alternatives for reducing the number of employees, with layoffs viewed as the last resort. It's much easier and simpler to encourage people to leave voluntarily through early retirement or other incentives. Following this approach, the downsizing effort is accompanied with a "buyout package" offered to certain classes of employees (e.g., those over 50 years old). The buyout package offers employees certain benefits and cash incentives if they voluntarily retire from the company. Other options are job sharing and transfers.

The charge of age discrimination is frequently associated with downsizing, as older workers are most often affected. As a result, job discrimination lawsuits have been filed against companies such as Pacific Telesis, AT&T, and IBM. To avoid costly lawsuits, employers need to develop and apply neutral, nondiscriminatory criteria for their staff reduction policies. For example, employees cannot be selected by a lottery nor can everyone over a certain age be downsized. Once an employer has established the criteria, they must be applied equally to all and with no exceptions. Otherwise, the employer runs the risk of a "disparate application" charge, which can lead to a costly judgment.

COMPETITIVE ADVANTAGE

competitive advantage
a significant and (ideally) long-term benefit to a company over its competition

A **competitive advantage** is a significant and (ideally) long-term benefit to a company over its competition. Establishing and maintaining a competitive advantage is complex, but a company's survival and prosperity depend on its success in doing so.

FACTORS THAT LEAD FIRMS TO SEEK COMPETITIVE ADVANTAGE

A number of factors can lead to the attainment of competitive advantage. Michael Porter, a prominent management theorist, suggested a now widely accepted **five-force model**. The five forces include (1) rivalry among existing competitors, (2) the threat of new entrants, (3) the threat of substitute products and services, (4) the bargaining power of buyers, and (5) the bargaining power of suppliers. The more these forces combine in any instance, the more likely firms will seek competitive advantage and the more dramatic the results of such an advantage will be.

five-force model
a widely accepted model that identifies five key factors that can lead to attainment of competitive advantage including (1) rivalry among existing competitors, (2) the threat of new entrants, (3) the threat of substitute products and services, (4) the bargaining power of buyers, and (5) the bargaining power of suppliers.

Rivalry among Existing Competitors

The rivalry among existing competitors is an important factor leading firms to seek competitive advantage. Typically, highly competitive industries are characterized by high fixed costs of entering or leaving the industry, low degrees of product differentiation, and many competitors. Although all firms are rivals with their competitors, industries with stronger rivalries tend to have more firms seeking competitive advantage.

Threat of New Entrants

The threat of new entrants is another important force leading an organization to seek competitive advantage. A threat exists when entry and exit costs to the industry are low and the technology needed to start and maintain the business is commonly available. For example, consider a small restaurant. The owner does not require millions of dollars to start the business, food costs do not go down substantially for large volumes, and food processing and preparation equipment is commonly available. When the threat of new market entrants is high, the desire to seek and maintain competitive advantage to dissuade new market entrants is usually high.

Threat of Substitute Products and Services

The more consumers are able to obtain similar products and services that satisfy their needs, the more likely firms are to try to establish competitive advantage. Such an advantage often creates a "new playing field" in which "substitute" products are no longer considered as such by the consumer. Consider the personal computer industry and the introduction of low-cost computers. A number of consultants and computer manufacturers made much of the high cost of ownership associated with personal computers in the mid-1990s. They introduced low-cost network computers with minimal hard disk space, slower CPUs, and less main memory than some consumers desired, but at half the cost of a standard workstation. There was considerable interest in these new machines for a while, but traditional personal computer manufacturers fought back. They developed a class of powerful workstations and implemented new pricing strategies to make them available at under $1,000. This eliminated the primary advantage of the stripped-down network computers and regained lost customers.

Bargaining Power of Customers and Suppliers

Large buyers tend to exert significant influence on a firm. This influence can be diminished if the buyers are unable to use the threat of going elsewhere. Suppliers can help an organization obtain a competitive advantage. In some cases, suppliers have entered into strategic alliances with firms. When they do so, suppliers act like a part of the company. Suppliers and companies can use telecommunications to link their computers and personnel to obtain fast reaction times and the ability to get the parts or supplies when they are needed to satisfy customers.

STRATEGIC PLANNING FOR COMPETITIVE ADVANTAGE

To be competitive, a company must be fast, nimble, flexible, innovative, productive, economical, and customer oriented. It must also align the information system strategy with general business strategies and objectives. Given the five market forces just mentioned, Porter proposed three general strategies to attain competitive advantage: altering the industry structure, creating new products and services, and improving existing product lines and services. Subsequent research into the use of information systems to help an organization achieve a competitive advantage has confirmed and extended Porter's original work to include additional strategies—such as forming alliances with other companies, developing a niche market, maintaining competitive cost, and creating product differentiation.[26]

Our fast-moving society is highly competitive. To maintain a competitive advantage, companies continually innovate and create new products. Think Outside was the first to market the Stowaway keyboard, a full-size keyboard that folds to pocket-size for use with handheld computers.

(Source: Courtesy of Think Outside, Inc.)

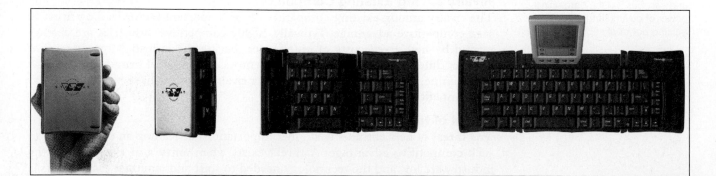

Altering the Industry Structure

Altering the industry structure is the process of changing the industry to become more favorable to the company or organization. This strategy can be accomplished by gaining more power over suppliers and customers. Some automobile manufacturers, for example, insist that their suppliers be located close to major plants and manufacturing facilities and that all business transactions be accomplished using electronic data interchange (EDI, direct computer-to-computer communications with minimal human effort). This system helps the automobile company control the cost, quality, and supply of parts and materials.

A company can also attempt to create barriers to new companies entering the industry. An established organization that acquires expensive new technology to provide better products and services can discourage new companies from getting into the marketplace. Creating strategic alliances may also have this effect. A **strategic alliance**, also called a **strategic partnership**, is an agreement between two or more companies that involves the joint production and distribution of goods and services.

strategic alliance (strategic partnership)
an agreement between two or more companies that involves the joint production and distribution of goods and services

Creating New Products and Services

Creating new products and services is always an approach that can help a firm gain a competitive advantage, and it is especially true of the computer industry and other high-tech businesses. If an organization does not introduce new products and services every few months, the company can quickly stagnate, lose market share, and decline. Companies that stay on top are constantly developing new products and services. Equifax, the largest credit-reporting agency in the United States, uses its information system to help it explore new products and services in new markets.[27] According to Equifax's Chief Financial Officer (CFO) Phil Mazzilli, "I'm involved in all strategic decisions, including how we use technology to get ourselves into new markets and improve efficiency, profitability, and shareholder value."

Improving Existing Product Lines and Services

Improving existing product lines and services is another approach to staying competitive. The improvements can be either real or perceived. Manufacturers of household products are always advertising new and improved products. In some cases, the improvements are more perceived than real refinements; usually, only minor changes are made to the existing product. Many food and beverage companies are introducing "Healthy" and "Light" product lines. A popular beverage company introduced "born on" dating for beer.

With soft sales in the PC market, Hewlett-Packard and Compaq agreed to a merger to make them more competitive. The strategic alliance was expected to help H-P compete in the larger computer server hardware market and in IS consulting services.

(Source: © AFP/CORBIS.)

Using Information Systems for Strategic Purposes

The first IS applications attempted to reduce costs and to provide more efficient processing for accounting and financial applications, such as payroll and general ledger. These systems were seen almost as a necessary evil—something to be tolerated to reduce the time and effort required to complete previously manual tasks. As organizations matured in their use of information systems, enlightened managers began to see how they could be used to improve organizational effectiveness and support the fundamental business strategy of the enterprise. Combining the improved understanding of the potential of information systems with the growth of new technology and applications has led organizations to use IS to gain a competitive advantage. In simplest terms, competitive advantage is usually embodied in either a product or service that has the most added value to consumers and that is unavailable from the competition or in an internal system that delivers benefits to a firm not enjoyed by its competition.

Although it can be difficult to develop information systems to provide a competitive advantage, some organizations have done so with success. A classic example is SABRE, a sophisticated computerized reservation system installed by American Airlines and one of the first CBISs recognized for providing competitive advantage. Travel agents used this system for rapid access to flight information, offering travelers reservations, seat assignments, and ticketing. The travel agents also achieved an efficiency benefit from the SABRE system. Because SABRE displayed American Airline flights whenever possible, it also gave the airline a long-term, significant competitive advantage.

Quite often, the competitive advantage a firm gains with a new information system is only temporary—competitors are quick to copy a good idea. So although the SABRE system was the first on-line reservation system, other carriers soon developed similar systems. However, SABRE has maintained a leadership position in the past because it was the first system available, has been aggressively marketed, and has had continual upgrades and improvements over time. Maintaining a competitive advantage takes effort and is not guaranteed. SABRE's competitive advantage, for example, is being challenged with the many Internet-based travel sites becoming popular with today's travelers.

TABLE 2.3

Competitive Advantage Factors and Strategies

Factors That Lead to Attainment of a Competitive Advantage	Alter Industry Structure	Create New Products and Services	Improve Existing Product Lines and Services
Rivalry among existing competitors	Blockbuster changes the industry structure with its chain of video and music stores.	Dell, Gateway, and other PC makers develop computers that excel at downloading Internet music and playing the music on high-quality speakers.	Food and beverage companies offer "healthy" and "light" product lines.
Threat of new entrants	H-P and Compaq merge to form a large Internet and media company.	Apple Computer introduces an easy-to-use iMac computer that can be used to create and edit home movies.	Starbucks offers new coffee flavors at premium prices.
Threat of substitute products and services	Ameritrade and other discount stockbrokers offer low fees and research on the Internet.	Wal-Mart uses technology to monitor inventory and product sales to determine the best mix of products and services to offer at various stores.	Cosmetic companies add sunscreen to their product lines.
Bargaining power of buyers	Ford, GM, and others require that suppliers locate near their manufacturing facilities.	Investors and traders of the Chicago Board of Trade (CBOT) put pressure on the institution to implement electronic trading.	Retail clothing stores require manufacturing companies to reduce order lead times and improve materials used in the clothing.
Bargaining power of suppliers	American Airlines develops SABRE, a comprehensive travel program used to book airline, car rental, and other reservations.	Intel develops SpeedStep, a chip for laptop computers, that operates at faster speeds when connected to an electrical outlet.	Hayworth, a supplier of office furniture, has a computerized-design tool that helps it design new office systems and products.

The extent to which companies are using computers and information technology for competitive advantage continues to grow. Many companies have even instituted a new position—chief knowledge officer—to help them maintain a competitive advantage. Forward-thinking companies must constantly update or acquire new systems to remain competitive in today's dynamic marketplace. In addition to using information systems to help a company achieve a competitive advantage internally, companies are increasingly investing in information systems to support their suppliers and customers. Investments in information systems that result in happy customers and efficient suppliers can do as much to achieve a competitive advantage as internal systems, such as payroll and billing. Table 2.3 lists several examples of how companies have attempted to gain a competitive advantage.

PERFORMANCE-BASED INFORMATION SYSTEMS

There have been at least three major stages in the business use of IS. The first stage started in the 1960s and was oriented toward cost reduction and productivity. This stage generally ignored the revenue side, not looking for opportunities to increase sales via the use of IS. The second stage started in the 1980s and was defined by Porter and others. It was oriented toward gaining a competitive advantage. In many cases, companies spent large amounts on ISs and ignored the

FIGURE 2.9

Three Stages in the Business Use of IS

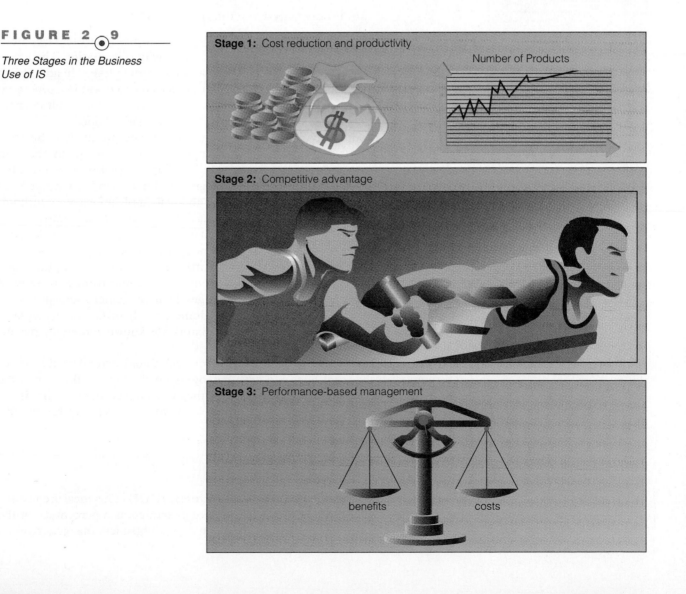

Stage 1: Cost reduction and productivity

Number of Products

Stage 2: Competitive advantage

Stage 3: Performance-based management

benefits   costs

costs. Today, we are seeing a shift from strategic management to performance-based management in many IS organizations. This third stage carefully considers both strategic advantage and costs. This stage uses productivity, return on investment (ROI), net present value, and other measures of performance. Figure 2.9 illustrates these stages.

PRODUCTIVITY

productivity
a measure of the output achieved divided by the input required

Developing information systems that measure and control productivity is a key element for most organizations.[28] **Productivity** is a measure of the output achieved divided by the input required. A higher level of output for a given level of input means greater productivity; a lower level of output for a given level of input means lower productivity. Consider a tax preparation firm, where productivity can be measured by the hours spent on preparing tax returns divided by the total hours the employee worked. For example, in a 40-hour week, an employee may have spent 30 hours preparing tax returns. The productivity is thus equal to 30/40, or 75 percent. With administrative and other duties, a productivity level of 75 may be excellent. The numbers assigned to productivity levels are not always based on labor hours—productivity may be based on factors like the amount of raw materials used, resulting quality, or time to produce the goods or service. In any case, what is important is not the value of the productivity number but how it compares with other time periods, settings, and organizations.

$$\text{Productivity} = (\text{Output/Input}) \times 100\%$$

Once a basic level of productivity is measured, an information system can monitor and compare it over time to see whether productivity is increasing. Then, corrective action can be taken if productivity drops below certain levels. In addition to measuring productivity, an information system can also be used within a process to significantly increase productivity. Thus, improved productivity can result in faster customer response, lower costs, and increased customer satisfaction.

Measuring productivity is important because improving productivity boosts a nation's standard of living. In an era of intense international competition, the need to improve productivity is critical to the well-being of any enterprise or country. If a company does not take advantage of technological and management innovation to improve productivity, its competitors will. The ability to apply information technology to improve productivity will separate successful enterprises from failures.

It is important to understand that information technology is not productive by itself. It takes well-managed, superbly trained, and motivated people—with or without information technology—to deliver measurable gains in output. Many people think that real improvements in productivity come from a synergy of information technology and sweeping changes in management and organizational structure that redefine how work gets done. Largely produced in response to increasing global competition, these overhauls are known loosely as reengineering, discussed previously.

Once work has been redefined, information technology can be used to move information to the front lines—to give employees on the factory floor or in the customer service department the knowledge they need to act quickly. That is the formula for a productivity explosion, which leads to an increase in the world's standard of living.

RETURN ON INVESTMENT AND THE VALUE OF INFORMATION SYSTEMS

return on investment (ROI)
one measure of IS value that investigates the additional profits or benefits that are generated as a percentage of the investment in information systems technology

One measure of IS value is **return on investment (ROI)**. This measure investigates the additional profits or benefits that are generated as a percentage of the investment in information systems technology. A small business that generates an

To compete successfully after the economic slowdown of the early 2000s and the September 11th tragedy, airlines are turning to technology to improve their productivity and customer service.

(Source: Stewart Cohen/Stone.)

additional profit of $20,000 for the year as a result of an investment of $100,000 for additional computer equipment and software would have a return on investment of 20 percent ($20,000/$100,000).

Because of the importance of ROI, many computer companies provide ROI calculators to potential customers. ROI calculators are typically found on a vendor's Web site and can be used to estimate returns in ten minutes or less. Plumtree Software, for example, provided Suncor Energy with an ROI calculator to help Suncor determine whether Plumtree's Internet software was worth the investment.[29] John Wharton, the e-business manager for Suncor, used the ROI calculator from Plumtree and decided to invest in the software.

Earnings Growth

Another measure of IS value is the increase in profit, or earnings growth, it brings. For instance, suppose a mail-order company, after installing an order processing system, had a total earnings growth of 15 percent compared with the previous year. Sales growth before the new ordering system was only about 8 percent annually. Assuming that nothing else affected sales, the earnings growth brought by the system, then, was 7 percent.

Market Share

Market share is the percentage of sales that one company's products or services have in relation to the total market. If installing a new on-line Internet catalog increases sales, it might help a company increase its market share by 20 percent. First Data Corp. embarked on a massive information systems upgrade to help it maintain its leading position in market share in automatic payment systems.[30] Denver-based First Data is the world's largest third-party transaction processor with over 300 million accounts.

Customer Awareness and Satisfaction

Although customer satisfaction can be difficult to quantify, about half of today's best global companies measure the performance of their information systems based on feedback from internal and external users. Some companies use surveys and questionnaires to determine whether the investment in information systems has increased customer awareness and satisfaction.

Total Cost of Ownership

total cost of ownership (TCO)
measurement of the total cost of owning computer equipment, including desktop computers, networks, and large computers

In addition to such measures as return on investment, earnings growth, market share, and customer satisfaction, some companies are also tracking total costs. One measure, developed by the Gartner Group, is the **total cost of ownership (TCO)**. This approach breaks total costs into such areas as the cost to acquire the technology, technical support, administrative costs, and end-user operations. Other costs in TCO include retooling and training costs. TCO can be used to get a more accurate estimate of the total costs for systems that range from small PCs to large mainframe systems. Market research groups often use TCO to compare different products and services.[31] In reviewing messaging products, Sara Radicati, president of a marketing research firm, said, "The market for messaging solutions is on the rise, but total cost of ownership continues to be a major issue for service providers."

The preceding are only a few measures that companies have used to plan for and maximize the value of their investments in information systems technology. In many cases, it is difficult to be accurate with ROI measures. For example, an increase in profits may be caused by an improved information system or by other factors, such as a new marketing campaign or a competitor that was late in delivering a new product to the market. Regardless of the difficulties, organizations must attempt to evaluate the contributions information systems make to be able to assess their progress and plan for the future. Information technology and personnel are too important to leave to chance.

JUSTIFYING INFORMATION SYSTEMS

Because information systems are so important to the work in organizations, businesses need to be sure that improvements or completely new systems are worthwhile. The process for reviewing IS changes involves justification that the change is necessary and will yield gains.

To avoid waste, each potential information systems project should be reviewed to ensure that the project meets an important business need, is consistent with corporate strategy, and leads to attainment of specific goals and objectives. A second check should be made to assess the degree of risk or uncertainty associated with each project. It is not unusual for IS departments to formally analyze and manage risk.

Risk can be assessed by answering questions such as:

1. How well are the requirements of the system understood?
2. To what degree does the project require pioneering effort in technology that is new to the firm?
3. Is there a risk of severe business repercussions if the project is poorly implemented?

The fundamental benefits for considering the project should be identified. Most IS projects fall into one of the following categories:

- **Tangible Savings.** Implementation of the project will result in hard dollar savings to the company that can be quantified (e.g., reduced staff, lowered operating costs, or increased sales).
- **Intangible Savings.** Implementation of the project will result in soft dollar savings to the company, the magnitude of which will be difficult to measure (e.g., help managers make better decisions or improve control over the operations of the business).
- **Legal Requirement.** Implementation of the project is required to meet a state or federal regulation (e.g., reporting information on the employment of handicapped or minorities).
- **Modernization.** Implementation of the project is needed to keep current with changing business requirements (e.g., systems changes needed due to a conversion from English to metric units of measure or conversion from many European currencies to the use of the Euro) or technology requirements (e.g., computer upgrade to improve work with new software).
- **Pilot Project.** Implementation of the project is required to gain experience in a new technology to the existing business (e.g., the use of portable computers by salespeople to enhance customer presentations).

Most organizations today realize that they must look at both sides of the equation—benefits as well as costs—in evaluating potential information system investments. Furthermore, determining return on investment can help the IS organization prove its contribution to the organization and ensure that its efforts are aligned with the company's overall business objectives.

CAREERS IN INFORMATION SYSTEMS

Realizing the benefits of any information system requires competent and motivated information systems personnel, and many companies offer excellent job opportunities.[32] *Computerworld* reported the 10 best places to work in IS based on training the company provides; recruiting, hiring, and retention practices; and day-to-day work environment (see Table 2.4).[33] Note the broad range of industries in which those companies operate—further proof of the widespread use of information systems and the value of an IS background whatever your career choice.

TABLE 2.4

The 10 Best Places to Work for IS

Rank	Company	Average Training Days	Average Cost of Training
1	The Home Depot	17	$9,200
2	Nationwide Insurance	15	$7,652
3	The Vanguard Group	15	$8,000
4	Forsythe Technology	15	$10,000
5	Avon Products	10	$11,000
6	FleetBoston Financial	7	$9,200
7	Towers Perrin	10	$10,000
8	PricewaterhouseCoopers	15	$7,907
9	Harrah's Entertainmen	14	$7,000
10	USAA	8	$7,200

Source: Data from Leslie Goff, "The Best Places to Work," *Computerworld*, June 25, 2001, p. 38.

Numerous schools have degree programs with such titles as information systems, computer information systems, and management information systems. These programs are typically in business schools and within computer science departments. Degrees in information systems have provided high starting salaries for many students after graduation from college. Demand for IS professionals has grown also in nonprofit organizations and in government.[34] In 2001, the federal government had about 60,000 people employed as IS professionals.

Many global opportunities in information systems exist. In addition, some companies seek skilled IS employees from foreign countries, including Russia and India.[35] "I'm finding the technical expertise in Russia is very high," says Ken Pocek, a manager for an Intel lab in Russia. The U.S. H-1B visa program is another approach to getting skilled employees from foreign lands to the United States under a special visa program.[36] Congress recently raised the cap on H-1B visas from 115,000 to 195,000 through 2003. But not everyone is happy with the H-1B program. Some companies may be firing U.S. workers and hiring less expensive workers under the H-1B program. Global opportunities in information systems exist for men and women.[37]

With any career, it can be hard to strike a balance between work life and home or personal life. This balance is important for employees and companies that want to keep their best people. To help achieve a good balance, many companies are turning to specialists, like Xylo, Inc., which is a leading provider of Web-based work/life solutions (see the "Ethical and Societal Issues" box).

ROLES, FUNCTIONS, AND CAREERS IN THE INFORMATION SYSTEMS DEPARTMENT

Information systems personnel typically work in an information systems department that employs Web developers, computer programmers, systems analysts, computer operators, and a number of other information systems personnel. They may also work in other functional departments or areas in a support capacity. In addition to technical skills, information systems personnel also need skills in written and verbal communication, an understanding of organizations and the way they operate, and the ability to work with people (the system users).[38] In general, information systems personnel are charged with maintaining the broadest perspective on organizational goals. For most medium- to large-sized organizations, information resources are typically managed through an IS department. In smaller businesses, one or more people may manage information resources, with support from outside services—outsourcing. As shown in Figure 2.10, the information systems organization has three primary responsibilities: operations, systems development, and support.

Xylo, Inc., Creates Work/Life Solutions

"The remarkable, and partly fortuitous, coming together of the technologies that make up what we label IT—information technologies—has begun to alter, fundamentally, the manner in which we do business and create economic value, often in ways that were not readily foreseeable even a decade ago," remarked Alan Greenspan, Chairman of the Federal Reserve System, to university students in Grand Rapids, Michigan. "As a consequence, growth in output per work hour has accelerated, elevating the standard of living of the average American worker."

Being more productive has been a morale booster for U.S. workers. A recent survey of U.S. adults over the age of 18 showed that 92 percent found meaning and purpose in their work. But as people become more involved with their careers, they often find it difficult to participate fully in their career and have enough time for their other life obligations. Work commitments may compete with life commitments, known as the work/life dilemma. A college student experiences this dilemma when he or she has to pass up an outing with friends to finish a class project due the next day. Working students have an even more complex dilemma by having to balance school commitments, work commitments, and a personal life—the work/student/life dilemma. As we take on full-time careers and full-time family responsibilities, the stakes in this balancing act become high. With the increase in dual-income homes and single working parents, employees are feeling the pull between work and home responsibilities more than ever. Businesses have recognized this problem and are investing in tools to help their employees successfully meet this challenge.

Xylo, Inc., is a leading provider of Web-based work/life solutions used by *Fortune* 500 and other leading companies to attract and retain employees. Xylo clients include Charles Schwab, EDS, Eddie Bauer, Hewlett-Packard, Microsoft, Nordstrom, Northwest Airlines, Sodexho, and many more. "These companies are gaining a competitive edge in the labor market and sending a powerful message to employees about their strong commitment to work/life," said Xylo Senior Vice President of Marketing Judy Meleliat.

Work/life solutions such as Xylo's provide the tools for employees to strike a balance between their work demands and personal lives, while allowing their companies to increase loyalty and reward their employees' commitment. Xylo provides its clients with a custom-designed, password-protected Web site tailored to fit the client's corporate objectives, culture, and values. The Web site offers employees direct access to a wide variety of services and information. It also includes a link to the company intranet for access from any computer with an Internet connection. The Web site is organized into three areas:

1. *Co-worker Connection:* Join a Company Team or Interest Group, Post a Classified Ad, Join a Carpool
2. *My Company:* Company News, Company Links, Employee Birthdays, Suggestion Box, On-line Surveys
3. *Discounts & Services:* The Mall, Travel, Entertainment, Financial Matters, Healthy Living, Family Matters

The Co-worker Connection area nurtures a community atmosphere among the members of the organization. The My Company area provides two-way communication between members and management and also can act as a gateway to the corporate intranet. Discounts & Services offers employees a convenient place to shop, make travel arrangements, and access helpful information. This area offers discounts from well-known Xylo marketing partners in travel, entertainment, and retail industries, and it can be edited to include local advertisements and discounts from the client company's own corporate partners.

Some might argue that allowing employees to shop on-line during work hours is distracting and counterproductive. But in the new work environment, with employees juggling so many responsibilities, many employers are finding that when they take an active role in helping employees handle work/life issues, employees become more content and focused, and the whole company benefits. Work/life solutions such as Xylo's exploit the benefits of modern-day telecommunications to build community, commitment, and contentment in the workforce.

Discussion Questions

1. How have work/life issues changed over the past 20 years? What role, if any, have information systems had on the way society views work?
2. What types of industry would benefit most from a work/life solution such as Xylo's? Is there any type of business in which this solution might be inappropriate?

Critical Thinking Questions

3. Many companies establish policies regarding appropriate use of the corporate intranet and the Internet. What policies would you create to accompany a product such as Xylo's?
4. Are we moving to a point where we no longer differentiate between a professional life and personal life? Are there any hazards in doing so?

Sources: "EDS Selects Xylo's Web-based Work/Life Solution," *Business Wire,* November 6, 2001; "Majority of Americans Find Work Meaningful and Purposeful," *Business Wire,* October 30, 2001; Xylo Web site, http://www.xylo.com.

Operations

The operations component of a typical IS department focuses on the use of information systems in corporate or business unit computer facilities. It tends to focus more on the *efficiency* of information system functions rather than their effectiveness.

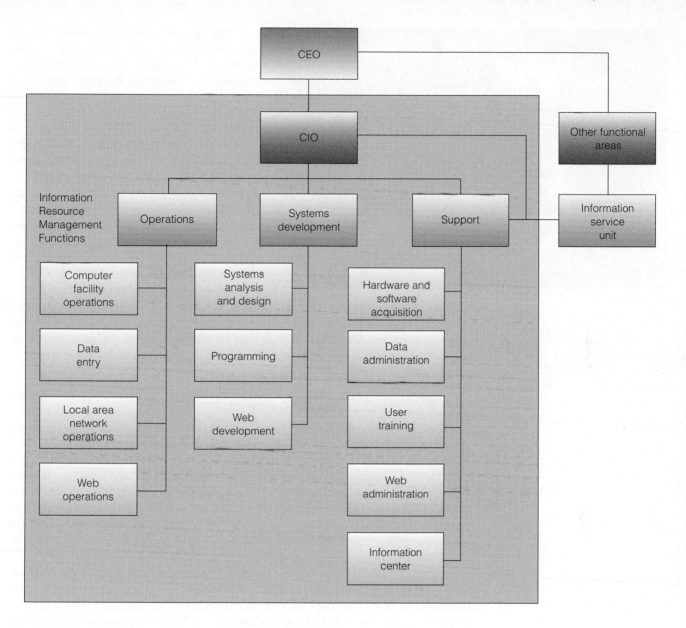

FIGURE 2•10

The Three Primary Responsibilities of Information Systems

Each of these elements—operations, systems development, and support—contains sub-elements critical to the efficient and effective operation of the organization.

The primary function of a system operator is to run and maintain IS equipment. System operators are responsible for starting, stopping, and correctly operating mainframe systems, networks, tape drives, disk devices, printers, and so on. System operators are typically trained at technical schools or through on-the-job experience. Other operations include logging, scheduling, hardware maintenance, and preparation of input and output. Data-entry operators convert data into a form the computer system can use. They may use terminals or other devices to enter business transactions, such as sales orders and payroll data. Increasingly, data entry is being automated—captured at the source of the transaction rather than being entered later. In addition, companies may have local area network and Web or Internet operators who are responsible for running the local network and any Internet sites the company may have.

Systems Development

The systems development component of a typical IS department focuses on specific development projects and ongoing maintenance and review. Systems analysts and programmers, for example, focus on these concerns.

Systems operators focus on the efficiency of information system functions rather than the effectiveness. Their primary function is to run and maintain IS equipment.

(Source: © 2002 PhotoDisc.)

The role of a systems analyst is multifaceted. Systems analysts help users determine what outputs they need from the system and construct the plans needed to develop the necessary programs that produce these outputs. Systems analysts then work with one or more programmers to make sure that the appropriate programs are purchased, modified from existing programs, or developed. The major responsibility of a computer programmer is to use the plans developed by the systems analyst to develop or adapt one or more computer programs that produce the desired outputs. The main focus of systems analysts and programmers is to achieve and maintain information system effectiveness.

With the dramatic increase in the use of the Internet, intranets, and extranets, many companies have Web or Internet developers who are responsible for developing effective and attractive Internet sites for customers, internal personnel, suppliers, stockholders, and others with a business relationship with the company.

Support

The support component of a typical IS department focuses on providing user assistance in the areas of hardware and software acquisition and use, data administration, user training and assistance, and Web administration. In many cases, the support function is delivered through an information center.

Because information systems hardware and software are costly, especially if mistakes are made, the acquisition of computer hardware and software is often managed by a specialized support group. This group sets guidelines and standards for the rest of the organization to follow in making purchases. Gaining and maintaining an understanding of available technology is an important part of the acquisition of information systems. Also, developing good relationships with vendors is important.

Firms may look to one outside source to supply part or all of their information systems needs—a single-vendor solution. There are advantages to this approach, such as potential cost savings and built-in compatibility. But using single vendors also involves risks, including lack of flexibility, vendor complacency due to lack of competitive bidding, and the possibility of missing out on new products from other vendors. Having an in-house specialist who focuses on the acquisition of information systems may also be wise when using the outsourcing approach.

A database administrator focuses on planning, policies, and procedures regarding the use of corporate data and information. For example, database administrators develop and disseminate information about the corporate databases for developers of information system applications. In addition, the database administrator is charged with monitoring and controlling database use.

User training is a key to get the most from any information system. The support area insures that appropriate training is available to users. Training can be provided by internal staff or from external sources. For example, internal support staff may train managers and employees in the best way to enter sales orders, to receive computerized inventory reports, and to submit expense reports electronically. Companies also hire outside firms to help train users in other areas, including the use of word processing, spreadsheets, and database programs.

Web administration is another key area of the support function. With the increased use of the Internet and corporate Web sites, Web administrators are sometimes asked to regulate and monitor Internet use by employees and managers to make sure that it is authorized and appropriate. Web administrators also are responsible for maintaining the corporate Web site. Keeping corporate Web sites accurate and current can require substantial resources.

The support component typically operates the information center. An **information center** provides users with assistance, training, application development, documentation, equipment selection and setup, standards, technical assistance, and troubleshooting. Although many firms have attempted to phase out information centers, others have changed the focus of this function from technical training to helping users find ways to maximize the benefits of the information resource.

information center
a support function that provides users with assistance, training, application development, documentation, equipment selection and setup, standards, technical assistance, and troubleshooting

information service unit
a miniature IS department

Information Service Units

An **information service unit** is basically a miniature IS department attached and directly reporting to a functional area. Notice the information service unit shown in Figure 2.10. Even though this unit is usually staffed by IS professionals, the project assignments and the resources necessary to accomplish these projects are provided by the functional area to which it reports. Depending on the policies of the organization, the salaries of IS professionals staffing the information service unit may be budgeted to either the IS department or the functional area.

The growth of information service units may be directly attributed to the increased number of users doing their own computing. The increasing use of networks has put computers at nearly every desk. Communication between IS personnel and users is more effective the closer they work together. When such information service units are not part of the formal organizational structure, they tend to arise informally in organizations. That is, a particular functional manager might establish and maintain informal groups of employees who are more proficient with IS than other users. As more employees become computer users, such cooperation must be considered to manage resources properly. It is probably more productive to support and provide training to these informal groups than to attempt to interfere with or thwart their activities.

TYPICAL IS TITLES AND FUNCTIONS

The organizational chart shown in Figure 2.10 is a simplified model of an IS department in a typical medium- or large-sized organization. Many organizations have even larger departments, with increasingly specialized positions such as librarian, quality assurance manager, and the like. Smaller firms often combine the roles depicted in Figure 2.10 into fewer formal positions.

The Chief Information Officer

The overall role of the chief information officer (CIO) is to employ an IS department's equipment and personnel to help the organization attain its goals. The CIO is usually a manager at the vice-president level concerned with the overall needs of the organization. He or she is responsible for corporatewide policy, planning, management, and acquisition of information systems. Some of the CIO's top concerns include integrating information systems operations with corporate strategies, keeping up with the rapid pace of technology, and defining and assessing the value of systems development projects in terms of performance, cost, control, and complexity. The high level of the CIO position is consistent with the idea that information is one of the organization's most important resources. This individual works with other high-level officers of an organization, including the chief financial officer (CFO) and the chief executive officer (CEO), in managing and controlling total corporate resources.

Increasingly, CIOs have to deal with international information systems. Chris Scalet, CIO for International Paper Co., has to manage information systems in the United States, Europe, South America, and Asia from his offices in New York.[39] According to Scalet, "It's a situation where you just have to communicate and recommunicate—you have to overcommunicate."

Depending on the size of the information systems department, there may be several people at senior IS managerial levels. Some of the job titles associated with information systems management are the CIO, vice president of information systems, and manager of information systems. A central role of all these individuals is to communicate with other areas of the organization to determine changing needs. Often these individuals are part of an advisory or steering committee that helps the CIO and other IS managers with their decisions about the use of information systems. Together they can best decide what information systems will support corporate goals. CIOs must work closely with advisory committees, stressing effectiveness and teamwork and viewing information systems as an integral part of the organization's business processes—not an adjunct to the organization. Thus, CIOs need both technical and business skills. Some companies, like discount broker Ameritrade, have co-CIOs.[40] One Ameritrade co-CIO has technical skills in developing computer programs, and the other has financial skills in developing planning tools for clients.

LAN Administrators

Local Area Network (LAN) administrators set up and manage the network hardware, software, and security processes. They manage the addition of new users, software, and devices to the network. They isolate and fix operations problems. LAN administrators are in high demand and often solve both technical and nontechnical problems. According to Chris Holmes of J.P. Morgan, "I transitioned into technology, where I am currently a LAN administrator and troubleshooter. I now have the opportunity to work directly with technology and solve problems, and that brings me much more satisfaction."[41]

Internet Careers

The recent bankruptcy of some Internet start-up companies, called the dot-gone era by some, has resulted in layoffs for some firms.[42] Some executives of these bankrupt start-up Internet companies lost hundreds of millions of dollars in a few months. Yet, the growth in the use of the Internet to conduct business continues and has caused a steady need for skilled personnel to develop and coordinate Internet usage. As seen in Figure 2.10, these careers are in the areas of Web operations, Web development, and Web administration. As with other areas in IS, there are a number of top-level administrative jobs related to the Internet. These career opportunities are with traditional companies and companies that specialize in the Internet.

Internet job sites such as Monster.com allow job hunters to browse job opportunities and post their resumes.

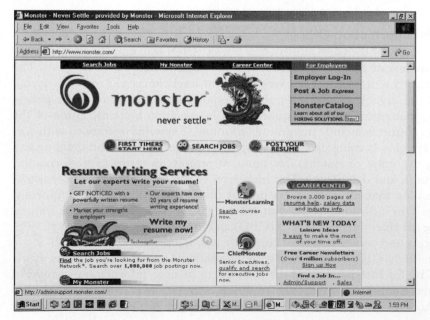

Internet jobs within a traditional company include Internet strategists and administrators, Internet systems developers, Internet programmers, and Internet or Web site operators. The Internet has become so important to some companies that some have suggested a new position, chief Internet officer, with responsibilities and salary similar to the CIO's.

In addition to traditional companies, there are many exciting career opportunities in companies that offer products and services over the Internet. These companies include Amazon.com, Yahoo!, eBay, and many others. Systest, for example, specializes in finding and eliminating digital bugs that could halt the operation of a computer system. According to Christopher Hardesty, chief financial officer, "The Internet has come in and really revolutionized the kind of performance testing we do."

A number of Internet sites post job opportunities for Internet careers and more traditional careers, such as Monster.com. These sites allow prospective job hunters to browse job opportunities, job locations, salaries, benefits, and other factors. In addition, some of these sites allow job hunters to post their resumes.

Quite often the people filling IS roles have completed some form of certification. **Certification** is a process for testing skills and knowledge resulting in an endorsement by the certifying authority that an individual is capable of performing a particular job. Certification frequently involves specific, vendor-provided or vendor-endorsed coursework. There are a number of popular certification programs, including Novell Certified Network Engineer, Microsoft Certified Professional Systems Engineer, Certified Project Manager, and others.[43] The Certified Information Systems Security Professional (CISSP) is also becoming increasingly important to companies.[44] The federal government is helping military personnel get IS certification. GI bill beneficiaries, for example, can now get reimbursed for technology certification through the Computing Technology Industry Association.[45]

One of the greatest fears of every IS manager is spending several thousand dollars to help an employee get certified and then to lose that person to a higher-paying position with a new firm. As a consequence, some organizations request a written commitment from individuals to stay a certain time after obtaining their certification. Needless to say, this requirement can create some ill will with the employee. To provide newly certified employees with incentives to remain, other organizations provide salary increases based on additional credentials acquired.

certification
process for testing skills and knowledge that results in a statement by the certifying authority that says an individual is capable of performing a particular kind of job

OTHER IS CAREERS

In addition to working for an information systems department in an organization, information systems personnel can work for one of the large consulting firms, such as Accenture, EDS, and others. These jobs often entail a large amount of travel, because consultants are assigned to work on various projects wherever the client is. Such roles require excellent people and project-management skills in addition to IS technical skills.

Another IS career opportunity is to be employed by a hardware or software vendor developing or selling products. Such a role enables the individual to work on the cutting edge of technology and can be extremely challenging and exciting!

SUMMARY

PRINCIPLE *The use of information systems to add value to the organization is strongly influenced by organizational structure, culture, and change.*

Value-added processes increase the relative worth of the combined inputs on their way to becoming final outputs of the organization. The value chain is a series (chain) of activities that includes (1) inbound logistics, (2) warehouse and storage, (3) production, (4) finished product storage, (5) outbound logistics, (6) marketing and sales, and (7) customer service.

Organizations use information systems to support organizational goals. Because information systems typically are designed to improve productivity, methods for measuring the system's impact on productivity should be devised.

An organization is a formal collection of people and various other resources established to accomplish a set of goals. The primary goal of a for-profit organization is to maximize shareholder value. Nonprofit organizations include social groups, religious groups, universities, and other organizations that do not have profit as the primary goal. Organizations are systems with inputs, transformation mechanisms, and outputs.

Organizational structure refers to organizational subunits and how they are related and tied to the overall organization. Several basic organization structures exist: traditional, project, team, multidimensional (also called *matrix structure*), and virtual organizational structure. A virtual organizational structure employs individuals, groups, or complete business units in geographically dispersed areas. The individuals, groups, or complete business units can involve people in different countries operating in different time zones with different cultures.

Organizational culture consists of the major understandings and assumptions for a business, corporation, or organization. Organizational change deals with how profit and nonprofit organizations plan for, implement, and handle change. Change can be caused by internal or external factors. The change model consists of these stages: unfreezing, moving, and refreezing. According to the concept of organizational learning, organizations adapt to new conditions or alter practices over time.

PRINCIPLE *Because information systems are so important, businesses need to be sure that improvements or completely new systems help lower costs, increase profits, improve service, or achieve a competitive advantage.*

Business process reengineering involves the radical redesign of business processes, organizational structures, information systems, and values of the organization to achieve a breakthrough in results. Continuous improvement constantly seeks ways to improve business processes to add value to products and services. The extent to which technology is used throughout an organization can be a function of technology diffusion, infusion, and acceptance. Technology diffusion is a measure of how widely technology is in place through an organization. Technology infusion is the extent to which technology is located in an area or department. The technology acceptance model (TAM) investigates factors, such as perceived usefulness of the technology, the ease of use of the technology, the quality of the information system, and the degree to which the organization supports the use of the information system, to predict information systems usage and performance. Total quality management consists of a collection of approaches, tools, and techniques that offers a commitment to quality throughout the organization. Outsourcing involves contracting with outside professional services to meet specific business needs. This approach allows the company to focus more closely on its core business and to target its limited resources to meet strategic goals. Downsizing involves reducing the number of employees to reduce payroll costs; however, it can lead to unwanted side effects.

Competitive advantage is usually embodied in either a product or service that has the most added value to consumers and that is unavailable from the competition or in an internal system that delivers benefits to a firm not enjoyed by its competition. A five-force model covers factors that lead firms to seek competitive advantage: rivalry among existing competitors, the threat of new market entrants, the threat of substitute products and services, the bargaining power of buyers, and the bargaining power of suppliers. Three strategies to address these factors and to attain competitive advantage include altering the industry structure, creating new products and services, and improving existing product lines and services.

The ability of the information system to provide or maintain competitive advantage should also be determined. Several strategies for achieving competitive advantage include enhancing existing products or services or developing new ones, as well as changing the existing industry or creating a new one.

The objectives of each potential information systems project are reviewed to ensure that the project meets an important business need, is consistent with corporate strategy, and leads to attainment of specific goals and objectives. A second check is made to assess the degree of risk or uncertainty associated with each project. The fundamental reason for considering the project should be identified. Developing information systems that measure and control productivity is a key element for most organizations. A useful measure of the value of an information system project is return on investment (ROI). This measure investigates the additional profits or benefits that are generated as a percentage of the investment in information systems technology. Total cost of ownership (TCO) can also be a useful measure. Most IT projects fall into one of the following categories: tangible savings, intangible savings, legal requirement, modernization, or pilot project.

PRINCIPLE *Information systems personnel are the key to unlocking the potential of any new or modified system.*

Information systems personnel typically work in an information systems department that employs a chief information officer, systems analysts, computer programmers, computer operators, and a number of other information systems personnel. The overall role of the chief information officer (CIO) is to employ an IS department's equipment and personnel to help the organization attain its goals. Systems analysts help users determine what outputs they need from the system and construct the plans needed to develop the necessary programs that produce these outputs. Systems analysts then work with one or more programmers to make sure that the appropriate programs are purchased, modified from existing programs, or developed. The major responsibility of a computer programmer is to use the plans developed by the systems analyst to build or adapt one or more computer programs that produce the desired outputs. Computer operators are responsible for starting, stopping, and correctly operating mainframe systems, networks, tape drives, disk devices, printers, and so on. LAN administrators set up and manage the network hardware, software, and security processes.

There is also an increasing need for trained personnel to set up and manage a company's Internet site, including Internet strategists, Internet systems developers, Internet programmers, and Web site operators. Information systems personnel may also work in other functional departments or areas in a support capacity. In addition to technical skills, information systems personnel also need skills in written and verbal communication, an understanding of organizations and the way they operate, and the ability to work with people (users). In general, information systems personnel are charged with maintaining the broadest enterprisewide perspective.

In addition to working for an information systems department in an organization, information systems personnel can work for one of the large information systems consulting firms, such as Accenture, EDS, and others. Another IS career opportunity is to be employed by a hardware or software vendor developing or selling products.

CHAPTER 2 SELF-ASSESSMENT TEST

The use of information systems to add value to the organization is strongly influenced by organizational structure, culture, and change.

1. Inbound logistics, warehouse and storage, production, and outbound logistics are all part of
   A. organization flow
   B. the value chain
   C. traditional organizational structure
   D. virtual organization structure

2. A (An) _____ is a formal collection of people and other resources established to accomplish a set of goals.

3. A virtual organizational structure is centered on major products or services. True False

4. The concept where organizations adapt to new conditions or alter their practices over time is called
   A. organizational learning
   B. organizational change
   C. continuous improvement
   D. reeengineering

Because information systems are so important, businesses need to be sure that improvements or completely new systems help lower costs, increase profits, improve service, or achieve a competitive advantage.

5. The idea of _____ is to constantly seek ways to improve the business processes to add value to products and services.

6. Today, quality means
   A. achieving production standards
   B. meeting or exceeding customer expectations
   C. maximizing total profits
   D. meeting or achieving design specifications

7. Technology diffusion is a measure of how widely technology is spread throughout an organization. True False

8. Reengineering is also called _____.

9. What is a measure of the output achieved divided by the input required?
   A. efficiency
   B. effectiveness
   C. productivity
   D. return on investment

10. _____ is a measure of the additional profits or benefits generated as a percentage of the investment in information systems technology.

Information systems personnel are the key to unlocking the potential of any new or modified system.

11. Who is involved in helping users determine what outputs they need and constructing the plans needed to produce these outputs?
    A. the CIO
    B. the applications programmer
    C. the systems programmer
    D. the systems analyst

12. The systems development component of a typical IS department focuses on specific development projects and ongoing maintenance and review. True False

13. The _____ is typically in charge of the information systems department or area in a company.

Chapter 2 Self-Assessment Test Answers

1. B, 2. organization, 3. False, 4. A, 5. continuous improvement, 6. B, 7. True, 8. process redesign, 9. C, 10. return on investment, 11. D, 12. True, 13. chief information officer (CIO).

KEY TERMS

certification, 71
change model, 51
competitive advantage, 57
continuous improvement, 53
culture, 50
downsizing, 56
empowerment, 47
five-force model, 57
flat organizational structure, 47
information center, 69
information service unit, 69

multidimensional organizational
   structure, 49
organization, 44
organizational change, 50
organizational culture, 50
organizational learning, 51
organizational structure, 47
outsourcing, 55
productivity, 62
project organizational structure, 48
quality, 54
reengineering (process redesign), 51

return on investment (ROI), 62
strategic alliance (strategic
   partnership), 59
team organizational structure, 48
technology acceptance model (TAM), 54
technology diffusion, 54
technology infusion, 54
total cost of ownership (TCO), 63
total quality management (TQM), 55
traditional organizational structure, 47
value chain, 45
virtual organizational structure, 49

REVIEW QUESTIONS

1. What is meant by organization structure?
2. What is the difference between a virtual organizational structure and a traditional organizational structure?
3. What is a value-added process? Give several examples.
4. What role does an information system play in the value-added processes of an organization?
5. What is reengineering? What are the potential benefits of performing a process redesign? What is the difference between reengineering and continuous improvement?
6. What is the technology acceptance model (TAM)?
7. What is quality? What is total quality management (TQM)?
8. What are organizational change and organizational learning?
9. List and define the four basic organizational structures.
10. Sketch and briefly describe the three-stage organizational change model.
11. What is downsizing? How is it different from outsourcing?
12. What are some general strategies employed by organizations to achieve competitive advantage?
13. What are the five common justifications for implementation of an information system?
14. Define the term *productivity*. Why is it difficult to measure the impact that investments in information systems have on productivity?
15. Briefly define technology diffusion and infusion.
16. What is the total cost of ownership?
17. What is an information systems unit?

DISCUSSION QUESTIONS

1. You have been hired to work in the IS area of a manufacturing company that is starting to use the Internet to order parts from its suppliers and offer sales and support to its customers. What types of Internet positions would you expect to see at the company?
2. What sort of information systems career would be most appealing to you—working as a member of an IS organization, working as a consultant, or working for a hardware or software vendor? Why?
3. What are the advantages of using a virtual organizational structure? What are the disadvantages?
4. As part of a TQM project initiated three months ago, you decided your company needed a new information system. The computer systems were brought in over the weekend. The first notice your employees received about the new information system was the computer located on each desk. How might the new system affect the culture of your organization? What types of behaviors might employees exhibit in response? As a manager, how should you have prepared the employees for the new system?

5. You have been asked to participate in the preparation of your company's strategic plan. Specifically, your task is to analyze the competitive marketplace using Porter's five-force model. Prepare your analysis, using your knowledge of a business you have worked for or have an interest in working for.

6. Based on the analysis you performed in Discussion Question 5, what possible strategies could your organization adopt to address these challenges? What role could information systems play in these strategies? Use Porter's strategies as a guide.

7. There are many ways to evaluate the effectiveness of an information system. Discuss each method and describe when one method would be preferred over another method.

8. Imagine that you are the CIO for a large, multinational company. Outline a few of your key responsibilities.

9. Discuss how the change model can be applied to breaking a bad habit—say, smoking or eating fatty foods. Some people have also related the stages in the change model to the changes one must go through to deal with a major life crisis—like divorce or the loss of a loved one. Explain.

PROBLEM-SOLVING EXERCISES

1. Locate a firm that uses the virtual organization structure. Then, using a presentation graphics package, draw the organizational structure for this firm. The firm should have more than 15 employees. Do the same for a firm that has another type of structure discussed in the text. Use your word processor to create a document that describes the differences between these two firms.

2. A new IS project has been proposed that will produce not only cost savings but also an increase in revenue. The initial costs to establish the system are estimated to be $500,000. The rest of the cash flow data is presented in the following table.

	Year 1	Year 2	Year 3	Year 4	Year 5
Increased Revenue	$0	$100	$150	$200	$250
Cost Savings	$0	$50	$50	$50	$50
Depreciation	$0	$75	$75	$75	$75
Initial Expense	$500				

Note: All amounts in 000's.

a. Using your spreadsheet program, calculate the return on investment for this project. Assume the cost of capital is 7 percent.

b. How would the rate of return change if the project were able to deliver $50,000 in additional revenue and generate cost savings of $25,000 in the first year?

TEAM ACTIVITIES

1. With your team, interview one or more people who were either outsourced or downsized from a position in the last few years. Find out how the process was handled and what justification was given for taking this action. Also try to get information from an objective source (financial reports, investment brokers, industry consultants) on how this action has affected the organization.

2. Have your team interview a company that recently introduced new technology. Write a brief report that describes the extent of technology infusion and diffusion.

WEB EXERCISES

1. This book emphasizes the importance of information. You can get information from the Internet by going to a specific address, such as http://www.ibm.com, http://www.whitehouse.gov, or http://www.fsu.edu. This will give you access to the home page of the IBM corporation, the White House, or Florida State University. Note that "com" is used for businesses or commercial operations, "gov" is used for governmental offices, and "edu" is used for educational institutions. Another approach is to use a search engine. Yahoo!, developed by two Tulane University students, was one of the first search engines on the Internet to find information. A search engine is a Web site that allows you to enter key words or phrases to find information. Lists or menus can also be used. The search engine will

return other Web sites (hits) that correspond to a search request. Using Yahoo! at http://www.yahoo.com, search for information about a company or topic discussed in Chapter 1 or 2. You may be asked to develop a report or send an e-mail message to your instructor about what you found.

2. Use the Internet to search for information about a company that you think would be a good career choice for you. You can use a search engine, like Yahoo!, or a database at your college or university. Write a brief report describing the company and why you would like to be employed by it.

CASES

CASE 1

GE Medical Systems Builds Nation's First All-Digital Hospital

If you've ever filled out a patient profile or medical history form for a family physician, you've had a good introduction to the complexities of medical record keeping. Such forms typically consist of several pages including dozens of questions about not only your own medical history but also your parents', siblings', and entire family tree's. As your life progresses and you experience the typical medical complications, your medical history becomes larger and more complex. Some of this information is added to your medical records by physicians; other information depends on your own recollections. If you switch doctors, you face the danger of having to start over from scratch.

Keeping up with patient records, along with continuously changing medical knowledge, is a considerable challenge for hospitals and the healthcare industry. This information management challenge has sparked the creation of a new area of specialized research known as *medical informatics*. The Department of Medical Informatics at Columbia University defines medical informatics as the scientific field that deals with the storage, retrieval, sharing, and optimal use of biomedical information, data, and knowledge for problem solving and decision making. In short, medical informatics deals with information systems for the medical community.

GE Medical Systems is an $8 billion global leader in medical imaging, healthcare services, and information technology, and it is a pioneer in the medical informatics industry. It is working on several fronts to provide solutions to the complexities of medical record keeping. The Indiana Heart Hospital in Indianapolis has partnered with GE Medical Systems to build the nation's first all-digital cardiac hospital. At this new paperless facility, doctors and nurses are able to access patient records and other medical information inside or outside the hospital from a Pocket PC or similar wireless, handheld computer. The new system does away with the need for nursing stations and medical records file rooms. This all-digital hospital has the highest degree of technology infusion and diffusion—all for efficiency and accuracy.

The system that GE has installed, the Centricity Information System, is an enterprisewide clinical information system that integrates patient information—including images, diagnostic readings, and medical history—from every area of the hospital into a single electronic record that can span a patient's entire lifetime. Bringing this variety of information into one easily accessible, centralized system will save healthcare professionals valuable time. "Current healthcare trends, including nursing shortages, make the all-digital concept crucial," David Veillette, CEO of the Indiana Heart Hospital, said. "The aging of the baby boomers means we have to find more efficient ways to take care of three times as many patients, with staffing levels that will be decreasing," he said. "The only way to do that is with information technology." With electronic records, hospital personnel won't have to struggle to read someone else's handwriting because data will be entered with a keyboard. Also, doctors and nurses won't have to search for paper files—reducing the possibility of errors, according to hospital officials. GE Medical Systems has an operations staff at the hospital to maintain the system and provide support to the medical staff.

GE Medical Systems has a vision of a massive healthcare network that can be accessed by any subscribing healthcare provider, where a patient's medical experience can be merged into a single electronic record that spans care given throughout the healthcare network. This practice echoes a common information management strategy used in all industries: digitize, centralize, and deliver. Digitize all data and information so that it can be stored electronically, store it centrally so that all information is accessed through one system, and then create easy access to that system. In most industries, creating effective and efficient information systems saves the company money and helps it gain a competitive advantage. In the healthcare industry, an effective and efficient information system saves lives.

Discussion Questions

1. How will the staffing of the Indiana Heart Hospital differ from that of a traditional hospital?
2. What type of privacy issues arise when developing a central healthcare network? What types of medical information might some patients want to keep private? What policies and procedures might be developed to safeguard private patient information?

Critical Thinking Questions

3. How will medical informatics affect the healthcare industry's culture? How will physicians and other healthcare providers who grew up prior to the computer revolution adapt to this new environment?
4. The vision of a centralized healthcare network seems to suggest an industry monopoly, with one company controlling the market. How would this benefit healthcare providers? Would such a monopoly inhibit innovation fostered by competition in the market?

Sources: Rick Barrett, "GE Medical Aids New All-Digital Cardiac Hospital in Indianapolis," *The Milwaukee Journal Sentinel*, January 24, 2002, http://www.jsonline.com; "Nation's First All-Digital Heart Hospital Uses Information Technology to Battle Heart Disease," *PR Newswire*, January 24, 2002; "New Patient Monitoring Technology from GE Provides Clinicians with Critical, Uninterrupted Flow of Patient Information for More Accurate Decision Making," *Business Wire*, November 12, 2001; "GE Medical Systems Agrees to Acquire the Business of MedicaLogic," *Business Wire*, January 24, 2002; http://www.gemedicalsystems.com; http://www.cpmc.columbia.edu/.

Delta and Tellabs Seek Higher Return on Investment

In these days of performance-based information systems, managers have to work hard to get their information system requests approved. Requests to develop a new information system or improve on existing systems are closely scrutinized by senior managers to assure that the investment is effectively supporting corporate goals and will bring in a quick return.

Delta Technology, the information technology arm of $16 billion Atlanta-based Delta Air Lines, presents a good example of this trend. "We have been carefully reviewing every project and every spend [expense] with approvals at the senior vice president level. Before, we delegated decisions to a lower level," says Curtis Robb, senior vice president and chief technology officer. "Finance is also much more actively involved in business cases that are developed [for IT projects]." In other words, Delta and many other companies have found it necessary to implement return on investment (ROI) standards and procedures for measuring return on information system investment.

Curtis Robb says there are critical issues that businesses must address to ensure ROI. The first is total cost of ownership. Each of Delta's business teams must develop plans that look ahead four years, he says. They look at not only the purchase price but also the "tail behind that purchase price"—hardware, software, maintenance, and support, Robb says. The second issue is finding the right level of support for the system once it is in place. Rightsizing maintenance contracts has helped Delta shed $10 million in expenses. Standardizing technology has also helped the company save on training and development costs. Rather than building new systems from scratch, Delta designs generic systems to allow portions of systems to be reused on new projects as they arise. The final issue is time to market. At Delta, "solution architects" are assigned to projects from the start to help create a blueprint and determine a timeline.

Once an information system project is under way, it is important to provide oversight to ensure that the project brings in a return. Some companies create technology review boards to provide monthly reviews of IS proposals. Projects are reviewed each month to make sure scope, costs, and time frames are on target.

Implementing a system such as Delta's often meets with a considerable amount of cultural resistance. Tellabs, a Naperville, Illinois–based communications equipment maker, has faced obstacles in implementing its new procedures for measuring return on IT investment. When information system proposals were reviewed for approval, CIO Cathy Kozik found a number of inaccuracies and a general lack of honesty. Managers and staffers were finding it hard to be objective due to concerns over budget cuts and worries about automating themselves out of their jobs. To overcome the honesty and accuracy problems, Kozik asked financial controllers from each unit to oversee the calculations of each proposal.

Implementing ROI standards must be a gradual process, Kozik warns. If Tellabs forced its ROI process on workers, "it would have collapsed under its own weight," she says. "Instead of going from 0 to 120, we're going from 0 to 30, 30 to 60."

The role of the CIO becomes all the more valuable to an organization when striving toward a high ROI. The CIO bridges the gap between top-level executives who may be technically naïve, and lower-level staff who may be more interested in preserving their jobs than saving the company money. Only the CIO can assure that the organization is getting the highest possible return on its information system investments to gain an advantage over the competition.

Discussion Questions

1. How might a CIO motivate the information system staff to assist in assessing return on investment and to overcome fears of job loss?
2. How does the trend of involving upper management in information system management decisions affect the balance of power within the organization? Does this undermine previous efforts to empower lower-level employees? Is it possible to have both a high ROI and empowered employees?

Critical Thinking Questions

3. How will the quest for leaner, meaner information systems affect innovation in the industry? Which IS employees are in danger of losing their jobs?
4. As Delta and its competitors strive for higher return on investment, what types of initiatives will give these companies a competitive advantage?

Sources: Melissa Solomon, "ROI: It's about People, Not Numbers," *Computerworld*, January 14, 2002, http://www.computerworld.com; Julia King, "ROI: Make It Bigger, Better, Faster," *Computerworld*, January 1, 2002, http://www.computerworld.com.

Pressplay: Defining the Internet Music Industry

The digitized world in which we now exist has brought with it wonderful conveniences, along with legal and ethical challenges. The digitization of data, information, and other media makes it possible to electronically store, access, and transfer a wide variety of material. This innovation has posed serious legal concerns over copyright infringement and ethical dilemmas for users, who find it possible to share copyrighted materials such as books, music, and video conveniently over the Internet. Sharing the intellectual and creative property of others deprives the copyright owners of the financial rewards they are due. Should this trend continue unchecked, some of the greatest creative minds in our society would no longer be able to make a living. In the year 2000 this problem exploded when the MP3 music file format was introduced, making it very easy to share music files over the Internet.

With every problem comes an opportunity. Here the challenge was to design a system that provides Internet users with a method for accessing music, while respecting copyrights, at an acceptable cost to compensate the record companies and musicians. The first company that could bring such a product to market would establish a solid competitive advantage in what could be a highly profitable industry.

The first companies to step up to the plate were Pressplay and Musicnet. Releasing their products within a week of each

WORLD VIEWS CASE

Information systems are used around the globe in a variety of business settings. For this edition of Principles of Information Systems, *we are pleased to include this and the other World Views Cases, which are written by people around the world, to demonstrate the international use of information systems. Our first case involves the use of the Internet to provide mortgage information and services.*

On-Line Presence Gives Mortgage Broker Global Reach

John Paynter
University of Auckland, New Zealand

Xiaohong Lu
University of Auckland, New Zealand

New Zealand is a small, diverse, geographically isolated country in the southwest Pacific with a population of 4 million people. Despite its size and seeming isolation, the country has a truly global focus. Recent immigration has aided population growth, and Kiwi expatriates work in places such as Japan, Hong Kong, the United Kingdom, and the United States. Traditionally, home ownership rates have been high. Yet, the standard of living has been eroding in world terms. Although the financial markets are unregulated, structural changes in the 1980s brought Reserve Bank control over the money rate, which is independent of government intervention. Despite economic contraction, inflation has been fueled by house price increases in Auckland, the largest city and the destination of most of the internal and external migration. This has caused interest rates to increase. In this tight market, mortgage lenders must offer competitive products. The last decade has seen the rise of mortgage brokers offering independent advice.

There are more than 40 mortgage brokerage sites in New Zealand. The first to move to on-line business was Kieran Trass, who founded mortgagenet (*www.amortgage.co.nz*) in 1996. He purchased his first computer and created a Web site to support his business. This effort took 6 weeks using Netscape Gold. Trass thought he could become a pure Internet player, so that he could run the business entirely through the Internet. But instead he found that "you have to have face-to-face to support the Internet presence. It's very difficult to do the entire process through the Internet, but we do because we have clients living overseas. We also build relationships with them by phone, fax, and e-mail. It's another channel, just like the yellow pages, but a little more interactive and powerful. You've still got to get people to visit." At first it cost him nothing to develop his site, but "after about 18 months, spending time and energy, I decided to get a dedicated Web site company."

The resulting Web site proved to be a problem, however. It didn't help Trass retain customers. "I had an initial influx of more people looking, but then I had less people on an ongoing basis. I couldn't understand, because I just spent $5,000." So he spent time surfing sites, looking for something different and thinking about his concept. He drew up icons and a home page and took them to his Web site company. "I said, 'Look, you've done a fantastic job of the database, but I'm not happy with the site. The design, layout, and information presentation are poor.'" So he went to a graphic artist and said, "I want you to create my Web site, the icons and how it drives and interacts." Once the artist was on board, he went to his initial developers and said, "I think you're really good at the back end and want to spend more money. I want you to develop my database further, and I'm using this guy to develop the front." He got the two of them together to develop the current format, which cost around $10,000.

Rather than charge people to access property investment information, Trass created a system whereby they could download articles after giving their e-mail addresses. "After five years I've got a database that's close to 5,000 subscribers. It's good target marketing; most are the property investors the site's designed to attract." He has also spent time researching what's happening overseas and locally to see what other people are doing. "Whose site looks stale or modern and innovative? In six years, I've spent around $50,000 [on the site]. I'm currently spending $1,000 to $2,000 monthly ... just keeping up to date, making a few changes."

One example of Trass's innovation is using text messaging in which the surfer can send a contact e-mail to his cell phone via the Internet. "It's been a support mechanism and a lead generator." In the future he may investigate video conferencing but is unsure whether the "power of the screen will be enough to replace a handshake."

Still, he explains, "I want to be different. It's better to be different than to be better. I like the simpleness of it; I get fantastic feedback. People do tell me they struggle to find things. I'm unsure how to address that without jeopardising my nice simple look."

When asked the main advantages of the site, he points to its global reach. "I've got clients worldwide—Sweden, Japan, USA, Philippines, France, England—I wouldn't get without the Internet. They are expatriate Kiwis owning properties in New Zealand, and they want to build a portfolio because when they come back one day, they would like to have a passive income and to secure a home base." He says that his clients are profitable because it's very cheap to deal with them via Internet, phone, and fax. And they are professionals who are accustomed to conducting business on-line. "They have to deal remotely with a lot of people here anyway. We can't spend money marketing. It's too hard for us to find them, so all we can do is have the presence, and they might come to us."

Discussion Questions
1. Describe the history and development of the mortgagenet Web site.
2. How does mortgagenet use technology for competitive advantage?

Critical Thinking Questions
3. How could other industries benefit from mortgagenet's lessons?
4. If you were Trass, what else would you do to keep your business competitive?

INFORMATION TECHNOLOGY CONCEPTS

Hardware: Input, Processing, and Output Devices

CHAPTER 3

PRINCIPLES

LEARNING OBJECTIVES

Assembling an effective, efficient computer system requires an understanding of its relationship to the information system and the organization. The computer system objectives are subordinate to, but supportive of, the information system and the needs of the organization.

- Describe how to select and organize computer system components to support information system objectives and business organization needs.

When selecting computer devices, you also must consider the current and future needs of the information system and the organization. Your choice of a particular computer system device should always allow for later improvements.

- Describe the power, speed, and capacity of central processing and memory devices.
- Describe the access methods, capacity, and portability of secondary storage devices.
- Discuss the speed, functionality, and importance of input and output devices.
- Identify popular classes of computer systems and discuss the role of each.

[Celera Genomics]

Supercomputers Are Key to Future Profitability

Celera Genomics is the company that recently made scientific history by mapping the genes of the human body. But after racing to finish work on the human genome in mid-2000, Celera surprised investors when it did not announce any profit-making products or services as a result of this breakthrough. As a result, its stock plunged from a high of nearly $270 in March 2000 to under $30 a year later.

Today Celera is building on its understanding of the human genome to branch profitably into proteomics, the study of the function, structure, and interactions of proteins in cells. Proteins regulate chemical reactions in the body as well as the activities of cells, tissues, and organs. Since proteins are directly involved in both normal and disease-associated biochemical processes, analyzing the proteins within a diseased cell can lead to a more complete understanding of the disease.

The combination of proteomics and genomics is expected to play a major role in biomedical research and lead to the development of exciting diagnostic and therapeutic products. Celera's new business strategy is to find isolated genes or proteins involved with a specific disease and then either sell those "drug targets" to major drug companies for a percentage of the profit or form a partnership with them. Celera itself is too small and inexperienced to bear the total cost of developing and marketing drugs on its own.

Although other companies have a head start in creating drugs from genomic data, Celera's knowledge of the human genome provides it with a competitive advantage in searching for leads that point to potential new medicines. To exploit that advantage, Celera began hiring world-class biologists and building a research lab to study proteins.

Celera's major investment in high technology to date includes separation devices, robotics, mass spectrometers, powerful computers, and other equipment for identifying and characterizing proteins and simulating disease progression. But this sophisticated modeling requires ever-more-powerful computer technology. In January 2001, Celera announced a four-year, $40 million project with Hewlett-Packard (H-P) and the Department of Energy to build a new supercomputer to analyze biological data. When fully operational, this computer will be capable of performing 100 trillion operations per second.

Celera already has one of the most powerful nongovernment supercomputing facilities in the world, featuring 800 interconnected H-P computer systems based on the 64-bit Alpha computer chip. Each of these computers is capable of performing more than 250 billion gene sequence comparisons per hour. Celera could not have sequenced and assembled the human genome so quickly without its supercomputer system. In fact, what Celera accomplished in just nine months would have taken hundreds of years using ordinary computers. In addition to powerful computers, Celera has more than 100 terabytes (about 100 trillion or 100 thousand billion bytes) of information stored on disk. That's larger than the data storage capacity at even classified government computing complexes.

Stock market analysts don't expect Celera to be profitable until 2004 or 2005. Fortunately, the company has plenty of cash in the bank and can afford to be patient in following its new business strategy. Celera went for a second round of financing while its stock was at an all-time high. Sales of additional stock raised more than $1 billion in cash, giving Celera unusual flexibility in launching its drug program.

As you read this chapter, consider the following:

- ◉ How are companies using hardware and technology to help them compete and achieve their mission?
- ◉ How must organizations integrate business strategy with technology, hardware, and systems to help them increase their sales and profits?

Today's use of technology is practical—intended to yield real business benefits, as seen with Celera Genomics. Employing information technology and providing additional processing capabilities can increase employee productivity, expand business opportunities, and allow for more flexibility. As we already discussed, a computer-based information system (CBIS) is a combination of hardware, software, database(s), telecommunications, people, and procedures—all organized to input, process, and output data and information. In this chapter, we concentrate on the hardware component of a CBIS. Hardware consists of any machinery (most of which uses digital circuits) that assists in the input, processing, storage, and output activities of an information system. The overriding consideration in making hardware decisions in a business should be how hardware can be used to support the objectives of the information system and the goals of the organization.

COMPUTER SYSTEMS: INTEGRATING THE POWER OF TECHNOLOGY

A computer system is a special subsystem of an organization's overall information system. It is an integrated assembly of devices—centered on at least one processing mechanism utilizing digital electronics—that are used to input, process, store, and output data and information.

Putting together a complete computer system, however, is more involved than just connecting computer devices. In an effective and efficient system, components are selected and organized with an understanding of the inherent trade-offs between overall system performance and cost, control, and complexity. For instance, in building a car, manufacturers try to match the intended use of the vehicle to its components. Racing cars, for example, require special types of engines, transmissions, and tires. The selection of a transmission for a racing car, then, requires not only consideration of how much of the engine's power can be delivered to the wheels (efficiency and effectiveness) but also how expensive the transmission is (cost), how reliable it is (control), and how many gears it has (complexity). Similarly, organizations assemble computer systems so that they are effective, efficient, and well suited to the tasks that need to be performed.

Because business needs and their importance vary at different companies, the information system solutions chosen can be quite different. Mike Meinz, Director of IT at General Mills, was facing a merger with Pillsbury and wanted to simplify the combined companies' computing infrastructure. He believed that the best way to do this was to minimize the number of hardware and software vendors, so he chose Hewlett-Packard as a key vendor and purchased three of its newest, most powerful computers. Homestore.com is a home improvement and real estate company that has seen a rapid increase in the number of visitors to its Web site. Scott Sullivan, VP for technology and operations, needed flexible technology that could be expanded quickly, easily, and inexpensively when business demanded. As a result, he elected to go with Dell computers because of Dell's ability to tailor computer systems to client needs. United Airlines needed powerful computers to plan flights based on passenger loads and to access large databases of passenger information. It chose IBM to provide the hardware to meet this need because of IBM's experience in large-capacity computer systems. Bell South's vice-president of enterprise data and infrastructure, Rich Liddell, needed outstanding service and support. He elected to go with Sun Microsystems because

United Airlines needed a powerful computer system to manage the large volume of flight and passenger data.

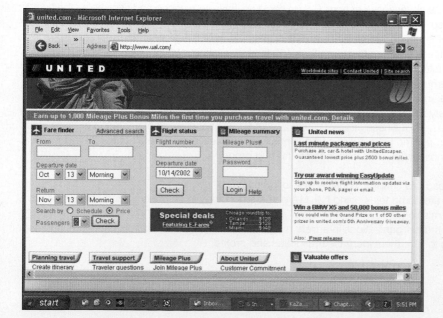

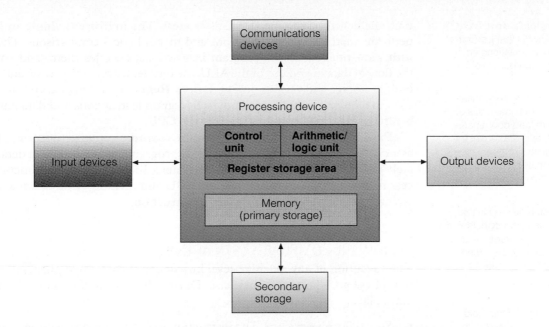

F I G U R E 3.1

Computer System Components

These components include input devices, output devices, communications devices, primary and secondary storage devices, and the central processing unit (CPU). The control unit, the arithmetic/logic unit (ALU), and the register storage areas constitute the CPU.

it had helped work out problems with software vendors in the past and also installed additional computing capacity over a weekend.[1]

As each of these examples demonstrates, assembling a computer subsystem requires an understanding of its relationship to the information system and the needs of the organization. Although we generally refer to the computer subsystem as simply a computer system, we must remember that the computer system objectives are subordinate to, but supportive of, the information system and the needs of the organization.

The components of all information systems—such as hardware devices, people, and procedures—are interdependent. Because the performance of one system affects the others, all of these systems should be measured according to the same standards of effectiveness and efficiency, given the constraints of cost, control, and complexity.

When selecting computer subsystem devices, you also must consider the current and future uses to which these systems will be put. Your choice of a particular computer system should always allow for later improvements in the overall information system. Reasoned forethought—a trait required for dealing with computer, information, and organizational systems of all sizes—is the hallmark of a true systems professional.

HARDWARE COMPONENTS

Computer system hardware components include devices that perform the functions of input, processing, data storage, and output (Figure 3.1). To understand how these hardware devices work together, consider an analogy from a paper-based office environment. Imagine a one-room office occupied by a single individual. The human (the processor) is capable of organizing and manipulating data. The person's mind (register storage) and the desk occupied by the human (primary storage) are places to temporarily store data. Filing cabinets fill the need for a more permanent form of storage (secondary storage). In this analogy, the incoming and outgoing mail trays can be understood as sources of new data (input) or as places to put the processed paperwork (output).

The ability to process (organize and manipulate) data is a critical aspect of a computer system, in which processing is accomplished by an interplay between one or more of the central processing units and primary storage. Each **central processing unit (CPU)** consists of three associated elements: the arithmetic/logic

central processing unit (CPU)
the part of the computer that consists of three associated elements: the arithmetic/logic unit, the control unit, and the register areas

arithmetic/logic unit (ALU)
portion of the CPU that performs mathematical calculations and makes logical comparisons

control unit
part of the CPU that sequentially accesses program instructions, decodes them, and coordinates the flow of data in and out of the ALU, the registers, primary storage, and even secondary storage and various output devices

register
high-speed storage area in the CPU used to temporarily hold small units of program instructions and data immediately before, during, and after execution by the CPU

primary storage (main memory; memory)
part of the computer that holds program instructions and data

instruction time (I-time)
the time it takes to perform the fetch-instruction and decode-instruction steps of the instruction phase

execution time (E-time)
the time it takes to execute an instruction and store the results

machine cycle
the instruction phase followed by the execution phase

pipelining
a form of CPU operation in which there are multiple execution phases in a single machine cycle

unit, the control unit, and the register areas. The **arithmetic/logic unit (ALU)** performs mathematical calculations and makes logical comparisons. The **control unit** sequentially accesses program instructions, decodes them, and coordinates the flow of data in and out of the ALU, the registers, primary storage, and even secondary storage and various output devices. **Registers** are high-speed storage areas used to temporarily hold small units of program instructions and data immediately before, during, and after execution by the CPU.

Primary storage, also called **main memory** or just **memory**, is closely associated with the CPU. Memory holds program instructions and data immediately before or immediately after the registers. To understand the function of processing and the interplay between the CPU and memory, let's examine the way a typical computer executes a program instruction.

HARDWARE COMPONENTS IN ACTION

The execution of any machine-level instruction involves two phases: the instruction phase and the execution phase. During the instruction phase, the following takes place:

- *Step 1: Fetch instruction.* The instruction to be executed is accessed from memory by the control unit.
- *Step 2: Decode instruction.* The instruction is decoded so the central processor can understand what is to be done, relevant data is moved from memory to the register storage area, and the location of the next instruction is identified.

Steps 1 and 2 are called the instruction phase, and the time it takes to perform this phase is called the **instruction time (I-time)**.

The second phase is the execution phase. During the execution phase, the following steps are performed:

- *Step 3: Execute the instruction.* The ALU does what it is instructed to do, such as making either an arithmetic computation or a logical comparison.
- *Step 4: Store results.* The results are stored in registers or memory.

Steps 3 and 4 are called the execution phase. The time it takes to complete the execution phase is called the **execution time (E-time)**.

After both phases have been completed for one instruction, they are again performed for the second instruction, and so on. The instruction phase followed by the execution phase is called a **machine cycle** (Figure 3.2). Some central processing units can speed up processing by using **pipelining**, whereby the CPU gets one instruction, decodes another, and executes a third at the same time. The

FIGURE 3●2

Execution of an Instruction

In the instruction phase, the computer's control unit fetches the instruction to be executed from memory (1). Then the instruction is decoded so the central processor can understand what is to be done (2). In the execution phase, the ALU does what it is instructed to do, making either an arithmetic computation or a logical comparison (3). Then the results are stored in the registers or in memory (4). The instruction and execution phases together make up one machine cycle.

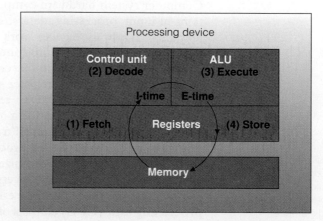

Pentium processor, for example, uses two execution unit pipelines. This gives the processing unit the ability to execute two instructions in a single machine cycle.

PROCESSING AND MEMORY DEVICES: POWER, SPEED, AND CAPACITY

The components responsible for processing—the CPU and memory—are housed together in the same box or cabinet, called the *system unit*. All other computer system devices, such as the monitor and keyboard, are linked either directly or indirectly into the system unit housing. As discussed previously, achieving information system objectives and organizational goals should be the primary consideration in selecting processing and memory devices. In this section, we investigate the characteristics of these important devices.

PROCESSING CHARACTERISTICS AND FUNCTIONS

Because having efficient processing and timely output is important, organizations use a variety of measures to gauge processing speed. These measures include the time it takes to complete a machine cycle and clock speed.

Machine Cycle Time

As we've seen, the execution of an instruction takes place during a machine cycle. The time in which a machine cycle occurs is measured in fractions of a second. Machine cycle times are measured in *microseconds* (one-millionth of one second) for slower computers to *nanoseconds* (one-billionth of one second) and *picoseconds* (one-trillionth of one second) for faster ones. Machine cycle time also can be measured in terms of how many instructions are executed in a second. This measure, called **MIPS**, stands for millions of instructions per second. MIPS is another measure of speed for computer systems of all sizes.

MIPS
millions of instructions per second

Clock Speed

Each CPU produces a series of electronic pulses at a predetermined rate, called the **clock speed**, which affects machine cycle time. The control unit portion of the CPU controls the various stages of the machine cycle by following predetermined internal instructions, known as **microcode**. You can think of microcode as predefined, elementary circuits and logical operations that the processor performs when it executes an instruction. The control unit executes the microcode in accordance with the electronic cycle, or pulses of the CPU "clock." Each microcode instruction takes at least the same amount of time as the interval between pulses. The shorter the interval between pulses, the faster each microcode instruction can be executed (Figure 3.3).

clock speed
a series of electronic pulses produced at a predetermined rate that affect machine cycle time

microcode
predefined, elementary circuits and logical operations that the processor performs when it executes an instruction

F I G U R E   3.3

Clock Speed and the Execution of Microcode Instructions

A faster clock speed means that more microcode instructions can be executed in a given time period.

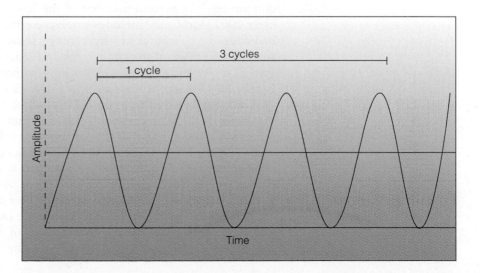

hertz
one cycle or pulse per second

megahertz (MHz)
millions of cycles per second

gigahertz (GHz)
billions of cycles per second

Clock speed is often measured in megahertz. As seen in Figure 3.3, a **hertz** is one cycle or pulse per second. **Megahertz (MHz)** is the measurement of cycles in millions of cycles per second, and **gigahertz (GHz)** stands for billions of cycles per second. The clock speed for personal computers can range from 200 MHz for computers bought in the mid-1990s to 2.2 GHz for the Intel Pentium 4 available in early 2002.[2]

Because the number of microcode instructions needed to execute a single program instruction—such as performing a calculation or printing results—can vary, there is no direct relationship between clock speed measured in megahertz and processing speed measures such as MIPS. Although widely touted, clock speed is only meaningful when making speed comparisons between computer chips in the same family from the same manufacturer—comparing one Intel Pentium 4 chip with another, for example.

Wordlength and Bus Line Width

bit
BInary digiT—0 or 1

wordlength
the number of bits the CPU can process at any one time

Data is moved within a computer system in units called bits. A **bit** is a binary digit—0 or 1. Another factor affecting overall system performance is the number of bits the CPU can process at one time, or the **wordlength** of the CPU. Early computers were built with CPUs that had a wordlength of 4-bits, meaning that the CPU was capable of processing 4 bits at one time. The 4 bits could be used to represent actual data, an instruction to be processed, or the address of data to be accessed. The 4-bit limitation was quite confining and greatly constrained the power of the computer. Over time, CPUs have evolved to 8-, 16-, 32-, and 64-bit machines with dramatic increases in power and capability. Computers with larger wordlengths can transfer more data between devices in the same machine cycle. They can also use the larger number of bits to address more memory locations and hence are a requirement for systems with certain large memory requirements.

Digital Equipment Corporation (since acquired by Compaq, which was acquired by Hewlett-Packard) was the first major chip manufacturer to produce a 64-bit chip (the Alpha chip) in 1994. Silicon Graphics, Sun Microsystems, Hewlett-Packard, IBM, and Intel have also come out with a 64-bit chip. Such a chip allows the CPU to directly address 16 quintillion (billion billion) unique address locations compared with 4.3 billion for a 32-bit processor. The ability to directly access a larger address space is critical for multimedia, imaging, and database applications; however, the computer's operating system and related application software must also support 64-bit technology to achieve the full benefit of the 64-bit architecture.

The Itanium processor is Intel's first microprocessor that is based on the 64-bit architecture called IA-64. It was codeveloped with Hewlett-Packard at a cost of over $1 billion. The Itanium will provide Intel's platform for 64-bit computing. The second member of the IA-64 family is McKinley, released in early 2002. McKinley includes extensive error-correcting code, which corrects bad code in a computer's main memory. It also can detect abnormal conditions such as the failure of the cooling fan and react by reducing the operation of the system to prevent it from overheating—without shutting it down completely.[3] Future IA-64 microprocessors have the code names Madison and Deerfield.

bus line
the physical wiring that connects the computer system components

Data is transferred from the CPU to other system components via **bus lines**, the physical wiring that connects the computer system components. The number of bits a bus line can transfer at any one time is known as bus line width. Bus line width should be matched with CPU wordlength for optimal system performance. It would be of little value, for example, to install a new 64-bit bus line if the system's CPU had a wordlength of only 16. Assuming compatible wordlengths and bus widths, the larger the wordlength, the more powerful the computer.

Because all these factors—machine cycle time, clock speed, wordlength, and bus line width—affect the processing speed of the CPU, comparing the speed of two different processors even from the same manufacturer can be confusing.

Although the megahertz rating has important consequences for the design of a PC system and is therefore important to the PC design engineer, it is not necessarily a good measure of processor performance, especially when comparing one family of processors with the next or comparing between manufacturers. Chip makers such as Intel, Advanced Micro Devices, and Sun Microsystems have developed a number of benchmarks for speed. To ensure objective comparisons, many people prefer to use general computer system benchmarks defined by non-chip manufacturers such as SYSmark, High End Winstone, and Business Winstone. In addition, there are benchmarks for specific applications, such as Internet access and multimedia. For less technical measures of performance, popular PC journals (such as *PC Magazine* and *PC World*) often rate personal computers on price, performance, reliability, service, and other factors.

Physical Characteristics of the CPU

CPU speed is also limited by physical constraints. Most CPUs are collections of digital circuits imprinted on silicon wafers, or chips, each no bigger than the tip of a pencil eraser. To turn a digital circuit within the CPU on or off, electrical current must flow through a medium (usually silicon) from point A to point B. The speed at which it travels between points can be increased by either reducing the distance between the points or reducing the resistance of the medium to the electrical current.

Reducing the distance between points has resulted in ever-smaller chips, with the circuits packed closer together. In the 1960s, shortly after patenting the integrated circuit, Gordon Moore, former chairman of the board of Intel (the largest microprocessor chip maker), formulated what is now known as **Moore's Law**. This hypothesis states that transistor (the microscopic on/off switches, or the microprocessor's brain cells) densities on a single chip will double every 18 months. Moore's Law has held up amazingly well over the years, although Moore himself thought that the industry could run into fundamental physical limits when the dimensions of the tiny transistors reached 0.25 micron (a micron is one millionth of a meter). This barrier was broken in 1998, and chips made for mass production in 2002 are at the 0.13 micron level. In June 2001, Intel reported that the company had successfully made a handful of silicon transistors 80 atoms wide and 3 atoms thick capable of switching on and off 1.5 trillion times a second. This breakthrough will make it possible to build a CPU with 1 billion transistors operating at 20 GHz by around 2007. Such a computer will have 25 times the number of transistors and be more than 10 times faster than the computers of 2002. Many researchers now forecast that it will be possible to continue doubling the number of transistors on a chip every 18 months until at least 2014.[4]

In addition to increased speeds, Moore's Law has had an impact on costs and overall system performance. As seen in Figure 3.4, the number of transistors on a chip continues to climb.

Researchers are taking many approaches to continue to improve the performance of computers. One approach is to substitute superconductive material for the silicon in computer chips. **Superconductivity** is a property of certain metals that allows current to flow with minimal electrical resistance. Traditional silicon chips create some electrical resistance that slows processing. Chips built from less resistant superconductive metals offer increases in processing speed. The use of materials other than silicon, including carbon and gallium arsenide (GaAs), is also being investigated and shows some promise.

Researchers at The Johns Hopkins University have developed a technique incorporating the use of thin layers of silicon placed on top of a layer of synthetic sapphire that allows the use of light, rather than electricity, to send data between microchips. The technique employs microscopic lasers and optical fibers to move data as much as 100 times faster while requiring less power than current chips.[5] Other companies are experimenting with chips called **optical processors** that

Moore's Law
a hypothesis that states that transistor densities on a single chip will double every 18 months

superconductivity
a property of certain metals that allows current to flow with minimal electrical resistance

optical processors
computer chips that use light waves instead of electrical current to represent bits

NUMBER OF
TRANSISTORS

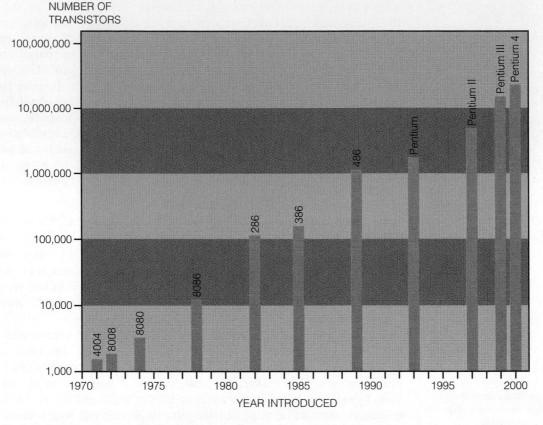

YEAR INTRODUCED

FIGURE 3.4

Moore's Law

(Source: Data from "Moore's Law:
Overview," Intel Web site,
www.intel.com/research/silicon/
mooreslaw.htm, accessed
July 22, 2002.)

use light waves instead of electrical current to represent bits. The primary advantage of optical processors is their speed. Optical processors have the potential to
be 500 times faster than traditional electronic circuits.

In August 2001, IBM announced that it had altered the structure of silicon in
a way that would increase semiconductor speed by 35 percent. This strained silicon technology is based on the tendency of atoms inside compounds to align
with one another, enabling a faster flow of electrons. This technology will be used
in high-end computers as early as 2003.[6]

Another advancement is the development of the carbon nanotube, so called
because it is a pure carbon tube made of hexagonal structures 1 to 3 nanometers
in diameter. The nanotubes can be used to form the tiny circuits for computer
components. The practical use of nanotubes may not occur until sometime in the
decade of 2010 because of manufacturing difficulties and other complications.[7]

Scientists at the Los Alamos National Laboratory have taken miniaturization
to the extreme. They are experimenting with radio waves to manipulate individual atoms into executing a simple computer program. Their goal is to be able to
manipulate thousands of atoms and build a computer many times smaller yet
more powerful than any computer currently in existence.[8]

Complex and Reduced Instruction Set Computing

**complex instruction set
computing (CISC)**
a computer chip design that
places as many microcode instructions into the central processor as
possible

**reduced instruction set
computing (RISC)**
a computer chip design based on
reducing the number of microcode
instructions built into a chip to an
essential set of common microcode instructions

Processors for many personal computers are designed based on **complex
instruction set computing (CISC)**, which places as many microcode instructions into the central processor as possible. In the mid-1970s John Cocke of IBM
recognized that most of the operations of a CPU involved only about 20 percent
of the available microcode instructions. This led to an approach to chip design
called **reduced instruction set computing (RISC)**, which involves reducing
the number of microcode instructions built into a chip to this essential set of
common microcode instructions. RISC chips are faster than CISC chips for processing activities that predominantly use this core set of instructions because
each operation requires fewer microcode steps prior to execution. Most RISC

RM7000 Family • 64-Bit MIPS-based Processor

PMC-Sierra's RM7000 family of MIPS RISC processors provides high-performance and low-power solutions for embedded applications such as networking, printing, workstation, and consumer devices.

(Source: Courtesy of PMC-Sierra, Inc.)

chips use pipelining, which, as mentioned earlier, allows the processor to execute multiple instructions in a single machine cycle. With less sophisticated microcode instruction sets, RISC chips are also less expensive to produce and are quite reliable.

The PowerPC chip is a RISC processor created by Motorola. By almost any benchmark, RISC processors run faster than Intel's Pentium processor. And because RISC chips have a simpler design and require less silicon, they are cheaper to produce. The PowerPC chip is designed to provide portable and desktop personal computers with the processing power normally associated with much more expensive computers. For example, the PowerPC has the ability to make functions such as voice recognition, dictation, pen input, and touch screens practical. Apple announced its Xserve servers based on dual 1-GHz PowerPC processors in July 2002.[9] Sun Microsystems' Sparc chip is another example of a RISC processor.

When selecting a CPU, organizations must balance the benefits of speed with the issues of cost, control, and complexity. CPUs with faster clock speeds and machine cycle times are usually more expensive than slower ones. This expense, however, is a necessary part of the overall computer system cost, for the CPU is typically the single largest determinant of the price of many computer systems. CPU speed can also be related to complexity. Having a less complex code, as in the case of RISC chips, not only can increase speed and reliability but can also reduce chip manufacturing costs.

MEMORY CHARACTERISTICS AND FUNCTIONS

Located physically close to the CPU (to decrease access time), memory provides the CPU with a working storage area for program instructions and data. The chief feature of memory is that it rapidly provides the data and instructions to the CPU.

Storage Capacity

Like the CPU, memory devices contain thousands of circuits imprinted on a silicon chip. Each circuit is either conducting electrical current (on) or not (off). By representing data as a combination of on or off circuit states, the data is stored in memory. Usually eight bits are used to represent a character, such as the letter *A*. Eight bits together form a **byte (B)**. Following is a list of storage capacity measurements. In most cases, storage capacity is measured in bytes, with one byte usually equal to one character.

byte (B)
eight bits together that represent a single character of data

Name	Abbreviation	Number of Bytes
Byte	B	1
Kilobyte	KB	1,024 Bytes
Megabyte	MB	1,024 Kilobytes (about 1 million)
Gigabyte	GB	1,024 Megabytes (about 1 billion)
Terabyte	TB	1,024 Gigabytes (about 1 trillion)
Petabyte	PB	1,024 Terabytes (about 1 quadrillion)

random access
memory (RAM)
a form of memory in which
instructions or data can be
temporarily stored

Types of Memory

There are several forms of memory, as shown in Figure 3.5. Instructions or data can be temporarily stored in **random access memory (RAM)**. RAM is temporary and volatile—RAM chips lose their contents if the current is turned off or disrupted (as in a power surge, brownout, or electrical noise generated by lightning or nearby machines). RAM chips are mounted directly on the computer's main circuit board or in other chips mounted on peripheral cards that plug into the computer's main circuit board. These RAM chips consist of millions of switches that are sensitive to changes in electric current.

RAM comes in many different varieties. The mainstream type of RAM is extended data out, or EDO, RAM, which is faster than older types of RAM. Another kind of RAM is called dynamic RAM (DRAM). SDRAM, or synchronous DRAM, needs high or low voltages at regular intervals—every two milliseconds (two one-thousandths of a second)—to retain its information. SDRAM can exceed EDO RAM in performance. SDRAM also has the advantage of a faster transfer speed between the microprocessor and the memory.

Over the past decade, microprocessor speed has doubled every 18 months, but memory performance has not kept pace. In effect, memory has become the principal bottleneck to system performance. Thus, microprocessor manufacturers are working with memory vendors to keep up with the performance of faster processors and bus architectures.

read-only memory (ROM)
a nonvolatile form of memory

Another type of memory, **ROM**, an acronym for **read-only memory**, is usually nonvolatile. In ROM, the combination of circuit states is fixed, and therefore its contents are not lost if the power is removed. ROM provides permanent storage for data and instructions that do not change, like programs and data from the computer manufacturer.

There are other types of nonvolatile memory as well. Programmable read-only memory (PROM) is a type in which the desired data and instructions—and hence the desired circuit state combination—must first be programmed into the memory chip. Thereafter, PROM behaves like ROM. PROM chips are used where the CPU's data and instructions do not change, but the application is so specialized or unique that custom manufacturing of a true ROM chip would be cost prohibitive. A common use of PROM chips is for storing the instructions to popular video games, such as those for Nintendo and Sega. Game instructions are programmed onto the PROM chips by the game manufacturer. Instructions and data can be programmed onto a PROM chip only once.

Erasable programmable read-only memory (EPROM) is similar to PROM except, as the name implies, the memory chip can be erased and reprogrammed. EPROM is used where the CPU's data and instructions change, but only infrequently. An automobile manufacturer, for example, might use an industrial robot

FIGURE 3.5

Basic Types of Memory Chips

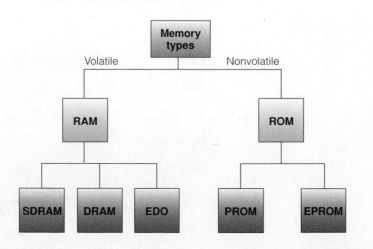

to perform repetitive operations on a certain car model. When the robot is performing its operations, the nonvolatility and rapid accessibility to program instructions offered by EPROM is an advantage. Once the model year is over, however, the EPROM controlling the robot's operation will need to be erased and reprogrammed to accommodate a different car model.

cache memory
a type of high-speed memory that a processor can access more rapidly than main memory

Cache memory is a type of high-speed memory that a processor can access more rapidly than main memory (Figure 3.6). Cache memory functions somewhat like a notebook used to record phone numbers. A person's private notebook may contain only 1 percent of all the numbers in the local phone directory, but the chance that the person's next call will be to a number in his or her notebook is high. Cache memory works on the same principle—frequently used data is stored in easily accessible cache memory instead of slower memory like RAM. Because there is less data in cache memory, the CPU can access the desired data and instructions more quickly than if it were selecting from the larger set in main memory. The CPU can thus execute instructions faster, and the overall performance of the computer system is raised. There are three types of cache memory present in the majority of systems shipped—the Level 1 (L1) cache is in the processor; the Level 2 (L2) cache memory is optional and found on the motherboard of most systems. In addition, the original Intel Itanium processor had a 4MB Level 3 (L3) cache on the motherboard that connected to the processor over a bus that runs as fast as the processor—800 MHz. The McKinley processor has a 3MB L3 cache integrated directly onto the processor. In addition, it has 256K of L2 cache and 32K L1 cache, also all on processor.

The main memory in your system that can move its information into your system's cache memory is called the *cacheable memory*. Memory in your system that is not cacheable performs as if your system is cacheless, moving information as needed directly to the processor without the ability to use the cache memory as a fast-retrieval storage bin. All systems have a main memory cacheable limit, typically 512 KB or greater.

Costs for memory capacity continue to decline. When considered on a megabyte-to-megabyte basis, memory is still considerably more expensive than most forms of secondary storage. Memory capacity can be important in the effective operation of a CBIS. The specific applications of a CBIS determine the amount of memory required for a computer system. For example, complex processing problems, such as computer-assisted product design, require more memory than simpler tasks like word processing. Also, because computer systems have different types of memory, other programs may be needed to control how memory is accessed and used. In other cases, the computer system can be configured to maximize memory usage. Before additional memory is purchased, all these considerations should be addressed.

FIGURE 3.6

Cache Memory

Processors can access this type of high-speed memory faster than main memory. Located near the CPU, cache memory works in conjunction with main memory. A cache controller determines how often the data is used and transfers frequently used data to cache memory, then deletes the data when it goes out of use.

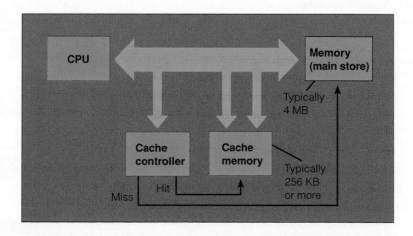

MULTIPROCESSING

multiprocessing
simultaneous execution of two or more instructions at the same time

coprocessor
part of the computer that speeds processing by executing specific types of instructions while the CPU works on another processing activity

A number of forms of **multiprocessing** involve the simultaneous execution of two or more instructions at the same time. One form of multiprocessing involves coprocessors. A **coprocessor** speeds processing by executing specific types of instructions while the CPU works on another processing activity. Coprocessors can be internal or external to the CPU and may have different clock speeds than the CPU. Each type of coprocessor best performs a specific function. For example, a math coprocessor chip can be used to speed mathematical calculations, and a graphics coprocessor chip decreases the time it takes to manipulate graphics.

Massively Parallel Processing

massively parallel processing
a form of multiprocessing that speeds processing by linking hundreds or thousands of processors to operate at the same time, or in parallel, with each processor having its own bus, memory, disks, copy of the operating system, and applications

Another form of multiprocessing, called **massively parallel processing**, speeds processing by linking hundreds and even thousands of processors to operate at the same time, or in parallel. Each processor includes its own bus, memory, disks, copy of the operating system, and application software. With parallel processing, a business problem (such as designing a new product or piece of equipment) is divided into several parts. Each part is "solved" by a separate processor. The results from each processor are then assembled to get the final output (Figure 3.7). Celera Genomics, mentioned at the beginning of the chapter, used this form of processing to decode the human genome.

Massively parallel processing systems can coordinate large amounts of data and access them with greater speed than was previously possible. The most frequent business uses for massive parallel processing include modeling, simulation, and the analysis of large amounts of data. In today's challenging marketplace, consumers are demanding increased product features and a whole array of new services. These consumer demands have forced companies to find more effective and insightful ways of gathering and analyzing information, not just about existing customers but also potential customers. Collecting and organizing this enormous amount of data is a difficult task. Massively parallel processing can access and analyze the data to create the information necessary to build an effective marketing program that can give the company a competitive advantage.

symmetrical multiprocessing (SMP)
another form of parallel processing in which multiple processors run a single copy of the operating system and share the memory and other resources of one computer

Symmetrical multiprocessing (SMP) is another form of parallel processing in which multiple processors run a single copy of the operating system and share the memory and other resources of one computer. Sharing resources creates more overhead than a single-processor system or the massively parallel processing system. As a result, the processing capability of SMP systems isn't proportionally greater than that of single-processor systems (i.e., the capability of an SMP processor with two processors is less than twice the speed of a single processor).

FIGURE 3.7

Massively Parallel Processing

Massively parallel processing involves breaking a problem into various subproblems or parts, then processing each of these parts independently. The most difficult aspect of massively parallel processing is not the simultaneous processing of the subproblems but the logical structuring of the problem into independent parts.

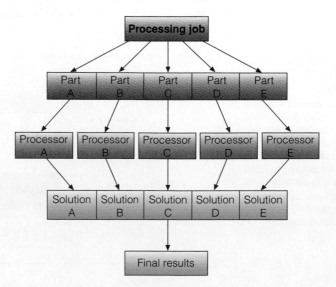

SMP has been implemented in the Sun Microsystems UltraSPARC and SPARCserver, IBM Alpha, Macintosh PowerPC, and Intel chips.

SECONDARY STORAGE

secondary storage (permanent storage)
devices that store larger amounts of data, instructions, and information more permanently than allowed with main memory

As we have seen, memory is an important factor in determining overall computer system power. However, main memory provides only a small amount of storage area for the data and instructions required by the CPU for processing. Computer systems also need to store larger amounts of data, instructions, and information more permanently than allowed with main memory. **Secondary storage**, also called permanent storage, serves this purpose.

Compared with memory, secondary storage offers the advantages of nonvolatility, greater capacity, and greater economy. As previously noted, on a megabyte-per-megabyte basis, most forms of secondary storage are considerably less expensive than memory (Figure 3.8). Because of the electromechanical processes involved in using secondary storage, however, it is considerably slower than memory. The selection of secondary storage media and devices requires an understanding of their primary characteristics—access method, capacity, and portability.

As with other computer system components, the access methods, storage capacities, and portability required of secondary storage media are determined by the information system's objectives. An objective of a credit card company's information system, for example, might be to rapidly retrieve stored customer data to approve customer purchases. In this case, a fast access method is critical. In other cases, such as salesforce automation via laptop computers, portability and storage capacity might be major considerations in selecting and using secondary storage media and devices.

sequential access
retrieval method in which data must be accessed in the order in which it is stored

direct access
retrieval method in which data can be retrieved without the need to read and discard other data

Storage media that provide faster access methods are generally more expensive than slower media. The cost of additional storage capacity and portability varies widely, but it is also a factor to consider. In addition to cost, organizations must address security issues to allow only authorized people access to sensitive data and critical programs. Because the data and programs kept in secondary storage devices are so critical to most organizations, all of these issues merit careful consideration.

ACCESS METHODS

Data and information access can be either sequential or direct. **Sequential access** means that data must be accessed in the order in which it is stored. For example, inventory data stored sequentially may be stored by part number, such as 100, 101, 102, and so on. If you want to retrieve information on part number 125, it is necessary to read and discard all the data relating to parts 001 through 124.

Direct access means that data can be retrieved directly, without the need to pass by other data in sequence. With direct access, it is possible to go directly to and access the needed data—say, part number 125—without having to read through parts 001 through 124. For this reason, direct access is usually faster

FIGURE 3.8

Cost Comparison for Various Forms of Data Storage

All forms of secondary storage cost considerably less per megabyte of capacity than RAM, although they have slower access times. A diskette costs about 35 cents per megabyte, while RAM can cost around $.90 per megabyte, 3 times more.

(Source: Data from CompUSA Direct Catalog, January 2002.)

Data Storage Media	Data Tape Cartridge	Data Tape Cartridge	Rewritable Optical Disk	Jaz Disk	Zip Disk	Zip Disk	Floppy Diskette	SDRAM
Capacity	30 GB	2 GB	2.6 GB	2 GB	200 MB	100 MB	1.44 MB	128 MB
Cost	$366.77	$40.45	$58.91	$98.34	$15.02	$9.45	$0.50	$111.93
Cost/MB	0.012	0.020	0.023	0.049	0.075	0.095	0.347	0.874

sequential access storage device (SASD)
device used to sequentially access secondary storage data

direct access storage device (DASD)
device used for direct access of secondary storage data

magnetic tape
common secondary storage medium, Mylar film coated with iron oxide with portions of the tape magnetized to represent bits

than sequential access. The devices used to sequentially access secondary storage data are simply called **sequential access storage devices (SASDs)**; those used for direct access are called **direct access storage devices (DASDs).**

DEVICES

The most common forms of secondary storage include magnetic tapes, magnetic disks, and optical disks. Some of these media (magnetic tape) allow only sequential access, while others (magnetic and optical disks) provide direct and sequential access. Figure 3.9 shows some different secondary storage media.

Magnetic Tapes

One common secondary storage medium is **magnetic tape**. Similar to the kind of tape found in audio- and videocassettes, magnetic tape is a Mylar film coated with iron oxide. Portions of the tape are magnetized to represent bits. Magnetic tape is an example of a sequential access storage medium. If the computer needs to read data from the middle of a reel of tape, all the tape before the desired piece of data must be passed over sequentially—one disadvantage of magnetic tape. When information is needed, it can take time for a tape operator to load the magnetic tape on a tape device and get the relevant data into the computer. Despite the falling prices of hard disk drives, tape storage is still a popular choice for low-cost data backup for off-site storage in the event of a disaster.

Technology is improving to provide tape drives with greater capacities and faster transfer speeds. There are three competing tape formats—Super Digital Linear Tape (SDLT), Advanced Intelligent Tape (AIT-3), and Linear Tape Open (LTO). SDLT from Quantum can store up to 110 GB per tape cartridge with a data transfer rate of 10 MB/second. Quantum forecasts that the SDLT capacity will ultimately increase to 1 TB per cartridge with a transfer rate of 100 MB/second. Sony Electronics Inc.'s AIT-3 cartridges offer a 100 GB capacity per cartridge, with a transfer rate of 11 MB/second. By 2007, it expects its AIT-6 product to have a capacity of 800 GB with a data transfer rate of 96 MB/second. LTO, sponsored by Hewlett-Packard, Seagate Technology, and IBM, stores up to 100 GB per cartridge at transfer speeds up to 15 MB/second. By 2003, LTO technology drives may be capable of storing up to 1.6 TB with a data transfer speed of 320 MB/second.[10]

FIGURE 3.9

Types of Secondary Storage

Secondary storage devices such as magnetic tapes and disks, optical disks, CD-ROMs, and DVDs are used to store data for easy retrieval at a later date.

(Source: Courtesy of Imation.)

FIGURE 3⊙10

Hard Disk

Hard disks give direct access to stored data. The read/write head can move directly to the location of a desired piece of data, dramatically reducing access times, as compared with magnetic tape.

(Source: Courtesy of Seagate Technology.)

magnetic disk
common secondary storage medium, with bits represented by magnetized areas

redundant array of independent/inexpensive disks (RAID)
method of storing data that generates extra bits of data from existing data, allowing the system to create a "reconstruction map" so that if a hard drive fails, it can rebuild lost data

disk mirroring
a process of storing data that provides an exact copy that protects users fully in the event of data loss

storage area network (SAN)
technology that provides high-speed connections between data storage devices and computers over a network using the Fibre Channel communications protocol

Magnetic Disks

Magnetic disks are also coated with iron oxide; they can be thin steel platters (hard disks; see Figure 3.10) or Mylar film (diskettes). As with magnetic tape, magnetic disks represent bits by small magnetized areas. When reading from or writing data onto a disk, the disk's read/write head can go directly to the desired piece of data. Thus, the disk is called a direct access storage medium. Although disk devices can be operated in a sequential mode, most disk devices use direct access. Because direct access allows fast data retrieval, this type of storage is ideal for companies that need to respond quickly to customer requests, such as airlines and credit card firms. For example, if a manager needs information on the credit history of a customer or the seat availability on a particular flight, the information can be obtained in a matter of seconds if the data is stored on a direct access storage device. If the data is stored on magnetic tape, it could take from a few minutes to over half an hour to load the tape and get the information.

Magnetic disk storage varies widely in capacity and portability. Standard diskettes are portable but have a slower access time and lower storage capacity (1.44 MB for some computers) than fixed hard disks. Hard disk storage, while more costly and less portable, has a greater storage capacity and quicker access time.

RAID

Companies' data storage needs are expanding rapidly. Today's storage configurations routinely entail many hundreds of gigabytes. However, putting the company's data on-line involves a serious business risk—the loss of critical business data can put a corporation out of business. The concern is that the most critical mechanical components inside a disk storage device—the disk drives, the fans, and other input/output devices—can break (like most things that move).

Organizations now require that their data storage devices be fault tolerant— the ability to continue with little or no loss of performance in the event of a failure of one or more key components. **Redundant array of independent/ inexpensive disks (RAID)** is a method of storing data that generates extra bits of data from existing data, allowing the system to create a "reconstruction map" so that if a hard drive fails, it can rebuild lost data. With this approach, data is split and stored on different physical disk drives using a technique called *stripping* to evenly distribute the data. Since being developed at the University of California, Berkeley in 1987, RAID technology has been applied to storage systems to improve system performance and reliability.

RAID can be implemented in several ways. In the simplest form, RAID subsystems duplicate data on drives. This process, called **disk mirroring**, provides an exact copy that protects users fully in the event of data loss. However, if full copies are always to be kept current, organizations need to double the amount of storage capacity that is kept on-line. Thus, disk mirroring is expensive. Other RAID methods are less expensive because they only partly duplicate the data, allowing storage managers to minimize the amount of extra disk space (or overhead) they must purchase to protect data.

SAN

A **storage area network (SAN)** is a special-purpose high-speed network that provides direct connections between data storage devices and computers (Figure 3.11). This built-in system redundancy makes it possible to add storage and computers without requiring the network to be brought down. In today's demanding 24/7 business operations, this is a significant advantage. Read the "IS Principles in Action" special-interest feature to learn more about the need for highly reliable data storage methods and how that need is met.

Currently, SANs use Fibre Channel, a communications protocol designed for high-speed information transfer for computer, storage, and network devices. This protocol supports data transmission at rates exceeding 100 MB/second over

FIGURE 3.11

Storage Area Network

SAN provides high-speed connections between data storage devices and computers over a network using the Fibre Channel communications protocol.

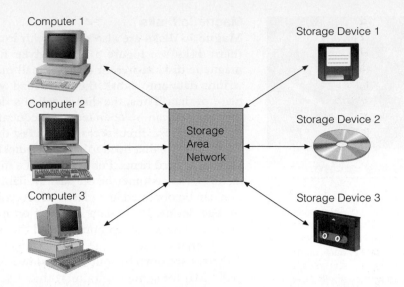

distances of a few miles. In the future, SANs may use other networking protocols including emerging Internet protocols. The use of a SAN enables different types of storage subsystems such as multiple RAID storage devices and magnetic tape backup systems to be integrated into a single system. SAN manufacturers include EMC and IBM.

Cox Communications Inc., a $3 billion cable TV and telecom services vendor in Atlanta, is adding a Dell storage area network, the Dell PowerVault SAN, so it can connect at least five storage systems of over 3.5 TB of data to all of its 300 Dell servers. The Fibre Channel–based SAN moves data at speeds of 100 MB per second, compared with the speeds of 20 MB per second Cox gets with standard connections.[11] USA Today implemented network-attached storage from Network Appliance Inc. as a way of replicating data over an IP network to a secure site in Maryland.[12]

Optical Disks

optical disk
a rigid disk of plastic onto which data is recorded by special lasers that physically burn pits in the disk

Another type of secondary storage medium is the **optical disk**. Similar in concept to a ROM chip, an optical disk is simply a rigid disk of plastic onto which data is recorded by special lasers that physically burn pits in the disk. Data is directly accessed from the disk by an optical disk device, which operates much like a stereo's compact disk player. This optical disk device uses a low-power laser that measures the difference in reflected light caused by a pit (or lack thereof) on the disk.

Each pit represents the binary digit 1; each unpitted area (called a *land*) represents the binary digit 0. Thus, the presence or lack of a pit determines the bit. Once a master optical disk has been created, duplicates can be manufactured using techniques similar to those used to produce music CDs.

compact disk read-only memory (CD-ROM)
a common form of optical disk on which data, once it has been recorded, cannot be modified

A common form of optical disk is called **compact disk read-only memory (CD-ROM)**. Once data has been recorded on a CD-ROM, it cannot be modified—the disk is "read only." CD-ROM disks and hardware have moved from being unique add-ons in the mid-1980s to being a standard feature of today's personal computers. **CD-writable (CD-W)** disks allow data to be written once to a CD disk. **CD-rewritable (CD-RW)** technology allows personal computer users to replace their diskettes with high-capacity CDs that can be written upon and edited over. The CD-RW disk can hold 740 MB of data—roughly 500 times the capacity of a 1.4-MB diskette.

CD-writable (CD-W) disk
an optical disk that can be written upon but only once

CD-rewritable (CD-RW) disk
an optical disk that allows personal computer users to replace their diskettes with high-capacity CDs that can be written upon and edited over

Magneto-Optical Disk

magneto-optical disk
a hybrid between a magnetic disk and an optical disk

A **magneto-optical disk** is a type of disk drive that combines magnetic disk technologies with CD-ROM technologies. Like magnetic disks, MO disks can be

Las Vegas Casinos Cover Their Bets with Highly Reliable Computers

Las Vegas casinos have mastered the art of gathering and using detailed customer information. They are especially interested in the data captured when guests use their loyalty cards—cards the casinos issue to frequent customers. A wealth of data is captured when players swipe their loyalty card at a table game or slot machine—how long they play, how much they win or lose, even what their betting strategy might be. The casino can compare recent statistics with previous visits and provide real-time hints to employees about how to treat a given customer based on how much earnings the guest may generate for the hotel and casino.

As they use their loyalty card to gamble, see shows, eat, or shop, guests are also racking up points that they can redeem for prizes, such as golf clubs, free hotel rooms, or tickets to popular shows. The Las Vegas casinos have perfected the art of providing complimentary items to encourage guests to return or spend more money. In 2000, the MGM Mirage gave out $286 million in such gifts while visitors spent $3.5 billion at its 10 properties. Harrah's has found such data to be so useful that it hasn't deleted any of the information it has gathered since 1995 on 23 million guests.

The casinos and hotels are always open. On the day that President Bush asked Americans to observe a national day of mourning, all the Las Vegas casinos owned by the giant MGM Mirage shut down their tables for one minute of silence. That brief moment on September 14 was the only time that action in Las Vegas casinos stopped entirely. Continuous computer availability is an absolute requirement in this $24 \times 7 \times 365$ industry. Even brief downtime for hardware maintenance or data backup can disrupt operations and cause disgruntled guests to leave for another casino—resulting in the loss of millions of dollars.

Casinos employ highly reliable computers that guarantee constant uptime. Disk mirroring or replicating data is commonly used to prevent downtime or data loss. Backup tapes are created each day and go out in armored trucks to a facility unknown to the casino and managed by a company that doesn't give its name out.

Some casinos completely mirror their main computers so that if one fails for any reason, its twin will be able to take over without skipping a beat. For example, the Venetian uses two IBM iSeries 830-2320 computers in a mirrored configuration.

Identical software is installed on each computer to run the hotel, casino, and slots plus support administrative functions such as reservations, finance, payroll, and time and attendance applications. As soon as a transaction has been processed on the primary system, it is passed to the backup system to be processed again, keeping the backup system in synch with the primary system. The goal is for the primary and backup systems to contain the same information. All of this duplication is expensive—each of the computers and its associated software costs several hundred thousand dollars.

Some casinos plan to add advanced storage area networks (SANs) to their hardware mix. The MGM Mirage is working with Dell Computer to install more flexible SANs so that all information systems can be switched to an alternate computer when the primary computer must be upgraded or brought down for maintenance. The goal is to allow the secondary computer to take over instantly and continue to update the SAN. Now, the SAN must come down during computer upgrades, and some data is lost.

Discussion Questions

1. Which of the three approaches to data protection do you think is most cost effective for a large casino such as the MGM Grand—disk mirroring, computer mirroring, or the use of a SAN? Why?
2. Why would a casino wish to keep secret the name of its data backup service provider as well as the location of the data backup facility?

Critical Thinking Questions

3. Identify at least three manufacturing or service-oriented companies that have similar needs for high-reliability data processing and data access. Explain why high reliability is important to each.
4. Identify at least three companies where high reliability of data processing and data access are *not* critical.

Sources: Adapted from Martin J. Garvey, "Casino CIOs Put Their Chips on IBM's eserver iseries," *Informationweek.com*, November 19, 2001, pp. 94–96; Mark Gimein, "Wish You Were Here," *Fortune*, October 15, 2001, accessed at http://www.fortune.com; Kim S. Nash, "Casinos Hit Jackpot with Customer Data," *Computerworld*, July 2, 2001, accessed at http://www.computerworld.com.

read from and written to. And like diskettes, they are removable. However, their storage capacity can exceed 5 GB, much greater than magnetic diskettes. In terms of data access speed, they are faster than diskettes but not as fast as hard disk drives. This type of disk uses a laser beam to change the molecular configuration of a magnetic substrate on the disk, which in turn creates visual spots. In conjunction with a photodetector, another laser beam reflects light off the disk

FIGURE 3.12

Digital Versatile Disk Player

DVDs look like CDs but have a much greater storage capacity and can transfer data at a much faster rate.

(Source: Courtesy of Sony Electronics.)

digital versatile disk (DVD)
storage medium used to store digital video or computer data

The PC memory card is like a portable hard disk that fits into any Type II PC Card slot and can store up to 5 gigabytes.

(Source: Courtesy of Kingston Technology.)

and measures the size of the spots; the presence or absence of a spot indicates a bit. The disk can be erased by demagnetizing the substrate, which in turn removes the spots, allowing the process to begin again.

A 2.3-GB MO disk format, nicknamed GigaMO, was jointly developed by Fujitsu and Sony in 2001. It can transfer data at the rate of 8.3 MB/second.[13] Meanwhile, Hewlett-Packard and Sony developed a 5.2-GB magneto-optical drive that costs around $2,000 with a single 5.2-GB WORM (write once, read many) disk costing nearly $75.

The primary advantage of optical disks is their huge storage capacities, compared with other secondary storage media. Optical disks can store large applications and programs that contain graphics and audio data. They also allow for storage of data that are not needed at a given moment but could possibly be useful later.

Digital Versatile Disk

A **digital versatile disk (DVD)** is a five-inch CD-ROM look-alike with the ability to store about 135 minutes of digital video or several gigabytes of data (Figure 3.12). Software programs, video games, and movies are common uses for this storage medium. At a data transfer rate of 1.25 MB/second, the access speed of a DVD drive is faster than the typical CD-ROM drive. From the former read-only DVD formats, demand for rewritable DVDs has driven their development.

The two main competing formats in the rewritable DVD market are DVD-RAM and DVD+RW. The DVD-RAM format is supported by Hitachi, Toshiba, and Matsushita/Panasonic and can store 2.6 GB of data on single-sided disks or 5.2 GB on double-sided disks. DVD-RAM drives from Hitachi have been shipping since 1998. The DVD+RW format was defined by Sony, Hewlett-Packard, and Phillips and allows for 3 GB of data on single-sided disks and 6 GB on double-sided disks. However, the three companies announced in May 2001 that they were adopting a variation of this format called DVD-R that lets users record disks once and rewrite compact disks (CDs) several times. The DVD-R format has a major advantage—it is compatible with almost all home DVD movie players. DVD-R disks cost $20 to $30 compared with $2 for a videocassette.[14]

Both DVD-RAM and DVD+R drives can read audio CDs, but there are a number of incompatibility issues. A disc created by a DVD-RAM drive won't work in a DVD+R drive and vice versa. And many of today's DVD-ROM (read-only memory, nonrewritable) and DVD-R drives cannot read disks created in the rewritable formats of DVD-RAM or DVD+RW. Much like the Beta versus VHS format battle for video tapes in the 1970s, these incompatibilities mean that if an organization adopts one format but the other format becomes the standard, the organization's DVD investment is greatly diminished.[15]

Memory Cards

A group of computer manufacturers formed the Personal Computer Memory Card International Association (PCMCIA) to create standards for a peripheral device known as a PC memory card. These PC memory cards are credit-card-size devices that can be installed in an adapter or slot in many personal computers. To the rest of the system, the PC memory card functions as though it were a fixed hard disk drive. Although the cost per megabyte of storage is greater than for traditional hard disk storage, these cards are less failure prone than hard disks, are portable, and are relatively easy to use. Software manufacturers often store the instructions for their program on a memory card for use with laptop computers.

Flash Memory

Flash memory is a silicon computer chip that, unlike RAM, is nonvolatile and keeps its memory when the power is shut off. Flash memory chips are small and can be easily modified and reprogrammed, which makes them popular in computers, cellular phones, and other products. Flash memory is also used in some handheld computers to store data and programs, in digital cameras to store photos, and in airplanes to store flight information in the cockpit. Compared with other types of secondary storage, flash memory can be accessed more quickly, consumes less power, and is smaller in size. The primary disadvantage is cost. Flash memory chips can cost almost three times more per megabyte than a traditional hard disk. Nonetheless, the market for flash memory has exploded in recent years.

Expandable Storage

Expandable storage devices use removable disk cartridges (Figure 3.13). When your storage needs to increase, you can use more removable disk cartridges. The storage capacity can range from less than 100 MB to several gigabytes per cartridge. In recent years, the access speed of expandable storage devices has increased. Some devices are about as fast as an internal disk drive.

Expandable storage devices can be internal or external. A few personal computers are now including internal expandable storage devices as standard equipment. Zip by Iomega is an example. Of course, a CD-RW drive by Hewlett-Packard, Iomega, and others can also be used for expandable storage. With so much critical data stored on your hard drive, it is wise to make frequent backups. However, use of the standard diskette would require over 200 disks and several hours to back up even a small 300-MB hard drive. So, expandable storage devices are ideal for backups. They can hold at least 80 times as much data as and operate five times faster than the existing 1.44-MB diskette drives. Although more expensive than fixed hard disks, removable disk cartridges combine hard disk storage capacity and diskette portability. Some organizations prefer removable hard disk storage for the portability and control it provides. For example, a large amount of data can be taken to any location, or it can be secured so that access is controlled.

The overall trend in secondary storage is toward direct-access methods, higher capacity, and increased portability. Organizations should select the specific type of storage based on their needs and resources. In general, the ability to store large amounts of data and information and access it quickly can increase organizational effectiveness and efficiency by allowing the information system to provide the desired information in a timely fashion. Table 3.1 lists the most common secondary storage devices and their capacities for easy reference.

FIGURE 3.13

Expandable Storage

Expandable storage drives allow users to add additional storage capacity by simply plugging in a removable disk or cartridge. The disks can be used to back up hard disk data or to transfer large files to colleagues.

(Source: Courtesy of Iomega.)

flash memory
a silicon computer chip that, unlike RAM, is nonvolatile and keeps its memory when the power is shut off

expandable storage devices
storage that uses removable disk cartridges to provide additional storage capacity

TABLE 3.1

Comparison of Secondary Storage Devices

Storage Device	Year First Introduced	Maximum Capacity
3.5-inch diskette	1987	1.44 MB
CD-ROM	1990	650 MB
Zip	1995	100–250 MB
DVD	1996	17 GB

INPUT AND OUTPUT DEVICES: THE GATEWAY TO COMPUTER SYSTEMS

A user's first experience with computers is usually through input and output devices. Through these devices—the gateways to the computer system—people provide data and instructions to the computer and receive results from it. Input and output devices are part of the overall user interface, which includes other hardware devices and software that allow humans to interact with a computer system.

As with other computer system components, the selection of input and output devices depends on organizational goals and information system objectives. For example, many restaurant chains use handheld input devices or computerized terminals that let waiters enter orders to ensure timely and accurate data input. These systems have cut costs by making inventory tracking more efficient and marketing to customers more effective.

CHARACTERISTICS AND FUNCTIONALITY

Rapidly getting data into a computer system and producing timely output is very important for today's organizations. The form of the output desired, the nature of the data required to generate this output, and the required speed and accuracy of the output and the input determine the appropriate output and input devices. Some organizations have very specific needs for output and input, requiring devices that perform specific functions. The more specialized the application, the more specialized the associated system input and output devices.

The speed and functions performed by the input and output devices selected and used by the organization should be balanced with their cost, control, and complexity. More specialized devices might make it easier to enter data or output information, but they are generally more costly, less flexible, and more susceptible to malfunction.

The Nature of Data

Getting data into the computer—input—often requires transferring human-readable data, such as a sales order, into the computer system. Human-readable data is data that can be directly read and understood by humans. A sheet of paper containing inventory adjustments is an example of human-readable data. By contrast, machine-readable data can be understood and read by computer devices (e.g., the universal bar code that grocery scanners read) and is typically stored as bits or bytes. Data on inventory changes stored on a diskette is an example of machine-readable data.

Data can be both human readable and machine readable. For example, magnetic ink on bank checks can be read by humans and computer system input devices. Most input devices require some human interaction, because people most often begin the input process by organizing human-readable data and transforming it into machine-readable data. Every keystroke on a keyboard, for example, turns a letter symbol of a human language into a digital code that the machine can understand.

Data Entry and Input

Getting data into the computer system is a two-stage process. First, the human-readable data is converted into a machine-readable form through a process called **data entry**. The second stage involves transferring the machine-readable data into the system. This is **data input**.

Today, many companies are using on-line data entry and input—the immediate communication and transference of data to computer devices directly connected to the computer system. On-line data entry and input places data into the computer system in a matter of seconds. Organizations in many industries require the instantaneous update offered by this approach. For example, an airline clerk may need to enter a last-minute reservation. On-line data entry and input is used to record the reservation as soon as it is made. Reservation agents at

data entry
process by which human-readable data is converted into a machine-readable form

data input
process that involves transferring machine-readable data into the system

other terminals can then access this data to make a seating check before they make another reservation.

Source Data Automation

Regardless of how data gets into the computer, it should be captured and edited at its source. **Source data automation** involves capturing and editing data where the data is originally created and in a form that can be directly input to a computer, thus ensuring accuracy and timeliness. For example, using source data automation, salespeople enter sales orders into the computer at the time and place they take the order. Any errors can be detected and corrected immediately. If any item is temporarily out of stock, the salesperson can discuss options with the customer. Prior to source data automation, orders were written on a piece of paper and entered into the computer later (often by someone other than the person who took the order). Often the handwritten information wasn't legible or, worse yet, got lost. If problems occurred during data entry, it was necessary to contact the salesperson or the customer to "recapture" the data needed for order entry, leading to further delays and customer dissatisfaction.

INPUT DEVICES

Literally hundreds of devices can be used for data entry and input. They range from special-purpose devices used to capture specific types of data to more general-purpose input devices. Some of the special-purpose data entry and input devices will be discussed later in this chapter. First, we will focus on devices used to enter and input more general types of data, including text, audio, images, and video for personal computers.

Personal Computer Input Devices

A keyboard and a computer mouse are the most common devices used for entry and input of data such as characters, text, and basic commands. Some companies are developing newer keyboards that are more comfortable, adjustable, and faster to use. These keyboards, such as the split keyboard by Microsoft and others, are designed to avoid wrist and hand injuries caused by hours of keyboarding. Using the same keyboard, you can enter sketches on the touchpad and text using the keys.

A computer mouse is used to "point to" and "click on" symbols, icons, menus, and commands on the screen. This causes the computer to take a number of actions, such as placing data into the computer system.

Voice-Recognition Devices

Another type of input device can recognize human speech. Called **voice-recognition devices**, these tools use microphones and special software to record and convert the sound of the human voice into digital signals. Speech recognition can be used on the factory floor to allow equipment operators to give basic commands to machines while they are using their hands to perform other operations. Voice recognition is also used by security systems to allow only authorized personnel into restricted areas. Voice recognition has been used in many industries, including automobiles. Japan's Nippon Telegraph and Telephone Corp. and Honda have collaborated to create a voice-activated vehicle information system that could let some Honda drivers check and respond to e-mail. The system relies on e-mail management software to prioritize and summarize messages. The in-car system uses both voice-recognition software that adapts to surrounding noises and voice-synthesis software to produce a high-quality natural tone when reading data requested by the driver.[16]

Voice-recognition systems now available on many makes of autos and trucks allow a driver to activate radio programs and CDs. It can even tell you the time. Asking "What time is it?" will get a response such as, "Eleven thirty-four."

source data automation
capturing and editing data where the data is originally created and in a form that can be directly input to a computer, thus ensuring accuracy and timeliness

voice-recognition device
an input device that recognizes human speech

An ergonomic keyboard is designed to be more comfortable to use.

(Source: Courtesy of Kinesis Corporation, www.kinesis-ergo.com.)

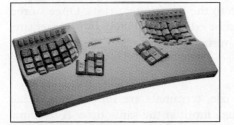

Voice-recognition devices analyze and classify speech patterns and convert them into digital codes. Some systems require "training" the computer to recognize a limited vocabulary of standard words for each user. Operators train the system to recognize their voices by repeating each word to be added to the vocabulary several times. Other systems allow a computer to understand a voice it has never heard. In this case, the computer must be able to recognize more than one pronunciation of the same word—for example, recognizing the use of the phrase "Please?" spoken by someone from Cincinnati as meaning the same as "Huh?" spoken by someone from the Bronx, or "I beg your pardon?" spoken by an English person.

Digital Computer Cameras

digital computer camera
input device used with a PC to record and store images and video in digital form

Some personal computers work with **digital computer cameras**, which record and store images and video in digital form. These cameras look very similar to a regular camera. When you take pictures, the images are electronically stored in the camera. A cable is then connected from the camera to a port on the computer, and the images can be downloaded. During the download, the visual images are converted into digital codes by a computer board. Once downloaded and converted into digital format, the images can be modified and included in other applications. For example, a photo of the company office recorded by a digital computer camera can be captured and then pasted into a word processing document used in a company brochure. You can even add sound and handwriting to the photo. Some digital cameras, like the Sony Mavica, can store images on diskettes. The diskettes can then be inserted into a computer to transfer the photos to a hard disk. Once on the hard disk, the images can be edited, sent to another location, pasted into another application, or printed. Some personal computers, as shown in Figure 3.14, have a video camera that records full-motion video. Fuji, Hewlett-Packard, Kodak, Minolta, Olympus, Sony, and Toshiba each offer at least one digital camera for about $300 with a standard eyepiece viewfinder, a color screen for previewing and reviewing shots, and a built-in flash.[17]

Digital camera manufacturers frequently promote the resolutions of their cameras, with 2, 3, and 5 megapixels resolution common today. Uninformed buyers assume that the number of pixels (or points of data) that a camera can capture is the definitive measure of image quality. However, resolution is simply the measure of the amount of data saved in the picture file. This relates to how large an image may be printed or displayed, but not how well the picture will appear. The key is to choose a camera with a resolution suited to your output purpose. A 1 megapixel resolution is good for a 5 × 7 inch print, but a 2 megapixel resolution is needed for a good 8 × 10 inch picture.

Digital cameras are available that rival cameras used by professional photographers for photo quality and such features as zoom, flash, exposure controls, special effects, and even video-capture capabilities. They support a wide variety of resolutions, taking either snapshot prints and images appropriate for the Web (such as those from VGA cameras) or large 16- × 20-inch prints (such as those from expensive 5 megapixel cameras).

FIGURE 3.14

A PC Equipped with a Computer Camera

Digital video cameras make it possible for people at distant locations to conduct videoconferences, thereby eliminating the need for expensive travel to attend physical meetings.

(Source: © Andreas Pollok/Stone.)

The number one advantage of digital cameras is saving time and money by eliminating the need to process film. Kodak is now allowing photographers to have it both ways. When Kodak print film is developed, Kodak offers the option of placing pictures on a CD in addition to the traditional prints. Once stored on the CD, the photos can be edited, placed on an Internet site, or sent electronically to business associates or friends around the world.

Terminals

Inexpensive and easy to use, terminals are input devices that perform data entry and data input at the same time. A terminal is connected to a complete computer system, including a processor, memory, and secondary storage. General commands, text, and other data are entered via a keyboard or mouse, converted into machine-readable form, and transferred to the processing portion of the computer system. Terminals, normally connected directly to the computer system by telephone lines or cables, can be placed in offices, in warehouses, and on the factory floor.

Scanning Devices

Image and character data can be input using a scanning device. A page scanner is like a copy machine. The page to be scanned is typically inserted into the scanner or placed face down on the glass plate of the scanner, covered, and scanned. With a handheld scanner, the scanning device is moved or rolled manually over the image to be scanned. Both page and handheld scanners can convert monochrome or color pictures, forms, text, and other images into machine-readable digits. It has been estimated that U.S. enterprises generate over one billion pieces of paper daily. To cut down on the high cost of using and processing paper, many companies are looking to scanning devices to help them manage their documents.

Optical Data Readers

A special scanning device called an *optical data reader* can also be used to scan documents. The two categories of optical data readers are optical mark recognition (OMR) and optical character recognition (OCR). OMR readers are used for test scoring and other purposes when test takers use pencils to fill in boxes on OMR paper, which is also called a "mark sense form." OMR is used in standardized tests, including SAT and GMAT tests. In comparison, most OCR readers use reflected light to recognize various characters. With special software, OCR readers can convert handwritten or typed documents into digital data. Once entered, this data can be shared, modified, and distributed over computer networks to individuals.

Magnetic Ink Character Recognition (MICR) Devices

In the 1950s, the banking industry became swamped with paper checks, loan applications, bank statements, and so on. The result was the development of magnetic ink character recognition (MICR), a system for reading this data quickly. With MICR, data is placed on the bottom of a check or other form using a special magnetic ink. Data printed with this ink using a character set is readable by both people and computers (Figure 3.15).

Point-of-Sale (POS) Devices

point-of-sale (POS) device
terminal used in retail operations to enter sales information into the computer system

Point-of-sale (POS) devices are terminals used in retail operations to enter sales information into the computer system. The POS device then computes the total charges, including tax. Many POS devices also use other types of input and output devices, like keyboards, bar code readers, printers, and screens. A large portion of the money that businesses spend on computer technology involves POS devices.

Automated Teller Machine (ATM) Devices

Another type of special-purpose input/output device, the automated teller machine (ATM), is a terminal used by most bank customers to perform withdrawals and other transactions with their bank accounts. The ATM, however, is no longer used only for cash and bank receipts. Companies use various ATM devices to support their specific business processes. Some can dispense tickets for airlines, concerts, and soccer games. Some colleges use them to output transcripts. For this reason, the input and output capabilities of ATMs are quite varied. Like POS devices, ATMs may combine other types of input and output devices. Unisys, for example, has developed an ATM kiosk that allows bank customers to make cash withdrawals, pay bills, and also receive advice on investments and retirement planning.[18]

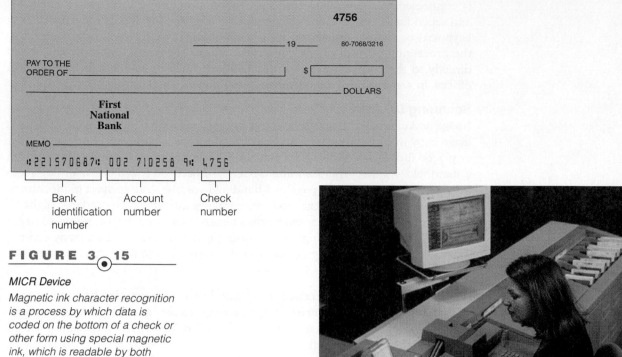

Bank
identification
number

Account
number

Check
number

FIGURE 3●15

MICR Device

Magnetic ink character recognition is a process by which data is coded on the bottom of a check or other form using special magnetic ink, which is readable by both computers and humans. For example, look at the bottom of a bank check or most utility bills.

(Source: Courtesy of NCR Corporation.)

Pen Input Devices

By touching the screen with a pen input device, it is possible to activate a command or cause the computer to perform a task, enter handwritten notes, and draw objects and figures. Pen input requires special software and hardware. Handwriting recognition software can convert handwriting on the screen into text. The Tablet PC from Microsoft and its various hardware partners will be able to transform handwriting into typed text and store the "digital ink" just the way a person writes it. Users will be able to use a pen to write and send e-mail, add comments to Word documents, mark up PowerPoint presentations, and even hand draw charts in a document. That data can then be moved, highlighted, searched, and converted into text. If successful, such an interface is likely to become widely used. Providing such a simple input method will be highly attractive to users who are uncomfortable using a keyboard. The keys to the success of this means of input are the accuracy at which handwriting can be read and translated into digital form and cost.

Light Pens

A light pen uses a light cell in the tip of a pen. The cell recognizes light from the screen and determines the location of the pen on the screen. Like pen input devices, light pens can be used to activate commands and place drawings on the screen.

Touch-Sensitive Screens

Advances in screen technology allow display screens to function as input as well as output devices. By touching certain parts of a touch-sensitive screen, you can execute a program or cause the computer to take an action. Touch-sensitive

Input on tablet PCs, such as this ViewPad 1000 from ViewSonic, is done either with the penlike device shown at bottom left or through a touch-sensitive virtual keyboard that appears on screen. The device uses handwriting recognition software for the pen input.

(Source: Courtesy of ViewSonic Corporation.)

screens are popular input devices for some small computers because they preclude the necessity of keyboard input devices that consume space in storage or in use. They are frequently used at gas stations for customers to select grades of gas and request a receipt, on photocopy machines to enable users to select various options, at fast-food restaurants for order clerks to enter customer choices, at information centers in hotels to allow guests to request facts about local eating and drinking establishments, and at amusement parks to provide directions to patrons. They also are used in kiosks at airports and department stores.

Bar Code Scanners

A bar code scanner employs a laser scanner to read a bar-coded label. This form of input is used widely in grocery store checkouts and in warehouse inventory control. Owens & Minor is a medical and surgical supply company that has equipped hospital operating rooms with supplies and labeled each item with a bar code. When an item is used during surgery, a wireless handheld computer scans its bar code to update a computerized inventory system. The system tracks how much inventory is left and automatically places an order for refills when an item falls below the reorder point.[19]

OUTPUT DEVICES

Computer systems provide output to decision makers at all levels of an organization to solve a business problem or capitalize on a competitive opportunity. In addition, output from one computer system can be used as input into another computer system within the same information system. The desired form of this output might be visual, audio, and even digital. Whatever the output's content or form, output devices function to provide the right information to the right person in the right format at the right time.

Display Monitors

The display monitor is a TV-screen-like device on which output from the computer is displayed. Because the monitor uses a cathode ray tube to display images, it is sometimes called a CRT. The monitor works much the same way a TV screen does—one or more electron beams are generated from cathode ray tubes. As the beams strike a phosphorescent compound (phosphor) coated on the inside of the screen, a dot on the screen called a **pixel** lights up. A pixel is a dot of color on a photo image or a point of light on a display screen. It can be in one of two modes: on or off. The electron beam sweeps back and forth across the screen so that as the phosphor starts to fade, it is struck again and lights up again.

pixel
a dot of color on a photo image or a point of light on a display screen

With today's wide selection of monitors, price and overall quality can vary tremendously. The quality of a screen is often measured by the number of horizontal and vertical pixels used to create it. A larger number of pixels per square inch means a higher resolution, or clarity and sharpness of the image. For example, a screen with a 1,024 × 768 resolution (786,432 pixels) has a higher sharpness than one with a resolution of 640 × 350 (224,000 pixels). The distance between one pixel on the screen and the next nearest pixel is known as *dot pitch*. The common range of dot pitch is from .25 mm to .31 mm. The smaller the number, the better the picture. A dot pitch of .28 mm or smaller is considered good. Greater pixel densities and smaller dot pitches yield sharper images of higher resolution.

A monitor's ability to display color is a function of the quality of the monitor, the amount of RAM in the computer system, and the monitor's graphics adapter card. The color graphics adapter (CGA) was one of the first technologies to display color images on the screen. Today, super video graphics array (SVGA) displays are standard, providing vivid colors and superior resolution.

CRT monitors are large and bulky in comparison with LCD monitors (flat displays).

(Source: Courtesy of ViewSonic Corporation.)

plotter

a type of hard-copy output device used for general design work

FIGURE 3.16

Laser Printer

Laser printers, available in a wide variety of speeds and price ranges, have many features, including color capabilities. They are the most common solution for outputting hard copies of information.

(Source: Courtesy of Epson America, Inc.)

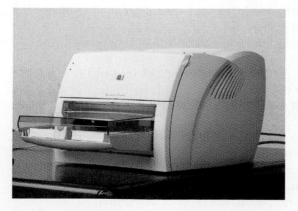

Liquid Crystal Displays (LCDs)

Because CRT monitors use an electron gun, there must be a distance of one foot between the gun and screen, causing them to be large and bulky. Thus, a different technology, flat-panel display, is used for portable personal computers and laptops. One common technology used for flat screen displays is the same liquid crystal display (LCD) technology used for pocket calculators and digital watches. LCD monitors are flat displays that use liquid crystals—organic, oil-like material placed between two polarizers—to form characters and graphic images on a back-lit screen. These displays are much easier on your eyes because they are flicker-free, far lighter, less bulky, and don't emit the type of radiation that makes some CRT users worry.

The primary choices in LCD screens are passive-matrix and active-matrix LCD displays. In a passive-matrix display, the CPU sends its signals to transistors around the borders of the screen, which control all the pixels in a given row or column. In an active-matrix display, each pixel is controlled by its own transistor attached in a thin film to the glass behind the pixel. Passive-matrix displays are typically dimmer, slower, but less expensive than active-matrix ones. Active-matrix displays are bright, clear, and have wider viewing angles than passive-matrix displays. Active-matrix displays, however, are more expensive and can increase the weight of the screen.

LCD technology is also being used to create thin and extremely high-resolution monitors for desktop computers. Although the screen may measure just 13 inches from corner to corner, the display's extremely high resolution—1,280 × 1,280 pixels—lets it show as much information as a conventional 20-inch monitor. And while cramming more into a smaller area causes text and images to shrink, you can comfortably sit much closer to an LCD screen than the conventional CRT monitor. Several manufacturers offer LCD displays at about $400.

Printers and Plotters

One of the most useful and popular forms of output is called *hard copy*, which is simply paper output from a printer. Printers with different speeds, features, and capabilities are available. Some can be set up to accommodate different paper forms such as blank check forms, invoice forms, and so forth. Newer printers allow businesses to create customized printed output for each customer from standard paper and data input using full color.

The speed of the printer is typically measured by the number of pages printed per minute (ppm). Like a display screen, the quality, or resolution, of a printer's output depends on the number of dots printed per inch. A 600-dpi (dots-per-inch) printer prints more clearly than a 300-dpi printer. A recurring cost of using a printer is the ink-jet or laser cartridge that must be replaced every few thousand pages of output. Figure 3.16 shows a laser printer.

Plotters are a type of hard-copy output device used for general design work. Businesses typically use these devices to generate paper or acetate blueprints, schematics, and drawings of buildings or new products onto paper or transparencies.

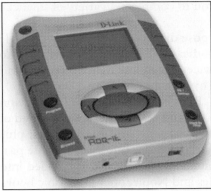

D-Link offers an MP3 player with a 10-GB storage capacity that can hold 150 hours of music.

(Source: Courtesy of D-Link Systems, Inc.)

music device
a device that can be used to download music from the Internet and play the music

multifunction device
a device that can combine a printer, fax machine, scanner, and copy machine into one device

Standard plot widths are 24 inches and 36 inches, and the length can be whatever meets the need—from a few inches to feet.

Computer Output Microfilm (COM) Devices

Companies that produce and store significant numbers of paper documents often use computer output microfilm (COM) devices to place data from the computer directly onto microfilm for future use. The traditional photographic phase of conversion to microfilm is eliminated. Once this is done, a standard microfilm reader can access the data. Newspapers and journals typically place their past publications on microfilm using COM, giving readers the ability to view past articles and news items.

Music Devices

Music devices, such as Diamond Multimedia's Rio 500 portable MP3 player, are about the size of a cigarette pack and can be used to download music from the Internet and other sources. These devices have no moving parts and can store hours of music. When you get tired of the music, you can always download new pieces. Music devices that use the MP3 standard can cost under $300. MP3 is an abbreviation for Motion Picture Experts Group Audio Layer 3, a popular music format for the Internet. This format requires about 1 MB of storage for every minute of music, which is about a tenth of the storage required for music on a standard CD. The MP3 standard allows the music to be compressed, so it takes less time to download from the Internet. A piece of music in this format also requires less storage space on the MP3 player or the hard drive of a computer. While MP3 is currently the leading digital audio format, it is facing competition from Microsoft's Windows Media, RealNetworks' Real Audio, and an upgraded version of MP3 called mp3PRO which will cut the storage space required in half.

A number of computer manufacturers—including Apple, Dell, Hewlett-Packard, NEC, and others—offer computers that make downloading and playing music from the Internet in the MP3 format easier with higher sound quality. These specialized computers offer easy downloading and superior speakers for playing music. A number of Internet sites allow people to share music using the MP3 format. Because musicians and production companies don't receive royalties from people sharing music in the MP3 format, there are also a number of legal and ethical issues that must be addressed.

SPECIAL-PURPOSE INPUT AND OUTPUT DEVICES

Many additional input and output devices are used for specialized or unique applications. A **multifunction device** can combine a printer, fax machine, scanner, and copy machine into one device. Multifunction devices are less expensive than buying these devices separately, and they take less space on a desktop compared with separate devices. For example, the Hewlett-Packard OfficeJet v40xi is a small, quiet, and efficient all-in-one unit that faxes black and white and color pages, supports Optical Character Recognition (OCR), prints photo-quality slicks, and copies documents at a cost under $250. Canon, Xerox, and others make similar devices.

Special-purpose hearing devices can be used to detect manufacturing or equipment problems. The Georgia Institute of Technology has developed a hardware device that can "listen to" equipment to detect worn or damaged parts. Voice-output devices, also called *voice-response devices,* allow the computer to send voice output in the form of synthesized speech over phone lines and other media. Some banks and financial institutions use voice recognition and response to give customers account information over the phone. The Smart Disk allows you to place a smart card, which is similar to a credit card in size and function, into the Smart Disk device. The Smart Disk is then inserted into a standard

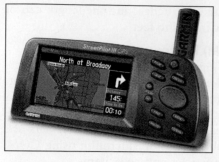

The GPS Street Pilot III provides voice prompts to guide you.

(Source: Courtesy of Garmin Corporation.)

diskette drive on a PC, which allows you to complete a variety of financial transactions using your smart card.

The e-book reader is a special-purpose display unit that allows people to download books, newspapers, periodicals, and other documents onto it for their reading enjoyment. The devices weigh less than 22 ounces and fit a paperback size of 5 inches wide by 7 inches high by 1 inch deep.[20] E-book readers are available from Rocket, Franklin, and Thomson Multimedia (under the RCA brand name) and cost in the neighborhood of $300. Readers benefit from built-in dictionaries and highlighters, plus e-books make it a cinch to bookmark and search for quotes or favorite passages. Once the book or document is read, another one can be downloaded into the device. Some bookstores, including Barnes and Noble, offer electronic books that can be downloaded into electronic book devices. Many electronic books allow you to quickly move up or back in the document and to search for key terms or themes. In addition, new technology, such as ClearType from Microsoft, makes text displayed on e-book screens easier to read and closer to the quality of a printed page. Many authors, including Stephen King, have experimented with releasing their books over the Internet to be read on a computer or using an e-book.

On-board computer-based navigation systems are an option on many luxury cars including the Lexus, Acura, Infiniti, and Cadillac. The system is able to pinpoint the position of the auto from location data received from a global positioning satellite (GPS) received through an onboard antenna. It uses vehicle speed information to determine how far you've traveled and a gyro sensor to tell when you've turned. Then the device compares all this information with the system's CD-ROM road map database to show your position within a few feet. The GPS Street Pilot III is a portable positioning system for your car that costs about $1,000. It provides turn-by-turn voice prompts that guide you to your destination. Just enter a street address or business location to access the shortest and fastest route directly to the door. Downloadable maps focus in on your region of choice.

COMPUTER SYSTEM TYPES, STANDARDS, SELECTING, AND UPGRADING

special-purpose computers
computers used for limited applications by military and scientific research groups

In general, computers can be classified as either special purpose or general purpose. **Special-purpose computers** are used for limited applications by military and scientific research groups such as the CIA and NASA. Other applications include specialized processors found in appliances, cars, and other products. Special-purpose computers are increasingly being used by businesses. For example, automobile repair shops connect special-purpose computers to your car's engine to identify specific performance problems.

general-purpose computers
computers used for a wide variety of applications

General-purpose computers are used for a variety of applications and are the most common. The computers used to perform business applications discussed in this book are general-purpose computer systems. General-purpose computer systems combine processors, memory, secondary storage, input and output devices, a basic set of software, and other components. These systems can range from inexpensive personal computers to expensive supercomputers. These systems display a wide range of capabilities. Table 3.2 shows general ranges of capabilities for various types of computer systems.

COMPUTER SYSTEM TYPES

Computer systems can range from desktop (or smaller) portable computers to massive supercomputers that require housing in large rooms. Let's examine the types of computer systems in more detail.

Type of computer	Typical Processor Speed	Weight	Cost	How Used	Example
Handheld	> 50 MHz	< .5 lb	< $500	Personal organizer	Palm
Notebook	> 500 MHz	< 4 lbs	< $1000	Improvement of individual worker's productivity	IBM
Laptop	> 500 MHz	< 7 lbs	< $2000	Improvement of individual worker's productivity	Apple iBook Hewlett-Packard
Network	> 200 MHz	< 15 lbs	< $750	Support for data entry and Internet connection	Oracle
Desktop	> 800 MHz	< 25 lbs	< $2000	Improvement of individual worker's productivity	Apple iMac Dell
Workstation	> 2 GHz	< 30 lbs	$4000 – $40,000	Engineering, CAD, software development	Sun Microsystems
Midrange	> 1 GHz	> 50 lbs	$20,000 – $250,000	Computing for a department or small company	IBM AS/400
Mainframe	> 300 MIPS	> 200 lbs	> $250,000	Computing for a large company	IBM Z/900
Supercomputer	> 2 Teraflops	> 200 lbs	> $1,000,000	Scientific applications, marketing, customer support, product development	Compaq Terascale

TABLE 3.2

Types of Computer Systems

personal computer (PC)
relatively small, inexpensive computer system, sometimes called a microcomputer

Personal Computers

As previously noted, **personal computers (PCs)** are relatively small, inexpensive computer systems, sometimes called *microcomputers*. Although personal computers are designed primarily for individual users, they are often tied in to larger computer and information systems as well. Personal computers can be purchased from retail stores or on-line. In 2000, 51 percent of the households in the U.S. had at least one personal computer and nearly 90 percent of children used personal computers in school.[21]

There are several types of personal computers. Named for their size (small enough to fit on an office desk), *desktop computers* are the most common personal computer system configuration. Increasingly, powerful desktop computers can provide sufficient memory and storage for most business computing tasks. Desktop PCs have become standard business tools; more than 30 million are in use in large corporations.

In addition to traditional PCs that use Intel processors and Microsoft software, there are other options. One of the most popular is the iMac by Apple Computer. In January 2002, Apple announced three models of its new iMac personal computer with a design that represents a radical departure from the previous version. Its base is a small, half sphere that measures 10.6 inches in diameter. A flat-panel monitor is attached to the base using a jointed chrome neck that can be adjusted to position the monitor. Some say that the new machine looks more like a desk lamp than a desktop computer. The models cost between $1,300 and $1,800, depending on the speed of the processor (700 MHz or 800 MHz), amount of main memory (128 or 256 MB of RAM), size of the hard drive (20, 40, or 60 GB), and option of including or not including Apple's all-in-one SuperDrive, which reads and writes both CDs and DVDs. The machines come loaded with its new Mac OS X version 10.1 operating system and a number of multimedia software applications including iDVD, which allows users to make DVD movies; iMovie, a video editing application; iPhoto, a new digital photo editing tool; and iTunes, which lets users convert CD music into MP3 files for use on Apple's iPod portable MP3 music player.[22]

Various smaller personal computers can be used for a variety of purposes. A *laptop computer* is a small, lightweight PC about the size of a briefcase. Apple, for example, has the iBook computer, which is a laptop system compatible with the iMac. The latest iBook, an all-white model, was announced by Apple Computer in May 2001. The machine is 11.2 inches wide, 9.1 inches deep, and 1.35 inches thick, and it weighs in at 4.9 pounds. It has a 500-MHz Power PC G3 processor, 256-KB cache, five-hour battery life, and a screen resolution of 1,024 by 768 pixels. There are four versions: a CD-ROM, DVD-ROM, CD-RW, and a combination model that includes both DVD-ROM and CD-RW drives. The base model includes 64 MB of RAM, while the other three models have 128 MB of RAM. The price ranges from $1,300 to $1,800. The iBook also comes "AirPort-ready" for wireless Internet access with two built-in antennas and a slot for an AirPort wireless communications card.[23]

Newer PCs include the even smaller and lighter *notebook* and *subnotebook* computers that provide similar computing power. Some notebook and subnotebook computers fit into docking stations of desktop computers to provide additional storage and processing capabilities. Ultraportable notebooks are the fastest-growing segment of the notebook class computer. Dell, Fujitsu, Hewlett-Packard, IBM, and Sharp all sell highly portable notebook computers that measure $1.0 \times 11.0 \times 9.0$ inches or less and weigh under 4 pounds. These computers come with a 600-MHz or faster processor, 128 to 256 MB of main memory, 20- to 30-GB hard drive, high-resolution ($1,024 \times 768$) 12.1-inch active-matrix displays, and a battery life exceeding 2 ½ hours. The keyboards are small, measuring just 18 mm from key center to key center, called the *pitch*. This size takes some getting used to for most desktop users, who are accustomed to a 19-mm pitch. The screens are small enough that the fonts and icons can be hard to read.[24]

Handheld (palmtop) computers are PCs that provide increased portability because of their smaller size—some are as small as a credit card. These systems often include a wide variety of software and communications capabilities. Palm was the company that invented the Palm Pilot organizer and its successors. Palm signed licensing agreements with Handspring, IBM, Sony, and many other manufacturers, permitting them to make what amounts to Palm clones. These computers are compatible with and able to communicate with desktop computers over wireless networks.

One of the shortcomings of palmtop computers is that they require lots of power relative to their size. The new Toshiba Genio E, Compaq iPaq, and HP Jornada can run for 8, 12, or 14 hours between battery charges, respectively. Each cost in the neighborhood of $600 when they were introduced in the fall of 2001 and came with a 200+ MHz processor, 64 MB of RAM, a speaker and a microphone for voice notes, a headphone jack for listening to digital music files and audio books, and a connection cradle that keeps the address book, calendar, and e-mail synchronized with data on the user's desktop computer.[25] Sony's Clié PEG-T415 is priced at $300 and is less than 0.4 inches thick, which makes it easy to carry at all times.[26]

The *Atlanta Journal-Constitution* expects to save over $250,000 a year on newspaper delivery services by using palmtop computers. The palmtops streamline the return of unsold papers at more than 15,000 locations around Atlanta.

Handheld (palmtop) computers such as this Handspring Treo 90 pack powerful features into small spaces. This device weighs only 4 ounces but has 16 MB of memory—enough to keep anyone organized on the go.

(Source: Courtesy of Handspring.)

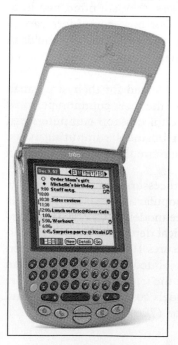

Network computers are stripped-down versions of PCs that are used primarily for data entry and Internet connection.

(Source: Courtesy of Wyse Technology.)

network computer
a cheaper-to-buy and cheaper-to-run version of the personal computer that is used primarily for accessing networks and the Internet

Each newspaper rack is given a unique bar code, and delivery people scan in the location and the number of newspapers that weren't purchased at that rack. This information is then relayed to corporate information systems to calculate the retailers' bill, something that took delivery workers hours to do by hand. Newspaper-delivery people are also using palmtop computers to streamline home delivery. Drivers get up-to-date delivery information in sequential order throughout their route via the palmtop computers rather than rely on printed address labels.[27]

Embedded computers are computers placed inside other products to add features and capabilities. In automobiles, embedded computers can help with navigation, engine performance, braking, and other functions. Household appliances, stereos, and some phone systems also use embedded computers. Raven Shoes from VectraSense Technologies feature embedded computers that monitor a jogger's level of activity and increase or decrease the support offered by the shoes.[28]

The **network computer** is a cheaper-to-buy and cheaper-to-run version of the personal computer that is used primarily for accessing the Internet and e-mail. These stripped-down versions of personal computers do not have the storage capacity or computing power of typical desktop computers, nor do they need it for the role they play. No hard disk means you never get viruses or a hard disk crash. Unlike personal computers, network computers—or thin clients—download software from a network when needed. This can make it much easier and less expensive to manage the support, distribution, and updating of software applications. Their primary market is for small business and education.

Advocates of network computers argue that they not only cost less to purchase compared with a standard desktop PC but also cost less to operate. Thus, it is not so much the low cost of purchasing a network computer that makes it so attractive but its low maintenance cost. However, the network computer's flexibility is extremely limited when compared with the personal computer. In addition, PC companies have reacted strongly to network computers by offering machines with lower prices and more features. Only about 150,000 such devices shipped in the United States in 2000 and it is debatable how successful this type of computer will be.[29]

The New Internet Computer (NIC) from Oracle has a starting cost of $200. It includes 64 MB of RAM, a 4 MB flash memory disk, CD-ROM drive, keyboard and mouse, speakers, and the ability to connect to the Internet or corporate networks. Sun Microsystems also sells a network computer of its own, known as the Sun Ray. IBM offers the Network Station Series 1000.

workstation
computer that fits between high-end personal computers and low-end midrange computers in terms of cost and processing power

Workstation computers provide high computing power and reliability. They are used to support engineering, design, and technical users.

(Source: Courtesy of IBM Corporation.)

Workstations are computers that fit between high-end personal computers and low-end midrange computers in terms of cost and processing power. Workstation manufacturers use the RISC rather than the CISC computer chip to provide high computing power and reliability. They cost from $4,000 to $40,000. Workstations are small enough to fit on an individual's desktop. A workstation may be dedicated to support a single user or a small group of users. High-end personal computers are approaching the computing power of a workstation.

Workstations are used to support engineering and technical users who perform heavy mathematical computing, computer-aided design (CAD), and other applications requiring a high-end processor. Such users need very powerful CPUs, large amounts of main memory, and extremely high-resolution graphic displays to meet their needs. Engineers use CAD programs to create two- and three-dimensional engineering drawings and product designs. Although initially creating a design with CAD software may take as long as, if not longer than, creating a design in the traditional manner, CAD is much faster when it comes time to revise. Instead of redrawing an entire plan, CAD allows the engineers to modify it with a few clicks of a mouse. They can also easily pan the design or zoom in to magnify a particular part. They can also rotate the view to examine it from different perspectives.

Web appliance
a device that can connect to the Internet, typically through a phone line

Many companies, including Compaq, Dell, Hitachi, and Samsung, are developing inexpensive **Web appliances**. A Web appliance is a device that can connect to the Internet, typically through a phone line. It can be used to check stock prices, check e-mail messages, search the Internet for information, and more. Web appliances come in a variety of configurations. Some have a keyboard, a passive-matrix display, a 200-MHz processor, and Web and e-mail software. These devices can cost less than $200 to purchase and about $20 per month for an Internet connection. Some Web appliances have the appearance of a cellular phone, with the capabilities of a standard phone and basic Internet connections. Other Web appliances are being attached to everyday products, such as TVs, stoves, and refrigerators. Once attached, the Web appliance will be able to get movie schedules, alert people when their stove may require maintenance, or advertise grocery specials. In the future, Web appliances may be built into many of the products we use every day.

Midrange Computers

midrange computer
formerly called minicomputer, a system about the size of a small three-drawer file cabinet that can accommodate several users at one time

Midrange computers (formerly called *minicomputers*) are systems about the size of a small three-drawer file cabinet that can accommodate several users at one time. These systems often have secondary storage devices with more capacity than workstation computers and can support a variety of transaction processing activities, including payroll, inventory control, and invoicing. Midrange computers often have excellent processing and decision-support capabilities. Many small to medium-size organizations—from manufacturers, to real estate companies, to retail operations—use midrange computers.

Mainframe Computers

mainframe computer
large, powerful computer often shared by hundreds of concurrent users connected to the machine via terminals

Mainframe computers are large, powerful computers often shared by hundreds of concurrent users connected to the machine via terminals. The mainframe computer must reside in an environment-controlled computer room or data center with special heating, venting, and air-conditioning (HVAC) equipment to control the temperature, humidity, and dust levels around the computer. In addition, most mainframes are kept in a secured data center with limited access to the room through some kind of security system. The construction and maintenance of such a controlled access room with HVAC can add hundreds of thousands of dollars to the cost of owning and operating a mainframe computer. Mainframe computers also require specially trained individuals (called *system engineers* and *system programmers*) to care for them. Mainframe computers can crunch numbers at a rate of over 300 million instructions per second and start at $250,000 for a fully configured system.

The traditional role of the mainframe computer was as the large, centrally located computer of a firm. Mainframes have been the cornerstone of computing in large corporations for many years. From the early 1950s until the mid-1970s, virtually all commercial computer processing was mainframe based. Mainframe computers were acquired by many companies to automate accounting and finance processes, such as payroll, general ledger, accounts receivable, and accounts payable. Order processing, billing, and inventory control were other early computer applications. IBM is now the sole manufacturer of mainframe computers.

Today the role of the mainframe is undergoing some remarkable changes as lower-cost midrange computers, workstations, and personal computers become increasingly powerful. Many computer jobs that used to run on mainframe computers have migrated onto these smaller, less expensive computers. This information processing migration is called *computer downsizing*. The new role of the mainframe is a large information processing and data storage utility for a corporation—running jobs too large for other computers, storing files and databases too large to be stored elsewhere, and storing backups of files and databases created elsewhere (these large stores of data are sometimes called *data warehouses*).

Midrange computers pack more capacity than workstations into systems the size of a three-drawer file cabinet. They can support several users at once for TPS and decision-support activities.

(Source: Courtesy of IBM Corporation.)

Mainframe computers have been the workhorses of corporate computing for more than 50 years. They can support hundreds of users simultaneously and handle all of the core functions of a corporation.

(Source: Courtesy of IBM Corporation.)

supercomputers

the most powerful computer systems, with the fastest processing speeds

ASCI White is a powerful supercomputer at Lawrence Livermore National Laboratory in Livermore, California. It is capable of 12.3 trillion operations per second, more than a person with a calculator could do in 10 million years and is used by the lab to simulate nuclear weapons tests.

(Source: AP/Wide World Photos.)

The mainframe is capable of handling the millions of daily transactions associated with airline, automobile, and hotel/motel reservation systems. It can process the tens of thousands of daily queries necessary to provide data to decision support systems. Its massive storage and input/output capabilities enable it to play the role of a video computer, providing full-motion video to multiple, concurrent users.

Wal-Mart upgraded its mainframes and associated disk storage at a cost of $50 million. The project replaced a number of IBM S/390 mainframes with IBM's newer z900 computers. The new 64-bit mainframes include a very fast coprocessor designed to support the complex operations used in communicating secure e-business applications over the Internet. The mainframes are being used to run core business applications such as invoicing, replenishing product inventories, and processing credit card transactions for over 4,000 stores.[30]

Supercomputers

Supercomputers are the most powerful computer systems, with the fastest processing speeds. Originally, supercomputers were used primarily by government agencies to perform the high-speed number crunching needed in weather forecasting and military applications. With recent improvements in the cost and performance (lower cost and faster speeds) of these machines, they are being used more broadly for commercial purposes today. Military and research organizations trying to solve complex problems use these machines, and some research universities and large, high-tech corporations also use supercomputers. For example, large oil companies use them to perform sophisticated data analysis to explore for oil.

Compaq's Terascale Computing System at the Pittsburgh Supercomputing Center was rated the world's most powerful supercomputer committed to unclassified research in November 2001. This system is used for a wide variety of research projects, including simulation of the blood flow in the human body, space weather modeling, virtual tests for therapeutic cancer drugs, global modeling of the earth's magnetic field, and simulation of shock waves and eddies in turbulent fluids. The National Science Foundation came up with the $45 million to buy the hardware and software and keep it running for three years.[31] The system has over 3 TB of total RAM, with a hard-disk array of 50 TB of primary storage, and an additional 300 TB of disk or tape storage available as needed. The system has a peak processing capability exceeding 6 trillion calculations per second (teraflops).

Scientists often use computer models to simulate the problems they are studying. These models typically omit certain details of the situation because they are not well understood or the effort required to include the details simply makes the model too difficult to create. Almost always, some degree of model completeness must be sacrificed to make the problem solvable in a reasonable amount of time. Lack of completeness leads to some loss of accuracy. For example, meteorologists studying how the oceans influence weather patterns know that ocean currents have a larger impact than the effect of sunlight reflected by the clouds. As a result, they may simplify their model to ignore the effect of the reflected sunlight to squeeze in more data on ocean currents. The degree of approximation is called the *granularity* of the computer model. Read the "Ethical and Societal Issues" special-interest feature to gain an appreciation of how even the most powerful computers and simulation models are sometimes still inadequate.

Ethical and Societal Issues

European Weather Forecasters Rely on Blue Storm

Few things affect us so profoundly as the weather. It affects our food supply and the way we live. It even inflicts untold damage when it turns savage. Although we can never control the weather, we can try to predict its changes and avoid loss of life and property. But predicting weather on a large scale takes some massive computing power. The European Center for Medium-range Weather Forecasts (ECMWF) is an international organization headquartered in the United Kingdom. Its purpose is to prepare medium-range (up to 10-day) weather forecasts for distribution to the weather services of 22 countries—Austria, Belgium, Croatia, the Czech Republic, Denmark, Finland, France, Germany, Greece, Hungary, Iceland, Ireland, Italy, the Netherlands, Norway, Portugal, Slovenia, Spain, Sweden, Switzerland, Turkey, and the United Kingdom. These countries use the forecasts to issue warnings of severe storms and floods, present daily television weather forecasts throughout Europe, and provide specialized services to their many commercial and governmental users (e.g., optimal routing of ships at sea). The ECMWF also conducts research to improve these forecasts.

The behavior of the Earth's atmosphere follows a set of physical laws that can be expressed in extremely complex mathematical equations. Using these equations, scientists can forecast how temperature, wind speed and direction, and humidity will change over time. They use actual weather conditions for initial starting values for computer modeling and then calculate the weather at each point throughout the model atmosphere. The ECMWF model has 21 million grid points distributed throughout the atmosphere between the surface and a height of 63 kilometers (39 miles, if you don't like metrics). The complete weather forecast is made in a series of short steps, each about 20 minutes ahead, with each intermediate forecast serving as the starting conditions for the next forecast step.

The ECMWF recently ordered a supercomputer nicknamed Blue Storm and a data storage network from IBM to help scientists improve their forecasting abilities. The system is an IBM supercomputer capable of performing 7 trillion calculations per second (7 teraflops). The supercomputer will be expanded in stages so that by 2004, the machine should run at more than 20 teraflops. At that point, the Blue Storm system will be roughly five times more powerful than ECMWF's current systems.

ECMWF employees will access Blue Storm using powerful IBM IntelliStation workstations, and researchers throughout Europe will be able to access the system via a network. Blue Storm is expected to contain a storage capacity of 1.5 PB (petabytes). Memory capacity is 4.1 TB.

Unfortunately, the goal of achieving precise, long-term forecasting (many weeks in advance) has proven difficult for several reasons. Small errors in the initial weather conditions and the approximations of atmospheric processes (e.g., variable cloud conditions) in the models are the two main sources of inaccurate forecasts. In addition, minor natural events that may seem trivial can result in larger-magnitude atmospheric changes, so accounting for those seemingly insignificant changes can be mind- and computer-boggling. In addition, gaps in our understanding of weather phenomena can confound even the best models. Much remains to be known about basic processes in our atmosphere. For example, scientists only recently discovered the existence of massive electrical discharges into the atmosphere above thunderstorms (called sprites and blue jets)—some of them many miles across. Still, Blue Storm is helping the ECMWF to advance weather forecasting by providing awesome computing power.

Discussion Questions

1. There is great room for improvement in weather forecasting by gathering more detailed and accurate weather data and conducting basic research into atmospheric processes. How can the ECMWF justify the expenditure of tens of millions of dollars for a faster computer with all these shortcomings?

2. Do weather forecasting organizations have a responsibility to do a better job of communicating the uncertainty associated with their forecasts? If so, how might this be done effectively?

Critical Thinking Questions

3. Conduct a little research of your own to find out what other forecasting models scientists use. How likely is it that problems exist in these models?

4. If Moore's Law continues in effect until at least 2012, what sort of computers will be available? Will these machines be capable of more accurate forecasts?

Sources: Adapted from Clint Boulton, "IBM Supercomputers Tabbed by Agencies," *Internet News,* November 9, 2001, accessed at http://www.internetnews.com; Clint Boulton, "IBM Commences Operation 'Blue Storm,'" *Internet News,* December 21, 2001, accessed at http://www.internetnews.com; Martin J. Garvey, "IBM to Build Supercomputer to Help Forecast Weather," *InformationWeek,* December 21, 2001, accessed at http://www.informationweek.com; John Marchese, "Forecast Hazy," *Discover,* June 21, 2001, p. 44–51; "Forecasting by Computer," accessed at the European Center for Medium-range Weather Forecasts at http://www.ecmwf.int, January 2, 2001.

Servers

Although all of the preceding computer system types can be used for general processing tasks, they can also be used to serve a specific and unique purpose, such as supporting Internet and network applications. A **computer server** is a computer designed for a specific task, such as network or Internet applications. Servers typically have large memory and storage capacities, along with fast and

computer server
a computer designed for a specific task, such as network or Internet applications

efficient communications abilities. They can range in size from a PC to a mainframe system, depending on the needs of the organization. A Web server is used to handle Internet traffic and communications. An Internet caching server stores Web sites that are frequently used by a company. An enterprise server is a computer containing programs that collectively serve the needs of an entire organization, rather than a single user, department, or specialized application. A file server, discussed in more detail in Chapter 6, stores and coordinates program and data files. A transaction server is used to store and process business transactions. As with general computers, there are benchmarks to help a company determine the performance of a server, such as ZD ServerBench, WebBench, and NetBench. High-end servers' performance can match or even exceed that of a mainframe computer by using such technologies as partitioning memory into separate units for different user purposes, sophisticated input/output and workload management, and better processor and memory support.

scalability
the ability to increase the capability of a computer system to process more transactions in a given period by adding more, or more powerful, processors

Servers offer great **scalability**, the ability to increase the processing capability of a computer system so that it can handle more transactions in a given period. Scalability is increased by adding more, or more powerful, processors. *Scaling up* adds more powerful processors, and *scaling out* adds many more equal (or even less powerful) processors to increase the total data processing capacity.

STANDARDS

With the wide variety of computer products now available, the importance of hardware standards cannot be overemphasized. They diminish the cost of integration, help a developer determine which devices will be compatible with the rest of the system, provide increased options, and make upgrading a system less complex. Several common standards are summarized in Table 3.3. Note that in some cases there are competing standards.

In addition to industry standards, many large corporations also set their own internal standards by selecting specific computer configurations from a small set of manufacturers. The goal is to reduce hardware support costs and increase the organization's flexibility. Business units within an organization that adopt different hardware complicate future corporate information system projects. For example, the installation of software is made much easier if similar equipment from the same manufacturer is involved, rather than new and different systems at each installation site.

SELECTING AND UPGRADING COMPUTER SYSTEMS

computer system architecture
the structure, or configuration, of the hardware components of a computer system

The structure, or configuration, of the hardware components of a computer system is called the **computer system architecture**. This architecture can include a mixture of components, including processing, memory, storage, input, and output devices. As discussed in Chapter 2, organizations are adaptive systems that must respond to changes in their environment. A computer system that was once effective may need to be enhanced or upgraded to support new business activities and a changing environment. The ability to upgrade a system can be an important factor in selecting the best computer hardware.

Computer systems can be upgraded by installing additional memory, additional processors (such as a math coprocessor), more hard disk storage, a memory card, or other devices. When upgrading or expanding an existing computer system, it is usually necessary to reconfigure the system.

Hard Drive Considerations

The optimal hard drive is a function of several overlapping features. Since its main role is to serve as a long-term data store, capacity is a big plus. Most mobile PCs today come equipped with 2.5-inch removable hard drives. Look for something between 20 GB and 40 GB, depending on the type of data you'll be storing. Other

Standard	How Used
MultiMedia Extension (MMX)	Multimedia standard that enables software and hardware vendors to build products that will work well together
Multimedia PC Council (MPC)	Alternative multimedia standard that enables software and hardware vendors to build products that will work well together
Musical Instrument Digital Interface (MIDI)	Standard system for connecting musical instruments and synthesizers to computers; defines codes for musical events, including the start of a note and its pitch, length, volume, and other attributes
Plug 'n' Play (PnP)	Hardware and software components that card, personal computer, and operating system manufacturers incorporate into their products to eliminate the need for manual configuration so that hardware can be installed and used immediately
Small Computer System Interface (SCSI)	Interface that ensures that any storage, input, or output device that meets this standard can be quickly added to a system
Fibre Channel	An alternative to SCSI for connecting devices to a computer; allows a greater distance between devices and offers faster performance than SCSI
Personal Computer Memory Card International Association (PCMCIA)	Standard that ensures compatibility between PC memory and communications cards
Peripheral Component Interface (PCI)	Standard for connecting personal computer system components including modems, printers, and add-in cards
Compact Peripheral Component Interface (sCPI)	Standard for connecting notebook computer components that takes into account the limited amount of space available

TABLE 3•3

Industry Standards in Common Use

considerations are access speed (look for a minimum of 10 to 12 milliseconds), RAM, and hard drive cache size. Access speeds should also be selected based on the type of data you'll be storing on your hard drive. Today's business software applications and large video, audio, and graphics files require several megabytes of storage.

Main Memory Considerations

Main memory stores software code while the processor reads and executes the code. Having more RAM main memory means you can run more software programs at the same time. The minimum capacity you'll need to run most mainstream business software is 64 MB. Systems with 256 MB are well suited to take advantage of today's advanced personal productivity software (word processing, spreadsheet, graphics, and database) and multimedia programs.

As discussed earlier, your system's processor, main memory, and cache memory depend heavily on each other to achieve optimal system functionality. The original manufacturer of your computer considers this interdependency when designing and choosing the parts for the system. If you plan to upgrade your system's main memory above 64 MB, you should consult your PC supplier to understand your system's main memory limits on size of cache and the implication of exceeding those limits.

Printer Considerations

Laser printers and inkjet printers are the two primary choices for printers, and the differences between the two are becoming smaller and smaller. While most inkjet printers are color and laser printers monochrome, there are also color laser printers. All produce sharp images, with resolutions of 600 × 600 dots per inch (dpi) to 1200 × 1200 dpi now common. The major differences are in price, color, and speed.

There are two cost factors to consider when purchasing a printer. The first is the purchase price of the printer. Prices for laser printers will range from $300 to over $2,000, while prices for inkjet printers range from $100 to $600. The second cost to consider is operating cost. Laser printers provide greater printing duty cycle (volume of pages printed per month) and longer life for the ink/toner products, giving them a much lower operating cost than inkjet printers. Laser printers typically have operating costs of from $.01 to $.04 per page. Inkjet printers can have operating costs of from $0.03 to $0.08 per page for black-and-white pages and $0.10 to $0.20 for color pages. The cost for printing photos on special paper can exceed $.50 per page for either type of printer because of the high cost of the paper.

Laser printers are generally faster than inkjet printers and capable of handling higher volume printing than the inkjet printers. Laser printers print 15 to 30 pages per minute (ppm), and inkjet printers print 5 to 10 ppm for black and white and 0.5 to 2 ppm for color.

For color printing, inkjet printers print vivid hues and cost much less than a color laser printer. Inkjet printers can produce high-quality banners, graphics, greeting cards, letters, text, and prints of photos. For most people that require color printing capability, inkjet printers will be the most cost-effective solution.

As mentioned throughout this chapter, a computer system's components and architecture should be chosen to support fundamental objectives, current business processes, and future needs of the organization and information system. Each computer system component—processing, memory, storage, input, and output devices—has a critical role in the successful operation of the computer system, the information system, and the organization. A thorough understanding of the broader system goals and the characteristics of the hardware as they relate to these goals will be an important guide for the future IS professional or business user.

SUMMARY

PRINCIPLE *Assembling an effective, efficient computer system requires an understanding of its relationship to the information system and the organization. The computer system objectives are subordinate to, but supportive of, the information system and the needs of the organization.*

Hardware includes any machinery (often using digital circuitry) that assists with the input, processing, and output activities of a computer-based information system (CBIS). A computer system is an integrated assembly of physical devices with at least one central processing mechanism; it inputs, processes, stores, and outputs data and information.

Computer system hardware should be selected and organized to effectively and efficiently attain computer system objectives. These objectives should in turn support information system objectives and organizational goals. Balancing specific computer system objectives in terms of cost, control, and complexity will guide selection.

Processing is performed by an interplay between the central processing unit (CPU) and memory. The CPU has three main components: the arithmetic/logic unit (ALU), the control unit, and register areas. The ALU performs calculations and logical comparisons. The control unit accesses and decodes instructions and coordinates data flow. Registers are temporary holding areas for instructions to be executed by the CPU.

PRINCIPLE *When selecting computer devices, you must also consider the current and future needs of the information system and the organization. Your choice of a particular computer system device should always allow for later improvements.*

Instructions are executed in a two-phase process. In the instruction phase, instructions are brought into the central processor and decoded. In the execution phase, the computer executes the instruction and stores the result. The completion of this two-phase process is a machine cycle. Processing speed is often measured by the time it takes to complete one machine cycle, which is measured in fractions of seconds.

Computer system processing speed is also affected by clock speed, which is measured in megahertz (MHz). Speed is further determined by a CPU's wordlength, the number of bits it can process at one time. (A bit is a binary digit, either 0 or 1.) A 64-bit CPU has a wordlength of 64 bits and will process 64 bits of data in one machine cycle.

Moore's Law is a hypothesis that states that the number of transistors on a single chip will double every 18 months. This hypothesis has held up amazingly well.

Processing speed is also limited by physical constraints, such as distance between circuitry points and circuitry materials. Advances in gallium arsenide (GaAs) and superconductive metals will result in faster CPUs. Many processors are complex instruction set computing (CISC) chips, which have many microcode instructions placed in them. With reduced instruction set computing (RISC) chips, only essential instructions are included, so processing is faster.

Primary storage, or memory, provides working storage for program instructions and data to be processed and provides them to the CPU. Storage capacity is measured in bytes. A common form of memory is random access memory (RAM). RAM is volatile—loss of power to the computer will erase its contents—and comes in many different varieties. The mainstream type of RAM is extended data out, or EDO, RAM, which is faster than older types of RAM. Two other variations of RAM include dynamic RAM (DRAM) and synchronous DRAM. SDRAM also has the advantage of a faster transfer speed between the microprocessor and the memory. DRAM chips need high or low voltages applied at regular intervals—every two milliseconds (two one-thousandths of a second) or so—if they are not to lose their information.

Read-only memory (ROM) is nonvolatile and contains permanent program instructions for execution by the CPU. Other nonvolatile memory types include programmable read-only memory (PROM) and erasable programmable read-only memory (EPROM). Cache memory is a type of high-speed memory that CPUs can access more rapidly than RAM.

Processing done using several processing units is called multiprocessing. One form of multiprocessing uses coprocessors; coprocessors execute one type of instruction while the CPU works on others. Parallel processing involves linking several processors to work together to solve complex problems.

Computer systems can store larger amounts of data and instructions in secondary storage, which is less volatile and has greater capacity than memory. The primary characteristics of secondary storage media and devices include access method, capacity, and portability. Storage media can implement either sequential access or direct access. Sequential access requires data to be read or written in sequence. Direct access means that data can be located and retrieved directly from any location on the media.

Common forms of secondary storage include magnetic tape, magnetic disk, compact disk, digital video disk, and optical disk storage. Other storage alternatives are flash memory chips—silicon chips with nonvolatile memory—and PC memory cards, removable credit-card-size storage devices that function like fixed hard disk drives. Redundant array of independent/inexpensive disks (RAID) is a method of storing data that generates extra bits of data from existing data, allowing the system to more easily recover data in the event of a hardware failure. Storage area network (SAN) uses computer servers, distributed storage devices, and networks to provide fast and efficient storage.

Input and output devices allow users to provide data and instructions to the computer for processing and allow subsequent storage and output. These devices are part of a user interface through which humans interact with computer systems. Input and output devices vary widely, but they share common characteristics of speed and functionality.

Data is thus placed in a computer system in a two-stage process: data entry converts human-readable data into machine-readable form; data input then transfers it to the computer. On-line data entry and input immediately converts and transfers data from devices to the computer system. Source data automation involves automating data entry and input so that data is captured close to its source and in a form that can be input directly to the computer.

Scanners are input devices that convert images and text into binary digits. Specialized scanners include magnetic ink character recognition (MICR) devices, optical mark recognition (OMR) devices, and optical character recognition (OCR) devices. Some input and output devices combine several into one. Point-of-sale (POS) devices are terminals with scanners that read and enter codes into computer systems. Automated teller machines (ATMs) are terminals with keyboards used for transactions.

Output devices provide information in different forms, from hard copy to sound to digital format. Display monitors are standard output devices; monitor quality is determined by size, color, and resolution. Other output devices include printers, plotters, and computer output microfilm. Printers are popular hard-copy output devices whose quality is measured by speed and resolution. Plotters output hard copy for general design work. Computer output microfilm (COM) devices place data from the computer directly onto microfilm.

Computers may be classified as special purpose or general purpose. General-purpose computers are used for numerous applications and can be classified by processing speed, RAM capacity, and size. The six computer system types are network computer, personal computer, workstation, midrange computer, mainframe computer, and supercomputer. The network computer is a diskless, inexpensive computer used for accessing server-based applications and the Internet. Personal computers (PCs) are small, inexpensive computer systems. Two major types of PCs are desktop and laptop computers. Workstations are advanced PCs with greater memory, processing, and graphics abilities. File cabinet–size minicomputers have greater secondary storage and support transaction processing. Even larger mainframes have higher processing capabilities, while supercomputers are extremely fast computers used to solve the most intensive computing problems. A computer server is a computer designed for a specific task, such as network or Internet applications. Servers typically have large memory and storage capacities, along with fast and efficient communications abilities. They can range in size from a PC to a mainframe system, depending on the needs of the organization.

The configuration of computer system hardware components is the computer system architecture. Computer systems can be upgraded by changing or adding memory, processors, printers, and other devices. Standards are being created to lower the cost and complexity of upgrades.

CHAPTER 3 SELF-ASSESSMENT TEST

Assembling an effective, efficient computer system requires an understanding of its relationship to the information system and the organization. The computer system objectives are subordinate to, but supportive of, the information system and the needs of the organization.

1. The overriding consideration in making hardware decisions in a business should be how hardware can be used to support the objectives of the information system and the goals of the organization. True False

2. The information system solutions chosen to be implemented at most manufacturing organizations are virtually identical. True False

3. _____ is a property of certain metals that lets current flow with minimal resistance.

4. One petabyte represents about 1,024 gigabytes. True False

5. A form of memory that loses its contents if power is lost is called
   A. ROM
   B. PROM
   C. CD-ROM
   D. RAM

6. Which of the following is a sequential access storage media?
   A. magnetic tape
   B. RAID
   C. magnetic disk
   D. optical disk

7. _____ involves capturing and editing data where the data is originally created and in a form that can be directly input to a computer, thus ensuring accuracy and timeliness.

When selecting computer devices, you also must consider the current and future needs of the information system and the organization. Your choice of a particular computer system device should always allow for later improvements.

8. On a megabyte-to-megabyte basis, memory is still more expensive than most forms of secondary storage. True False

9. _____ include devices that perform the functions of input, processing, data storage, and output.

10. The relative clock speed of two CPUs from different manufacturers is a good indicator of their relative processing speed. True False

11. _____ is the ability to increase the capability of a computer system to process more transactions in a given period by adding more, or more powerful, processors.

KEY TERMS

arithmetic/logic unit (ALU), 88
bit, 90
bus line, 90
byte (B), 93
cache memory, 95
CD-rewritable (CD-RW) disk, 100
CD-writable (CD-W) disk, 100
central processing unit (CPU), 87
clock speed, 89
compact disk read-only memory (CD-ROM), 100
complex instruction set computing (CISC), 92
computer server, 118
computer system architecture, 119
control unit, 88
coprocessor, 96
data entry, 104
data input, 104
digital computer camera, 106
digital versatile disk (DVD), 102
direct access, 97
direct access storage device (DASD), 98
disk mirroring, 99
execution time (E-time), 88
expandable storage devices, 103

flash memory, 103
general-purpose computers, 112
gigahertz (GHz), 90
hertz, 90
instruction time (I-time), 88
machine cycle, 88
magnetic disk, 99
magnetic tape, 98
magneto-optical disk, 100
mainframe computer, 116
massively parallel processing, 96
megahertz (MHz), 90
microcode, 89
midrange computer, 116
MIPS, 89
Moore's Law, 91
multifunction device, 111
multiprocessing, 96
music device, 111
network computer, 115
optical disk, 100
optical processors, 91
personal computer (PC), 113
pipelining, 88
pixel, 109
plotters, 110
point-of-sale (POS) device, 107

primary storage (main memory; memory), 88
random access memory (RAM), 94
read-only memory (ROM), 94
reduced instruction set computing (RISC), 92
redundant array of independent/ inexpensive disks (RAID), 99
register, 88
scalability, 119
secondary storage (permanent storage), 97
sequential access, 97
sequential access storage device (SASD), 98
source data automation, 105
special-purpose computers, 112
storage area network (SAN), 99
supercomputers, 117
superconductivity, 91
symmetrical multiprocessing (SMP), 96
voice-recognition device, 105
Web appliance, 116
wordlength, 90
workstation, 115

REVIEW QUESTIONS

1. What is a computer system and what is the role of hardware in the system?
2. Describe a storage area network (SAN) system.
3. Why is it said that the components of all information systems are interdependent?
4. Explain the two-phase process for executing instructions.
5. Identify the three components of the CPU and explain the role of each.
6. What are the properties of flash memory and how is such a device typically used?
7. What is Moore's Law?
8. What is the difference between CISC and RISC instruction sets?

9. What is granularity when speaking about computer models?
10. Describe the various types of memory.
11. What is the difference between sequential and direct access of data?
12. Describe various types of secondary storage media in terms of access method, capacity, and portability.
13. Identify and briefly describe the various classes of personal computers.
14. What is the difference between CD-R and CD-RW?
15. What is the difference between cache memory and main memory?
16. What is source data automation?

17. What is the overall trend in secondary storage devices?
18. What are the computer system types? How do these types differ?
19. Discuss the methods of upgrading a computer system.

20. Discuss the role standards play in making it easier to use computer hardware.
21. Describe three special-purpose devices.
22. What is microcode?

DISCUSSION QUESTIONS

1. What are the implications of Moore's Law—that is, continuing the trend of increased computing power at lower costs? Use Moore's Law to forecast the personal computer computing power that could be available in three years. What sort of applications could benefit from that level of computer power?

2. Imagine that you are a recruiter for a university and want to use access to the university's powerful supercomputer as an incentive to attract scientists and researchers. What information would you need to communicate to candidates to entice them?

3. Identify and briefly describe the leading formats in the rewritable DVD market. Why is it important which format becomes recognized as the standard?

4. Identify and briefly describe three forms of multiprocessing.

5. Imagine that you are the business manager for your university. What type of computer would you recommend for broad deployment in the university's computer labs—a standard desktop personal computer or a network computer? Why?

6. Discuss fully: What advantages does a 64-bit processor have over a 32-bit processor? Are there any disadvantages?

7. If cost were not an issue, describe the characteristics of your ideal laptop computer.

8. What if you discovered that your favorite recording group composes, edits, and records all its music using multimedia computer technology? How do you feel? Does the use of computer technology to create original works of art or music diminish or enhance the accomplishment? Should such artists be considered as great as others who do not use computer technology?

PROBLEM-SOLVING EXERCISES

1. Do research (read various trade journals and search the Internet) on companies that make rewritable DVD devices. Use your word processing program to write a short report summarizing your findings. Make sure to include the speed, features, and price of the various devices. Be sure to discuss the compatability of each device with movie DVD devices. Develop a simple spreadsheet to compare the features and costs of the rewritable DVD devices you found.

2. Over the upcoming year, your department is expected to add eight people to its staff. You will need to acquire eight personal computer systems and two additional printers for the new employees to share. Standard office computers have a Pentium (2 GHz) processor with 128 MB of RAM, 15-inch SVGA color monitor, and a minimum of 20 GB hard disk drive.

At least four of the new people will use their computers more than three hours per day. You would like to provide larger monitors and special ergonomic keyboards for these people—if it fits within your budget. You are not sure whether you want to upgrade the machines to 256 MB of RAM and 40 GB hard drive.

Your department budget will allow a maximum of $20,000 for computer hardware purchases this year, and you want to select only one vendor for all of the hardware. A price list from three vendors appears in the following table, with prices for a single unit of each component. Use a spreadsheet to find the department's best solution; write a short memo explaining your rationale. Specify which vendor to choose and which items to be ordered as well as the total cost.

Component	Expert Solutions Ltd.	Business Processing Enterprises	Super Systems Inc.
2 GHz Pentium with 128 MB RAM			
20 GB hard drive	$1,245	$1,275	$1,200
Upgrade to 256 MB RAM	250	225	245
Upgrade to 40 GB hard drive	190	215	205
15-inch .28 dpi SVGA monitor	200	210	225
15-inch flat screen monitor	400	420	450
17-inch .28 dpi SVGA monitor	425	400	415
Ergonomic keyboard	55	50	50
12 ppm color ink-jet printer	120	120	110
Surge protector/ power strip	35	32	35
Three-year warranty (parts and labor)	340	300	320

TEAM ACTIVITIES

1. With two of your classmates, visit a major computer retail store (e.g., Gateway, CompUSA, MicroCenter). Spend a couple of hours during which each of you concentrates on identifying the latest developments in storage, input, and output devices. Write a brief report summarizing your findings.

2. With two or three of your classmates, visit the main computer facility of your college or university. Find out the manufacturer and model number as well as the specifications (speed of CPU, amount of main memory, disk drive capacity, etc.) of a mainframe or midrange computer. How long has it been in use? How much longer does the university plan to

use it before replacing it with something different? What business processes have changed that spurred this alteration? Will your college or university upgrade the computer or buy a new one?

3. Identify a manager of a computer center and arrange a meeting to discuss the process followed to evaluate new equipment and vendors. Find out how the organization is kept abreast of new developments, how they decide it is time to phase out one piece of equipment in favor of a new one, and what ongoing process they have to evaluate and monitor vendors.

WEB EXERCISES

1. Visit the Web sites of Intel, Sun Microsystems, Hewlett-Packard, IBM, and other computer manufacturers. Identify as many as possible industry standards that these manufacturers are following in the development of their products. Also, identify other standardization efforts in which they are involved.
2. Search the Web for companies that make secondary storage devices and systems, including disk, tape, RAID, SAN, and others. Summarize your findings using your word processing device.
3. Visit the Web sites of three manufacturers of handheld computers. Make a list of features and software applications available from each. Develop a spreadsheet that compares features and prices. Which system and manufacturer do you prefer? Why?

CASES

CASE 1

Australia Experiments with Electronic Voting

Government exists at three levels in Australia—Commonwealth (federal), state, and local. Australia is divided into six self-governing states (New South Wales, Queensland, South Australia, Tasmania, Victoria, and Western Australia) and two mainland territories (Australian Capital Territory and Northern Territory). Canberra, the capital of Australia, is situated in the Australian Capital Territory (ACT) and has a population of around 313,000. Federal Parliament passed the Australian Capital Territory (Self-Government) Act in 1988 that established self-government in the ACT. The ACT Legislative Assembly has 17 elected members.

In October 2001, roughly 215,000 Canberrans went to the polls to elect their local 17-member Legislative Assembly. At eight booths across the city, 16,500 voters elected to participate in a pilot test of a new electronic voting system (at an estimated cost of $400,000) rather than use the traditional paper-and-pencil method. These electronic voters were issued a barcode that gave them secure, once-only access to the electronic voting system. They then viewed lists of candidates on a screen-based ballot recording system and selected their preferred candidates using a keypad. Votes were confirmed by another barcode swipe, and then the barcode was dropped into the ballot box.

The Australian Capital Territory uses a complex balloting system called Hare-Clark. Under this method of balloting, voters express their first, second, third, etc. preferences for each candidate for each office. A candidate is elected if he or she gains 50 percent of the votes plus one (an absolute majority). But if no candidate achieves such a majority, the lowest vote-getter is excluded and that candidate's second preferences are distributed on a proportional basis to the remaining candidates as full votes. This process repeats until one of the candidates gains the required absolute majority.

All 16,500 electronic votes were recorded and stored on a secure, standalone computer and then transferred to a zip disk for transport to the central database on election night. It took a full week to read the hand-marked paper ballots, perform the necessary data entry to capture the votes in an electronic format, and then transfer the results to the central computer system. Twelve days were required to complete the count, including a built-in delay of six days to allow for postal votes to come in. The accuracy of the count eliminated the need for time-intensive hand recounts. A recount in 1998 resulted in a delay of 22 days before a result was known.

Members of Australia's other electoral commissions across the country watched the pilot of the new system closely. While they were impressed, they were not eager to expand use of the electronic system. They thought that Canberra, with its high number of computer technology users, was atypical of the rest of the country and that unequal access to technology would inhibit broad-scale acceptance of electronic voting. Also, Canberra's relatively small geographic area means fewer ballot locations and thus less computer hardware is required to implement an electronic system. The electoral commissioners also raised concerns about security.

Discussion Questions

1. What percentage of Canberrans voted in this election? What percentage of the voters tried the new electronic system? Do you find these percentages surprising? Why?
2. The electoral commissioners observing the pilot raised several issues. Which of these issues do you think are most significant? Which issues were raised that were not so significant or that can be easily overcome?

Critical Thinking Questions

3. What suggestions for improvement would you make for conducting a second pilot—either in the voting and counting process itself or in the technology used?
4. Do you think such a voting system would be successful in the U.S.? Why or why not?

Sources: Adapted from Selina Mitchell, "Canberra Leads Way in E-Voting," *Australian IT,* October 9, 2001, accessed at http://www.AustralianIT. news.com.AU; Selina Mitchell, "States Say E-Vote Cost Is Too High," *Australian IT,* November 13, 2001, accessed at http://www.AustralianIT. news.com.AU; "Legislative Council and Periodic Elections," from the Tasmanian Parliamentary Library Web site at http://www.parliament. tas.gov.au, accessed January 3, 2002.

CASE 2

Information Systems Support Future Soldiers

The Land Warrior is a state-of-the-art weapons system for the modern foot soldier. The system components include a self-contained computer and radio system, a global positioning system (satellite) receiver, a helmet-mounted LCD display, and thermal and video sights plus laser range sensors for the soldier's M4 carbine or M16A2 rifle. The goal of all this high-tech equipment is to increase the effectiveness of the soldier and the likelihood of his survival by providing improved communications, mobility, navigation, and situational awareness.

At the heart of the Land Warrior system is a computer that integrates inputs from all the other subsystems and enables the soldier to control them. The computer is carried in a vest and at the small of the back. It is equivalent to a 166-MHz Pentium with 236 MB of RAM and 800 MB of flash disk for storage. The system uses a standard computer peripheral interface to enable the soldier to connect his weapons, navigation, and communications systems.

Maps, graphics, overlays, orders, and tactical aids can be stored in the computer. Time-sensitive information critical to the mission at hand can be loaded onto a flash disk prior to each mission. Wireless technology is used to relay messages and updates to the soldier's computer over a range of up to three miles. As a safeguard against critical information falling into enemy hands, the computer system, including all data storage devices, self-destructs if tampered with or a destruct button is pushed.

The Land Warrior computer contains navigation software and hardware as a backup to the GPS system. This backup system comes in handy in areas where satellite signals can't reach the GPS, such as under thick jungle canopy, or in city streets between tall buildings. The Land Warrior system also includes a voice-activated radio to keep members of the combat group in contact. Soldiers can send or receive coded messages using voice commands, leaving their hands free to fire weapons or handle other tasks.

Traditionally equipped Marines carry anywhere from 75 to 90 pounds of basic equipment into the field. The initial weight goal of 75 pounds for the Land Warrior was achieved in spite of the many batteries that must be carried to run all the electronic equipment. The computer and batteries together weigh 16 pounds, but there are plans to cut that weight in half. The Land Warrior runs on lithium-ion batteries that provide enough power for a 12-hour mission. It is expected that some 35,000 soldiers will be outfitted with the Land Warrior at a cost of $20,000 per soldier over the next few years.

Discussion Questions
1. What are some technical and practical issues that could limit the effectiveness of the Land Warrior system? How might these be overcome?
2. Can you identify additional features that could be useful in the Land Warrior system?

Critical Thinking Questions
3. What impact might the use of the Land Warrior system have on the recruiting and training of soldiers?
4. The high cost of the Land Warrior system will limit the use of this equipment to a relatively small percentage of combat troops. How should the military decide to whom this equipment is deployed? What issues might arise from soldiers sent into war without benefit of the Land Warrior equipment?

Sources: Adapted from Patricia Daukantas, "Future Soldiers May Wear Their Computers," *Newsbytes News Network*, June 4, 2001, accessed at http://www.findarticles.com; Max Hawkins, "Future Soldier Takes Shape," *Australian IT*, November 30, 2001, accessed at http://australianit.news.com.au; Nancy Beth Jackson, "Palmtop Makers Take Aim at the Military Market," *The New York Times on the Web*, November 8, 2001, accessed at http://www.nytimes.com; Stephan J. Mraz, "21st Century Soldier," *Machine Design*, March 1, 2001, accessed at http://www.findarticles.com.

CASE 3

Saving Data Center Space with Smaller Servers

Floor space within a data center is usually very limited and comes at a premium price—in the neighborhood of $300 per square foot. Hardware manufacturers are developing creative ways of packing more servers into less space while still meeting the data processing needs of their customers.

Servers are typically mounted onto a chassis that provides power and I/O connectivity. The standard data center rack is 42 U high (U is a measurement unit equaling 1.75 inches) and can hold 14 standard chassis, each 3U high. New ultraslim server architectures reduce the server height to 2U or 1U. This enables a rack to hold 21 or 42 servers—increasing the processing power by 50 to 100 percent in the same floor space.

The ultrathin 1U servers include one or two processors, up to 4 GB of RAM, three 20 GB hard drives, and at least two network ports. It is possible to pack 42 of these into a single rack to provide lots of processing power in a minimal space.

Although the ultraslim servers are designed to minimize power consumption, a rack holding 42 servers demands a lot of power. It may be necessary to rewire the data center, upgrade the entire power supply, or move to a new data processing facility to meet the increased power needs. In addition, a rack with 42 two-processor servers generates a lot of heat, which must be removed to prevent damaging the components. This increased air-conditioning capacity can be quite expensive.

Another approach to reducing floor space is the use of *blades,* a complete server on a single circuit board. Up to twelve blades can be inserted onto a chassis sharing a single bus. Freed of the physical bulk and components associated with traditional servers, blades slide into slots on racks. In most cases, blade servers consist of processing and storage components housed in a rack unit that provides network and external storage connections, reducing both cabling and space requirements. Blade servers take up less space, generate less heat, use less power, and don't have the environmental requirements of air conditioning or raised flooring, as larger servers. As a result blade servers cost 30 percent to 50 percent less than traditional rack-mounted servers, with the biggest savings derived from their smaller size and low power-consumption costs. One blade can handle one or two processors and includes as many as three 20 GB hard drives, 512 MB of RAM, and two network connections. The use of blades provides for a maximum of 336 processors

(12 blades/chassis × 2 processors/blade × 14 chassis/rack) in a single rack, a large number of processors in a small amount of space.

Blades can be specialized for particular tasks, such as Web hosting, media streaming, handling e-mail, file storage, and the management of other servers, depending on their hardware design or the software installed. Blades are in use in telecommunications, military, medical imaging, and industrial applications. Internet service providers and major corporations are also beginning to use blades.

Blade technology provides improved flexibility and performance over traditional servers and even the ultraslim servers. The blade architecture makes it easier to switch out servers and replace parts—you can just plug something in or pull it out. You don't have to unscrew parts of a standard rack-mount server or deal with a tangle of cascading cords running down the back.

While blades clearly deliver even greater data processing capacity than 1U servers, the same considerations of power consumption, heat dissipation, storage connection, cable handling, and management all apply, but to a vastly greater population of servers. In addition, the software available to manage and reconfigure racks of server blades dynamically may be inadequate to meet your needs. It may be necessary for you to write your own code to do this. Additional training will be required in the new skills required to perform design, maintenance, systems management, and tech support on these new devices.

Discussion Questions
1. What are the major advantages and disadvantages of the ultra slim servers over traditional servers?
2. What are the major advantages and disadvantages of server blades versus traditional servers? Blades versus ultra slim servers?

Critical Thinking Questions
3. Imagine that you are the CIO for a Fortune 500 manufacturing company. Outline the steps you would recommend to successfully implement server blade technology at your firm.
4. Imagine that you are the product development manager for a manufacturer of servers. What additional new features or services do you think are potentially worth exploring?

Sources: Adapted from Edmund X. DeJesus, "Server Size Matters," *Computerworld*, September 24, 2001, accessed at http://www.computerworld.com; Paul McDougal, "Intel and IBM Spin Server Design New Ways," *InformationWeek*, November 19, 2001, accessed at http://www.informationweek.com; and Ashlee Vance and Peter Sayer, "HP Takes Lead with Blade Servers," *Computerworld*, December 4, 2001, accessed at http://www.computerworld.com.

NOTES

Sources for the opening vignette on p.85: Adapted from Geoffrey Colvin, "Celera and Money," *Fortune*, March 1, 2001, accessed at http://www.fortune.com; Leslie Jaye Goff, "Helping to Map the Code of Life," *Computerworld*, October 22, 2001, accessed at http://www.computerworld.com; Brian O'Keefe, "Post-Genome, Celera Now Shoots for Profits," *Fortune*, February 19, 2001, accessed at http://www.fortune.com; Aaron Ricadela, "IT at the Edge of Science," *Informationweek.com*, August 13, 2001, pp. 30–32 and 71–76.

1. Paul McDougall, "Lowdown on the High-End," *InformationWeek*, June 11, 2001, accessed at http://www.informationweek.com.
2. "Intel and AMD Set to Unveil New Chips on Monday," *The New York Times of the Web*, January 4, 2002, accessed at http://www.nytimes.com.
3. Douglas F. Gray, "Intel Looks Beyond Clock Speeds," *Computerworld*, August 28, 2001, accessed at http://www.computerworld.com.
4. John Markoff, "Intel Makes an Ultra-Tiny Chip," *The New York Times on the Web*, June 10, 2001, accessed at http://www.nytimes.com.
5. Sam Costello, "Scientists Use Optics to Speed Data Transfer on Chips," *Computerworld*, January 2, 2002, accessed at http://www.computerworld.com.
6. Steve Ulfelder, "Transistor Triumphs," *Computerworld*, August 13, 2001, accessed at http://www.computerworld.com.
7. Steve Ulfelder, "Transistor Triumphs," *Computerworld*, August 13, 2001, accessed at http://www.computerworld.com.
8. George Johnson, "Computing, One Atom at a Time," *The New York Times on the Web*, March 27, 2001, accessed at http://www.nytimes.com.
9. Mark Hall, "Xserve Grabs the Spotlight," *Computerworld*, July 1, 2002, accessed at http://www.computerworld.com.
10. Howard Millman, "The Terabyte Tapes," *Computerworld*, February 26, 2001, accessed at http://www.computerworld.com.
11. Martin J. Garvey, "Cox Adds Dell to Its Storage Lineup," *InformationWeek*, September 3, 2001, accessed at http://www.informationweek.com.
12. Martin J. Garvey, "A New Game Plan," *InformationWeek*, October 29, 2001, accessed at http://www.informationweek.com.
13. Martyn Williams, "Fujitsu Begins Sampling 2.3 GB Magneto-Optical Disk Drive," *Computerworld*, July 5, 2001, accessed at http://www.computerworld.com.
14. Greg Wright, "Make Your Own DVD Movies," *Cincinnati Enquirer*, May 30, 2001, p. G5.

15. John C. Dvorak, "The Moving Target of Rewritable DVD," *PC Magazine*, November 8, 2001, accessed at http://www.pcmag.com.
16. Tischelle George, "Could You Send My Car an E-mail About That?" *InformationWeek*, October 16, 2001, accessed at http://www.informationweek.com.
17. David Pogue, "Digital Cameras for Less: How Much Will $300 Buy?" *The New York Times on the Web*, December 20, 2001, accessed at http://www.nytimes.com.
18. Unisys Web site, http://www.unisys.com, accessed January 16, 2002.
19. David M. Ewalt, "Staying Healthy with Wireless," *InformationWeek*, September 17, 2001, accessed at http://www.informationweek.com.
20. eBook Web site accessed at http://www.gemstar-ebook.com/ebcontent/devices/ on January 22, 2002.
21. "Report Counts Computers in Majority of U.S. Homes," *The New York Times on the Web*, September 7, 2001, accessed at http://www.nytimes.com.
22. Sumner Lemon and Matt Berger, "Jobs Unveils New iMac, Calls Apple's 'i' Products a Success," IDG News Service, January 7, 2002, accessed at http://www.computerworld.com.
23. Douglas F. Gray, IDG News Service, "Apple Unveils Thinner, Lighter iBook, OS Update," *Computerworld*, May 1, 2001, accessed at http://www.computerworld.com.
24. Bill Howard, "Ultrauseful Ultraportables," November 27, 2001, *PC Magazine*, accessed at http://www.pcmag.com.
25. David Pogue, "A New Crop of Palmtops with Roots," *The New York Times on the Web*, October 4, 2001, accessed at http://www.nytimes.com.
26. David Pogue, "An Elegant New Sony Handhled," *The New York Times on the Web*, November 29, 2001, accessed at http://www.nytimes.com.
27. Aishia M. Williams, "Doing Business Without Wires," *InformationWeek*, January 15, 2001, accessed at http://www.informationweek.com.
28. Matt Hamblen, "The Wireless Geek Can Be Tres Chic," *Computerworld*, March 20, 2002, accessed at http://www.computerworld.com.
29. Ian Fried, "Ellison's NIC Co. to Team with Sun," *Cnet News.com*, June 25, 2001, accessed at http://news.com.com.
30. Lucas Mearian, "Wal-Mart Deal Boosts IBM in Storage Wars," *Computerworld*, October 22, 2001, accessed at http://www.computerworld.com.
31. Mark K. Anderson, "Public Computing on a Super Scale," *Wired News*, October 4, 2001, accessed at http://www.wirednews.com.

Software: Systems and Application Software

PRINCIPLES	LEARNING OBJECTIVES
When selecting an operating system, you must consider the current and future needs for application software to meet the needs of the organization. In addition, your choice of a particular operating system must be consistent with your choice of hardware.	• Identify and briefly describe the functions of the two basic kinds of software. • Outline the role of the operating system and identify the features of several popular operating systems.
Do not develop proprietary application software unless doing so will meet a compelling business need that can provide a competitive advantage.	• Discuss how application software can support personal, workgroup, and enterprise business objectives. • Identify three basic approaches to developing application software and discuss the pros and cons of each.
Choose a programming language whose functional characteristics are appropriate to the task at hand, taking into consideration the skills and experience of the programming staff.	• Outline the overall evolution of programming languages and clearly differentiate among the five generations of programming languages.
The software industry continues to undergo constant change; users need to be aware of recent trends and issues to be effective in their business and personal life.	• Identify several key issues and trends that have an impact on organizations and individuals.

[Flextronics]

Global Deployment of Software Improves Key Business Process

Flextronics is a multinational manufacturer that designs products, manufactures them, and delivers them to its client companies in twenty-eight countries on four continents. Flextronics has headquarters in both San Jose, California, and Singapore and has established a network of facilities in key markets to provide top service to its customers while efficiently controlling its operations. As a contract manufacturer, Flextronics builds other companies' products—everything from Microsoft's new Xbox video-game system to Ericsson's cell phones.

Flextronics specializes in networking and telecommunications gear, computers, consumer electronics, and medical instruments. Key customers include Cisco Systems, Ericsson, Hewlett-Packard, Microsoft, Nokia, and Philips. Its major competitors are Solectron, SCI Systems, and Celestica. Recent annual revenue exceeded $12 billion, and its employees number more than 70,000.

Operating as a contract manufacturer requires lots of quoting and bidding on the various components for each job. To streamline the complex bidding process, Flextronics began using an on-line quoting software package called QuoteWin to generate its quotes electronically. As the company became more familiar with the software program, Flextronics management was so impressed with the results that they decided to implement the system globally. In addition to simplifying and streamlining the quoting and bidding process, they hoped to maintain consistency across all divisions of the company.

The QuoteWin software system helps simplify and reduce the time required to create quotes by passing information back and forth over networks to its suppliers and to clients once a bid has been compiled. In fact, it takes only minutes to gather contract pricing data. As a result, Flextronics was able to cut the average time to prepare a quote from three weeks to two. The QuoteWin system also maintains an accurate database of all quotes associated with each job. The availability of historical data coupled with good reporting tools in the software enables business managers to do a thorough "what-if" analysis of each quote (e.g., managers can ask, "If this part is eliminated, how much money can we save the customer?").

Flextronics hopes to convert all its suppliers to the new system; however, many smaller suppliers are concerned with the cost. Also, some suppliers have been less than enthusiastic about adopting the QuoteWin software because they work with other manufacturers—who use a variety of quote preparation systems. The suppliers would naturally prefer to see some standardization in the software to help streamline the process for everyone.

Flextronics currently has six global "quote hubs," which manage its bidding and quoting activities. It evaluated the use of QuoteWin in each region and now has over 60 users on the system. The global rollout of the software package was scheduled hub by hub based on the number of quotes generated. Europe was the first region to get up and running because that hub generated the most quotes.

In the future, Flextronics plans to enhance the bid and quote process even further with the implementation of a custom-built software package called FlexDesign. This system will compile company rules and guidelines about product design and components. Engineers involved in the bid and quote process can use this information to choose components based on proven business practices and technical expertise, including such factors as whether the component is from an approved supplier or whether it's been tested elsewhere in the company.

As you read this chapter, consider the following:

◉ What are the various kinds of software and how are they used?
◉ What are the sources for acquiring software and what are their pros and cons?
◉ What are the issues in adopting software to be used in supporting a global operation?

In the 1950s, when computer hardware was relatively rare and expensive, software costs were a comparatively small percentage of total information systems costs. Today, the situation has dramatically changed. Software can represent 75 percent or more of the total cost of a particular information system for three major reasons: advances in hardware technology have dramatically reduced hardware costs, increasingly complex software requires more time to develop and so is more costly, and salaries for software developers have increased because, over the long term, the demand for these workers far exceeds the supply. In the future, as suggested in Figure 4.1, software is expected to make up an even greater portion of the cost of the overall information system. The critical functions software serves, however, make it a worthwhile investment.

AN OVERVIEW OF SOFTWARE

computer programs
sequences of instructions for the computer

documentation
text that describes the program functions to help the user operate the computer system

systems software
the set of programs designed to coordinate the activities and functions of the hardware and various programs throughout the computer system

computer system platform
the combination of a particular hardware configuration and systems software package

application software
programs that help users solve particular computing problems

One of software's most critical functions is to direct the workings of the computer hardware. As we saw in Chapter 1, software consists of computer programs that control the workings of the computer hardware. **Computer programs** are sequences of instructions for the computer. **Documentation** describes the program functions to help the user operate the computer system. The program displays some documentation on screen, while other forms appear in external resources, such as printed manuals. There are two basic types of software: systems software and application software.

SYSTEMS SOFTWARE

Systems software is the set of programs designed to coordinate the activities and functions of the hardware and various programs throughout the computer system. A particular systems software package is designed for a specific CPU design and class of hardware. The combination of a particular hardware configuration and systems software package is known as a **computer system platform**.

APPLICATION SOFTWARE

Application software consists of programs that help users solve particular computing problems. Both systems and application software can be used to meet the needs of an individual, a group, or an enterprise. Application software can support individuals, groups, and organizations to help them realize business objectives.

FIGURE 4.1

The Importance of Software in Business

Since the 1950s, businesses have greatly increased their expenditures on software as compared with hardware.

Application software has the greatest potential to affect processes that add value to a business because it is designed for specific organizational activities and functions.

(Source: © Charlie Westerman/Stone.)

sphere of influence
the scope of problems and opportunities addressed by a particular organization

personal sphere of influence
sphere of influence that serves the needs of an individual user

personal productivity software
software that enables users to improve their personal effectiveness, increasing the amount of work they can do and its quality

workgroup
two or more people who work together to achieve a common goal

workgroup sphere of influence
sphere of influence that serves the needs of a workgroup

TABLE 4.1

Classifying Software by Type and Sphere of Influence

Application software has the greatest potential to affect the processes that add value to a business because it is designed for specific organizational activities and functions, as we saw in the case of Flextronics. The effective implementation and use of application software can provide significant internal efficiencies and support corporate goals. Before an individual, a group, or an enterprise decides on the best approach for acquiring application software, goals and needs should be analyzed carefully.

SUPPORTING INDIVIDUAL, GROUP, AND ORGANIZATIONAL GOALS

Every organization relies on the contributions of individuals, groups, and the entire enterprise to achieve business objectives. And conversely, the organization also supports individuals, groups, and the entire enterprise with specific application software and information systems. As the power and reach of information systems expand, they promise to reshape every aspect of our lives: how we work and play, how we are educated, how we interact with others, how our businesses and governments conduct their work, and how scientists perform research. Information systems are changing virtually every method for capturing, storing, transmitting, and analyzing knowledge, including books, newspapers, magazines, movies, television, phone calls, musical recordings, and architectural drawings. One useful way of classifying the many potential uses of information systems is to identify the scope of the problems and opportunities addressed by a particular organization. This is called the **sphere of influence**. For most companies, the spheres of influence are personal, workgroup, and enterprise, as shown in Table 4.1.

Information systems that operate within the **personal sphere of influence** serve the needs of an individual user. These information systems enable their users to improve their personal effectiveness, increasing the amount of work they can do and its quality. Such software is often referred to as **personal productivity software**. There are many examples of such applications operating within the personal sphere of influence—a word processing application to enter, check spelling, edit, copy, print, distribute, and file text material; a spreadsheet application to manipulate numeric data in rows and columns for analysis and decision making; a graphics application to perform data analysis; and a database application to organize data for personal use.

A **workgroup** is two or more people who work together to achieve a common goal. A workgroup may be a large, formal, permanent organizational entity such as a section or department, or a temporary group formed to complete a specific project. The human resource department of a large firm is an example of a formal workgroup. It consists of several people, is a formal and permanent organizational entity, and appears on a firm's organization chart. An information system that operates in the **workgroup sphere of influence** supports a workgroup in the attainment of a common goal. Users of such applications are operating in an environment where communication, interaction, and collaboration are critical

Software	Personal	Workgroup	Enterprise
Systems software	Personal computer and workstation operating systems	Network operating systems	Midrange computer and mainframe operating systems
Application software	Word processing, spreadsheet, database, graphics	Electronic mail, group scheduling, shared work	General ledger, order entry, payroll, human resources

Lotus Notes is an application that enables a workgroup to schedule meetings and coordinate activities.

(Source: Courtesy of Lotus Development Corporation.)

to the success of the group. Applications include systems that support information sharing, group scheduling, group decision making, and conferencing. These applications enable members of the group to communicate, interact, and collaborate.

Information systems that operate within the **enterprise sphere of influence** support the firm in its interaction with its environment. The surrounding environment includes customers, suppliers, shareholders, competitors, special-interest groups, the financial community, and government agencies. Every enterprise has many applications that operate within the enterprise sphere of influence. The input to these systems is data about or generated by basic business transactions with someone outside the business enterprise. These transactions include customer orders, inventory receipts and withdrawals, purchase orders, freight bills, invoices, and checks. One of the results of processing transaction data is that the records of the company are updated. For example, the processing of employee time cards updates individuals' payroll records used to generate their checks. The order entry, finished product inventory, and billing information systems are examples of applications that operate in the enterprise sphere of influence. These applications support interactions with customers and suppliers.

SYSTEMS SOFTWARE

Controlling the operations of computer hardware is one of the most critical functions of systems software. Systems software also supports the application programs' problem-solving capabilities. Different types of systems software include operating systems and utility programs.

OPERATING SYSTEMS

enterprise sphere of influence
sphere of influence that serves the needs of the firm in its interaction with its environment

operating system (OS)
a set of computer programs that controls the computer hardware and acts as an interface with application programs

An **operating system (OS)** is a set of computer programs that controls the computer hardware and acts as an interface with application programs (Figure 4.2). The operating system, which plays a central role in the functioning of the complete computer system, is usually stored on disk. After a computer system is started, or "booted up," portions of the operating system are transferred to memory as they are needed. The collection of programs, collectively called the operating system, executes a variety of activities, including:

- Perform common computer hardware functions
- Provide a user interface
- Provide a degree of hardware independence
- Manage system memory
- Manage processing tasks
- Provide networking capability
- Control access to system resources
- Manage files

kernel
the heart of the operating system which controls the most critical processes

The **kernel**, as its name suggests, is the heart of the operating system and controls the most critical processes. The kernel ties all of the components of the operating system together and regulates other programs.

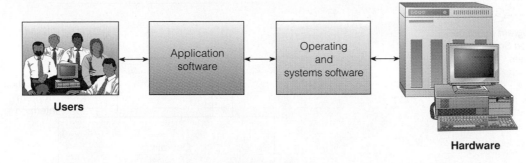

Users

Hardware

FIGURE 4.2

The role of the operating system and other systems software is as an interface or buffer between application software and hardware.

Common Hardware Functions

All application programs must perform certain tasks. For example:

- Get input from the keyboard or some other input device
- Retrieve data from disks
- Store data on disks
- Display information on a monitor or printer

Each of these basic functions requires a more detailed set of instructions to complete. The operating system converts a simple, basic instruction into the set of detailed instructions required by the hardware. In effect, the operating system acts as intermediary between the application program and the hardware. The typical OS performs hundreds of such functions, each of which is translated into one or more instructions for the hardware. The OS will notify the user if input/output devices need attention, if an error has occurred, and if anything abnormal occurs in the system.

User Interface

user interface
element of the operating system that allows individuals to access and command the computer system

command-based user interface
a user interface that requires that text commands be given to the computer to perform basic activities

One of the most important functions of any operating system is providing a **user interface**. A user interface allows individuals to access and command the computer system. The first user interfaces for mainframe and personal computer systems were command based. A **command-based user interface** requires that text commands be given to the computer to perform basic activities. For example, the command ERASE 00TAXRTN would cause the computer to erase or delete a file called 00TAXRTN. RENAME and COPY are other examples of commands used to rename files and copy files from one location to another. Many mainframe computers use a command-based user interface. In some cases, a specific job control language (JCL) is used to control how jobs or tasks are to be run on the computer system.

graphical user interface (GUI)
an interface that uses icons and menus displayed on screen to send commands to the computer system

icon
picture

A **graphical user interface (GUI)** uses pictures (called **icons**) and menus displayed on screen to send commands to the computer system. Many people find that GUIs are easier to use because users intuitively grasp the functions. Today, the most widely used graphical user interface is Windows by Microsoft. Alan Kay and others at Xerox PARC (Palo Alto Research Center, located in California) were pioneers in investigating the use of overlapping windows and icons as an interface. As the name suggests, Windows is based on the use of a window, or a portion of the display screen dedicated to a specific application. The screen can display several windows at once. The use of GUIs has contributed greatly to the increased use of computers because users no longer need to know command-line syntax to accomplish tasks.

Hardware Independence

application program interface (API)
interface that allows applications to make use of the operating system

The applications make use of the operating system by making requests for services through a defined **application program interface (API)**, as shown in Figure 4.3. Programmers can use APIs to create application software without having to understand the inner workings of the operating system.

FIGURE 4.3

Application Program Interface Links Application Software to the Operating System

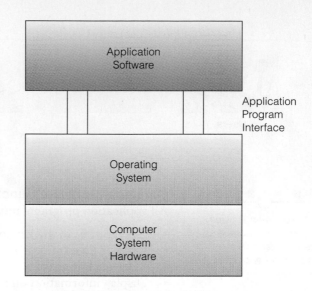

Application Software

Application Program Interface

Operating System

Computer System Hardware

Suppose a computer manufacturer designs new hardware that can operate much faster than before. If the same operating system for which an application was developed can run on the new hardware, minimal (or no) changes are needed to the application to enable it to run on the new hardware. If APIs did not exist, the application software developers might have to completely rewrite the application program to take advantage of the new, faster hardware.

Memory Management

The purpose of memory management is to control how memory is accessed and to maximize available memory and storage. The memory management feature of many operating systems allows the computer to execute program instructions effectively and speed processing.

Controlling how memory is accessed allows the computer system to efficiently and effectively store and retrieve data and instructions and to supply them to the CPU. Memory management programs convert a user's request for data or instructions (called a *logical view* of the data) to the physical location where the data or instructions are stored. A computer understands only the physical view of data—that is, the specific location of the data in storage or memory and the techniques needed to access it. This concept is described as logical versus physical access. For example, the current price of an item, say, a Texas Instruments BA-35 calculator with an item code of TIBA35, might always be found in the logical location "TIBA35$." If the CPU needed to fetch the price of TIBA35 as part of a program instruction, the memory management feature of the operating system would translate the logical location "TIBA35$" into an actual physical location in memory or secondary storage (Figure 4.4).

Memory management is important because memory can be divided into different segments or areas. With some computer chips, memory is divided into conventional, upper, high, extended, and expanded memory. In addition, some computer chips provide "rings" of protection. An operating system can use one or more of these rings to make sure that application programs do not penetrate an area of memory and disrupt the functioning of the operating system, which could cause the entire computer system to crash. Memory management features of today's operating systems are needed to make sure that application programs can get the most from available memory without interfering with other important functions of the operating system or with other application programs.

Most operating systems support **virtual memory**, which allocates space on the hard disk to supplement the immediate, functional memory capacity of RAM. Virtual memory works by swapping programs or parts of programs between

virtual memory
memory that allocates space on the hard disk to supplement the immediate, functional memory capacity of RAM

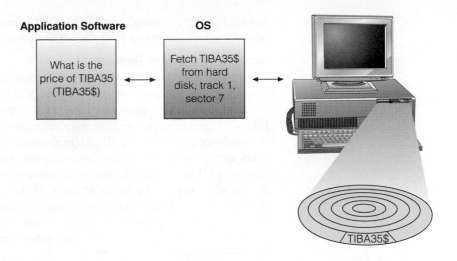

FIGURE 4.4

An Example of the Operating System Controlling Physical Access to Data

The user prompts the application software for specific data. The operating system translates this prompt into instructions for the hardware, which finds the data the user requested. Having successfully completed this task, the operating system then relays the data back to the user via the application software.

paging
process of swapping programs or parts of programs between memory and one or more disk devices

memory and one or more disk devices—a concept called **paging**. This reduces CPU idle time and increases the number of jobs that can run in a given time span.

Processing Tasks

Managing all processing activities is accomplished by the task management features of today's operating systems. Task management allocates computer resources to make the best use of each system's assets. Task management software can permit one user to run several programs or tasks at the same time (multitasking) and allow several users to use the same computer at the same time (time-sharing).

multitasking
capability that allows a user to run more than one application at the same time

An operating system with **multitasking** capabilities allows a user to run more than one application at the same time. Without having to exit a program, you can work in one application, easily pop into another, and then jump back to the first program, picking up where you left off. Better still, while you're working in the *foreground* in one program, one or more other applications can be churning away, unseen, in the *background,* sorting a database, printing a document, or performing other lengthy operations that otherwise would monopolize your computer and leave you staring at the screen unable to get other work done. Multitasking can save users a considerable amount of time and effort.

time-sharing
capability that allows more than one person to use a computer system at the same time

Time-sharing allows more than one person to use a computer system at the same time. For example, 15 customer service representatives may be entering sales data into a computer system for a mail-order company at the same time. In another case, thousands of people may be simultaneously using an on-line computer service to get stock quotes and valuable business news.

Time-sharing works by dividing time into small CPU processing time slices, which can be a few milliseconds or less in duration. During a time slice, some tasks for the first user are done. The computer then goes from that user to the next. During the next time slice, some tasks for the next user are completed. This process continues through each user and cycles back to the first user. Because the CPU processing time slices are small, it appears that all jobs or tasks for all users are being completed at the same time. In reality, each user is sharing the time of the computer with other users.

The ability of the computer to handle an increasing number of concurrent users smoothly is called scalability. This feature is critical for systems expected to handle a large number of users such as a mainframe computer or a Web server. Because personal computer operating systems usually are oriented toward single users, the management of multiple-user tasks often is not needed.

Networking Capability

The operating system can provide features and capabilities that aid users in connecting to a computer network. For example, Apple computer users have built-in network access through the AppleShare feature, and the Microsoft Windows operating systems come with the capability to link users to the Internet.

Access to System Resources

Computers often handle sensitive data that can be accessed over networks. The operating system needs to provide a high level of security against unauthorized access to the users' data and programs. Typically, the operating system establishes a log-on procedure that requires users to enter an identification code and a matching password. If the identification code is invalid or if the password does not go with the identification code, the user cannot gain access to the computer. The operating system also requires that user passwords be changed frequently—say, every 20 to 40 days. If the user is successful in logging on to the system, the operating system records who is using the system and for how long. In some organizations, such records are also used to bill users for time spent using the system. The operating system also reports any attempted breaches of security.

File Management

The operating system performs a file management function to ensure that files in secondary storage are available when needed and that they are protected from access by unauthorized users. Many computers support multiple users who store files on centrally located disks or tape drives. The operating system keeps track of where each file is stored and who may access it. The operating system must be able to resolve what to do if more than one user requests access to the same file at the same time. Even on stand-alone personal computers with only one user, file management is needed to keep track of where files are located, what size they are, when they were created, and who created them.

PERSONAL COMPUTER OPERATING SYSTEMS

Early operating systems for personal computers were very basic. In the last several years, however, more advanced operating systems have been developed, incorporating some features previously available only with mainframe operating systems. Table 4.2 classifies a number of current operating systems by sphere of influence. This section reviews selected popular personal computer operating systems.

Microsoft PC Operating Systems

Ever since a then-small company called Microsoft developed PC-DOS and MS-DOS to support the IBM personal computer introduced in the 1970s, there has been a continuous and steady evolution of personal computer operating systems. Each new version of operating system has improved the ease of use, processing capability, reliability, and ability to support new computer hardware devices. *PC-DOS* and *MS-DOS* had command-driven interfaces that were difficult to learn and use. An operating system called *DOS with Windows* converted PCs to a graphical user interface with a desktop metaphor that showed files as icons within folders (directories).

Windows 95 evolved from these early operating systems. Windows 95 included communications software to simplify connection to the Internet and allow users to send and receive electronic mail and faxes. It also enabled multitasking, which improved users' abilities to complete tasks more quickly. Windows 95 also introduced plug-and-play capabilities, whereby end users could add new hardware devices (e.g., a printer or scanner) to their system with minimal effort. *Windows 98* let an organization's IS department install and configure the operating system and all applications on one machine and then copy the image to each end user's machine. System start-up and shutdown were also speeded up.

The *Windows New Technology (NT) Workstation* operating system was designed to take advantage of 32-bit processors, and it featured multitasking and advanced networking capabilities. NT can also run programs written for other operating

Personal	Workgroup	Enterprise	Consumer
Windows 98			
Windows NT	Windows NT Server	Windows NT Server	
Windows 2000	Windows 2000 Server		
		Windows Advanced Server, Limited Edition	
Windows ME			
Windows XP	Windows XP	Windows XP	Windows XP Embedded
MAC OS			
MAC OS X	MAC OS Server		
Unix	Unix	Unix	
Solaris	Solaris	Solaris	
Linux	Linux	Linux	Linux
Red Hat Linux	Red Hat Linux	Red Hat Linux	
	Netware		
	IBM OS/390	IBM OS/390	
	IBM z/OS	IBM z/OS	
	HP MPE/iX	HP MPE/iX	
			Windows CE.Net
			Pocket PC
			Handheld PC
			Palm OS

TABLE 4.2

Popular Operating Systems across All Three Spheres of Influence

systems. NT supports symmetric multiprocessing, the ability to make simultaneous use of multiple processors. The many features and capabilities of NT made it very attractive for use on many computers. Microsoft renamed the next release of the Windows NT line of operating systems *Windows 2000*. This operating system, with 30 million lines of code, took four years to complete and cost Microsoft more than $1 billion to develop.[1] Microsoft designed Windows 2000 to be easy to operate, and it contained high-level security and significant enhancements for laptop users. The operating system was also designed to be highly reliable.

Windows Millennium Edition (ME) was designed for home use and enables even novice computer users to organize photos, make home movies and records, and play music, as well as the usual computer tasks such as accessing the Internet, playing games, and performing word processing. *Windows XP* (XP reportedly stands for the wonderful e*xp*erience that you will have with your personal computer) was released in fall 2001. Previous consumer versions of Windows were notably unstable and crashed frequently, requiring frustrating and time-consuming reboots. With XP, Microsoft hopes to bring the reliability of Windows 2000 to the consumer. The operating system is enormous—requiring more than 2 GB of hard drive space and more than an hour to install. It only works well on personal computers with at least 128 MB of RAM and a 400 MHz or faster processor.[2] Its redesigned icons, task bar, and window borders make for more pleasant viewing. The Start menu is two columns wide with recently used programs in the left column and everything else (e.g., My Documents, My Computer, and Control Panel) in the right column. It comes with Internet Explorer 6 browser software, which boasts improved security and reliability features including a one-way firewall that blocks hacker invasions coming in from the Internet.

Apple Computer Operating Systems

While IBM system platforms traditionally use Intel microprocessors and one of the Windows operating systems, Apple computers typically use non-Intel microprocessors designed by Apple, IBM, and Motorola and a proprietary Apple operating system—the Mac OS. Although IBM and IBM-compatible computers hold the largest share of the business PC market, Apple computers are also quite popular, especially in the fields of publishing, education, graphic arts, music, movies, and media. The Apple operating systems have also evolved over a number of years and often provide features not available from Microsoft. For example, the *Mac OS 9* operating system provided a Sherlock 2 feature as your personal search detective and personal shopper searching the Internet to locate multiple sources for products you request and compare prices and availability. The Mac OS 9 also had a Multiple Users feature that allowed you to safely share your Macintosh computer with other people. To keep your computer running smoothly, Auto Updating delivers the latest software updates and system enhancements directly from Apple's Web site.

The *Mac OS X ("Ten")* operating system is a completely new implementation of the Mac operating system that Apple began working on in 1996. Starting in July 2001, the Mac OS X came installed on all new Macs. It includes an entirely new user interface called "Aqua," which provides a new visual appearance for users—including luminous and semitransparent elements such as buttons, scroll bars, windows, and fluid animation to enhance the user's experience. One goal of the new software is to provide an even more stable computing environment than the Mac OS 9 operating system. It also comes with new features like automatic networking and an instant wake-from-sleep capability for portable computers. The new, more modular structure of the OS X programming code will permit easier changes and faster improvements.[3] Unfortunately, programs developed to run in earlier Mac operating system environments will not run in OS X. They need to run in what Apple calls the Classic environment. When you open one of the programs, you wait while this environment is established and essentially run the older operating system within the new OS X. Cline, Davis & Mann Inc., a New York advertising agency, specializes in healthcare industry campaigns with major pharmaceutical clients such as Pfizer (Viagra); GlaxoSmithKline (Serevent, an asthma inhaler); and Janssen Pharmaceutica (Risperdal for schizophrenia and Reminyl for Alzheimer's disease). About 90 percent of the computers for the ad agency's 360 employees are Macs, which are used mainly for graphics and creative work. The firm is anxious to move to Mac OS X to take advantage of the improved reliability and new features.[4]

Linux

Linux is an operating system developed under the GNU General Public License, and its source code is freely available to everyone. This doesn't mean, however, that Linux and its assorted distributions are free—companies and developers may charge money for it as long as the source code remains available. Linux is actually only the *kernel* of an operating system, the part that controls hardware, manages files, separates processes, and so forth. Several combinations of Linux are available, with sets of capabilities and applications to form a complete operating system. Each of these combinations is called a *distribution* of Linux.

WORKGROUP OPERATING SYSTEMS

To keep pace with today's high-tech society, the technology of the future must support a world in which network usage, data storage requirements, and data processing speeds increase at a dramatic rate. This rapid increase in communications and data processing capabilities pushes the boundaries of computer science and physics. Powerful and sophisticated operating systems are needed to run the servers that meet these business needs for workgroups.

Windows 2000 Server

Microsoft designed *Windows 2000 Server* to do a host of new tasks that are vital for Web sites and corporate Web applications on the Internet. Besides being more reliable than Windows NT, this operating system is capable of handling extremely demanding computer tasks such as order processing. It can be tuned to run on machines with up to 32 microprocessors—satisfying the needs of all but the most demanding of Web operators. Four machines can be clustered together to prevent service interruptions, which are disastrous for Web sites. With Windows 2000, Microsoft introduced Active Directory, which lets corporations keep track of every employee, computer, software package, and even scrap of data in one place.

Microsoft *Windows Advanced Server, Limited Edition,* was the first 64-bit version of the Windows Server family. Introduced in August 2001, it was designed to run on the 64-bit Itanium processor from Intel (also known as the IA64). This operating system enables Microsoft to begin competing with rival Linux vendors (Red Hat, Caldera, SuSE, and Turbolinux), which already have 64-bit Itanium versions of their Linux distributions. In addition, Sun Microsystems and IBM have had 64-bit Unix operating systems for years. Microsoft also plans a 64-bit version of Windows XP Professional, the desktop version of the operating system, and Windows 2000 DataCenter, the high-end server platform that supports up to 32 processors.[5]

Unix

Unix is a powerful operating system originally developed by AT&T for minicomputers. Unix can be used on many computer system types and platforms, from personal computers to mainframe systems. Unix also makes it much easier to move programs and data among computers or to connect mainframes and personal computers to share resources. Unix is considered to have a complex user interface with strange and arcane commands, so software developers have provided shells such as Motif from Open Systems Foundation and Open Look from Sun Microsystems. These shells provide a graphical user interface and shield the users from the complexity of the underlying operating system. There are many variants of Unix—including HP/UX from Hewlett-Packard, AIX from IBM, UNIX SystemV from UNIX Systems Lab, Solaris from Sun Microsystems, and SCO from Santa Cruz Operations.

Solaris is the Sun Microsystems variation of the Unix operating system, and Solaris is the current server operating system of choice for large Web sites. Solaris is highly reliable and handles the most demanding tasks. It can supervise servers with as many as 64 microprocessors, and eight such computers can be clustered together to work as one. The Solaris operating system can run on Sun's Sparc family of microprocessors, as well as computers with Intel microprocessors. An example of the unique features of Solaris is fault detection and analysis that lets IS administrators establish policies for problematic conditions. For example, if a processor gets too hot, the capability may instruct a system to shut down the processor and reboot. Another feature, the reconfiguration coordination

Red Hat Linux 7.2 is a network operating system from RedHat Software.

(Source: Box shot copyright © 2002 Red Hat, Inc. All rights reserved. Reprinted with permission from Red Hat, Inc.)

manager, lets administrators write policies that automatically redistribute system capacity.

Cleveland-based paint manufacturer and retailer Sherwin-Williams is deploying IBM PCs running the Turbolinux operating system to replace existing Unix-based systems. Under the multimillion-dollar deal, IBM is installing nearly 10,000 desktop personal computers—along with monitors, printers, cash drawers, and related products—in 2,500 Sherwin-Williams stores across North America. The new Turbolinux-equipped systems run all store functions, including customer transactions, inventory management, and software applications used for paint mixing and tinting. The company's IT staff is working with up to 1,000 IBM employees who are assisting in the deployment, which began in July 2002 and is scheduled for completion in one year.[6]

NetWare

NetWare is a network operating system sold by Novell that can support end users on Windows, Macintosh, and Unix platforms. NetWare provides directory software to track computers, programs, and people on a network, making it easier for large companies to manage complex networks. NetWare users can log in from any computer on the network and still get their own familiar desktop with all their applications, data, and preferences.

Red Hat Linux

Red Hat Software offers a Linux network operating system that taps into the talents of tens of thousands of volunteer programmers who generate a steady stream of improvements for the Linux operating system. The *Red Hat Linux* network operating system is very efficient at serving up Web pages and can manage a cluster of up to eight servers. Burlington Coat Factory needed a new operating system on more than 1,250 of its personal computers to support back-office functions such as shipping, receiving, and order processing at 250 stores. Red Hat Linux was chosen because the operating system is inexpensive and runs on standard industry hardware so that any risks were minimal. If Red Hat Linux didn't work as expected, Burlington had the option of keeping the hardware and buying another operating system. In addition, Linux environments typically have fewer virus and security problems than other operating systems. Red Hat Linux has proven to be a very stable and efficient operating system—it doesn't crash.[7]

Mac OS X Server

The *Mac OS X Server* is the first modern server operating system from Apple Computer. It provides Unix-style process management. Protected memory puts each service in its own well-guarded chunk of dynamically allocated memory, preventing a single process from going awry and bringing down the system or other services. Under preemptive multitasking, a computer operating system uses some criteria to decide how long to allocate to any one task before giving another task a turn to use the operating system. Preempting is the act of taking control of the operating system from one task and giving it to another. A common criterion for preempting is simply elapsed time. In more sophisticated operating systems, certain applications can be given higher priority than other applications, giving the higher priority programs longer processing times. Preemptive multitasking ensures that each process gets the right amount of CPU time and the system resources it needs for optimal efficiency and responsiveness.

ENTERPRISE OPERATING SYSTEMS

A few years ago, computer industry pundits were predicting the end of large-scale computing systems. The future of enterprise computing, they claimed, was in networks of servers that were less expensive to buy, more flexible, and more adaptable to the changing demands facing businesses in today's marketplace. In fact, Windows NT, Unix, Solaris, and other workgroup operating systems have versions that are designed to operate with extremely powerful network servers to meet enterprise computing needs.

In spite of this vision, there has been a renewed appreciation for the traditional strengths of mainframe computers. The new generation of mainframe computers provides the computing and storage capacity to meet massive data processing requirements and provide a large number of users with high performance and excellent system availability, strong security, and scalability. In addition, a wide range of application software has been developed to run in the mainframe environment, making it possible to purchase software to address almost any business problem. As a result, mainframe computers remain the computing platform of choice for mission-critical business applications for many companies. OS/390 and z/OS from IBM and MPE/iX from Hewlett-Packard are three examples of mainframe operating systems.

OS/390

In the mid-1990s, IBM introduced a new operating system, called *OS/390*, which is an evolution of an early IBM mainframe operating system called MVS first developed in the 1960s. OS/390 makes it possible for S/390 computers to manage and share information and transactions across a wide range of platforms and multivendor networks. Within OS/390 is a full implementation of the Unix operating system, so developers can use off-the-shelf Unix packages, develop Unix applications, or link them from other Unix environments. LAN Services, another integrated function of the OS/390 base system, extends S/390's strengths in data and systems management to an enterprise's workstations, whether they use the Windows, Macintosh, Unix, or AIX (an IBM variant of Unix) operating system.

z/OS

The *z/OS* is IBM's first 64-bit mainframe operating system. It supports IBM's z900 and z800 lines of mainframes that can come with up to 16 64-bit processors. (The z stands for zero downtime.) It provides several new capabilities to make it easier and less expensive for users to run large mainframe computers. The IBM zSeries mainframe, like previous generations of IBM mainframes, lets users subdivide a single computer into multiple smaller servers, each of which is capable of running a different application. Users are able to set priorities for each application. The Intelligence Resource Director (IRD) component of z/OS manages the partitions and is able to dynamically redirect resources from a low-priority application and reassign them to a high-priority application as the need arises.

In recognition of the widespread popularity of a competing operating system, z/OS allows partitions to run a version of the Linux operating system. Russell Corporation, an apparel maker in Atlanta, recently upgraded from an IBM S/390 mainframe to a z800 mainframe computer. The company was attracted by the z800's faster processor speed and support for up to 32 GB of memory. In addition, Russell can use the z800 to run the Linux operating system if it decides to adopt the open-source operating system because of the virtual partitioning technology that IBM ships with its zSeries mainframes.[8] Z/OS also makes it safer to run Internet and intranet applications by providing new intrusion-detection services and support for services that are frequently used in e-commerce applications.[9]

MPE/iX

Multiprogramming Executive with integrated POSIX (MPE/iX) is the Internet-enabled operating system for the Hewlett-Packard e3000 family of computers using RISC processing. MPE/iX is a robust operating system designed to handle a variety of business tasks, including on-line transaction processing and Web applications. It runs on a broad range of HP e3000 servers—from entry-level to workgroup and enterprise servers within the data centers of large organizations.

Linux

Red Hat Software announced the availability of *Red Hat Linux for IBM* mainframe computers in December 2001. The Linux mainframe operating system is optimized for IBM's S/390 Parallel Enterprise and Multiprise 3000 servers, but it can also run on the IBM's zSeries 800 and 900 servers. This new version of Red Hat Linux means that the company now has Linux versions for everything from handheld devices to the largest enterprise mainframes.[10]

CONSUMER APPLIANCE OPERATING SYSTEMS

New operating systems and other software are changing the way we interact with personal digital assistants (PDAs), mobile phones, digital cameras, TVs, and other appliances. Here are some of the more popular operating systems for such devices.

Windows CE.Net

Windows CE.Net is a key step in taking Microsoft closer to its vision of anywhere, anytime access to Web-based content and services. It is an embedded operating system for use in mobile devices, such as smart phones and PDAs, and can also be used in a variety of other devices, such as digital cameras, thin clients, TV set-top boxes, and automotive computers. PDAs with Windows CE try to bring as much of the functionality of a desktop PC as possible to a handheld device. Such a PDA is a programmable computer that performs most of the functions of a dedicated device.[11]

Windows CE.Net is broken down into software components. Hardware developers are able to pick and choose among these components to customize the operating system for the specific application they are developing. Microsoft hopes to entice hardware developers away from using rival operating systems such as Linux by providing both tools to customize Windows CE.Net to a specific device and specific software applications. If successful, this dual capability could enable hardware developers to get their products to market faster.[12]

Windows XP Embedded

The *Windows XP Embedded* operating system is used in devices such as handheld computers, TV set-top boxes, and automated industrial machines. It is based on Microsoft's Windows XP Professional desktop operating system and includes more than 10,000 software components, including such features as a built-in chat feature that is Microsoft's answer to America Online Inc.'s popular Instant Messaging and support for several network variations. In addition, it is designed so that applications run in their own memory spaces so that they don't interfere with or corrupt one another. As with Windows CE.Net, hardware developers can choose the pieces of the operating system they need for certain devices.[13]

Handheld PC

Handheld PC 2000 is a Microsoft operating system designed to manage a wide range of Windows-powered mobile devices including the Pocket PC and Handheld PCs. The operating system supports forms-based applications used for data collection, connects the user to a server to run desktop applications, and enables connections to the Internet.

Pocket PC

Pocket PC 2002 is the second version of Microsoft's Pocket PC operating system for handheld computers. The new version provides several improvements, including handwriting recognition, the ability to beam information to devices running either Pocket PC or Palm Inc.'s competing operating system, Microsoft's instant messaging technology, and support for more secure

FIGURE 4.5

Not only is the new Nokia 7650 a phone, it's also an integrated imaging device. Point, use the color display as a viewfinder, and snap a picture. Images can be stored on the device and sent to a friend.

(Source: Courtesy of Nokia.)

Utility programs can provide detailed information about your entire system and can help improve your computer's performance by scanning for errors and defragmenting your hard drive.

Internet connections.[14] Casio, Hewlett-Packard, Symbol Technologies, and Toshiba have announced new products that will run Pocket PC 2002.[15] In June 2002, China Merchants Bank and Legend Group Ltd. began offering on-line banking services for Chinese users of Legend's Tianji XP personal digital assistant (PDA) as part of an effort to improve the bank's services. The PDA uses Microsoft's Pocket PC 2002 software and provides the ability to check account balances, transfer funds, and trade securities online. The service can be accessed over any Internet connection.[16]

Palm OS

The strategy Palm has taken with its *Palm OS* operating system and Palm PDA is to extend these devices from single-purpose schedule managers to more general-purpose devices. The company has added features to allow better integration with desktop PCs and enabled users to add applications to the device. Such flexibility has enabled the Palm to remain relatively easy to use while adding some expandability and capability as a general-purpose computing platform.

Palm has licensed its operating system to major chip makers including Intel, Motorola, and Texas Instruments for use in all kinds of mobile devices, from handheld computers to cell phones and even wristwatches. The top two cell phone manufacturers—Nokia and Motorola—plan to release PDA smart phones that run on the Palm OS.[17] Palm is also allowing parts of its operating system to be built into microprocessors based on the ARM architecture (a technology from U.K.-based chip designer ARM Limited), which has become a chip standard for complex wireless communications applications. This move makes it easier for developers to create more powerful and smarter applications for Palm OS–based mobile devices.[18]

Office Depot completed a year-long nationwide rollout of wireless devices containing the Palm OS to its 2,000 truck drivers. Drivers use the system to scan shipments, create an electronic manifest as deliveries are loaded on a truck, and capture customer signatures electronically. Future enhancements include enabling customers to track the status of their orders via its Web site.[19]

UTILITY PROGRAMS

Utility programs are used to merge and sort sets of data, keep track of computer jobs being run, compress files of data before they are stored or transmitted over a network (thus saving space and time), and perform other important tasks. Utility programs often come installed on computer systems; a number of utility programs can also be purchased.

Norton Utilities from Symantec is a useful collection of software utilities that perform many different functions. It includes a Windows hardware utility that checks the status of all parts of the PC including hard disk, memory, modems, speakers, and printers. Its Disk Doctor utility checks the hard disk's boot sector, file allocation tables, and directories and analyzes them to ensure the hard disk wasn't tampered with. Its SpeedDisk utility optimizes placement of files on a crowded disk.

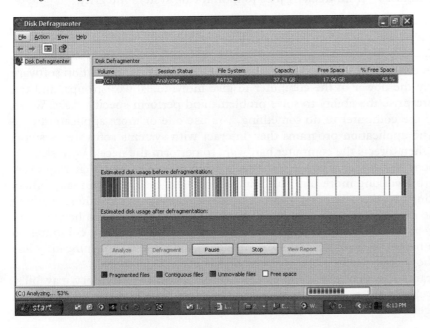

Personal	Workgroup	Enterprise
Software to compress data so that it takes less hard disk space	Software to provide detailed reports of workgroup computer activity and status of user accounts	Software to archive contents of a database by copying data from disk to tape
Screen saver	Software that manages an uninterruptable power source to do a controlled shutdown of the workgroup computer in the event of a loss of power	Software that compares the content of one file with another and identifies any differences
Virus detection software	Software that reports unsuccessful user log-on attempts	Software that reports the status of a particular computer job

TABLE 4.3

Examples of Utility Programs

utility programs
programs used to merge and sort sets of data, keep track of computer jobs being run, compress data files before they are stored or transmitted over a network, and perform other important tasks

A broad range of network and systems management utility software is available to monitor hardware and network performance and trigger an alert when a Web server is crashing or a network problem occurs. Although these general management features are helpful, what is needed is a way to pinpoint the cause of the problem. Topaz from Mercury Interactive is an example of software called an *advanced Web-performance monitoring utility*. It is designed not only to set off an alarm when there are problems but to let network administrators isolate the most likely causes of the problems. Its Auto RCA (root-cause analysis) module uses statistical analysis with built-in rules to measure system and Web performance. Actual performance data is compared with the rules, and the results can help pinpoint where trouble originated—in the application software, database, server, network, or the security features.[20]

IBM has created systems management software that allows individual support people to monitor the growing number of desktop computers in a business. With this software, the support people can sit at their personal computers and check or diagnose problems, such as a hard-disk failure on individual computers on a network. The support people can even repair individual systems anywhere on the organization's network, often without having to leave their desks. The direct benefit is to the system manager, but the business gains from having a smoothly functioning information system. Utility programs can meet the needs of an individual, a workgroup, or an enterprise, as listed in Table 4.3. They perform useful tasks—from tracking jobs to monitoring system integrity.

APPLICATION SOFTWARE

As discussed earlier in this chapter, the primary function of application software is to apply the power of the computer to give individuals, workgroups, and the entire enterprise the ability to solve problems and perform specific tasks. When you need the computer to do something, you use one or more application programs. The application programs then interact with systems software. Systems software then directs the computer hardware to perform the necessary tasks.

Suppose a manager is concerned that too many employees are getting overtime pay by working more than 40 hours each week, even though many others are working less than 40 hours per week. She would like to have those working below the 40-hour threshold replace those over the threshold, and hence avoid the time-and-a-half overtime pay rate. The manager can enlist a computer to print the names of all employees working significantly more or significantly less than 40 hours per week on average over the last three months.

Programs like this that complete sales orders, control inventory, pay bills, write paychecks to employees, and provide financial and marketing information to managers and executives are examples of application software. Most of the computerized business jobs and activities discussed in this book involve the use of application software.

TYPES AND FUNCTIONS OF APPLICATION SOFTWARE

The key to unlocking the potential of any computer system is application software. A company can either develop a one-of-a-kind program for a specific application (called **proprietary software**) or purchase and use an existing software program (sometimes called **off-the-shelf software**). It is also possible to modify some off-the-shelf programs, giving a blend of off-the-shelf and customized approaches. These different sources of software are represented in Figure 4.6. The relative advantages and disadvantages of proprietary software and off-the-shelf software are summarized in Table 4.4.

Proprietary Application Software

Software to solve a unique or specific problem is called *proprietary application software*. This type of software is usually built, but it can also be purchased from an outside company. If the organization has the time and IS talent, it may opt for **in-house development** for all aspects of the application programs. Alternatively, an organization may obtain customized software from external vendors. For example, a third-party software firm, often called a *value-added software vendor,* may develop or modify a software program to meet the needs of a particular industry or company. A specific software program developed for a particular company is called **contract software**.

Off-the-Shelf Application Software

Software can also be purchased, leased, or rented from a software company that develops programs and sells them to many computer users and organizations. Software programs developed for a general market are called *off-the-shelf software packages* because they can literally be purchased "off the shelf" in a store. Many companies use off-the-shelf software to support business processes. Key criteria for selecting off-the-shelf software include the following: (1) Will the software run on the operating system and hardware you have selected? (2) Does the software meet the essential business requirements that have been defined? (3) Is the software manufacturer financially solvent and reliable? (4) Does the total cost of purchasing, installing, and maintaining the software versus the business benefits to be achieved make this worthwhile? Most organizations form a small selection

proprietary software
a one-of-a-kind program for a specific application

off-the-shelf software
existing software program

in-house development
development of application software using the company's resources

contract software
software developed for a particular company

FIGURE 4.6

Sources of Software: Proprietary and Off-the-Shelf
Some off-the-shelf software may be modified to allow some customization.

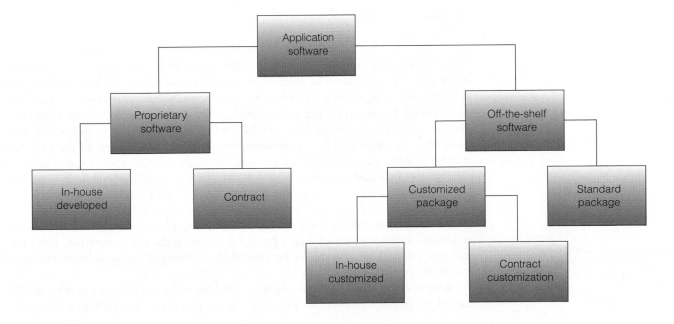

Proprietary Software		Off-the-Shelf Software	
Advantages	**Disadvantages**	**Advantages**	**Disadvantages**
You can get exactly what you need in terms of features, reports, and so on.	It can take a long time and significant resources to develop required features.	The initial cost is lower since the software firm is able to spread the development costs over a large number of customers.	An organization might have to pay for features that are not required and never used.
Being involved in the development offers a further level of control over the results.	In-house system development staff may become hard pressed to provide the required level of ongoing support and maintenance because of pressure to get on to other new projects.	There is a lower risk that the software will fail to meet the basic business needs—you can analyze existing features and the performance of the package.	The software may lack important features, thus requiring future modification or customization. This can be very expensive because users must adopt future releases of the software as well.
There is more flexibility in making modifications that may be required to counteract a new initiative by one of your competitors or to meet new supplier and/or customer requirements. A merger with another firm or an acquisition also will necessitate software changes to meet new business needs.	There is more risk concerning the features and performance of the software that has yet to be developed.	Package is likely to be of high quality since many customer firms have tested the software and helped identify many of its bugs.	Software may not match current work processes and data standards.

TABLE 4.4

A Comparison of Proprietary and Off-the-Shelf Software

committee consisting of end users, business managers, and IS personnel to first define the essential business requirements, identify a large number of potential suppliers, screen the potential suppliers down to a small number of "best qualified" candidates, and then make a final selection based on actual demonstrations of the software and input from current customers.

Customized Package

In some cases, companies use a blend of external and internal software development. That is, off-the-shelf software packages are modified or customized by in-house or external personnel. For example, a software developer may write a collection of programs to be used in an auto body shop that includes features to generate estimates, order parts, and process insurance. Body shops of all types have these needs. Designed properly—and with provisions for minor tailoring for each user—the same software package can be sold to many users. However, since each body shop has slightly different requirements, some modifications to the software may be needed. As a result, software vendors often provide a wide range of services, including installation of their standard software, modifications to the software required by the customer, training of the end users, and other consulting services.

Some software companies encourage their customers to make changes to their software. In some cases, the software company supplying the necessary software will make the necessary changes for a fee. Other software companies, however, will not allow their software to be modified or changed by those purchasing or leasing it.

Another approach to obtaining a customized software package is through the use of an application service provider. An **application service provider** is a

application service provider
a company that provides both end-user support and the computers on which to run the software from the user's facilities

company that can provide software, end-user support, and the computer hardware on which to run the software from the user's facilities. They can also take a complex corporate software package and simplify it for the users so it is easier to set up and manage. They provide contract customization of off-the-shelf software, and they speed deployment of new applications while helping IS managers avoid implementation headaches, reducing the need for many skilled IS staff members and reducing project start-up expenses. Such an approach allows companies to devote more time and resources to more important tasks. The use of an application service provider makes the most sense for relatively small, fast-growing companies with limited information system resources. It is also a good strategy for companies looking to deploy a single, functionally focused application quickly, such as setting up an e-commerce Web site or supporting expense reporting. Contracting with an application service provider may make less sense, however, for larger companies with major systems that have their technical infrastructure already in place.

Ingersoll-Rand Co. is an industrial and commercial equipment manufacturer with recent annual sales in excess of $9 billion. In May 2002, it rolled out a hosted version of Oracle software to 1,000 of its shared services employees, with the help of application service provider Corio, Inc. These employees manage finances, human resources, and procurement for several departments within the company. Under the terms of the five-year contract, Corio runs the application from one of its data centers in San Jose, California, and provides access to Ingersoll-Rand shared services employees worldwide.[21]

There is a long tradition of service companies managing a client's computing systems—the so-called outsourcers like IBM and Electronic Data Services. But what's different with application service providers is the use of the Internet. Through application service providers, software can be made available quickly. Users can gather and update information from their applications via simple Web browsers, so even users who recoil from dealing with complicated software can feel comfortable and be trained quickly. Another improvement over the old outsourcing model is that the data traffic is carried over public networks instead of far more expensive private networks. The software that companies are using is changing, too. Application service providers employ standard software packages, replacing, in many cases, custom programs that require small armies of programmers to support. That means application service providers can also devise simple methods for handling all their clients' applications, driving costs even lower.

Using an application service provider is not without risks—sensitive information could be compromised in a number of ways, including unauthorized access by employees or computer hackers, the application service provider being unable to keep its computers and network up and running as consistently as is needed, or a disaster disabling the application service provider's data center, temporarily putting an organization out of business. These are legitimate concerns that the application service provider must address.

Perhaps the biggest advantage of employing an application service provider is that it frees in-house corporate resources from staffing and managing complex computing projects so that they can focus on more important things.

PERSONAL APPLICATION SOFTWARE

There are literally hundreds of computer applications that can help individuals at school, home, and work. Personal application software includes general-purpose tools and programs that can support a number of individual needs. For example, a graphics program can be purchased to help a sales manager develop an attractive presentation to give to the sales force at its annual meeting. A spreadsheet program allows a financial executive to test possible investment outcomes. The primary programs are word processing, spreadsheet analysis, database, graphics, and on-line services. Advanced software tools—like project management, financial management, desktop publishing, and creativity software—are finding more and more use in business. The features of personal application software are summarized in Table 4.5. In addition to these general-purpose programs, there are literally thousands of other personal computer applications to perform specialized tasks: to help you do your taxes, get in shape, lose weight, get medical advice, write wills and other legal documents, make repairs to your computer, fix your car, write music, and edit your pictures and videos (see Figures 4.7 and 4.8). This type of software, often called *user software* or *personal productivity software*, includes the general-purpose tools and programs that support individual needs.

TABLE 4.5

Examples of Personal Productivity Software

Type of Software	Explanation	Example	Vendor
Word processing	Create, edit, and print text documents	Word WordPerfect	Microsoft Corel
Spreadsheet	Provide a wide range of built-in functions for statistical, financial, logical, database, graphics, and data and time calculations	Excel Lotus 1-2-3 Quattro Pro	Microsoft Lotus/IBM Originally developed by Borland
Database	Store, manipulate, and retrieve data	Access Approach FoxPro dBASE	Microsoft Lotus/IBM Microsoft Borland
On-line information services	Obtain a broad range of information from commercial services	America Online CompuServe Prodigy	America Online CompuServe Prodigy
Graphics	Develop graphs, illustrations, and drawings	Illustrator FreeHand	Adobe Macromedia
Project management	Plan, schedule, allocate, and control people and resources (money, time, and technology) needed to complete a project according to schedule	Project for Windows On Target Project Schedule Time Line	Microsoft Symantec Scitor Symantec
Financial management	Provide income and expense tracking and reporting to monitor and plan budgets (some programs have investment portfolio management features)	Managing Your Money Quicken	Meca Software Intuit
Desktop publishing (DTP)	Works with personal computers and high-resolution printers to create high-quality printed output, including text and graphics; various styles of pages can be laid out; art and text files from other programs can also be integrated into "published" pages	QuarkXPress Publisher PageMaker Ventura Publisher	Quark Microsoft Adobe Corel
Creativity	Helps generate innovative and creative ideas and problem solutions. The software does not propose solutions, but provides a framework conducive to creative thought. The software takes users through a routine, first naming a problem, then organizing ideas and "wishes," and offering new information to suggest different ideas or solutions	Organizer Notes	Macromedia Lotus

FIGURE 4.7

TurboTax

Tax preparation programs can save hours of work and are typically more accurate than doing a tax return by hand. Programs can check for potential problems and give you help and advice about what you may have forgotten to deduct.

(Source: Courtesy of Intuit.)

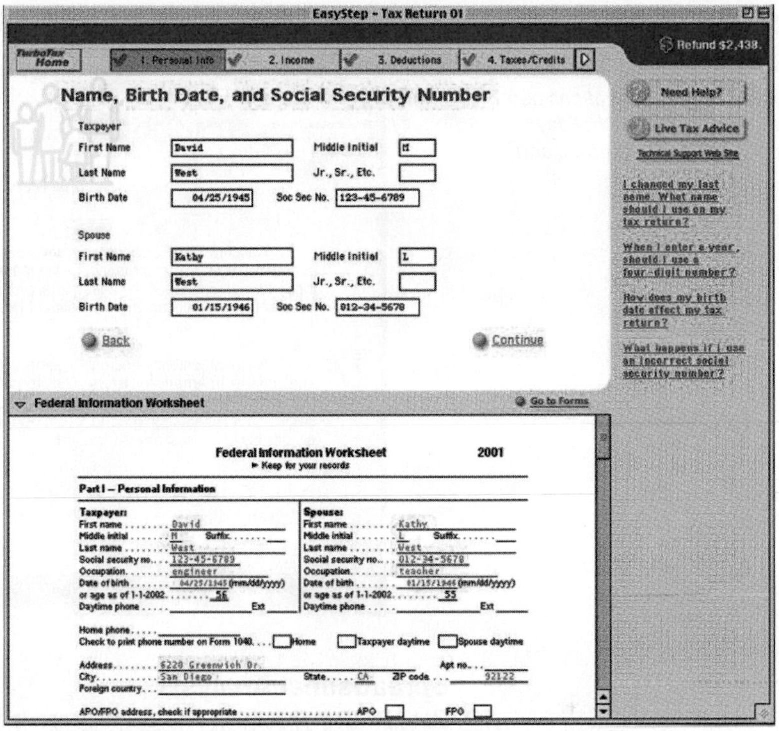

Word Processing

If you write reports, letters, or term papers, word processing applications can be indispensable. The majority of personal computers in use today have word processing applications installed. Such applications can be used to create, edit, and print documents. Most come with a vast array of features, including those for checking spelling, creating tables, inserting formulas, creating graphics, and much more (see Figure 4.9). This book (and most like it) was entered into a word processing application using a personal computer.

FIGURE 4.8

Quicken

Off-the-shelf financial management programs are useful for paying bills and tracking expenses.

(Source: Courtesy of Intuit.)

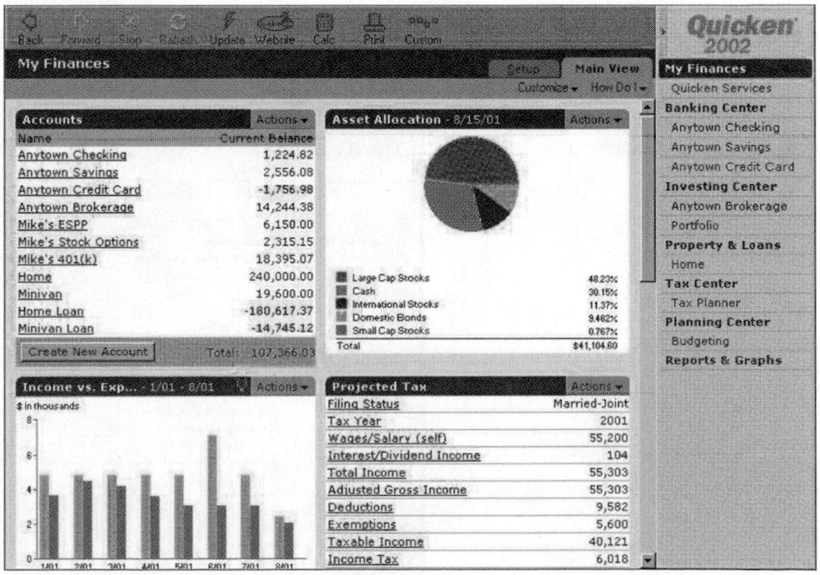

FIGURE 4-9

Word Processing Program

Word processing applications can be used to write letters, holiday greeting cards, work reports, and term papers.

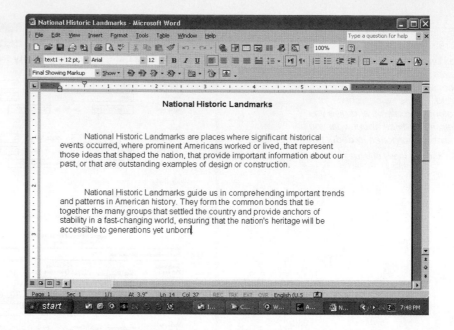

Spreadsheet Analysis

People use spreadsheets to prepare budgets, forecast profits, analyze insurance programs, summarize income tax data, and analyze investments. Whenever numbers and calculations are involved, spreadsheets should be considered. Features of spreadsheets include graphics, limited database capabilities, statistical analysis, built-in business functions, and much more (see Figure 4.10).

Database Applications

Database applications are ideal for storing, manipulating, and retrieving data. These applications are particularly useful when you need to manipulate a large amount of data and produce reports and documents. Database manipulations include merging, editing, and sorting data. The uses of a database application are varied. You can keep track of a CD collection, the items in your apartment, tax records, and expenses using a database application. A student club can use a database to store names, addresses, phone numbers, and dues paid. In business, a

FIGURE 4-10

Spreadsheet Program

Spreadsheet programs should be considered when calculations are required.

database application can help process sales orders, control inventory, order new supplies, send letters to customers, and pay employees. A database can also be a front end to another application. For example, a database application can be used to enter and store income tax information; the stored results can then be exported to other applications, such as a spreadsheet or tax preparation application (see Figure 4.11).

Graphics Program

It is often said that a picture is worth a thousand words. With today's graphics programs, it is easy to develop attractive graphs, illustrations, and drawings. Graphics programs can be used to develop advertising brochures, announcements, and full-color presentations. If you are asked to make a presentation at school or work, you can use a graphics program to develop and display slides while you are making your talk. A graphics program can be used to help you make a presentation, a drawing, or an illustration (see Figure 4.12).

On-Line Information Services

On-line services allow you to connect a personal computer to the outside world through phone lines. Using an on-line service, you can get investment information, make travel plans, and check news from around the world. You can also get prices and features for most consumer items, learn about companies, send e-mail to friends and family, learn about degree programs offered by colleges and universities around the world, and search for job openings in your area (see Figure 4.13).

Software Suite

software suite
a collection of single-application software packages in a bundle

A **software suite** is a collection of single-application software packages in a bundle. Software suites can include word processors, spreadsheets, database management systems, graphics programs, communications tools, organizers, and more. There are a number of advantages to using a software suite. The software programs have been designed to work similarly, so once you learn the basics for one application, the other applications are easy to learn and use. Buying software in a bundled suite is cost-effective; the programs usually sell for a fraction of what they would cost individually.

Microsoft Office, Corel's WordPerfect Office, Lotus SmartSuite, and Sun Microsystems's StarOffice are examples of popular general-purpose software suites

FIGURE 4.11

Database Program
Once entered into a database application, information can be manipulated and used to produce reports and documents.

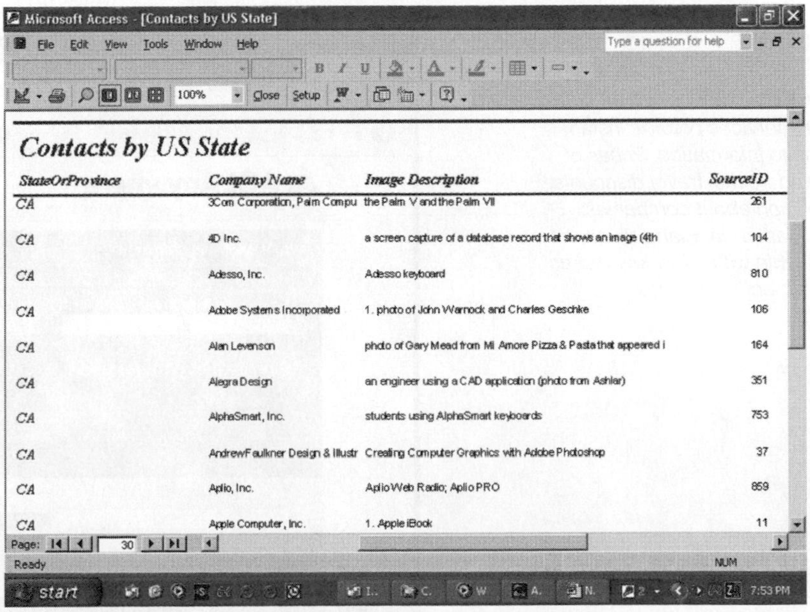

F I G U R E 4 ⦿ 12

Graphics Program

F I G U R E 4 ⦿ 12

Graphics Program

Graphics programs can help you make a presentation at school or work. They can also be used to develop attractive brochures, illustrations, drawings, and maps.

(Source: Courtesy of Adobe Systems Incorporated.)

for personal computer users (see Figure 4.14). Each of these software suites includes a spreadsheet program, word processor, database program, and graphics package with the ability to move documents, data, and diagrams among them (see Table 4.6). Thus, a user can create a spreadsheet and then cut and paste that spreadsheet into a document created using the word processing application.

There are more than 120 million users worldwide of the Microsoft Office software suite, with Office XP representing the sixth version of the decade-old productivity software.[22] XP aims to reveal more of the product's full functionality; a frequent rule of thumb is that most users are aware of only 20 percent of a software suite's capabilities. To help users, a new Smart Tags icon appears periodically and offers a list of features that might come in handy for performing whatever current operation you are executing. The resulting short list of commands may let you get driving directions (from the Web) for an address you've just typed or offer to undo (or turn off) the program's autocapitalization or other autoformatting features. A Task Pane interface helps you perform common editing functions. It's a narrow window that appears every so often at the right edge of the screen, exposing relevant controls. For example, if you use the Find command, the Task

F I G U R E 4 ⦿ 13

On-Line Services

On-line services provide instant access to information. Prices of cars and trucks, travel discounts, information about companies, stock market data, and much more is available with a few keystrokes using an on-line service.

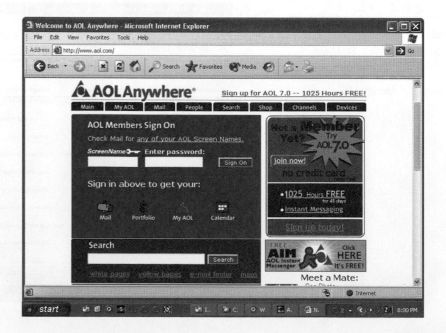

FIGURE 4.14

Software Suite

A software suite, such as Corel WordPerfect Office 2002, offers a collection of powerful programs, including word processing, spread-sheet, database, graphics, and other programs. The programs in a software suite are designed to be used together. In addition, the commands, icons, and procedures are the same for all programs in the suite.

(Source: Box shot reprinted with permission from Corel Corporation.)

groupware
software that helps groups of people work together more efficiently and effectively

collaborative computing software
software that helps teams of people to work together toward a common goal

TABLE 4.6

Major Components of Leading Software Suites

Pane shows options for finding files on your hard drive. Office XP employs memory-protection technology, with each application confined to its own portion of main memory to prevent a crash in one application from causing a crash in any other XP program or the operating system. If an application does crash, an error report is automatically e-mailed to Microsoft to help the company identify what went wrong. Office X is Microsoft's version of Office XP for the Mac OS X operating system.

Microsoft Office goes beyond its role as a mainstream package of ready-to-run applications with the extensive custom development facilities of Visual Basic for Applications (VBA)—a built-in facility that is part of every Office application. VBA provides a means for users to enhance off-the-shelf applications to tailor the programs for special tasks.

Since one or more applications in a suite may not be as desirable as the others, some people still prefer to buy separate packages. Another issue with the use of software suites is the large amount of main memory required to run them effectively. For example, many users find that they must spend hundreds of dollars for additional internal memory to upgrade their personal computer to be able to run a software suite. Continual debates rate one vendor's spreadsheet superior to another vendor's, and yet a third vendor may have the best word processing package. Thus, some users prefer using individual software packages from different vendors rather than a software suite from a single vendor.

WORKGROUP APPLICATION SOFTWARE

The software class known as **groupware** cannot be concretely defined but is, in general, software that helps groups of people work together more efficiently and effectively. **Collaborative computing software**, an only slightly better term, at least conveys the sense that teams are working toward a common goal. Collaborative computing software can support a team of managers working on the same production problem, letting them share their ideas and work via connected computer systems. Examples of such software include group scheduling software, electronic mail, and other software that enables people to share ideas. Read the "IS Principles in Action" special feature to learn more about collaborative computing software.

Lotus Notes

Lotus, a division of IBM, has defined *knowledge management* as the ability to provide individuals and groups of users with a method to find, access, and deliver valuable information in a coherent fashion. Its Lotus Notes product is an attempt to provide this ability. Lotus Notes gives companies the capability of using one software package, and one user interface, to integrate many business processes. For example, it can allow a global team to work together from a common or shared set of documents, have electronic discussions using common threads of discussion, and schedule team meetings. A key design feature is to make Notes 5 very easy to use.

Personal Productivity Function	Microsoft Office XP	Lotus SmartSuite Millennium Edition	Corel WordPerfect Office 2002	Sun Microsystems
Word Processing	Word	WordPro	WordPerfect	Writer
Spreadsheet	Excel	Lotus 1-2-3	Quattro Pro	Calc
Presentations	PowerPoint	Freelance Graphics	Presentations	Impress
Database	Access	Lotus Approach	Paradox	

Honda Uses Collaboration Software to Speed Product Development

Honda is the world's largest motorcycle maker and Japan's third largest automaker behind Toyota and Nissan. It also has a power products division that manufactures lawn mowers, snow blowers, portable generators, and outboard motors. The company's car models include the Accord, Acura, Civic, Legend, Prelude, and Insight—a gasoline-electric hybrid.

The automobile industry is intensely competitive, with manufacturers facing cutthroat global competition. They are under pressure to reduce costs to the minimum and pass the savings along to customers. They must also reduce the time to produce new models that appeal to customers in many parts of the world while tailoring automobiles to customer specifications. To overcome these challenges, manufacturers are turning over an increasing portion of the design, development, production, and customer-support activities to multiple suppliers and other business partners. Such collaborative product development is changing the way Honda brings products to market.

The research and development activities for Honda uses software called eMatrix from MatrixOne to support collaborative product development. This Internet-based system helps document and manage the product's bill of materials (BOM)— a list that describes all components of the finished product, such as part numbers, quantity, measurements, and labor and material costs. An accurate and complete BOM is critical to the manufacturing process—any mistake can cost the manufacturer in loss of labor and materials and valuable time-to-market.

eMatrix enables Honda research and design engineers to share design information from the early stages of product development so that the accuracy of the engineering BOM is greatly improved. The system supports engineers as they manage and configure all the details that go into a product: basic car design, options, features, local variations, components and larger assembled parts, specifications, supporting documentation, costs, and engineering changes. The single, accurate view of the engineering BOM across all organizations involved in the product ensures that most design and development issues are identified and resolved earlier in the design and manufacturing cycle. Such coordination leads to the early detection and resolution of many problems—saving time and resources. The bottom line is higher quality, decreased costs, and the ability to bring automobiles to market faster.

The eMatrix system is being deployed in multiple stages with the ultimate goal to link Honda's global research and design activities wherever they are located. The first phase of the rollout was completed in October 2001 and connected several thousand design engineers throughout Japan, the United States, the United Kingdom, and Thailand. The software vendor, MatrixOne, partnered with Fujitsu to deploy the eMatrix platform and integrate it with Honda's other key information systems.

Discussion Questions

1. Product design has always been a collaborative process. How has Honda brought collaboration to a new level?
2. Why is the BOM such a critical element of collaborative product development?

Critical Thinking Questions

3. Why did Honda resort to a software package to support this critical business process rather than develop a custom application to meet its needs?
4. What issues do you anticipate Honda must overcome in trying to implement this standard software to support its worldwide research and design activities?

Sources: Adapted from Antone Gonsalves, "It's the Tiger in Their Tanks," *InformationWeek,* September 17, 2001, accessed at http://www.informationweek.com; Demir Sarias, "Honda's E-Business Strategy," December 11, 2001, *Line56,* accessed at http://www.line56.com; Honda Motor Co., Ltd,: Company Report, accessed at http://www.moneycentral.msn.com on December 28, 2001; "Automotive Solutions," accessed at the MatrixOne Web site at http://www.matrixone.com on December 28, 2001; and "MatrixOne Solutions Drive New Paradigm for Automobile Design at Honda," accessed at About Us, Press Box at MatrixOne Web site at http://www.matrixone.com on December 28, 2001.

As Lotus Notes matured, Lotus added services to it and renamed it Domino (Lotus Notes is now the name of the e-mail package), and now an entire third-party market has emerged to build collaborative software based on Domino. These products, which include Changepoint's Involv, remove the burden of Notes administration and broaden the application's scope to better support the Internet. For example, Domino.Doc is a Domino-based document management application with built-in workflow and archiving capabilities. Its "life cycle" feature tracks a document through the review, approval, publishing, and archiving processes. Similarly, the workflow integration adds support for multiple roles, log tracking, and distributed approval.

The NHL has developed an automated workflow process based on Lotus Notes and Domino software to support its annual player draft. On draft day, the league's 30 teams, NHL officials, and journalists use some 60 workstations that are connected to two Notes servers. Using this system, a team sends a request for a player as a draft pick in a Notes e-mail message to the central scouting desk. Requests for players preapproved by the NHL are automatically forwarded to a central registry desk. If the draft pick is approved there, the name goes to the podium, where there is also a workstation, and NHL officials post the name on a large display board. The playing histories of those who haven't been vetted are compiled from scouting reports and local news coverage. This information is available in a database containing information on all prospective draft picks. NHL officials review that material before they approve or disapprove the draft pick.[23]

Group Scheduling

Group scheduling is another form of groupware, but not all software schedulers approach their tasks in the same way. Some schedulers, known as personal information managers (PIMs), tend to focus on personal schedules and lists, as opposed to coordinating the schedules and meetings of a team or group. Schedulers do not suit everyone's needs, and if they are not truly required, they could impede efficiency. The "Three Cs" rule for successful implementation of groupware is summarized in Table 4.7.

ENTERPRISE APPLICATION SOFTWARE

Software that benefits the entire organization can also be developed or purchased. A fast-food chain, for example, might develop a materials ordering and distribution program to make sure that each fast-food franchise gets the necessary raw materials and supplies during the week. This materials ordering and distribution program can be developed internally using staff and resources in the IS department or purchased from an external software company. Table 4.8 lists a number of applications that can be addressed with enterprise software.

Many organizations are moving to integrated enterprise software that supports supply chain management (movement of raw materials from suppliers through shipment of finished goods to customers), as shown in Figure 4.15.

Organizations can no longer respond to market changes using nonintegrated information systems based on overnight processing of yesterday's business transactions, conflicting data models, and obsolete technology. As a result, many corporations are turning to enterprise resource planning (ERP) software, a set of integrated programs that manage a company's vital business operations for an entire multisite, global organization. Thus, an ERP system must be able to support multiple legal entities, multiple languages, and multiple currencies. Although the scope of an ERP system may vary from vendor to vendor, most ERP systems provide integrated software to support manufacturing and finance. In addition to these core business processes, some ERP systems may be capable of supporting additional business functions such as human resources, sales, and distribution.

TABLE 4.7

Ernst & Young's "Three Cs" Rule for Groupware

Convenient	If it's too hard to use, it doesn't get used; it should be as easy to use as the telephone.
Content	It must provide a constant stream of rich, relevant, and personalized content.
Coverage	If it isn't close to everything you need, it may never get used.

TABLE 4 8

Examples of Enterprise Application Software

Accounts receivable	Sales ordering
Accounts payable	Order entry
Airline industry operations	Payroll
Automatic teller systems	Human resource management
Cash-flow analysis	Check processing
Credit and charge card administration	Tax planning and preparation
Manufacturing control	Receiving
Distribution control	Restaurant management
General ledger	Retail operations
Stock and bond management	Invoicing
Savings and time deposits	Shipping
Inventory control	Fixed asset accounting

Successful linking of these business functions means that business requirements must come first and technology constraints second. Software will need to be rethought and reconfigured to reflect that reality. Unfortunately, the effort to implement ERP software is great and the path to success is full of hazards. Software vendors that provide integrated enterprise software are listed in Table 4.9.

Most ERP vendors specialize in software that addresses the needs of well-defined markets, such as automotive, semiconductor, petrochemical, and food/beverage manufacturers, with solutions targeted to meet specific needs in those industries. ERP has become one of the most lucrative segments of the software market as companies continue to switch from old information systems to more efficient client/server-based systems.

Increased global competition, new executive management needs for control over the total cost and product flow through their enterprises, and more customer

FIGURE 4 15

Use of Integrated Supply Chain Management Software

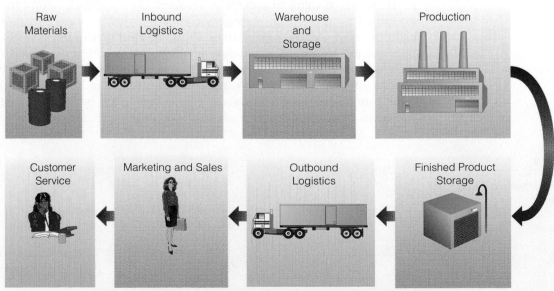

Integrated Enterprise Software to Support Supply Chain Management

TABLE 4.9

Selected Enterprise Resource Planning Vendors

SAP	Baan
Oracle	SSA
PeopleSoft	Marcam
Dun & Bradstreet	QAD
J.D. Edwards	Ross Systems

interactions are driving the demand for enterprisewide access to current business information. ERP offers integrated software from a single vendor that helps meet those needs. The primary benefits of implementing ERP include eliminating inefficient systems, easing adoption of improved work processes, improving access to data for operational decision making, standardizing technology vendors and equipment, and enabling the implementation of supply chain management.

Hoffman Enclosures is a manufacturer of construction and industrial enclosures. It is implementing ERP software from J.D. Edwards to help it assess its own resources and those of its suppliers in real time so it can better respond to customer requests. The company will implement Buyer Workstation, the first of several new supplier-relationship management tools to be released in 2003 as part of the J.D. Edwards upgrade to the company's OneWorld ERP suite. This software will combine strategic planning and fulfillment functions into a single real-time application used to manage the supply chain. The new tools will integrate capabilities ranging from design collaboration and supply-chain planning and execution to fulfillment and settlement capabilities.[24]

PROGRAMMING LANGUAGES

programming languages
sets of keywords, symbols, and a system of rules for constructing statements by which humans can communicate instructions to be executed by a computer

Both operating systems and application software are written in coding schemes called *programming languages*. The primary function of a programming language is to provide instructions to the computer system so that it can perform a processing activity. IS professionals work with **programming languages**, which are sets of keywords, symbols, and a system of rules for constructing statements by which humans can communicate instructions to be executed by a computer. Programming involves translating what a user wants to accomplish into a code that the computer can understand and execute. Like writing a report or a paper in English, writing a computer program in a programming language requires the programmer to follow a set of rules. Each programming language uses a set of symbols that have special meaning. Each language also has its own set of rules, called the **syntax** of the language. The language syntax dictates how the symbols should be combined into statements capable of conveying meaningful instructions to the CPU.

syntax
a set of rules associated with a programming language

Programming languages were developed to help solve particular problems. Since they were each designed for different problems, they contain different attributes. Each of the attributes in Table 4.10 represents two extremes, with most languages falling somewhere between these extremes.

THE EVOLUTION OF PROGRAMMING LANGUAGES

The desire to use the power of information processing efficiently in problem solving has pushed development of newer programming languages. The evolution of programming languages is typically discussed in terms of generations of languages.

TABLE 4.10

Programming Language Attributes

Extreme 1	Extreme 2
Supports programming of batch processing systems with data collected into a set and processed at one time.	Supports programming of real-time systems with each data transaction processed when it occurs.
Requires programmer to write procedure-oriented code, describing step by step each action the computer must take.	Enables a programmer to write non-procedure-oriented code, describing the end result desired without having to specify how to accomplish it.
Supports business applications that require the ability to store, retrieve, and manipulate alphanumeric data and process large files.	Supports sophisticated scientific computations.
Programmers write code with a relatively high level of errors.	Programmers write code with a relatively low level of errors.
Programmers are less productive and able to create only a small amount of code per unit of time.	Programmers are more productive and are able to create a large amount of code per unit time.

First Generation

The first generation of programming languages is *machine language*, which requires the use of binary symbols (0s and 1s). Because this is the language of the CPU, text files translated into binary sets can be read by almost every computer system platform.

Second Generation

Developers of programming languages attempted to overcome some of the difficulties inherent in machine language by replacing the binary digits with symbols programmers could more easily understand. These second-generation languages use codes like A for add, MVC for move, and so on. Another term for these languages is *assembly language*, which comes from the programs (called *assemblers*) used to translate it into machine code. Systems software programs such as operating systems and utility programs are often written in an assembly language.

Third Generation

Third-generation languages continued the trend toward greater use of symbolic code and away from specifically instructing the computer how to complete an operation. BASIC, COBOL, C, and FORTRAN are examples of third-generation languages that use Englishlike statements and commands. This type of language is easier to learn and use than machine and assembly languages because it more closely resembles everyday human communication and understanding.

With third-generation and higher level programming languages, each statement in the language translates into several instructions in machine language. A special software program called a **compiler** converts the programmer's source code into the machine language instructions consisting of binary digits, as shown in Figure 4.16. A compiler creates a two-stage process for program execution. First, it translates the program into a machine language; second, the CPU executes that program.

compiler
a special software program that converts the programmer's source code into the machine language instructions consisting of binary digits

Fourth Generation

Fourth-generation programming languages emphasize what output results are desired rather than how programming statements are to be written. As a result, many managers and executives with little or no training in computers and programming are using fourth-generation languages (4GLs). Languages for accessing information in a database are often fourth-generation languages. Prime examples include PowerBuilder, Delphi, Essbase, Forte, Focus, Powerhouse, SAS, and

FIGURE 4 ⊙ 16

How a Compiler Works
A compiler translates a complete program into a complete set of binary instructions (Stage 1). Once this is done, the CPU can execute the converted program in its entirety (Stage 2).

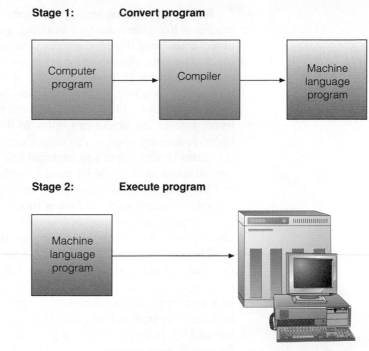

Stage 1:   Convert program

Computer program → Compiler → Machine language program

Stage 2:   Execute program

Machine language program →

Program execution

many others. Another popular fourth-generation language is called Structured Query Language (SQL), which is often used to perform database queries and manipulations.

When catalog retailer Blair Corp. relaunched its Web site in December 2001, it wanted a software tool to analyze the behavior of its Web customers to determine which areas within the site attracted the most customers and exactly why those browsing did or didn't buy. Blair chose IntelliVisor, an on-line service based on the SAS fourth-generation language to analyze such things as which product categories suffered the most shopping-cart abandonment (customers ending their shopping process without completing the order) and which generated the highest conversion rates (turning browsers into buyers.)[25]

Fifth Generation

A fifth-generation programming language uses a visual or graphical development interface to create source language that is usually compiled with a 3GL or 4GL language compiler. Microsoft, Borland, IBM, and other companies make 5GL visual programming products for developing applications in Java, for example. Visual Basic, PC COBOL, and Visual C++ are examples of visual programming languages. Microsoft's Visual Studio 7, now known as Visual Studio.Net, allows 20-some programming languages—APL, COBOL, C++, Perl, Smalltalk, C#, JScript, Visual Basic, Transact-SQL, Visual FoxPro, and even Java—to share a single GUI. The layout and behavior of Visual Studio.Net requires some getting used to for Visual Basic developers.

OBJECT-ORIENTED PROGRAMMING LANGUAGES

The preceding programming languages separate data elements from the procedures or actions that will be performed on them, but another type of programming language ties them together into units called *objects*. An object consists of data and the actions that can be performed on the data. For example, an object could be data about an employee and all the operations (such as payroll calculations) that might be performed on the data. Programming languages that are based on objects are called *object-oriented programming languages*.

Building programs and applications using object-oriented programming languages is like constructing a building using prefabricated modules or parts. The object containing the data, instructions, and procedures is a programming building block. The same objects (modules or parts) can be used repeatedly. One of the primary advantages of an object is that it contains *reusable code*. In other words, the instruction code within that object can be reused in different programs for a variety of applications, just as the same basic prefabricated door can be used in two different houses. An object can relate to data on a product, an input routine, or an order-processing routine. An object can even direct a computer to execute other programs or to retrieve and manipulate data. So, a sorting routine developed for a payroll application could be used in both a billing program and an inventory control program. By reusing program code, programmers are able to write programs for specific application problems more quickly (see Figure 4.17). By combining existing program objects with new ones, programmers can easily and efficiently develop new object-oriented programs to accomplish organizational goals.

There are several object-oriented programming languages; some of the most popular include Smalltalk, C++, Java, and C#. Java is an Internet programming language from Sun Microsystems. One of the key advantages of Java is the ability of a Java application to run on a variety of computers and operating systems, including Unix, Windows, and Macintosh operating systems. An increasing number of U.S. colleges are using Java as their first programming language. C# is a Microsoft programming language that the software giant hopes will unseat Java. J2EE is an industry-standard version of Java used by many software development companies including IBM and BEA Systems.

The various languages have distinguishing characteristics that make them appropriate for particular types of problems or applications. Among the third-generation languages, COBOL has excellent file- and database-handling capabilities for manipulating large volumes of business data, while FORTRAN is better suited for scientific applications. Java is an obvious choice for people doing Web development. End users will choose one of the fourth- or fifth-generation languages to develop programs. Although many programming languages are used to write new business applications, there are more lines of code written in COBOL in existing business applications than any other programming language.

FIGURE 4.17

Reusable Code in Object-oriented Programming

By combining existing program objects with new ones, programmers can easily and efficiently develop new object-oriented programs to accomplish organizational goals. Note that these objects can be either commercially available or designed internally.

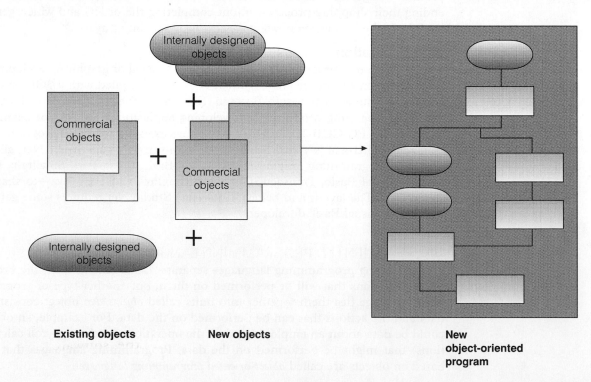

Existing objects   **New objects**   **New object-oriented program**

SELECTING A PROGRAMMING LANGUAGE

Selecting the best programming language to use for a particular program involves balancing the functional characteristics of the language with cost, control, and complexity issues.

Machine and assembly languages provide the most direct control over computer hardware. For this reason, many vendors of popular application software programs take the time and effort to code portions of their leading programs in assembly language to maximize their speed. When a programmer requires a high degree of control over how various hardware components are used, these languages should be used. In selecting any programming language, the amount of direct control that is needed over the operation of the hardware can be an important factor to consider.

More recent programming languages are typically more complex than earlier programming languages. Although these newer languages appear to be simpler because they are more Englishlike, each command can drive complex routines and functions that operate behind the scenes. It takes less time to develop computer programs using higher-level languages than with lower-level languages. This means that the cost to develop computer programs can be substantially less with these more recent programming languages. Although training programmers to use these higher-level programming languages may produce high up-front costs, using higher-level languages can reduce the total costs to develop computer programs in the long run.

C++ and Java both have advantages and disadvantages, but Java may be the future of programming. Java is far easier to learn, and as a result, people become productive much sooner. Programmers who learn C++ must spend a lot of time debugging rather than learning software engineering techniques. An increasing number of U.S. colleges are using Java as their first programming language. Java is also more portable—with the ability to run on more operating systems and hardware. However, C++ will not disappear anytime soon. A large base of C++ programs is installed, and there are a large number of users because Microsoft uses it for programming. The ANSI and ISO standards committees have also been working on C and C++ since 1990, and it is apparent that people will continue to develop C++ in or outside of a Microsoft environment.

SOFTWARE ISSUES AND TRENDS

Because software is such an important part of today's computer systems, issues such as software bugs, licensing, upgrades, and global software support have received increased attention. We highlight several major software issues and trends in this section: software bugs, open source software, opensourcing, antitrust issues, software licensing, software upgrades, and global software support.

SOFTWARE BUGS

software bug
a defect in a computer program that keeps it from performing in the manner intended

A **software bug** is a defect in a computer program that keeps it from performing in the manner intended. Some software bugs are obvious and cause the program to terminate unexpectedly. Other bugs are subtler and allow errors to creep into your work. Computer and software vendors tell us that since humans design and program hardware and software, bugs are inevitable. In fact, according to the Pentagon and the Software Engineering Institute at Carnegie Mellon University, there are typically 5 to 15 bugs in every 1,000 lines of code—the software instructions that make sense only to computers and programmers.

Most software bugs arise because manufacturers release new software as early as possible instead of waiting until all bugs are identified and removed. They are under intense pressure from customers to deliver the software they have announced and from shareholders to begin selling the new product to increase sales. Meanwhile, the software manufacturer's quality-assurance people fight a losing battle for more testing time to identify and remove bugs. Although the decision of when to release new software is based on a fine line, the industry clearly favors releasing software early and with defects. After all, software companies make money on upgrades, so there is little incentive to achieve a perfect first release.

In 2002, Bill Gates announced a companywide shift in Microsoft's emphasis to developing high-quality code that is reliable and secure—even if it comes at the expense of adding new features. Many software customers and IT professionals call the commitment long overdue, but not just from Microsoft.[26] Software bugs have plagued most new releases of all computer software, with the most damaging bugs affecting software used by major industries, the government, and the military. And those bugs have been costly.

A programming error at drug maker Eli Lilly and Co. led the company's e-mail system to disclose the addresses of 600 patients who were taking the manufacturer's Prozac antidepressant medication. The incident brought a complaint from the American Civil Liberties Union (ACLU) that claimed that the disclosure sets "a dangerous precedent" and constitutes an unfair and deceptive trade practice by violating the terms of the privacy policy posted on Lilly's Web site.[27] Toshiba Corp. had a software bug in its Version 1.7 BIOS for its 5005 Series Satellite notebooks that caused the unit's 1.1-GHz Pentium III desktop processor to slow dramatically during intense computing. The problem rendered the performance notebook nearly useless for some tasks.[28]

A programming glitch temporarily converted thousands of pay phones in Ontario and Quebec into a toll-free service. For six days people who punched in a sequence of numbers—10–10–620—on any of Bell Canada's Millennium pay phones were suddenly able to call anyone, anywhere, with no charge.[29]

NTT DoCoMo had to suspend sales of a Sony handset featuring an advanced version of its popular "i-mode" Internet access service due to a software glitch. About 420,000 units of the Sony phone model SO503i had already been sold and will be replaced if customers request it. The sales suspension came at the worst possible time—just as the electronics giant had accelerated its drive in the handset market by forming an alliance with Ericsson.[30]

Once a company has gained a reputation for releasing buggy software, it takes a long time and lots of effort to overcome people's negative perceptions. Nineteen months after the initial release of its 11i E-Business applications suite failed to meet users' expectations, Oracle users were still expressing concerns about the quality of its software applications.[31] Table 4.11 summarizes tips for reducing the impact of software bugs.

TABLE 4.11

How to Deal with Software Bugs

Register all software so that you receive bug alerts, fixes, and patches.
Check the manual or read-me files for workarounds.
Access the support area of the manufacturer's Web site for patches.
Install the latest software updates.
Before reporting a bug, make sure that you can re-create the circumstances under which it occurs.
Once you can re-create the bug, call the manufacturer's tech support line.
Avoid buying the latest release of software for several months or a year until the software bugs have been discovered and removed.

OPEN SOURCE SOFTWARE

open source software
software that is freely available to anyone in a form that can be easily modified

Open source software is software that is freely available to anyone in a form that can be easily modified. Users can download the source code and build the software themselves, or the software's developers can make executable versions available along with the source. Open source software development is a collaborative process—developers around the world use the Internet to keep in close contact via e-mail and to download and submit new software. Major software changes can occur in days rather than weeks or months. Open source software is at the heart of many of the Internet's most popular services, including e-mail and the Web. A number of open source software packages are widely used, including the Linux system; Free BSD, another operating system; Apache, the most popular Web server in the world; Sendmail, a program that delivers e-mail for most systems on the Internet; and Perl, a programming language used to develop Internet application software.

Why would an organization run its business using software that's free? How can something that's given away over the Internet be stable or reliable or sufficiently supported to place at the core of a company's day-to-day operations? The answer is surprising—open source software is often *more* reliable than commercial software. How can this be? First, by making a program's source code readily available, users can fix any problems they discover. A fix is often available within hours of the problem's discovery. Second, with the source code for a program accessible to thousands of people, the chances of a bug being discovered and fixed before it does any damage are much greater than with traditional software packages.

The question of software support is the biggest stumbling block to the acceptance of open source software at the corporate level. Getting support for traditional software packages is easy—you call a company's toll-free support number or access its Web site. But how do you get help if an open source package doesn't work as expected? Since the open source community lives on the Internet, you look there for help. Through use of Internet discussion areas, you can communicate with others who use the same software, and you may even reach someone who helped develop it. Users of popular open source packages can get correct answers to their technical questions within a few hours of asking for help on the appropriate Internet forum. Another approach is to contact one of the many companies emerging to support and service such software—for example, Red Hat for Linux, C2Net for Apache, and Sendmail, Inc., for Sendmail. These companies offer high-quality, for-pay technical assistance.

OPENSOURCING

opensourcing
extending software development beyond a single organization by finding others who share the same problem and involving them in a common development effort

The use of the Internet to spur development of open source software has led to **opensourcing**, or extending software development beyond a single organization by finding others who share the same problem and involving them in a common development effort. Such an approach spreads development costs across multiple organizations and increases the opportunity for people with highly specialized expertise to contribute. It increases the number of developers and users who can identify and eliminate bugs in software. It also fosters the growth of a community of people who can support the software.

ANTITRUST ISSUES FOR SOFTWARE

An ongoing debate in the software industry is the issue of competition—is the field level for a variety of manufacturers to provide products, or does one company have a stranglehold on the industry? Microsoft is one of the largest and most powerful companies in the world, and its impact on the software industry is unequaled. It has been the subject of much legal action to limit what some people see as its abuses.

In the most significant lawsuit against the company, the U.S. Justice Department and several states brought an antitrust lawsuit against Microsoft. After almost three years of court proceedings, a U.S. District Court ruled in April 2000 that Microsoft violated federal antitrust law by abusing its monopoly position in the PC operating systems market. As of July 2002, the courts are still trying to decide what remedy to impose on the software giant. A list of possible remedies focuses on changing Microsoft business practices to require that Microsoft fully disclose Windows technical information and that Microsoft give PC makers flexibility in deciding which operating system and options to include in new systems while barring any retaliation from the software maker.[32]

This proposed settlement has come under heavy criticism by groups claiming that it should have tougher terms. The attorneys general of nine states and the District of Columbia have declined to participate in the proposed settlement. They use Microsoft's Passport and Wallet services to drive home their point. Passport allows people to sign on to a number of Web sites with one password. Microsoft Wallet lets people fill in billing information once and shop on many e-commerce sites. Under the proposed settlement, Microsoft is free to include these services as elements of the Windows operating system, just as it did with its Internet Explorer browser, the action that provoked the government to launch its antitrust case in the first place. Meanwhile, rival Web service providers such as AOL Time Warner must pay PC makers to include their services on their PCs. Opponents of the proposed settlement argue that the agreement fails to protect consumer choice and promote competition; Microsoft is free to continue to abuse its monopoly.

In a closely related case, Netscape Communications Corp., a subsidiary of AOL Time Warner, filed a lawsuit against Microsoft in January 2002 alleging that the software maker harmed Netscape with anticompetitive practices associated with the Windows operating system. Netscape is seeking an injunction against Microsoft that would restrict how Microsoft would be able to sell Windows to consumers and PC makers. Netscape is also seeking an award of three times any damages set by the court—which, if granted, could be worth billions of dollars.[33]

In June 2002, InterTrust Technologies, Inc., alleged that Microsoft products—including Windows XP, Office XP, Microsoft.Net-based products and services, and Windows Media Player—infringe on its patents. Burst.com, Inc., also filed a suit in June 2002 accusing Microsoft of illegally employing Burst's video delivery technologies while creating its upcoming Corona digital video playback technology.[34]

How these various lawsuits will affect Microsoft in particular and the software industry in general remain to be seen. However, it is clear that software rivals are increasingly taking legal action as part of their corporate strategy.

SOFTWARE LICENSING

In general, software manufacturers want to license their software to lock in a steady, predictable stream of revenue from customers. Software manufacturers also want to encourage customers to move to the latest releases of their software products to minimize the effort required to support out-of-date products. There are numerous types of software licenses to help accomplish these objectives.

Usage-based License

In this arrangement, software fees are based on the amount of actual usage of the manufacturer's products. Licensees are charged in much the same way that utility firms charge their customers—increasing fees for increased use of power or water.

Capacity-based License

With capacity-based licenses, the fees paid to the software manufacturer are based on the processing power of the computer on which the software is installed. Users who run their software on a more powerful processor pay more for the software. The fees charged do not relate to the actual use of the products.

Software-as-a-Network Service

When software is accessed as a network service, the software manufacturer makes its products available through the Internet. The advantages of this form of usage is that the software manufacturer automatically offers users bug fixes, enhancements, and other updates over the Web and charges a subscription fee for the software and associated services.

Subscription Licensing

With subscriptions, user companies sign a multiyear deal with a manufacturer for individual products or a collection of products and then pay annual subscription fees based on how many PCs they have.

In July 2002, Microsoft put into effect its Version 6 volume licensing program. Under this program, companies that enroll in Microsoft's new Software Assurance program are entitled to receive the latest versions of Microsoft products during its contract term. The annual cost is 25 percent of the volume license fee for server software products and 29 percent for desktop products. To be eligible for this program, the company must already be running the most current version of the product. After July 31, 2002, the company must purchase new licenses to be eligible for Software Assurance. Many users and companies viewed this program as a way to ensure they stay on current software releases but at a substantial increase in their software budget.[35]

Read the "Ethical and Societal Issues" special-interest box to gain a deeper insight into the implications of a software manufacturer that changed its software licensing strategy.

SOFTWARE UPGRADES

Software companies revise their programs and sell new versions periodically. In some cases, the revised software offers new and valuable enhancements. In other cases, the software uses complex program code that offers little in terms of additional capabilities. In addition, revised software can contain bugs or errors. Deciding whether to purchase the newest software can be a problem for corporations and individuals with a large investment in software. Should the newest version be purchased when it is released? Some organizations and individuals do not always get the most current software upgrades or versions, unless there are significant improvements or capabilities. Instead, they may upgrade to newer software only when there are vital new features. Software upgrades usually cost much less than the original purchase price.

GLOBAL SOFTWARE SUPPORT

Large, global companies have little trouble persuading vendors to sell them software licenses for even the most far-flung outposts of their company. But can those same vendors provide adequate support for their software customers in all locations? Taking into account the support requirements of local operations is one of the biggest challenges information systems teams face when putting together standardized, companywide systems. They must ensure that the vendor has sufficient support for local operations. In slower technology growth markets, such as Eastern Europe and Latin America, there may be no vendor presence at all. Instead, large vendors such as Sybase, IBM, and Hewlett-Packard typically contract out support for their software to local providers.

Oracle Tries to Recover from Pricing Miscue

In mid-2000, Oracle introduced a complex pricing scheme called Universal Power Unit (UPU) for its database management software. The UPU pricing was based on the number and speed of processors in the server on which the Oracle software was run. As soon as it was instituted, many Oracle users immediately criticized the UPU pricing as exorbitant. Some financial analysts even placed part of the blame for Oracle's lower-than-expected database sales on the UPU approach. By June 2001, Oracle realized its mistake, and CEO Larry Ellison eliminated the controversial capacity-based licensing approach and changed to the standard per-processor pricing used by all its competitors.

The enterprise edition of the Oracle database software was set at $40,000 per processor, with the standard edition set at $15,000 per CPU. With the new pricing, it became crystal clear that Oracle's per-processor price was twice as expensive as IBM's DB2 database management software. But Ellison was quick to explain that the Oracle software comes with more features than IBM's DB2 and that once users start adding in the cost of features that IBM charges extra for, IBM's program is actually more expensive.

Responses to the switch in pricing varied. For some users that had been paying for software under the UPU performance measurement, the new per-processor scheme cut prices by as much as 40 percent. For example, Eastman Chemical Company claimed it paid $2.1 million for an Oracle license under UPU pricing but would have paid $800,000 less under the new pricing policy. Other users claimed that the per-processor pricing resulted in higher costs for users with low-end servers with multiple processors. Overall, financial analysts predicted that the pricing change would likely have a positive impact on Oracle's database sales.

According to Jacqueline Woods, vice president of global practices at Oracle, the pricing system was changed in response to user complaints. It was also aimed at minimizing haggling over license fees. While large users, in particular, stood to benefit from the new pricing approach, Oracle wanted to be neutral on the hardware companies used by charging the same per-processor price no matter what kind of CPU a customer uses.

Ellison said he wanted to counter a persistent marketing blitzkrieg by IBM that tried to show that Oracle's databases were much more expensive than IBM's. Thus, one goal of the new pricing scheme was to make cost comparisons between Oracle and IBM as easy as possible.

Discussion Questions

1. What do you see as the pros and cons of Oracle's new pricing scheme for Oracle database users? For Oracle?
2. Why did Oracle convert to the new pricing scheme if it made it even more clear that the Oracle database management software was more expensive than IBM's?

Critical Thinking Questions

3. Do you think Larry Ellison's comment about Oracle being cheaper than DB2 helps users accurately compare prices between the competing vendors? Why or why not?
4. How should Oracle deal with a customer who agreed to a license based on the old pricing scheme but who would save much money if it could convert to the new pricing scheme?

Sources: Adapted from Marc Songini, "Opinion Split on New Oracle Pricing," *Computerworld,* December 17, 2001, accessed at http://www.computerworld.com; Dan Verton, "Oracle Reverses Course on Database Pricing," *Computerworld,* June 15, 2001, accessed at http://www.computerworld.com; and Rick Whiting, "CFO: Market Hit Bottom Last Quarter," *InformationWeek,* December 17, 2001, accessed at http://www.informationweek.com.

One approach that has been gaining acceptance in North America is to outsource global support to one or more third-party distributors. The software user company may still negotiate its software license with the software vendor directly, but it then hands over the global support contract to a third-party supplier. The supplier acts as a middleman between software vendor and user, often providing distribution, support, and invoicing. This is how American Home Products Corporation handles global support for both Novell NetWare and Microsoft Office applications in the 145 countries in which it operates. American Home Products, a pharmaceutical and agricultural products company, negotiated the agreements directly with the vendors for both purchasing and maintenance, but fulfillment of the agreement is handled exclusively by Philadelphia-based Softsmart, an international supplier of software and services.

In today's computer systems, software is an increasingly critical component. Whatever approach individuals and organizations take to acquire software, it is important for everyone to be aware of the current trends in the industry. Informed users are wiser consumers, and they can make better decisions.

SUMMARY

PRINCIPLE *When selecting an operating system, you must consider the current and future needs for application software to meet the needs of the organization. In addition, your choice of a particular operating system must be consistent with your choice of hardware.*

Software consists of programs that control the workings of the computer hardware. There are two main categories of software: systems software and application software. Systems software is a collection of programs that interacts between hardware and application software. Systems software includes utility programs and operating systems. Application software enables people to solve problems and perform specific tasks. Application software may be proprietary or off-the-shelf.

An operating system (OS) is a set of computer programs that controls the computer hardware to support users' computing needs. The OS functions by converting an instruction from an application into a set of instructions needed by the hardware. The OS also serves as an intermediary between application programs and hardware, allowing hardware independence. Memory management involves controlling storage access and use by converting logical requests into physical locations and by placing data in the best storage space, perhaps expanded or virtual memory.

Task management allocates computer resources through multitasking and time-sharing. With multitasking, users can run more than one application at a time. Time-sharing allows more than one person to use a computer system at the same time.

The ability of a computer to handle an increasing number of concurrent users smoothly is called *scalability,* a feature critical for systems expected to handle a large number of users.

An OS also provides a user interface, which allows users to access and command the computer. A command-based user interface requires text commands to send instructions; a graphical user interface (GUI), like Windows, uses icons and menus.

Software applications make use of the operating system by making requests for services through a defined application program interface (API). Programmers can use APIs to create application software without having to understand the inner workings of the operating system. APIs also provide a degree of hardware independence so that the underlying hardware can change without necessarily requiring a rewrite of the software applications.

Over the years, several popular operating systems have been developed. These include several proprietary operating systems used primarily on mainframes. MS-DOS is an early OS for IBM-compatibles. Older Windows operating systems are GUIs used with DOS. Newer versions such as Windows 95, Windows 98, Windows NT, and Windows XP, are fully functional operating systems that do not need DOS. Apple computers use proprietary operating systems like the Mac OS and Mac OS X. Unix is a powerful operating system that can be used on many computer system types and platforms, from personal computers to mainframe systems. Use of the Unix operating system makes it easy to move programs and data among computers or to connect mainframes and personal computers to share resources. Linux is the kernel of an operating system whose source code is freely available to everyone. Several variations of Linux are available, with sets of capabilities and applications to form a complete operating system; for example, Red Hat Linux. OS/390, z/OS, and MPE iX are operating systems for mainframe computers. A number of operating systems have been developed to support consumer appliances such as Windows CE.Net, Windows XP Embedded, Pocket PC, Palm OS, and variations of Linux.

Utility programs are used to perform many useful tasks and often come installed on computers along with the operating system. This software is used to merge and sort sets of data, keep track of computer jobs being run, compress files of data, protect against harmful computer viruses, and monitor hardware and network performance.

PRINCIPLE *Do not develop proprietary application software unless doing so will meet a compelling business need that can provide a competitive advantage.*

Application software applies the power of the computer to solve problems and perform specific tasks. One useful way of classifying the many potential uses of information systems is to identify the scope of problems and opportunities addressed by a particular organization— its sphere of influence. For most companies, the spheres of influence are personal, workgroup, and enterprise.

User software, or personal productivity software, includes general-purpose programs that enable users to improve their personal effectiveness, increasing the amount of work that can be done and its quality. Software that helps groups work together is often referred to as groupware. Examples of such software include group scheduling software, electronic mail, and other software that enables people to share ideas. Enterprise software that benefits the entire organization can also be developed or purchased. Many organizations are turning to enterprise resource planning software, a set of integrated programs that manage a company's vital business operations for an entire multisite, global organization.

Three approaches to developing application software are as follows: build proprietary application software, buy existing programs off the shelf, or use a combination of customized and off-the-shelf application software. Building proprietary software (in-house or contracting out) has the following advantages: the organization will get software that more closely matches its needs; by being involved with the development, the organization has further control over the results; and the organization has more flexibility in making changes. The disadvantages include the following: it is likely to take longer and cost more to develop; the in-house staff will be hard pressed to provide ongoing support and maintenance; and there is a greater risk that the software features will not work as expected or that other performance problems will occur.

Purchasing off-the-shelf software has its advantages. The initial cost is lower; there is a lower risk that the software will fail to work as expected; and the software is likely to be of higher quality than proprietary software. Some of the disadvantages are that the organization may pay for features in the software it does not need; the software may lack important features requiring expensive customization; and the system may work in such a way that work process reengineering is required.

Application software packages are typically chosen by a small selection committee that takes into account the operating environment in which the software must run, the business requirements that must be met, the viability of the supplier, and the costs versus benefits.

Some organizations have taken a third approach—customizing software packages. This approach can combine all of the above advantages and disadvantages and must be carefully managed.

Application service providers are companies that can provide software, end-user support, and the computer hardware on which to run the software from the user's facilities. They provide contract customization of off-the-shelf software, and they speed deployment of new applications while helping IS managers avoid implementation headaches, reducing the need for many skilled IS staff members and reducing project start-up expenses.

Although there are literally hundreds of computer applications that can help individuals at school, home, and work, the primary ones are word processing, spreadsheet analysis, database, graphics, and on-line services. A software suite, such as SmartSuite, WordPerfect, StarOffice, or Office, offers a collection of powerful programs.

PRINCIPLE *Choose a programming language whose functional characteristics are appropriate to the task at hand, taking into consideration the skills and experience of the programming staff.*

All software programs are written in coding schemes called programming languages, which provide instructions to a computer to perform some processing activity. There are several classes of programming languages, including machine, assembly, high-level, query and database, object-oriented, and visual programming languages.

Programming languages have gone through changes since their initial development in the early 1950s. In the first generation, computers were programmed in machine language, or binary code, a series of statements written in 0s and 1s. The second generation of languages was termed assembly languages; these languages support the use of symbols and words rather than 0s and 1s. The third generation consists of many high-level programming languages that use Englishlike statements and commands. They also must be converted to machine language by special software called a *compiler* but are easier to write than assembly or machine language code. These languages include BASIC, COBOL, FORTRAN, and others. A fourth-generation language is less procedural and more Englishlike than third-generation languages. The fourth-generation languages include database and query languages like SQL. Fifth-generation programming languages combine rules-based code generation, component management, visual programming techniques, reuse management, and other advances. These languages offer the greatest ease of use yet.

Object-oriented programming languages—like Smalltalk, C++, Java, and C#—use groups of related data, instructions, and procedures called *objects*, which serve as reusable modules in various programs. These languages can reduce program development and testing time. Java can be used to develop applications on the Internet.

Selecting the best programming language to use for a particular program involves balancing the functional characteristics of the language with cost, control, and complexity issues.

PRINCIPLE *The software industry continues to undergo constant change; users need to be aware of recent trends and issues to be effective in their business and personal life.*

Software bugs, open source software, opensourcing, antitrust issues, software licensing, software upgrades, and global software support are all important software issues.

A software bug is a defect in a computer program that keeps it from performing in the manner intended. Software bugs are common, even in key pieces of business software.

Open source software is software that is freely available to anyone in a form that can be easily modified. Open source software development and maintenance is a collaborative process with developers around the world using the Internet to keep in close contact via e-mail and to download and submit new software.

Opensourcing is the process of extending software development beyond a single organization by finding others who share the same business problem and involving them in a common development effort.

An ongoing debate in the software industry is the issue of competition—is the field level for a variety of manufacturers to provide products, or does one company have a stranglehold on the industry? Microsoft is one of the largest and most powerful companies in the world, and its impact on the software industry is unequaled. It has been the subject of much legal action to limit what some people see as its abuses. How these various lawsuits will affect Microsoft in particular and the software industry in general remain to be seen. However, it is clear that software rivals are increasingly taking legal action as part of their corporate strategy.

Software manufacturers are developing new approaches to licensing their software to lock in a steady, predictable stream of revenue from their customers. Some of these new approaches include usage-based licensing, capacity-based licensing, software-as-a-network service, and subscription licensing.

Software upgrades are an important source of increased revenue for software manufacturers and can provide useful new functionality and improved quality for software users.

Global software support is an important consideration for large, global companies putting together standardized, companywide systems. A common solution is outsourcing global support to one or more third-party software distributors.

KEY TERMS

application program interface (API), 135
application service provider, 148
application software, 132
collaborative computing software, 155
command-based user interface, 135
compiler, 160
computer programs, 132
computer system platform, 132
contract software, 147
documentation, 132
enterprise sphere of influence, 134
graphical user interface (GUI), 135

groupware, 155
icon, 135
in-house development, 147
kernel, 134
multitasking, 137
off-the-shelf software, 147
open source software, 165
opensourcing, 165
operating system (OS), 134
paging, 137
personal productivity software, 133
personal sphere of influence, 133
programming languages, 159

proprietary software, 147
software bug, 163
software suite, 153
sphere of influence, 133
syntax, 159
systems software, 132
time-sharing, 137
user interface, 135
utility programs, 146
virtual memory, 136
workgroup, 133
workgroup sphere of influence, 133

CHAPTER 4 SELF-ASSESSMENT TEST

When selecting an operating system, you must consider the current and future needs for application software to meet the needs of the organization. In addition, your choice of a particular operating system must be consistent with your choice of hardware.

1. What is the heart of the operating system that controls the most critical processes?
   A. platform
   B. instruction set
   C. kernel
   D. CPU

2. Multitasking and time sharing are essentially the same thing. True False

3. _____ is the process of swapping programs or parts of programs between memory and disk.

4. The file manager component of the operating system controls how memory is accessed and maximizes available memory and storage. True False

Do not develop proprietary application software unless doing so will meet a compelling business need that can provide a competitive advantage.

5. The primary function of application software is to apply the power of the computer to give individuals, workgroups, and the entire enterprise the ability to solve problems and perform specific tasks. True False

6. Software that enables users to improve their personal effectiveness, increasing the amount of work they can do and its quality is called
   A. personal productivity software
   B. operating system software
   C. utility software
   D. graphics software

7. Which of the following is NOT a characteristic of ERP software?
   A. ability to support a multisite, global organization
   B. ability to support multiple legal entities, multiple languages, and multiple currencies
   C. ability to integrate software to support manufacturing and finance functions
   D. ability to let individual users set up their own systems and work practices the way they like it

8. Software used to solve a unique or specific problem that is usually built, but can also be purchased from an outside company is called
   _____.

9. Software that has the greatest potential to affect the processes that add value to a business because it is designed for specific organizational activities and functions.
   A. personal productivity software
   B. operating system software

   C. utility software
   D. applications software

Choose a programming language whose functional characteristics are appropriate to the task at hand, taking into consideration the skills and experience of the programming staff.

10. A built-in scripting facility that is part of every Microsoft Office application and provides a means of enhancing off-the-shelf applications to allow users to tailor the programs.
    A. Visual Basic
    B. Smalltalk
    C. Norton Utilities
    D. Java

11. A class of applications software that helps groups work together and collaborate is called
    _____.

12. Each language also has its own set of rules, called the _____ of the language.

13. A special software program called a *compiler* performs the conversion from the programmer's source code into the machine language instructions consisting of binary digits. True/False

Chapter 4 Self-Assessment Test Answers

1. c, 2. False, 3. Paging, 4. False, 5. True, 6. a, 7. d, 8. proprietary software, 9. d, 10. a, 11. groupware or collaborative computing software, 12. syntax, 13. True

---

REVIEW QUESTIONS

1. State three reasons the relative cost of hardware and software has shifted so dramatically.
2. Give four examples of personal productivity software.
3. How do software bugs arise?
4. Identify and briefly discuss two types of user interface provided by the operating system.
5. What are the two basic types of software? Briefly describe the role of each.
6. Name four operating systems that support the personal, workgroup, and enterprise spheres of influence.
7. Define the term *utility software* and give two examples.
8. Identify the two primary sources for acquiring application software.

9. What is an application service provider? What issues arise in considering the use of one?
10. What is open source software? What is the biggest stumbling block with the use of open source software?
11. What does the acronym API stand for? What is the role of an API?
12. Briefly discuss the advantages and disadvantages of frequent software upgrades.
13. Describe the term *enterprise resource planning (ERP) system*. What functions does such a system perform?
14. Identify and briefly discuss four different types of software licenses.

DISCUSSION QUESTIONS

1. Assume that you must take a computer programming course next semester. What language do you think would be best for you to study? Why? Do you think that a professional programmer needs to know more than one programming language? Why or why not?

2. Identify the three spheres of influence and briefly discuss the software needs of each.

3. Identify the three fundamental types of applications software. Discuss the advantages and disadvantages of each type.

4. You are using a new release of an application software package. You think that you have discovered a bug. Outline the approach that you would take to confirm that it is indeed a bug. What actions would you take if it truly were a bug?

5. How can application software improve the effectiveness of a large enterprise? What are some of the benefits associated with implementation of an enterprise resource planning system? What are some of the issues that could keep the use of enterprise resource planning software from being successful?

6. Define the term *application service provider*. What are some of the advantages and disadvantages of employing an ASP? What precautions might you take to minimize the risk of using one?

7. Briefly outline the evolution of programming languages. Use your imagination and creativity to develop a brief description of the sixth generation of programming languages. How would they work? What sort of features might be included?

8. Why was an antitrust lawsuit filed against Microsoft? What are some of the possible implications for Microsoft and the technology industry?

9. What is opensourcing? If you were the IT manager for a large manufacturing company, what issues might you have with the use of opensource software? What advantages might there be for use of such software?

10. Identify four types of software license frequently used. Which approach does the best job of ensuring a steady, predictable stream of revenue from customers? Which approach is most fair for the small company that makes infrequent use of the software?

PROBLEM-SOLVING EXERCISES

1. Choose a programming language of interest to you and develop a six-slide presentation of its history, current level of usage, typical applications, ease of use, etc.

2. Use a spreadsheet package to prepare a simple monthly budget and forecast your cash flow—both income and expenses for the next six months (make up numbers rather than use actual). Now use a graphics package to plot the total monthly income and monthly expenses for six months. Cut and paste both the spreadsheet and the graph into a word processing document that summarizes your financial condition.

TEAM ACTIVITIES

1. Form a group of three or four classmates. Find articles from business periodicals, search the Internet, or interview people on the topic of software bugs. How frequently do they occur, and how serious are they? What can software users do to encourage defect-free software? Compile your results for an in-class presentation or a written report.

2. Form a group of three or four classmates. Identify and contact an information systems professional with a local firm. Interview the individual to discover the relative level of use of custom software versus standard application packages. If you uncover examples of the use of custom software, find out why the software was developed this way. Write a brief report summarizing your findings.

WEB EXERCISES

1. Microsoft, IBM/Lotus, Sun, and Corel are the four main providers of personal productivity software suites. Do research to assess the relative success of these three products in terms of sales of their software suites. Do you think it is possible that Microsoft will become the only provider of such software? Would this be good or bad? Why? Write a brief report summarizing your findings and conclusions.

2. Do research on the Web and develop a two-page report summarizing the latest consumer appliance operating systems. Which one seems to be gaining the most widespread usage? Why do you think this is the case?

3. Do research to document the current status of the Microsoft antitrust lawsuit. Summarize the opinion of at least three commentators on the impact that this lawsuit has had on the industry.

CASES

CASE 1

Infosys Uses Offshore Outsourcing Model to Provide Contract Software

Infosys is one of India's best-known technology companies. It has been in operation over 20 years and has 10,000 employees in 30 offices worldwide, providing software and services to large companies. Much of the work that information systems professionals do can be done anywhere—on a company's premises or thousands of miles away in a foreign country. But to make this process work well, someone must provide strong project management, ensure good communications, provide high-quality products, meet deadlines, and stay within budget constraints. Infosys has gained such a reputation with its clients.

Infosys's strategy is to set up world-class software development campuses to ensure itself a good supply of well-trained employees. For example, Infosys signed an agreement with the local Indian government to establish a software development campus in Hyderabad, India. Eventually this campus will provide training facilities for 2,500 software professionals.

Infosys has also been highly successful in forming partnerships with major U.S. companies to provide programming expertise. Three of its customers are Aetna, Kansas City Southern, and SunAmerica. Aetna, one of the leading U.S. providers of healthcare and related financial services, is using more than 500 Infosys employees to create on-line financial and retirement benefit services to complement Aetna's growing line of "E-health" initiatives. Aetna retains control of the overall system design and overall architecture, but it uses Infosys to program, test, and deliver the various system components. Aetna turned to Infosys for part of its software needs because the cost of hiring in-house programming staff in the U.S. is seven times higher than in India. As an additional benefit, time-zone differences mean that work can be done around the clock—one shift of workers in the U.S. and two shifts of workers from Infosys at locations around the world keep the projects moving. With the shortage of U.S. programmers, the use of offshore outsourcing was an alternative to hiring inexperienced college graduates or enticing seasoned IT veterans out of retirement.

Infosys worked with Kansas City Southern to build a $50 million management control system to help the railway guarantee on-time shipments and cut freight cycle times. Kansas City Southern (KCS) carries automotive and computer parts from Canada and Chicago to Mexico City and brings finished Volkswagen Beetles and PCs back north. The new system helps customers order and track freight shipments online, replaces paper records used by train engineers and customer service agents, and provides business managers with a more effective way to track costs such as those of leased freight cars. When Infosys came on board, software development occurred round-the-clock, in tandem with developers from Infosys in Madras, India, with some 1.5 million worker hours expended against the effort.

Infosys has contracted with SunAmerica, a financial services company, to transform its old policy administration system into a new Web-based, thin client system. The new system will improve the policy administration process for SunAmerica and redesign its output systems to print policies for some 130,000 users.

Discussion Questions

1. Briefly describe the offshore outsourcing model for delivering contract software. Why have many major U.S. companies turned to this approach to develop their key applications?

2. What are some of the potential negatives associated with offshore outsourcing?

Critical Thinking Questions

3. What unique risks are associated with the execution of an offshore outsourcing project?

4. What specific actions can be taken to reduce these risks?

Sources: Adapted from Business Editors/High-Tech Writers, "Infosys Signs MoU with the Government of Andhra Pradesh to Establish a Software Development Campus in Hyderabad," *BusinessWire*, January 11, 2001; Matt Hamblen, "The Little Engine That Might," *Computerworld*, August 6, 2001, accessed at http://www.computerworld.com; Saritha Rai, "World Business Briefing/Asia: India: High-Technology Optimism," *The New York Times*, June 5, 2001, late edition—final, section W, p. 1; "SunAmerica Selects Infosys to Transform Core Policy Systems," *The New York Times on the Web*, December 4, 2001, accessed at http://www.nytimes.com; and Business Editors/High-Tech Writers, "Infosys Expands Strategic Relationship with Aetna," *BusinessWire*, March 14, 2001, accessed at http://www.businesswire.com.

CASE 2

Crystal Flash Opts for Application Service Provider

Crystal Flash is a supplier of energy-related products and services for the state of Michigan. It employs 600 workers and is owned by Heritage, one of the largest oil producers in the United States. Crystal Flash's retail food and fuel stores, with their bright yellow canopies, are highly visible and instantly recognizable in western Michigan. Its fleet of 100 trucks delivers propane and heating oil to thousands of homes in rural areas plus gasoline and diesel fuel to trucking companies, construction firms, and farms.

In 2000, Crystal Flash decided to revamp its outdated sales practices that led to customers being called on by more than one salesperson, creating heavy administrative workloads. A committee consisting of representatives from sales, marketing, and information systems was formed to determine how to establish a more uniform and efficient set of sales processes. The group focused on implementing a sales management software application. After six months of work, they decided that a custom sales management application would be too expensive due to the required consulting, hardware, ongoing maintenance, and licensing fees. They spent the next six months reviewing and evaluating existing sales management software packages. The solution they chose was Salesnet Sales Force Automation, a software application that runs on server hardware owned and operated by Salesnet, Inc., a sales software application service provider.

The Salesnet Process Builder software module enables sales organizations to define and build their own sales processes. Crystal Flash was able to define a set of standard sales processes that will reinforce effective selling and closing behaviors among all its sales reps. No longer are Crystal Flash's sales reps spending time completing unnecessary paperwork. They are able to use the software's calendar, scheduling, and contact management features to support greater teamwork. In addition, use of the software enables Crystal Flash managers to access real-time information about sales team activities and to obtain sales reports, forecasts, and customer information.

Sales reps can access the Internet-based application through desktop PCs, by dialing up through notebook computers or wirelessly on smaller devices.

Because Salesnet is a hosted application, there is no upfront capital investment in software, hardware, IT resources, or ongoing maintenance fees. As a result, Crystal Flash saved up to $100,000 over other solutions. The standard version of Salesnet costs $59 per user per month.

Discussion Questions

1. In addition to economic factors, why would a small company such as Crystal Flash find the use of an application service provider especially attractive?
2. What are the biggest benefits of this system for Crystal Flash?

Critical Thinking Questions

3. Check out the Salesnet Web site at http://www.salesnet.com. Can you identify any limitations of their software that might make its use inappropriate for a large sales organization?
4. Can you identify any potential problems or risks with Crystal Flash's choice to use an application service provider?

Sources: Adapted from Linda Formichelli, "Sovereign Tracks Processes, Prospects, with Salesnet," *CRM White Paper*, June 18, 2002, accessed at http://searchcrm.techtarget.com; Stacy Crowley, "Salesnet Preps 'Extended' Version, Offline App," *Computerworld*, June 24, 2002, accessed at http://www.computerworld.com; Eric M. Zeman, "A Flash of Brilliance," *Field Force Automation*, May 2002, accessed at http://www.ffa.com; Crystal Flash Web page "About Us" at http://www.crystalflash.com, accessed on July 28, 2002.

CASE 3

Microsoft Takes Action to Eliminate Software Bugs

High-quality software systems function safely and dependably and meet our needs for continuous and secure operations. Reliable software has long been required in air traffic control, atomic power plant operation, automobile safety, electronic commerce, healthcare, Internet communication, military and defense systems, and space exploration. Increasingly, individual users are demanding quality in the basic software used to support factory and office workers. Poor-quality software has resulted in crashed systems, lost work, and security holes through which intruders are increasingly spreading viruses, stealing data, and shutting down Web sites. In short, users don't necessarily want new features, they want products that work and are secure.

Often the root cause of the poor software is that quality is not designed into the software from the very start. To do so, software developers must define and follow good engineering principles and be committed to learn from past mistakes. In addition, they must consider the environment within which their systems will operate and design systems that are more immune to human error.

Even if a system is well designed, programmers often make mistakes when they translate design specifications into lines of code. It is estimated that even an experienced programmer unknowingly injects about one defect into every 10 lines of code. Even if 99.9 percent of those mistakes are caught, that's still 1 bug per 10,000 lines of code. Consider the Microsoft Windows operating system, which is composed of hundreds of millions of lines of code bundled with products made by other Microsoft divisions. The complexity of the software makes it highly likely that there are thousands of bugs in this commonly used software.

In recognition of this reality, Microsoft plans to build tools into Windows to report software problems back to Microsoft and, ultimately, fix them automatically. Bill Gates has focused Microsoft employees on a companywide effort to develop high-quality code, even it comes at the expense of adding new features to its products. In the first month of the quality program, 7,000 Microsoft programmers were ordered to drop what they were doing and devote their efforts to eliminating flaws in Windows 2000, the operating system used by corporate computer servers, and Windows XP used by desktop PCs. Before beginning this task, Windows programmers and managers were sent to two days of training based on the book

Writing Secure Code by Microsoft security specialists Michael Howard and David Leblane.

Skepticism runs high about Microsoft's new-found interest in high-quality software. Microsoft became the world's biggest software maker but in the process gained a reputation for rushing feature-packed software to market—and fixing bugs later. To build more secure products, Microsoft must transform its deeply ingrained, highly bureaucratic culture. Experts doubt that two days of seminars and a couple weeks of bug fixing will have a significant effect. Also, Microsoft lacks a plan to make systemic changes involving other key products such as Office, SQL database, and Exchange e-mail. Many point out that diverting programmers to bug fixing came at a good time. Any delay in the release of its next product, a Windows 2000 upgrade, won't hurt too much because corporations have reduced their software budgets.

Obviously, Microsoft is not the only software manufacturer guilty of turning out poor quality software. The National Infrastructure Protection Center's recent summary of software vulnerabilities is 70 pages long, representing companies from Abode to Zendown.

Discussion Questions

1. Identify several examples where poor-quality software has had a negative impact on your life.
2. What are the some of the reasons why a software manufacturer will sacrifice quality in a rush to get a new product to market?

Critical Thinking Questions

3. Why are experts skeptical that Microsoft's actions will have lasting effect?
4. What fundamental changes must be made for Microsoft to significantly improve the quality of its software?

Sources: Adapted from "Groups Vow to Stamp Out Computer Bugs," *USA Today*, January 19, 2001, accessed at http://www.usatoday.com; George V. Hulme, "Software's Challenge," *InformationWeek*, January 21, 2002, accessed at http://www.informationweek.com; Byron Acohido, "Microsoft Does Security Sweep Workers Drop Everything to Check Windows for Holes," *USA Today*, accessed at http://www.usatoday.com; Matt Richtel, "Computer Security Experts Warn of Internet Vulnerability," *The New York Times on the Web*, February 13, 2002, accessed at http://www.nytimes.com.

NOTES

Sources for the opening vignette on p. 131: Adapted from Corporate Information—Flextronics Web page accessed at http://www.flextronics.com on December 27, 2001; PolyDyne Software Inc. Web page accessed at http://www.polydyne.com on December 27, 2001; David Hannon, "Contract Manufacturer Moves to Online Quoting," *Purchasing*, August 23, 2001, p. 23; Jeff Sweat, "Customer Collaboration Counts—Flextronics Depends on Collaborative Tools to Work with Clients and Connect Co-Workers," *InformationWeek*, December 10, 2001, accessed at http://www.information week.com.

1. Mary Jo Foley and Steven J. Vaughan-Nicols, "Microsoft Trims Windows 2000," *PC Week*, September 20, 1999, p. 18.
2. David Pogue, "Windows XP: Microsoft's New Look for Fall, in Size XXL," September 6, 2001, *The New York Times on the Web*, accessed at http://www.nytimes.com.
3. David Pogue, "A New Face (and Heart) for the Mac," March 29, 2001, *The New York Times*, accessed at http://www.nytimes.com.
4. Leslie Jaye Goff, "Supporting Creativity," *Computerworld*, December 10, 2001, accessed at http://www.computerworld.com.
5. John Fontana, "Microsoft Releases 64-bit Windows," *Network World Fusion*, August 28, 2001, accessed at http://www.mwfusion.com.
6. Todd R. Weiss, "Sherwin-Williams Brushes SCO Unix Aside, Adopts Linux," *Computerworld*, May 27, 2002, accessed at http://www.computer world.com.
7. Todd R. Weiss, "Burlington Coat Factory Warehouse Corp," *Computerworld*, March 11, 2002, accessed at http://www.computerworld.com.
8. Jaikumar Vijayan, "Mainframe Users Turn to IBM's z800 for Cost Savings," *Computerworld*, June 24, 2002, accessed at http:// www.computerworld.com.
9. Jaikumar Vijayan, "What's New in z/OS?," *Computerworld*, November 19, 2001, accessed at http://www.computerworld.com.
10. Todd R. Weiss, "Red Hat Linux Now Available for IBM S/390 Mainframes," *Computerworld*, December 18, 2001, accessed at http://www.computerworld.com.
11. Larry Mittag, "Palm OS or Windows CE?" *Computerworld*, May 10, 2001, accessed at http://www.computerworld.com.
12. Sumner Lemon, IDG News Service, "Windows CE.Net Ready for January Launch," *Computerworld*, December 27, 2001, accessed at http://www.computeroworld.com.

13. Matt Berger, IDG News Service, "Microsoft Offers Test Version of Windows XP Embedded," *Computerworld*, September 24, 2001, accessed at http://www.computerworld.com.
14. Douglas F. Gray, IDG News Service, "Vendors Unleash New Handhelds for Pocket PC 2002," *Computerworld*, October 4, 2001.
15. Bob Brewin and Matt Hamblen, "Microsoft Debuts Pocket PC Operating System with Enterprise Features," *Computerworld*, September 6, 2001.
16. Sumner Lemon, "China Merchant's Bank, Legend Offer PDA Banking Service," *Computerworld*, June 25, 2002, accessed at http://www.computerworld.com.
17. Joris Evers, IDG News Service, "Palm: Chip Makers Ready Palm OS for ARM Chips," *Computerworld*, July 24, 2001.
18. "Palm on Your Wristwatch," *Wired*, July 24, 2001, accessed at http://www.wired.com.
19. Bob Brewin, "Office Depot Hedges Its Handheld Bets," *Computerworld*, October 15, 2001, accessed at http://www.computerworld.com.
20. Martin J. Garvey, "Mercury Interactive Adds Statistical-Analysis Engine with Built-in Rules to Topaz Software," *InformationWeek*, December 24, 2001, accessed at http://www.informationweek.com.
21. Larry Greenemeier, "Ingersoll-Rand and Corio Get Cozier on ASP Front," *InformationWeek*, May 1, 2002, accessed at http://www. informationweek.com.
22. Aaron Ricadela, "Microsoft Touts Office XP's Ease of Use, Net Integration," *InformationWeek*, June 4, 2001, p. 28.
23. Jennifer Disabatino, "NHL Scores with Database on Draft Day," *Computerworld*, July 9, 2001, accessed at http://www.computerworld.com.
24. Steve Konicki, "Hoffman Enclosures Wants to Keep It Real," *InformationWeek*, June 12, 2002, accessed at http://www. informationweek.com.
25. Robert L. Scheier, "Finding Pearls in an Ocean of Data," *Computerworld*, July 23, 2001, accessed at http://www.computerworld.com.
26. George V. Hulme, "Software's Challenge," *InformationWeek*, January 21, 2002, pp. 22–24, accessed at http://www.informationweek.com.
27. John Rendleman, "Eli Lilly under Fire for E-Mail Glitch," *InformationWeek*, July 9, 2001, accessed at http://www. informationweek.com.

28. Tom Mainelli, "Toshiba Fixes Flawed Notebooks, Again," *Computerworld*, July 18, 2002, accessed at http://www.computerworld.com.

29. Michelle Delio, "Canada's Free Phone Call Frenzy," *Wired*, May 21, 2001, accessed at http://www.wired.com.

30. "DoCoMo Hits a Snag," *Wired News Report*, May 11, 2001, accessed at http://www.wired.com.

31. Steve Konicki and Jennifer Maselli, "After Woes, Oracle Woos Business with New Tools," *InformationWeek*, January 21, 2002, p. 24.

32. Patrick Thibodeau, "States Alter Remedy List at Microsoft Trial," *Computerworld*, June 24, 2002, accessed at http://www.computerworld.com.

33. Patrick Thibodeau, "Microsoft Faces More Perils in New Lawsuit," *Computerworld*, January 28, 2002, accessed at http://www.computerworld.com.

34. Todd R. Weiss, "Vendor Expands Patent Infringement Allegations Against Microsoft," *Computerworld*, June 24, 2002, accessed at http://www.computerworld.com.

35. Carol Sliwa, "License Tracker Launched as Microsoft Deadline Looms," *Computerworld*, July 15, 2002, accessed at http://www.computerworld.com.

Organizing Data and Information

PRINCIPLES	LEARNING OBJECTIVES
The database approach to data management provides significant advantages over the traditional file-based approach.	• Define general data management concepts and terms, highlighting the advantages and disadvantages of the database approach to data management. • Name three database models and outline their basic features, advantages, and disadvantages.
A well-designed and well-managed database is an extremely valuable tool in supporting decision making.	• Identify the common functions performed by all database management systems and identify three popular end-user database management systems.
The number and types of database applications will continue to evolve and yield real business benefits.	• Identify and briefly discuss current database applications.

[Valio]

Building Brand Loyalty with Better Decision-Making Data

Valio is the largest dairy company in Finland, generating annual sales in excess of $1.5 billion. Founded in 1905, Valio is a farmer's cooperative that supplies nearly 80 percent of Finland's fresh milk. The 33 owner dairies collect milk from some 17,200 dairy farms, and Valio processes the milk into 800 dairy products. Valio then markets the products at home and in 60 other countries under various brand names.

Valio's probiotic dairy products and food supplements are especially profitable. (Probiotics are food supplements or dairy products containing live bacteria that are beneficial to the health of the consumer.) Valio owns the worldwide patent for Lactobacillus GG, one of the leading probiotic bacterial strains, and it allows other companies to use the bacteria in their products through licensing agreements. Fermented dairy products, "sweet milk," fresh cheeses, ripened cheeses, infant foods, dairy and nondairy drinks, and food supplements all can contain the bacteria.

Long known for its high-quality products, Valio must now compete in the global marketplace for shelf space with aggressively priced rival products. As part of its strategy to remain successful, Valio built a huge database with information from multiple sources that feeds a number of smaller databases with subsets of its data. For example, one smaller database contains information on domestic markets and another has data on export markets. Company decision makers can access the data from anywhere in the world via the Internet. They then use software to analyze the data to track brand and product performance by location and customer and to track sales, profits, and inventory. These tools and the data enable employees to focus on customer needs, cut costs, and increase revenue.

Data consistency is a key benefit derived from use of the large database. All employees—including sales managers, sales reps, and marketing teams—work with the same data, so they start with a common understanding of the current state of the company.

The system was designed for ease of use, so even workers who are new to PCs can use the system with minimal training. Standard formatted reports enable managers to quickly view the latest business developments. If necessary, workers can drill down into the data to obtain more in-depth analyses.

Valio made a strategic decision to offer its customers Internet access to a special database that contains market data—its own and suppliers'—as well as raw materials forecasts. Customers can then modify their orders, and Valio can alter its production schedules to match consumer buying habits. This collaboration enables Valio and its customers to minimize their inventories and develop a more competitive supply chain. Valio's efforts to build a larger and more sophisticated database than its rivals has provided it with a competitive edge and strengthened consumer loyalty to its brands.

As you read this chapter, consider the following:

⊙ How can databases be used to support business objectives?
⊙ What are some of the issues associated with compiling and managing massive amounts of data?

The bane of modern business is too much data and not enough information. Computers are everywhere, accumulating gigabytes galore. Yet it seems to get harder to find the forest for the trees—that is, to extract significance from the blizzard of numbers, facts, and statistics. Like other components of a computer-based information system, the overall objective of a database is to help an organization achieve its goals. A database can contribute to organizational success in a number of ways, including the ability to provide managers and decision makers with timely, accurate, and relevant information based on data. As we saw in the case of Valio, a database can help companies organize data to learn from this valuable resource. Databases also help companies generate information to reduce costs, increase profits, track past business activities, and open new market opportunities. Indeed, the ability of an organization to gather data, interpret it, and act on it quickly can distinguish winners from losers in a highly competitive marketplace. It is critical to the success of an organization that database capabilities be aligned with the company's goals. Because data is so critical to an organization's success, many firms develop databases to help them access data more efficiently and use it more effectively. In this chapter, we will investigate the development and use of different types of databases.

database management system (DBMS)
a group of programs that manipulate the database and provide an interface between the database and the user of the database and other application programs

As we saw in Chapter 1, a database is a collection of data organized to meet users' needs. Throughout your career, you will be directly or indirectly accessing a variety of databases, ranging from a simple roster of departmental employees to a fully integrated corporatewide database. You will probably access these databases using software called a **database management system (DBMS)**. A DBMS consists of a group of programs that manipulate the database and provide an interface between the database and the user of the database and other application programs. A database, a DBMS, and the application programs that utilize the data in the database make up a database environment. Understanding basic database system concepts can enhance your ability to use the power of a computerized database system to support IS and organizational goals.

DATA MANAGEMENT

Without data and the ability to process it, an organization would not be able to successfully complete most business activities. It would not be able to pay employees, send out bills, order new inventory, or produce information to assist managers in decision making. As you recall, data consists of raw facts, like employee numbers and sales figures. For data to be transformed into useful information, it must first be organized in a meaningful way.

THE HIERARCHY OF DATA

Data is generally organized in a hierarchy that begins with the smallest piece of data used by computers (a bit) and progresses through the hierarchy to a database. As discussed in Chapter 3, a bit (a binary digit) represents a circuit that is either on or off. Bits can be organized into units called bytes. A byte is typically eight bits. Each byte represents a **character**, which is the basic building block of information. A character may consist of uppercase letters (A, B, C, ... , Z), lowercase letters (a, b, c, ... , z), numeric digits (0, 1, 2, ... , 9), or special symbols (.![+][-]/ ...).

character
basic building block of information, consisting of uppercase letters, lowercase letters, numeric digits, or special symbols

field
typically a name, number, or combination of characters that describes an aspect of a business object or activity

record
a collection of related data fields

Characters are put together to form a field. A **field** is typically a name, number, or combination of characters that describes an aspect of a business object (e.g., an employee, a location, a truck) or activity (e.g., a sale). A collection of related data fields is a **record**. By combining descriptions of various aspects of an object or activity, a more complete description of the object or activity is obtained. For instance, an employee record is a collection of fields about one employee. One field would be the employee's name, another her address, and still others her phone number, pay rate, earnings made to date, and so forth. A collection of related records is a **file**—for example, an employee file is a collection of all company employee records. Likewise, an inventory file is a collection of all inventory records for a particular company or organization. PC database software often refers to files as tables.

At the highest level of this hierarchy is a *database*, a collection of integrated and related files. Together, bits, characters, fields, records, files, and databases form the **hierarchy of data** (Figure 5.1). Characters are combined to make a field, fields are combined to make a record, records are combined to make a file, and files are combined to make a database. A database houses not only all these levels of data but the relationships among them.

DATA ENTITIES, ATTRIBUTES, AND KEYS

Entities, attributes, and keys are important database concepts. An **entity** is a generalized class of people, places, or things (objects) for which data is collected, stored, and maintained. Examples of entities include employees, inventory, and customers. Most organizations organize and store data as entities.

An **attribute** is a characteristic of an entity. For example, employee number, last name, first name, hire date, and department number are attributes for an employee (Figure 5.2). Inventory number, description, number of units on hand, and the location of the inventory item in the warehouse are examples of attributes for items in inventory. Customer number, name, address, phone number, credit rating, and contact person are examples of attributes for customers. Attributes are usually selected to capture the relevant characteristics of entities like employees or customers. The specific value of an attribute, called a **data item,** can be found in the fields of the record describing an entity.

As discussed, a collection of fields about a specific object is a record. A **key** is a field or set of fields in a record that is used to identify the record. A **primary key** is a field or set of fields that uniquely identifies the record. No other record can have the same primary key. The primary key is used to distinguish records so

file
a collection of related records

hierarchy of data
bits, characters, fields, records, files, and databases

entity
generalized class of people, places, or things for which data is collected, stored, and maintained

attribute
a characteristic of an entity

data item
the specific value of an attribute

key
a field or set of fields in a record that is used to identify the record

primary key
a field or set of fields that uniquely identifies the record

FIGURE 5.1

The Hierarchy of Data

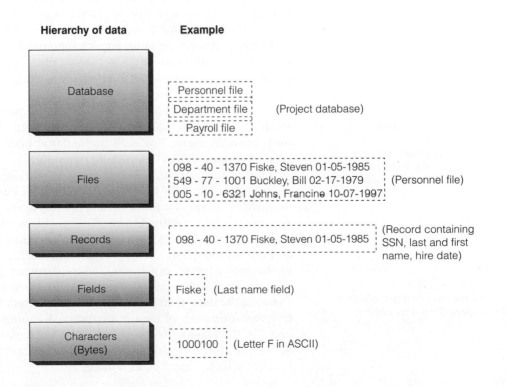

Hierarchy of data	Example
Database	Personnel file / Department file / Payroll file (Project database)
Files	098 - 40 - 1370 Fiske, Steven 01-05-1985 / 549 - 77 - 1001 Buckley, Bill 02-17-1979 / 005 - 10 - 6321 Johns, Francine 10-07-1997 (Personnel file)
Records	098 - 40 - 1370 Fiske, Steven 01-05-1985 (Record containing SSN, last and first name, hire date)
Fields	Fiske (Last name field)
Characters (Bytes)	1000100 (Letter F in ASCII)

FIGURE 5•2

Keys and Attributes

The key field is the employee number. The attributes include last name, first name, hire date, and department number.

Employee #	Last name	First name	Hire date	Dept. number
005-10-6321	Johns	Francine	10-07-1997	257
549-77-1001	Buckley	Bill	02-17-1979	632
098-40-1370	Fiske	Steven	01-05-1985	598

Entities (records)

Key field

Attributes (fields)

that they can be accessed, organized, and manipulated. For an employee record such as the one shown in Figure 5.2, the employee number is an example of a primary key.

Locating a particular record that meets a specific set of criteria may require the use of a combination of secondary keys. For example, a customer might call a mail-order company to place an order for clothes. If the customer does not know his primary key (such as a customer number), a secondary key (such as last name) can be used. In this case, the order clerk enters the last name, such as Adams. If there are several customers with a last name of Adams, the clerk can check other fields, such as address, first name, and so on, to find the correct customer record. Once the correct customer record is obtained, the order can be completed and the clothing items shipped to the customer.

THE TRADITIONAL APPROACH VERSUS THE DATABASE APPROACH

The Traditional Approach

Organizations are adaptive systems with constantly changing data and information needs. For any growing or changing business, managing data can become quite complicated. One of the most basic ways to manage data is via files. Because a file is a collection of related records, all records associated with a particular application (and therefore related by the application) can be collected and managed together in an application-specific file. At one time, most organizations had numerous application-specific data files; for example, customer records often were maintained in separate files, with each file relating to a specific process completed by the company, such as shipping or billing. This approach to data management, whereby separate data files are created and stored for each application program, is called the **traditional approach to data management**. For each particular application, one or more data files is created (Figure 5.3).

One of the flaws in this traditional file-oriented approach to data management is that much of the data—for example, customer name and address—is duplicated in two or more files. This duplication of data in separate files is known as **data redundancy**. The problem with data redundancy is that changes to the data (e.g., a new customer address) might be made in one file and not another. The order-processing department might have updated its file to the new address, but the billing department is still sending bills to the old address. Data redundancy, therefore, conflicts with **data integrity**—the degree to which the data in any one file is accurate. Data integrity follows from the control or elimination of data redundancy. Keeping a customer's address in only one file decreases the possibility that the customer will have two different addresses stored in different locations. The efficient operation of a business requires a high degree of data integrity.

traditional approach to data management
an approach whereby separate data files are created and stored for each application program

data redundancy
duplication of data in separate files

data integrity
the degree to which the data in any one file is accurate

Data	Files	Application programs	Users

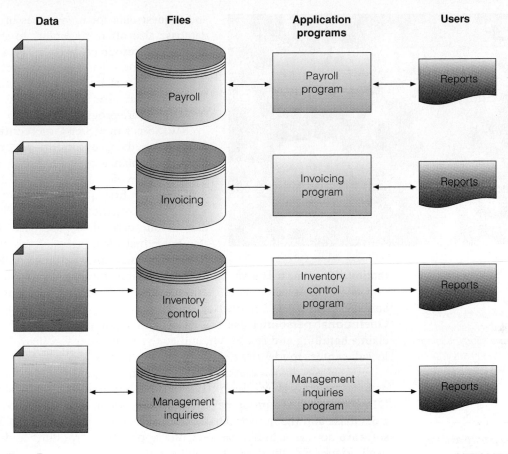

FIGURE 5.3

The Traditional Approach to Data Management

With the traditional approach, one or more data files is created and used for every application. For example, the inventory control program would have one or more files containing inventory data, such as the inventory item, number on hand, and item description. Likewise, the invoicing program can have files on customers, inventory items being shipped, and so on. With the traditional approach to data management, it is possible to have the same data, such as inventory items, in several different files used by different applications.

program-data dependence
concept according to which programs and data developed and organized for one application are incompatible with programs and data organized differently for another application

database approach to data management
an approach whereby a pool of related data is shared by multiple application programs

In many database systems based on the traditional file approach, the data is organized for a particular application program (say, billing). These applications have **program-data dependence**—that is, programs and data developed and organized for one application are incompatible with programs and data organized differently for another application. For example, one programmer might develop a billing program that stores ZIP code data in one format with five numbers, while another programmer might develop a separate order-processing program that stores ZIP code data in a nine-number format. In a file-based environment, all the programs that access this ZIP code data would need to be changed. Bridging the gap between two files with different program-data dependencies is often difficult and expensive.

Despite the drawbacks of using the traditional file approach in database systems, some organizations continue to use it. For these firms, the cost of converting to another approach is too high.

The Database Approach

Because of the problems associated with the traditional approach to data management, many managers wanted a more efficient and effective means of organizing data. The result was the **database approach to data management**. In a database approach, a pool of related data is shared by multiple application programs. Rather than having separate data files, each application uses a collection of data that is either joined or related in the database.

The database approach offers significant advantages over the traditional file-based approach. For one, by controlling data redundancy, the database approach can use storage space more efficiently and increase data integrity. The database approach can also provide an organization with increased flexibility in the use of data. Because data once kept in two files is now located in the same database, it is easier to locate

and request data for many types of processing. A database also offers the ability to share data and information resources. This can be a critical factor in coordinating organizationwide responses across diverse functional areas of a corporation. In sharing data, however, some consistency should exist among software programs.

SWISS is a new Swiss intercontinental airline formed from the Crossair, Swissair, and SairGroup merger. Its planes began flying in March 2002. The airline, which served almost 2 million passengers in its first quarter, is using the database approach to organize and store all its customer data in one centralized database. The database also holds historical flight information, sales and marketing data, and customer preference infor-

mation. As such, it is a valuable source of information for the entire organization. SWISS sales and marketing employees use the data to segment its customers and develop targeted marketing campaigns to appeal to each resulting group. Operational personnel use the data to monitor postflight activities, such as claims handling and resolution, and analyze customer feedback data received at its call centers to identify necessary improvements.[1]

To use the database approach to data management, additional software—a database management system (DBMS)—is required. As previously discussed, a DBMS consists of a group of programs that can be used as an interface between a database and the user or the database and application programs. Typically, this software acts as a buffer between the application programs and the database itself. Figure 5.4 illustrates the database approach.

An enterprisewide database enables Best Buy to reduce inventory costs and provide customers with the products they demand. The database pulls information from 350 sources across the enterprise.

(Source: AP/Wide World Photos.)

FIGURE 5.4

The Database Approach to Data Management

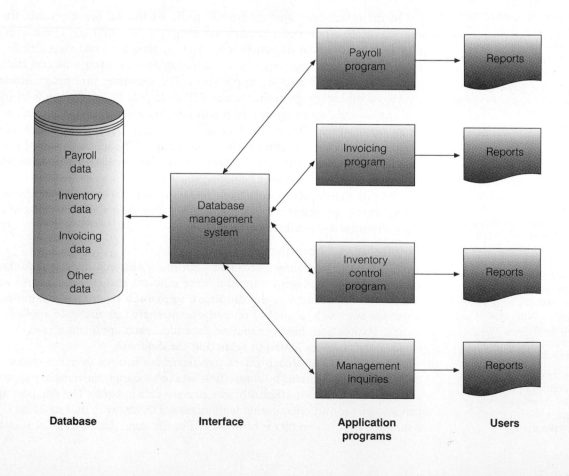

Advantages	Explanation
Improved strategic use of corporate data	Accurate, complete, up-to-date data can be made available to decision makers where, when, and in the form they need it.
Reduced data redundancy	The database approach can reduce or eliminate data redundancy. Data is organized by the DBMS and stored in only one location. This results in more efficient utilization of system storage space.
Improved data integrity	With the traditional approach, some changes to data were not reflected in all copies of the data kept in separate files. This is prevented with the database approach because there are no separate files that contain copies of the same piece of data.
Easier modification and updating	With the database approach, the DBMS coordinates updates and data modifications. Programmers and users do not have to know where the data is physically stored. Data is stored and modified once. Modification and updating is also easier because the data is stored at only one location in most cases.
Data and program independence	The DBMS organizes the data independently of the application program. With the database approach, the application program is not affected by the location or type of data. Introduction of new data types not relevant to a particular application does not require the rewriting of that application to maintain compatibility with the data file.
Better access to data and information	Most DBMSs have software that makes it easy to access and retrieve data from a database. In most cases, simple commands can be given to get important information. Relationships between records can be more easily investigated and exploited, and applications can be more easily combined.
Standardization of data access	A primary feature of the database approach is a standardized, uniform approach to database access. This means that the same overall procedures are used by all application programs to retrieve data and information.
A framework for program development	Standardized database access procedures can mean more standardization of program development. Because programs go through the DBMS to gain access to data in the database, standardized database access can provide a consistent framework for program development. In addition, each application program need address only the DBMS, not the actual data files, reducing application development time.
Better overall protection of the data	The use of and access to centrally located data are easier to monitor and control. Security codes and passwords can ensure that only authorized people have access to particular data and information in the database, thus ensuring privacy.
Shared data and information resources	The cost of hardware, software, and personnel can be spread over a large number of applications and users. This is a primary feature of a DBMS.

TABLE 5.1

Advantages of the Database Approach

The database approach to data management involves a combination of hardware and software. Tables 5.1 and 5.2 list some of the primary advantages and disadvantages of the database approach and explore some of these issues.

Because of the many advantages of the database approach, most businesses use databases to store data on customers, orders, inventory, employees, and suppliers. This data is used as the input to the various information systems throughout an organization. For example, the transaction processing system can use the data to support daily business processes like billing, inventory tracking, and ordering. This same data can be processed by a management information system to create reports or a decision support system to provide information to aid managerial decision making.

Many modern databases are enterprisewide, encompassing much of the data of the entire organization. Often, distinct yet related databases are linked to provide enterprisewide databases. Much planning and organization go into the development of such databases. For example, Best Buy is a specialty retailer of consumer electronics, personal computers, entertainment software, and appliances. It operates

Disadvantages	Explanation
Relatively high cost of purchasing and operating a DBMS in a mainframe operating environment	Some mainframe DBMSs can cost hundreds of thousands of dollars.
Increased cost of specialized staff	Additional specialized staff and operating personnel may be needed to implement and coordinate the use of the database. However, some organizations have been able to implement the database approach with no additional personnel.
Increased vulnerability	Even though databases offer better security because security measures can be concentrated on one system, they also make more data accessible to the trespasser if security is breached. In addition, if for some reason there is a failure in the DBMS, multiple application programs are affected.

TABLE 5.2

Disadvantages of the Database Approach

nearly 2,000 retail stores and commercial Web sites under the names Best Buy, Magnolia Hi-Fi, Media Play, On Cue, Sam Goody, and Suncoast. Best Buy uses information about the business and its customers to tailor the product mix to its customer base, minimize the time items are held in inventory to reduce costs, and respond quickly to customer needs. At the center of this strategic information is a database, which consolidates information from about 350 different sources across the enterprise.[2]

Databases are in wide use in the public sector as well as the private. After decades of complaints about the use of excessive force, false arrests, and racial profiling, the Los Angeles Police Department agreed in a November 2000 consent decree to implement a database to track the conduct and performance of its 10,000 police officers. The database records information about officers' use of force, search and seizure, and citizen complaints, as well as criminal charges or civil lawsuits filed against them. Police officials check the system at least once each day to identify any inappropriate behavior by an officer.[3]

DATA MODELING AND DATABASE MODELS

Because there are so many elements in today's businesses, it is critical to keep data organized so that it can be used effectively. A database should be designed to store all data relevant to the business and provide quick access and easy modification. Moreover, it must reflect the business processes of the organization. When building a database, an organization must carefully consider these questions:

- *Content:* What data should be collected and at what cost?
- *Access:* What data should be provided to which users and when?
- *Logical structure:* How should data be arranged so that it makes sense to a given user?
- *Physical organization:* Where should data be physically located?

DATA MODELING

Key considerations in organizing data in a database include determining what data is to be collected in the database, who will have access to it, and how they might wish to use the data. Based on these determinations, a database can then be created. Building a database requires two different types of designs: a logical design and a physical design. The logical design of a database shows an abstract model of how the data should be structured and arranged to meet an organization's information needs. The logical design of a database involves identifying relationships among the different data items and grouping them in an orderly fashion. Because databases provide both input and output for information systems throughout a business, users from all functional areas should assist in creating the logical design to ensure that their needs are identified and addressed. Physical database design starts from the logical database design and fine-tunes it for performance and cost considerations (e.g., improved response time, reduced storage space, lower operating cost). The person who fine-tunes the physical design must have an in-depth knowledge of the DBMS to implement the database. For example, the logical database design may need to be altered so that certain data entities are combined, summary totals are carried in the data records rather than calculated from elemental data, and some data attributes are repeated in more than one data entity. These are examples of **planned data redundancy**, which is done to improve the system performance so that user reports or queries can be created more quickly.

One of the tools database designers use to show the logical relationships among data is a data model. A **data model** is a diagram of entities and their relationships. Data modeling usually involves understanding a specific business problem and analyzing the data and information needed to deliver a solution. When done at the level of the entire organization, this is called **enterprise data modeling**. Enterprise data modeling is an approach that starts by investigating the general data and information needs of the organization at the strategic level and then examining more specific data and information needs for the various functional areas and departments within the organization. Various models have been developed to help managers and database designers analyze data and information needs. An entity-relationship diagram is an example of such a data model.

Entity-relationship (ER) diagrams use basic graphical symbols to show the organization of and relationships between data. In most cases, boxes are used in ER diagrams to indicate data items and entities, and connecting lines show relationships between data items and entities.

ER diagrams help ensure that the relationships among the data entities in a database are correctly structured so that any application programs developed are consistent with business operations and user needs. In addition, ER diagrams can serve as reference documents once a database is in use. If changes are made to the database, ER diagrams help design them. Figure 5.5 shows an ER diagram for an order database. In this database design, one salesperson services many customers. This is an example of a one-to-many relationship, as shown by the one-to-many symbol ("crow's foot") shown in Figure 5.5. The ER diagram also shows that each customer can place one-to-many orders, each order includes one-to-many line items, and many line items can specify the same product (a many-to-one relationship). There can also be one-to-one relationships. For example, one order generates one invoice.

DATABASE MODELS

The structure of the relationships in most databases follows one of three logical database models: hierarchical, network, and relational. While hierarchical and network models were used to build older databases, most new databases are built based on the relational database model. It is important to remember that the records represented in the models are actually linked or related logically to one another. These links dictate the way users can access data with application programs. Because the different models involve different links between data, each model has unique advantages and disadvantages.

Hierarchical (Tree) Models

In many situations, data follows a hierarchical, or treelike, structure. In a **hierarchical database model**, the data is organized in a top-down, or inverted tree, structure. For example, data about a project for a company can follow this

planned data redundancy
a way of organizing data in which the logical database design is altered so that certain data entities are combined, summary totals are carried in the data records rather than calculated from elemental data, and some data attributes are repeated in more than one data entity to improve database performance

data model
a diagram of data entities and their relationships

enterprise data modeling
data modeling done at the level of the entire enterprise

entity-relationship (ER) diagrams
data models that use basic graphical symbols to show the organization of and relationships between data

hierarchical database model
a data model in which data is organized in a top-down, or inverted tree, structure

An Entity-Relationship (ER) Diagram for a Customer Order Database

Development of this type of diagram helps ensure the logical structuring of application programs that are able to serve users' needs and are consistent with the data relationships in the database.

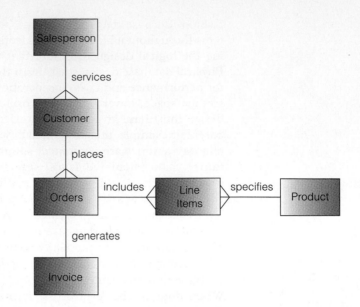

network model

an expansion of the hierarchical database model with an owner-member relationship in which a member may have many owners

relational model

a database model that describes data in which all data elements are placed in two-dimensional tables, called *relations,* that are the logical equivalent of files

type of model, as shown in Figure 5.6. The hierarchical model is best suited to situations in which the logical relationships between data can be properly represented with the one-to-many approach.

Network Models

A **network model** is an expansion of the hierarchical model. Instead of having only various levels of one-to-many relationships, however, the network model is an owner-member relationship in which a member may have many owners (Figure 5.7).

Databases structured according to either the hierarchical model or the network model suffer from the same deficiency: once the relationships are established between data elements, it is difficult to modify them or to create new relationships.

Relational Models

Relational models have become the most popular database models, and use of these models will increase in the future. The relational model describes data using a standard tabular format. In a database structured according to the **relational model,**

A Hierarchical Database Model

Project 1 is the top, or root, element. Departments A, B, and C are under this element, with Employees 1 through 6 beneath them as follows: Employees 1 and 2 under Department A, Employees 3 and 4 under Department B, and Employees 5 and 6 under Department C. Thus, there is a one-to-many relationship among the elements of this model.

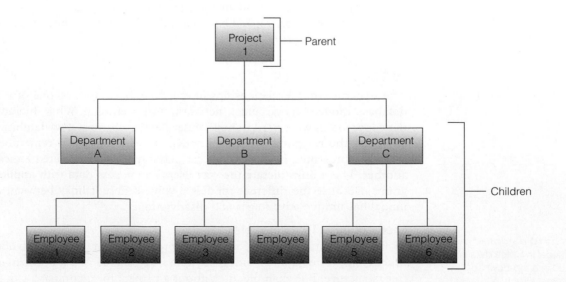

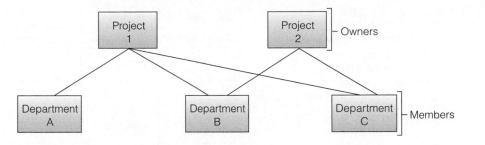

FIGURE 5.7

A Network Database Model

In this network model, two projects are at the top. Departments A, B, and C are under Project 1; Departments B and C are under Project 2. Thus, the elements of this model represent a many-to-many relationship.

all data elements are placed in two-dimensional tables, called *relations,* that are the logical equivalent of files. The tables in relational databases organize data in rows and columns, simplifying data access and manipulation. It is normally easier for managers to understand the relational model (Figure 5.8) than the hierarchical and network models.

In the relational model, each row of a table represents a data entity, with the columns of the table representing attributes. Each attribute can take on only certain values. The allowable values for these attributes are called the **domain**. The domain for a particular attribute indicates what values can be placed in each of the columns of the relational table. For instance, the domain for an attribute such as gender would be limited to male or female. A domain for pay rate would not include negative numbers. Defining a domain can increase data accuracy. For example, a pay rate of –$5.00 could not be entered into the database because it is a negative number and not in the domain for pay rate.

The process of taking a complex set of data and converting it into a set of simple two-dimensional tables is called **data normalization**. During this process, redundant data elements and cumbersome many-to-many data relationships are eliminated. When data has been normalized, it is easy to use for reporting and data updating purposes. You can recognize whether the data in a table has been normalized. If so, every attribute for each entity in the table is dependent on the unique identifier for that entity (the key), the whole key, and nothing else but the key.

Once data has been placed into a relational database, users can make inquiries and analyze data. Basic data manipulations include selecting, projecting, and joining. **Selecting** involves eliminating rows according to certain criteria. Suppose a project table contains the project number, description, and department number for all projects being performed by a company. The president of the company might want to find the department number for Project 226, a sales manual project. Using selection, the president can eliminate all rows but the one for Project 226 and see that the department number for the department completing the sales manual project is 598.

Projecting involves eliminating columns in a table. For example, we might have a department table that contains the department number, department name, and social security number (SSN) of the manager in charge of the project. The sales manager might want to create a new table with only the department number and the social security number of the manager in charge of the sales manual project. Projection can be used to eliminate the department name column and create a new table containing only department number and SSN.

Joining involves combining two or more tables. For example, we can combine the project table and the department table to get a new table with the project number, project description, department number, department name, and social security number for the manager in charge of the project.

As long as the tables share at least one common data attribute, the tables in a relational database can be **linked** to provide useful information and reports

domain
the allowable values for data attributes

data normalization
the process of taking a complex set of data and converting it into a set of simple two-dimensional tables

selecting
data manipulation that eliminates rows according to certain criteria

projecting
data manipulation that eliminates columns in a table

joining
data manipulation that combines two or more tables

linking
data manipulation that combines two or more tables using common data attributes to form a new table with only the unique data attributes

FIGURE 5.8

A Relational Database Model

In the relational model, all data elements are placed in two-dimensional tables, or relations. As long as they share at least one common element, these relations can be linked to output useful information.

Data table 1: Project table

Project number	Description	Dept. number
155	Payroll	257
498	Widgets	632
226	Sales Manual	598

Data table 2: Department table

Dept. number	Dept. name	Manager SSN
257	Accounting	005-10-6321
632	Manufacturing	549-77-1001
598	Marketing	098-40-1370

Data table 3: Manager table

SSN	Last name	First name	Hire date	Dept. number
005-10-6321	Johns	Francine	10-07-1997	257
549-77-1001	Buckley	Bill	02-17-1979	632
098-40-1370	Fiske	Steven	01-05-1985	598

(see Figure 5.9). Being able to link tables to each other through common data attributes is one of the keys to the flexibility and power of relational databases. Suppose the president of a company wants to find out the name of the manager of the sales manual project and the length of time the manager has been with the company. The president would make the inquiry to the database, perhaps via a desktop personal computer. The DBMS would start with the project description and search the project table to find out the project's department number. It would then use the department number to search the department table for the department manager's social security number. The department number is also in the department table and is the common element that allows the project table and the department table to be linked. The DBMS then uses the manager's social security number to search the manager table for the manager's hire date. The manager's social security number is the common element between the department table and the manager table. The final result: the manager's name and hire date are presented to the president as a response to the inquiry.

One of the primary advantages of a relational database is that it allows tables to be linked, as shown in Figure 5.9. This linkage is especially useful when information is needed from multiple tables, as in our example. The manager's social security number, for example, is maintained in the manager table. If the social security number is needed, it can be obtained by linking to the manager table.

The relational database model is by far the most widely used. It is easier to control, more flexible, and more intuitive than the others because it organizes

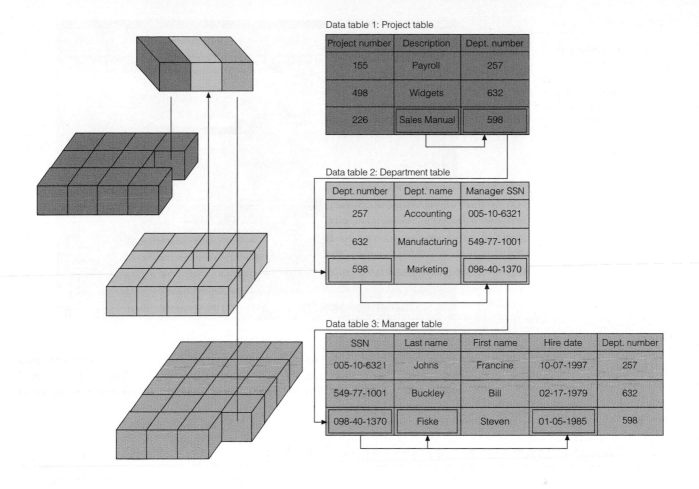

Data table 1: Project table

Project number	Description	Dept. number
155	Payroll	257
498	Widgets	632
226	Sales Manual	598

Data table 2: Department table

Dept. number	Dept. name	Manager SSN
257	Accounting	005-10-6321
632	Manufacturing	549-77-1001
598	Marketing	098-40-1370

Data table 3: Manager table

SSN	Last name	First name	Hire date	Dept. number
005-10-6321	Johns	Francine	10-07-1997	257
549-77-1001	Buckley	Bill	02-17-1979	632
098-40-1370	Fiske	Steven	01-05-1985	598

FIGURE 5.9

Linking Data Tables to Answer an Inquiry

In finding the name and hire date of the manager working on the sales manual project, the president needs three tables: project, department, and manager. The project description (Sales Manual) leads to the department number (598) in the project table, which leads to the manager's SSN (098-40-1370) in the department table, which leads to the manager's name (Fiske) and hire date (01-05-1985) in the manager table.

data cleanup
the process of looking for and fixing inconsistencies to ensure that data is accurate and complete

data in tables. As seen in Figure 5.10, a relational database management system, such as Access, provides a number of tips and tools for building and using database tables. This figure shows the database displaying information about data types and indicating that additional help is available. The ability to link relational tables also allows users to relate data in new ways without having to redefine complex relationships. Because of the advantages of the relational model, many companies use it for large corporate databases, such as in marketing and accounting. The relational model can be used with personal computers and mainframe systems. Galileo International is one of four major worldwide reservation systems that handle travel industry bookings. In 2002, it implemented a new fare-pricing system based on the use of relational database technology. The millions of daily queries from on-line travel companies, such as Expedia, Travelocity, and Orbitz, plus the need to adjust pricing on the fly for those sites, require the most efficient database management technology possible. With the new system, Galileo can also accept lists of airline airfares automatically from ATPco, the industry-funded company that publishes airfares.[4]

Data Cleanup

As discussed in Chapter 1, the characteristics of valuable data include that the data is accurate, complete, economical, flexible, reliable, relevant, simple, timely, verifiable, accessible, and secure. The purpose of **data cleanup** is to develop data with these characteristics. A 4 percent error rate may not sound like much, but a multibillion-dollar corporation could lose millions of dollars if bad information caused the firm to bill all but 4 percent of its orders. When a database is created with data from multiple sources, those disparate sources may store different values

**Building and Modifying a
Relational Database**

*Relational databases provide
many tools, tips, and tricks to simplify the process of creating and
modifying a database.*

for the same customer due to spelling errors, multiple account numbers, and
address variations. The purpose of data cleanup is to look for and fix these and
other inconsistencies that can result in duplicate or incorrect records ending up in
the database.

DATABASE MANAGEMENT SYSTEMS (DBMSs)

Creating and implementing the right database system ensures that the database will
support both business activities and goals. But how do we actually create, implement, use, and update a database? The answer is found in the database management
system. As discussed, a database management system (DBMS) is a group of programs used as an interface between a database and application programs or a database and the user. DBMSs are classified by the type of database model they
support. For example, a relational database management system follows the relational model. Access by Microsoft is a popular relational DBMS for personal
computers. Popular mainframe relational DBMSs include DB2 and Informix
(acquired for $1 billion by IBM in July 2001) by IBM, Oracle, and Sybase. A
number of open source database management systems are also available, including MySQL, PostgreSQL, and Berkeley DB. These DBMSs come with open
source licenses so that they are free to obtain. Companies have a number of options
for support as new firms try to undercut the major proprietary competitors. Yahoo
and Slashdot are two Web sites that rely on open source databases to store articles
and comments.[5]

All DBMSs share some common functions, like providing a user view, physically storing and retrieving data in a database, allowing for database modification, manipulating data, and generating reports. These DBMSs are capable of
handling the most complex of data processing tasks. For example, the Mayo
Clinic and IBM are using IBM's DB2 DBMS to develop an all-inclusive database containing patient records, physician notes, demographic data, genetic
data, and proteomic (protein-related) data for some 4 million patients. The
database is expected to grow to several petabytes (PB). It will be used to
uncover links between specific diseases and patient behavior, diagnose illnesses
faster, and prescribe treatments to meet the needs of the half-million patients
the facility treats annually.[6]

PROVIDING A USER VIEW

Because the DBMS is responsible for access to a database, one of the first steps in installing and using a database involves telling the DBMS the logical and physical structure of the data and relationships among the data in the database. This description is called a **schema** (as in schematic diagram). A schema can be part of the database or a separate schema file. The DBMS can reference a schema to find where to access the requested data in relation to another piece of data.

schema
a description of the entire database

A DBMS also acts as a user interface by providing a view of the database. A user view is the portion of the database a user can access. To create different user views, subschemas are developed. A **subschema** is a file that contains a description of a subset of the database and identifies which users can view and modify the data items in that subset. While a schema is a description of the entire database, a subschema shows only some of the records and their relationships in the database. Normally, programmers and managers need to view or access only a subset of the database. For example, a sales representative might need only data describing customers in her region, not the sales data for the entire nation. A subschema could be used to limit her view to data from her region. With subschemas, the underlying structure of the database can change, but the view the user sees might not change. For example, even if all the data on the southern region changed, the northeast region sales representative's view would not change if she accessed data on her region.

subschema
a file that contains a description of a subset of the database and identifies which users can view and modify the data items in the subset

A number of subschemas can be developed for different users and the various application programs. Typically, the database user or application will access the subschema, which then accesses the schema (Figure 5.11). Subschemas can also provide additional security because programmers, managers, and other users are typically allowed to view only certain parts of the database.

CREATING AND MODIFYING THE DATABASE

Schemas and subschemas are entered into the DBMS (usually by database personnel) via a data definition language. A **data definition language (DDL)** is a collection of instructions and commands used to define and describe data and

data definition language (DDL)
a collection of instructions and commands used to define and describe data and data relationships in a specific database

FIGURE 5.11

The Use of Schemas and Subschemas

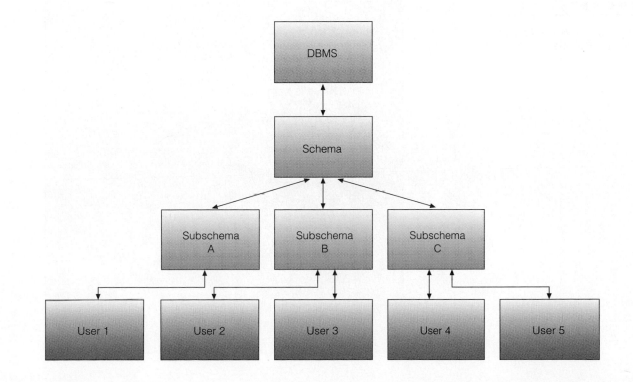

data relationships in a specific database. A DDL allows the database's creator to describe the data and the data relationships that are to be contained in the schema and the many subschemas. In general, a DDL describes logical access paths and logical records in the database. Figure 5.12 shows a simplified example of a DDL used to develop a general schema. The *Xs* in Figure 5.12 reveal where specific information concerning the database is to be entered. File description, area description, record description, and set description are terms the DDL defines and uses in this example. Other terms and commands can be used, depending on the particular DBMS employed.

data dictionary
a detailed description of all the data used in the database

Another important step in creating a database is to establish a **data dictionary**, a detailed description of all data used in the database. The data dictionary contains the name of the data item, aliases or other names that may be used to describe the item, the range of values that can be used, the type of data (such as alphanumeric or numeric), the amount of storage needed for the item, a notation of the person responsible for updating it and the various users who can access it, and a list of reports that use the data item. Figure 5.13 shows a typical data dictionary entry.

For example, the information in a data dictionary for the part number of an inventory item can include the name of the person who made the data dictionary entry (D. Bordwell), the date the entry was made (August 4, 2002), the name of the person who approved the entry (J. Edwards), the approval date (October 13, 2002), the version number (3.1), the number of pages used for the entry (1), the part name (PARTNO), other part names that may be used (PTNO), the range of values (part numbers can range from 100 to 5000), the type of data (numeric), and the storage required (four positions are required for the part number). Following are some of the typical uses of a data dictionary.

- *Provide a standard definition of terms and data elements.* This standardization can help in programming by providing consistent terms and variables to be used for all programs. Programmers know what data elements are already "captured" in the database and how they relate to other data elements.

FIGURE 5.12

Using a Data Definition Language to Define a Schema

```
SCHEMA DESCRIPTION
SCHEMA NAME IS XXXX
AUTHOR      XXXX
DATE        XXXX
FILE DESCRIPTION
     FILE NAME IS XXXX
        ASSIGN XXXX
     FILE NAME IS XXXX
        ASSIGN XXXX
AREA DESCRIPTION
     AREA NAME IS XXXX
RECORD DESCRIPTION
     RECORD NAME IS XXXX
     RECORD ID IS XXXX
     LOCATION MODE IS XXXX
     WITHIN XXXX AREA FROM XXXX THRU XXXX
SET DESCRIPTION
     SET NAME IS XXXX
     ORDER IS XXXX
     MODE IS XXXX
     MEMBER IS XXXX
     .
     .
     .
```

FIGURE 5.13

A Typical Data Dictionary Entry

NORTHWESTERN MANUFACTURING

PREPARED BY:	D. BORDWELL
DATE:	04 AUGUST 2002
APPROVED BY:	J. EDWARDS
DATE:	13 OCTOBER 2002
VERSION:	3.1
PAGE:	1 OF 1

DATA ELEMENT NAME:	PARTNO
DESCRIPTION:	INVENTORY PART NUMBER
OTHER NAMES:	PTNO
VALUE RANGE:	100 TO 5000
DATA TYPE:	NUMERIC
POSITIONS:	4 POSITIONS OR COLUMNS

- *Assist programmers in designing and writing programs.* Programmers do not need to know which storage devices are used to store needed data. Using the data dictionary, programmers specify the required data elements. The DBMS locates the necessary data. More important, programmers can use the data dictionary to see which programs already use a piece of data and, if appropriate, can copy the relevant section of the program code into their new program, thus eliminating duplicate programming efforts.

- *Simplify database modification.* If for any reason a data element needs to be changed or deleted, the data dictionary would point to specific programs that utilize the data element that may need modification.

A data dictionary helps achieve the advantages of the database approach in these ways:

- *Reduced data redundancy.* By providing standard definitions of all data, it is less likely that the same data item will be stored in different places under different names. For example, a data dictionary would reduce the likelihood that the same part number would be stored as two different items, such as PTNO and PARTNO.
- *Increased data reliability.* A data dictionary and the database approach reduce the chance that data will be destroyed or lost. In addition, it is more difficult for unauthorized people to gain access to sensitive data and information.
- *Faster program development.* With a data dictionary, programmers can develop programs faster. They don't have to develop names for data items because the data dictionary does that for them.
- *Easier modification of data and information.* The data dictionary and the database approach make modifications to data easier because users do not need to know where the data is stored. The person making the change indicates the new value of the variable or item, such as part number, that is to be changed. The database system locates the data and makes the necessary change.

STORING AND RETRIEVING DATA

As just described, one function of a DBMS is to be an interface between an application program and the database. When an application program needs data, it requests that data through the DBMS. Suppose that to calculate the total price of a new car, an auto dealer pricing program needs price data on the engine option—six cylinders instead of the standard four cylinders. The application program thus requests this data from the DBMS. In doing so, the application program follows a logical access path. Next, the DBMS, working in conjunction with various system software programs, accesses a storage device, such as disk or tape, where the data is stored. When the DBMS goes to this storage device to retrieve the data, it follows a path to the physical location (physical access path) where the price of this option is stored. In the pricing example, the DBMS might go to a disk drive to retrieve the price data for six-cylinder engines. This relationship is shown in Figure 5.14.

This same process is used if a user wants to get information from the database. First, the user requests the data from the DBMS. For example, a user might give a command, such as LIST ALL OPTIONS FOR WHICH PRICE IS GREATER THAN 200 DOLLARS. This is the logical access path (LAP). Then the DBMS might go to the options price sector of a disk to get the information for the user. This is the physical access path (PAP).

When two or more people or programs attempt to access the same record in the same database at the same time, there can be a problem. For example, an inventory control program might attempt to reduce the inventory level for a product by ten units because ten units were just shipped to a customer. At the same time, a purchasing program might attempt to increase the inventory level for the same product by 200 units because more inventory was just received. Without proper database control, one of the inventory updates may not be correctly made, resulting in an inaccurate inventory level for the product. **Concurrency control** can be used to avoid this potential problem. One approach is to lock out all other application programs from access to a record if the record is being updated or used by another program.

Users increasingly need to be able to access and update databases via the Internet. Many software vendors are incorporating this capability into their products, including Microsoft, Macromedia, Inline Internet Systems, and Netscape Communications. Such databases allow companies to create an Internet-accessible catalog, which is nothing more than a database of items, descriptions, and prices.

concurrency control
a method of dealing with a situation in which two or more people need to access the same record in a database at the same time

FIGURE 5.14

Logical and Physical Access Paths

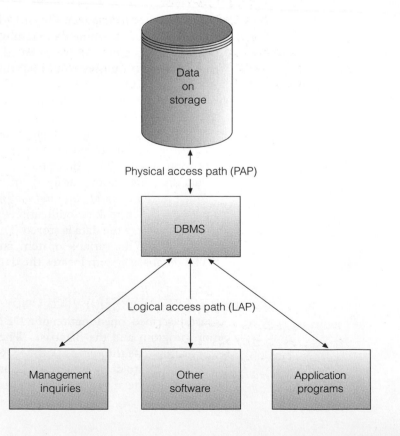

MANIPULATING DATA AND GENERATING REPORTS

Once a DBMS has been installed, the system can be used by all levels of employees via specific commands in various programming languages. For example, COBOL commands can be used in simple programs that will access or manipulate certain pieces of data in the database. Here's another example of a DBMS query: SELECT * FROM EMPLOYEE WHERE JOB_CLASSIFICATION = "C2". The * tells the program to include all columns from the EMPLOYEE table. In general, the commands that are used to manipulate the database are part of the **data manipulation language (DML)**. This specific language, provided with the DBMS, allows managers and other database users to access, modify, and make queries about data contained in the database to generate reports. Again, the application programs go through subschemas, schemas, and the DBMS before actually getting to the physically stored data on a device such as a disk.

In the 1970s, D. D. Chamberlain and others at the IBM Research Laboratory in San Jose, California, developed a standardized data manipulation language called *Structured Query Language (SQL)*, pronounced like the word *sequel* or simply spelled out as *SQL*. The EMPLOYEE query shown earlier is written in SQL. In 1986, the American National Standards Institute (ANSI) adopted SQL as the standard query language for relational databases. Since ANSI's acceptance of SQL, interest in making SQL an integral part of relational databases on both mainframe and personal computers has increased.

SQL lets programmers learn one powerful query language and use it on systems ranging from PCs to the largest mainframe computers (Figure 5.15). Programmers and database users also find SQL valuable because SQL statements can be embedded into many programming languages, such as the widely used C and COBOL languages. Because SQL uses standardized and simplified procedures for retrieving, storing, and manipulating data in a database system, the popular database query language can be easy to understand and use.

Once a database has been set up and loaded with data, it can produce desired reports, documents, and other outputs (see Figure 5.16). These outputs usually appear in screen displays or hard-copy printouts. The output-control features of a database program allow you to select the records and fields to appear in reports. You can also complete calculations specifically for the report by manipulating database fields. Formatting controls and organization options (like report headings) help you to customize reports and create flexible, convenient, and powerful information-handling tools.

A database program can produce a wide variety of documents, reports, and other outputs that can help organizations achieve their goals. The most common reports select and organize data to present summary information about some aspect of company operations. For example, accounting reports often summarize financial

data manipulation language (DML)

the commands that are used to manipulate the data in a database

---

Structured Query Language

SQL has become an integral part of most relational database packages, as shown by this screen from Microsoft Access.

Database Output

A database application offers sophisticated formatting and organization options to produce the right information in the right format.

data such as current and past-due accounts. Many companies base their routine operating decisions on regular status reports that show the progress of specific orders toward completion and delivery. Increasingly, companies are using databases to provide improved customer services.

Exception, scheduled, and demand reports, first discussed in Chapter 1, highlight events that require urgent management attention. Database programs can produce literally hundreds of documents and reports. A few examples include these:

- Form letters with address labels
- Payroll checks and reports
- Invoices
- Orders for materials and supplies
- A variety of financial performance reports

DATABASE ADMINISTRATOR

database administrator (DBA)

a highly skilled and trained systems professional who directs or performs all activities related to maintaining a successful database environment

A **database administrator (DBA)** is a highly skilled and trained systems professional who directs or performs all activities related to maintaining a successful database environment. The DBA's responsibilities include designing, implementing, and maintaining the database system; establishing policies and procedures pertaining to the management, security, maintenance, and use of the database management system; and training employees in database management and use.

A DBA is expected to have a clear understanding of the fundamental business of the organization, be proficient in the use of selected database management systems, and stay abreast of emerging technologies and new design approaches. Typically, a DBA has either a degree in computer science or management information systems and some on-the-job training with a particular database product or more extensive experience with a range of database products.

The DBA works with users to decide the content of the database—exactly what entities are of interest and what attributes are to be recorded about those entities. Using this data, the DBA will create a logical database model that satisfies the needs of its users. The DBA then works with users to learn the specifics of how the database will be used—the number and frequency of different kinds of requests for data, the frequency of updating and deleting of records, and the people who will need update and/or read access to the various data entities and attributes. The DBA also helps define the types of data-edit changes and controls needed for the database. All of this information is compared with the features of a specific database management system to define the physical structure of the database that will satisfy users' needs.

The DBA works with programmers as they build applications that will use the database to ensure that their programs comply with database management system standards and conventions. Once the database is built and operating, the DBA monitors operations logs for database security violations. Database performance is also monitored to ensure that the system's response time meets users' needs and that it operates efficiently. If there is a problem, the DBA attempts to correct it before it gets serious.

Some organizations have created a position called *data administrator*. The **data administrator** is a nontechnical, but important role that ensures that data is managed as an important organizational resource. The data administrator is responsible for defining and implementing a consistent set of principles for a variety of data issues, including setting data standards and data definitions that apply across the many databases that an organization may have. For example, this would ensure that a term such as "customer" is defined and treated consistently in all corporate databases. The data administrator also works with business managers to identify who should have read and/or update access to certain databases and to selected attributes within those databases. This information is then communicated to the database administrator for implementation.

data administrator
a non-technical, but important role that ensures that data is managed as an important organizational resource

POPULAR DATABASE MANAGEMENT SYSTEMS

The latest generation of database management systems makes it possible for end users to build their own database applications. End users are using these tools to address everyday problems such as how to manage a mounting pile of information on employees, customers, inventory, or sales and fun stuff such as wine lists, CD collections, and video libraries. These database management systems are an important personal productivity tool along with word processing, spreadsheet, and graphics software.

A key to making DBMSs more usable for some databases is the incorporation of "wizards" that walk you through how to build customized databases, modify ready-to-run applications, use existing record templates, and quickly locate the data you want. These applications also include powerful new features such as help systems and Web-publishing capabilities. For example, users can create a complete inventory system and then instantly post it to the Web, where it does double duty as an electronic catalog. Some of the more popular DBMSs for end users include Corel's Paradox, FileMaker's FileMaker Pro, Microsoft's Access, and Lotus's Approach.

The complete database management software market encompasses software used by professional programmers and that runs on midrange, mainframe, and supercomputers. The entire market generates $10 billion per year in revenue, with IBM, Oracle, and Microsoft, the leaders (Figure 5.17). Although Microsoft rules in desktop PC software, its share of database software on bigger computers is small.

SELECTING A DATABASE MANAGEMENT SYSTEM

Selecting the best database management system begins by analyzing database needs and characteristics. The information needs of the organization affect the type of data that is collected and the type of database management system that is used. Important characteristics of databases include the size of the database, number of concurrent users, performance, the ability of the DBMS to integrate with other systems, the features of the DBMS, vendor considerations, and the cost of the database management system.

FIGURE 5.17

*Worldwide Database Market
Share, 2001—Based on
Percentage of Worldwide New
License Revenue from DBMSs*

(Source: Data from James Niccolai,
"Gartner: IBM Steals Database Crown
from Oracle," *Computerworld*, May 7,
2002, accessed at
http://www.computerworld.com.

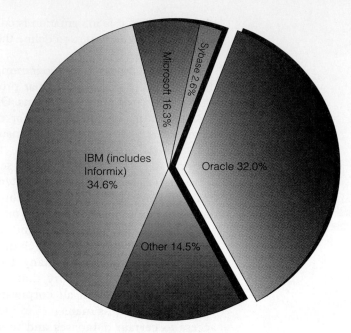

Database Size

The database size depends on the number of records or files in the database. The size determines the overall storage requirement for the database. Most database management systems can handle relatively small databases of less than 100 million bytes; fewer can manage terabyte-size databases.

Number of Concurrent Users

The number of simultaneous users that can access the contents of the database is also an important factor. Clearly, a database that is used by a large workgroup must be able to support a number of concurrent users; if it cannot, then the efficiency of the members of the workgroup will be lowered. The term *scalability* is sometimes used to describe how well a database performs as the size of the database or the number of concurrent users is increased. A highly scalable database management system is desirable to provide flexibility. Unfortunately, many companies make a poor DBMS choice in this regard and then later are forced to convert to a new DBMS when the original does not meet expectations.

Performance

How fast the database is able to update records can be the most important performance criterion for some organizations. Credit card and airline companies, for example, must have database systems that can update customer records and check credit or make a plane reservation in seconds, not minutes. Other applications, such as payroll, can be done once a week or less frequently and do not require immediate processing. If an application demands immediacy, it also demands rapid recovery facilities in the event the computer system shuts down temporarily. Other performance considerations include the number of concurrent users that can be supported and how much main memory is required to execute the database management program.

Integration

A key aspect of any database management system is its ability to integrate with other applications and databases. A key determinant here is what operating systems it can run under—such as Unix, Windows NT, and Windows 2000. Some companies use several databases for different applications at different locations. A manufacturing company with four plants in three different states might have a separate database at each location. The ability of a database program to import data from and export data to other databases and applications can be a critical consideration.

A database management system used by on-line stores must be able to support a large number of concurrent users by quickly checking a customer's credit card and processing his or her order for merchandise.

Features

The features of the database management system can also make a big difference. Most database programs come with security procedures, privacy protection, and a variety of tools. Other features can include the ease of use of the database package and the availability of manuals and documentation to help the organization get the most from the database package. Additional features such as wizards and ready-to-use templates help improve the product ease of use and are very important. In fact, because of the pressure to reduce information systems budgets and the scarcity of experienced database administrators, organizations are demanding database software that comes with features that simplify database management tasks. In response, IBM's DB2 version 8 comes with the capability to alert database administrators by E-mail, pager, or mobile device when a problem is developing, such as an application running out of memory or a user query taking an excessive amount of time. The DB2 software then recommends the specific steps that must be taken to resolve the problem. In some cases, the DB2 software can diagnose the problem and take remedial action on its own.[7] In another approach aimed at minimizing the database management workload, Oracle enables its customers to outsource the database management activities. Customers can choose to have the software installed and run on Oracle systems in an Oracle data center or have Oracle manage the software installed on the customer's systems at the customer's location.[8]

The Vendor

The size, reputation, and financial stability of the vendor should also be considered in making any database purchase. Some vendors are well respected in the information systems industry and have a large support staff to give assistance, if necessary. A well-established and financially secure database company is more likely to remain in business than others.

The ability of the vendor to provide global support for large, multinational companies or companies outside the U.S. is becoming increasingly important. CERN is the European Organization for Nuclear Research located near Geneva, Switzerland. It is building a particle accelerator that will give scientists new insights into the structure of matter. When the Large Hadron Collider begins operation in 2006, it will generate a prodigious amount of information—between 5 and 20 PB (petabytes) of raw data each year. CERN information systems people are considering using Oracle9i rather than the Objectivity database software that has so far been an unofficial standard among nuclear research labs. The primary reason is that Oracle has extensive European operations while Objectivity does not and is headquartered thousands of miles away in Mountain View, California.[9]

Cost

Database packages for personal computers can cost a few hundred dollars, but large database systems for mainframe computers can cost hundreds of thousands of dollars. In addition to the initial cost of the database package, monthly operating costs should be considered. Some database companies rent or lease their database software. Monthly rental or lease costs, maintenance costs, additional hardware and software costs, and personnel costs can be substantial.

DATABASE APPLICATIONS

The types of data and information that managers need change as business processes change. A number of effective database applications can help managers meet their needs, among them linking the company databases to the Internet, setting up data warehouses and marts, using databases for strategic business intelligence, allowing organizations to place data at different locations, using on-line processing and open connectivity standards for increased productivity, developing databases with the object-oriented approach, and searching for and using unstructured data such as graphics and video.

LINKING THE COMPANY DATABASE TO THE INTERNET

Customers, suppliers, and company employees must be able to access corporate databases through the Internet, intranets, and extranets to meet various business needs. For example, when they are shopping, Internet customers need to access the corporate product database to obtain product information, including size, color, type, and price details. Suppliers use the Internet and corporate extranets to view inventory databases to check levels of raw materials and the current production schedule to determine when and how much of their products must be delivered to support just-in-time inventory management. Company employees need to be able to access corporate databases to support decision making even when they are off site. In such cases, they may use laptop computers and access the data via the Internet or company intranet.

In enabling access of corporate databases via the Internet, intranets, and extranets, it is important that any software installation required at the user end be extremely simple. In addition, only authorized users should be able to access the databases. As a result, organizations are adopting the use of **application servers**—software packages, often written in the Java programming language for use on computers running the Windows NT operating system, that connect end users to the databases holding the information they need to access. Application servers manage the process of connecting users to that data by setting up an application session for each user, checking each user's identification and password, fetching requested information from the appropriate database, and building the data into a Web page for display to the users. Application server software packages, offered by more than three dozen vendors, also provide application management services such as monitoring system performance to identify any system bottlenecks.

Many companies accept data from customers in developing their production forecast. Customers enter the amount of each product they will order each week for the next several weeks. The company then aggregates this data by week and knows how much product it must have on hand to meet future needs. Here are the steps users follow to connect to a company's demand forecast database using the Internet and their Web browser software:

1. A customer points his or her Web browser to the company's demand forecast Web page and is connected to an application server.
2. The application server establishes a user session and verifies the user's identification and password information against a database.
3. A Web page is created by the application server, enabling the user to view the company's demand forecast and input the data for his or her company.
4. All demand forecast inputs are sent to the application server, which establishes a link to the database and updates the data.
5. The application server collects the data and builds a new Web page for the user.

application servers
Software packages, often written in the Java programming language for use on computers running the Windows NT operating system, that connect end users to the databases holding the information they need to access

DATA WAREHOUSES, DATA MARTS, AND DATA MINING

The raw data necessary to make sound business decisions is stored in a variety of locations and formats—hierarchical databases, network databases, flat files, and spreadsheets, to name a few. This data is initially captured, stored, and managed by transaction processing systems that are designed to support the day-to-day operations of the organization. For decades, organizations have collected operational, sales, and financial data with their on-line transaction processing (OLTP) systems.

Traditional OLTP systems are designed to put data into databases very quickly, reliably, and efficiently. These systems are not good at supporting meaningful analysis of the data. In fact, tuning a system to provide excellent performance for OLTP often renders rapid data retrieval for data analysis nearly impossible. Furthermore, data stored in OLTP databases is inconsistent and constantly changing. The database contains the current transactions required to operate the business, including errors, duplicate entries, and reverse transactions, which get in the way of a business analyst, who needs stable data. Historical data is missing from the OLTP database, which makes trend analysis impossible. Because of the application orientation of the data, the variety of nonintegrated data sources, and the lack of historical data, companies were limited in their ability to access and use the data for other purposes. So, although the data collected by OLTP systems doubles every two years, it does not meet the needs of the business decision maker—they are data rich but information poor.

Data Warehouses

data warehouse
a database that collects business information from many sources in the enterprise, covering all aspects of the company's processes, products, and customers

A **data warehouse** is a database that holds business information from many sources in the enterprise, covering all aspects of the company's processes, products, and customers. The data warehouse provides business users with a multidimensional view of the data they need to analyze business conditions. A data warehouse is designed specifically to support management decision making, not to meet the needs of transaction processing systems. The data warehouse provides a specialized decision support database that manages the flow of information from existing corporate databases and external sources to end-user decision support applications. A data warehouse stores historical data that has been extracted from operational systems and external data sources (Figure 5.18). This operational and external data is "cleaned up" to remove inconsistencies and integrated to create a new information database that is more suitable for business analysis.

Data warehouses typically start out as very large databases, containing millions and even hundreds of millions of data records. As this data is collected from the various production systems, a historical database is built that business analysts can use.

Sears's data warehouse is enormous—and growing. When completed, it will contain roughly enough information to fill 20 million four-drawer filing cabinets.

(Source: © Syracuse Newspapers/Dennis Nett/The Image Works.)

To remain fresh and accurate, the data warehouse receives regular updates. Old data that is no longer needed is purged from the data warehouse. Updating the data warehouse must be fast, efficient, and automated, or the ultimate value of the data warehouse is sacrificed. It is common for a data warehouse to contain from 3 to 10 years of current and historical data. Data cleaning tools can merge data from many sources into one database, automate data collection and verification, delete unwanted data, and maintain data in a database management system. The amount of data the average business collects and stores is doubling each year. If that holds true at a company such as Sears, Roebuck and Co., which is combining its customer and inventory data warehouses to create

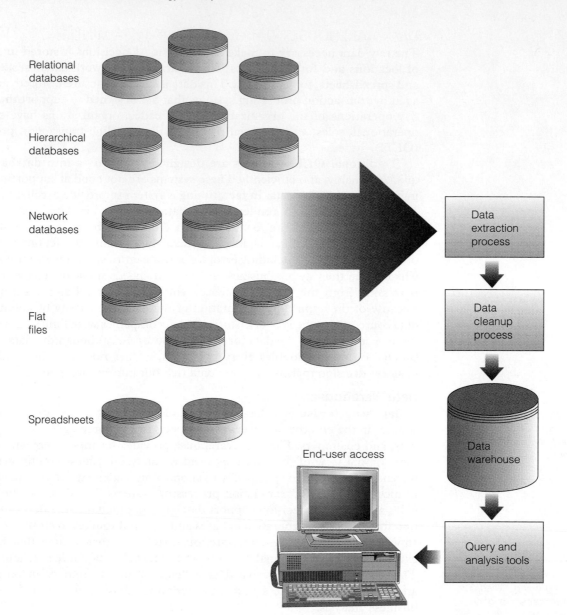

FIGURE 5.18

Elements of a Data Warehouse

a 70-TB system, the retailer will hit the 1-PB threshold (approximately 1,000 TB) within four years. A petabyte of data is the equivalent of 250 billion pages of text, enough to fill 20 million four-drawer filing cabinets.[10]

The primary advantage of data warehousing is the ability to relate data in new, innovative ways. However, a data warehouse can be extremely difficult to establish, with the typical cost exceeding $2 million. Table 5.3 compares on-line transaction processing to data warehousing.

Data Mart

data mart
a subset of a data warehouse

A **data mart** is a subset of a data warehouse. Data marts bring the data warehouse concept—on-line analysis of sales, inventory, and other vital business data that has been gathered from transaction processing systems—to small and medium-sized businesses and to departments within larger companies. Rather than store all enterprise data in one monolithic database, data marts contain a subset of the data for a single aspect of a company's business—for example, finance, inventory, or personnel. In fact, there may even be more detailed data for a specific area in a data mart than what a data warehouse would provide.

Characteristic	OLTP Database	Data Warehousing
Purpose	Support transaction processing	Support decision support
Source of data	Business transactions	Multiple files, databases—data internal and external to the firm
Data access allowed users	Read and write	Read only
Primary data access mode	Simple database update and query	Simple and complex database queries with increasing use of data mining to recognize patterns in the data
Primary database model employed	Relational	Relational
Level of detail	Detailed transactions	Often summarized data
Availability of historical data	Very limited—typically a few weeks or months	Multiple years
Update process	On-line, ongoing process as transactions are captured	Periodic process, once per week or once per month
Ease of update	Routine and easy	Complex, must combine data from many sources; data must go through a data cleanup process
Data integrity issues	Each individual transaction must be closely edited	Major effort to "clean" and integrate data from multiple sources

TABLE 5.3

Comparison of OLTP and Data Warehousing

data mining
an information analysis tool that involves the automated discovery of patterns and relationships in a data warehouse

Data marts are most useful for smaller groups who want to access detailed data. A warehouse is used for summary data that can be used by an entire company. Because data marts typically contain tens of gigabytes of data, as opposed to the hundreds of gigabytes in data warehouses, they can be deployed on less powerful hardware with smaller secondary storage devices, delivering significant savings to an organization. Although any database software can be used to set up a data mart, some vendors deliver specialized software designed and priced specifically for data marts. Already, companies such as Sybase, Software AG, Microsoft, and others have announced products and services that make it easier and cheaper to deploy these scaled-down data warehouses. The selling point: data marts put targeted business information into the hands of more decision makers.

Data Mining

Data mining is an information analysis tool that involves the automated discovery of patterns and relationships in a data warehouse. Data mining represents the next step in the evolution of decision support systems. It makes use of advanced statistical techniques and machine learning to discover facts in a large database, including databases on the Internet. Unlike query tools, which require users to formulate and test a specific hypothesis, data mining uses built-in analysis tools to automatically generate a hypothesis about the patterns and anomalies found in the data and then from the hypothesis predict future behavior.

Data mining's objective is to extract patterns, trends, and rules from data warehouses to evaluate (i.e., predict or score) proposed business strategies, which in turn will improve competitiveness, improve profits, and transform business processes. It is used extensively in marketing to improve customer retention; cross-selling opportunities; campaign management; market, channel, and pricing analysis; and customer segmentation analysis (especially one-to-one marketing).

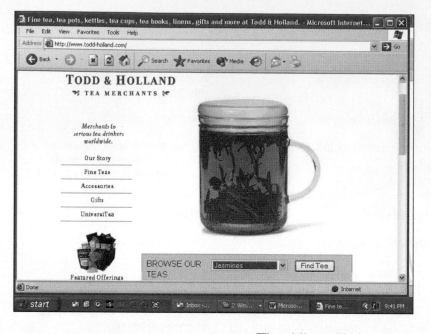

Todd & Holland Tea Merchants used predictive analysis data mining software to redefine its target market—to include professional women from 25 to 35 years of age. Previously, the company had overlooked that customer segment.

predictive analysis
a form of data mining that combines historical data with assumptions about future conditions to predict outcomes of events such as future product sales or the probability that a customer will default on a loan

business intelligence
the process of gathering enough of the right information in a timely manner and usable form and analyzing it to have a positive impact on business strategy, tactics, or operations

In short, data mining tools help end users find answers to questions they never even thought to ask.

Jiffy Lube uses data mining to attract new customers. The company keeps records on millions of vehicles, tracking how often the owners come in and for what services. Using this data, it has profiled its best customers so its service centers can custom design direct mailings to locals who match its best customers' profiles but haven't yet visited Jiffy Lube.[11] United Airlines built a data warehouse that eventually will hold 6 TB of customer, reservation, and flight data. The cost of the project is estimated to have exceeded $10 million. United will use data mining to better manage customer data, improve its frequent-flier programs, study flight data, and analyze reservation patterns. The airline will also use the data to identify high-value customers who were inconvenienced by delays and offer them some sort of compensation.[12]

E-commerce presents another major opportunity for effective use of data mining. Attracting customers to on-line Web sites is tough; keeping them can be next to impossible. For example, when on-line retail Web sites launch deep-discount sales, they cannot easily figure out how many first-time customers are likely to come back and buy again. Nor do they have a way of understanding which customers acquired during the sale are price sensitive and more likely to jump on future sales. As a result, companies are gathering data on user traffic through their Web sites into databases. This data is then analyzed using data mining techniques to personalize the Web site and develop sales promotions targeted at specific customers.

Predictive analysis is a form of data mining that combines historical data with assumptions about future conditions to predict outcomes of events such as future product sales or the probability that a customer will default on a loan. Retailers use predictive analysis to upgrade occasional customers into frequent purchasers by predicting what products they will buy if offered an appropriate incentive. Genalytics, Magnify, NCR Teradata, SAS Institute, Sightward, SPSS, and Quadstone have developed predictive analysis tools. Todd & Holland Tea Merchants sells specialty tea over the Internet and in a retail shop. The company used predictive analysis software from SPSS to analyze the company's customer list and a year's worth of sales data. The resulting recommendation was that Todd & Holland should market the tea to professional women between 25 and 35—a market segment that had gone unrecognized by the owners.[13]

Traditional DBMS vendors are well aware of the great potential of data mining. Thus, companies like Oracle, Sybase, Tandem, and Red Brick Systems are all incorporating data mining functionality into their products. Table 5.4 summarizes a few of the most frequent applications for data mining. Read the "Ethical and Societal Issues" special-interest box to see how predictive analysis and data mining can be used along with database management technology to help make the skies safer for all of us.

BUSINESS INTELLIGENCE

Closely linked to the concept of data mining is use of databases for business intelligence purposes. **Business intelligence** is the process of gathering enough of the right information in a timely manner and usable form and analyzing it so that it can have a positive impact on business strategy, tactics, or operations. Business intelligence involves turning data into useful information that is then distributed throughout an enterprise. Companies use this information to make improved strategic decisions about which markets to enter, how to select and manage key customer relationships, and how to select and effectively promote products to increase profitability and market share.

AXA Financial, a member of Paris-based AXA Group, is one of the world's largest providers of insurance and financial services with over 4.5 million customers. It subdivided its large DB2-based data warehouse into data marts, with each focused on a different business area. This approach enables workers to use business intelligence tools to gain an in-depth understanding of AXA's customers—from the profits they generate, their rate of retention, and the opportunities they offer to cross-sell the company's products.[14] Employees can obtain the data needed to zero in on problem areas, obtain a detailed picture of the profitability of any customer, and see which products are selling and which are not. If sales decline, they can track those declines to specific offices or even to individual sales reps to pinpoint problems and take immediate steps to remedy them.

Competitive intelligence is one aspect of business intelligence and is limited to information about competitors and the ways that knowledge affects strategy, tactics, and operations. Effective **competitive intelligence** is a continuous process, involving the legal and ethical collection of information, analysis that doesn't avoid unwelcome conclusions, and controlled dissemination of that information to decision makers. Are you ahead of your competitors or are you racing to catch up? To stay ahead in the marketplace, you must be able to integrate competitive intelligence into your company's strategic plans and decisions. Competitive intelligence is a critical part of a company's ability to see and respond quickly and appropriately to the changing marketplace.

Competitive intelligence is not espionage—the use of illegal means to gather information. In fact, almost all the information a competitive intelligence professional needs can be collected by examining published information sources, conducting interviews, and using other legal, ethical methods. Using a variety of analytical tools, a skilled competitive intelligence professional can by deduction fill the gaps in information already gathered.

competitive intelligence
a continuous process involving the legal and ethical collection of information, analysis, and controlled dissemination of information to decision makers

TABLE 5.4

Common Data Mining Applications

Application	Description
Branding and positioning of products and services	Enable the strategist to visualize the different positions of competitors in a given market using performance (or importance) data on dozens of key features of the product in question and then to condense all that data into a perceptual map of just two or three dimensions
Customer churn	Predict current customers who are likely to go to a competitor
Direct marketing	Identify prospects most likely to respond to a direct marketing campaign such as telephone solicitation or direct mailing
Fraud detection	Highlight transactions most likely to be deceptive or illegal
Market basket analysis	Identify products and services that are most commonly purchased at the same time (e.g., nail polish and lipstick)
Market segmentation	Group customers based on who they are or on what they prefer
Trend analysis	Analyze how key variables (e.g., sales, spending, promotions) vary over time

Database Technology Used to Improve Airport Security

The Transportation Security Administration is evaluating various approaches to harness database technology in its efforts to improve airport security. Unfortunately, any such system may require airlines to invest in additional information systems technology—at a time when they are suffering from a lack of revenue. One major airline has already declared bankruptcy.

One idea is to develop a database system that links every airline reservation system in the country with a number of private and government databases. Data mining and predictive analysis would be used to sort through personal travel histories, the backgrounds of passengers aboard particular flights, and a wealth of other data to assign numerical threat ratings to individuals. Warnings would be sent electronically to workers at airport screening locations to inspect individuals with high threat ratings more closely.

Another approach would allow prescreened "trusted travelers" to pass through airport security checkpoints quickly, avoiding long lines and congestion. This system would devote more time and resources to screening other travelers whose level of risk is higher or unknown. Those who apply for the "trusted traveler" program would have to pass a background check using data from a number of state and federal databases. Once at the airport, "trusted traveler" passengers would be identified, perhaps by scanning their fingerprints or retinas or requiring some form of identification card. (The federal government is considering developing a security ID card for airline passengers that would rely on biometric identification and be linked to government databases). The system would also cross-check the passenger's identification with the FBI's watch list database and a federal passenger profiling system known as Computer-Assisted Passenger Screening. Provided everything was clear, the passenger could then proceed to his or her airplane using expedited security check-in procedures.

Discussion Questions

1. Which approach is best in terms of improving airport security: assigning a threat rating to individuals or prescreening individuals to identify "trusted travelers"? Why do you think this approach is best?

2. Identify specific data that could be used to assign numeric threat ratings to individuals. Describe how the system would work.

Critical Thinking Questions

3. Briefly discuss the data integrity or data privacy issues associated with the two approaches outlined in the box.

4. Identify other technical and economic issues that may make it difficult to implement either approach. Should the federal government help pay the cost of implementing these safeguards?

Sources: Adapted from Dan Verton, "Feds Mulling New Airline Surveillance," *Computerworld*, February 1, 2002, accessed at www.computerworld.com; Matt Berger, "'Trusted Traveler' Aims to Streamline Flight Security," *Computerworld*, March 18, 2002, accessed at www.computerworld.com; Brian Sullivan, "EPIC Files Suit Against the Bush Administration," *Computerworld*, April 2, 2002, accessed at www.computerworld.com; Larry Greenemeier, "Security Technology Modeled on Israeli Example," *InformationWeek*, March 11, 2002, accessed at www.informationweek.com.

Larger businesses often have staff who can do most of the data gathering and analysis. They often benefit, though, by seeking an outside perspective during analysis. Smaller businesses almost never have the internal resources to do effective business intelligence. The exception to this rule is when there is a person (often the CEO) who does the business intelligence work because he or she likes it. Smaller businesses and independent business professionals can usually benefit from outside help with information gathering and with analysis and recommendation.

The term **counterintelligence** describes the steps an organization takes to protect information sought by "hostile" intelligence gatherers. One of the most effective counterintelligence measures is to define "trade secret" information relevant to the company and control its dissemination.

Knowledge management is the process of capturing a company's collective expertise wherever it resides—in computers, on paper, or in people's heads—and distributing it wherever it can help produce the biggest payoff. The goal of knowledge management is to get people to record knowledge (as opposed to data) and then share it. Although a variety of technologies can support it, knowledge management is really about changing people's behavior to make their experience and expertise available to others.[15] Knowledge management had its start in large consulting firms and has expanded to nearly every industry. Pharmaceutical companies must have access to various databases from different biotechnology companies to ensure they make informed decisions. Kyowa Pharmaceutical

counterintelligence
the steps an organization takes to protect information sought by "hostile" intelligence gatherers

knowledge management
the process of capturing a company's collective expertise wherever it resides—in computers, on paper, in people's heads—and distributing it wherever it can help produce the biggest payoff

depends on access to multiple databases to facilitate the flow of information throughout the clinical trials of its new products. It uses IBM Life Sciences' DiscoveryLink software to connect information from distributed databases and makes the data available to biotechnology researchers via the Internet.[16]

DISTRIBUTED DATABASES

Distributed processing involves placing processing units at different locations and linking them via telecommunications equipment. A **distributed database**—a database in which the data may be spread across several smaller databases connected via telecommunications devices—works on much the same principle. A user in the Milwaukee branch of a clothing manufacturer, for example, might make a request for data that is physically located at corporate headquarters in Milan, Italy. The user does not have to know where the data is physically stored. The user makes a request for data, and the DBMS determines where the data is physically located and retrieves it (Figure 5.19).

Distributed databases give corporations more flexibility in how databases are organized and used. Local offices can create, manage, and use their own databases, and people at other offices can access and share the data in the local databases. Giving local sites more direct access to frequently used data can improve organizational effectiveness and efficiency significantly.

Despite its advantages, distributed processing creates additional challenges in maintaining data security, accuracy, timeliness, and conformance to standards. Distributed databases allow more users direct access at different sites; thus, controlling who accesses and changes data is sometimes difficult. Also, because distributed

distributed database
a database in which the data may be spread across several smaller databases connected via telecommunications systems

FIGURE 5.19

The Use of a Distributed Database
For a clothing manufacturer, computers may be located at corporate headquarters, in the research and development center, in the warehouse, and in a company-owned retail store. Telecommunications systems link the computers so that users at all locations can access the same distributed database no matter where the data is actually stored.

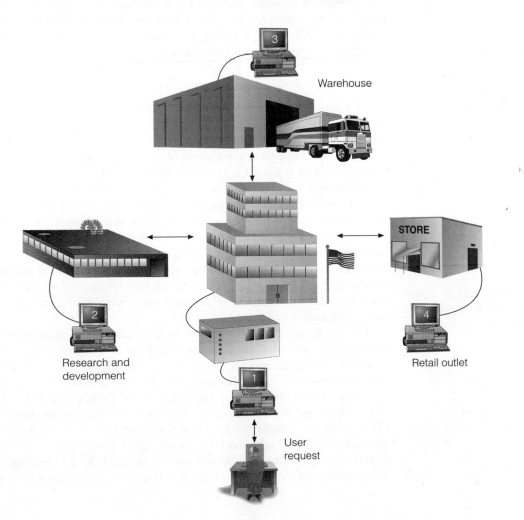

Warehouse

Research and development

Retail outlet

STORE

User request

replicated database
a database that holds a duplicate set of frequently used data

databases rely on telecommunications lines to transport data, access to data can be slower. To reduce telecommunications costs, some organizations build a replicated database. A **replicated database** holds a duplicate set of frequently used data. At the beginning of the day, the company sends a copy of important data to each distributed processing location. At the end of the day, the different sites send the changed data back to update the main database.

Swiss Federal Railways (SBB) carries more than 85 percent of Switzerland's passenger traffic. It uses replicated databases to maintain its reputation for punctuality, safety, and reliability. An Oracle database is stored in Bern, Switzerland, and holds data on SBB's passenger service operation including data on fares, routes, and other essential information. The database is continually replicated and downloaded from the master database to 176 satellite servers across the country. The satellite locations can only read the data. Then they send back the latest figures on ticket sales and reservations for high-speed and InterCity services.[17]

Another challenge created by distributed databases involves integrating the various databases. For example, some organizations use many database management systems. A hierarchical database management system may be used at the headquarters of a large manufacturing company, and different relational DBMSs may be used by the various regional offices. Businesses must develop a solution that enables them to access data in these different DBMSs. IBM, Oracle, and Microsoft all are pursuing approaches that would enable a user to access all of a company's computerized data, no matter where it is stored (file servers, e-mail servers, corporate databases, and employees) with a single query. IBM calls its approach the *federated data concept*, and it relies on special software to seek out the data wherever it might be (including on non-IBM databases) and return the results to the user. Oracle calls for users to combine all data into a few central databases.[18] Microsoft is taking an approach similar to IBM and creating tools and mechanisms that would allow a user to go get the data wherever it might be.[19]

ON-LINE ANALYTICAL PROCESSING (OLAP)

Most industry surveys today show that the majority of data warehouse users rely on spreadsheets, reporting and analysis tools, or their own custom applications to retrieve data from warehouses and format it into business reports and charts. In general, these approaches work fine for questions that can be answered when the amount of data involved is relatively modest and can be accessed with a simple table lookup.

For nearly two decades, multidimensional databases and their analytical information display systems have provided flashy sales presentations and trade show demonstrations. All you have to do is ask where a certain product is selling well, for example, and a colorful table showing sales performance by region, product type, and time frame automatically pops up on the screen. Called **on-line analytical processing (OLAP)**, these programs are now being used to store and deliver data warehouse information. OLAP allows users to explore corporate data from a number of different perspectives.

on-line analytical processing (OLAP)
software that allows users to explore data from a number of different perspectives

OLAP servers and desktop tools support high-speed analysis of data involving complex relationships, such as combinations of a company's products, regions, channels of distribution, reporting units, and time periods. Speed is essential in a booming economy, as businesses grow and accumulate more and more data in their operational systems and data warehouses. Long popular with financial planners, OLAP is now being put in the hands of other professionals. The leading OLAP software vendors include Cognos, Comshare, Hyperion Solutions, Oracle, MineShare, WhiteLight, and Microsoft. Blockbuster Inc. is considering a half-dozen OLAP projects that could save it as much as $30 million in operational costs during the next several years. Blockbuster currently uses Hyperion's Essbase to extract budgeting and planning data from its enterprise systems and analyze how the weather or given movie titles affect sales in its stores. It also uses the software to help in planning how to exploit peak rental times.[20]

Access to data in multidimensional databases can be very quick because they store the data in structures optimized for speed, and they avoid SQL and index processing. But multidimensional databases can take a great deal of time to update; in very large databases, update times can be so great that they force updates to be made only on weekends. Despite this flaw, multidimensional databases have continued to prosper because of their great retrieval speed. Some software providers are attempting to counteract this flaw through the use of partitioning and calculations-on-the-fly capabilities.

Consumer goods companies use OLAP to analyze the millions of consumer purchase records captured by scanners at the checkout stand. This data is used to spot trends in purchases and to relate sales volume to promotions and store conditions, such as displays, and even the weather. OLAP tools let managers analyze business data using multiple dimensions, such as product, geography, time, and salesperson. The data in these dimensions, called *measures*, is generally aggregated— for example, total or average sales in dollars or units, or budget dollars or sales forecast numbers. Rarely is the data studied in its raw, unaggregated form. Each dimension also can contain some hierarchy. For example, in the time dimension, users may examine data by year, by quarter, by month, by week, and even by day. A geographic dimension may compile data from city, state, region, country, and even hemisphere. Read the "IS Principles in Action" special feature to learn more about the use of data warehouses and OLAP tools.

The value of data ultimately lies in the decisions it enables. Powerful information-analysis tools in areas such as OLAP and data mining, when incorporated into a data warehousing architecture, bring market conditions into sharper focus and help organizations deliver greater competitive value. OLAP provides top-down, query-driven data analysis; data mining provides bottom-up, discovery-driven analysis. OLAP requires repetitive testing of user-originated theories; data mining requires no assumptions and instead identifies facts and conclusions based on patterns discovered. OLAP, or multidimensional analysis, requires a great deal of human ingenuity and interaction with the database to find information in the database. A user of a data mining tool does not need to figure out what questions to ask; instead, the approach is, "Here's the data, tell me what interesting patterns emerge." For example, a data mining tool in a credit card company's customer database can construct a profile of fraudulent activity from historical information. Then, this profile can be applied to all incoming transaction data to identify and stop fraudulent behavior, which may otherwise go undetected.

Table 5.5 compares the OLAP and data mining approaches to data analysis.

TABLE 5.5

Comparison of OLAP and Data Mining

Characteristic	OLAP	Data mining
Purpose	Supports data analysis and decision making	Supports data analysis and decision making
Type of analysis supported	Top-down, query-driven data analysis	Bottom-up, discovery-driven data analysis
Skills required of user	Must be very knowledgeable of the data and its business context	Must trust in data mining tools to uncover valid and worthwhile hypothesis

Sears Competes Using Data Warehouse and OLAP Tools

Sears, Roebuck and Co. is completing a project that will allow it to combine information about customer buying trends with inventory and sales data. As part of the project, Sears is buying more EMC Symmetrix disk arrays, a new WorldMark Unix server, and a Teradata database from NCR. The additional hardware and data storage devices will allow Sears to combine its inventory and sales data warehouse with a customer data warehouse onto a single server and provide access to 140 TB of data. Sears is also working with EMC to build a 25 TB storage area network (SAN) to link additional data from Unix and Windows NT servers for use in product assortment planning, human resources, and enterprise resource planning. The new hardware costs in the neighborhood of $15 million.

The recent downturn in retailing as well as competitive pricing and merchandising have made Sears recognize that it must use the huge amounts of data it gathers more wisely. Consolidating its inventory and sales data with its customer information data will enable Sears's employees to analyze customer buying habits, inventory levels, and sales data. The goal is to gain a better understanding of customer shopping habits through market-basket analysis, thus improving its ability to market goods and control inventory. To remain competitive, Sears must ensure that its customers find the merchandise and service they want in its stores, while eliminating what they don't want—faster than the competition.

Prior to the data consolidation, Sears could track sales down to basic store levels, such as how many size 36 walking shorts it sold in any of its stores on a particular day. But the $41 billion retailer couldn't link the data to customer purchases—to see whether any of the buyers of those shorts also bought a swimsuit that day, indicating that they might be taking a vacation. The consolidated data will support such analysis, and with that kind of knowledge, Sears can target its promotional mailings on other products, such as offering discounts on golf shirts to those it thinks are planning vacations. Sears sees many other opportunities arising from the merger of its inventory and sales data with customer data. Executives recently used the data to determine that Sears wasn't making money on cosmetics and bicycles, two product lines it has decided to drop.

Sears uses online analytical processing (OLAP) software to run complex queries, create different scenarios, and detect patterns in sales and financial data. Hyperion's Essbase OLAP software is used by workers at headquarters to manipulate financial data about store profitability and costs. The data can be analyzed many different ways using the OLAP software—by geography, lines of business, store size, and other factors. Sears eventually will provide store managers the ability to access the data using the Essbase software. The holdup is a lack of training and cultural resistance—many store managers prefer to remain on the store floor rather than crunch numbers on a computer.

Discussion Questions

1. How can analyzing sales by customer purchases help Sears improve its ability to market goods and control inventory?
2. Identify three additional examples of how the new sales data could be used in decision making.

Critical Thinking Questions

3. What can be done to overcome store managers' resistance to the use of data and analytical tools? Is their resistance to analysis necessarily a bad thing?
4. How might Sears try to quantify the benefits derived from the use of the consolidated data and analysis tools?

Sources: Adapted from Lucas Mearian, "Sears Triples Its Storage Capacity," *Computerworld*, January 28, 2002, accessed at http://www.computerworld.com; Lucas Mearian, "Sears to Build Huge Storage Network for CRM," *Computerworld*, January 24, 2002, accessed at http://www.computerworld.com; Marc L. Songini, "Firms Face Barriers in Push for Data Analysis," *Computerworld*, April 30, 2001, accessed at http://www.computerworld.com; Rick Whiting, "Tower of Power," *InformationWeek*, February 11, 2002, accessed at http://www.informationweek.com.

OPEN DATABASE CONNECTIVITY (ODBC)

open database connectivity (ODBC)
standards that ensure that software can be used with any ODBC-compliant database

To help with database integration, many companies rely on **open database connectivity (ODBC)** standards. ODBC standards help ensure that software can be used with any ODBC-compliant database, making it easier to transfer and access data among different databases. For example, a manager might want to take several tables from one database and incorporate them into another database that uses a different database management system. In another case, a manager might want to transfer one or more database tables into a spreadsheet program. If all this software meets ODBC standards, the data can be imported, exported, or linked to other applications (see Figure 5.20). For example, a table in an Access database can be exported to a Paradox database or a spreadsheet. Tables and data can also be imported using ODBC. For example, a table in a Paradox database or an Excel spreadsheet can be imported into an Access database. Linking allows an application to use data or an object stored in another application without actually importing the data or object into the application. The Access database, for example, can link to a table in the Lotus 1-2-3 spreadsheet or the FileMaker Pro database. Applications that follow the ODBC standard can use these powerful ODBC features to share data between different applications stored in different formats.

ODBC-compliant products do suffer from their all-purpose nature, however. Their overall performance is usually less efficient than that of products designed for use with a specific database. Yet, more and more vendors are building ODBC-compliant products as businesses increasingly use distributed databases. Many organizations are using such tools to allow their workers and managers easier access to a variety of databases and data sources. ODBC standards also make it easier for growing companies to integrate existing databases, to connect more users into the same database, and to move application programs from PC-oriented databases to larger, workstation-based databases, and vice versa.

OBJECT-RELATIONAL DATABASE MANAGEMENT SYSTEMS

Many of today's newer application programs require the ability to manipulate audio, video, and graphical data. Conventional database management systems are not well suited for this, because these types of data cannot easily be stored in rows or tables. Manipulation of such data requires extensive programming so that the DBMS can translate data relationships. An **object-relational database management system (ORDBMS)** provides a complete set of relational database capabilities plus the ability for third parties to add new data types and operations to the database. These new data types can be audio, images, unstructured text, spatial, or time series data that require new indexing, optimization, and retrieval features.

In such a database, these types of data are stored as objects, which contain both the data and the processing instructions needed to complete the database transaction. The objects can be retrieved and related by an ORDBMS. Businesses can then mix and match these elements in their daily search for clues and information.

object-relational database management system (ORDBMS)
a DBMS capable of manipulating audio, video, and graphical data

FIGURE 5.20

Advantages of ODBC
ODBC can be used to export, import, or link tables between different applications.

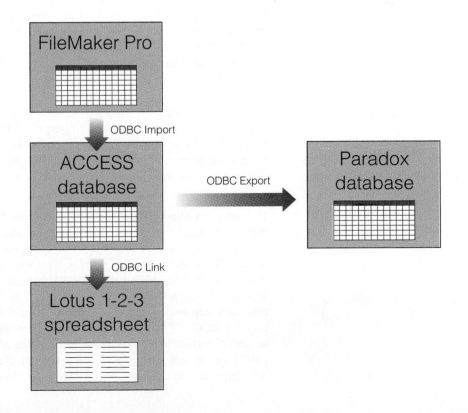

For example, by clicking on a picture of a red Corvette, a market analyst at General Motors might be able to call up a profile of red Corvette buyers. If he wants to break that up by geographic region, he might click on a portion of a map. He might also view a GM sales-training film to see whether the sales pitch is appropriate, given the most recent market trends. Chances are, the analyst will do all this using an Internet site tied into a database. As another example, MasterCard is interested in object-oriented technology to combine transactional data with cardholder fingerprints to prevent fraud.

Each of the vendors offering ORDBMS facilities provides a set of application programming interfaces to allow users to attach external data definitions and methods associated with those definitions into the database system. They are essentially offering a standard socket into which users can plug special instructions. DataBlades, Cartridges, and Extenders are the names applied by Oracle and IBM to describe the plug-ins to their respective products. Other plug-ins serve as interfaces to Web servers.

Web-based applications increasingly require complex object support to link graphical and other media components back to the database. These systems make sense for developers of systems that are highly dependent on complex data types, particularly Web and multimedia applications. Because it supports so many applications, an ORDBMS is also called a universal database server.

An increasing amount of data that organizations use is in the form of images, which can be stored in object-relational databases. Credit card companies, for example, input pictures of charge slips into an image database using a scanner. The images can be stored in the database and later sorted by customer, printed, and sent to customers along with their monthly statements. Image databases are also used by physicians to store X rays and transmit them to clinics away from the main hospital. Financial services, insurance companies, and government branches are also using image databases to store vital records and replace paper documents.

Image data has some disadvantages, one of which is the increased secondary storage requirements. It is also sometimes difficult to locate the desired data in the database. Retrieval of image data from databases can be made easier by using the object-oriented approach. Other ways to index and cross-reference data are being developed.

TimeWare, Inc., provides hardware and software products to meet the labor management needs for many industries. Its flagship product, TimeWare RMS, includes a software module to capture time and attendance data using a "one touch" single-step process using its BioScreen terminal. The BioScreen terminal is a computerized employee self-service station that provides accurate attendance information through biometric, fingerprint identification, and touch screen technology. The TimeWare RMS software uses the ObjectStore object relational database management system from eXcelon to store and retrieve the biometric data-fingerprint images.[21]

Hypermedia

Hypermedia allows businesses to search and manipulate multimedia forms of data—graphics, sound, video, and alphanumeric data. A marketing manager, for example, might store notes about the competition and new trends in the marketplace. These notes could include written material on new markets, product images, advertising brochures, and TV commercials used by competitors. Using a hypermedia database management system, the marketing manager could organize this data into nodes and define the relationships among them. For example, she could link all TV commercials and written brochures about new products from several competitors. With the hypermedia database approach, many types of data can be organized into a web of nodes connected by links established by the user. The hypermedia database approach is also used to enable surfers to access topics

Spatial data technology is used by NASA to store data from satellites and earth stations. Location-specific information can be accessed and compared.

(Source: Courtesy of NASA.)

on a Web site in any order that makes sense to them. The hypertext links from the on-screen page to other related Web pages are highlighted in color. Putting the cursor on a hypertext link and clicking takes the user directly to this new page.

Spatial Data Technology

Spatial data technology involves the use of an object-relational database to store and access data according to the locations it describes and to permit spatial queries and analysis. MapExtreme is spatial technology software from MapInfo that extends a user's database so it can store, manage, and manipulate location-based data. New York City police use the software to bring together crime data and map the data visually so that patterns are easier to analyze. Commanders can select and work with spatial data at a specified location, within a rectangle, a given radius, or a polygon such as a precinct. For example, a commander can request a list of all liquor stores within a two-mile radius of the precinct.[22]

Builders and insurance companies use spatial data to make decisions related to natural hazards. Spatial data can even be used to improve financial risk management with information stored by investment type, currency type, interest rates, and time.

SUMMARY

PRINCIPLE *The database approach to data management provides significant advantages over the traditional file-based approach.*

Data is one of the most valuable resources a firm possesses. It is organized into a hierarchy that builds from the smallest element to the largest. The smallest element is the bit, a binary digit. A byte (a character such as a letter or numeric digit) is made up of eight bits. A group of characters, such as a name or number, is called a *field* (an object). A collection of related fields is a *record*; a collection of related records is called a *file*. The database, at the top of the hierarchy, is an integrated collection of records and files.

An entity is a generalized class of objects for which data is collected, stored, and maintained. An attribute is a characteristic of an entity. Specific values of attributes—called *data items*—can be found in the fields of the record describing an entity. A data key is a field within a record that is used to identify the record. A primary key uniquely identifies a record, while a secondary key is a field in a record that does not uniquely identify the record.

The traditional approach to data management has been from a file perspective. Separate files are created for each application. This approach can create problems over time: as more files are created for new applications,

data that is common to the individual files becomes redundant. Also, if data is changed in one file, those changes might not be made to other files, reducing data integrity.

Traditional file-oriented applications are often characterized by program-data dependence, meaning that they have data organized in a manner that cannot be read by other programs. To address problems of traditional file-based data management, the database approach was developed. Benefits of this approach include reduced data redundancy, improved data consistency and integrity, easier modification and updating, data and program independence, standardization of data access, and more efficient program development.

Potential disadvantages of the database approach include the relatively high cost of purchasing and operating a DBMS in a mainframe operating environment, specialized staff required to implement and coordinate the use of the database, and increased vulnerability if security is breached and there is a failure in the DBMS.

When building a database, careful consideration must be given to content and access, logical structure, and physical organization. One of the tools database designers use to show the relationships among data is a data model. A data model is a map or diagram of entities and their relationships. Enterprise data modeling involves analyzing the data and information needs of the entire organization. Entity-relationship (ER) diagrams can be employed to show the relationships between entities in the organization. Entities may have a one-to-one (1:1), one-to-many (1:N), or many-to-one relationship (N:1).

Databases typically use one of three common models: hierarchical (tree), network, and relational. The hierarchical model has one main record type at the top, with subordinate records below. The network model, an expansion of the hierarchical structure, involves an owner-member relationship in which each member may have more than one owner. The newest, most flexible structure is the relational model. Instead of a hierarchy of predefined relationships, data is set up in two-dimensional tables. Tables can be linked by common data elements, which are used to access data when the database is queried. Each row represents a record. Columns of the tables are called attributes, and allowable values for these attributes are called the domain. Basic data manipulations include selecting, projecting, and joining.

The relational model, the most widely used database model, is easier to control, more flexible, and more intuitive than the other models because it organizes data in tables. The process of taking a complex set of data and converting it into a set of simple two-dimensional tables is called data normalization.

PRINCIPLE   *A well-designed and well-managed database is an extremely valuable tool in supporting decision making.*

A DBMS is a group of programs used as an interface between a database and application programs. When an application program requests data from the database, it follows a logical access path. The actual retrieval of the data follows a physical access path. Records can be considered in the same way: a logical record is what the record contains; a physical record is where the record is stored on storage devices. Schemas are used to describe the entire database, its record types, and their relationships to the DBMS.

A database management system provides four basic functions: providing user views, creating and modifying the database, storing and retrieving data, and manipulating data and generating reports.

Subschemas are used to define a user view, the portion of the database a user can access and manipulate. Schemas and subschemas are entered into the computer via a data definition language, which describes the data and relationships in a specific database. Another tool used in database management is the data dictionary, which contains detailed descriptions of all data in the database.

Once a DBMS has been installed, the database may be accessed, modified, and queried via a data manipulation language. A more specialized data manipulation language is the query language, the most common being Structured Query Language (SQL). SQL is used in several popular database packages today and can be installed on PCs and mainframes.

A database administrator (DBA) is a highly skilled and trained systems professional who directs or performs all activities related to maintaining a successful database environment. The DBA's responsibilities include designing, implementing, and maintaining the database system; establishing policies and procedures pertaining to the management, security, maintenance, and use of the database management system; and training employees in database management and use. The data administrator is a nontechnical, but important role that ensures that data is managed as an important organizational resource.

Popular end-user DBMSs include Corel's Paradox, FileMaker's FileMaker Pro, Microsoft's Access, and Lotus's Approach. These DBMSs provide "wizards" to help end users create a new database and load data and to perform many other functions. IBM, Oracle, and Microsoft are the leading DBMS vendors.

Selecting a database management system begins by analyzing the information needs of the organization. Important characteristics of databases include the size of the database, number of concurrent users, performance, the ability of the DBMS to integrate with other systems, the features of the DBMS, vendor considerations, and the cost of the database management system.

PRINCIPLE   *The number and types of database applications will continue to evolve and yield real business benefits.*

Customers, suppliers, and company employees must be able to access corporate databases through the Internet, intranets, and extranets to meet various business needs. As a result, organizations are adopting the use of application servers—software packages, often written in the Java programming language for use on computers running the Windows NT operating system, that connect end users to the databases holding the information they need to access. Application servers manage the process of connecting users to that data by setting up an application session for each user, checking each user's identification and password, fetching requested information from the appropriate database, and building the data into a Web page for display to the users.

Traditional on-line transaction processing (OLTP) systems put data into databases very quickly, reliably, and efficiently, but they do not support the types of data analysis needed today. So, organizations are building data warehouses, which are relational database management systems specifically designed to support management decision making. Data marts are subdivisions of data warehouses, which are commonly devoted to specific purposes or functional business areas.

Data mining, which is the automated discovery of patterns and relationships in a data warehouse, is emerging as a practical approach to generate a hypothesis about the patterns and anomalies in the data that can be used to predict future behavior.

Predictive analysis is a form of data mining that combines historical data with assumptions about future conditions to predict outcomes of events such as future product sales or the probability that a customer will default on a loan.

Business intelligence is the process of getting enough of the right information in a timely manner and usable form and analyzing it so that it can have a positive impact on business strategy, tactics, or operations. Competitive intelligence is one aspect of business intelligence limited to information about competitors and the ways that information affects strategy, tactics, and operations. Competitive intelligence is not espionage—the use of illegal means to gather information. Counterintelligence describes the steps an organization takes to protect information sought by "hostile" intelligence gatherers. Knowledge management is the process of capturing a company's collective expertise wherever it resides—in computers, on paper, or in people's heads—and distributing it wherever it can help produce the biggest payoff. The goal of knowledge management is to get people to record knowledge (as opposed to data) and then share it.

With the increased use of telecommunications and networks, distributed databases, which allow multiple users and different sites access to data that may be stored in different physical locations, are gaining in popularity. To reduce telecommunications costs, some organizations build replicated databases, which hold a duplicate set of frequently used data.

Multidimensional databases and on-line analytical processing (OLAP) programs are being used to store data and allow users to explore the data from a number of different perspectives. Open database connectivity (ODBC) standards allow different database applications to share information.

An object-relational database management system (ORDBMS) provides a complete set of relational database capabilities, plus the ability for third parties to add new data types and operations to the database. These new data types can be audio, video, and graphical data that require new indexing, optimization, and retrieval features.

Hypermedia allows organizations to access and manipulate all types of data. Spatial data technology involves the use of an object-relational database to store and access data according to the locations it describes and to permit spatial queries and analysis.

⊙

CHAPTER 5 SELF-ASSESSMENT TEST

The database approach to data management provides significant advantages over the traditional file-based approach.

1. _____ is a group of programs that manipulate the database and provide an interface between the database and the user of the database and other application programs.
   A. GUI
   B. operating system
   C. DBMS
   D. productivity software

2. A characteristic of an entity is called a(n) _____.

3. A primary key is a field or set of fields that uniquely identifies the record. True False

4. The duplication of data in separate files is known as
   A. data redundancy
   B. data integrity
   C. data relationships
   D. data entities

5. _____ is a data modeling approach that starts by investigating the general data and information needs of the organization at the strategic level and then examining more specific data and information needs for the various functional areas and departments within the organization.

6. The structure of the relationships in most databases follows one of three logical database models:
   A. hierarchical, network, and relational
   B. enterprise, departmental, distributed
   C. normalized, unnormalized, standard
   D. tactical, strategic, global

A well-designed and well-managed database is an extremely valuable tool in supporting decision making.

7. The process of taking a complex set of data and converting it into a set of simple two-dimensional tables is called _____.

8. Once data has been placed into a relational database, users can make inquiries and analyze data. Basic data manipulations include selecting, projecting, and normalization. True False

9. Because the DBMS is responsible for access to a database, one of the first steps in installing and using a database involves telling the DBMS the logical and physical structure of the data and relationships among the data in the database. This description is called a(n) _____.

10. The commands that are used to access and report information from the database are part of the:
   A. data definition language
   B. data manipulation language
   C. data normalization process
   D. subschema

11. Three popular DBMSs for end users are
   _____, _____,
   _____.

12. The ability of a vendor to provide global support for large, multinational companies or companies outside the U.S. is becoming increasingly important. True False

The number and types of database applications will continue to evolve and yield real business benefits.

13. A(n) _____ holds business information from many sources in the enterprise, covering all aspects of the company's processes, products, and customers.

14. An information analysis tool that involves the automated discovery of patterns and relationships in a data warehouse is called
   A. data mart
   B. data mining
   C. predictive analysis
   D. business intelligence

15. _____ is a continuous process involving the legal and ethical collection of information, analysis that doesn't avoid unwelcome conclusions, and controlled dissemination of that information to decision makers.

Chapter 5 Self-Assessment Test Answers

1. c; 2. attribute; 3. True; 4. a; 5. Enterprise data modeling; 6. a; 7. data normalization; 8. False; 9. schema; 10. b; 11. any of three of the following: Corel's Paradox, FileMaker's FileMaker Pro, Microsoft's Access, and Lotus's Approach; 12. True; 13. data warehouse; 14. b; 15. Competitive intelligence

KEY TERMS

application servers, 202
attribute, 181
business intelligence, 206
character, 180
competitive intelligence, 207
concurrency control, 196
counterintelligence, 208
data administrator, 199
data cleanup, 191
data definition language (DDL), 193
data dictionary, 194
data integrity, 182
data item, 181
data manipulation language (DML), 197
data mart, 204
data mining, 205
data model, 187
data normalization, 189
data redundancy, 182
data warehouse, 203
database administrator (DBA), 198

database approach to data management, 183
database management system (DBMS), 180
distributed database, 209
domain, 189
enterprise data modeling, 187
entity, 181
entity-relationship (ER) diagrams, 187
field, 180
file, 181
hierarchical database model, 187
hierarchy of data, 181
joining, 189
key, 181
knowledge management, 208
linking, 189
network model, 188
object-relational database management system (ORDBMS), 213

on-line analytical processing (OLAP), 210
open database connectivity (ODBC), 212
planned data redundancy, 187
predictive analysis, 206
primary key, 181
program-data dependence, 183
projecting, 189
record, 180
relational model, 188
replicated database, 210
schema, 193
selecting, 189
subschema, 193
traditional approach to data management, 182

REVIEW QUESTIONS

1. What is an attribute? How is it related to an entity?
2. Define the term *database*. How is it different from a database management system?
3. What is the purpose of data cleanup?
4. How would you describe the traditional approach to data management? How does it differ from the database approach?
5. What is data modeling? What is its purpose? Briefly describe three commonly used data models.
6. What is a database schema and what is its purpose?
7. Identify important characteristics in selecting a database management system.
8. What is the difference between a data definition language (DDL) and a data manipulation language (DML)?

9. What is a distributed database system?
10. What advantages does the open database connectivity (ODBC) standard offer?
11. What is a data warehouse, and how is it different from a traditional database used to support OLTP?
12. What is data mining? What is OLAP? How are they different?
13. What is an ORDBMS? What kind of data can it handle?
14. What is business intelligence? How is it used?
15. What is spatial data technology? How might it be used?
16. What is predictive analysis? How might it be used?

DISCUSSION QUESTIONS

1. You have been selected to represent the student body on a project to develop a new student database for your school. What actions might you take to fulfill this responsibility to ensure that the project met the needs of students and was successful?
2. Your company is releasing a major new product. To announce the product, you, the advertising manager in charge of the new product launch, need to develop advertising and other promotions. What counterintelligence initiatives might you undertake?
3. What is a data model, and what is data modeling? Why is data modeling an important part of strategic planning?
4. You are going to design a database for your cooking club to track its recipes. Identify the database characteristics most important to you in choosing a DBMS. Which of the database management systems described in this chapter would you choose? Why? Is it important for you to know what sort of computer the database will run on? Why or why not?
5. Distinguish OLAP from predictive analysis. Under what conditions would you use one technique over the other?

6. Make a list of the databases in which data about you exists. How is the data in each database captured? Who updates each database and how often? Is it possible for you to request a printout of the contents of your data record from each database? What data privacy concerns do you have?
7. You are the vice president of information technology for a large, multinational, consumer packaged goods company (e.g., Procter & Gamble, Unilever, or Gillette). You must make a presentation to persuade the board of directors to invest $5 million to establish a competitive intelligence organization—including people, data gathering services, and software tools. What key points do you need to make in favor of this investment? What arguments can you anticipate others might make?
8. Briefly discuss what impact data privacy legislation may have on the building and use of customer and employee data warehouses.

PROBLEM-SOLVING EXERCISES

1. Develop a simple data model for an end-user database management system to record your personal items and home/apartment furnishings so that you have a log of all valuable items for insurance purposes in case of theft, fire, or natural disaster. For each item, what attributes should you capture? What will be the unique key for the records in your database?

2. A video movie rental store is using a relational database to store information on movie rentals to answer customer questions. Each entry in the database contains the following items: Movie ID No. (primary key), Movie Title, Year Made, Movie Type, MPAA Rating, Number of Copies on Hand, and Quantity Owned. Movie types are comedy, family, drama, horror, science fiction, and western. MPAA ratings are G, PG, PG-13, R, X, and NR (not rated). Use an end-user database management system to build a data entry screen to enter this data. Build a small database with at least 10 entries.

3. To improve service to their customers, the salespeople at the video rental store have proposed a list of changes being considered for the database in the previous exercise. From this list, choose two database modifications and modify the data entry screen to capture and store this new information. Proposed changes:
A. Add the date that the movie was first available to help locate the newest releases.
B. Add the director's name.
C. Add the names of three primary actors in the movie.
D. Add a rating of one, two, three, or four stars.
E. Add the number of Academy Award nominations.

TEAM ACTIVITIES

1. In a group of three or four classmates, do research to identify the largest existing data warehouse. Write a brief paragraph about the data warehouse and how it is used. Try to identify any unique problems due to the size of the data warehouse.

2. As a team of three or four classmates, interview business managers from three different businesses that use databases to help them in their work. What data entities and data attributes are contained in each database? How do they access the database to perform analysis? Have they received training in any query or reporting tools? What do they like about their database and what could be improved? Do any of them use data mining or OLAP techniques? Weighing the information obtained, select one of these databases as being most strategic for the firm and briefly present your selection and the rationale for the selection to the class.

3. Imagine that you and your classmates are a research team developing an improved process for evaluating auto loan applicants. The goal of the research is to predict which applicants will become delinquent or forfeit their loan. Those who score well on the application will be accepted; those who score exceptionally well will be considered for lower-rate loans. Prepare a brief report for your instructor addressing these questions:
A. What data do you need for each loan applicant?
B. What data might you need that is not typically requested on a loan application form?
C. From where might you get this data?
Take a first cut at designing a database for this application. Using the chapter material on designing a database, show the logical structure of the relational tables for this proposed database. In your design, include the data attributes you believe are necessary for this database, and show the primary keys in your tables. Keep the size of the fields and tables as small as possible to minimize required disk drive storage space. Fill in the database tables with the sample data for demonstration purposes (10 records). Once your design is complete, implement it using a relational DBMS.

WEB EXERCISES

1. Use a Web search engine to find information on one of the following topics: business intelligence, knowledge management, predictive analysis. Find a definition of the term, an example of a company using the technology, and three companies that provide such software. Cut graphics and text material from the Web pages and paste them into a word processing document to create a two-page report on your selected topic. At the home page of each software company, request further information from the company about its products.

2. Use a Web search engine to find three companies that provide competitive intelligence services. How are the services that they provide similar? How are they different? Which companies seem to be the most ethical? Why?

CASES

CASE 1

VNU, International Media and Information Company

VNU NV in 1994 was a Netherlands-based owner of Dutch newspapers and television stations and European magazines. Over the next seven years, it spent a total of $8 billion acquiring other companies to transform itself into a leading provider of information about media and consumers for use by media companies and advertisers. Today VNU owns 145 trade newspapers and magazines; Soundscan, which tracks retail sales of recorded music; Nielsen Media Research, which rates television shows; and ACNielsen, which compiles market data on the sales of packaged goods. With operations in over 100 countries and employing 35,000 people, VNU provides millions of people around the world with business intelligence.

Its marketing information activities are built on the data collected, edited, cleaned, and stored by its ACNielsen subsidiary. The basic data collection process begins with in-store scanning of product codes at the checkout and observations of professional auditors during store visits. This data is gathered from retail outlets wherever food, household, health and beauty, durables, confectionery, and beverage products are sold. The data can be used to track sales volume, selling price, and effectiveness of promotions and merchandising—not just of your own company's products, but of competitors as well. The data can also be used to measure overall product performance, assess the extent of product distribution in area stores, quantify the effectiveness of special promotions such as sales or advertisements, and measure consumers' reactions to changes in price.

Recognizing the need to support global decision making, ACNielsen introduced an Internet service to deliver its information products in 1996. This expansion on-line allows people within client organizations to obtain ACNielsen information, regardless of their location.

In addition to its information products, ACNielsen markets a broad range of advanced decision support software that helps more than 9,000 clients spread across 100 countries to obtain large volumes of information, evaluate it, make judgments about their growth opportunities, and plan future marketing and sales campaigns. The software includes tools that perform sophisticated multidimensional reporting, data mining, analytical modeling, graphical presentations, and expert systems. The goal is to enable decision makers to make sound recommendations to optimize their business: price levels, promotion methods, media spending, product portfolio mix, retail category optimization, and other choices. If made correctly, these key decisions can generate increased sales and profits.

Discussion Questions

1. What issues might be involved in collecting, editing, cleaning, and storing the ACNielsen data? Identify some specific issues related to data privacy.
2. What concerns might be associated with providing customers around the world with access to the ACNielsen data via the Internet?

Critical Thinking Questions

3. What ethical issues might arise for VNU in providing marketing data, tools, and services to two major multinational companies that are competitors (e.g., Procter & Gamble and Unilever)? How might these issues be minimized?
4. Which do you think is more important to VNU customers—good data or good tools for the analysis of data? Why?

Sources: Adapted from Geraldine Fabrikant, "MEDIA; Big Makeover for Dutch Media Data Company," *The New York Times*, August 13, 2001, Late Edition-Final, Section C, p. 9; "About VNU," VNU Web site at http://www.vnu.com, accessed on February 25, 2002; "Sybase Solution Analyzes Television Viewing Trends for Nielsen Research," Sybase press release, October 22, 2001.

CASE 2

Wells Fargo Bank Uses Predictive Analysis to Improve Profits

Banks are major investors in information systems. In fact, the largest banks spend 20 to 25 percent of their overall budgets in this area. Much of this spending is driven by the ongoing consolidation of banks and the intense pressures to remain profitable. One area of spending is for predictive analysis to improve profitability and increase revenue.

Wells Fargo Home Mortgage is a subsidiary of Wells Fargo & Co. in San Francisco. The company is one of the top providers of home loans in the United States, handling 1 of every 12 U.S. mortgages. Wells Fargo built a data warehouse that contains more than 200 million payment records on 12 million borrowers and 5 million home loans. It uses this data and predictive analysis software to forecast the performance of its mortgage loan portfolio and individual mortgage loans. The

goal is to manage the risk of loan defaults. Too tight a credit policy means the bank loses revenue by refusing loans that would be paid back. Too loose a credit policy means the bank loses money on defaulted loans.

A credit-scoring model helps loan officers evaluate new applications and predict the likelihood of default. Completing such an analysis gives loan officers a better sense of the potential risk up front so that they can price the loan accordingly. Other models predict the performance of the entire loan portfolio to help managers assess the potential loss from bad loans. Loan portfolio analysis looks at the payment history of millions of customers over two or three years—often examining between 40 million and 50 million payment records, plus quarterly credit reports for each loan. With this knowledge, the company can then price loans appropriately and market various loans to different segments of consumers. Such thorough analysis of its home mortgage portfolio has helped Wells Fargo extend credit to customers who might not otherwise be eligible for a home loan.

Consolidating data from multiple sources to build a single, companywide data warehouse has helped Wells Fargo ensure accurate and consistent results. This increased reliability has led to savings in human and computer resources by cutting the turnaround time on data requests from many days to just a few hours or minutes.

Wells Fargo used the data warehouse and predictive analysis tools to carefully analyze loan performance over time. What they learned was that their loan default rates were half as much as rating services such as Moody's and Standard & Poor's had predicted. Based on the new, lower default rate, Wells Fargo was able to renegotiate the loans and assume more risk, thus saving $250,000 per month in interest expenses.

The system also helped Wells Fargo enter into new partnerships with mortgage insurance companies and generate more than $30 million a year in additional revenues. Without the system, Wells Fargo would not be able to meet stringent legal reporting requirements for these partnerships.

Discussion Questions

1. What benefits has Wells Fargo achieved through the use of a data warehouse and predictive analysis tools?

2. Why is it important to have multiple years of consolidated data for all types of customers when performing predictive analyses? Why wouldn't the analysis be as worthwhile if it were done with a representative subset of the data?

Critical Thinking Questions

3. How important is data completeness and quality to the predictive analysis process? What actions might Wells Fargo take to ensure the accuracy and completeness of its data?

4. What other predictive analyses or models would be useful to Wells Fargo besides those already mentioned?

Sources: Adapted from "Success Stories—Wells Fargo Manages Credit Risk of 12 Million Borrowers with SAS," from SAS Institute Web site at http://www.sas.com, accessed February 27, 2002; Lucas Mearian, "Study: IT Spending at Banks to Increase in 2002," *Computerworld*, January 14, 2002, http://www.computerworld.com; Rick Whiting, "Companies Boost Sales Efforts with Predictive Analysis," *InformationWeek*, February 25, 2002, http://www.informationweek.com.

CASE 3

J. Crew Turns to Data Warehousing to Increase On-line Sales

J. Crew is a global retailer and catalog merchant of prestige fashions with headquarters in New York City. It offers a wide range of men's, women's, and children's apparel, shoes, accessories, and personal care products through its fast-expanding retail network of 133 U.S. stores and 76 licensed stores in Japan.

When J. Crew set up its jcrew.com site in June 1997, it was one of the first apparel sites on the Web. Today the Web site is J. Crew's fastest-growing distribution channel. Part of the reason for its huge success is that J. Crew uses a data warehouse and software tools to identify for on-line shoppers what J. Crew clothes, shoes, and accessories customers frequently purchase together. That information is fed to applications running the Web site so that when on-line shoppers click on an item, the Web site recommends complementary products that the customer might be interested in buying. Delivering dynamic, relevant product recommendations to shoppers has increased the average order size and raised customer satisfaction and loyalty.

J. Crew uses DigiMine's Enterprise Analytics data-mining software to analyze sales data from its Web site, retail stores, and catalog sales operation. All of this data is collected and stored in a 500-GB data warehouse running on a Microsoft SQL Server database, which took J. Crew and DigiMine six months to develop. J. Crew combines the data generated by visitors clicking on its Web pages with product sales data from corporate systems that process order data from catalog and retail operations. The consolidation of all this data gives J. Crew a complete view of its customers' preferences and enables it to analyze sales trends, build customer profiles, and generate product recommendations for e-mail marketing campaigns. It also uses this pooled data to pair products and advise shoppers which shoes customers most often buy with which slacks. As a result, every shopper at jcrew.com can view compelling apparel, shoe, or accessory suggestions, based on their browsing and purchasing behavior.

Discussion Questions

1. Visit the J. Crew Web site at http://www.jcrew.com and shop for some apparel. Be alert for the site's attempts to increase your order size through recommendations of complementary apparel, shoe, or accessories.

2. In serving their on-line shoppers, most retailers do not combine data from the brick-and-mortar stores and catalog purchases with browsing and shopping data from their Web sites. J. Crew does. How does this provide J. Crew a competitive advantage?

Critical Thinking Questions

3. Identify some potential data privacy issues that might arise with the capture, storage, and analysis of customers' on-line shopping data from the J. Crew Web site.

4. Imagine that J. Crew is considering offering a new service for on-line customers—help in shopping for gifts for friends and family members. Interested customers would register friends and family members, along with their birth dates, anniversaries, and other pertinent information such as sizes and color preferences. J. Crew would send an e-mail reminding the customer that a friend's birth date or other special occasion was near and suggest that they visit the Web site for help in choosing a gift. What are the pros and cons of such a service? What additional information would need to be captured in the data warehouse to support this service? Would you be in favor of this service? Why or why not?

Sources: Adapted from Ann Bednarz, "Cents and Retail Sensibility," *Computerworld*, January 7, 2002, http://www.computerworld.com; Mark Hall, "Finding Answers in Data Haystacks," *Computerworld*, April 23, 2001, http://www.computerworld.com; Rick Whiting, "Retailer Seeks Success in How Customers Dress," *InformationWeek*, November 26, 2001, http://www.informationweek.com; "J. Crew Deploys digiMine's Data Mining Solutions," digimine company press release, November 13, 2001, http://www.digimine.com; J. Crew Web site at http://www.jcrew.com.

NOTES

Sources for the opening vignette on p.179: Adapted from "Finland's Largest Dairy Company Farms Profitable Pastures," *Success Stories,* accessed at Compaq Web site, http://www.compaq.com, February 20, 2002; "Company and Functional Products," Valio Web site, http://valio.com, accessed February 20, 2002; "Carton Finds Its Thrills … With a Blueberry-Flavored Milk Drink," *Packaging Digest,* April 2001, accessed at http://www.findarticles.com; and "Valio's Profit Surges in First Eight Months," *Eurofood,* November 8, 2001, accessed at http://www.findarticles.com.

1. Jennifer Maselli, "Swiss Hopes to Reach New Heights," *InformationWeek*, June 24, 2002, accessed at *www.informationweek.com.*
2. Kelli Wiseth, "Find Meaning," *Oracle Magazine*, September 2001, accessed at http://www.oracle.com.
3. Lucas Mearian and Linda Rosencrance, "Police Pleased with Data Mining Engines," *Computerworld*, April 2, 2001, accessed at http://www.computerworld.com.
4. Jennifer DiSabatino, "Galileo Moves Fare Pricing onto Unix-Based System," *Computerworld*, March 1, 2002, accessed at http://www.computerworld.com.
5. Peter Wayner, "Open Source Databases Bloom," *Computerworld*, September 10, 2001, accessed at *www.computerworld.com.*
6. Rick Whiting, "Patient Care Is Goal of Super-Size Database," *InformationWeek*, April 1, 2002, accessed at *www.informationweek.com.*
7. Rick Whiting, "Database Management Made Manageable," *InformationWeek*, July 29, 2002, accessed at *www.informationweek.com.*
8. Joris Evers, "Oracle Wants Users to Hand Over App Management," *Computerworld*, March 22, 2002, accessed at *www.computerworld.com.*
9. Rick Whiting, "CERN Project Will Collect Hundreds of Petabytes of Data," *InformationWeek*, February 11, 2002, accessed at http:/www.informationweek.com.
10. Rick Whiting, "Tower of Power," *InformationWeek*, February 11, 2002, accessed at http://www.informationweek.com.
11. Bill Miles, "Slick," *Darwin,* June 21, 2001, accessed at http://www.darwinmag.com.
12. Rick Whiting, "United Building 6-Terabyte Warehouse," *InformationWeek*, April 1, 2002, accessed at *www.informationweek.com.*
13. Rick Whiting, "Companies Boost Sales Efforts With Predictive Analysis," *InformationWeek*, February 25, 2002, accessed at http://www.informationweek.com.
14. Mary Brandel, "Masters of Business Intelligence," *Computerworld*, February 26, 2001, accessed at http://www.computerworld.com.
15. Pete Loshin, "Knowledge Management," *Computerworld*, October 22, 2001, accessed at http://www.computerworld.com.
16. Larry Greenemeier, "The IT Prescription for Faster Drug Delivery," *Computerworld*, February 25, 2002, accessed at http://www.computerworld.com.
17. "Oracle Enterprise Manager Helps SBB Reduce Costs by Enabling a Seamless Operation of a Replicated Environment," Oracle Web site, accessed at *www.oracle.com* on August 1, 2002.
18. Gary H. Anthes, "Agreeing to Disagree," *Computerworld*, August 5, 2002, accessed at *www.computerworld.com.*
19. Gary Anthes, "Database Horizons," *Computerworld*, August 5, 2002, accessed at www.computerworld.com.
20. Marc L. Songini, "Firms Face Barriers in Push for Data Analysis," *Computerworld*, April 30, 2001, accessed at http://www.computerworld.com.
21. Press release, "eXcelon's Objectstore—Powers Timeware Inc's Labor Management Suite," March 19, 2002, accessed at eXcelon's Web site at www.exln.com.
22. Linda Rosencrance, "NYPD Selects MapInfo for Citywide Crime Analysis," *Computerworld*, January 4, 2002, accessed at http://www.computerworld.com.

Telecommunications and Networks

CHAPTER 6

PRINCIPLES	LEARNING OBJECTIVES
Effective communication is essential to organizational success.	• Define the terms *communication* and *telecommunications* and describe the components of a telecommunications system.
An unmistakable trend of communications technology is that more people are able to send and receive all forms of information over greater distances at a faster rate.	• Identify three basic types of communications media and discuss the basic characteristics of each. • Identify several types of telecommunications hardware devices and discuss the role that each plays. • Identify the benefits associated with a telecommunications network. • Name three distributed processing alternatives and discuss their basic features. • Define the term *network topology* and identify five alternatives.
The effective use of telecommunications and networks can turn a company into an agile, powerful, and creative organization, giving it a long-term competitive advantage.	• Identify and briefly discuss several telecommunications applications.

[DHL]

Implementing a Global Network

Adrian Dalsey, Larry Hillblom, and Robert Lynn founded DHL Worldwide Express as a service shuttling bills of lading (receipts for shipped goods) between San Francisco and Honolulu in 1969. The company grew rapidly, expanding both the locations served and the services offered. Soon DHL was providing international door-to-door express shipments to the Philippines, Japan, Hong Kong, Singapore, and Australia. Steady expansion continued in the 1970s and 1980s as DHL initiated service to Europe, Latin America, the Middle East, Africa, the Eastern Bloc countries, and the People's Republic of China.

DHL's amazing growth mirrors the rapid increase in the globalization of trade. As more and more companies move into international markets, DHL continually expands and upgrades its information network of 5,900 offices and 36 central communication hubs. Today DHL provides service to 120,000 destinations in more than 228 countries and territories. Its 70,000 employees and fleet of 254 aircraft worldwide provide the flexibility to use the fastest possible means of transportation to any given destination.

DHLNet is an on-line telecommunications network used to transmit package location data and summary status and billing information to DHL customers around the world. DHLNet links the Internet to the company's internal information systems to allow customers immediate access to accurate data about their shipments. DHL agents anywhere in the world also use DHLNet to access the company's mainframe computers and databases and develop rapid price quotes for upcoming shipments. DHLNet can be accessed by personal computer, telephone, and even PDA.

The successful and reliable operation of DHLNet is so essential to the firm that DHL is implementing a trio of data centers to provide round-the-clock network management. Support personnel at each data center will manage both the airfreight carrier's information systems and its DHLNet telecommunications network. The newest data center—the Americas Information Services Center—is located in a 106,000 square foot facility in Scottsdale, Arizona. It cost the company $250 million and took five years to build. The company employs about 350 workers at the facility, mixing new hires with more experienced workers who are transferred from other DHL sites in the United States, Canada, and Latin America.

This new facility is linked via DHLNet to the company's existing data centers in London and Kuala Lumpur, Malaysia. The three data centers work in tandem to manage the company's entire network infrastructure, 24 hours a day, seven days a week. Each data center has primary responsibility for eight hours and then passes control to the next facility. The centers manage information systems operations in nine-hour shifts, covering the eight hours of network support plus a one-hour overlap for the transfer of control.

Network specialists at the data centers use network monitoring tools and fault-reporting systems to detect any computing or network problems quickly. In the event of a network failure, backup systems and network circuits have been programmed to take over automatically. Keeping operations running smoothly, even in the face of problems, is a high priority as DHL strives to provide top-notch support and services to more than 1 million customers worldwide.

As you read this chapter, consider the following:
- ⊙ Why is it so critical for companies like DHL to have an efficient and effective telecommunications system?
- ⊙ What are some specific telecommunications applications used by organizations to gain a competitive advantage?
- ⊙ What measures do companies take to ensure the reliable and uninterrupted operations of their networks?

In today's high-speed business world, effective communication is critical to organizational success, as it is with DHL. Often, what separates good management from poor management is the ability to identify problems and solve them with available resources. Efficient communications is one of the most valuable of these resources, because it enables a company to keep in touch with its operating divisions, customers, suppliers, and stockholders. For example, when auto supplier Johnson Controls took on the job of building the dashboard controls, overhead computer console, interior lights, and seats for the 2002 Jeep Liberty, it needed ideas from its own designers and from a worldwide network of 35 major suppliers and dozens of smaller ones. Managers at Johnson Controls relied on a worldwide telecommunications network to work with DaimlerChrysler as they designed part of the vehicle and with its own supplier network to engineer and build it.[1] Companies hope to save billions of dollars, reduce time to market, and enable collaboration with their business partners through the use of telecommunications systems.

Memos, notices on bulletin boards, and presentations are all obvious examples of continual communication within a business organization. Other not-so-obvious examples include policy and procedure manuals and even salaries (they communicate the company's perception of the value of the contribution of the person being paid). Communication also exists in other forms—for instance, warning lights from a computer system that monitors manufacturing processes and signals from a building management system that monitors temperature, humidity, lighting, and security of a building. Communication is any process that permits information to pass from a sender to one or more receivers. Communications of all types form a major part of any business system. Therefore, managers must gain an appreciation of communication concepts, media, and devices—as well as an understanding of how these factors may best be employed to develop effective and efficient business systems.

AN OVERVIEW OF COMMUNICATIONS SYSTEMS

COMMUNICATIONS

Communications is the transmission of a signal by way of a medium from a sender to a receiver (Figure 6.1). The signal contains a message composed of data and information. The signal goes through some communications medium, which is anything that carries a signal between a sender and receiver. In human speech, the sender transmits a signal through the transmission medium of the air. In telecommunications, the sender transmits a signal through a transmission medium such as a cable.

The components of communication can easily be recognized if you consider human communication (Figure 6.2). When we talk to one another face to face, we send messages to each other. A person may be the sender at one moment and the receiver a few seconds later. The same entity, a person in this case, can be a sender, a receiver, or both. This process is typical of two-way communication. The signals we use to convey these messages are our spoken words—our language. For communication to be effective, both sender and receiver must understand the signals and agree on the way they are to be interpreted. For example, if the sender in Figure 6.2 is speaking in a language the receiver does not understand, or if the sender believes a particular word has one meaning and the receiver believes the word has some other meaning, effective communication will not occur.

F I G U R E   6.1

Overview of Communications

The message (data and information) is communicated via the signal. The transmission medium "carries" the signal.

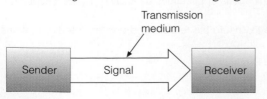

Transmission medium

Sender   Signal   Receiver

FIGURE 6.2

Communication and Telecommunication

In human speech, the sender transmits a signal through the transmission medium of the air. In telecommunication, the sender transmits a signal through a cable or other telecommunication medium.

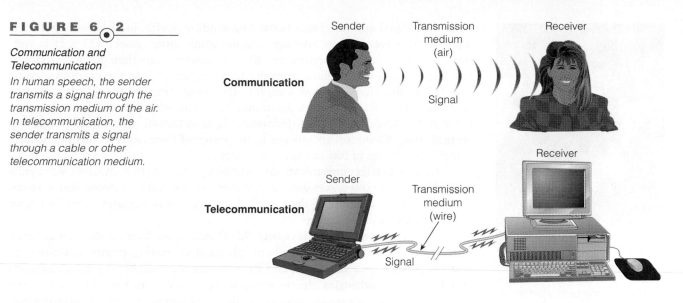

synchronous communications
communication in which the receiver gets the message instantaneously

asynchronous communications
communication in which the receiver gets the message minutes, hours, or days after it is sent

data communications
a specialized subset of telecommunications that refers to the electronic collection, processing, and distribution of data—typically between computer system hardware devices

telecommunications medium
anything that carries an electronic signal and interfaces between a sending device and a receiving device

In addition to the flow of communications, shown in Figure 6.2, communications can be synchronous or asynchronous. With **synchronous communications**, the receiver gets the message instantaneously, when it is sent. Voice and phone communications are examples of synchronous communications. With **asynchronous communications,** the receiver gets the message later—sometimes hours or days after the message is sent. Sending a letter through the post office or e-mail over the Internet are examples of asynchronous communications. Academic researchers are actively investigating the impact of both synchronous and asynchronous communications on effectiveness, performance, and other organizational measures. Both types of communications are important in business, regardless of whether the communication is done manually or electronically.

TELECOMMUNICATIONS

Telecommunications refers to the electronic transmission of signals for communications, including such means as telephone, radio, and television. Telecommunications has the potential to create profound changes in business because it lessens the barriers of time and distance.

Telecommunications not only is changing the way businesses operate but also is altering the nature of commerce itself. As networks are connected with one another and information is transmitted more freely, a competitive marketplace is making excellent quality and service imperative for success. **Data communications**, a specialized subset of telecommunications, refers to the electronic collection, processing, and distribution of data—typically between computer system hardware devices. Data communications is accomplished through the use of telecommunications technology.

Figure 6.3 shows a general model of telecommunications. The model starts with a sending unit (1), such as a person, a computer system, a terminal, or another device, that originates the message. The sending unit transmits a signal (2) to a telecommunications device (3). The telecommunications device performs a number of functions, which can include converting the signal into a different form or from one type to another. A telecommunications device is a hardware component that allows electronic communication to occur or to occur more efficiently. The telecommunications device then sends the signal through a medium (4). A **telecommunications medium** is anything that carries an

Telecommunications technology enables businesspeople to communicate with co-workers and clients from remote locations.

(Source: Stone/Terry Vine.)

computer network
the communications media, devices, and software needed to connect two or more computer systems and/or devices

FIGURE 6.3

Elements of a Telecommunications System

Telecommunications devices relay signals between computer systems and transmission media.

electronic signal and interfaces between a sending device and a receiving device. The signal is received by another telecommunications device (5) that is connected to the receiving computer (6). The process can then be reversed and another message can go back from the receiving unit (6) to the original sending unit (1). In this chapter, we will explore the components of the telecommunications model shown in Figure 6.3. An important characteristic of telecommunications is the speed at which information is transmitted, measured in bits per second (bps). Common speeds are in the range of thousands of bits per second (Kbps) to millions of bits per second (Mbps).

Advances in telecommunications technology allow us to communicate rapidly with clients and co-workers almost anywhere in the world. Telecommunications also reduces the amount of time needed to transmit information that can drive and conclude business actions.

Celanese Chemicals employs over 5,000 employees located globally in seven different countries. It produces chemicals used in making paints and coatings, textiles, plastics, and adhesives, and its products are used heavily in the housing and automotive industries. Previously, sales representatives had to make extraordinary efforts to respond to customer calls. They had no access to current information and had to call a home-office customer service rep who did have network access. After retrieving the necessary product information, the customer service rep would call the sales rep back, who would then call the customer with the information. The whole process took from four hours to over a day. Now, thanks to a revamping of its telecommunications network, sales reps can use their handheld computers linked wirelessly to corporate networks, the Internet, and a Web-based enterprise resource planning system to get real-time information and provide it to customers on the spot. This system empowers the sales representative and often results in faster, higher-quality customer service. Celanese expects customer satisfaction to rise as a result of the fast response times and greatly improved one-on-one service.[2]

In addition to external communications, telecommunications technology also helps businesses coordinate activities and integrate various departments to increase operational efficiency and support effective decision making. The far-reaching developments of telecommunications are having and will continue to have a profound effect on business information systems and on society in general.

NETWORKS

A **computer network** consists of communications media, devices, and software needed to connect two or more computer systems and/or devices. Once connected, computers can share data, information, and processing jobs. More and more businesses are linking computers in networks to streamline work processes and allow employees to collaborate on projects. The effective use of networks can turn a company into an agile, powerful, and creative organization, giving it a long-term competitive advantage. Networks can be used to share hardware, programs,

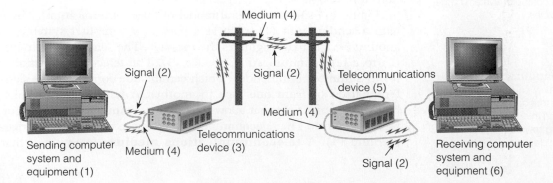

Sending computer
system and
equipment (1)

Signal (2)

Medium (4)

Telecommunications
device (3)

Medium (4)

Signal (2)

Medium (4)

Signal (2)

Telecommunications
device (5)

Receiving computer
system and
equipment (6)

and databases across the organization. They can transmit and receive information to improve organizational effectiveness and efficiency. They enable geographically separated workgroups to share documents and opinions, which fosters teamwork, innovative ideas, and new business strategies.

TELECOMMUNICATIONS

In today's global, fast-paced business environment, the use of telecommunications can help businesses solve problems and maximize opportunities. But using telecommunications effectively requires careful analysis of telecommunications media, devices, and carriers and services.

BASIC COMMUNICATIONS CHANNEL CHARACTERISTICS

A communications channel is the transmission medium that carries a message from the source of the message to its receivers. In a communications network, the channel is frequently a telephone line; however, there are many other transmission media. Communications channel can be classified as simplex, half-simplex, or full duplex.

A **simplex channel** can transmit data in only one direction. It is seldom used for business telecommunications. Doorbells and radio and TV broadcasting operate using a simplex channel. A **half-duplex channel** can transmit data in either direction, but not simultaneously. For example, A can begin transmitting to B over a half-duplex line, but B must wait until A is finished to transmit to A. Personal computers are usually connected to a remote computer over a half-duplex channel. A **full-duplex channel** permits data transmission in both directions at the same time, so a full-duplex channel is like two simplex lines. Private leased lines or two standard phone lines are required for full-duplex transmission.

CHANNEL BANDWIDTH AND INFORMATION-CARRYING CAPACITY

In addition to the directions that telecommunications can travel, businesses must consider the speed at which signals can be transmitted. Speed depends on the channel bandwidth. **Bandwidth** is the range of frequencies that an electronic signal occupies on a given transmission medium. Claude Elwood Shannon, a U.S. mathematician, spent the bulk of his 30-year career doing research at Bell Labs, now part of Lucent Technologies. In the 1940s, Shannon developed mathematical equations relating the rate of data transmission to channel bandwidth. **Shannon's fundamental law of information theory** states that the information-carrying capacity of a channel is directly proportional to its bandwidth—the broader the bandwidth, the more information that can be carried. In general, **broadband** refers to telecommunications in which a wide band of frequencies is available to transmit information, allowing more information to be transmitted in a given amount of time. Related terms are *wideband* (a synonym), *baseband* (a one-channel band), and *narrowband* (meaning just wide enough to carry voice data).

Telecommunications professionals consider the capacity of the channel when they recommend media for particular business needs. In general, today's organizations need more bandwidth for increased transmission speed to carry out their daily functions. Let's take a look at the different types of telecommunications media that are available.

TYPES OF MEDIA

Each type of communications media exhibits its own characteristics, including transmission capacity and speed. In developing a telecommunications system, the selection of media depends on the purpose of the overall information and organizational systems, the purpose of the telecommunications subsystems, and

---

simplex channel
a communications channel that can transmit data in only one direction

half-duplex channel
a communications channel that can transmit data in either direction, but not simultaneously

full-duplex channel
a communications channel that permits data transmission in both directions at the same time, thus the full-duplex channel is like two simplex lines

bandwidth
the width of the range of frequencies that an electronic signal occupies on a given transmission medium

Shannon's fundamental law of information theory
the law of telecommunications that states that the information-carrying capacity of a channel is directly proportional to its bandwidth—the broader the bandwidth, the more information can be carried

broadband
telecommunications in which a wide band of frequencies is available to transmit information, allowing more information to be transmitted in a given amount of time

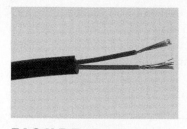

FIGURE 6.4

Twisted-Pair Wire Cable

(Source: Fred Bodin.)

the characteristics of the media. As with other system components, the media should be chosen to support the goals of the information and organizational systems at the least cost and to allow for possible modification of system goals over time. The proper media will help a company link subsystems to maximize effectiveness and efficiency.

Twisted-Pair Wire Cable

Twisted-pair wire cable is, as you might expect, a cable consisting of pairs of twisted wires (Figure 6.4). A typical cable contains two or more twisted pairs of wire, usually copper. Proper twisting of the wire keeps the signal from "bleeding" into the next pair and creating electrical interference. Because the twisted-pair wires are insulated, they can be placed close together and packaged in one group. Hundreds of wire pairs can be grouped into one large wire cable.

There are two kinds of twisted-pair wire cable: shielded and unshielded. Shielded twisted-pair wire cable has a special conducting layer within the normal insulation. This conducting layer makes the cable less prone to electrical interference, or "noise." Unshielded twisted-pair (UTP) wire cable lacks this special insulation shield. UTP cables have historically been used for telephone service and to connect computer systems and devices. Newer types of cable, however, have begun to replace UTP cable in both businesses and homes. Communications over unshielded twisted-pair generates considerable electromagnetic radiation, making it relatively easy for determined individuals to "listen in" on such communications undetected.

Twisted-pair cables come in different categories (Category 1, 2, 3, 4, 5, 5E, and 6). The lower categories are used primarily in homes. Higher categories are used as a cheaper alternative to coaxial cable for smaller networks. Category 1 is traditional telephone cable. Category 3 is the most common type of existing cable found in corporate settings, and it normally contains four pairs of wire. Category 5 is frequently installed in new buildings and is capable of carrying data at speeds faster than 1 gigabit/second.

Coaxial Cable

Figure 6.5 shows a typical coaxial cable, similar to that used in cable television installations. A coaxial cable consists of an inner conductor wire surrounded by insulation, called the *dielectric*. The dielectric is surrounded by a conductive shield (usually a layer of foil or metal braiding), which is in turn covered by a layer of nonconductive insulation, called the *jacket*. When used for data transmission, coaxial cable falls in the middle of the cabling spectrum in terms of cost and performance. The cable itself is more expensive than twisted-pair wire cable but less so than fiber-optic cable (discussed next). However, the cost of installation and other necessary communications equipment makes it difficult to compare the total costs of using each medium. Coaxial cable offers cleaner and crisper data transmission (less noise) than twisted-pair wire cable. It also offers a higher data transmission rate.

One of the main deterrents to deploying broadband to hotel guest rooms is the limitation of single twisted-pair wiring used in most properties. Recently, a number of telecommunications companies began offering a technology called *long-reach Ethernet (LRE)* that supports broadband transmission over hotel phone wiring. This breakthrough makes it unnecessary to string more sophisticated twisted-pair wiring or coaxial cable to each hotel room. Rewiring or pulling cable is labor intensive and costly, and hotels frequently have to block off rooms during installation. Wyndham Hotels is one chain that implemented broadband Internet connectivity for its guest rooms using the LRE technology in 2002.[3]

Fiber-Optic Cable

Fiber-optic cable, consisting of many extremely thin strands of glass or plastic bound together in a sheathing (a jacket), transmits signals with light beams

FIGURE 6.5

Coaxial Cable

(Source: Fred Bodin.)

FIGURE 6●6

Fiber-Optic Cable

(Source: Stone/Greg Pease.)

(Figure 6.6). These high-intensity light beams are generated by lasers and are conducted along the transparent fibers. These fibers have a thin coating, called *cladding*, which effectively works like a mirror, preventing the light from leaking out of the fiber.

Fiber-optic technology is dramatically improving. Optisphere Networks and WorldCom conducted a month-long test sending data at a rate of 3.2 terabits per second (Tbps) over a total distance of 150 miles using fiber networks. The test was the equivalent of sending more than 41 million telephone calls at the same time on one fiber.[4] Bell Labs has demonstrated the capability to send 2.56 Tbps over a distance of 2,500 miles.[5]

The much smaller diameter of fiber-optic cable makes it ideal in situations where there is not room for bulky copper wires—for example, in crowded conduits, which can be pipes or spaces carrying both electrical and communications wires. In such tight spaces, the smaller fiber-optic telecommunications cable is very effective. Because fiber-optic cables are immune to electrical interference, signals can be transmitted over longer distances with fewer expensive repeaters to amplify or rebroadcast the data. Fiber-optic information is also difficult to steal—a special plus for security. In fact, with the right equipment installed, it is virtually impossible to tap into fiber-optic cable without being detected. Fiber-optic cable and associated telecommunications devices are more expensive to purchase and install than the twisted-pair wire, although the cost is coming down.

Microwave Transmission

Microwave transmissions are sent through the atmosphere and space. Although these transmission media do not entail the expense of laying cable, the transmission devices needed to utilize this medium are quite expensive. Microwave is a high-frequency radio signal that is sent through the air (Figure 6.7). Microwave transmission is line-of-sight, which means that the straight line between the transmitter and receiver must be unobstructed. Typically, microwave stations are placed in a series—one station will receive a signal, amplify it, and retransmit it to the next microwave transmission tower. Such stations can be from to 30 to 70 miles apart (depending on the height of the towers) before the curvature of the earth makes it impossible for the towers to "see one another." Microwave signals can carry thousands of channels at the same time.

FIGURE 6●7

Microwave Communications

Because they are line-of-sight transmission devices, microwave dishes must be placed in relatively high locations such as atop mountains, towers, and tall buildings.

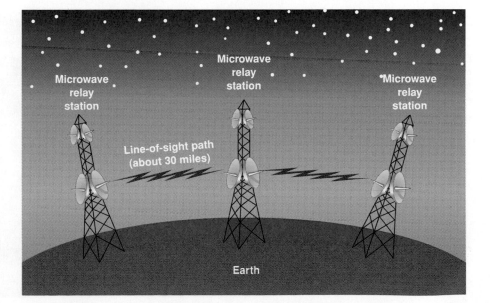

A communications satellite is basically a microwave station placed in outer space (Figure 6.8). The satellite receives the signal from the earth station, amplifies the relatively weak signal, and then rebroadcasts it at a different frequency. The advantage of satellite communications is the ability to receive and broadcast over large geographic regions. Such problems as the curvature of the earth, mountains, and other structures that block the line-of-sight microwave transmission make satellites an attractive alternative. Geostationary and low-earth-orbit satellites are the two most common forms of communications satellites.

A *geostationary satellite* orbits the earth directly over the equator, approximately 22,000 miles above the earth. At this altitude, one complete trip around the earth takes 24 hours so that the satellite remains over the same spot on the earth's surface at all times. The satellite thus stays fixed in the sky relative to any point on the earth from which it can be seen. Three such satellites, spaced at equal intervals (120 angular degrees apart), can provide coverage of the entire world. A geostationary satellite can be accessed using a dish antenna aimed at the spot in the sky where the satellite hovers.

A *low-earth-orbit (LEO) satellite* system employs a large number of satellites, each in a circular orbit at a constant altitude of a few hundred miles. Each orbit takes the satellites over the geographic poles with one orbit taking roughly 90 minutes. The satellites are spaced so that, from any point on the earth at any time, at least one satellite is on a line of sight. The entire system operates similar to the way a cellular telephone functions except that the wireless receivers/transmitters are moving, rather than fixed, and are in space rather than on the earth. LEO service subscribers can access the satellites using an antenna a little more sophisticated than old-fashioned television "rabbit ears."

Most of today's communications satellites are owned by companies that rent or lease satellite communications capacity to other companies. However, several large companies are now using their own satellites for internal telecommunications. Some large retail chains, such as Sears and Wal-Mart, use satellite transmission to connect their main offices to retail stores and warehouses throughout the country or the world. Bob Evans Farms uses a satellite network to connect its 459 restaurants and six food production plants. Prior to the satellite network, the restaurant chain experienced unacceptable delays processing credit card authorizations, especially on busy weekend mornings. Now the average time to authorize a credit-card purchase is about three seconds, including printing the customer's receipt.

FIGURE 6.8

Satellite Communications

Communications satellites are relay stations that receive signals from one earth station and rebroadcast them to another.

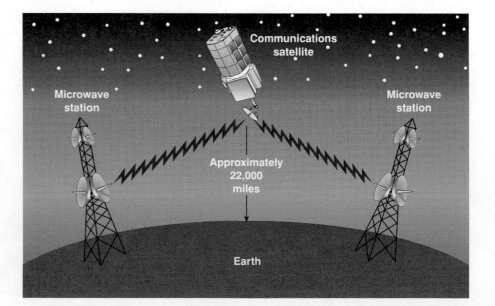

The effective use of telecommunications and networks can turn a company into an agile, powerful, and creative organization, giving it a long-term competitive advantage.

Penske Uses Satellite Communications to Meet Customer Needs

Penske Logistics provides service across three continents using the company's technology, engineering, and operational expertise for global transportation management, distribution management, and integrated logistics. In fall 2001, Penske installed wireless computers in its 4,000-vehicle fleet. Each truck's cab computer features a wireless network and satellite connection, effectively turning the cab into a communications hub. The wireless networks are based on a pioneering technology called Code Division Multiple Access (CDMA). That technology powers wireless networks and handsets all over the world. Qualcomm, based in San Diego, developed CDMA, which makes very efficient use of the radio frequency spectrum and allows more people to share the airwaves at the same time—without cross-talk, static, or interference.

The satellite connections for Penske truck drivers are made possible by a consortium of international telecommunications companies formed to deliver satellite communications through a network of service providers called Globalstar. Members include Aerospazio, Alcatel, Alenia, China Telecom, DACOM, DaimlerChrysler Aerospace, Elsacom, France Telecom, Hyundai, Loral Space & Communications, Qualcomm, and Vodafone. The Globalstar system provides high-quality satellite-based voice, fax, and data transmission services to a broad range of customers using Globalstar's set of 48 low-earth-orbit (LEO) satellites. Globalstar sells access to its satellite communications system to regional and local telecommunications service providers around the world.

At the company's distribution centers, the Penske driver captures information about cargo deliveries and pickups using a bar-code scanner. The scanner comes equipped with a card allowing the driver to relay data to the cab computer, a Qualcomm MVPc CE unit running the Windows CE operating system. The cab computer is programmed to immediately relay critical information—such as a change in the number of items delivered to a customer—to Penske corporate distribution systems and to the affected customer. This information is sent over the Globalstar satellite system. So, the customer knows exactly what items and what quantities are being shipped as soon as they are loaded on the truck. Penske's central operators are also able to send route alerts to its drivers about traffic jams or changes in schedules over the satellite communications system.

Routine shipment information—such as records of all items dropped off at a particular location—is stored in the onboard computer until the truck gets within close range of a Penske distribution terminal. At that point, the computer in the truck's cab senses the approach of the distribution terminal and starts downloading its data at 11 Mbps.

The cab computers also run software from Xata, a supplier of onboard computing systems for transportation companies. The software enables Penske to automate driver logs, monitor state line crossings, and calculate fuel tax. The wireless and satellite-based telecommunications system enables Penske to deliver superior customer service while reducing costs and enabling it to comply with all distribution regulations.

Discussion Questions

1. Why does Penske use the satellite link only for transmission of critical information, while transmitting routine information over the short-range link?
2. Why is the information concerning exactly what was loaded on the truck deemed to be critical enough to relay to the customer immediately?

Critical Thinking Questions

3. With this basic infrastructure in place, what additional applications can you identify that would further reduce costs and add customer value?
4. What components of this system could fail? Which ones need some sort of backup system? What would you recommend to provide backup capability at a reasonable cost?

Sources: Adapted from Bob Brewin, "Penske Outfits Fleet with Wireless Terminals," *Computerworld*, June 11, 2001, http://www.computerworld.com; "Penske Logistics Signs License Agreement with Xata for Xatanet Software," Penske press release, February 26, 2001, http://www.penske.com; "Qualcomm's FleetAdvisor Offers Private Fleets Affordable Solutions for Improved Productivity and Customer Service," Qualcomm press release, February 5, 2001, http://www.qualcomm.com.

Each night the stores' point-of-sale (POS) systems are polled for financial data, and this data is transmitted over the satellite network. E-mail and on-line manuals on restaurant procedures, facilities, and physical plant maintenance are also sent over the network. On-line inventory management and electronic ordering applications are also being implemented.[6]

In addition to standard satellite stations, small mobile satellite systems allow people and businesses to communicate. These portable systems have a dish that is a few feet in diameter and can operate on battery power anywhere in the world. This capability is important for news organizations that need to transmit news stories from remote locations. Many people are also investing in direct satellite dish technology to receive TV and send and receive computer communications.

Read the "IS Principles in Action" special-interest box to learn more about how one company is using satellite communications to gain a long-term competitive advantage.

Cellular Transmission

With cellular transmission, a local area, such as a city, is divided into cells. As a car or vehicle with a cellular device, such as a mobile phone, moves from one cell to another, the cellular system passes the phone connection from one cell to another (Figure 6.9). The signals from the cells are transmitted to a receiver and integrated into the regular phone system. Cellular phone users can thus connect to anyone that has access to regular phone service, like a child at home or a business associate in London. They can also contact other cellular phone users. Because cellular transmission uses radio waves, it is possible for people with special receivers to listen to cellular phone conversations, so they are not secure.

Cellular transmission is used for much more than making phone calls from remote locations. Combining cellular transmission with other devices opens up the power of networks, communications, and the Internet. For example, combining cellular transmission with some phone devices allows people to read and respond to their e-mails, check stock prices, get news on their favorite topics, and more. With cell phone rates dropping, some people are disconnecting their old, wired phones in their homes and offices and using cellular phones for all of their calling needs. New devices, called *net phones*, combine cellular transmission with access to networks and the Internet. Handheld computers are now using wireless communications to enhance their power. The Palm VII, for example, can use wireless communications to check appointments from a central computer, download information from corporate computers, and browse the Internet.

Infrared Transmission

Another mode of transmission, called *infrared transmission*, sends signals through the air via light waves. Infrared transmission requires line-of-sight transmission and short distances—under a few hundred yards. Infrared transmission can be used to connect various small devices and computers. For example, infrared transmission has been used to allow handheld computers to transmit data and information to larger computers within the same room. Infrared transmission can also be used to connect a display screen, a printer, and a mouse to a computer. Some special-purpose phones can also use infrared transmission. This means of transmission can be used to establish a wireless network with the advantage that devices can be moved, removed, and installed without expensive wiring and network connections.

FIGURE 6.9

A Typical Cellular Transmission Scenario

Using a cellular car phone, the caller (1) dials the number. The signal is sent from the car's antenna to the low-powered cellular antenna located in that cell (2). The signal is sent to the regional cellular phone switching office, also called the mobile telephone subscriber office (MTSO) (3). The signal is switched to the local telephone company switching station located nearest the call destination (4). Now integrated into the regular phone system, the call is automatically switched to the number originally dialed (5), all without the need for operator assistance.

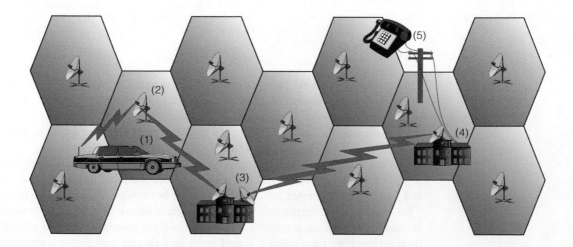

CONNECTORS

Each type of telecommunications media uses different types of connectors to link them to devices. Twisted-pair cables can have many different types of connectors. The two most common are Registered Jack (RJ) 11 and RJ45. RJ11 connectors are commonly used in U.S. telephones and have four or six contact points. RJ45 connectors are similar but wider, with eight contact points, and are generally used in heavy-duty computing environments. Coaxial network cables generally use Bayonet Neill-Concelman (BNC) twist-lock connectors. The most common fiber-optic cable connectors are ST (a twist-lock device), FC/PC (which screws on), and the snap-in SC. It is important when ordering or installing network equipment to specify the correct connector type.

DEVICES

A telecommunications device is one of various hardware devices that allow electronic communication to occur or to occur more efficiently. Almost every telecommunications system uses one or more of these devices to transmit or receive signals.

Modems

analog signal
a continuous, curving signal

digital signal
a signal represented by bits

In data telecommunications, it is not uncommon to use transmission media of differing types and capacities at various stages of the communications process. If a typical telephone line is used to transfer data, it can only accommodate an **analog signal** (a continuous, curving signal). Because a computer generates a **digital signal** represented by bits, a special device is required to convert the digital signal to an analog signal, and vice versa (Figure 6.10). Translating data from digital to analog is called *modulation*, and translating data from analog to digital is called *demodulation*. Thus, these devices are modulation/demodulation devices, or **modems**. Penril/Bay Networks, Hayes, Microcom, Motorola, and U.S. Robotics are examples of modem manufacturers.

modem
a device that translates data from digital to analog and analog to digital

Modems can automatically dial telephone numbers, originate message sending, and answer incoming calls and messages. Modems can also perform tests and checks on how they are operating. Some modems are able to vary their transmission rates, commonly measured in bits per second. The V.90 modem protocol was adopted by the International Telecommunications Union in 1998 and provides a common framework for 56 Kbps data transmission. The V.92 Modem on Hold protocol allows a user to suspend a modem session to answer an incoming voice call or to place an outgoing call while engaged in a modem session.

Special-Purpose Modems

Various types of special-purpose modems are available. Cellular modems are placed in laptop personal computers to allow people on the go to communicate with

FIGURE 6.10

How a Modem Works
Digital signals are modulated into analog signals, which can be carried over existing phone lines. The analog signals are then demodulated back into digital signals by the receiving modem.

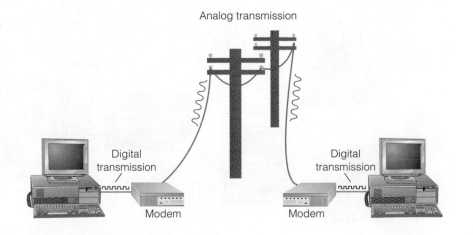

Analog transmission

Digital transmission

Digital transmission

Modem

Modem

A cable modem can deliver network and Internet access up to 500 times faster than a standard modem and phone line.

(Source: Courtesy of Linksys.)

multiplexer
a device that allows several telecommunications signals to be transmitted over a single communications medium at the same time

front-end processor
a special-purpose computer that manages communications to and from a computer system

common carriers
long-distance telephone companies

value-added carriers
companies that have developed private telecommunications systems and offer their services for a fee

switched line
a communications line that uses switching equipment to allow one transmission device to be connected to other transmission devices

other computer systems and devices. With a cellular modem, you can connect to other computers while in your car, on a boat, or in any area that has cellular transmission service. Expansion slots used for PC memory cards can also be used for standardized credit-card-size PC modem cards, which work like standard modems. PC modems are becoming increasingly popular with notebook and portable computer users.

Cable companies are promoting the cable modem, which has a low initial cost and transmission speeds up to 10 Mbps. A cable modem can deliver network and Internet access up to 500 times faster than a standard modem and phone line. In addition, a cable modem is always on, so you can be on the Internet 24 hours a day, 7 days a week. Fees from $30 to $50 per month usually include unlimited service, local news, and an e-mail account. Cable service may cost only an additional $10 per month if you have existing Internet access through an Internet provider and want to upgrade from phone to cable service. @Home and AOL Time Warner are leading companies using cable TV to bring the Internet to homes and businesses.

Multiplexers

Because media and channels are expensive, devices that allow several signals to be sent over one channel have been developed. A **multiplexer** is one of these devices. A multiplexer allows several telecommunications signals to be transmitted over a single communications medium at the same time (Figure 6.11).

Front-End Processors

Front-end processors are special-purpose computers that manage communications to and from a computer system. Like a receptionist handling visitors at an office complex, communications processors direct the flow of incoming and outgoing jobs. They connect a midrange or mainframe computer to hundreds or thousands of communications lines. They poll terminals and other devices to see if they have any messages to send. They provide automatic answering and calling, as well as perform circuit checking and error detection. Front-end processors also develop logs or reports of all communications traffic, edit data before it enters the main processor, determine message priority, automatically choose alternative and efficient communications paths over multiple data communications lines, and provide general data security for the main system CPU. Because front-end processors perform all these tasks, the midrange or mainframe computer is able to process more work (Figure 6.12).

FIGURE 6.11

Use of a Multiplexer to Consolidate Data Communications onto a Single Communications Link

CARRIERS AND SERVICES

Telecommunications carriers provide the telephone lines, satellites, modems, and other communications technology used to transmit data from one location to another. They also provide many types of services. Telecommunications carriers

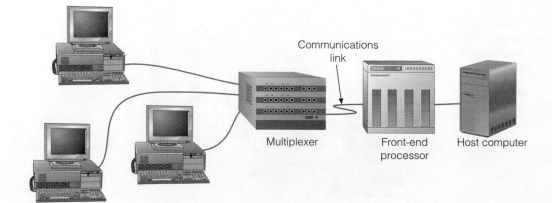

Communications link

Multiplexer   Front-end processor   Host computer

Telecommunications networks require state-of-the-art computer software technology to continuously monitor the flow of voice, data, and image transmission over billions of circuit miles worldwide.

(Source: Stone/Roger Tully.)

dedicated line

a communications line that provides a constant connection between two points; no switching or dialing is needed, and the two devices are always connected

are classified as either common carriers or other special-purpose carriers. The **common carriers** are primarily the long-distance telephone companies. American Telephone & Telegraph (AT&T), one of the largest companies providing communications media and services, is a common carrier for long-distance service and a special-purpose carrier for other services. MCI WorldCom, Sprint, and others make up a significant part of the telecommunications industry as well. **Value-added carriers** are companies that have developed private telecommunications systems and offer their services for a fee. Some value-added carriers that offer communications services include SprintNet and Telenet (developed by GTE) and Tymnet.

Switched and Dedicated Lines

Common carriers typically provide the use of standard telephone lines, called **switched lines**. These lines use switching equipment to allow one transmission device (e.g., your telephone) to be connected to other transmission devices (e.g., the telephones of your friends and relatives). A switch is a special-purpose circuit that directs messages along specific paths in a telecommunications system. When you make a phone call, the local telephone service provider's switching equipment connects your phone to the phone of the person you're calling. Fees for a switched business line (versus residential line) can range from $25 to $100 or more per month. A **dedicated line**, also called a *leased line*, provides a constant connection between two points. No switching or dialing is needed; the two devices are always connected. Many firms with high data transfer requirements between two points—say, sharing headquarters between East Coast and West Coast offices—utilize dedicated lines. The high initial cost of purchasing or leasing such a line is offset by eliminating long-distance charges

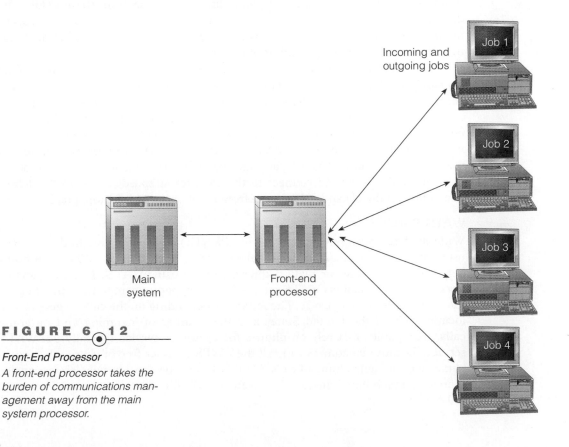

Incoming and outgoing jobs

Job 1

Job 2

Job 3

Job 4

Main system

Front-end processor

FIGURE 6.12

Front-End Processor

A front-end processor takes the burden of communications management away from the main system processor.

incurred with a switched line. Monthly fees for a dedicated line can range from $50 to $500 or more, but there is no additional charge for use.

Private Branch Exchange

Traditional phone service (sometimes called *POTS* for "plain old telephone service") connects your home or small business to a telephone company's central office using twisted pairs of copper wires. The central office has equipment that can switch calls locally or to long-distance carrier phone offices. (The term *public exchange* is used outside the United States in place of *central office.*) The wired connection between your home or small business and the central office is called the *local loop.* POTS was originally designed only for voice transmission using analog transmission technology on a single voice channel. If digital devices are tied to the voice channel, modems are required to make the conversion between analog and digital signals.

private branch exchange (PBX)
a communications system that can manage both voice and data transfer within a building and to outside lines

A **private branch exchange (PBX)** is a communications system that can manage both voice and data transfer within a building and to outside lines. In a PBX system, switching equipment routes phone calls and messages within the building. PBXs can be used to connect hundreds of internal phone lines to a few phone company lines. For example, an organization might have five phone lines coming in from the outside phone company. These five lines may be connected to 50 phones within the organization. Any of the 50 phones can use one of the five phone lines to make calls outside the organization. These same five lines may also be used for incoming calls. Furthermore, it is usually possible for any of the 50 phones to connect to another internal phone on an intercom system.

Not only can PBXs store and transfer calls, but they can also serve as connections between different office devices. With PBX technology, a manager could connect his computer to the PBX via a modem and then send instructions to a copy machine through his PC. Another advantage of a PBX system is that it requires a business to have fewer phone lines coming in from the outside. The disadvantage is that the company has to purchase, rent, or lease the PBX equipment. Thus, there is a trade-off between the expense of the PBX equipment and the savings in the reduced number of incoming phone lines.

The Royal Canadian Pacific operates summer-only excursion trains through the Canadian Rockies. It retrofitted its 1920s-era cars to provide its passengers Internet access by installing a standard telephone with a modem jack in the train's lounge area and each stateroom. The phones are connected via ordinary twisted-pair copper wiring to an onboard private branch exchange (PBX) that is wired to six rack-mounted cellular telephones, which are in turn connected to antennas mounted on the roof of the railcar. When a passenger plugs his or her laptop into the modem jack, he or she is able to connect to the Internet at speeds of up to 9.6 Kbps, depending on the distance of the train from a cell tower and the topography.[7]

WATS Service

Wide-area telecommunications service (WATS) is a billing method for heavy users of voice services. When you dial a company at a toll-free 800 or 888 number to place an order or make a query, you are using WATS. The company or organization you call via WATS pays a fee to the phone company, depending on the level of service and usage. The fee varies depending on the caller's geographic location within the United States and the number of incoming and outgoing calls. Companies that rely on phones for customer service typically use WATS services because customers can call the WATS number free of charge. For companies with a high volume of calls, WATS can also be substantially less expensive than a normal billing schedule. It is even possible for individuals to get a personal toll-free number.

Phone and Dialing Services

Common carriers are beginning to provide more and more phone and dialing services to home and business users. Automatic number identification (ANI), or caller ID, equipment can be installed on a phone system to identify and display the number of an incoming call. In a business setting, ANI can be used to identify the caller and link that caller with information stored in a computer. For example, when a customer calls Federal Express, the customer service rep uses the ANI to identify the name and address of the customer, thus saving time when handling a request for a pickup. The ANI can be very useful in helping people screen calls before they are answered. Unwanted phone calls from other people and businesses can be identified before the phone is ever answered. ANI, however, doesn't always work well with different carriers. Common carriers offer even more services to extend the capabilities of the typical phone system. Even with all the advances in computers and telecommunications, common carrier services remain important. Some of these services are listed here:

- The ability to integrate personal computers so that the telephone number of the caller is automatically captured and used to look up information in a database about a customer
- Use of access codes to screen out junk calls, wrong numbers, and other unwanted phone calls
- Call screening priorities (e.g., only certain calls are transmitted during certain times of the day, such as from 10:00 P.M. to 7:00 A.M.)
- The ability to use one number for a business phone, home phone, personal computer, fax, etc.
- Intelligent dialing (when a busy signal is received, the phone redials the number when your line and the line of the party you are trying to reach are both free).

Telemarketers use many of these dialing services to capture your telephone number. Read more about telemarketers and what you can do to protect your privacy in the "Ethical and Societal Issues" special-interest box.

Digital Subscriber Line

digital subscriber line (DSL)
a communications line that uses existing phone wires going into today's homes and businesses to provide transmission speeds exceeding 500 Kbps at a cost of $20 or more per month

A **digital subscriber line (DSL)** transmits digital data to your computer directly as digital data and enables the phone company to use a much wider bandwidth for transmitting it to you. If you wish, the signal can be separated so that some of the bandwidth is used to transmit an analog signal so that you can use your telephone and computer on the same line and at the same time. A DSL uses existing phone wires going into today's homes and businesses to provide transmission speeds exceeding 500 Kbps at a cost of $20 or more per month. This speed means faster Internet access and downloads compared with standard phone lines. DSL are not available everywhere, but their use is growing rapidly.

In early 2002, the Federal Communications Commission (FCC) decided to reclassify high-speed Internet access services offered over telephone facilities as information services rather than the more heavily regulated telecommunications services. The reclassification applies to DSL services offered by local phone companies. The FCC's rationale for the reclassification was to increase the availability of DSL to all Americans, to spur competition, and to boost investment and innovation by ensuring the services are minimally regulated.[8]

A special modem costing a few hundred dollars is required for use of DSL. DSL modems follow standards established by both North American and European services. In general, the maximum range for DSL without a repeater is about 3.4 miles. The closer the DSL connection is to the telephone company office, the faster the data access. Another factor affecting transmission rates is the

Telemarketers—Please Do *Not* Call!

All of us get telemarketing calls—and usually at the most inconvenient times! While our meal gets cold, marketers rapidly recite their sales pitch and don't even allow us to get a word in edgewise about a product or service in which we have absolutely no interest.

Telemarketers can get your phone number in various ways. Of course, it's easy enough for them to look up your number in the phone book. If you have an unlisted number, they can use a Criss-Cross Directory that shows addresses, names, and telephone numbers for nearly everyone. Many people have their phone number preprinted on their checks, where it is readily available to be copied and input into a telemarketer's database. If you sign up for a contest or drawing, your name also ends up on a telemarketer's call list. In fact, collecting lots of names and phone numbers is the real purpose of many promotions. When you call 800, 888, 877, and 900 numbers, your phone number can be captured through Automatic Number Identification and then matched to computerized lists and street address directories. Telemarketers purchase lists of telephone numbers from reputable companies who refuse to keep such information private. Telemarketers don't even need to know your phone number; they also use automatic dialing devices that call every possible phone number.

Two federal laws regulate telemarketing. In addition, many states have laws regarding the use of telephone lines for telemarketing. The Telephone Consumer Protection Act of 1991 requires telemarketers to follow strict guidelines for solicitation calls, including the creation and use of do not call (DNC) lists. Direct response marketers must maintain an updated database of DNC consumers to avoid federal and state fines. The Telemarketing and Consumer Fraud Abuse Prevention Act also requires telemarketers to keep DNC lists, and under certain conditions, you can sue telemarketers in federal court if they do not comply. Telemarketers are restricted in the hours they may call (from 8:00 A.M. to 9:00 P.M.) and must make certain disclosures about the nature of the call and the goods or services they represent. They cannot make false or misleading statements or misrepresent any information, including the total cost and the quality of any goods or services they are offering.

Software manufacturers have developed products to help telemarketers conform to these acts. For example, Gryphon Networks has introduced a patented technology for telemarketers that automatically blocks them from making calls to DNC list registrants or placing calls during call curfew hours.

The service also allows telemarketers to add consumers to their DNC lists, at the consumer's request, by simply pushing two keys on their phone at the time of the call.

Numerous products are also available for consumers. Easy HangUp and Phone Butler get telemarketers off the line with the push of a button. A recorded message tells them to stop calling and ends the call. Telemarketer Stopper and Call Me Not play a "disconnected" tone to callers. Computer dialers detect the tone and remove your number from their database because they think it is dead.

The Direct Marketing Association (DMA) offers a Telephone Preference Service used by national telemarketers. You can write to the Direct Marketing Association (DMA) to request that your phone number be added to their "don't call" list. Unfortunately, not all telemarketers participate in the DMA program.

The Consumer Protection Association of America will actively pursue any complaints on your behalf with telemarketing companies or other organizations that continue to call you after you've requested to be taken off their list. You can file a complaint with CPAA on-line at www.consumerpro.com.

Discussion Questions

1. What are some of the ways mentioned here that telemarketers might use to get your phone number? Which of these strike you as being unethical?
2. Visit the Web site of the Consumer Protection Association and write a brief paragraph about the information you find on telemarketers.

Critical Thinking Questions

3. The Telemarketing and Consumer Fraud Abuse Prevention Act does not apply to nonprofit organizations. How do you feel about this?
4. The Telemarketer Stopper and Call Me Not products play a "disconnected" tone to callers. Computer dialers detect the tone and remove your number from their database because they think it is dead. Are there any potential negative repercussions from the use of such products?

Sources: Adapted from "Reduce Telemarketing Calls," accessed at http://amerishop.com, April 3, 2002; "Gryphon Networks: Protecting Consumers and Telemarketing Firms; Company Works to Help Telemarketers Comply with Do-Not-Call Legislation," *PRNewswire*, June 28, 2001, accessed at http://www.prnewswire.com; "How to Hang Up on Telemarketers for Good," Consumer Protection Association of America Web site at http://www.consumerpro.com, accessed April 3, 2002.

---

gauge of the copper wire. The heavier 24 gauge wire carries the same data rate farther than 26 gauge wire. If you live beyond the 3.4 mile range, you may still be able to have DSL access if your phone company has extended the local loop with fiber-optic cable.

ISDN

integrated services digital network (ISDN)
a technology that uses existing common-carrier lines to simultaneously transmit voice, video, and image data in digital form

Many telephone companies now offer **integrated services digital network (ISDN)** service, a technology that uses existing common-carrier lines to simultaneously transmit voice, video, and image data in digital form. ISDN also offers high rates of transmission: the digital service has the capacity to send a 22-page document in about a second! With ISDN, communications devices require a special ISDN board. These data communications systems use an ISDN network switch—a digital switch that allows different communications services to be connected to the system. For example, ISDN allows long-distance services, video and voice services, facsimile devices, telephones, and private branch exchanges to be integrated into one telecommunications system (Figure 6.13). ISDN digital networks are typically faster (from 64 Kbps to 2 Mbps) than standard phone lines and can carry more signals than analog networks. They also allow for easier sharing of image, multimedia, and other complex forms of data across telephone lines.

T1 Carrier

T1 carrier

a line or channel developed by AT&T and used in North America to increase the number of voice calls that can be handled through existing cables

The **T1 carrier** was developed by AT&T to increase the number of voice calls that could be handled through the existing cables. For digital communications, T1 is the carrier used in North America. T1 is also suitable for data and image transmissions. Large companies frequently purchase T1 lines to develop an integrated telecommunications network that can carry voice, data, and images. T1 has a speed of 1.544 Mbps developed from two dozen 64-Kbps channels, together with one 8-Kbps channel for carrying control information. T1 services are quite expensive, with subscribers paying a monthly fee based on the distance (several dollars per mile in some cases), and there is also a high installation fee.

With all of these telecommunications options, the problem for individuals and businesses is how to choose the best. Each option has its own cost, speed, and reliability to consider. Table 6.1 shows some of the costs, advantages, and disadvantages of different lines and services offered by communications carriers.

FIGURE 6.13

ISDN Network Switching

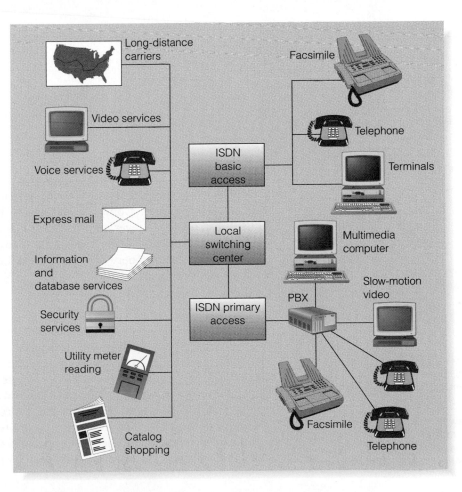

Line/Service	Speed	Cost per Month	Advantages	Disadvantages
Standard phone service	56 Kbps	$10–$40	Low cost and broadly available	Too slow for video and downloads of large files
ISDN	64 Kbps–128 Kbps	$50–$150	Fast for video and other applications	Higher costs and not available everywhere
DSL	500 Kbps–1.544 Mbps	$20–$120 in addition to standard phone service	Fast, and the service comes over standard phone lines	Slightly higher costs and not available everywhere
Cable modem	Receive at up to 500 Kbps and send at speeds up to 1.544 Mbps	$20–$120	Fast and uses existing cable that comes into the home	Slightly higher costs and not available everywhere
T1	1.544 Mbps	$600–$1,200	Very fast broadband service, typically used by corporations and universities	Very expensive, high installation fee, and users pay a monthly fee based on distance
Satellite	300 Kbps–800 Kbps downstream; 60 Kbps–300 Kbps upstream	$30–$120	Can be used where cable and DSL are not available. Data services can come bundled with TV service.	Installation fees can be several hundred dollars

TABLE 6.1

Costs, Advantages, and Disadvantages of Several Line and Service Types

NETWORKS AND DISTRIBUTED PROCESSING

Businesses link their personnel and equipment to enable people to work quicker and more efficiently. Computer networks allow organizations flexibility—to accomplish work wherever and whenever it is most beneficial. To take full advantage of networks and distributed processing, professionals in all areas must understand strategies, network concepts and considerations, network types, and related network topics.

BASIC PROCESSING STRATEGIES

When an organization needs to use two or more computer systems, one of three basic processing strategies may be followed: centralized, decentralized, or distributed. With **centralized processing**, all processing occurs in a single location or facility. This approach offers the highest degree of control. For example, centralized processing is useful for financial institutions that require a high degree of security. With **decentralized processing**, processing devices are placed at various remote locations. The individual computer systems are isolated and do not communicate with each other. Decentralized systems are suitable for companies that have independent operating divisions. Some drug store chains, for example,

centralized processing
processing alternative in which all processing occurs in a single location or facility

decentralized processing
processing alternative in which processing devices are placed at various remote locations

operate each location as a completely separate entity; each store has its own computer system that works independently of the computers at other stores. With **distributed processing**, computers are placed at remote locations but connected to each other via telecommunications devices. Consider a manufacturing company with plants in Milwaukee, Chicago, and Atlanta and a corporate headquarters in New York. Each location has its own computer system. By connecting all the computer systems into a distributed processing system, all the locations can share data and programs. Distributed processing also allows each plant to perform its own processing (say, for example, inventory) while the New York computer system coordinates and processes other applications, like payroll.

One benefit of distributed processing is that processing activity can be allocated to the location(s) where it can most efficiently occur. For example, the New York headquarters may have the largest computer system, but the Atlanta office might have hundreds of employees to input the data. The system's output may be most needed in Chicago, the location of the warehouse. With distributed processing, each of these offices can organize and manipulate the data to meet its specific needs, as well as share its work product with the rest of the organization. The distribution of the processing across the organizational system ensures that the right information is delivered to the right individuals, maximizing the capabilities of the overall information system by balancing the effectiveness and efficiency of each individual computer system.

The September 11 terrorist attacks caused many companies to distribute their workers, operations, and systems much more widely, a reversal of the recent trend toward centralization. The goal is to minimize the consequences of a catastrophic event at one location while ensuring uninterrupted systems availability. Empire Blue Cross Blue Shield, New York's largest health insurer, had its headquarters in one of the World Trade Center towers. Despite the destruction of the facility, most of Empire's services continued uninterrupted because of the planning and effort the insurer had put into building a fully redundant technology architecture. Part of the planning included redundant communications facilities with wireless and satellite communications augmenting traditional communications.[9]

NETWORK CONCEPTS AND CONSIDERATIONS

Networks that link computers and computer devices provide for flexible processing. Building networks involves two types of design: logical and physical. A logical model shows how the network will be organized and arranged. A physical model describes how the hardware and software in the network will be physically and electronically linked.

Network Topology

The number of possible ways to logically arrange the nodes, or computer systems and devices on a network, may seem limitless. Actually, there are only five major types of **network topologies**—logical models that describe how networks are structured or configured. These types are ring, bus, hierarchical, star, and hybrid (Figure 6.14).

The **ring network** contains computers and computer devices placed in a ring, or circle. With a ring network, there is no central coordinating computer. Messages are routed around the ring from one device or computer to another in one direction. A **bus network** consists of computers and computer devices on a single line. Each device is connected directly to the bus and can communicate directly with all other devices on the network. The bus network is one of the most popular types of personal computer networks. The **hierarchical network** uses a treelike structure. Messages are passed along the branches of the hierarchy until they reach their destination. Like a ring network, a hierarchical network does not require a centralized computer to control communications. Hierarchical networks are easier to repair

---

distributed processing
processing alternative in which computers are placed at remote locations but connected to each other via telecommunications devices

network topology
logical model that describes how networks are structured or configured

ring network
a type of topology that contains computers and computer devices placed in a ring, or circle; there is no central coordinating computer; messages are routed around the ring from one device or computer to another

bus network
a type of topology that contains computers and computer devices on a single line; each device is connected directly to the bus and can communicate directly with all other devices on the network; one of the most popular types of personal computer networks

hierarchical network
a type of topology that uses a treelike structure with messages passed along the branches of the hierarchy until they reach their destination

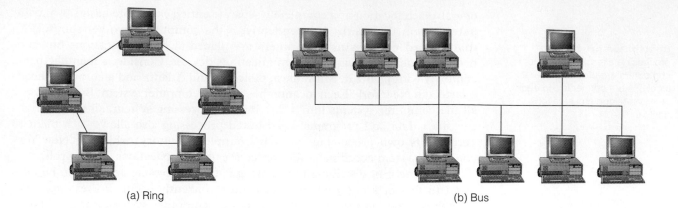

(a) Ring   (b) Bus

(c) Hierarchical   (d) Star

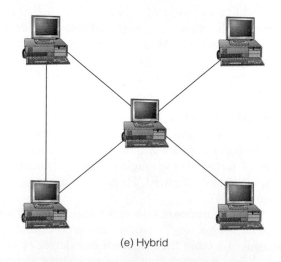

FIGURE 6⊙14

The four basic types of network topology are (a) ring, (b) bus, (c) hierarchical, and (d) star. In addition, a hybrid configuration (e) can be formed from elements of any of these four topologies.

(e) Hybrid

star network
a type of topology that has a central hub or computer system, and other computers or computer devices are located at the end of communications lines that originate from the central hub or computer

than other topologies because you can isolate and repair one branch without affecting the others. A **star network** has a central hub or computer system. Other computers or computer devices are located at the end of communications lines that originate from the central hub or computer system. The central computer of a star network controls and directs messages. If the central computer breaks down, it results in a breakdown of the entire network. Many organizations

hybrid network
a network topology that is a combination of other network types

use a **hybrid network**, which is simply a combination of two or more of the four topologies just discussed. The exact configuration of the network depends on the needs, goals, and organizational structure of the company involved.

NETWORK TYPES

Depending on the physical distance between nodes on a network and the communications and services provided by the network, networks can be classified as local area, wide area, or international. Local area networks tie together equipment in a building or local area; international networks are used to communicate between countries. Wide area networks operate over a broad geographic area.

Local Area Networks

local area network (LAN)
a network that connects computer systems and devices within the same geographic area

A network that connects computer systems and devices within the same geographic area is a **local area network (LAN)**. A local area network can be a ring, bus, star, hierarchical, or hybrid network. Typically, local area networks are wired into office buildings and factories (Figure 6.15). While unshielded twisted-pair wire cable is the most widely used media with LANs, other media—including fiber-optic cable—is also popular. They can be built around powerful personal computers, minicomputers, or mainframe computers. When a personal computer is connected to a local area network, a network interface card (NIC) is usually required. A network interface card is a card or board that is placed in a computer's expansion slot, usually provided by the PC manufacturer when you obtain your PC, to allow it to communicate with the network. A wire or connector from the network is plugged directly into the network interface card. For example, a salesperson whose notebook computer has an interface card can establish a

FIGURE 6.15

A Typical LAN in a Bus Topology
All network users within an office building can connect to each other's devices for rapid communication. For instance, a user in research and development could send a document from her computer to be printed at a printer located in the desktop publishing center.

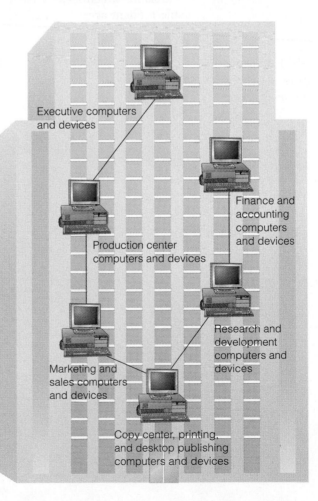

Executive computers and devices

Finance and accounting computers and devices

Production center computers and devices

Research and development computers and devices

Marketing and sales computers and devices

Copy center, printing, and desktop publishing computers and devices

link to the corporate LAN. He or she can access the network while at the office and download data needed for the next sales call.

Another basic LAN is a simple peer-to-peer network that might be used for a very small business to allow the sharing of files and hardware devices such as printers. In a peer-to-peer network, each computer is set up as an independent computer, except that other computers can access specific files on its hard drive or share its printer. These types of networks have no server. Instead, each computer uses a network interface card and cabling to connect it to the next machine. Examples of peer-to-peer networks include Windows for Workgroups, Windows 98, Windows NT, Windows 2000, and AppleShare. Performance of the computers on a peer-to-peer network is usually slower because one computer is actually sharing the resources of another computer. These networks, however, are a good beginning network from which small businesses can grow. The software cost is minimal, and network cards can be used if the company decides to enlarge the system. In addition, peer-to-peer networks are becoming cheaper, faster, and easier to use for home-based businesses.

Wide Area Networks

wide area network (WAN)
a network that ties together large geographic regions using microwave and satellite transmission or telephone lines

A **wide area network (WAN)** ties together large geographic regions using microwave and satellite transmission or telephone lines. When you make a long-distance phone call, you are using a wide area network. AT&T, MCI WorldCom, and others are examples of companies that offer WAN services to the public. Companies also design and implement WANs. These WANs usually consist of computer equipment owned by the user, together with data communications equipment provided by a common carrier (Figure 6.16).

Ireland, which lacks a single commercial DSL provider and has almost no cable modem access, is tied with Greece in last place for domestic high-speed

FIGURE 6.16

A Wide Area Network

Wide area networks are the basic long-distance networks used by organizations and individuals around the world. The actual connections between sites, or nodes (shown by dashed lines), may be any combination of satellites, microwave, or cabling. When you make a long-distance telephone call, you are using a WAN.

North America

connectivity in Western Europe. The Irish government will invest 300 million euros to build 50,000 kilometers of high-speed, fiber-optic Internet access rings around 123 of its towns and cities. The country hopes that the rings, funded 90 percent by the Irish government and 10 percent by local authorities, will help raise the country out of last place. A national public access network is also in the works and would ultimately string together all the fiber rings.[10]

International Networks

international network
a network that links systems between countries

Networks that link systems between countries are called **international networks**. However, international telecommunications comes with special problems. In addition to requiring sophisticated equipment and software, global area networks must meet specific national and international laws regulating the electronic flow of data across international boundaries, often called *transborder data flow*. Some countries have strict laws limiting the use of telecommunications and databases, making normal business transactions such as payroll costly, slow, or even impossible. Other countries have few laws concerning telecommunications and database use. These countries, sometimes called *data havens*, allow other governments and companies to avoid their country's laws by processing data within their boundaries. International networks in developing countries can have inadequate equipment and infrastructure that can cause problems and limit the usefulness of the network.

Despite the obstacles, numerous private and public international networks exist. United Parcel Service, for example, has invested in an international network, called UPSnet. UPSnet allows drivers to use handheld computers to send real-time information about pickups and deliveries to central data centers. The huge UPS network allows data to be retrieved by customers to track packages or to be used by the company for faster billing, better fleet planning, and improved customer service. (The Internet, which we will discuss in Chapter 7, is the largest public international network.)

Home and Small Business Networks

With more people working at home, connecting home computing devices and equipment together into a unified network is on the rise. Small businesses are also connecting their systems and equipment. With a home or small business network, computers, printers, scanners, and other devices can be connected. A person working on one computer, for example, can use data and programs stored on another computer's hard disk. In addition, a single printer can be shared by several computers on the network.

To make home and small business networking a reality, a number of companies are offering standards, devices, and procedures. Home Phoneline Network Alliance, for example, has developed HomePNA 2. This system allows 10-Mbps speeds over existing phone lines, without interfering with existing phone service (Figure 6.17).

Companies such as Gigafast, Linksys, NetGear, SMC, and Phonex also offer home networking products based on a new networking standard called *Powerline* that runs on your home's existing electrical wiring system. Powerline is more convenient than phone-line based systems because a typical house has several electrical outlets in every room, but phone outlets in just a few rooms. A sandwich-size Powerline adapter is required for each computer (currently about $135 per adapter, with costs expected to drop). The far end of each adapter plugs in to any electrical outlet to provide an instant high-speed network at 14 Mbps. The Powerline signal travels about 1,000 feet and doesn't slow down as you approach the limit. Nor is it susceptible to disruption to the use of other appliances.[11]

TERMINAL-TO-HOST, FILE SERVER, AND CLIENT/SERVER SYSTEMS

If an organization chooses distributed information processing, it can connect computers in several ways. Most common are terminal-to-host, file server, and client/server architecture.

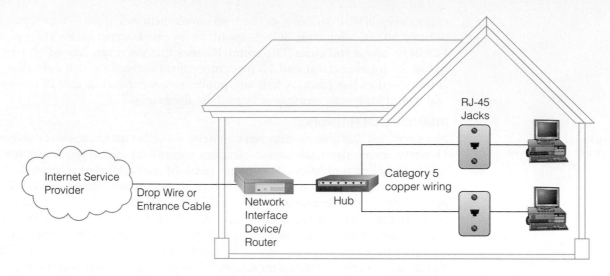

FIGURE 6.17

*Connecting Computing Devices
Using a Home Network*

(Source: Courtesy of Access
Computing, Corning, New York.
http://www.accesscomputing.com/
access/pages/homenetworks.htm,
accessed August 28, 2002.)

terminal-to-host

an architecture in which the application and database reside on one
host computer, and the user interacts with the application and data
using a "dumb" terminal

file server

an architecture in which the application and database reside on the
one host computer, called the file
server

client/server

an architecture in which multiple
computer platforms are dedicated
to special functions such as database management, printing, communications, and program
execution

FIGURE 6.18

Terminal-to-Host Connection

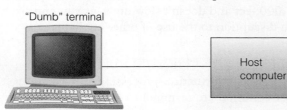

Terminal-to-Host

With **terminal-to-host** architecture, the application and database reside on one
host computer, and the user interacts with the application and data using a
"dumb" terminal. (Even if you use a PC to access the application, you run terminal emulation software on the PC to make it act as if it were a dumb terminal
with no processing capacity.) Since a dumb terminal has no data processing
capability, all computations, data accessing and formatting, and data display are
done by an application that runs on the host computer (Figure 6.18).

File Server

In **file server** architecture, the application and database reside on the one host
computer, called the *file server*. The database management system runs on the
end user's personal computer or workstation. If the user needs even a small subset of the data that resides on the file server, the file server sends the user the
entire file that contains the data requested, including a lot of data the user does
not want or need. The downloaded data can then be analyzed, manipulated, formatted, and displayed by a program that runs on the user's personal computer
(Figure 6.19).

Client/Server

In **client/server** architecture, multiple computer platforms are dedicated to special functions such as database management, printing, communications, and program execution. These platforms are called *servers*. Each server is accessible by all
computers on the network. Servers can be computers of all sizes; they store both
application programs and data files and are equipped with operating system software to manage the activities of the network. The server distributes programs and
data files to the other computers (clients) on the network as they request them. An
application server holds the programs and data files for a particular application,
such as an inventory database. Processing can be done at the client or server.

A client is any computer (often an end user's personal computer) that sends
messages requesting services from the servers on the network. A client can con-
verse with many servers concurrently. A user at a personal computer initiates a
request to extract data that resides in a database somewhere on the network. A
data request server intercepts the request and determines
on which data server the data resides. The server then
formats the user's request into a message that the database server will understand. Upon receipt of the message,
the database server extracts and formats the requested
data and sends the results to the client. Only the data
needed to satisfy a specific query is sent—not the entire

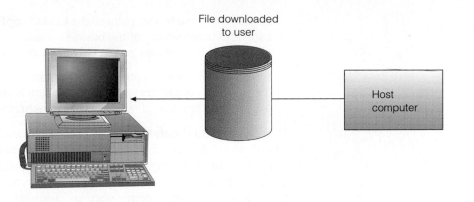

File downloaded to user

FIGURE 6.19

File Server Connection

The file server sends the user the entire file that contains the data requested. The downloaded data can then be analyzed, manipulated, formatted, and displayed by a program that runs on the user's personal computer.

file (Figure 6.20). As with the file server approach, once the downloaded data is on the user's machine, it can then be analyzed, manipulated, formatted, and displayed by a program that runs on the user's personal computer.

There are several advantages of the client/server approach over both the terminal-to-host and file server approaches: reduced cost, improved performance, and increased security. IBM Global Services signed a five-year, $50 million contract with The American Cancer Society, giving it responsibility for running the nonprofit organization's data center in Austin, Texas, helping implement an enterprise resource planning system, and installing a client/server computing environment that would replace the organization's mainframe computers. The client/server environment is complex, and IBM is providing the expertise to operate and support the new systems. In addition, the client/server environment is expected to help the Cancer Society reduce operating costs, improve relationships with donors, and help it make better use of donor contributions.[12]

Reduced cost potential The functionality achieved with client/server computing can exceed that provided by a traditional minicomputer or even a mainframe-based computer system at a lower cost. With client/server computing, a powerful workstation costing less than $25,000 may replace much of the function provided by a midrange computer costing over $100,000. In addition, vendor contracts for workstation software and hardware support are cheaper than for midrange and mainframe computers. Thus, many organizations view the migration of applications from mainframe computers and terminal-to-host architecture to client/server architecture as a significant cost-saving opportunity. This downsizing (or, as some call it, "rightsizing") can yield significant savings in reduced hardware and software support costs.

FIGURE 6.20

Client/Server Connection

Multiple computer platforms, called servers, are dedicated to special functions such as database management, data storage, printing, communications, network security, and program execution. Each server is accessible by all computers on the network. A server distributes programs and data files to the other computers (clients) on the network as they request them. The client requests services from the servers, provides a user interface, and presents results to the user. Once data is moved from a server to the client, the data may be processed on the client.

Improved performance The most important difference between the file server and client/server architecture is that the latter much more efficiently minimizes traffic on the network. With client/server computing, only the data needed to satisfy a user query is moved from the database to the client device, whereas the entire file is sent in file server computing. The smaller amount of data being sent over the network also greatly reduces the amount of time needed for the user to receive a response.

Increased security Security mechanisms can be implemented directly on the database server through the use of stored procedures. These procedures execute faster than the password protection and data validation rules attached to individual applications on a file server. They can also be shared across multiple applications.

The type of application most appropriate for client/server architecture is one that uses large data files, requires fast response time, and needs strong security and recovery

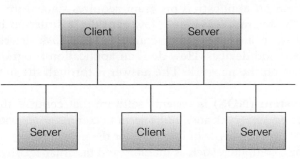

options. All these factors point to the kind of applications that are central to the operation and management of the business. On-line transaction processing and decision support applications are particularly good candidates for client/server computing.

Although client/server systems have much to offer in terms of practical benefits, some problems are associated with such systems: increased cost, loss of control, and complexity of vendor environment.

Increased cost potential   If all costs associated with client/server computing are accounted for, expected savings may fail to materialize. Moving to client/server architecture is a major two- to five-year conversion process. Over that time period, considerable costs are incurred for hardware, software, communications equipment and links, data conversion, and training. Costs are even higher for multiple-site companies converting to client/server computing. These expenses are difficult for the IS organization to track because they are often paid by the end users directly. Thus, the move to client/server architecture may be much more expensive than the IS organization realizes.

Loss of control   Controlling the client/server environment to prevent unauthorized use, invasion of privacy, and viruses is also difficult. Despite these concerns, many companies expect to gain long-term efficiency and effectiveness by moving away from large mainframe systems. The overall use of mainframe computers may not decrease, however, because mainframes are often reconfigured to become primary servers for large-scale client/server systems.

Complex multivendor environment   Implementation of client/server architecture leads to operating in a multivendor environment with, in many cases, relatively new and immature products. Situations such as these make it likely that problems will arise. Often such problems are difficult to identify and isolate to the appropriate vendor.

Nevertheless, the dominance of single-vendor environments and terminal-to-host architecture is fading fast as corporations move into the much more complex client/server environment with multiple vendors for networks, hardware, and software. Open systems are essential to implementing a client/server architecture so that managers are free to choose clients and servers and be assured that their combinations will be able to communicate with one another.

COMMUNICATIONS SOFTWARE AND PROTOCOLS

Communications Software

communications software
software that provides a number of important functions in a network, such as error checking and data security

Communications software provides a number of important functions in a network. Most communications software packages provide error checking and message formatting. In some cases, when there is a problem, the software can indicate what is wrong and suggest possible solutions. Communications software can also maintain a log listing all jobs and communications that have taken place over a specified period of time. In addition, data security and privacy techniques are built into most packages.

In Chapter 4, you learned that all computers have operating systems that control many functions. When an application program requires data from a disk drive, it goes through the operating system. Now consider a situation in which a computer is attached to a network that connects large disk drives, printers, and other equipment and devices. How does an application program request data from a disk drive on the network? The answer is through the network operating system.

network operating system (NOS)
systems software that controls the computer systems and devices on a network and allows them to communicate with each other

A **network operating system (NOS)** is systems software that controls the computer systems and devices on a network and allows them to communicate with each other. An NOS performs the same types of functions for the network as operating system software does for a computer, such as memory and task management and coordination of hardware. When network equipment (such as printers, plotters, and disk drives) is required, the network operating system makes sure that these resources are correctly used. In most cases, companies that produce and sell networks provide the NOS. For example, NetWare is the NOS from Novell, a popular network environment for personal computer systems and equipment. Windows NT and Windows 2000 are other commonly used network operating systems.

network management software

software that enables a manager on a networked desktop to monitor the use of individual computers and shared hardware (like printers), scan for viruses, and ensure compliance with software licenses

Software tools and utilities are available for managing networks. With **network management software**, a manager on a networked desktop can monitor the use of individual computers and shared hardware (like printers), scan for viruses, and ensure compliance with software licenses. Network management software also simplifies the process of updating files and programs on computers on the network—changes can be made through a communications server instead of being made on individual computers. Network management software also protects software from being copied, modified, or downloaded illegally and performs error control to locate telecommunications errors and potential network problems. Some of the many benefits of network management software include fewer hours spent on routine tasks (like installing new software), faster response to problems, and greater overall network control.

Network management is one of the most important tasks of IS managers. In fact, poor management of the network can cause a whole company to suffer. With networks now being used to communicate with customers and business partners, network outages or slow performance can even mean a loss of business. Network management includes a wide range of technologies and processes used to automate infrastructure monitoring and help IS staffs identify and address problems before they affect customers, business partners, or employees.

Fault detection and *performance management* are the two types of network management products. Both employ the *Simple Network Management Protocol* (SNMP) to obtain key information from individual network components. SNMP is the standard management protocol used on most Internet networks. It allows virtually anything on the network—including switches, routers, firewalls, hubs, and even operating systems and server products and utilities—to communicate with management software about its current operations and state of health. SNMP can also be used to control these devices and products, telling them to redirect traffic, change traffic priorities, or even shut down.

Fault management alerts IS staff in real-time when a device is failing. Equipment vendors place traps (code in a software program for handling unexpected or unallowable conditions) on their hardware to identify the occurrence of problems. In addition, the IS staff can place agents—automated pieces of software—on networks to monitor different functions. When a device exceeds a given performance threshold, the agent sends an alarm to the company's IS fault management program. For example, if a CPU registers that it is more than 80 percent busy, an alarm may be generated.

Performance management software sends messages to the various devices (i.e., polls them) to sample their performance and to determine whether they are operating within acceptable levels. The devices reply to the management system with performance data that the system stores in a database. This real-time data is automatically correlated to historical trends and displayed graphically so that the IS staff can identify any unusual variations.

Today, most IS organizations use a combination of fault management and performance management to ensure that their network remains up and running and that every network component and application is performing acceptably. With the two technologies the IS staff can identify and resolve fault and performance issues before they affect customers and service. The latest network management technology even incorporates automatic fixes—the network management system identifies a problem, notifies the IS manager, and automatically corrects the problem before anyone outside the IS department notices it.

Ipswitch's network management software, WhatsUp Gold, uses SNMP to communicate with network devices—polling them to check their status and responding to messages they send. If there is a problem, the software is programmed to notify the designated network manager via e-mail, automated voice mail, or pager. The software offers numerous reporting tools, including maps and charts that can be understood easily, even by nonexperts. Georgia State University uses the WhatsUp Gold network management software to support approximately 4,400 nodes connected by numerous networking technologies.[13]

Communications Protocols

Communications protocols make communications possible. A number of communications **protocols** are used by companies and organizations of all sizes. Just as standards are important in building computer and database systems, established protocols help ensure communications among computers of different types and from different manufacturers.

OSI Many protocols have layers of standards and procedures. The **Open Systems Interconnection (OSI) model** serves as a standard model for network architectures and is endorsed by the International Standards Committee. The OSI model divides data communications functions into seven distinct layers to promote the development of modular networks that simplify the development, operation, and maintenance of complex telecommunications networks. These layers are described in Figure 6.21.

TCP/IP In the 1970s, the U.S. government pioneered development of the **Transmission Control Protocol/Internet Protocol (TCP/IP)** to link its defense research agencies. The government has adopted OSI standards to replace TCP/IP, but TCP/IP remains the major network protocol used by schools and businesses. It is the primary communications protocol of the Internet. The most recent version available before publication of this book was TCP/IP 4.0.

SNA IBM has also developed a communications protocol, called Systems Network Architecture (SNA), which is a protocol used for IBM systems. Because of the popularity of IBM systems, many other computer manufacturers and communications companies have made their systems compatible with the SNA protocol.

Ethernet Ethernet is the most widely installed local area network (LAN) technology. Specified in a standard, IEEE 802.3, Ethernet was originally developed by Xerox and then developed further by Xerox, Digital Equipment Corporation, and Intel. An Ethernet LAN typically uses coaxial cable or special grades of twisted-pair wires. The most commonly installed Ethernet systems are called 10BASE-T and provide transmission speeds up to 10 Mbps. Devices are connected to the cable and compete for access using a Carrier Sense Multiple Access with Collision Detection (CSMA/CD) protocol. Following this protocol, devices on the LAN "listen" before and during transmitting to detect whether another device on the LAN also transmitted, thus disrupting their original message. If such a "collision" is detected, the device retransmits the original message after waiting a random time interval.

Fast Ethernet or 100BASE-T provides transmission speeds up to 100 Mbps and is typically used for LAN systems, supporting workstations with 10BASE-T cards. Gigabit Ethernet provides an even higher level of support at 1000 Mbps—equal to 1 gigabit (Gb), or 1 billion bits per second.

ATM ATM (asynchronous transfer mode) is a switching technology that relies on dedicated connections and organizes digital data into 53-byte cell units. These cells are then transmitted over a physical medium using digital signal technology. Individually, a cell is processed asynchronously relative to other related cells and is queued before being multiplexed over the transmission path. ATM is designed to

protocols
rules that ensure communications among computers of different types and from different manufacturers

Open Systems Interconnection (OSI) model
a standard model for network architectures that divides data communications functions into seven distinct layers to promote the development of modular networks that simplify the development, operation, and maintenance of complex telecommunications networks

Transmission Control Protocol/Internet Protocol (TCP/IP)
the primary communications protocol of the Internet, originally developed to link defense research agencies

FIGURE 6.21

The Seven Layers of the OSI Model

This Open Systems Interconnection (OSI) model is designed to permit communication among different computers from different manufacturers using different operating systems—as long as each conforms to the OSI model.

(Source: *Information Systems for Managers, 3/e*, pp. 134–135, by George Reynolds, 1995, West Publishing. Reprinted with permission from Course Technology.)

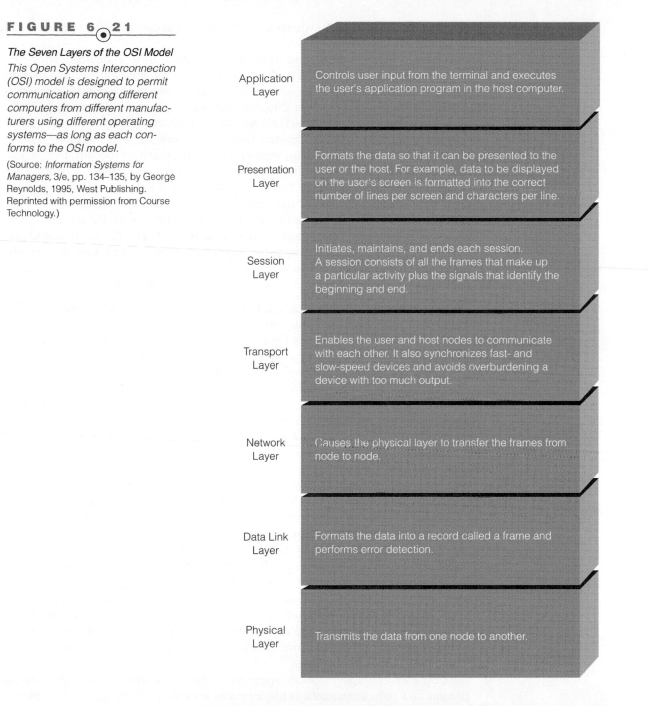

Application Layer — Controls user input from the terminal and executes the user's application program in the host computer.

Presentation Layer — Formats the data so that it can be presented to the user or the host. For example, data to be displayed on the user's screen is formatted into the correct number of lines per screen and characters per line.

Session Layer — Initiates, maintains, and ends each session. A session consists of all the frames that make up a particular activity plus the signals that identify the beginning and end.

Transport Layer — Enables the user and host nodes to communicate with each other. It also synchronizes fast- and slow-speed devices and avoids overburdening a device with too much output.

Network Layer — Causes the physical layer to transfer the frames from node to node.

Data Link Layer — Formats the data into a record called a frame and performs error detection.

Physical Layer — Transmits the data from one node to another.

be easily implemented in hardware rather than software, so that faster processing and switch speeds are possible. Speeds on ATM networks can reach 10 Gbps.

Bluetooth Named after Harald Bluetooth, a Danish king of the 10th century, Bluetooth is a communications standard defined by researchers at telecommunications giant Ericsson in 1994. Its goal is to simplify communications among cell phones, handheld computers, and other wireless devices by streaming information back and forth using short-range radio waves.[14] A few Bluetooth products have been developed, including cell phones that connect wirelessly to earpieces, computers that can beam PowerPoint presentations directly to a projector, and medical gear that stream a patient's vital statistics onto a display screen. Luxury automobiles Jaguar and Infiniti used Bluetooth and voice-recognition software to create a state-of-the-art car phone system that can detect a phone anywhere in the car, mute the car radio when calls come in, and allow the caller to place a call by speaking the phone number.

FedEx couriers track and scan package data into a device called the SuperTracker, a handheld bar-code scanner. Currently, they must return to their vans to upload data from the scanner to a radio device that communicates with FedEx information systems for shipping rates and mailing restrictions. But FedEx is testing a new PowerPad portable wireless device that incorporates Bluetooth wireless technology. With it, couriers instantly retrieve and send shipping data to the FedEx information systems, without having to return to the van. FedEx also plans to equip its couriers with Bluetooth-enabled portable printers to print shipping labels and with smart phones to upload data to the wireless area network.[15]

Palm Inc. released a software development kit that enables developers to write Bluetooth-compatible programs for the Palm operating system. Palm is also working with hardware manufacturers, including Red-M, Northstar Systems, and TDK Systems, to build compatible hardware development kits. Bluetooth-enabled Palm-powered devices will be able to communicate with a wide range of other devices, including cell phones, computers, and LAN access points.[16]

802.11 There are a number of 802.11 technical specifications for wireless LANs developed by a working group of the Institute of Electrical and Electronics Engineers (IEEE). All use the Ethernet protocol and CSMA/CA (carrier sense multiple access with collision avoidance) for path sharing.

The first 802.11 standard was originally developed and patented in 1942. It allowed vendors to use either Frequency Hopping Spread Spectrum (FHSS) or Direct Sequence Spread Spectrum (DSSS). FHSS "hops" among 79 1-MHz-wide channels of the 2.4-GHz band, while DSSS spreads its signals across the band, with three 22-MHz-wide DSSS channels that are available.

The 802.11a specification applies to wireless ATM systems and high-speed switching devices. 802.11a operates at frequencies between 5 GHz and 6 GHz, with a maximum data transmission speed of 54 Mbps, but most commonly, communications takes place at 6 Mbps, 12 Mbps, or 24 Mbps. Mercedes-Benz has developed a C320 sedan that incorporates the 802.11a wireless LAN technology in a high-speed wireless communications system capable of sending and receiving data in very quick bursts. Unlike regular cell-phone networks that provide constant coverage, the Mercedes DriveBy InfoFueling system sends data in a quick burst as vehicles pass transceiver sites with limited range. A car would send and receive data as it sped past a transceiver at a base station. It would not be able to send or receive again until it passed another base station. The system can be used to download updated maps, real-time traffic data, and digital music or video. It also can send data from the moving car to a person connected to a conventional LAN or the Internet.[17]

The 802.11b specification operates at 2.4 GHz and is used by wireless phones and remote controls. It has a maximum transmission speed of 11 Mbps. The 802.11b standard, also called Wi-Fi, is considered a competitor to the Bluetooth standard. Wi-Fi allows faster transmissions than Bluetooth and can support connections across longer distances, up to 300 feet compared with just 30 feet for Bluetooth. However, Bluetooth chips are less expensive to make and consume less power than Wi-Fi chips. They are also easier to build into small devices that run on batteries, such as cell phones and palmtop computers. 802.11b equipment is not compatible with 802.11a because it operates in different frequencies. People who have 802.11b, or Wi-Fi gear, and want additional bandwidth to surf the Web at the faster 802.11a speed must buy all new equipment or upgrade their current networks with dual-radio modems or PC cards that accommodate both standards.

Hofstra's School of Law implemented an in-building wireless data network based on the 802.11b standard for wireless LANs. Students and teachers can sit in a courtyard, study area, or classroom and use notebook computers and wireless modems to link to the network to conduct research, work on joint projects, and communicate via instant messaging and e-mail. Students must have their own laptop or notebook computer and either buy a wireless 802.11b card ($126) or borrow one free from the school to access the wireless network. The wireless network cost the school only about $12,000 to install—meaning each of the 30 access points cost only around $400.[18]

WiFi or 802.11b is another way to provide wireless network connections for the computers in your home. This system works as cordless phones do—you need a base station (about $180), which communicates with antenna cards ($80 to $100) in each personal computer. You can move your laptop computer through the house and remain online without plugging in to anything. A drawback is that the wireless laptop's connection slows down as it moves farther from the base station, and at about 150 feet, it disconnects completely. Another problem is that 802.11b wireless networks are disrupted by 2.4-gigahertz cordless phones and microwave ovens. Security and privacy are also an issue—if you live next door to a family of WiFi users, it's possible for their wireless computer to get a free ride to the Internet on your cable modem.[19]

The most recently approved standard, 802.11g, offers wireless transmission over relatively short distances at up to 54 Mbps and operates in the 2.4 GHz range and is thus compatible with 802.11b, or Wi-Fi.

Glendale, California, is using wireless LAN technology to provide high-speed data service to its police, fire, and public works departments. Glendale uses 802.11 wireless LAN equipment that provides 1 Mbps throughput. While this is not as fast as the 802.11b standard that supports 11 Mbps, it does provide a much wider range. The high cost and slow throughput (19.2 Kbps) of Cellular Digital Packet Data (CDPD) services available from cellular carriers eliminated that as an option. Higher throughput is needed to support certain applications, such as the transmission of mug shots to patrol cars and data from a geographic information system used by the public works department. Glendale's solution is an indication that wireless LANs are starting to crack into markets that cellular carriers, which offer wide-area service, previously dominated.[20]

MMDS Multichannel Multipoint Distribution System (MMDS) allows phone companies to deliver T1 speeds to the homes of individual users over rooftop antennas mounted anywhere within a 35-mile radius of a powerful transmission tower. Spike Broadband Systems is using this technology to build a wireless broadband-access system for Denmark's second largest mobile-phone company, Sonofon. Spike claims they can build the infrastructure at a fraction of the cost required to build cable-modem or DSL networks. In addition, the system can also provide local telephone service.[21]

Gorman Uniform Service, a uniform rental business in Houston, Texas, contracted with Sprint Corp. to provide MMDS broadband wireless. Gorman chose MMDS wireless over DSL and T1 services based on costs—the MMDS service cost $149 per month compared with a T1 line that cost $700 per month and with DSL at $300 to $400 per month.[22]

BRIDGES, ROUTERS, GATEWAYS, AND SWITCHES
Many LANs have hardware and software devices that allow them to communicate with other networks employing different transmission media or protocols (Figure 6.22).

Bridge
A **bridge** connects two or more networks at the media access control portion of the data link layer. The two networks must use the same communications protocol.

bridge
connection between two or more networks at the media access control portion of the data link layer; the two networks must use the same communications protocol

Two Approaches to Electronic Data Interchange

Many organizations now insist that their suppliers operate using EDI systems. Often the EDI connection is made directly between vendor and customer (a); alternatively, the link may be provided by a third-party clearinghouse, which provides data conversion and other services for the participants (b).

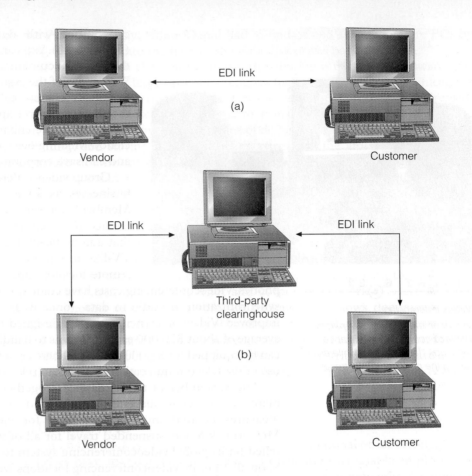

Vendor   Customer

EDI link

(a)

EDI link   EDI link

Third-party
clearinghouse

(b)

Vendor   Customer

activities will continue to evolve. Processes as simple as billing and ordering will change and new industries will emerge to help build the networks needed to support EDI.

PUBLIC NETWORK SERVICES

public network services
systems that give personal computer users access to vast databases and other services, usually for an initial fee plus usage fees

Public network services give personal computer users access to vast databases, the Internet, and other services, usually for an initial fee plus usage fees. Public network services allow customers to book airline reservations, check weather forecasts, get information on TV programs, analyze stock prices and investment information, communicate with others on the network, play games, and receive articles and government publications. Fees, based on the services used, can range from under $15 to over $500 per month (Figure 6.25). Providers of public network services include Microsoft, America Online, and Prodigy. These companies provide a vast array of services, including news, electronic mail, and investment information. AOL is the number one provider of public network services in terms of size.

DISTANCE LEARNING

Telecommunications can be used to extend the classroom. Northrop Grumman Information Technology needed to provide consistent training to its 5,000 employees around the world. The federal contractor chose SmartForce PLC's Enterprise System, with over 1,500 on-line courses, at a total annual cost of $67 per employee. In one year, employees completed a total of 4,000 courses and 8,000 hours of training. Employee surveys have been "overwhelmingly positive," and 37 percent of employees have taken at least one class or used a module from a class as a just-in-time learning tool.[31] Travel costs are being dramatically cut with the new system, and the quality of the training is expected to greatly improve. Often called

distance learning
the use of telecommunications to
extend the classroom

distance learning or cyberclasses, these electronic classes are likely to thrive in the future.

With distance learning software and systems, instructors can easily create course home pages on the Internet. Students can access the course syllabus and instructor notes on the Web page. E-mail mailing lists can be established so students and the instructor can easily e-mail one another as a means of turning in homework assignments or commenting and asking questions about material presented in the course. It is also possible to form chat groups so that students can work together as a "virtual team," which meets electronically to complete a group project.

SPECIALIZED SYSTEMS AND SERVICES

In addition to the applications just discussed, there are a number of other specialized telecommunications systems and services. For example, with millions of personal computers in businesses across the country, interest in specialized and regional information services is increasing. Specialized services, which can be expensive, include professional legal, patent, and technical information. For example, investment companies can use systems such as Quotron and Shark to get up-to-the-minute information on stocks, bonds, and other investments.

Regional services, also called metropolitan services, include local electronic bulletin boards and electronic mail facilities that offer information regarding local club, school, and government activities. An electronic bulletin board is a message center that displays messages in electronic form, much like a bulletin board displays paper messages in schools and offices. An electronic bulletin board can be accessed by subscribers with personal computers, network equipment, and software. In addition to regional bulletin boards, national and international bulletin boards are available for people and groups with special interests or needs. These types of bulletin boards exist for many users, such as users of certain software packages and users with certain hobbies. Many public network services, including Prodigy, America Online, and CompuServe, provide access to hundreds of different bulletin boards on a variety of topics and interest areas.

Global positioning systems (GPSs) are other types of specialized telecommunications services. They have long been used by the military to find locations of troops, equipment, and the enemy—within yards in some cases. Today, GPS is being used by companies to survey land and buildings and by individuals to locate their positions while camping or exploring. Steve Wozniak, who along with Steve Jobs founded Apple Computer Inc. in 1976, has launched a new business whose goal is to make GPS devices cheap, plentiful, and indispensable to businesses and individuals. Recent advances in GPS and antenna technology, coupled with the declining cost of processing power and two-way networking, have created many new possibilities for small, inexpensive, and easy-to-use location devices. The thinking is that millions of people would buy such devices to put on their pets' collars or their children's clothing, and companies could use them to track deliveries.[32] Some auto companies have placed GPS systems in their cars to assist travelers in need. The auto systems and some advanced GPSs combine basic GPS features with a cell phone.

FIGURE 6.25

Public network services provide users with the latest information required to remain competitive. Yahoo, for example, enables registered users to obtain up-to-the-minute stock quotes.

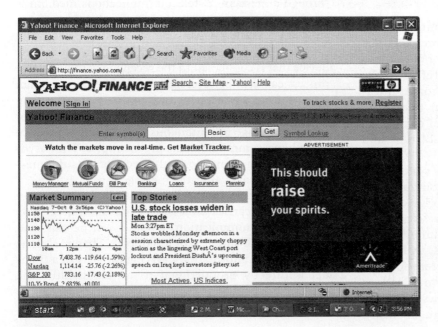

The Burlington Northern Santa Fe Corp. (BNSF) railroad has implemented a satellite-based control system to enable it to monitor and control the temperature settings in "smart" refrigerator cars. A satellite communications transmitter/receiver is installed in the refrigeration unit of each car, and status data is sent via a satellite link. If a refrigeration unit breaks down, a microchip controller card is programmed to send a message alerting the shipper and the railroad, which can dispatch a technician to fix the problem. Each smart refrigerator car is also equipped with a GPS receiver that allows the car's location to be determined within a few feet. The system helps eliminate the spoilage of perishable goods such as fruit, fruit juice, and cheese and thus has helped increase the number of shipments from customers like Tropicana and Kraft Foods.[33]

The use of pagers is also on the increase. Today's pagers contain many features, including two-way paging. Airline companies have placed phone and Internet services on many of their aircraft, allowing people to stay in touch at 30,000 feet. Six Flags is partnering with vendor Lo-Q PLC to rent Q-bot text pager devices to customers at 9 of its 38 theme parks. Instead of having to wait in long lines, park visitors can pay an extra $10 or so to register at a special kiosk for each ride. The devices will then notify them when they can get on the ride without having to wait. About 1,200 of the devices will be offered at the Six Flags Over Georgia park.[34]

With all these telecommunications systems and services, it is no wonder that managers and workers are able to conduct business from remote locations. Often called *virtual workers*, these employees can conduct business at any time and at any place. Today, new telecommunications systems and services are being introduced every month. In the future, we can expect even more innovations to dramatically alter how businesses and individuals stay connected and in touch.

SUMMARY

PRINCIPLE   *Effective communications is essential to organizational success.*

Communications is any process that permits information to pass from a sender to one or more receivers. Communications of all types form a major part of any business system. Telecommunications refers to the electronic transmission of signals for communications, including telephone, radio, and television. Telecommunications is creating profound changes in business because it lessens the barriers of time and distance.

The elements of a telecommunications system start with a sending unit, such as a person, a computer system, a terminal, or another device, that originates the message. The sending unit transmits a signal to a telecommunications device. The telecommunications device performs a number of functions, which can include converting the signal into a different form or from one type to another. A telecommunications device is a hardware component that allows electronic communication to occur or to occur more efficiently.

The telecommunications device then sends the signal through a medium. A telecommunications medium is anything that carries an electronic signal and interfaces between a sending device and a receiving device. The signal is received by another telecommunications device that is connected to the receiving computer. The process can then be reversed, and another message can go back from the receiving unit to the original sending unit. With synchronous communications, the receiver gets the message instantaneously, when it is sent. Voice and phone communications are examples. With asynchronous communications, the receiver gets the message hours or days after the message is sent.

A communications channel is the transmission medium that carries a message from the source to its receivers. Communications channel can be classified as simplex, half-simplex, or full duplex. Shannon's Law states that the information-carrying capacity of a channel is directly proportional to its bandwidth—the broader the bandwidth, the more information can be carried.

PRINCIPLE *An unmistakable trend of communications technology is that more people are able to send and receive all forms of information over greater distances at a faster rate.*

The telecommunications media that physically connect data communications devices are twisted-pair wire cable, coaxial cable, and fiber-optic cable. Twisted-pair cable consists of pairs of twisted wires, either shielded or unshielded. A coaxial cable consists of an inner conductor wire surrounded by insulation (the dielectric) and a nonconductive insulating shield (the jacket). Fiber-optic cable consists of thousands of extremely thin glass or plastic strands bound together in a sheathing (a jacket) for transmitting signals via light. Fiber-optic cables are faster and more reliable than other types, but they are expensive to install. Microwave transmission consists of a high-frequency radio signal sent through the air. Other transmission options that give organizations portability and flexibility in transmitting data are cellular and infrared transmission.

Each type of telecommunications media uses different types of connectors to connect them to devices. Twisted-pair cables frequently use Registered Jack (RJ) 11 and RJ45. Coaxial network cables generally use Bayonet Neill-Concelman (BNC) twist-lock connectors. The most common fiber-optic cable connectors are ST (a twist-lock device), FC/PC (which screws on), and the snap-in SC.

There are several types of telecommunications hardware devices. Four types of modems are internal modems, external modems, cellular modems, and cable modems. The V.90 modem protocol provides a common framework for 56-Kbps data transmission. The V.92 Modem on Hold protocol allows a user to suspend a modem session to answer an incoming voice call or to place an outgoing call while engaged in a modem session. Two additional hardware devices include multiplexers and communications processors.

Modems convert signals from digital to analog for transmission, then back to digital. Modems can be either internal or external. Cellular modems placed in portable computers allow communication with other computer systems and devices, and cable modems are designed for use with coaxial cable media. A multiplexer allows several signals to be transmitted over a single communications medium at the same time. A communications processor connects to a large number of communications lines and performs a number of tasks, including polling, providing automatic answering and calling, performing circuit checking and error detection, developing logs or reports of all communications traffic, editing basic data entering the main processor, determining message priority, choosing alternative and efficient communications paths over multiple data communications lines, and providing general data security for the main system.

Traditional phone service (sometimes called *POTS* for "plain old telephone service") connects your home or small business to a telephone company central office using twisted pairs of copper wires. The central office has switching equipment that can switch calls locally or to long-distance carrier phone offices. Switched lines use switching equipment to allow one transmission device to be connected to other transmission devices. A dedicated line, also called a *leased line,* provides a constant connection between two points. No switching or dialing is needed; the two devices are always connected.

The effective use of networks can turn a company into an agile, powerful, and creative organization, giving it a long-term competitive advantage. Networks can be used to share hardware, programs, and databases across the organization. They can transmit and receive information to improve organizational effectiveness and efficiency. They enable geographically separated workgroups to share documents and opinions, which fosters teamwork, innovative ideas, and new business strategies.

When an organization needs to use two or more computer systems, one of three basic data processing strategies may be followed: centralized, decentralized, or distributed. With centralized processing, all processing occurs in a single location or facility. This approach offers the highest degree of control. With decentralized processing, processing devices are placed at various remote locations. The individual computer systems are isolated and do not communicate with each other.

With distributed processing, computers are placed at remote locations but connected to each other via telecommunications devices. The September 11 terrorist attacks caused many companies to distribute their workers, operations, and systems much more widely, a reversal of the recent trend toward centralization. The goal is to minimize the consequences of a catastrophic event at one location while ensuring uninterrupted systems availability. Three distributed processing alternatives include terminal-to-host, file server, and client/server.

With terminal-to-host architecture, the application and database reside on the same host computer, and the user interacts with the application and data using a "dumb" terminal. Since a dumb terminal has no data processing capability, all computations, data access and formatting, and data display are done by an application that runs on the host computer.

In the file server approach, the application and database reside on the same host computer, called the *file server.* The database management system runs on the end user's personal computer or workstation. If the user needs even a small subset of the data that resides on the file server, the file server sends the user the entire file that contains the data requested, including a lot of data the user does not want or need. The downloaded data can then be analyzed, manipulated, formatted, and displayed by a program that runs on the user's personal computer.

A client/server system is a network that connects a user's computer (a client) to one or more host computers (servers). A client is often a PC that requests services from the server, shares processing tasks with the server, and displays the results. Many companies have reduced their use of mainframe computers in favor of client/server systems using midrange or personal computers to achieve cost savings, provide more control over the desktop, increase flexibility, and become more responsive to business changes. The start-up costs of these systems can be high, and the systems are more complex than a centralized mainframe computer.

Network topology refers to the manner in which devices on the network are physically arranged. Communications networks can be configured in numerous ways, but five designs are most prevalent: bus, hierarchical, star, ring, and hybrid (hybrid networks combine the basic designs of the four other topologies to suit the specific communication needs of an organization).

Ethernet is the most widely installed local area network (LAN) technology. Specified in a standard, IEEE 802.3, an Ethernet LAN typically uses coaxial cable or special grades of twisted-pair wires. The most commonly installed Ethernet systems are called 10BASE-T and provide transmission speeds up to 10 Mbps. Devices are connected to the cable and compete for access using a Carrier Sense Multiple Access with Collision Detection (CSMA/CD) protocol. ATM (asynchronous transfer mode) is a switching technology that relies on dedicated connections and organizes digital data into 53-byte cell units. These cells are then transmitted over a physical medium using digital signal technology. Bluetooth is a communications standard whose goal is to simplify communications among cell phones, hand-held computers, and other wireless devices by streaming information back and forth using short-range radio waves. 802.11 is a family of technical specifications for wireless local area networks developed by a working group of the Institute of Electrical and Electronics Engineers (IEEE). All use the Ethernet protocol and CSMA/CA (Carrier Sense Multiple Access with Collision Avoidance) for path sharing. Multichannel Multipoint Distribution System (MMDS) allows phone companies to deliver T1 speeds to the homes of individual users over rooftop antennas mounted anywhere within a 35-mile radius of a powerful transmission tower.

The physical distance between nodes on the network determines whether it is called a local area network (LAN), a wide area network (WAN), or an international network. The major components in a LAN are a network interface card, a file server, and a bridge and/or gateway. WANs tie together large geographic regions using microwave and satellite transmission or telephone lines. Value-added networks (VANs) are special WANs with additional services that permit more economical and faster communications. International networks involve communications between countries, linking systems together from around the world. These networks are also called *global area networks*. The electronic flow of data across international and global boundaries is often called *transborder data flow*.

PRINCIPLE *The effective use of telecommunications and networks can turn a company into an agile, powerful, and creative organization, giving it a long-term competitive advantage.*

Many applications of telecommunications exist today, including the following: personal computer to mainframe links, voice mail, electronic document distribution, electronic software distribution, telecommuting, videoconferencing, electronic data interchange, public network services, specialized and regional information services, and distance learning. Personal computer to mainframe links enable people to upload and download data. Voice mail users can leave, receive, and store messages from other people around the world. Electronic document distribution allows organizations to transmit documents without the use of paper, thus cutting costs and saving time. Electronic software distribution involves installing software on a computer by sending programs over a network so they can be downloaded into individual computers. Telecommuting uses information technology to enable employees to work away from the office. Videoconferencing brings groups together in voice, video, and audio calls. Electronic data interchange (EDI), another rapidly growing area, enables customers, suppliers, and manufacturers to exchange data electronically. EDI reduces the need for manual paper systems while speeding up the rate at which business can be transacted. Public network services give users access to vast databases and services, usually for an initial fee plus usage fees. Distance learning is a way to support education of students who are unable to meet frequently with their instructor. Specialized services, which are more expensive, include legal, patent, and technical information. Regional services include local electronic bulletin boards that offer e-mail facilities and information regarding local activities.

CHAPTER 6 SELF-ASSESSMENT TEST

Effective communications is essential to organizational success.

1. The electronic collection, processing, and distribution of data—typically between computer system hardware devices—is called:
   A. asynchronous communications
   B. synchronous communications
   C. data communications
   D. telecommunications

2. A (An) _____ consists of communications media, devices, and software needed to connect two or more computer systems and/or devices.

3. Communication channels can be classified as simplex, half-duplex, or full-duplex. True False

4. The translation of data from a digital signal into an analog form is called:
   A. multiplexing
   B. simplex communications
   C. modulation
   D. front-end processing

An unmistakable trend of communications technology is that more people are able to send and receive all forms of information over greater distances at a faster rate.

5. _____ states that the information-carrying capacity of a channel is directly proportional to its bandwidth—the broader the bandwidth, the more information can be carried.

6. Data has successfully been transmitted using fiber-optic network over a distance of 2,500 miles at rates faster than 2.5 terabits/second. True False

7. A(An) _____ allows several telecommunications signals to be transmitted over a single communications medium at the same time, thus reducing data communications costs.

8. Switching equipment that routes phone calls and messages within a building and connects internal phone lines to a few phone company lines is called a PBX. True False

9. With _____, the data transmission rate actually increases the closer you are located to the local telephone company's central office.

The effective use of telecommunications and networks can turn a company into an agile, powerful, and creative organization, giving it a long-term competitive advantage.

10. When an organization needs to use two or more computer systems, one of three basic processing strategies may be followed: _____, _____, or _____.

11. Which of the following is NOT one of the four basic types of network topology:
    A. ring
    B. bus
    C. hierarchical
    D. relational

12. A network that connects computer systems and devices within the same geographic area is a (an) _____.

13. In _____ architecture, multiple computer platforms are dedicated to special functions such as database management, printing, communications, and program execution.
    A. client/server
    B. file server
    C. terminal-to-host
    D. database server

14. A (An) _____ is systems software that controls the computer systems and devices on a network and allows them to communicate with each other.

15. The primary communications protocol of the Internet.
    A. OSI
    B. SNA
    C. TCP/IP
    D. Ethernet

16. The 802.11b communications protocol standard is also called Wi-Fi and is a competitor to Bluetooth. True False

17. Electronic document distribution reduces the use of paper and can be a major cost and time saver. True False

18. Installing software on a file server for users to share by signing onto the network and requesting that the software be downloaded onto their computers over a network is called:
    A. electronic document distribution
    B. voice mail
    C. electronic software distribution
    D. telecommuting

Chapter 6 Self-Assessment Test Answers:

1. c; 2. computer network; 3. True; 4. c; 5. Shannon's law; 6. True; 7. multiplexer; 8. True; 9. DSL; 10. centralized, decentralized, and distributed; 11. d; 12. LAN; 13. a; 14. network operating system; 15. c; 16. True; 17. True; 18. c.

KEY TERMS

analog signal, 235
asynchronous communications, 227
bandwidth, 229
bridge, 255
broadband, 229
bus network, 243
centralized processing, 242
client/server, 248
common carriers, 236
communications software, 250
computer network, 228
data communications, 227
decentralized processing, 242
dedicated line, 237
digital signal, 235
digital subscriber line (DSL), 239
distance learning, 261
distributed processing, 243
electronic data interchange (EDI), 259
electronic document distribution, 258
electronic software distribution, 257

file server, 248
front-end processor, 236
full-duplex channel, 229
gateway, 256
half-duplex channel, 229
hierarchical network, 243
hybrid network, 245
integrated services digital network (ISDN), 240
international network, 247
local area network (LAN), 245
modem, 235
multiplexer, 236
network management software, 251
network operating system (NOS), 250
network topology, 243
Open Systems Interconnection (OSI) model, 252
private branch exchange (PBX), 238
protocols, 252
public network services, 260

ring network, 243
router, 256
Shannon's fundamental law of information theory, 229
simplex channel, 229
star network, 244
switch, 256
switched line, 236
synchronous communications, 227
T1 carrier, 241
telecommunications medium, 227
telecommuting, 258
terminal-to-host, 248
Transmission Control Protocol/Internet Protocol (TCP/IP), 252
value-added carriers, 236
videoconferencing, 258
voice mail, 257
wide area network (WAN), 246

REVIEW QUESTIONS

1. What is a communications channel? Identify three types.
2. Describe the elements and steps involved in the telecommunications process.
3. What is Shannon's law?
4. Identify three telecommunications media that physically connect to telecommunications devices.
5. What is the role of a connector? What kinds are there?
6. Name and briefly describe two modem protocols.
7. What is POTS? What is the local loop?
8. Define the term *computer network*.
9. What advantages and disadvantages are associated with the use of client/server computing?
10. What is a T1 line? How might it be used?
11. What is a network operating system? What is its role?

12. What is network management software? Identify the two types of network management software and briefly explain how they work.
13. Identify four basic types of network topologies.
14. What role do the bridge, router, gateway, and switch play in a network?
15. Describe a local area network and its various components.
16. What is a wide area network? What is a value-added network?
17. What is EDI? Why are companies using it?
18. What is telecommuting? Why is it being used by many companies?
19. What are some of the factors that are causing an increase in the use of videoconferencing?

DISCUSSION QUESTIONS

1. Why is effective communications critical to organizational success?
2. Identify and briefly discuss three distributed processing alternatives. How does distributed processing help to minimize the consequences of a

catastrophic event at one location while ensuring uninterrupted systems availability?
3. What issues would you expect to encounter in establishing an international network for a large, multinational company?

4. Briefly discuss the pros and cons of e-mail versus voice mail. Under what circumstances would you would use one and not the other?

5. Compare and contrast the Bluetooth and the Wi-Fi communications protocols. Which one do you think will be the most successful? Why?

6. If it were available and you could afford it, which would you rather have in your home—T1, ISDN, or DSL? Why?

7. Consider an industry that you are familiar with through work experience, coursework, or a study of industry performance. How could electronic data interchange be used in this industry? What limitations would EDI have in this industry?

8. What is telecommuting? What are the advantages and disadvantages of telecommuting? Do you anticipate that you will telecommute in your career?

9. Discuss the pros and cons of conducting this course as a distance learning course.

10. Identify at least seven telecommunications protocols. Why do you think that there are so many protocols? Will there always be that many?

PROBLEM-SOLVING EXERCISES

1. You have been hired as a telecommunications consultant for a small but growing automobile parts supplier. The company wishes to develop a telecommunications system to link itself to its major customers, primarily U.S. auto manufacturers. There are many options, including EDI, use of a third party, and the Internet. The company has hired you to review its needs for the new system and to select a telecommunications solution. How would you proceed? What questions need to be asked? Use word processing software to prepare a list of at least 10 questions that you need to answer to evaluate this project. Make some assumptions about the answers and write your opinion of this project.

Embed a spreadsheet that details the approximate cost to set up the telecommunications connection between the company and its suppliers.

2. Your company is considering investing in the installation of videoconferencing equipment for each of its eight locations around the United States. Use PowerPoint or similar software to make a convincing presentation to management to adopt such a program. Your presentation must address such questions as what the benefits and disadvantages are to the company, what the costs of the hardware and software are, and how any potential environmental, technical, or competitive issues will be addressed.

TEAM ACTIVITIES

1. With a group of your classmates, visit one of the following: a cellular phone company, the college computing center, a phone or cable company, a police department, or any other interesting organization that relies heavily on telecommunications. Prepare a report on how the organization is planning to use telecommunications to enhance access to information—both yours and its. Find out what kind of telecommunications media and devices it uses currently and what changes the organization might make to improve data and information access.

2. Form a team to identify the public locations (airport, public library, Starbucks, etc.) in your area where the capability of wireless LAN connections is available. Visit at least two locations and write a brief paragraph discussing your experience at each location.

WEB EXERCISES

1. Network monitoring equipment is available from numerous vendors. Do research on the Web to identify three major vendors and write a paragraph summarizing each vendor's primary products. Try to identify at least one major customer for each vendor.

2. There are a number of on-line job-search companies, including Monster.com. Investigate one or more of these companies and research the positions available in the telecommunications industry, including the Internet. You may be asked to summarize your findings for your class in a written or verbal report.

CASES

CASE 1

Starbucks Deploys Wireless LANs

Starbucks is the number one U.S. specialty coffee retailer, and it operates more than 5,000 coffee shops in office buildings, shopping centers, airport terminals, and supermarkets in some 20 countries worldwide. Starbucks sells coffee drinks and beans, pastries, and other food items and beverages, as well as mugs, coffeemakers, coffee grinders, and storage containers. The company also sells its beans to restaurants, businesses, airlines, and hotels, and it offers mail-order and online catalogs. Starbucks has expanded into coffee ice cream (with Dreyer's) and makes Frappuccino, a bottled coffee drink (with PepsiCo).

In January 2001, Starbucks and Microsoft announced an agreement to jointly offer high-speed Internet connections in all 3,000 Starbucks stores throughout North America. Starbucks chose Compaq Computer as the main hardware supplier for this project, now part of Hewlett-Packard. H-P is providing iPaq handheld computers, which Starbucks customers can use to access broadband content and services.

The project is part of an overall Starbucks strategy to upgrade its operations including the introduction of Starbucks customer cards, which are intended to speed up orders and eventually allow customers to order drinks over mobile phones. Starbucks also believes that giving customers the ability to access a 10-Mbps wireless LAN while enjoying a cup of coffee on a comfortable couch will lure more customers into its coffee shops during off-peak post-breakfast hours—the company currently does 80 percent of its business in the morning. Starbucks has 15 million customers a week, and 90 percent of them use the Internet. So, providing Internet access may be a benefit to many of its customers. Starbucks offers the service at fees ranging from $2.50 for 15 minutes to $59.95 per month for unlimited access.

MobileStar Network Corporation was picked to install an industry-standard 802.11b LAN complete with wireless radios and antennae in Starbucks stores. One LAN is capable of supporting 20 to 40 concurrent users. Later, MobileStar pulled the plug on its wireless network services, so Starbucks partnered with the wireless subsidiaries of Deutsche Telekom, I-Mobile International, and VoiceStream Wireless. Customers access the network through a portal that Microsoft developed for Starbucks on its MSN online service. Access to that portal carries no charge, but once customers surf beyond it or check their e-mail, they will either have to enter a user number or sign up on-line. The first installations were completed in spring 2001 in the Pacific Northwest. Starbucks planned to expand its Internet access for customers to all 3,000 North American outlets and 70 percent of its stores worldwide by the end of 2003.

Starbucks is weighing several options to encourage use of the wireless LAN technology, including the sale of wireless modem cards in its stores—provided it does not detract from the coffee experience. Starbucks has also started discussions with major laptop computer manufacturers to provide incentives for customers to purchase the wireless LAN devices.

Some industry analysts believe that Starbucks has positioned itself too far ahead of the technology curve by launching this service so early. After all, not that many people have a wireless modem today. A writer for *Fortune* magazine tested access to the Internet from Starbucks outlets and complained that he had to visit five stores before he could get it to work. He also pointed out the total lack of publicity for the wireless networks—there were no signs, no brochure, no one available to offer customer service, nobody in the stores who knew anything about the wireless LAN.

Rollout plans were delayed when MobileStar Network Corporation shut down its offices in October 2001. By February 2002, Starbucks confirmed that VoiceStream, the North American affiliate of T-Mobile International, had acquired MobileStar and will be the service provider for the high-speed Internet access that Starbucks is continuing to roll out to company-owned stores.

Discussion Questions

1. Visit a local Starbucks store (in an off-peak time period) and ask the workers about the wireless LAN service. Has the service been installed there yet? If so, do you notice any customers taking advantage of it? Ask for a demo. If it has not been installed yet, find out why not.

2. What evidence might support critics' comments that Starbucks has positioned itself too far ahead of the technology curve?

Critical Thinking Questions

3. Make a list of actions that are required to complete the rollout of this new technology in a manner that will provide an effective and useful service to customers.

4. Discuss the potential risk of "pushing the technology so hard" that it takes away from the coffee experience.

Sources: Adapted from Elisa Batista, "Bluetooth Fakes Lack Proper Bite," *Wired*, October 24, 2001, accessed at http://www.wired.com; Bob Brewin, "Starbucks Takes Wireless Leap," *Computerworld*, January 8, 2001, accessed at http://www.computerworld.com; A. Lee Graham, "MobileStar Cites Starbucks Project as Its Grounds," *The Dallas Morning News*, February 20, 2001, accessed at MobileStar Web site at http://www.mobilestar.com; David Lidsky, "Shortcut to Hell," *Fortune*, March 1, 2002, accessed at http://www.fortune.com; Jaikumar Vijayan, "Starbucks Picks Compaq as Hardware Supplier for Wireless Access in Stores," *Computerworld*, May 2, 2001, accessed at http://www.computerworld.com; "Starbucks Annual Shareholders Meeting to Highlight New Initiatives," Starbucks press release, February 26, 2002, http://www.starbucks.com.

CASE 2

Bear, Stearns & Co. Builds Redundant Telecommunications Network

Bear, Stearns & Company is a leading investment banking and securities trading and brokerage firm that serves governments, corporations, institutions, and individuals worldwide. The company headquarters is in New York City, and it has 10,500 employees with additional offices in several U.S. cities and foreign countries. Through Bear, Stearns Securities Corporation, it processes approximately 10 percent of the New York Stock Exchange volume cleared daily through the National Securities Clearing Corp. Bear, Stearns is also one of the largest securities borrowers in the world.

After three years of planning and three and one-half years of construction, Bear, Stearns moved into a new seven-story building on Madison Avenue in New York in 2002. Bear, Stearns saw the relocation of its New York headquarters building as the perfect opportunity to provide innovations that would enable its employees to work more efficiently and more effectively.

High on its list of priorities was the need to upgrade the computers and the network its traders rely on for market data and trade execution. The new building houses 2,000 traders, each with multiple, state-of-the-art desktop computers. The typical trader uses two PCs, but some use as many as four. Flat panel displays were a requirement to reduce the amount of desk space required.

The company identified the network interface card (NIC) on the traders' PCs as critical to ensuring complete fault tolerance. The NIC is a computer circuit board or card that is installed in a computer so that it can be connected to a network. Personal computers and workstations on a local area network typically contain an NIC specifically designed for the LAN transmission technology. All Bear, Stearns PCs were outfitted with two network interface cards for redundancy. In addition, Bear, Stearns ran 12 Category 5 communications lines connecting each trader's PC to Bear, Stearns's 100-Mbps Ethernet LAN. Two fiber-optic cables also provide additional redundancy. The goal is to eliminate any downtime—traders must be able to continue business even in the event of a network failure.

IP Metrics Software was contracted to provide the NICs. The firm also provided high-availability/high-reliability software and consulting services to increase the performance and reliability of the network infrastructure and eliminate all single points of failure on the network. The current network design employs Cisco Systems 6509 routers in a four-tier architecture. The headquarters LAN connects to a three-node optical metropolitan area network that carries data at 6 Gbps among the headquarters, a site in downtown Brooklyn, and Bear, Stearns's data center in Whippany, New Jersey. This network uses Nortel Networks' Optera equipment and runs alongside a synchronous optical network–asynchronous transfer mode network that's there as a backup.

Discussion Questions
1. Draw a rough sketch showing the equipment and network connections available to the typical Bear, Stearns trader.
2. Is there any single point of failure in the Bear, Stearns network?

Critical Thinking Questions
3. Why is it so critical to Bear, Stearns to avoid a business interruption?
4. How could you quantify the cost of a business interruption at Bear, Stearns?

Sources: Adapted from "Investor Relations," Bear, Stearns Web site at http://www.bearstearns.com, accessed March 11, 2002; Marianne Kolbasuk McGee, "IT Leadership Put to the Test," *InformationWeek*, July 2, 2001, http://www.informationweek.com; and John Randleman, "No Downtime, No Matter What," *InformationWeek*, March 11, 2002, http://www.informationweek.com.

CASE 3

Networks Support Collaborative Product Design

Manufacturers are involving business partners and suppliers in their product-development phase to reduce time to market and achieve significant business advantages. But to do so, they must save product information in a secure area where it can be managed while still allowing selected business partners and suppliers to access and update it via telecommunications networks.

Such a collaborative product design process avoids many of the problems of managing and sharing the frequently changing information associated with the design, development, and manufacturing of a product. Significant business benefits are achieved when suppliers are allowed to comment and even make changes to the product design. For example, a supplier might suggest a cheaper component, or a fabricator might suggest design changes that would reduce the weight or increase the durability of the product. Of course, all such changes must be managed through version control.

Without a collaborative product design process, designers and engineers must communicate via couriers, faxes, face-to-face meetings, and e-mail. Each of these methods has its own drawbacks, including security issues, problems with version control, and juggling large electronic design files. Using telecommunications to support collaborative design, when the change is posted to the network, every party is immediately informed and can access the latest version. Telecommunications and collaborative product design enables team members from around the world to attend virtual meetings, share product designs and drawings, and gain access to information. It also eliminates distance and time barriers by enabling designers to use networks to exchange files so they can work together on a project. For example, a designer at a Malaysian supplier can work during the day and then leave a marked-up design document for U.S.-based designers to use when their workday starts.

Collaborative product design tools typically combine information management, version control, configuration management, workflow management, and collaborative tools, such as the ability to view and mark up documents and drawings. Market leaders include Parametric Technical Corp.'s Windchill design-collaboration tool and IBM's Envoia. Other significant players include Agile Software, Alventive, Dassault Systèmes, Framework Technologies, Matrix One, SDRC, and Unigraphics.

Ingersoll-Rand is a global manufacturer with a diverse product lineup including Schlage locks and security solutions, Thermo King transport temperature control equipment, Hussmann refrigeration equipment, Bobcat construction equipment, Club Car golf cars and utility vehicles, Torrington bearings and components, and Power Works microturbines. As a global, diversified industrial enterprise, it must bring together the talent, energy, and enthusiasm of all Ingersoll-Rand people. To that end, the company has implemented a design and manufacturing strategy that it calls "design anywhere, build anywhere." The company relies on a global network and Parametric's network-based Windchill software to link designers and factories around the globe to work on new products. Windchill is an Internet-based suite of collaborative software applications for product development. It provides a secure environment for product lifecycle management that allows enterprises to integrate business processes and product data with dispersed divisions, partners, and customers. For example, Ingersoll-Rand designers in China, Europe, India, and the United States are collaborating on the design of air compressors, refrigeration units, paving equipment, and rock drills. The goal is to lower design and engineering costs while leveraging global buying power.

One of the biggest challenges in the successful use of collaborative product design is a cultural issue, not a technical one. Companies aren't accustomed yet to sharing information for collaborative efforts. Most design engineers don't welcome critical input from production managers or component buyers. Also, few companies support sharing closely held information with outsiders, especially new product-design details.

Discussion Questions

1. What are some of the operational requirements for a network that forms the communications infrastructure for a company like Ingersoll-Rand, which must compete in today's 24×7×365 environment?

2. In what ways does the use of a network to support collaborative product design dictate the need for improved network management and redundant systems?

Critical Thinking Questions

3. What are some of the potential issues associated with the sharing of early product design details with business partners and suppliers? What are the implications of this on the design of the network?

4. What network features could a company implement to reduce the primary cultural barrier that limits the use of collaborative product design?

Sources: Adapted from Business Editors, "Brembo Chooses CATIA and ENOVIA Solutions for E-collaboration," *Business Wire*, September 3, 2001, accessed at http://www.businesswire.com; Steve Konicki, "Toyota Builds Global Collaborative Design Network," *InformationWeek*, March 26, 2002, http://www.informationweek.com; Steve Konicki, "Groupthink Gets Smart," *InformationWeek*, January 14, 2002, http://www.informationweek.com; Alan Radding and Gina Roos, "Online Design Collaboration Gets Real," *EE Times*, July 16, 2001, http://www.eetimes.com.

NOTES

Sources for the opening vignette on p.225: Adapted from "About DHL" from the DHL Worldwide Express Web site at http://www.dhl-usa.com, accessed March 27, 2002; Bob Brewin, "UPS Takes Wireless Application to Asia," *Computerworld*, February 11, 2002, accessed at http://www.computerworld.com; Linda Rosencrance, "DHL Builds U.S. Data Center to Help Manage Global Net," *Computerworld*, March 4, 2002, http://www.computerworld.com; Linda Rosencrance, "DHL to Open IT Services Center in Arizona," *Computerworld*, February 27, 2002, http://www.computerworld.com.

1. Steve Konicki, "Groupthink Gets Smart," *InformationWeek*, January 14, 2002, accessed at http://www.informationweek.com.

2. "The List of Wireless 25 Innovators," *Computerworld*, September 24, 2001, accessed at http://www.computerworld.com.

3. James Cope, "Ethernet Offers Faster Way to Wire Hotels," *Computerworld*, March 5, 2001, accessed at www.computerworld.com.

4. Rick Perera, "Companies Claim 3.2 T Bit/Sec Data Transmission," *Computerworld*, March 16, 2001, accessed at http://www.computerworld.com.

5. Reuters, "Bell Labs Says It Shatters Data Delivery Record," *The New York Times on the Web*, March 22, 2002, accessed at http://www.nytimes.com.

6. Sami Lais, "Satellites Link Bob Evans Farms," *Computerworld*, July 2, 2001, accessed at www.computerworld.com.

7. Bob Brewin, "E-Mail, Web Access Arrive on Rail Service," *Computerworld*, May 21, 2001, accessed at www.computerworld.com.

8. John Rendleman, "FCC Reclassifies DSL Service to Spur Competition," *InformationWeek*, February 25, 2002, accessed at http://www.informationweek.com.

9. Jaikumar Vijayan, "Sept. 11 Attacks Prompt Decentralization Moves," *Computerworld*, December 17, 2001, accessed at http://www.computerworld.com.

10. Karlin Lillington, "Irish Fitted for Broadband Rings," *Wired*, March 19, 2002, accessed at http://www.wired.com.

11. David Pogue, "Instant Home Computer Networks," *The New York Times on the Web*, April 4, 2002, accessed at http://www.nytimes.com.

12. Alorie Gilbert, "IBM Cuts Deals with Cancer Society Shanghai Telecom," *InformationWeek*, August 9, 2001, accessed at http://www.informationweek.com.

13. Tom Krazit, "Ipswitch Upgrades Network Monitoring Tool," *Computerworld*, December 3, 2001, accessed at http://www.computerworld.com.

14. Chris Gaither, "Bluetooth Defies Obituaries," *The New York Times on the Web*, December 20, 2001, accessed at http://www.nytimes.com.

15. Tischelle George, "FedEx to Use AT&T's GPRS Network," *InformationWeek*, March 19, 2002, accessed at http://www.informationweek.com.

16. David M. Ewait, "Palm Releases Bluetooth Developer's Kit," *InformationWeek*, January 14, 2002, accessed at http://www.informationweek.com.

17. John Rendleman, "Mercedes Speeds Up Wireless Transmission," *InformationWeek*, November 14, 2001, accessed at http://www.informationweek.com.

18. John Rendlemann, "Wireless Network Helps Lower Law School's Infrastructure Costs," *InformationWeek*, July 31, 2001, accessed at http://www.informationweek.com.

19. David Pogue, "Instant Home Computer Networks," *The New York Times on the Web*, April 4, 2002, accessed at http://www.nytimes.com.

20. Bob Brewin, "California City Plans Wireless LAN for Critical Communications," *Computerworld*, February 18, 2002, accessed at http://www.computerworld.com.

21. "Broadband's Next Wave: Wireless?," *The New York Times on the Web*, May 17, 2001, accessed at http://www.nytimes.com.

22. James Cope, "Bridging the Long Last Mile ," *Computerworld*, September 17, 2001, accessed at http://www.computerworld.com.

23. Jim Duffy, "Cisco Finally Rachets Up to 10 G Bits," *Computerworld*, January 31, 2001, accessed at http://www.computerworld.com.

24. Mary Brandel, "Teamwork Buoys Big Audio at Bose," *Computerworld*, February 11, 2002, accessed at http://www.computerworld.com.

25. Michael Meehan, "Home Depot Seeks Remote Control of the Desktop," *Computerworld*, January 7, 2002, accessed at http://www.computerworld.com.

26. Solutions Overview at Mobius Web site at http://www.mobius-inc.com, accessed on March 25, 2002.

27. Matt Hamblen, "Avoiding Travel, Users Turn to Communications Technology," *Computerworld*, September 24, 2001, accessed at http://www.computerworld.com.

28. James Cope, "Videoconferencing Getting Easier, Cheaper," *Computerworld*, April 23, 2001, accessed at http://www.computerworld.com.

29. Matt Hamblen, "Avoiding Travel, Users Turn to Communications Technology," *Computerworld*, September 24, 2001, accessed at http://www.computerworld.com.

30. Linda Rosencrance, "Brief: Covisint Launches Web-Based EDI Tool," *Computerworld*, March 15, 2002, accessed at http://www.computerworld.com.

31. Kathleen Melymuka, "Executive Education on a Shoestring," *Computerworld*, March 11, 2002, accessed at http://www.computerworld.com.

32. Todd R. Weiss, "Apple Co-Founder Hatches GPS Company," *Computerworld*, January 24, 2002, accessed at http://www.computerworld.com.

33. Bob Brewin and Linda Rosencrance, "Smart Boxcars Give Rail Shippers Control," *Computerworld*, July 22, 2002, accessed at www.computerworld.com.

34. Todd R. Weiss, "Wireless Devices to Help Cut Visitor Waits at Nine Six Flags Parks," *Computerworld*, February 1, 2002, accessed at http://www.computerworld.com.

The Internet, Intranets, and Extranets

PRINCIPLES

The Internet is like many other technologies—it provides a wide range of services, some of which are effective and practical for use today, others are still evolving, and still others will fade away from lack of use.

Originally developed as a document-management system, the World Wide Web is a menu-based system that is easy to use for personal and business applications.

Before the Internet and the World Wide Web become universally used and accepted for business use, management issues; service and speed issues; and fraud, security, and unauthorized Internet sites must be addressed and solved.

LEARNING OBJECTIVES

- Briefly describe how the Internet works, including alternatives for connecting to it and the role of Internet service providers.
- Identify and briefly describe the services associated with the Internet.

- Describe the World Wide Web and how it works.
- Explain the use of Web browsers, search engines, and other Web tools.

- Identify who is using the Web to conduct business and discuss some of the pros and cons of Web shopping.
- Outline a process for creating Web content.
- Describe Java and discuss its potential impact on the software world.
- Define the terms *intranet* and *extranet* and discuss how organizations are using them.
- Identify several issues associated with the use of networks.

Attracting Customers with High-Tech Internet Services

Some merchandise is easier to sell on-line than others. Items that don't need to be held and touched prior to the purchase, such as books, music CDs, and software, do well. Brand-name electronics and computers sell well on the Web because customers are familiar with the products from past experience or by reputation and use the Web to find the best deal. In fact, electronics and PC manufacturers credit the Web for more than 25 percent of total sales. But of all the items that sell well on the Web, clothing is not one; Web sales make up less than 2 percent of all apparel sales. What's more, 30 percent of all apparel bought on-line is returned, minimizing the profit margins even more. This lackluster performance has led some clothing merchants to all but forsake the Web as a tool for selling their wares. A notable exception is Lands' End.

As many companies experienced serious financial losses resulting from the downturn in the economy, Lands' End, the well-known direct merchant, experienced record earnings. President and Chief Executive Officer David F. Dyer credits the company's success to the business principles to which it adheres. Roughly stated, they include the following:

1. Do everything possible to make better products
2. Price products fairly and honestly
3. Accept any return for any reason at any time
4. Ship products faster than anyone
5. What is best for the customer is best for all of us

Lands' End uses technology extensively to support its business principles. Its streamlined information and communications systems allow it to quickly fill orders, shipping them within 36 hours even during its busiest holiday rush. The company keeps Inventory levels low, avoiding, as much as possible, the necessity of liquidation sales. Lands' End was an "early adopter" of the Internet, launching its Web site in 1995. Today, landsend.com is the world's largest apparel Web site in business volume, maintaining Web sites for seven countries in six languages. It has also been a pioneer in enhancing the shopping experience and fostering one-on-one relationships with its customers.

Lands' End was the first apparel company to offer "My Virtual Model," an innovative program that runs from a company's Web site, allowing customers to create a 3-D model of themselves by providing critical measurements. Once a virtual model is created, the customer can use it to "try on" items and outfits to see how they will look in real life.

Lands' End was also one of the first E-tailers to provide instant personal help on-line using a chat utility that links a customer directly to a human representative. Typing a question in the chat window results in a customer service representative's reply on screen moments later. The representative can even take control of a customer's browser window to show customers Web pages of merchandise they may want to buy. Lands' End "Shop with a friend" Web service also allows two shoppers sitting at two different computers to browse the site together, chat with each other, and add items to a single shopping basket.

Another recent Lands' End innovation is called "Lands' End Custom," which allows customers to order personally tailored clothes. After a customer supplies his or her physical measurements, a software system calculates that person's weight distribution and the clothing measurements conforming to it. A single item of clothing is then cut and assembled to match the measurements. For this service Lands' End struck deals with new manufacturers in Mexico that could handle the challenge of custom-cut clothing with a quick turnaround time, rather than its usual Asian manufacturers. Customers pay a bit more for this service and must wait three weeks for delivery, but initial response has shown that there is a market for this service. Although Lands' End won't disclose the profit margin on this service, it is clear that profit margin per unit is not the company's only consideration. Lands' End hopes that this service will further reduce excess inventory and merchandise returns. Also, the fact that Lands' End is first to attempt this new service gives it a jump on the competition.

Through technical innovation, Lands' End has been able to streamline its operations to reduce overhead and provide customers with an on-line shopping experience that it hopes will be easier and more pleasant than shopping at the mall.

As you read this chapter, consider the following:

◉ What services can a brick-and-mortar clothing store offer that are difficult to duplicate on the Web? Can any services be offered by a Web-based clothing merchant that would be difficult to offer in a brick-and-mortar store?

◉ What types of issues may arise while shopping at landsend.com that might frustrate or concern customers? How do these issues compare with those of a brick-and-mortar business? How can Land's End overcome these challenges?

To speed communications and share information, businesses are linking employees, branch offices, and global operations via networks, whether they set up their own or use outside services. As seen with Lands' End in the opening vignette, companies are also using the Internet to provide customized products and services. If everyone is talking about the Internet and companies are increasingly using it for competitive advantage, then what exactly is it? The Internet is the world's largest computer network. Actually, the Internet is a collection of interconnected networks, all freely exchanging information (Figure 7.1). Research firms, colleges, and universities have long been part of the Internet, and now businesses, high schools, elementary schools, and other organizations are joining up as well. Nobody knows exactly how big the Internet is because it is a collection of smaller computer networks with no single place where all the connections are registered.

FIGURE 7.1

Routing Messages over the Internet

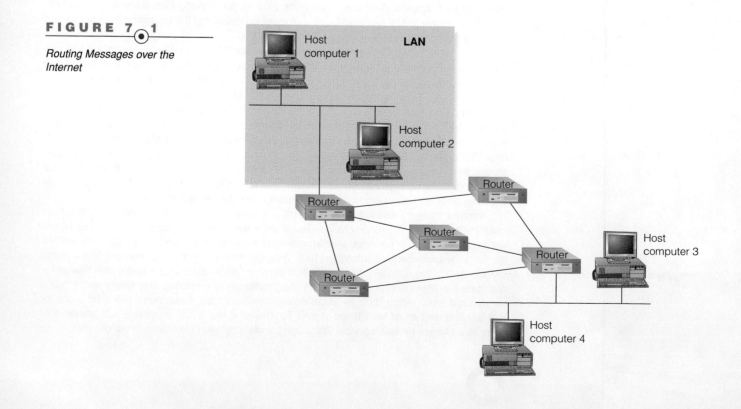

USE AND FUNCTIONING OF THE INTERNET

The Internet is truly international in scope, with users on every continent—including Antarctica. However, the United States has the most usage, by far. According to a Pew Internet and American Life Project study, an estimated 104 million people, or about 56 percent of U.S. adults, have used the Internet.[1] Older and poorer Americans, however, are less likely to use the Internet compared with younger and wealthier people. Figure 7.2 shows the use of the Internet in the United States by age, gender, region, and income.[2]

Although the United States still claims more Web activity than other countries, the Internet is expanding around the globe but at differing rates for different countries. In Africa, Internet connectivity is limited in every country except South Africa. But even there, its use is very limited, since even a slow 14.4-Kbps modem costs more than a month's salary for most people. In Russia, using the Internet's e-mail capabilities provides a timely mail service; it may take weeks for an airmail letter to reach the United States. International use of the Internet is expected to continue its growth. One study estimates that international Internet usage could surge to over a billion users by 2005.[3]

The Internet is increasingly going wireless. Jim Shelby, for example, developed a wireless Internet system for himself and others in the Aspen, Colorado area.[4] His service is mostly for personal use. Shelby's homemade system provides fast, free Internet connection to people living in the Aspen area. "I've got Aspen nailed," says Mr. Shelby. In addition to land-based systems, the Internet is also becoming available at sea and in the air. Going on-line while aboard a cruise ship is becoming a reality. Royal Caribbean, with the help of IBM, is installing Internet appliances that allow crew members to access the Internet to get news and send and receive e-mails.[5] Boeing recently received an FCC license to provide Internet service on its airplanes.[6] The license allows Boeing to operate up to 800 broadband connections on flights.

Many believe that the Internet will eventually become so pervasive that using it will be as simple as pushing a button on a remote control.[7] Hewlett-Packard,

FIGURE 7.2

U.S. Internet Usage by Age, Gender, Region, and Income

(Source: Reprinted by permission of *The Wall Street Journal*, Copyright © 2001 Dow Jones & Company, Inc. All Rights Reserved Worldwide. License number 597221359195.)

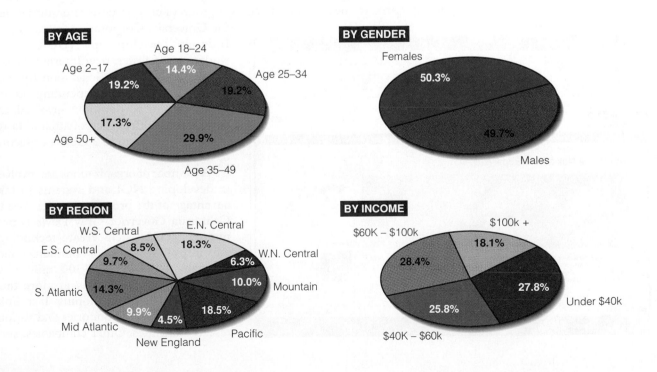

for example, has started its Internet project, CoolTown, to develop products for the Internet. The project is investigating the feasibility of Internet-linked computers, cell phones, personal digital assistants, stereo systems, kitchen appliances, and other appliances. These devices will use infrared signals to connect to the Internet from every room of a house or office building.

The ancestor of the Internet was the **ARPANET**, a project started by the U.S. Department of Defense (DOD) in 1969. The ARPANET was both an experiment in reliable networking and a means to link DOD and military research contractors, including a large number of universities doing military-funded research. (*ARPA* stands for the Advanced Research Projects Agency, the branch of Defense in charge of awarding grant money. The agency is now known as DARPA—the added *D* is for *Defense*.) The ARPANET was highly successful, and every university in the country wanted to sign up. This wildfire growth made it difficult to manage the ARPANET, particularly the large and rapidly growing number of university sites on it. It was decided to break the ARPANET into two networks: MILNET, which included all military sites, and a new, smaller ARPANET, which included all the nonmilitary sites. The two networks remained connected, however, through use of the **Internet protocol (IP)**, which enabled traffic to be routed from one network to another as needed. All the networks connected to the Internet speak IP, so they all can exchange messages.

Unlike a corporate network with a centralized infrastructure, the Internet is nothing more than an ad hoc linkage of many networks that adhere to basic standards. Since these networks are constantly changing and being improved, the Internet itself is perpetually evolving. However, since the Internet is such a loose collection of networks, nothing prevents some participants from using outdated or slow equipment.

Today, several people, universities, and companies are attempting to make the Internet faster and easier to use. Robert Kahn, who managed the early development of the ARPANET, is one individual who wants to take the Internet to the next level.[8] He is president of the nonprofit organization National Research Initiatives that provides guidance and funding for the development of a national information infrastructure. The organization is looking into using "digital objects" that allow programs and data to be used and shared on all types of computer systems. To speed Internet access, a group of corporations and universities, the University Corporation for Advanced Internet Development (UCAID), is working on a faster, new Internet. Called Internet2 (I2), Next Generation Internet (NGI), and Abilene, depending on the universities or corporations involved, the new Internet offers the potential of faster Internet speeds, up to 2 Gbits per second or more.

A number of organizations are involved in developing NGI and systems to take advantage of the promise of faster speeds. California Governor Gray Davis is promoting NGI to help create technology jobs in California.[9] One communications company is investing $300 million to increase Internet capacity at more than 190 universities, developing fast links between national laboratories and various research facilities. Other companies, such

ARPANET
project started by the U.S. Department of Defense (DOD) in 1969 as both an experiment in reliable networking and a means to link DOD and military research contractors, including a large number of universities doing military-funded research

Internet protocol (IP)
communication standard that enables traffic to be routed from one network to another as needed

There are more than 414 million Internet users around the world, and the number is expected to triple by the end of 2005 to nearly 1.7 billion.

as FedEx, are starting to build global systems to take advantage of NGI or I2 when it arrives.

HOW THE INTERNET WORKS

The Internet transmits data from one computer (called a *host*) to another (see Figure 7.1). If the receiving computer is on a network to which the first computer is directly connected, it can send the message directly. If the receiving computer is not on a network to which the sending computer is connected, the sending computer relays the message to another computer that can forward it. The message may be sent through a router (see Chapter 6) to reach the forwarding computer. The forwarding host, which presumably is attached to at least one other network, in turn delivers the message directly if it can or passes it to yet another forwarding host. It is quite common for a message to pass through a dozen or more forwarders on its way from one part of the Internet to another.

The various networks that are linked to form the Internet work pretty much the same way—they pass data around in chunks called *packets*, each of which carries the addresses of its sender and its receiver. The set of conventions used to pass packets from one host to another is known as the Internet protocol (IP), which operates at the network layer of the seven-layer OSI model discussed in Chapter 6. Many other protocols are used in connection with IP. The best known is the **transport control protocol (TCP)**, which operates at the transport layer. TCP is so widely used as the transport layer protocol that many people refer to TCP/IP, the combination of TCP and IP used by most Internet applications. Adhering to the same technical standards allows the more than 100,000 individual computer networks owned by governments, universities, nonprofit groups, and companies to constitute the Internet. Once a network following these standards links to a **backbone**—one of the Internet's high-speed, long-distance communications links—it becomes part of the worldwide Internet community. The Finland Post, for example, uses an IP that allows data and phone calls to be transmitted over the same infrastructure.[10] As a result, the cost of telecommunications equipment has dropped over 40 percent and Finnish postal employees can now use their PCs to make phone calls. According to Kari Haaha, the Finland Post's business-development manager, "IP has left a measurable impact on our performance."

Each computer on the Internet has an assigned address called its **uniform resource locator,** or **URL**, to identify it from other hosts. The URL gives those who provide information over the Internet a standard way to designate where Internet elements such as servers, documents, newsgroups, etc. can be found. Let's look at the URL for Course Technology, http://www.course.com.

The "http" specifies the access method and tells your software to access this particular file using the HyperText Transport Protocol. This is the primary method for interacting with the Internet.

The "www" part of the address signifies that the address is associated with the World Wide Web service discussed later. The "course.com" part of the address is the domain name that identifies the Internet host site. The domain names must adhere to strict rules. They always have at least two parts separated by dots (periods). For all countries except the United States, the rightmost part of the domain name is the country code (au for Australia, ca for Canada, dk for Denmark, fr for France, jp for Japan, etc.). Within the United States, the country code is replaced with a code denoting affiliation categories (Table 7.1 contains a few popular categories). The leftmost part of the domain name identifies the host network or host provider, which might be the name of a university or business.

transport control protocol (TCP)
widely used transport layer protocol that is used in combination with IP by most Internet applications

backbone
one of the Internet's high-speed, long-distance communications links

uniform resource locator (URL)
an assigned address on the Internet for each computer

TABLE 7.1

U.S. Top-Level Domain Affiliations

Affiliation ID	Affiliation
com	business organization
edu	educational sites
gov	government sites
net	networking organizations
org	organizations

Herndon, Virginia–based Network Solutions Inc. (NSI) was the sole company in the world with the direct power to register addresses using .com, .net, or .org domain names. But this government contract ended in October 1998, as part of the U.S. government's move to turn management of the Web's address system over to the private sector. Today, other companies, called *registrars*, can register domain names, and additional companies are seeking accreditation to register domain names from the Internet Corporation for Assigned Names and Numbers (ICANN). Some registrars are concentrating on large corporations, where the profit margins may be higher, compared with small businesses or individuals.

There are hundreds of thousands of registered domain names. Some people, called *cyber-squatters*, have registered domain names in the hope of selling the names to corporations or people at a later date. The domain name Business.com, for example, sold for $7.5 million. In one case, a federal judge ordered the former owner of the Sex.com Web site to pay the person who originally registered the domain name $40 million in compensatory damages and an additional $25 million in punitive damages.[11] But some companies are fighting back, suing people who register domain names in hopes of trying to sell them to companies. Today, the Internet Corporation for Assigned Names and Numbers has the authority to resolve domain name disputes.[12] Under new rules, if an address is found to be "confusingly similar" to a registered trademark, the owner of the domain name has no legitimate interest in the name. The rule was designed in part to prevent cyber-squatters.

ACCESSING THE INTERNET

There are three ways to connect to the Internet (Figure 7.3). Which method is chosen is determined by the size and capability of the organization or individual. As seen in the "Ethical and Societal Issues" box, there is a digital divide between people who have access to the Internet and those who do not.

Connect via LAN Server

This approach requires the user to install on his or her PC a network adapter card and Open Datalink Interface (ODI) or Network Driver Interface Specification (NDIS) packet drivers. These drivers allow multiple transport protocols to run on one network card simultaneously. LAN servers are typically connected to the Internet at 56 Kbps or faster. In addition, the higher cost of this service can be shared among several dozen LAN users to get to a reasonable cost per user. Additional costs associated with a LAN connection to the Internet include the cost of the software mentioned at the beginning of this section.

FIGURE 7·3

Three Ways to Access the Internet

There are three ways to access the Internet—using a LAN server, dialing into the Internet using SLIP or PPP, or using an on-line service with Internet access.

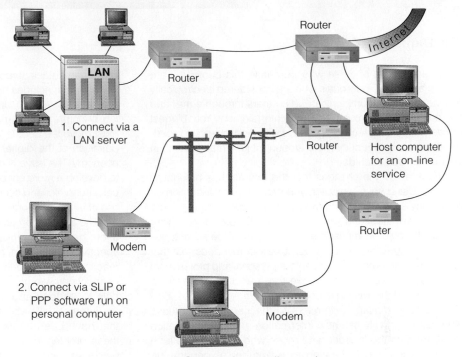

1. Connect via a LAN server

2. Connect via SLIP or PPP software run on personal computer

3. Connect via an on-line service

Connect via SLIP/PPP

serial line Internet protocol (SLIP)

a communications protocol that transmits packets over telephone lines

point-to-point protocol (PPP)

a communications protocol that transmits packets over telephone lines

This approach requires a modem and the TCP/IP protocol software plus **serial line Internet protocol (SLIP)** or **point-to-point protocol (PPP)** software. SLIP and PPP are two communications protocols that transmit packets over telephone lines, allowing dial-up access to the Internet. If you are running Windows, you will also need Winsock. Users must also have an Internet service provider that lets them dial into a SLIP/PPP server. SLIP/PPP accounts can be purchased for $30 a month or less from regional providers. With all this in place, a modem is used to call into the SLIP/PPP server. Once the connection is made, you are on the Internet and can access any of its resources. The costs include the cost of the modem and software, plus the service provider's charges for access to the SLIP/PPP server. The speed of this Internet connection is limited to the slower of your computer's modem and the speed of the modem of the SLIP/PPP server to which you connect.

Connect via an On-Line Service

This approach requires nothing more than what is required to connect to any of the on-line information services, such as a modem, standard communications software, and an on-line information service account. Increasingly, on-line services are offering DSL, satellite, and cable connection to the Internet, offering faster speeds. These technologies were discussed in Chapter 6. There is normally a fixed monthly cost for basic services, including e-mail. Additional fees usually apply for DSL, satellite, or cable access to the Internet, although these costs are falling.[13] Cox Communications, for example, offers high-speed Internet access in Las Vegas for slightly more than the slower modem connection offered by many on-line services. The on-line information services provide a wide range of services, including e-mail and the World Wide Web. America Online, Microsoft Network, and Prodigy are examples of such services.

INTERNET SERVICE PROVIDERS

Internet service provider (ISP)

any company that provides individuals or organizations with access to the Internet

An **Internet service provider (ISP)** is any company that provides individuals and organizations with access to the Internet. ISPs do not offer the extended

The Digital Divide

If you are typical of those who purchase this book, you are among a group of individuals who are considered electronically "connected." You communicate with others through e-mail and rely on the Web for information and entertainment. Your biggest concern regarding technology may be finding low-price broadband service and reducing the amount of junk mail you find in your e-mail box each day.

Think about the changes that the Internet has brought to your life. Using your computer, you can access information on nearly any subject and communicate with everyone from your family, classmates, and your teacher, to the president of your college. It would be impossible to list here all of the advantages that the Internet provides for us. It is a tremendous tool that assists us in defining and meeting our personal and professional goals. But, where does that leave those who cannot afford a computer and the training to use it?

The term *digital divide* refers to the gap between those who can effectively use new information and communication tools, such as the Internet, and those who cannot. This term can be discussed in relation to a number of geographic focuses: local, national, and international.

The digital divide in the U.S., although a significant concern, has been shrinking over the last few years. Low-income Internet users (those earning less than $25,000) soared 46 percent in 2001, making them the fastest-growing segment of Internet users. According to a 2002 study by the U.S. Department of Commerce (DOC), more than half of the nation is now on-line, with 2 million new Internet users per month. The Commerce Department credits the school system for leveling the playing field, stating that "computers at schools substantially narrow the gap in computer usage rates for children from high and low income families."

A less-optimistic view exists for the *global* digital divide. The World Economic Forum Web site states that industrialized countries, with only 15 percent of the world's population, are home to 88 percent of all Internet users. Finland alone has more Internet users than all of Latin America. A recent study by Nielsen//NetRatings found that:

- The United States has more computers than the rest of the world combined
- 41 percent of the global on-line population is in the United States and Canada
- 27 percent of the on-line population lives in Europe, the Middle East, and Africa (25 percent of European households are on-line)
- 20 percent of the on-line population logs on from Asia Pacific countries (33 percent of all Asian households are on-line)
- Only 4 percent of the world's on-line population is in South America

To build a global community through the Internet, more affluent countries cannot ignore the needs of the less fortunate.

Dr. Kenan Patrick Jarboe with Athena Alliance believes that this commitment includes more than just passing out PCs. He says, "We must move from 'divide' to 'inclusion' as the central organizing principle of our analysis and actions. In other words, we must move the debate from access to the Internet—the original definition of the digital divide—to inclusion in the information economy. The issue is not simply the utilization of IT. Our task is to develop a more encompassing description of the technological, economic, and social aspects of the revolution in IT and the rise of a new economy."

The information and communication technology revolution offers genuine potential, and if we do not take a global perspective, a significant portion of the world may lose out. As we reach out to the world with fiber-optic cable and high-speed connections, we must make sure that the rest of the world is able to respond and participate. In creating a smaller world, the Internet calls attention to social and economic problems that have been neglected. It is up to us now to recognize these problems and develop solutions that will help to eliminate them. Only then can we attain the ideal global community to which we aspire.

Discussion Questions

1. A new neighbor moves in next door. He or she is the same age, gender, and race as you, with one major difference: the new neighbor has never touched a computer. What professional advantages do you have over your new neighbor? What personal advantages do you have over your new neighbor? Are you inclined to help this person?
2. Name three significant benefits of investing in underdeveloped countries to build an inclusive global economy.

Critical Thinking Questions

3. Is it possible to include all nations in the new economy? What about those that shun technology as being an evil Western tool? What will their role be in the new economy?
4. You have been assigned to assist a developing country to participate in the new economy. The citizens have knowledge of spoken and written English but know nothing about science and technology. You've been given $1 million and ten years to bring this country to the point where it can conduct commerce over the Internet. Present an outline of your plan.

Sources: Thea Williams, "Leapfrog the Digital Divide," http://australianit.news.com.au, February 23, 2002; John Surmacz, "Five Thoughts about the Digital Divide," *Darwin*, September 13, 2001, www.darwinmag.com; Michael Pastore, "Global Digital Divide Still Very Much in Existence," Cyberatlas Web site, http://cyberatlas.internet.com, accessed March 1, 2002; Digital Divide Network staff, "Digital Divide Basics Fact Sheet," Benton Foundation, http://www.digitaldividenetwork.org, accessed March 1, 2002; "A Nation Online: How Americans Are Expanding Their Use of the Internet," a report by the National Telecommunications and Information Administration (NTIA) accessed at http://www.ntia.doc.gov, March 1, 2002.

To use an ISP like MSN, you must have an account with the service provider and software that allows a direct link via TCP/IP.

informational services offered by commercial on-line services such as America Online or Prodigy. There are literally thousands of Internet service providers, ranging from universities making unused communications line capacity available to students and faculty to major communications giants such as AT&T and Sprint. To use this type of connection, you must have an account with the service provider and software that allows a direct link via TCP/IP.

In choosing an Internet service provider, the important criteria are cost, reliability, security, the availability of enhanced features, and the service provider's general reputation. Reliability is critical because if your connection to the ISP fails, it interrupts your communications with customers and suppliers.

Among the value-added services ISPs provide are electronic commerce, networks to connect employees, networks to connect with business partners, host computers to establish your own Web site, Web transaction processing, network security and administration, and integration services. Many corporate IS managers welcome the chance to turn to ISPs for this wide range of services because they do not have the in-house expertise and cannot afford the time to develop such services from scratch. In addition, when organizations go with an ISP-hosted network, they can also tap the ISP's national infrastructure at minimum cost. That's important when a company has offices spread across the country.

In most cases, ISPs charge a monthly fee that can range from $15 to $30 for unlimited Internet connection through a standard modem. The fee normally includes e-mail. Some ISPs, however, are experimenting with no-fee Internet access. But there are strings attached to the no-fee offers in most cases. Some free ISPs require that customers provide detailed demographic and personal information. In other cases, customers must put up with extra advertising banners on every Web site. Table 7.2 identifies several corporate Internet service providers.

Many ISPs and on-line services offer broadband Internet access through digital subscriber lines (DSLs), cable, or satellite transmission. In a survey of broadband users, over 70 percent were extremely or very satisfied with their service.[14] About 90 percent of broadband users paid $50 or less per month for unlimited service. These technologies were discussed in Chapter 6.

TABLE 7.2

A Representative List of Internet Service Providers

Internet Service Provider	Web address
AT&T's WorldNet Service	www.att.net
BellSouth	www.bellsouth.com
EarthLink	www.earthlink.net
Sprint	www.sprint.com

INTERNET SERVICES

The types of Internet services available are vast and ever expanding. These services are discussed next and summarized in Table 7.3.

E-MAIL AND INSTANT MESSAGING

Electronic mail, or e-mail, has been used internally in business networks for years, but with the spread of Internet use, it is now commonly used for national and international communications. With the increased awareness of possible terrorist threats such as the anthrax scare, e-mail and electronic document transfers are receiving increased attention as a safer alternative to traditional "snail" mail sent through the U.S. Post Office.[15] The U.S. Department of

TABLE 7.3

Summary of Internet Services

Service	Description
E-mail	Enables you to send text, sound, and images to others.
Telnet	Enables you to log on to another computer and access its public files. Users can log on to a work computer from an offsite location.
FTP	Enables you to copy a file from another computer to your computer.
Usenet and newsgroups	An on-line discussion group that focuses on a particular topic.
Chat rooms	Enables two or more people to carry on on-line text conversations in real-time.
Internet phone	Enables you to communicate with other Internet users around the world who have equipment and software compatible to yours.
Internet video conferencing	Supports simultaneous voice and visual communications.
Content streaming	Enables you to transfer multimedia files over the Internet so that the data stream of voice and pictures plays more or less continuously.
Instant messaging	Allows two or more people to communicate instantly on the Internet.
Shopping on the Web	Allows people to purchase products and services on the Internet.
Web auctions	Lets people bid on products and services.
Music, radio, and video on the Internet	Lets users play or download music, radio, and video.
Office on the Web	Allows people to have access to important files and information through a Web site.
Internet sites in 3-D	Allows people to view products and images at different angles in what appears to be three dimensions.
Free software and services	Allows people to obtain a wealth of free software, advice, and information on the Internet. Unwanted advertising and false information are potential drawbacks.
Additional Internet services	Provides a variety of other services to individuals and companies.

Transportation (DOT), for example, encourages files to be submitted electronically. The DOT's Web site says, "Those persons making filings in DOT dockets are encouraged to file electronically by using the DOT DMS Web site."

E-mail is no longer limited to simple text messages. Depending on your hardware and software and the hardware and software of the recipient, you can embed sound and images in your message and attach files that contain text documents, spreadsheets, graphs, or executable programs. E-mail travels through the systems and networks that make up the Internet. Gateways can receive e-mail messages from the Internet and deliver them to users on other networks.

Many of these networks have agreements with the Internet and with each other to exchange e-mail, just as countries exchange regular mail across their borders. Similarly, an e-mail message may pass through a series of intermediate networks to reach the destination address. Since not all networks use the same e-mail format, a gateway translates the format of the e-mail message into one that the next network can understand. Each gateway reads the "To" line of the e-mail message and routes the message closer to the destination mailbox. Thus, you can send e-mail messages to literally anyone in the world if you know that person's e-mail address and if you have access to the Internet or another system that can send e-mail.

E-mail has changed the way people communicate. It improves the efficiency of communications by reducing interruptions from the telephone and unscheduled personal contacts. Also, messages can be distributed to multiple recipients easily and quickly without the inconvenience and delay of scheduling meetings. Because past messages can be saved, they can be reviewed, if necessary. And because messages are received at a time convenient to the recipient, the recipient has time to respond more clearly and to the point. For large organizations whose operations span a country or the world, e-mail allows people to work around the time zone changes. Some users of e-mail estimate that they eliminate two hours of verbal communications for every hour of e-mail use. But the person at the other end still must check the mailbox to receive messages. Table 7.4 lists some abbreviations commonly used in personal e-mail messages. These abbreviations are normally not appropriate for business correspondence.[16]

With its popularity and ease of use, some people feel they are drowning in too much e-mail, however.[17] According to International Data Corporation, 1.4 trillion e-mail messages were sent from businesses in North America in 2001. This staggering number is up from 40 billion e-mail messages in 1995. Companies and individuals are taking a number of steps to help them manage and cope.

TABLE 7.4

Some Common Abbreviations Used in Personal E-Mail

Expressions	Abbreviations
;-) Smile with a wink	AAMOF As a matter of fact
;-(Frown with a wink	AFAIK As far as I know
:-# My lips are sealed	BTW By the way
:-D Laughing	CUL8R See you later
:-0 Shocked	F2F Face to face
:-] Blockhead	LOL laughing out loud
:-@ Screaming	OIC Oh, I see
:-& Tongue-tied	TIA Thanks in advance
%-) Brain-dead	TTFN Ta-Ta for now

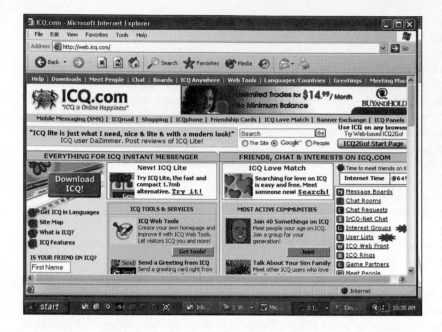

ICQ is a client program that informs you who's on-line and enables you to contact them and chat with them in real-time.

instant messaging
a method that allows two or more individuals to communicate on-line using the Internet

Some e-mail services scan for possible junk or bulk mail and delete it or place it in a separate file. Software products, such as EchoMail, help companies, individuals, and even some U.S. senators sort and answer large amounts of e-mail.[18] The software has the ability to recognize key words and phrases and respond to them. In other cases, some companies are using software to avoid legal problems from e-mail sent by managers or employees by deleting it after a certain amount of time.[19] VCNMail 2, for example, allows a sender to set a date when sent e-mail is to be deleted from the recipient's disk, set the number of times e-mail can be viewed, and prevent printing and forwarding of e-mail.

Instant messaging is on-line, real-time communication between two or more people who are connected to the Internet. With instant messaging, two or more screens open up. Each screen displays what one person is typing. Because the typing is displayed on the screen in real-time, it is like talking to someone using the keyboard.

A number of companies offer instant messaging, including America Online, Yahoo!, and Microsoft. America Online is one of the leaders in instant messaging, with about 40 million users of its Instant Messenger and about 50 million people using its client program ICQ. In addition to being able to type messages on a keyboard and have the information instantly displayed on the other person's screen, some instant messaging programs are allowing voice communication or connection to cell phones. A wireless service provider announced that it has developed a technology that can detect when a person's cell phone is turned on. With this technology, someone on the Internet can use instant messaging to communicate with someone on a cell phone anywhere in the world.

Instant messaging services often use a *buddy list* that alerts people when their friends or buddies are also on-line.[20] This feature makes instant messaging even more useful. Instant messaging is so popular that it helps Internet service providers and on-line services draw new customers and keep old ones.

TELNET AND FTP

telnet
a terminal emulation protocol that enables users to log on to other computers on the Internet to gain access to public files

Telnet is a terminal emulation protocol that enables you to log on to other computers on the Internet to gain access to their publicly available files. Telnet is particularly useful for perusing library card files and large databases. It is also called *remote logon.*

file transfer protocol (FTP)
a protocol that describes a file transfer process between a host and a remote computer and allows users to copy files from one computer to another

File Transfer Protocol (FTP) is a protocol that describes a file transfer process between a host and a remote computer. Using FTP, you can copy a file from another computer to your computer. FTP is often used to gain access to a wealth of free software on the Internet.

USENET AND NEWSGROUPS

usenet
a system closely allied with the Internet that uses e-mail to provide a centralized news service; a protocol that describes how groups of messages can be stored on and sent between computers

Usenet is a system closely allied with the Internet that uses e-mail to provide a centralized news service. It is actually a protocol that describes how groups of messages can be stored on and sent between computers. Following the usenet protocol, e-mail messages are sent to a host computer that acts as a usenet server. This server gathers information about a single topic into a central place for messages. A user sends e-mail to the server, which stores the messages. The user can then log on to the server to read these messages or have software on the computer log on and automatically download the latest messages to be read at leisure. Thus, usenet forms a virtual forum for the electronic community, and this forum is divided into newsgroups.

Newsgroups make up usenet, a worldwide discussion system classified by subject. Articles or messages are posted to newsgroups using newsreader software and are then broadcast to other interconnected computer systems via a wide variety of networks. A newsgroup is essentially an on-line discussion group that focuses on a particular topic. Newsgroups are organized into various hierarchies by general topic, and within each topic there can be many subtopics. On the Internet, there are tens of thousands of newsgroups, covering topics from astrology to zoology (Table 7.5). Discussions take place via e-mail, which is sent to the newsgroup's address. A newsgroup may be moderated or unmoderated. If a newsgroup is moderated, e-mail is automatically routed to the moderator, a person who screens all incoming e-mail to make sure it is appropriate before posting it to the newsgroup. Some people who are frequent newsgroup users invest in a newsgroup reader that makes reading and posting messages easier. People posting messages on newsgroups should be careful. The search engine Google put 700 million newsgroup messages that go back to 1981 on-line for anyone to see.[21]

Newsgroup servers around the world host newsgroups that share information and commentary on predefined topics. Each group takes the form of a large bulletin board where members post and reply to messages, creating what is called a *message thread*. The open nature of newsgroups encourages participation, but the discussions often become rambling and unfocused. As a result of so much active participation, newsgroups can evolve into tight communities where certain members tend to dominate the discussions.

Here are some tips to consider when accessing newsgroups. When you join a newsgroup, first check its list of Frequently Asked Questions, or FAQs (pronounced "facks"), before submitting any questions to the newsgroup. The FAQ list will have answers to common questions the group receives. It is considered impolite to waste the group's time by asking common questions when FAQs are available. Most new users just read messages without responding at first. Many newsgroups include members from around the world, and in the interest of courtesy, you should pick up some sense of the audience and its culture before jumping in with questions and opinions. It is impolite to jump into the middle of a conversation. You may raise points and issues long since discussed and abandoned. Be concerned about what you say and the feelings of others. Remember, a person is receiving your messages. Do not use extreme words or repeat rumors (you could risk libel or defamation lawsuits). Do not post copyrighted material, and be careful how you use copyrighted material downloaded to your computer. Protect yourself by not offering personal information such as home address, employer, or phone number. Remember that this global on-line community has

newsgroups
on-line discussion groups that focus on specific topics

TABLE 7.5

Selected Usenet Newsgroups

alt.airline	alt.sports.basketball.college
alt.aol	alt.current-events.net-abuse.spam
alt.books	alt.politics
alt.fan	alt.hackers
alt.sports.baseball.atlanta-braves	alt.music.gossip
alt.sports.basketball.nba.la-lakers	alt.politics

fragmented into thousands of different groups for a reason—to maintain the focus of each conference. Respect the specific subject matter of the group.

CHAT ROOMS

chat room
a facility that enables two or more people to engage in interactive "conversations" over the Internet

A **chat room** is a facility that enables two or more people to engage in interactive "conversations" over the Internet. When you participate in a chat room there may be dozens of participants from around the world. Multiperson chats are usually organized around specific topics, and participants often adopt nicknames, also called handles, to maintain anonymity. One form of chat room, Internet Relay Chat (IRC), requires participants to type their conversation rather than speak. Voice chat is also an option, but you must have a microphone, sound card and speakers, a fast modem, and voice-chat software compatible with the other participants.

INTERNET PHONE AND VIDEOCONFERENCING SERVICES

Internet phone service enables you to communicate with other Internet users around the world who have equipment and software compatible to yours. This service is relatively inexpensive and can make sense for international calls. With some services, it is possible to make a call from someone using the Internet to someone using a standard phone. The cost to make a call can be as low as 1 cent a minute for calls in the United States. Low rates are available for calling outside the United States. Voice mail and fax capabilities are also available.

voice-over-IP (VOIP)
technology that enables network managers to route phone calls and fax transmissions over the same network they use for data

Using **voice-over-IP (VOIP)** technology, network managers can route phone calls and fax transmissions over the same network they use for data—which means no more phone bills. Gateways installed at both ends of the communications link convert voice IP packets into voice and back. With the advent of widespread, low-cost Internet telephony services, traditional long-distance providers are being pushed to either respond in kind or trim their own long-distance rates.

Here's how VOIP works (Figure 7.4). Voice travels over the corporate intranet or Internet rather than the circuit-switched public network. Most corporate-class IP telephony applications use gateways that sit between the PBX and a router to convert calls into IP packets and shunt them onto the network. Using packets allows multiple parties to share digital lines so data transmission is much more efficient than traditional phone conversations, each of which requires a line. When the packets hit the destination gateway, the message is depacketized, converted back into voice, and sent out via local phone lines. With newer multiVOIP, the PBX, seen in Figure 7.4, is not needed. Phones are directly connected to a multiVOIP box. This arrangement can be cheaper than standard VOIP, making the technology more attractive to small businesses.

What is especially interesting about VOIP is the promise of new ways for merging voice with video and data communications over the Web or a company's data network. In the long run, it's not the cost savings that will boost the market, it's the multimedia capabilities it gives us and the smart call-management capabilities. Travel agents could use voice and video over the Internet to discuss travel plans; Web merchants could use it to show merchandise and take orders; customers could show suppliers problems with their products.

A codec (*compression-dec*ompression) device prepares the voice transmission by squeezing the recorded sound data and slicing it into packets for transfer over the Internet. On the receiving end, a codec reassembles and decompresses the data for playback. Different codecs are optimized for different uses and conditions, and the characteristics of a specific codec can affect voice quality.

Internet videoconferencing, which supports both voice and visual communications, is another emerging service. Hardware and software are available to support a two-party conferencing system. The key here is a video codec to convert visual images into a stream of digital bits and translate back again. The ideal video product will support multipoint conferencing in which multiple users appear simultaneously on the multiple screens.

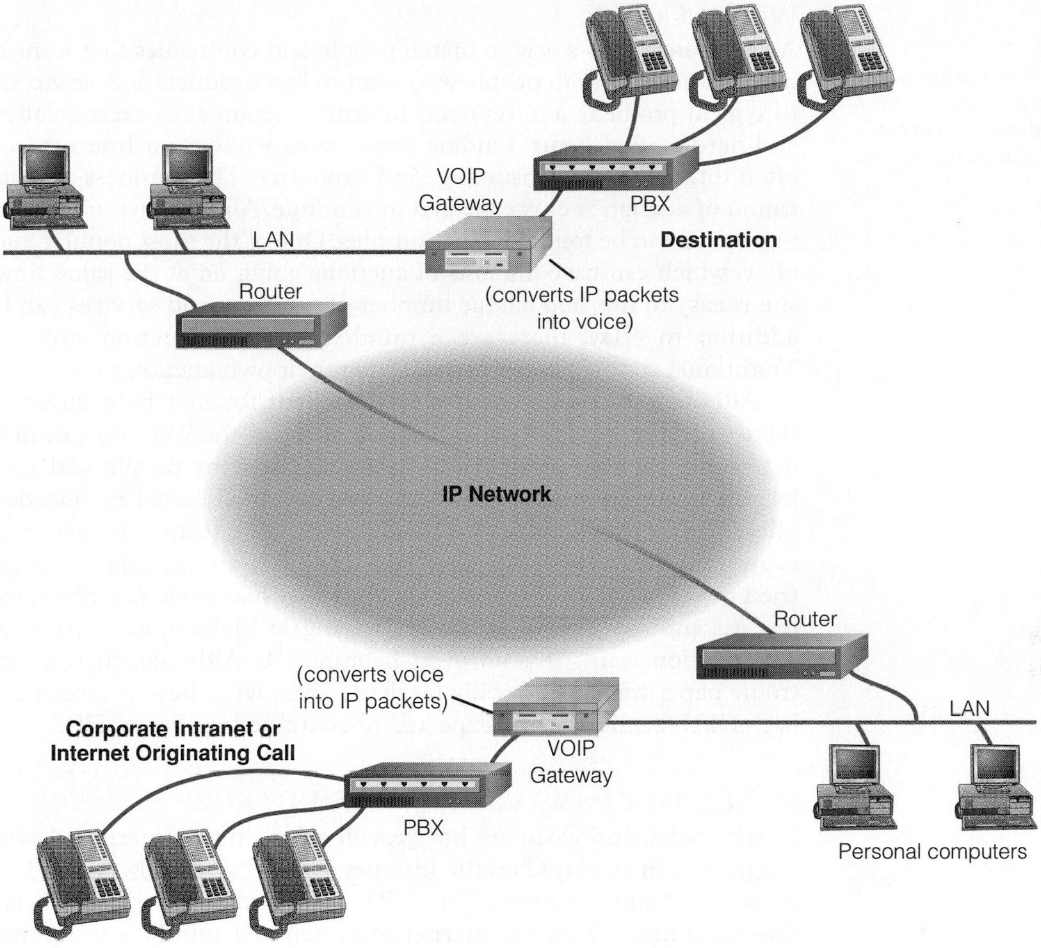
VOIP
Gateway
PBX
LAN
Destination
Router
(converts IP packets
into voice)
IP Network
Router
(converts voice
into IP packets)
LAN
Corporate Intranet or
Internet Originating Call
VOIP
Gateway
PBX
Personal computers

FIGURE 7.4

How Voice-Over-IP (VOIP) Works

content streaming
a method for transferring multimedia files over the Internet so that the data stream of voice and pictures plays more or less continuously without a break, or very few of them; enables users to browse large files in real time

CONTENT STREAMING

Content streaming is a method for transferring multimedia files over the Internet so that the data stream of voice and pictures plays more or less continuously, without a break, or very few of them. It also enables users to browse large files in real time. For example, rather than wait the half-hour it might take for an entire 5-MB video clip to download before they can play it, users can begin viewing a streamed video as it is being received.

SHOPPING ON THE WEB

Shopping on the Web for books, clothes, cars, drugs, and even medical advice can be convenient and easy. Some Internet shoppers are often loyal to a few familiar Internet sites, even if it is slightly more expensive to buy the same items there.[22] One study completed at the Columbia Business School found that only 10 percent of people buying books on the Internet looked at more than one Internet site before making a purchase. Some believe that on-line shoppers are lazy. Others believe that on-line shoppers have strong preferences for a few Internet sites and are reluctant to switch or look for other sites—good news for well-established and popular Internet sites.

Increasingly, people are using bots to help them search for information or shop on the Internet. A **bot** is a software tool that searches the Web for information, products, prices, etc. A bot, short for a robot, can find the best prices or features from multiple Web sites. Hotbot.lycos.com, for example, searches a number of Internet sites using keywords.

bot
a software tool that searches the Web for information, products, prices, etc.

Web auction
an Internet site that matches people who want to sell products and services with people who want to purchase these products and services

WEB AUCTIONS

A **Web auction** is a way to match people and companies that want to sell products and services with people who want to buy products and services. In addition to typical products and services, Internet auction sites excel at offering unique and hard-to-find items. Finding these items without an Internet auction site is often difficult, time consuming, and expensive. This business-to-business application of auction sites is expected to continue. Almost anything you may want to buy or sell can be found on auction sites. One of the most popular auction sites is eBay, which can have millions of auctions going on at the same time. The eBay site is easy to use, and a large number of products and services can be found. In addition to eBay, there are a number of other auction sites on the Web. Traditional companies are even starting their own auction sites.

Although auction Web sites are excellent for matching buyers and sellers, there are potential problems. Auction sites on the Web are not always able to determine whether products and services listed by people and companies are legitimate. In addition, some Web sites have had illegal or questionable items offered. eBay has an aggressive fraud investigation team to prevent and help prosecute fraudulent use of its site.[23] In one case, an individual was arrested who tried to sell a Porsche Roadster on eBay that had been assembled from a crash-test car and spare parts. According to Angela Malacri, manager of eBay's fraud investigation team, "It's an overwhelming job. Although there's often an electronic paper trail, it's basically faceless." Even with these potential problems, the use of Web auction sites is expected to continue to grow rapidly.

MUSIC, RADIO, AND VIDEO ON THE INTERNET

Music, radio, and video are hot growth areas on the Internet. Audio and video programs can be played on the Internet or files can be downloaded. Using music players and music formats like MP3, discussed in Chapter 3, it is possible to download music from the Internet and listen to it anywhere using small, portable music players. A number of companies are now offering music over the Internet. According to one AOL Time Warner executive, "We have the opportunity to create a personal jukebox in the house and the car." Another key executive of AOL said, "This is the takeoff point for the music business." Music on the Internet is not without controversy, however. The Recording Industry Association of America won a legal battle against Napster, preventing the company from allowing free copies of music to be shared over the Internet.[24] But a number of other companies have surfaced that offer music sharing over the Internet. "Stopping the services will be a long process," says Cary Sherman, general counsel for the Recording Industry Association of America.

The Internet is also being used to form music collaborations that would be difficult or impossible otherwise.[25] Internet sites such as RocketNetwork.com and Tonos.com are allowing musicians to record music from long distances. Todd Shoemaker, a musician and studio engineer for Elton John, uses the Internet to produce music for a collaboration called NegativePositive. According to Shoemaker, "Usually, techno musicians are single individuals who sit in a room with a lot of keyboards and equipment, doing their thing. It's not easy to collaborate live, partly because techno musicians all use the same instruments and have to take turns at the controls."

It is also possible to listen to radio broadcasts over the Internet or to download radio programs. WorldClassRock, for example, entered into a partnership with Clear Channel to broadcast over the Internet using an Internet streaming program.[26] The Web radio program makes a profit and charges Microsoft and RealNetworks a fee to link to Internet sites where their Internet streaming software can be downloaded. Entire audio books can also be downloaded for later listening, using devices like the Audible Mobile Player. This technology is similar to the popular books-on-tape media, except you don't need a cassette tape or a tape player. Worldstream Communications is now offering interactive talk shows on the Internet, with a format that resembles the popular TV talk shows. Topics range from politics, to economics, to news. A typical program has a reporter interviewing a guest. On the Internet, you can see pictures of the guest and hear the live talk and audience reactions.

Some corporations have also started to use Internet video to broadcast corporate messages or to advertise on the Web. Victoria's Secret, for example, used Internet video to advertise its lingerie line. The video was so popular that 1.5 million viewers jammed the Internet site. In addition to advertising, some companies are now investigating the use of Internet video to broadcast stockholder meetings, statements to the public from top-level executives, and other messages. Doctors can also use Internet video to monitor and even control surgical operations that take place thousands of miles from them. As mentioned in Chapter 6, Internet video is also being used successfully for teleconferencing, which can connect employees, managers, and corporate executives around the world in private conversations. Using Internet video, it is also possible to receive TV programs from an Internet site.

OFFICE ON THE WEB

After the terrorist attacks on the World Trade Center and Pentagon, travel for business has been drastically reduced. Aside from the use of video and audio conferences to connect employees who are widely spread, employers are also offering employees the option of telecommuting to their jobs. From home offices, co-workers can connect to the workplace via a virtual office on the Web. For example, you can receive a phone call from your boss, who wants you to send a financial document to a co-worker immediately. You may also want to set up conference calls or track upcoming appointments.

To help solve these problems, you can set up an Internet office.[27] An Internet office is a Web site that contains files, phone numbers, e-mail addresses, an appointment calendar, and more. Using a standard Web browser, you can access important files and information. FleetBoston offers an Internet safe-deposit box service that allows you to safeguard important files on its Internet site and share them with others.[28] You can download a draft of a sales report to your laptop computer or send documents from the Web site to co-workers. You can also search your online appointment book.

An Internet office allows your desktop computer, phone books, appointment schedulers, and other important information to be with you wherever you are. For those who do still travel for their work, products such as Hotoffice.com, eRoom.com, and Quickplace.com allow them to travel light and never leave important information behind. GoToMyPC.com allows you to connect to an office or home PC while traveling using the Internet.[29] It is also possible to print documents while traveling.[30] PrintMe Networks allows you to print from the Internet to printers at popular hotels, shops, airports, and other locations. For individuals and employees who travel, these services can be invaluable.

INTERNET SITES IN THREE DIMENSIONS

Some Web sites offer three-dimensional views of places and products. For example, a 3-D Internet auto showroom allows people to get different views of a car, simulating the experience of walking around in a real auto showroom. When looking at a 3-D real estate site on the Web, people can tour the property, go into different rooms, look at the kitchen appliances, and even take a virtual walk in the garden. Some 3-D sites allow people to interact with the product they are viewing on the Web. Sony, for example, is experimenting with this technology to allow people to open and close the lid of a Sony laptop using a mouse. It is likely that 3-D Internet sites will become common in the future.

FREE SOFTWARE AND SERVICES

The Internet has always been known as a source for free software, advice, and services. The software for many of the services just discussed can be downloaded from the Internet free of charge. ICQ, the instant messaging system discussed earlier, for example, can be downloaded at no charge. Some e-mail services, such as those offered by Yahoo! and Hotmail, are also free. In addition, there is a wealth of information and advice on the Internet. Using a search engine, it is possible to obtain free information on almost any topic, ranging from investments to dating. Table 7.6 lists some popular and useful free software and services.[31] Note that free services can change or even stop their offerings with little or no warning.

The disadvantages to free services can be many, however. There is a wealth of free software and services on the Internet, but many of the sites bombard the user with annoying advertising. In addition, the information and advice may not always be truthful or helpful. For example, some people have posted false information on investment chat rooms, hoping to manipulate the price of a stock and make a profit. In some cases, groups or sites on the Internet have a bias. They can appear to be helpful but are actually trying to advance their own agenda by posting false or misleading information on the Internet. Thus, great care must be exercised when obtaining free software or services from the Internet.

OTHER INTERNET SERVICES

Other Internet services are constantly emerging. The Mars Odyssey space vehicle, for example, included a transceiver that could eventually allow the use of the Internet on the distant planet.[32] The transceiver would permit different Mars probes to communicate with each other and Earth stations. Eventually, this technology could allow e-mail when man visits Mars. The Internet can be essential to provide critical information during times of disaster or terrorism.[33] During the anthrax scare of 2001, critical medical information was transmitted over the Internet. According to Dr. Cordell, Professor of Emergency Medicine at Indiana University, "Never before in my medical career have I had a more urgent need for just-in-time information, and less time to obtain it than now."

With interest rate declines of the early 2000s, many Internet lending sites have also emerged.[34] People wanting to consolidate their credit card debt or to

TABLE 7.6

Free Internet Services

Site	Description
www.freeservers.com	Gives a limited amount of free Web site space.
www.freemerchant.com	Gives a free on-line storefront that uses your domain name or company name.
www.companysleuth.com	Keeps track of press releases, job sites, investor forums, SEC documents, and other information on companies you specify.
www.software602.com	Free word processor and spreadsheet program.
www.namedemo.com	One free domain name and e-mail address.
www.netzero.com and www.bluelight.com	Free Internet service providers. You have to endure ads on your screen.
www.games.msn.com and www.heat.net	Free games on the Internet.

obtain lower payments on their existing home mortgages have turned to sites such as Quicken Loan, E-Loan, and LendingTree for help. These and similar sites loaned more than $2 billion during a three-month period in 2001, up almost 200 percent from the previous year.

The Internet also facilitates distance learning, which has dramatically increased in the last several years. Many colleges and universities now allow students to take courses without ever visiting campus. Businesses are also taking advantage of distance learning through the Internet.[35] Dan Thomas, Senior Director of Education Products and Services at Mortgage Bankers Association of America (MBAA), is in charge of Campus-MBAA, a distance learning facility for the association. More than 11,000 students are taking distance-learning courses over the Internet in real estate through MBAA.

⊙ THE WORLD WIDE WEB

World Wide Web (WWW, or W3)
a collection of tens of thousands of independently owned computers that work together as one in an Internet service

The World Wide Web was developed by Tim Berners-Lee at CERN, the European Organization for Nuclear Research in Geneva. He originally conceived of it as an internal document-management system. This server can be located at http://welcome.cern.ch/welcome/gateway.html. From this modest beginning, the **World Wide Web** (the Web, WWW, or W3) has grown to a collection of tens of thousands of independently owned computers that work together as one in an Internet service. These computers, called Web servers, are scattered all over the world and contain every imaginable type of data. Thanks to the high-speed Internet circuits connecting them and some clever cross-indexing software, users are able to jump from one Web computer to another effortlessly—creating the illusion of using one big computer. Because of its ability to handle multimedia objects, including linking multimedia objects distributed on Web servers around the world, the Web has become the most popular means of information access on the Internet today.

The Web is a menu-based system that uses the client/server model. It organizes Internet resources throughout the world into a series of menu pages, or screens, that appear on your computer. Each Web server maintains pointers, or links, to data on the Internet and can retrieve that data. However, you need the right hardware and telecommunications connections, or the Web can be painfully slow. Traditionally, graphics and photos have taken a long time to materialize on the screen, and an ordinary phone line connection may not always provide sufficient speed to use the Web effectively. Serious Web users need to connect via the LAN server, SLIP/PPP, DSL, cable, or other approaches discussed earlier.

home page
a cover page for a Web site that has graphics, titles, and text

hypermedia
tools that connect the data on Web pages, allowing users to access topics in whatever order they wish

Data can exist on the Web as ASCII characters, word processing files, audio files, graphic and video images, or any other sort of data that can be stored in a computer file. A Web site is like a magazine, with a cover page called a **home page** that has color graphics, titles, and text. All the highlighted type (sometimes underlined) is hypertext, which links the on-screen page to other documents or Web sites. **Hypermedia** connects the data on pages, allowing users to access topics in whatever order they wish. As opposed to a regular document that you read linearly, hypermedia documents are more flexible, letting you explore related documents at your own pace and navigate in any direction. For example, if a document mentions the Egyptian pharaohs, you can choose to see a picture of the pyramids, jump into a description of the building of the pyramids, and then jump back to the original document. Hypertext links are maintained using URLs. Table 7.7 lists some interesting Web sites.[36] Many PC and business magazines also publish interesting and useful Web sites, and Web sites are often evaluated and reviewed in print media and on-line.

Site	Description	URL
Monster	This is a job-hunting site. You can search for a job by type or company, list your résumé, and perform basic company research. One feature, Talent Market, allows people to put their skills up for bid.	www.monster.com
Centers for Disease Control (CDC)	This government site provides a wealth of information on a wide variety of health topics.	www.cdc.gov
ICQ	This chat facility offers free chat services for two or more people.	www.icq.com
NASA Human SpaceFlight	This site from NASA gives information about past and present missions into space.	www.spaceflight.nasa.gov
MSN MoneyCentral	This Microsoft site offers a large range of financial and investment information.	Moneycentral.msn.com
Britannica	This site provides the popular encyclopedia on-line.	www.britannica.com
Yahoo Maps	This service offers street addresses and driving directions.	www.maps.yahoo.com
eBay	This is a popular auction site on the Internet.	www.ebay.com
Amazon.com	This popular site sells books, videos, music, furniture, and much more.	www.amazon.com
Travelocity	This large site offers travel information and bargains.	www.travelocity.com
WebMD	This site provides medical information and advice.	www.webmd.com

T A B L E 7 ⊙ 7

Several Interesting Web Sites

hypertext markup language (HTML)
the standard page description language for Web pages

HTML tags
codes that let the Web browser know how to format text—as a heading, as a list, or as body text—and whether images, sound, and other elements should be inserted

Extensible Markup Language (XML)
markup language for Web documents containing structured information, including words, pictures, and other elements

Hypertext Markup Language (HTML) is the standard page description language for Web pages. One way to think about HTML is as a set of highlighter pens in different colors that you use to mark up plain text to make it a Web page—red for the headings, yellow for bold, and so on. The **HTML tags** let the browser know how to format the text: as a heading, as a list, or as body text. HTML also tells whether images, sound, and other elements should be inserted. Users mark up a page by placing HTML tags before and after a word or words. For example, to turn a sentence into a heading, you place the <H1> tag at the start of the sentence. At the end of the sentence, you place the closing tag </H1>. When you view this page in your browser, the sentence will be displayed as a heading. So, a Web page is made up of two things: text and tags. The text is your message, and the tags are codes that mark the way words will be displayed. All HTML tags are encased in a set of less than (<) and greater than (>) arrows, such as <H2>. The closing tag has a forward slash in it, such as </B> for closing bold. Figure 7.5 shows a simple document and its corresponding HTML tags.

A number of newer Web standards are gaining in popularity, including Extensible Markup Language (XML), Extensible Hypertext Markup Language (XHTML) cascading style sheets (CSS), and Dynamic HTML (DHMTL).[37] **Extensible Markup Language (XML)** is a markup language for Web documents containing structured information, including words and pictures. The Vancouver Police Department, for example, used XML to develop a secure Web site to help it sniff out stolen property.[38] With the new approach, second-hand stores can enter items they might buy into the Internet Web site to determine whether the item was stolen. What used to take weeks of manually shuffling paper containing stolen property lists can now be done instantly. In the field of

FIGURE 7.5

Sample Hypertext Markup Language

Shown at the left on the screen is a document, and at the right are the corresponding HTML tags.

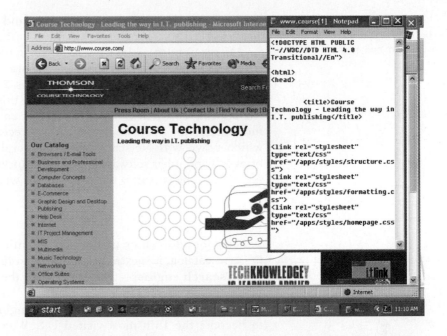

finance, Fidelity Investments decided to convert its corporate data to the XML format.[39] The purpose of the conversion was to simplify the interaction between the company's Web applications and its traditional, back-office operations.

With XML, there is no predefined tag set. With HTML, for example, the <H1> tag always means a first-level heading. With XML, tags and relationships between them can be defined. XML Web documents consist of content and markup. For example, <?xml version="1.0"?> is a processing markup that identifies a document as an XML document. <!ELEMENT oldjoke (burns +, allen, applause?)> is an example of an element declaration markup. XML includes the capabilities to define and share document information over the Web. CSS improves Web page presentation, and DHTML provides dynamic presentation of Web content. These standards move more of the processing for animation and dynamic content to the browser and provide quicker access and displays. XML documents contain tags that pertain specifically to the information you requested. Thus, you are able to navigate to desired data more quickly. However, entire industries will need to agree on sets of tags to achieve this goal. Fortunately, software is available to help people interested in Web authoring in HTML, XML, and other markup languages.

WEB BROWSERS

Web browser

software that creates a unique, hypermedia-based menu on a computer screen, providing a graphical interface to the Web

A **Web browser** creates a unique, hypermedia-based menu on your computer screen that provides a graphical interface to the Web. The menu consists of graphics, titles, and text with hypertext links. The hypermedia menu links you to Internet resources including text documents, graphics, sound files, and newsgroup servers. As you choose an item or resource, or move from one document to another, you may be jumping between computers on the Internet without knowing it, while the Web handles all the connections. The beauty of Web browsers and the Web is that they make surfing the Internet fun. Just clicking with a mouse on a highlighted word or graphic whisks you effortlessly to computers halfway around the world. Most browsers offer basic features such as support for backgrounds and tables, the ability to view a Web page's HTML source code, and a way to create hot lists of your favorite sites.

By 1996, Netscape Communications' Navigator had become the most widely used Web browser. It was setting the pace of browser development by embracing the HTML 3.0 standard and other hot new technologies. Navigator enables Net

applet
small program embedded in Web pages

surfers to view more complex graphics and 3-D models as well as audio and video material and runs small programs embedded in Web pages called **applets**.

Microsoft released Internet Explorer in the summer of 1995 to compete with Netscape. Some questioned whether it was appropriate for Explorer to come bundled with Windows 98 and Windows 2000. The battle between the U.S. Department of Justice and Microsoft over this issue appears to be slowing down. An appeals court in 2001 rejected the charge that Microsoft attempted to monopolize the browser market. More recently, however, AOL Time Warner, which now owns Netscape, filed a federal lawsuit against Microsoft about browsers and other software issues.[40] According to John Buckley, a spokesperson for AOL, "The lawsuit is to get justice for what Microsoft did to Netscape."

SEARCH ENGINES

Looking for information on the Web is a little like browsing in a library—without the card catalog, it is extremely difficult to find information. Web search tools—called **search engines**—take the place of the card catalog. Most search engines, like Yahoo.com and Google.com, are free. They make money by charging advertisers to put ad banners on their search engines. Google has grown dramatically to become the 15th most visited U.S. Web site, processing over 120 million searches every day.[41]

search engine
a Web search tool

The Web is a huge place, and it gets bigger with each passing day, so even the largest search engines do not index all Internet pages. Even if you do find a search site that suits you, your query might still miss the mark. So when searching the Web, you may wish to try more than one search engine to expand the total number of potential Web sites of interest. Another option is to use a meta-search engine. A **meta-search engine** submits keywords to several individual search engines and returns the results from all search engines queried. Ixquick (*www.ixquick.com*) and ProFusion (*www.profusion.com*) are examples of meta-search engines. However, meta-search engines do not query all search engines and may not permit more powerful searches.

meta-search engine
a tool that submits keywords to several individual search engines and returns the results from all search engines queried

Once you find a document that comes close to your goal, you can usually find related material by following the highlighted entries that take you to other Web pages when you click on them. And if you come across something you think you'll want to return to, you can add it to the "hot list" or "favorites" list on your Web browser to save time in the future.

Google is one of the most popular search engines on the Web.

Search engines that use keyword indexes produce an index of all the text on the sites they examine. Typically, the search engine reads at least the first few hundred words on a page, including the title, the HTML "alt text" coded into Web-page images, and any keywords or descriptions that the author has built into the page structure. The search engine throws out words such as "and," "the," "by," and "for." The search engine assumes whatever words are left are valid page content; it then alphabetizes these words (with their associated sites) and places them in an index where they can be searched and retrieved.

This type of search engine usually does no content analysis per se but uses word placement and frequency to determine how a page ranks among other pages containing the same or similar words. For example, when someone searches for the word *alien,* a page with "alien" in its title will appear higher in the search results than a site that doesn't mention "alien" in the title. Likewise, a page with 20 mentions of "alien" in the body text will rank higher than a page with one instance of the word.

Keyword indexes tend to be fast and broad; you'll typically get search results in seconds (faster than other kinds of engines). But unless you're careful about how you construct your query, you're likely to be overwhelmed with data.

Subject directories operate like a card catalog: They assign sites to specific topic categories based on the site's content. The advantage of this approach is that sites are grouped and easier to browse than those in a raw keyword index. Some sites use both keyword and subject directories. There are a number of Web search tools to choose from, as summarized in Table 7.8.[42]

JAVA

Java
an object-oriented programming language from Sun Microsystems based on C++ that allows small programs (applets) to be embedded within an HTML document

Java is an object-oriented programming language from Sun Microsystems based on the C++ programming language that allows small programs—the *applets* mentioned earlier—to be embedded within an HTML document. When the user clicks on the appropriate part of the HTML page to retrieve it from a Web server, the applet is downloaded onto the client workstation environment, where it begins executing.

Java lets software writers create compact "just-in-time" programs that can be dispatched across a network such as the Internet. On arrival, the applet automatically loads itself on a personal computer and runs—reducing the need for computer owners to install huge programs anytime they need a new function. And unlike other programs, Java software can run on any type of computer. Java is used by programmers to make Web pages come alive, adding splashy graphics, animation, and real-time updates. Java-enabled Web pages are more interesting than plain Web pages. Caterpillar Financial Services Corporation, for example, used Java to develop ExpressTrack, a Web-based financial system.[43] The system took three years to build and cost several million dollars. General Motors Corporation used Java to add a large number of new features to its Web sites in 40 countries.[44] According to Steve Hannah, director of business development for e-GM, "We're going from a patchwork quilt to a single blanket with this new approach."

TABLE 7.8

Popular Search Engines

Search Engine	Web Address
Altavista	http://www.altavista.com
Ask Jeeves	http://www.ask.com
Google	http://google.com
HotBot	http://www.hotbot.lycos.com
Infoseek	http://infoseek.go.com
Northern Light	http://www.northernlight.com
Yahoo!	http://www.yahoo.com

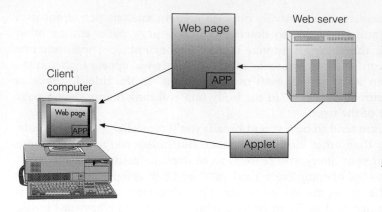

FIGURE 7.6

Downloading an Applet from a Web Server

The user accesses the Web page from a Web server. If the user clicks on the APP rectangle to execute the Java application, the client's computer checks for a copy on its local hard drive. If the applet is not present, the client requests that the applet be downloaded.

The relationship among Java applets, a Java-enabled browser, and the Web is shown in Figure 7.6. To develop a Java applet, the author writes the code for the client side and installs that on the Web server. The user accesses the Web page and pulls it down to his personal computer, which serves as a client. The Web page contains an additional HTML tag called APP, which refers to the Java applet. A rectangle on the page is occupied by the Java application. If the user clicks on the rectangle to execute the Java application, the client computer checks to see if a copy of the applet is already stored locally on the computer's hard drive. If it is not, the computer accesses the Web server and requests that the applet be downloaded. The applet can be located anywhere on the Web. If the user's Web browser is Java enabled (e.g., Sun's HotJava browser or Netscape's Navigator product), then the applet is pulled down into the user's computer and is executed within the browser environment.

The Web server that delivers the Java applet to the Web client is not capable of determining what kind of hardware or software environment the client is running on, and the developer who creates the Java applet does not want to worry about whether it will work correctly on Windows, Unix, and MacOS. Java is thus often described as a "cross-platform" programming language.

The development of Java has had a major impact on the software industry. Sun Microsystems's strategy is to open up Java to any and all. Any software vendors and individual developers—from development tool vendors, language compiler developers, database management system vendors, and client/server application vendors to small businesses—can then use Java to create Internet-capable, run-anywhere applications and services (Figure 7.7). As a result, the Java community is becoming broader every day, encompassing some of the world's biggest independent software vendors, as well as users ranging from corporate CIOs, programmers, multimedia designers, and marketing professionals to educators, managers, film and video producers, and hobbyists.

FIGURE 7.7

Web Page Providing Java Applets

Free Java Applets can be downloaded from this site for use on your own Web site.

Java could change the economics of paying for software—most of the software industry today is based on the concept of entire applications delivered to the marketplace for a fixed, one-time cost, for which the customer is given a license that allows him or her to use the software forever on a single computer platform. Java makes it possible to sell one-time usage of a piece of software. This usage could be defined for a single transaction or for a single session during which the user is connected to a Web server.

PUSH TECHNOLOGY

push technology
automatic transmission of information over the Internet rather than making users search for it with their browsers

Push technology is used to send information automatically over the Internet rather than make users search for it with their browsers. Frequently the information, or "content," is customized to match an individual's needs or profile. The use of push technology is also frequently referred to as "Webcasting." Most push systems rely on HTTP (HyperText Transport Protocol) or Java technology to collect content from Web sites and deliver it to employees' or users' desktops. A business, for example, could use push technology to deliver important marketing information to its sales force.

Before they can be "pushed," employees or users must download and install software that acts like a TV antenna, capturing transmitted content. As with any new technology, the people paying for push have yet to venture beyond rudimentary applications. Most are focusing on improving communications with employees, customers, and business partners.

A number of companies are using push technology to deliver critical information over the Internet. SAP, for example, uses push technology to deliver critical enterprise resource planning software over the Internet. Enterprise resource planning software is used by many large corporations to streamline operations. Convene, an enterprise training and performance company, uses push technology to deliver critical business information.[45] Other companies, including oil service supplier Schlumberger and Carlson Wagonlit Travel, also use push technology to make priority deliveries of information over the Internet.

There are drawbacks to the use of push technology. One issue, of course, is information overload. Another is the volume of data being broadcast is so great that push technology can clog up the Internet communications links with traffic.

CVS.com enables users to order prescriptions and buy health and beauty supplies on-line.

BUSINESS USES OF THE WEB

In 1991, the Commercial Internet Exchange (CIX) Association was established to allow businesses to connect to the Internet. Since then, businesses have been using the Internet for a number of applications. Electronic mail is a major application for most companies. Many companies display products over the Internet, including catalogs and sample texts. Customers can place orders by keying in payment information and shipping addresses.

By linking buyers and sellers electronically on the Web, businesses are able to establish new and ongoing relationships with customers, allowing them access to information or products whenever it suits them. Businesses can use the

Web as a tool for marketing, sales, and customer support. The Web can also serve as a low-cost alternative to fax, express mail, and other communications channels. It also can eliminate paperwork and drive down the cost per business transaction.

The Internet's business potential has just begun to be tapped. As more and more people gain access to the World Wide Web, its functions are changing drastically. We discuss a couple of these applications, corporate intranets and extranets after a discussion of developing Web content. Chapter 8 will cover e-commerce in more detail.

DEVELOPING WEB CONTENT

Web authors work with several standards to create their pages. The HTML standards were created by a committee of various people involved in the Web. Anyone can create tags, and others may adopt them, modify them, or reject them. Thus, the HTML standards are evolving. HTML 1.0 was the standard in 1994 and is now obsolete. HTML 2.0 introduced a forms feature for allowing users to enter data and became the standard in 1995. HTML 3.0 allows banners, centering, right text alignment, tables, mathematical formulas, and image alignment. Netscape has added a number of new tags to this standard, such as <BLINK>, which causes blinking words that work only within Netscape. Thus, not all browsers will work the same way when used to view the same Web page. For some browsers, a tag will work wonderfully. On others, that tag won't do anything at all, or, worse, it may cause problems. Web authors need to keep these inconsistencies in mind when they develop pages. The art of Web design involves getting around the technical limitations of the Web and using a limited set of tools to make appealing designs. Following are tips to create a Web page.

1. Your computer must be linked to a Web server, which can deliver Web pages to other browsers.
2. You will need a Web browser program to look at HTML pages you create.
3. The actual design can take one of the following approaches: (a) Write your copy with a word processor, then use an HTML converter to convert the page into HTML format complete with tags so the browser knows how it should format the page. (b) Use an HTML editor to write text and add HTML tags at the same time. (c) Edit an existing HTML template (with all the tags ready to use) to meet your needs. (d) Use an ordinary text editor and type in the start and end tags for each item.
4. Open the page with the browser and see the result. You can correct mistakes by correcting the tags.
5. Add links to your home page to allow your readers to click on a word and be taken to a related home page. The new page may be either a part of your Web site or a home page on a different Web site.
6. To add pictures, you must first store them as a file on your hard drive. This can be done in one of several ways: draw them yourself using a graphics software package, copy pictures from other Web pages, buy a disk of clip art, scan photos, or use a digital camera.
7. You can add sound by using a microphone connected to your computer to record a sound file; adding links to the page will enable those who access your Web page to hear it.
8. Upload the HTML file to your Web site using e-mail or FTP.
9. Review the Web page to make sure that all links are correctly established to other Web sites.
10. Advertise your Web page to others and encourage them to stop, take a look, and send feedback by e-mail.

After Web content development, the next step is to place the content on a Web site or home page. Popular options include ISPs, free sites, and Web hosting.

Some Internet service providers include limited Web space, typically 1 to 6 MB, as part of their monthly fee. If more disk space is needed, there are additional charges. Free sites offer limited space for an Internet site. In return, free sites often require the user to view advertising or agree to other terms and conditions. A Web host is another option. A Web host can charge $15 or more per month, depending on services. Some Web hosting sites include domain name registration, Web authoring software, and activity reporting and monitoring of the Web site.

A number of products make developing and maintaining Web content easier, including Web services. The "IS Principles in Action" box provides more details about Web services, but in short, these products can greatly simplify creation of a Web page. Microsoft, for example, has introduced a Web development platform called .NET.[46] The .NET platform allows different programming languages to be used and executed. It also includes a rich library of programming code to help build XML Web applications. Once a Web site has been constructed, a *content management system (CMS)* can keep the Web site running smoothly.[47] CMS consists of both software and support. Companies that provide CMS can charge from $15,000 to over $500,000 annually, depending on the complexity of the Web site being maintained and the services being performed. Leading CMS vendors include BroadVision, Documentum, EBT, FileNet, Open Market, and Vignette.

INTRANETS AND EXTRANETS

An intranet is an internal corporate network built using Internet and World Wide Web standards and products. Employees of an organization use it to gain access to corporate information. After getting their feet wet with public Web sites that promote company products and services, corporations are seizing the Web as a swift way to streamline—even transform—their organizations. These private networks use the infrastructure and standards of the Internet and the World Wide Web. A big advantage of using an intranet is that many people are already familiar with the Internet and the Web, so they need little training to make effective use of their corporate intranet.

Most companies already have the foundation for an intranet—a network that uses the Internet's TCP/IP protocol. Computers using Web server software can store and manage documents built on the Web's HTML format. With a Web browser on your PC, you can call up any Web document—no matter what kind of computer it is on.

An intranet is an inexpensive yet powerful alternative to other forms of internal communications, including conventional computer setups. One of an intranet's most obvious virtues is its ability to slash the need for paper. Because Web browsers run on any type of computer, the same electronic information can be viewed by any employee. That means that all sorts of documents (such as internal phone books, procedure manuals, training manuals, and requisition forms) can be inexpensively converted to electronic form on the Web and be constantly updated. An intranet provides employees with an easy and intuitive approach to access information that was previously difficult to obtain. For example, it is an ideal solution to provide information to a mobile sales force that needs access to a lot of rapidly changing information. Intranets can also do something far more important. By presenting information in the same way to every computer, they can do what computer and software makers have frequently promised but never actually delivered: pull all the computers, software, and databases that dot the corporate landscape into a single system that enables employees to find information wherever it resides.

The Internet is like many other new technologies—it provides a wide range of services, some of which are effective and practical for use today, others are still evolving, and still others will fade away from lack of use.

Web Services Take the Web to a New level

The term *Web services* refers to an emerging technology that promises to revolutionize the way we use the Web for business and personal productivity. The Web has traditionally been used to find and display information. By using a Web browser, such as Netscape or Internet Explorer, we request pages of information from Web servers that locate the requested material. In technical terms this is referred to as a client/server system—the Web browser is the client program that requests information from a server program running on a computer maintained by a business or organization.

The technology behind Web services utilizes programs that can communicate with each other over the Web. Computer programs communicating with each other over a network to work together to accomplish tasks is called *distributed computing.* Distributed computing is not new, but until now, there hasn't been a framework for systems developers to easily create such applications for the Web. Web services provide that framework.

Through Web services, systems developers can automate trivial or repetitive tasks that traditionally required human interaction. For example, Microsoft has developed a calendar service that allows users to show their appointment books to others on the Web. Using this service, you could easily make appointments with your dentist, hair stylist, or mechanic through your Web browser. The use of Web services is quickly spreading:

- Dollar Rent A Car created a Web services interface to allow existing and potential business partners to access its mainframe-based reservation system.
- Expedia.com is transforming travel itineraries into communication centers with Web services—allowing travelers to pick distinct notification settings for different members of their integrated buddy list.
- Web services has helped CertifiedMail.com create secure messaging that embraces open standards so that the company can broaden its customer base and expand into lucrative new markets, such as the healthcare industry.

With a little imagination, it is easy to think up many applications that could benefit from this new ability to "program the Web."

The key to Web services is XML (Extensible Markup Language). XML is a standard for describing data on the Web. Just as HTML (Hypertext Markup Language) was developed as a standard for formatting Web content into Web pages, XML is used within a Web page to describe and transfer data between Web service applications. Besides XML, three other components are used in Web service applications:

1. SOAP (Simple Object Access Protocol) is a specification that defines the XML format for messages
2. WSDL (Web Services Description Language) provides a way for a Web service application to describe its interfaces in enough detail to allow a user to build a client application to talk to it

3. UDDI (Universal Discovery Description and Integration) is used to register Web service applications so that potential users can easily find them

There are strong indications that XML Web services technology is more than just a passing technical fad. Three major forces in technology, Microsoft, IBM, and Sun Microsystems, are all investing heavily in Web services development. The Web Services Interoperability (WS-I) Organization has been established to help Web services run on different platforms, operating systems, and programming languages. WS-I members include dozens of big technology companies, including Hewlett-Packard and Oracle. By 2005 it is anticipated that Web services technology and interfaces will be a standard for application integration and that commercial Web services will be common in many organizations.

Although Web services promise to improve the way we do business on the Web, they are not without their share of problems. Amid the fanfare of Microsoft's launch of Visual Studio .Net, a popular Web services programming platform, news services reported a security flaw within the application that could leave the Web service server open to attack by hackers. Microsoft worked quickly to seal the security hole, but when it comes to Internet security, it seems that just as one hole is patched, another is found.

A larger concern is over authentication—the confirmation that incoming data and messages are being sent by the advertised source. For example, if you use a Web service to instruct First National Bank to transfer funds to your Discover Card, can First National verify that it is really you who sent the instruction? Through digital signature technology, such verification can take place. The World Wide Web Consortium's (W3C) XML Signature recommendation, developed in conjunction with the Internet Engineering Task Force (IETF), provides a standard way of signing XML documents so that recipients can verify the identity of the sender and the integrity of the data.

Web services acceptance and implementation hinges on developments in Internet security and privacy. It was no coincidence that within a week of the release of Visual Studio .Net, Bill Gates sent out a well-publicized memo to all Microsoft employees stating that security was to be Microsoft's new focus. As security and privacy issues are resolved to satisfy the public and advocacy groups, we will no doubt rely more heavily on Web services to handle many of our on-line errands. Web services could develop into an electronic extension of ourselves, anticipating our needs and delivering information and services as we need them, freeing up time for more interesting endeavors.

Discussion Questions

1. What types of Web services might be offered by the following businesses and organizations? Travel agents, realtors, restaurants, the IRS. What other industries could benefit from Web services?

2. What specific security and privacy concerns might revolve around Web services offered by the preceding industries and organizations?

Critical Thinking Questions

3. What does Microsoft hope to gain by being the first to offer a Web services programming environment? Why might the company consider this area of development more important than its Windows operating system product and Office Suite?

4. How might Web services change the way we use our home PC in terms of software distribution and storage?

Sources: David Coursey, "Why Web Services Will Be the Next Big Thing," CNet.com, http://techupdate.cnet.com, accessed February 25, 2002; Daniel Sholler, "What Are Web Services, Anyway?" October 1, 2001, Metagroup Web site, http://www.metagroup.com, accessed February 25, 2002; Margaret Kane, "Oops! Security Flaw Found in VS.Net," *ZDNet News*, http://techupdate.cnet.com, accessed February 25, 2002; Paul Festa, "W3C Backs XML-Based Digital Signature," *ZDNet News*, http://zdnet.com, February 14, 2002; Microsoft Web site, http://msdn.microsoft.com, accessed February 25, 2002; WS-I Web site, http://www.ws-i.org, accessed April 5, 2002.

Universal reach is what made the Internet grow so rapidly. But Internet enthusiasts tended to focus on how to link far-flung people and businesses. When the Internet caught on, people were not considering it as a tool for running their business—but that is what is happening, with amazing speed. Just as the simple act of putting millions of computers around the world on speaking terms initiated the Internet revolution, so connecting all the islands of information in a corporation is sparking unprecedented collaboration. Corporate intranets are breaking down the walls within corporations.

More advanced use of the corporate intranet supports what has come to be known as workgroup computing. Workgroup computing involves many aspects, but basically it is an approach to support people working together in teams. One of the key aspects is the ability to store and share information in any form—text, video, sound, graphics, handwritten memos, or hand-drawn figures—which is often called a *knowledge base*. The key feature of workgroup computing is being able to organize and retrieve all this data simply. Group calendaring and scheduling allows an employee to check others' schedules and set up meetings. Another advantage of intranets is support for real-time meetings with people linked over networks, instead of making them travel to one place. Workgroup computing also supports work flow processes, tracking the status of documents—who has them, who is behind or ahead of schedule, and who gets them next.

A rapidly growing number of companies have advanced beyond the workgroup stage to offer limited network access to selected customers and suppliers. Such networks are referred to as extranets, which connect people who are external to the company. An extranet is a network that links selected resources of the intranet of a company with its customers, suppliers, or other business partners. Again, an extranet is built based on Web technologies.

Security and performance concerns are different for an extranet than for a Web site or network-based intranet. Authentication and privacy are critical on an extranet so that information is protected. Obviously, performance must be good to provide quick response to customers and suppliers. Table 7.9 summarizes the differences between users of the Internet, intranets, and extranets.

Secure intranet and extranet access applications usually require the use of a virtual private network (VPN). A **virtual private network (VPN)** is a secure connection between two points across the Internet. VPNs transfer information by encapsulating traffic in IP packets and sending the packets over the Internet, a practice called **tunneling**. Most VPNs are built and run by Internet service providers. Companies that use a

virtual private network (VPN)
a secure connection between two points across the Internet

tunneling
the process by which VPNs transfer information by encapsulating traffic in IP packets over the Internet

---

An intranet is an internal corporate network used by employees to gain access to company information.

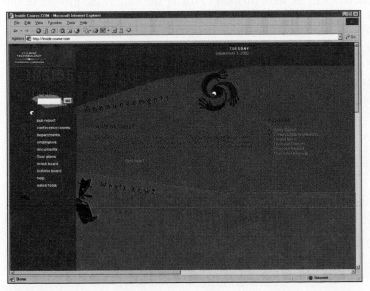

TABLE 7.9

Summary of Internet, Intranet, and Extranet Users

Type	Users	Need for User ID and Password
Internet	Anyone	No
Intranet	Employees and managers	Yes
Extranet	Business partners	Yes

VPN from an Internet service provider have essentially outsourced their networks to save money on wide-area network equipment and personnel. In using a VPN, a user sends data from his or her personal computer to the company's firewall, discussed later in the chapter, which also converts the data into a coded form that cannot be easily read by an interceptor. The coded data is then sent via an access line to the company's Internet service provider. From there, the data is transmitted through tunnels across the Internet to the recipient's Internet service provider and then over an access line to the receiving company's firewall, where it is decoded and sent to the receiver's personal computer (Figure 7.8).

Companies and governmental agencies are big users of VPNs. NASA, for example, uses a VPN to transfer data between the space shuttles and earth. The technology makes the shuttle appear as a node on the Internet. This allows NASA specialists on earth to control experiments being done in space. It also allows people to monitor shuttle operations using a standard Web browser.

NET ISSUES

The topics raised in this chapter apply not only to the Internet and intranets but also to LANs, private WANS, and every type of network. Control, access, hardware, and security issues affect all networks, so it is important to mention some of these management issues.

MANAGEMENT ISSUES

Although the Internet is a huge, global network, it is managed at the local level; no centralized governing body controls the Internet. Although the U.S. federal government provided much of the early direction and funding for the Internet, the government does not own or manage the Internet. The Internet Society and the Internet Activities Board (IAB) are the closest the Internet has to a centralized governing body. These societies were formed to foster the continued growth of the Internet. The IAB oversees a number of task forces and committees that deal with Internet issues. One of the main functions of the IAB is to manage the network protocols used by the Internet, including TCP/IP. Some universities and government agencies are investigating how the Internet can be controlled to prevent sensitive information and pornographic material from being placed on the Internet.

FIGURE 7.8

Virtual Private Network

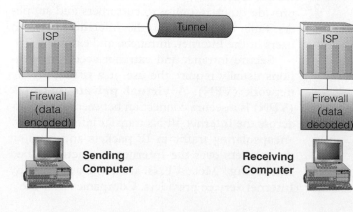

SERVICE AND SPEED ISSUES

Service and speed issues on the Internet are a function of the volume of traffic and more sophisticated Web sites. The growth in Internet traffic continues to be phenomenal. Traffic volume on company intranets is growing even faster than the Internet. Companies setting up an Internet or intranet Web site often underestimate the amount of computing power and communications capacity they need to serve all the "hits" (requests for pages) they get from Web cruisers. Web server computers can be overwhelmed with thousands of hits per hour. In addition, Web sites are becoming more sophisticated with video and audio clips and other features that require faster Internet speeds.

Connection agreements exist among the backbone companies to accept one another's traffic and provide a certain level of service. Some Internet providers do a good job and provide a high level of quality and service. Others have not done such a good job, creating wide variations in the quality of the Internet. As a result, a shakeout in ISPs has occurred, as with other Web-based companies and services. Most providers lease their lines from phone companies, and the cost of even a medium-speed line to connect California with Australia is over $1.2 million per year—ten times the cost of a New York–to–San Francisco link, so some providers scrimp. This leads to inadequate capacity, which slows transmissions.

Routers, the specialized computers that send packets down the right network pathways, can also become bottlenecks. For each packet, every router along the way must scan a massive address book of about 40,000 area destinations (akin to Internet ZIP codes) to pick the right one. These routers can get overloaded and lose packets. The TCP/IP protocol compensates for this by detecting a missing packet and requesting the sending device to resend the packet. However, this leads to a vicious circle, as the network devices continually try to resend lost packets, further taxing the already overworked routers. This leads to long response times or loss of the connection to the network.

Several actions are opening up the bottleneck. Various backbone providers have been upgrading their backbone links, installing bigger, faster "pipes" and converting to newer transmission technology, such as asynchronous transfer mode (ATM), which can send a message down the right path more quickly than standard packet-switching technology. Each ATM transmission is pre-addressed with its own route, so routing addresses do not have to be looked up and the packet can zip right through an ATM switch. Also, router manufacturers are working to develop improved models with increased hardware capacity and more efficient software to provide quick access to addresses. Yet a third solution is to prioritize traffic. Today, all network traffic travels through the same big backbone pipes. There is no way to make sure that your urgent message is not stalled behind someone downloading a magazine page. With prioritized service, customers could pay more for guaranteed delivery speed, much like an overnight package costs more than second-day delivery. If implemented, this solution could also affect the cost of network services that generate a lot of traffic, such as Internet phone and videoconference services. Finally, the increased availability of faster DSL, satellite, and cable connections, discussed in Chapter 6, can also be used to speed Internet access. The cost of these faster services are also starting to fall.

PRIVACY, FRAUD, SECURITY, AND UNAUTHORIZED INTERNET SITES

As the use of the Internet grows, privacy, fraud, and security issues become even more important. People and companies are reluctant to embrace the Internet unless these issues are successfully addressed.

Privacy

From a consumer perspective, the protection of individual privacy is essential. Yet, many people use the Internet without realizing their privacy may be in jeopardy. A number of companies, including Jupiter Media Metrix and Nielsen//NetRatings, help other companies monitor visits to their Internet sites.[48] These companies often call a random sample of Internet users to gain insights into the habits and desires of Internet users. Some companies, such as Hallmark Cards, hire people to visit chat rooms on the Internet to get important marketing information.[49] Hallmark rewards participants who help it collect information on the Internet with Hallmark gifts and products.

cookie
a text file that an Internet company can place on the hard disk of a computer system

Many Internet sites use cookies to gather information about people who visit their sites. A **cookie** is a text file that an Internet company can place on the hard disk of a computer system. These text files keep track of visits to the site and the actions people take. To help prevent this potential problem, some companies are developing software to help prevent these files from being placed on computer systems. CookieCop, for example, allows Internet users to accept or reject cookies by Internet site. Microsoft's Internet Explorer 6 browser also has the ability to screen Web sites according to their privacy policy.[50] Using the Platform for Privacy Preferences (P3P), Internet Explorer 6 can summarize the privacy policy for Web sites and prevent information from being transmitted from your computer to a Web site that doesn't meet certain criteria.

Fraud

Internet fraud is another important issue. Some people have received false messages that seem to be from their Internet service providers asking them to update their personal information, including social security numbers and credit card information.[51] But instead of going to the Internet service provider, the information is captured and used by on-line thieves. The possibility of Internet fraud has prevented many people from using the Internet. Local, state, and federal agencies are actively pursuing Internet fraud. Federal law enforcement officers, in a major sting operation, brought charges against 90 individuals and companies involved in Internet schemes that may have bilked more than 50,000 people out of $117 million.[52] As law enforcement agencies crack down on fraud, public confidence in using the Internet should increase.

Security with Encryption and Firewalls

When it comes to security on the Internet, it is essential to remember two things. First, there is no such thing as absolute security. Second, plenty of clever people consider it great sport to try to breach any security measures—the better your security, the greater the challenge to them.

cryptography
the process of converting a message into a secret code and changing the encoded message back to regular text

encryption
the conversion of a message into a secret code

From a corporate strategy perspective, security of data is essential. Such approaches as cryptography can help. **Cryptography** is the process of converting a message into a secret code and changing the encoded message back to regular text. The original conversion is called **encryption**. The unencoded message is called *plaintext*. The encoded message is called *ciphertext*. Decryption converts ciphertext back into plaintext (Figure 7.9). For much of the Cold War era, cryptography was the province of military and intelligence agencies; uncrackable codes were reserved for people with security clearance only.

Widespread deployment of cryptography requires additional hardware and software but is becoming increasingly necessary to support electronic commerce, copyright management, and electronic delivery of services. Without cryptography, people will not trust that electronic financial transactions, secret or private data, and valuable intellectual property will remain confidential across networks.

A cryptosystem is a software package that uses an algorithm, or mathematical formula, plus a key to encrypt and decrypt messages. The algorithm is calculated with the key and converts every character of the plaintext into other coded characters, thus creating the ciphertext. Only someone with the correct key should be able to decode the ciphertext. Good ciphertext appears to be nothing more than random characters. Encryption makes information useless to hackers and thieves.

The Data Encryption Standard (DES), adopted as a federal standard in 1977 to protect unclassified communications and data, was designed by IBM and modified by the National Security Agency. It uses 56-bit keys, meaning a user must employ precisely the right combination of 56 1s and 0s to decode information correctly. Other technologies offer a range of key lengths; in the case of RC5, up to 2,048 bits. The RSA protocol has no limit on key length, but it can slow transmission, since it uses separate keys for encryption and decryption. Many products mix technologies: They use a fast algorithm like DES for the actual encryption but send the DES key through a more secure method like RSA.

FIGURE 7 • 9

Cryptography is the process of converting a message into a secret code and changing the encoded message back into regular text.

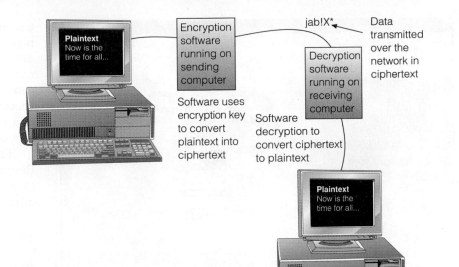

U.S. banks and brokerage houses use the federal government's DES algorithm to protect the integrity and confidentiality of fund transfers totaling some $2.3 trillion a day worldwide. Organizations encrypt the words and videos of their teleconferencing sessions. Individuals encode their electronic mail. And researchers use encryption to hide information about new discoveries from prying eyes.

Encryption is not just for keeping secrets. It can also be used to verify who sent a message and to tell whether the message was tampered with en route. A **digital signature** is a technique used to meet these critical needs for processing on-line financial transactions. Digital signatures involve a complicated technique that combines the public-key encryption method with a "hashing" algorithm that prevents reconstructing the original message. The hashing algorithm provides further encoding by using rules to convert one set of characters to another set (e.g., the letter *s* is converted to a *v*, 2 is converted to 7, etc.). Thus, encryption also can prevent electronic fraud by authenticating senders' identities with digital signatures.

The most popular method of preventing unauthorized access to corporate computer data is to construct what is known as a firewall between company computers and the Internet. An Internet **firewall** is a device that sits between your internal network and the Internet. Its purpose is to limit access into and out of your network based on your organization's access policy. A firewall can be anything from a set of filtering rules set up on the router to an elaborate application gateway consisting of one or more specially configured computers that control access. BellSouth, for example, uses a hardware firewall by SonicWall to provide security for its FastAccess Internet service.[53]

Firewalls permit desired services on the outside, such as e-mail, to pass. In addition, most firewalls allow access to the Web from inside protected networks. But firewalls deny other, unwanted access. For example, you may be able to use the telnet utility to log in to systems on the Internet, but users on remote systems cannot log in to your local system because the firewall prevents it.

A firewall can be set up to allow or to prevent access to and from specific hosts and networks. In addition, security personnel can assign different levels of access to hosts; a preferred host may have full access, and a secondary host may only have access to certain portions of the directory structure.

For a higher level of security, companies can install an assured pipeline, which uses more sophisticated methods to prevent access. A firewall looks at only

digital signature
encryption technique used to verify the identity of a message sender for processing on-line financial transactions

firewall
a device that sits between an internal network and the Internet, limiting access into and out of a network based on access policies

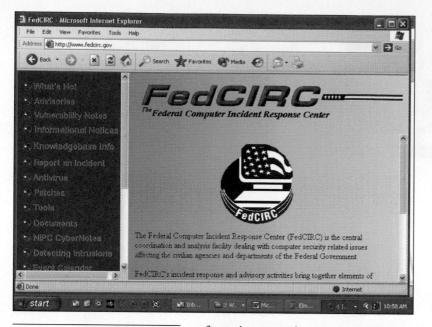

The header of a packet, but an assured pipeline looks at the entire request for data and then determines whether the request is valid. Inappropriate requests can be routed away, and files that meet specific criteria (such as those that contain the word *confidential*) can never be sent over the Internet.

The Federal Computer Incident Response Capability (FedCIRC) responds to virus attacks, network intrusions, and other threats. It also provides security training and consulting services to individual agencies. Vulnerability and threat information is shared with the public at this group's Web site (http://www.fedcirc.gov).

Unauthorized Sites

Unauthorized and unwanted Internet sites are also problems some companies face. A competitor or an unhappy employee can create an Internet site with an address similar to a company's. When someone searches for information about the company, he or she may find an unauthorized site instead. In some cases, the site may appear to be the legitimate, official corporate site. In others, it is obvious that the site is not sponsored or authorized by the company.

The Federal Computer Incident Response Capability assists civilian government agencies with computer security incidents.

Some unauthorized or unwanted sites contain damaging information about a company. Sometimes the information is true, and sometimes it is false or misleading. A fired employee, for example, could post stories about his boss or former employer that may not be entirely true. A competitor or an environmental watch group could also post information about why a customer should not do business with a company. As people travel the Internet, they should be aware that not all sites they see are endorsed by companies or organizations, so a wise Net surfer uses caution and some skepticism in accepting posted information as the whole truth. Unauthorized and unwanted Internet sites can be very troublesome, and it is not unusual for companies to sue those who post these sites. Network management issues will take an increasing amount of time for IS personnel, but any user needs to be aware of the basics to function effectively in business. Communications, service, and daily work are all at stake.

SUMMARY

PRINCIPLE   *The Internet is like many other technologies— it provides a wide range of services, some of which are effective and practical for use today, others are still evolving, and still others will fade away from lack of use.*

The Internet started with ARPANET, a project started by the U.S. Department of Defense (DOD). Today, the Internet is the world's largest computer network. Actually, it is a collection of interconnected networks, all freely exchanging information. The Internet transmits data from one computer (called a *host*) to another. The set of conventions used to pass packets from one host to another is known as the Internet protocol (IP). Many other protocols are used in connection with IP. The best known is the Transport Control Protocol (TCP). TCP is so widely used that many people refer to TCP/IP, the combination of TCP and IP used by most Internet applications. Each computer on the Internet has an assigned address to identify it from other hosts, called its uniform resource locator (URL). There are three ways to connect to the Internet: via a LAN whose server is an Internet host, via SLIP or PPP, and via an on-line service that provides Internet access.

An Internet service provider is any company that provides individuals or organizations with access to the Internet. To use this type of connection, you must have an account with the service provider and software that allows a direct link via TCP/IP. Among the value-added services ISPs provide are electronic commerce, intranets and extranets, Web site hosting, Web transaction processing, network security and administration, and integration services.

Internet services include (1) e-mail, (2) telnet, (3) FTP, (4) usenet and newsgroups, (5) chat rooms, (6) Internet phone, (7) Internet videoconferencing, (8) content streaming, (9) instant messaging, (10) shopping on the Web, (11) Web auctions, (12) music, radio, and video, (13) office on the Web, (14) 3-D Internet sites, (15) free software, and (16) other services. E-mail is used to send messages. Telnet enables you to log on to remote computers. FTP is used to transfer a file from another computer to your computer. Usenet supports newsgroups, which are on-line discussion groups focused on a particular topic. Chat rooms let you talk to dozens of people at one time, who can be located all over the world. Internet phone service enables you to communicate with other Internet users around the world who have equipment and software compatible with yours. Internet videoconferencing enables people to conduct virtual meetings. Content streaming is a method of transferring multimedia files over the Internet so that the data stream of voice and pictures plays continuously. Instant messaging allows people to communicate in real-time using the Internet. Shopping on the Web is popular for books, videos, music, and a host of other items and services. Items that may require style and fit, such as clothes, are not as popular on the Web. Web auctions are a way to match people looking for products and services with people selling these products and services. The Web can also be used to download and play music, radio, and video programs. With office on the Web, it is possible to store important files and information on the Internet. When telecommuting or traveling, these files and information can be downloaded or sent to other people. Some Internet sites are three-dimensional, allowing people to manipulate the site to see different views of products and images on the Internet. A wealth of free software and services are available through the Internet. Some of the free information, however, may be misleading or even false. Some of the other Internet services include information about space exploration, fast information transfer, obtaining a home loan, and distance learning.

The Web is a collection of independently owned computers that work together as one in an Internet service. High-speed Internet circuits connect these computers, and cross-indexing software is employed to enable users to jump from one Web computer to another effortlessly. Because of its ability to handle multimedia objects and hypertext links between distributed objects, the Web is emerging as the most popular means of information access on the Internet today.

PRINCIPLE *Originally developed as a document-management system, the World Wide Web is a menu-based system that is easy to use for personal and business applications.*

A Web site is like a magazine, with a cover page called a *home page* that has graphics, titles, and black and highlighted text. Web pages are loosely analogous to chapters in a book. Hypertext links are maintained using URLs (uniform resource locator), a standard way of coding the locations of the HTML (Hypertext Markup Language) documents. In addition to HTML, a number of newer Web standards are gaining in popularity, including Extensible Markup Language (XML), Extensible Hypertext Markup Language (XHTML), cascading style sheets (CSS), and Dynamic HTML (DHTML).

The client communicates with the server computer according to a set of rules called HTTP (Hypertext Transport Protocol), which retrieves the document and presents it to the users. HTML is the standard page description language for Web pages. The HTML tags let the browser know how to format the text: as a heading, as a list, or as body text. HTML also tells whether images, sound, and other elements should be inserted.

A Web browser reads HTML and creates a unique, hypermedia-based menu on your computer screen that provides a graphical interface to the Web. The browser uses data about links to accomplish this; the data is stored on the Web server. The hypermedia menu links you to other Internet resources, not just text documents, graphics, and sound files. Internet Explorer and Netscape are examples of Web browsers. A search engine helps find information on the Internet. Popular search engines include Yahoo and Google. A meta-search engine submits keywords to several individual search engines and returns the results. Push technology is used to send information automatically over the Internet rather than make users search for it with their browsers.

PRINCIPLE *Before the Internet and the World Wide Web become universally used and accepted for business use, management issues; service and speed issues; and privacy, fraud, security and unauthorized Internet sites must be addressed and solved.*

Java is an object-oriented programming language from Sun Microsystems based on C++ that allows small programs—applets—to be embedded within an HTML document. When the user clicks on the appropriate part of the HTML page to retrieve it from a Web server, the applet is downloaded onto the client workstation, where it begins executing. The development of Java has had a major impact on the software industry and could change the economics of paying for software. Java makes it possible to sell one-time usage of a piece of software.

A rapidly growing number of companies are doing business on the Web and enabling shoppers to search for and buy products online. The travel, entertainment, gift, greetings, book, and music businesses are experiencing the fastest growth on the Web. For many people, it is easier to shop on the Web than search through catalogs or trek to the shopping mall. However, some shoppers are concerned about the potential for credit card numbers to be stolen over the Internet.

The steps to creating a Web page include getting space on a Web server; getting a Web browser program; writing your copy with a word processor, using an HTML editor, editing an existing HTML document, or using an ordinary text editor to create your page; opening the page using a browser, viewing the result, and correcting any tags; adding links to your home page to take viewers to another home page; adding pictures and sound; uploading the HTML file to your Web site; reviewing the Web page to make sure that all links are working correctly; and advertising your Web page. There are a number of products that make developing and maintaining Web content easier, such as Microsoft's .NET. Once a Web site has been constructed, a content management system (CMS) can used to keep the Web site running smoothly.

An intranet is an internal corporate network built using Internet and World Wide Web standards and products. It is used by the employees of the organization to gain access to corporate information. Computers using Web server software store and manage documents built on the Web's HTML format. With a Web browser on your PC, you can call up any Web document—no matter what kind of computer it is on. Because Web browsers run on any type of computer, the same electronic information can be viewed by any employee. That means that all sorts of documents can be converted to electronic form on the Web and constantly updated.

An extranet is a network that links selected resources of the intranet of a company with its customers, suppliers, or other business partners. It is built based on Web technologies. Security and performance concerns are different for an extranet than for a Web site or network-based intranet. Authentication and privacy are critical on an extranet. Obviously, performance must be good to provide quick response to customers and suppliers.

Management issues and service bottlenecks are issues that affect all networks. No centralized governing body controls the Internet. Also, because the amount of Internet traffic is so large, service bottlenecks often occur. Before the use of the Internet will be accepted by everyone, privacy, fraud, and security issues must be addressed and solved. Cryptography techniques and firewalls are required to combat information thieves and provide as much security as possible. Some unauthorized and unwanted Internet sites contain damaging information about companies. These sites can be placed on the Internet by unhappy employees, competitors, or other individuals and groups.

CHAPTER 7 SELF-ASSESSMENT TEST

The Internet is like many other new technologies—it provides a wide range of services, some of which are effective and practical for use today, others are still evolving, and still others will fade away from lack of use.

1. The _____ was the ancestor of the Internet. It was developed by the U.S. Department of Defense.

2. What enables traffic to flow on the Internet from one network to another?
   A. Internet Protocol
   B. ARPANET
   C. Uniform Resource Locator
   D. LAN Server

3. Each computer on the Internet has an address called the transport control protocol. True False

4. Which of the following is NOT a way to gain access to the Internet?
   A. LAN Server
   B. Usenet
   C. Point-to-point protocol, or PPP
   D. On-line Service

5. _____ is a protocol that describes a file transfer process between a host and remote computer.

6. Usenet provides a central news service through e-mail. True False

7. What allows two or more people to engage in on-line, interactive "conversation" over the Internet?
   A. Content streaming
   B. Chat rooms
   C. Newsgroups
   D. Usenet

8. _____ can be used to route phone calls over networks and the Internet.

9. What is the standard page description language for Web pages?
   A. Home page language
   B. Hypermedia Language
   C. Java
   D. Hypertext markup language (HTML)

Before the Internet and World Wide Web become universally used and accepted for business use, management issues; service and speed issues; and privacy, fraud, security, and unauthorized Internet sites must be addressed and solved.

10. A (An) _____ is a network based on Web technology that links customers, suppliers, and others to the company.

11. A cookie is a text file that an Internet company can place on the hard drive of a computer system. True False

12. What sits between an internal network or computer and the Internet to prevent unauthorized access to a computer system?
    A. Digital signature
    B. Firewall
    C. Extranet
    D. Internet

13. _____ is used to send information automatically over the Internet.

Chapter 7 Self-Assessment Test Answers

1. ARPANET, 2. A, 3. False, 4. B, 5. File transfer protocol (FTP), 6. True, 7. B, 8. Voice-over IP (VOIP), 9. D, 10. extranet, 11. True, 12. B. 13, Push technology.

KEY TERMS

applet, 294
ARPANET, 276
backbone, 277
bot, 287
chat room, 286
content streaming, 287
cookie, 304
cryptography, 304
digital signature, 305
encryption, 304
Extensible Markup Language (XML), 292
file transfer protocol (FTP), 284

firewall, 305
home page, 291
HTML tags, 292
hypermedia, 291
hypertext markup language (HTML), 292
instant messaging, 284
Internet service provider (ISP), 279
Internet protocol (IP), 276
Java, 295
meta-search engine, 294
newsgroups, 285
point-to-point protocol (PPP), 279

push technology, 297
search engine, 294
serial line Internet protocol (SLIP), 279
telnet, 284
transport control protocol (TCP), 277
tunneling, 301
uniform resource locator (URL), 277
usenet, 284
virtual private network (VPN), 301
voice-over-IP (VOIP), 286
Web auction, 288
Web browser, 293
World Wide Web, 291

REVIEW QUESTIONS

1. What is the Internet? Who uses it and why?
2. What is TCP/IP? How does it work?
3. Explain the naming conventions used to identify Internet host computers.
4. What is a domain name?
5. Briefly describe three different ways to connect to the Internet. What are the advantages and disadvantages of each approach?

6. What is an Internet service provider? What services do they provide?
7. What are the advantages and disadvantages of e-mail?
8. What is a newsgroup? How would you use one?
9. What are telnet and FTP used for?
10. What is an Internet chat room?
11. What is content streaming?

12. What is instant messaging?
13. Briefly describe a Web auction.
14. What is the Web? Is it another network like the Internet or a service that runs on the Internet?
15. What is hypermedia?
16. What is a URL and how is it used?
17. What is HTML and how is it used?
18. What is a Web browser? How is it different from a Web search engine?

19. What is push technology?
20. What is an intranet? Provide three examples of the use of an intranet.
21. What is an extranet? How is it different from an intranet?
22. What is cryptography?
23. What are firewalls? How are they used?

DISCUSSION QUESTIONS

1. Instant messaging is being widely used today. Describe how this technology could be used in a business setting. Are there any drawbacks or limitations in using instant messaging in a business setting?
2. How would you protect yourself from unwanted e-mail?
3. Briefly describe how the Internet phone service operates. Discuss the potential for this service to impact traditional telephone services and carriers.
4. The U.S. federal government is against the export of strong cryptography software. Discuss why this may be so. What are some of the pros and cons of this policy?
5. Identify three companies with which you are familiar that are using the Web to conduct business. Describe their use of the Web.
6. What is voice-over IP (VOIP) and how could it be used in a business setting?
7. Outline a process to create a Web page. What computer hardware and software do you need if you

wish to create a Web home page containing both sound and pictures?
8. One of the key issues associated with the development of a Web site is getting people to visit it. If you were developing a Web site, how would you inform others about it and make it interesting enough that they would return and also tell others about it?
9. Getting music, radio, and video programs from the Internet is getting easier, but some companies are still worried that people will illegally obtain copies of this programming without paying the artists and producers royalties. If you were an artist or producer, what would you do?
10. How could you use the Internet if you were a traveling salesperson?
11. How has the Java programming language changed the software industry?
12. Briefly summarize the differences in how the Internet, a company intranet, and an extranet are accessed and used.

PROBLEM-SOLVING EXERCISES

1. Do research on the Web to find several popular Web auction sites. After researching these sites, use a word processor to write a report on the advantages and potential problems of using a Web auction site to purchase a product or service. Also discuss the advantages and potential problems of selling a product or service on a Web auction site.

2. Develop a brief proposal for creating a business Web site. Describe how users will interact with the Web site. How will you

get people to visit your site? Develop a simple spreadsheet to analyze the income you need to cover your Web site and other business expenses.

3. You are a manager of a small company with a new Web site. How would you avoid privacy, fraud, and security problems? Develop a brief report describing what you would do. Using a graphics program, diagram how you would protect your Internet site from outside hackers.

TEAM ACTIVITIES

1. Identify a company that is making effective use of a company extranet. Find out all you can about its extranet. Try to speak with one or more of the customers or suppliers who use the extranet and ask what benefits it provides from their perspective.

2. Have each team member use a different search engine to find information about MP3 players. Meet as a team and decide which search engine was the best for this task. Write a brief report to your instructor summarizing your findings.

WEB EXERCISES

1. This chapter covers a number of powerful Internet tools, including Internet phones, search engines, browsers, e-mail, newsgroups, Java, intranets, and much more. Pick one of these topics and get more information from the Internet. You may be asked to develop a report or send an e-mail message to your instructor about what you found.

2. You can download free software using the Internet. Explore free software and write a brief report describing what you found. What are the advantages and disadvantages of free software?

CASES

CASE 1

Google Uses New Approach to Web Advertising

Banner ads have become a familiar, and sometimes barely tolerable, part of the Web-browsing experience. Go to your favorite Web portal, on-line news service, or any number of sites, and you need to be patient as the banner ad graphics are downloaded before you can view the page content. The irony of the situation is that we have become so accustomed to these ads that we no longer pay them any attention. Can you recall the content of any banner ad that you've seen recently?

On average, only 0.5 percent of visitors to a Web site click and follow a banner ad link. With such a dismally low click-through rate, Web marketing experts have moved to more creative ways to capture the user's attention. Welcome to pop-up ads, lay-under ads, blanket ads, ads that fly around the display while you're trying to view the content underneath, along with a wide assortment of other multimedia gimmicks. Pop-up ads open in additional browser windows on top of a Web site, while lay-under ads open in additional browser windows under a site, surprising you when you close the primary window. Although such ads are more effective for catching the valuable eye of the user, they also push the limits of what users will tolerate. Each additional window is an inconvenience. They slow download times and make the user manually close more windows. How far can marketers go before they reach the sheesh response? "Sheesh! I'm not visiting this Web site again!"

Google (www.google.com), the very popular Web search engine, is working to find new ways of marketing products and

services to its users. Google is renowned for its clean and obnoxious-ad-free design. In early 2002 Google banned the use of pop-up ads on its Web site. Google also refrains from using any form of banner ad. So how does Google make money? Through an approach it calls "AdWords."

AdWords is a unique, text-only, approach to Web marketing. You won't see any advertisements on Google's home page. It's only when the results of your search are displayed that you see advertisements—that is, if you look closely. In a narrow column on the right edge of the page is a list of text-only advertisements under the heading "Sponsored Links." The key selling feature of these advertisements is that they are directly related to the search keywords entered by the user. So, in a manner of speaking, they are a service to the user.

Businesses interested in advertising on Google with AdWords are asked to buy keywords that relate to their product or service. When those keywords are included in a search, the company's ad will be displayed. Pricing for AdWords is based on the position in the list of sponsors, first position being most costly. Google offers an alternative package named AdWords Select that uses cost-per-click (CPC) pricing; the company pays only when a customer clicks on the ad, regardless of how many times it is shown. In this scenario the click-through rate and CPC together determine where an ad is shown, so better ads rise to the top and no ads can be locked out of the top position.

Google offers one other advertising option: Premium Sponsorship Advertising. Premium sponsors pay premium rates to have links to their products or services appear in the first two places of the actual search results. These premium advertisements are the closest Google comes to providing fuel to the anti-Web-advertisement contingent.

None of the advertising options that Google offers slows the delivery of search results, nor do they consume a high percentage of screen real estate. Because the advertisements are tied to search keywords, they are more likely to yield results to the sponsors. Overall, Google has developed a winning Web marketing strategy that assists both its sponsors and its users.

Discussion Questions

1. Google charges more for popular keywords than for less-popular ones. How else might Google profit by recording the keywords that are entered by users and tracking the search result links that are followed?
2. How much advertising is too much? Everyone has a different threshold of tolerance for advertising. What is your limit? Can you tolerate one banner ad per Web page? More? How about pop-up advertisement windows?

Critical Thinking Questions

3. What other types of Web sites, besides search engines, might be able to offer user-targeted advertising like AdWords?
4. Is the future of the Web more likely to resemble network television or public TV? What exterior forces can be applied to sway the Web to be less commercial?

Sources: James Lewin, "The Future of Web Advertising," February 5, 2002, http://www.itworld.com; Chris Sherman, "No, Google Hasn't Sold Out." March 12, 2002, http://siliconvalley.internet.com; Danny Sullivan, "Up Close With Google AdWords," March 4, 2002, http://searchenginewatch.com, Brian Sullivan, "Intrusive Ads a Sign of Online Advertising Evolution, Analysts Say," February 19, 2002, http://www.computerworld.com; http://www.google.com, accessed March 25, 2002.

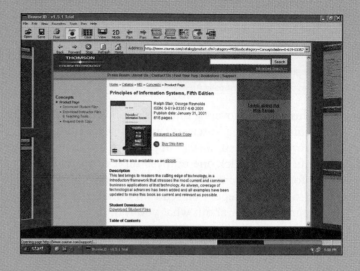

The Web in Three Dimensions

If you think about it, Web browsers haven't changed much over the past six years. Mosaic was the first graphical Web browser, followed shortly by Netscape, and then Internet Explorer. Since those early days, hundreds of Web browsers have been developed and marketed, none able to compete successfully with Netscape and Internet Explorer, but all following the same general framework as the original: start at your home page, click hyperlinks to move to other Web pages or type in a URL, click the Back or Forward buttons to progress through Web pages that you have visited, click the Home button to return to your home page. We all know the routine. There has been little innovation in Web browsers—until recently, when a small company in Virginia released a product they call Browse3D.

President and CEO of Browse3D Corp, David Shuping, got bored with the Internet, so he decided to make it more interesting. "You can view one page at a time," Shuping said. "It's like the old saying, 'You move one step forward and take two steps back.' We were doing the same on the Web." So Shuping and a former colleague, Bill Johnson, along with four consultants, spent 18 months developing the new way to browse the Web.

The result of their effort is a browser that allows you to view Web pages in a virtual 3D room. In the center of your display is the primary wall that works much like traditional Web browsers. What's different about this Web browser are the left and right walls that appear to surround you on either side. The left wall holds nine screenshots of the Web pages you have visited, while the right wall holds nine screenshots of the Web pages that are referenced in hyperlinks on the page you are currently viewing. The left wall is your Web-browsing past, and

the right wall is your possible future. The benefit is that you are able to actually see the pages, rather than just their text links. The left and right walls can be flipped to display nine more pages either further back in your history, or further ahead in your possible future.

Browse3D offers numerous other features. For instance, you can make a page "sticky" so that it is saved on a fourth wall for future reference during your browsing session. You can save collections of sticky pages for future reference either as static pages or active pages. Static pages are preserved in their current state so that if the page is updated on the Web, your saved static page remains unchanged. Active pages call up the current version of the page on the Web. Browse3D offers tools to rotate the virtual room, zoom in and out, or switch to a standard 2-D browser.

Browse3D offers a way for users to become more immersed in the Web-browsing experience by graphically viewing their past and future and quickly selecting the correct path to traverse. The beta version of this desktop product debuted at COMDEX in November 2001, where it was named "Best Internet Software" and generated interest from thousands of potential customers. Will Browse3D be the product that finally gives Internet Explorer and Netscape a run for their money? Perhaps it will threaten the big browser companies enough to be bought out and become a feature in the next version of Internet Explorer or Netscape. Or, maybe it will fail to capture the attention of the public and simply fade away. See whether Browse3D has survived its first two years by visiting http://www.browse3d.com.

Discussion Questions

1. Browse3D allows you to view multiple Web pages simultaneously on your display. Can this be accomplished with a traditional Web browser? How?

2. After reading about how Browse3D works, what do you think its limitations and shortcomings are? What type of computer and Internet connection would be ideal for this software?

Critical Thinking Questions

3. Why do you think that Netscape and Internet Explorer have been hesitant to make innovative changes to the traditional Web browser interface?

CASE 3

Hummingbird Provides Leading Enterprise Information Portal (EIP) Solutions

In this era of information overload, it is often challenging to put a finger on the exact document that is needed. In a typical work environment, an employee locates and accesses thousands of documents, including saved e-mail messages; documents containing corporate policies, procedures, and forms; external information such as news and stock market figures; and information from the corporate database on customers, inventory, transactions, and other business-related information. Finding relevant information in the workplace can be difficult. Because much of this information is private, accessing it off-site may be impossible.

To address this issue, many businesses are turning to Hummingbird to provide them with an Enterprise Information Portal (EIP). Hummingbird, based in Toronto, Canada, assists businesses in organizing and delivering corporate documents and information across intranets, extranets, and the Internet. Hummingbird has more than 40,000 customers—including 75 percent of the *Fortune* 500 companies—and over 5 million users worldwide.

Web portals such as yahoo.com, msn.com, and aol.com provide convenient access to popular Web resources such as news, weather, and shopping. Most Web portals also allow you to personalize the Web page (for example, my.yahoo.com) so that it only lists items of personal and local interest, such as the weather in your town, your stock quotes, your horoscope, and news categories of interest. Operating under similar principles, an Enterprise Information Portal is a user interface that provides one-click access to documents, applications, and information that are used most frequently in day-to-day work. Enterprise Information Portals provide customizable interfaces for private and secure business documents and information, as well as applications, e-mail, and public information.

Hummingbird claims to be the only vendor that can supply an end-to-end EIP solution. They do so by providing:

- Document and knowledge management
- EIP integration technologies
- Business intelligence
- Intranet connectivity

Hummingbird is particularly well known for its success in tying together a variety of document types residing on varying systems and platforms. Using the Hummingbird EIP, an employee could access data stored in a mainframe database, a Unix server, or Windows PC—whether it is structured data stored in various databases, applications, or legacy systems or unstructured data such as e-mail, documents, multimedia, or other files—and then act on that information using related enterprise applications.

4. What other improvements would you like to see to the traditional Web Browser interface?

Sources: Ellen McCarthy, "Browse3D, a More Visual Way to Surf the Web," *The Washington Post*, March 4, 2002, p. E5; Harry McCracken, "Comdex: Web Browsing Revisits Three Dimensions," November 15, 2001, http://www.pcworld.com; "Browse3D Releases Next Generation Browser; Desktop Product First Major Improvement in Browser Technology in Years," *Business Wire*, January 22, 2002; http://www.browse3d.com, accessed March 28, 2002.

The Hummingbird EIP translates documents and data into HTML and XML for convenient access with a Web browser. The user interface typically contains several frames of content. One frame might contain a directory tree, as used in Internet Explorer, for easy access to documents, another could hold an e-mail inbox, still other frames could contain a stock ticker, the current weather, company news, etc. The majority of the screen is used as work space, where your applications are opened and used. The interface can include several tabs for switching between different screens and a search utility to assist in finding information. Often corporate information system specialists will design the interface so that some areas are customizable. This way, items on the desktop can be rearranged for optimal convenience. Other portions of the desktop may be configured with items that are important to all users of the system.

Since corporate data and documents are translated to HTML and XML, it is relatively simple to deliver the content over the corporate intranet, extranet, or the Internet. Hummingbird secures the system over the Internet by providing login and password protection. They also encrypt the data as it is being transmitted so that if it is intercepted while moving over the Internet, the bytes would be unintelligible.

Hummingbird's EIP saves employees time by connecting them with the information and knowledge they need almost instantaneously. By creating a single portal for access to all information, Hummingbird creates a virtual office, so knowledge workers can do their job wherever they can connect to the Internet. EIPs use the strength of the Internet to offer the best that information systems can offer.

Discussion Questions

1. List three ways that an EIP can help a business save money.
2. What do you think is the most time-consuming part of setting up an EIP?

Critical Thinking Questions

3. What types of policies might be needed when allowing employees to customize the EIP with applications, images, and other information?
4. What do you think would be the primary concern of a business that allows its employees to access secure information over the Internet with their user name and password?

Sources: "Fasken Martineau Chooses Hummingbird Solutions to Increase Technology Advantage," *Canada NewsWire*, January 16, 2002; "Hummingbird and STG, Inc. to Present Network-Based Electronic Records Management at FOSE 2002," *PR Newswire*, March 18, 2002; Susan Levi Wallach, "DOD Certifies Hummingbird's Electronic Records Management Software," November 12, 2001, http://www.itworld.com; http://www.hummingbird.com, accessed March 28, 2002.

NOTES

Sources for the opening vignette on p. 273: "Lands' End to Beat Goal," *The New York Times*, February 15, 2002, Section C, p. 16; "E-Commerce Report; Selling Made-to-Order Clothing On-Line without Sending the Customer to a Tailor," *The New York Times*, November 5, 2001, Section D, p. 7; Tom Daykin, "Lands' End Earnings Soar 173 Percent in Third Quarter," *The Milwaukee Journal Sentinel*, November 7, 2001, www.msjonline.com; Company Web site, www.landsend.com, accessed February 18, 2002.

1. WSJ.com, "Internet Access Grows among U.S. Adults," *The Wall Street Journal,* February 20, 2001, p. B19.

2. Reagan, Brad, "Who Goes There," *The Wall Street Journal*, October 29, 2001, p. R4, Source: Jupiter Media Metrix.

3. Ewalt, David, "Report Predicts a Billion Web Users by 2005," *InformationWeek Online*, February 2, 2001.

4. Tam, Pui-Wing, "In Digital Revolution, Mr. Selby Is a Kind of Wireless Guerrilla," *The Wall Street Journal*, December 7, 2001, p. A1.

5. Nelson, Matthew, "Royal Caribbean Connects Crew to Internet," *InformationWeek Online*, May 15, 2001.

6. Mearian, Lucas, "Boeing Gets FCC License for Internet Service on Planes," *Computerworld*, January 12, 2002, p. 17.

7. Scott, Karyl, "The Omnipresent Web," *InformationWeek Online*, December 10, 2001.

8. Anthes, Gary, "Reinventing the Internet," *Computerworld*, August 27, 2001, p. 52.

9. Marceau, Colby, "Governor Davis Participates in Next Generation World Wide Webcast," *Business Wire*, January 25, 2002.

10. Latour, Almar, "IP Sales," *The Wall Street Journal*, December 6, 2001, p. B6.

11. Sinrod, Eric, "The Sex.com Saga," *Computerworld*, April 16, 2001, p. 34.

12. Angwin, Julia, "Are Domain Panels the Hanging Judges of Cyberspace Court?" *The Wall Street Journal*, August 20, 2001, p. B1.

13. Weber, Thomas, "More Trouble for AOL," *The Wall Street Journal*, April 22, 2002, p. B1.

14. Grimes, Brad, "Ditch Your Dial Up," *PC World*, February 2002, p. 68.

15. Disabatino, Jennifer, "Anthrax Prompts Shift in Fed's E-Mail Policy," *Computerworld*, October 29, 2001, p. 14.

16. Kay, Russell, "Emotions and Internet Shorthand," *Computerworld*, January 14, 2002, p. 42.

17. Gubernick, Lisa, "Help! I'm Drowning in E-Mail!" *The Wall Street Journal*, January 10, 2002, p. B1.

18. Bulkeley, William, "EchoMail Can Sort, Answer Deluge of E-Mails," *The Wall Street Journal*, November 15, 2001, p. B10.

19. Kontzer, Tony, "This E-Mail Will Self-Destruct in Five Seconds," *InformationWeek Online,* July 23, 2001.

20. Angwin, Julia, "Talk Is Cheap," *The Wall Street Journal*, March 23, 2001, p. A1.

21. Fleishman, Glenn, "Usenet Archive Is Posted on Web," *The New York Times,* December 12, 2001.

22. Fry, Jason, "Why Shoppers' Loyalty to Familiar Web Sites Isn't So Crazy," *The Wall Street Journal*, August 13, 2001, p. B1.

23. Goff, Robert, "eBay's Cop," *Forbes*, June 25, 2001, p. 42.

24. Leonard, Devin, "Don't Call Them Napster," *Fortune*, June 25, 2001, p. 44.

25. Ratliff, Ben, "Virtual Jam," *Forbes*, June 25, 2001, p. 48.

26. Muto, Sheila, "Kicked Off the Air, It Becomes a Web Model," *The Wall Street Journal*, February 6, 2001, p. B1.

27. Weber, Thomas, "Updating Address Books Via the Web," *The Wall Street Journal*, July 9, 2001, p. B1.

28. Colkin, Eileen, "FleetBoston Takes Safe-Deposit Boxes to a New Level," *Information Week Online*, March 19, 2001.

29. Mossberg, Walter, "Here's How to Run Your PCs Remotely by Using the Internet," *The Wall Street Journal*, September 6, 2001, p. B1.

30. Clark, Don, "Laptop Users Will Soon Print While Traveling," *The Wall Street Journal*, October 22, 2001, p. B5.

31. Lake, Matt, "Best Free Stuff Online," *PC World*, March 2001, p. 92.

32. Jesdanun, Anick, "E-Mail on Mars? Space Mission Could Be Giant Step for Tech," *The Summit Daily News*, June 4, 2001, p. 11.

33. Landro, Laura, "Web Can Offer Public the Information Needed to Respond to Bioterror," *The Wall Street Journal*, November 9, 2001, p. B1.

34. Mullaney, Timothy, "Don't Expect Miracles," *Business Week*, April 16, 2001, p. EB8.

35. Dragan, Richard and Behr, Mary, "Help Employees Stay Knowledgeable through Virtual Education," *PC Magazine*, July 2001, p. 172.

36. Behr, Mary et al., "Top 100 Undiscovered Web Sites," *PC Magazine*, February 26, 2002, p. 87.

37. Holzschlag, Molly, "The Fear of X," *PC Magazine*, August 2001, p. IP01.

38. Gross, Neil, "Software That Sniffs Out Stolen Property," *Business Week*, December 24, 2001, p. 75.

39. Mearian, Lucas, "Fidelity Makes Big XML Conversion," *Computerworld*, October 1, 2001, p. 12.

40. Bray, Hiawatha, "AOL Takes Fight with Microsoft to Courts," *The Wall Street Journal*, February 11, 2002, p. 12B.

41. Elgin, Ben, "Why They're Agog over Google," *Business Week*, September 24, 2001, p. 83.

42. Gowan, Michael and Spanbauer, Scott, "Find Everything Faster," *PC World.*

43. Copeland, Lee, "Caterpillar Digs into Agile Development," *Computerworld*, January 7, 2002, p. 14.

44. Wallace, Bob, "Truckload of Features Added to GM Sites," *InformationWeek Online*, March 4, 2001.

45. McKenna, Susan, "Convene Releases Update of Convene Learning Platform," *Business Wire*, February 19, 2002.

46. Floyd, Michael, "Inside Visual Studio.NET," *PC Magazine*, September 4, 2001, p. IP01.

47. Paul, Fredric, "Choosing the Right Content Management System," *PC World*, July 2001, p. 142.

48. Snel, Ross, "Start-Ups Try New Ways to Track Web Users," *The Wall Street Journal*, April 5, 2001, p. B7.

49. Keenan, Faith, "Friendly Spies on the Net," *Business Week*, July 9, 2001, p. EB26.

50. Weber, Thomas, "A New Privacy Tool Is at Your Disposal," *The Wall Street Journal*, September 10, 2001, p. B1.

51. Bridges, Tony, "Users, Beware: AOL Scam on the Loose," *The Tallahassee Democrat*, July 2, 2001.

52. Thibodeau, Patrick, "Feds Charge 90 in Net Fraud Crackdown," *Computerworld*, June 4, 2001, p. 17.

53. "SonicWall Provides Security for Bell South DSL Subscribers," *RBOC Update*, March 2002.

Ecotourism is important for the economy of many countries, including New Zealand. From its Maori beginnings to the setting for the Lord of the Rings *films, this country is blessed with colorful creatures in beautiful environments. This World Views Case examines the use of the Internet to entice visitors from around the world to watch whales in its pristine waters. The case shows how an effective Web site can be developed, used, and maintained by employing a variety of Internet tools.*

From Railway Station to Whaleway Station

Peter Blakey
Massey University, New Zealand

On the South Island of New Zealand lies Kaikoura, an area steeped in Maori history and legend. Off the coast of the village is a marine environment so rich in nutrients that it attracts some of the most magnificent creatures with which we share our planet. Among them is the giant sperm whale, which can grow up to 20 metres in length and has the largest brain of any animal alive.

An icon of New Zealand's industrial past has become a centrepiece for the future of Whale Watch® Kaikoura, an ecotourism business that has helped regenerate the town. The town's central railway station underwent a $750,000 refurbishment to be more attractive for Whale Watch for tourists arriving in Kaikoura. "With everything from state-of-the-art computer systems, to a restaurant and café/bar, and outside areas for enjoying the ocean, the Whaleway Station is another major investment by the company in the future of our business. It's also a vote of confidence in our future growth," said Whale Watch Chief Executive Wally Stone.

Whale Watch is owned and operated by Kaikoura's community of indigenous people, part of a Maori tribe known as Ngai Tahu. The company's dividends are dedicated to providing education and training for Ngai Tahu adults and young people. Whale Watch was formed in 1987 and is still growing steadily. Booking numbers currently exceed 120,000 per year, although unfavourable weather or sea conditions reduce actual trip numbers to about 72,000 annually. About 90 percent of its tourists are from outside New Zealand.

There is a high demand for tours; therefore, advance booking is essential. The company recommends 7– to 10–day trips during the high season (from November through April) and 3– to 4–day trips during off peak times (from May through October). Reservations can be made through the Web site (http://www.whalewatch.co.nz). Visitors can make reservations for lodging, restaurants, and other tours from this site.

The Web site was designed and is maintained for Whale Watch by Aslan Ltd of Christchurch, New Zealand (http://aslan.co.nz). The site is hosted by Digiweb, also of Christchurch (http://www.digiweb.co.nz). It was built using Microsoft FrontPage. Occasionally, minor problems occur with the site, but they are usually fixed by reloading the software. Other software products used for the site include Ulead and Adobe for working with images and Microsoft Word for the text. The Web site development language is HTML, with some Java for the animations. There are no secure areas or cookies at the site.

The intention of the site at all times is to reflect the ocean habitat of the main attraction, the whales and dolphins. The banner is composed of pictures of the Whale Watch boats, a huge whale's tale, and the lovely blue-green of the ocean. The use of colour and movement invites bookings on whale-watching tours, encourages tourists to visit Kaikoura's other attractions, recommends a visit to the café at the Whaleway Station, and provides an opportunity to browse the database on recent whale activity. As an organization whose business is ecotourism, Whale Watch very wisely has employed professional consultants to develop and manage the Web site. The consultants are not on site (Christchurch and Kaikoura are about 100 kilometres apart on the east coast of the South Island), but through e-mail and the World Wide Web, with some visits by the consultants, communication is not a problem.

Recent visitors to the Web who then actually visited Kaikoura and the Whaleway Station, and took a whale-watching trip, include Steve, a geologist from Washington State. He commented, "In New Zealand, all the different geology unfolds before your eyes in such a confined area. It's a must for all geologists, and Kaikoura offers a magnificent place to start." Gary from Kaikoura's Hog Holidays®, another tourism company, welcomes bikers from all over the world on his polished Harley-Davidson. Kaikoura and the whale-watching tours are always on Hog Holidays' itinerary. "The Web site makes it so easy for overseas riders to experience what they are going to see when they arrive and book their trip on the Whale Watch boats just when they want it."

Visit the Whale Watch Web site and then answer the following questions.

Discussion Questions

1. Why do you think that the presence of other links on this Web site increases the number of visitors to the site?
2. Explore the Internet and try to find other tourist sites in New Zealand.

Critical Thinking Questions

3. Discuss how visitors who participated in the whale-watching tours might be encouraged to add their personal experiences to the site. In your opinion, would this be a good idea?
4. Would there be any advantages to carrying advertisements for other businesses on the site to reduce its costs? Any disadvantages?

PART

3

BUSINESS INFORMATION SYSTEMS

Electronic Commerce

CHAPTER 8

PRINCIPLES	LEARNING OBJECTIVES
E-commerce is a new way of conducting business, and as with any other new application of technology, it presents both opportunities for improvement and potential problems.	• Identify several advantages of e-commerce. • Outline a multistage model that describes how e-commerce works. • Identify some of the major challenges companies must overcome to succeed in e-commerce. • Identify several e-commerce applications.
E-commerce requires the careful planning and integration of a number of technology infrastructure components.	• Outline the key components of technology infrastructure that must be in place for e-commerce to succeed. • Discuss the key features of the electronic payments systems needed to support e-commerce.
Users of the new e-commerce technology must take safeguards to protect themselves.	• Identify the major issues that represent significant threats to the continued growth of e-commerce.
Organizations must define and execute a strategy to be successful in e-commerce.	• Outline the key components of a successful e-commerce strategy.

[Tesco]

Grocer Implements Successful E-Commerce Business Model

As the death of many dot.coms has shown, conducting business on-line involves much more than simply setting up a Web site and raking in profits. On-line groceries offer a good example of why pure e-commerce operations can encounter serious new business problems compared with Web sites that add an on-line component to existing brick-and-mortar sales (sales out of physical stores). Solid sales and operations experience must be combined with good technical knowledge to create a successful e-commerce Web site.

A number of existing grocery chains and dot.com start-ups have tried to enter the on-line grocery business—unsuccessfully. One on-line grocer that is profitable is UK–based Tesco. Through its Web site, Tesco.com, the company serves 1 million registered customers in the United Kingdom, handles more than 85,000 orders per week with an average order of $133, and generates nearly $500 million in annual sales. Tesco.com operates in four countries—the UK, Republic of Ireland, South Korea, and the United States.

Tesco has been the most successful of the on-line grocers largely because net margins for grocers in England are close to 8 percent versus 2 percent for U.S.–based grocers. Another reason for its success is that it has kept things simple. When shoppers log on to Tesco.com, they key in their postal code, and their order is routed to their local Tesco store. Once received, workers pick items from the store shelves to fill the order. Tesco has not invested in multimillion-dollar automated distribution facilities or fancy conveyor belts that zip products through warehouses.

Albertsons (U.S.) and Royal Ahold (Netherlands) have followed this "keep things simple" model for e-business and fill orders from their own brick-and-mortar stores. Filling on-line orders directly out of a local store instead of from a warehouse results in lower costs for the company and faster service for customers.

On the other side of the coin, numerous on-line grocers have failed with different business models. After fewer than eight months of its on-line shopping trial, Safeway plc in the United Kingdom shut down its on-line service in November 2000. The chain did not even deliver to customers; they ordered via the Web and picked up their groceries at a store at a prearranged time. Webvan is another notable failure. It filed for bankruptcy in July 2001 after spending more than $800 million in two years. HomeRuns.com, in business since 1996, also shut down in July 2001. Both the Webvan and HomeRuns.com business models were based on establishing expensive warehouses in many locations and required costly fleets of delivery personnel and equipment. The companies simply couldn't recover the heavy start-up costs until their customer bases grew much larger. Furthermore, since neither operated brick-and-mortar stores, they relied on a major behavior change in the way that people buy groceries to become profitable.

As you read this chapter, consider the following:
- How are organizations using e-commerce to provide improved customer service, become more productive, and remain competitive?
- What are the key issues that must be addressed to ensure effective implementation of e-commerce?

AN INTRODUCTION TO ELECTRONIC COMMERCE

business-to-business (B2B) e-commerce
a form of e-commerce in which the participants are organizations

business-to-consumer (B2C) e-commerce
a form of e-commerce in which customers deal directly with the organization, avoiding any intermediaries

consumer-to-consumer (C2C) e-commerce
a form of e-commerce in which the participants are individuals, with one serving as the buyer and the other as the seller

Dot.com companies take many forms. Early e-commerce news profiled start-ups that used Internet technology to compete with the traditional players in an industry. For example, Amazon.com challenged well-established booksellers Waldenbooks and Barnes and Noble. Like Amazon, Tesco (the grocer discussed in the opening vignette) provides an example of **business-to-consumer (B2C) e-commerce**, in which customers deal directly with an organization and avoid any intermediaries. Other types of e-commerce are **business-to-business (B2B) e-commerce**, in which the participants are organizations, and **consumer-to-consumer (C2C) e-commerce**, which involves consumers selling directly to other consumers. Neoforma.com is a B2B e-commerce company that received more than $80 million in financing to create an Internet marketplace to take on the $140 billion hospital supply industry. eBay is an example of a C2C e-commerce site; customers buy and sell items directly to each other through the site.

Aside from the major categories of e-commerce, companies are also using Internet technologies to enhance their current operations, such as inventory control and distribution. But whatever model is used, successful implementation of e-business requires significant changes to existing business processes and substantial investment in information systems technology.

Over the past few years, we have learned a lot about the practical limitations of e-commerce. It has become painfully clear that before companies can achieve profits, they must understand their business, their consumers, and the constraints of e-commerce. Although it once seemed so, selling cheap consumer goods on-line in a virtual storefront may not always be a great way to compete. And inventing a new use for cutting-edge technology isn't necessarily enough to guarantee a successful business. Starting up a dot.com company, taking it public, and selling shares at inflated stock prices before the company has earned a profit doesn't work anymore either—investors have become wary of flimsy schemes.

Still, e-commerce is not dead; it is maturing and evolving, with the focus currently shifted from B2C to B2B. E-commerce is a useful tool for connecting business partners in a virtual supply chain to cut resupply times and reduce costs. Noted research firm International Data Corporation (IDC) predicts B2B e-commerce will have a $5.3 trillion impact on the worldwide economy by 2005.[1] But even B2C e-commerce is experiencing rapid growth. The U.S. Department of Commerce reported that on-line sales reached $10 billion in the final quarter of 2001, a 13.1 percent increase over the fourth quarter of 2000. Yet traditional retail sales figures still dwarf on-line sales-total retail sales for the same quarter of 2001 were estimated at $861 billion.[2]

Businesses and individuals use e-commerce to reduce transaction costs, speed the flow of goods and information, improve the level of customer service, and enable close coordination among manufacturers, suppliers, and customers. E-commerce also enables consumers and companies to gain access to worldwide markets. E-commerce is not limited to use by manufacturing firms; many service firms have also implemented successful e-commerce projects. Blue Cross/Blue Shield's Federal Employee Program moved onto the Web, enabling the health insurer to provide real-time claims processing and interactive customer service. The company processes medical, dental, and pharmacy claims for half the government employees across the country. The new system gives the plan's customers, health-care providers, and claims processors access to pertinent information quickly over the Internet.[3]

Business processes that are strong candidates for conversion to e-commerce are those that are paper based and time consuming and those that can make business more convenient for customers. Thus, it comes as no surprise that the first business processes that companies converted to an e-commerce model were those related to buying and selling. For example, after Cisco Systems, the maker of Internet routers and other telecommunications equipment, put its procurement operation on-line in 1998, the company reported that it halved cycle times and saved an additional $170 million in material and labor costs. Similarly, Charles Schwab & Co. slashed transaction costs by as much as 80 percent by shifting brokerage transactions from traditional channels like retail and phone centers to the Internet.[4]

Some companies, such as those in the automotive and aerospace industries, have been conducting e-commerce for decades through the use of electronic data interchange (EDI), which involves application-to-application communications of business data (invoices, purchase orders, etc.) between companies in a standard data format. Many companies have now gone beyond simple EDI-based applications to launch e-commerce initiatives with suppliers, customers, and employees to address business needs in new areas.

Because of the costs involved in buying new technology, the EDI capabilities of most small businesses are nonexistent or extremely limited. A few major retailers and manufacturers have enlisted the help of third parties to bring smaller firms into their EDI supply chain. For example, SPS Commerce specializes in hooking up small businesses like American Outdoor Products (25 employees) to big supply chains like Recreational Equipment Incorporated (REI). SPS built an Internet-based application that translates EDI ordering and shipping requirements so that workers can access them through a Web browser on their PC.[5]

MULTISTAGE MODEL FOR E-COMMERCE

A successful e-commerce system must address the many stages consumers experience in the sales life cycle. At the heart of any e-commerce system is the user's ability to search for and identify items for sale; select those items and negotiate prices, terms of payment, and delivery date; send an order to the vendor to purchase the items; pay for the product or service; obtain product delivery; and receive after-sales support. Figure 8.1 shows how e-commerce can support each of these stages. Product delivery may involve tangible goods delivered in a traditional form (e.g., clothing delivered via a package service) or goods and services delivered electronically (e.g., software downloaded over the Internet).

A multistage model for purchasing over the Internet includes search and identification, selection and negotiation, purchasing, product or service delivery, and after-sales support, as shown in Figure 8.1.

Search and Identification

An employee ordering parts for a storeroom at a manufacturing plant would follow the steps shown in Figure 8.1. Such a storeroom stocks a wide range of office supplies, spare parts, and maintenance supplies. The employee prepares a list of needed items—for example, fasteners, piping, and plastic tubing. Typically, for each item carried in the storeroom, a corporate buyer has already identified a preferred supplier based on the vendor's price competitiveness, level of service, quality of products, and speed of delivery. The employee then logs on to the Internet and goes to the Web site of the preferred supplier.

From the supplier's home page, the employee can access a product catalog and browse until finding the items that meet the storeroom's specifications. The employee fills out a request-for-quotation form by entering the item codes and quantities needed. When the employee completes the quotation form, the supplier's Web application prices the order with the most current prices and shows the additional cost for various forms of delivery—overnight, within two working days, or next week. The employee may elect to visit other suppliers' Web home pages and repeat this process to search for additional items or obtain competing prices for the same items. As mentioned in Chapter 7, bots are software programs that can follow a user's instructions; they can also be used for search and identification.

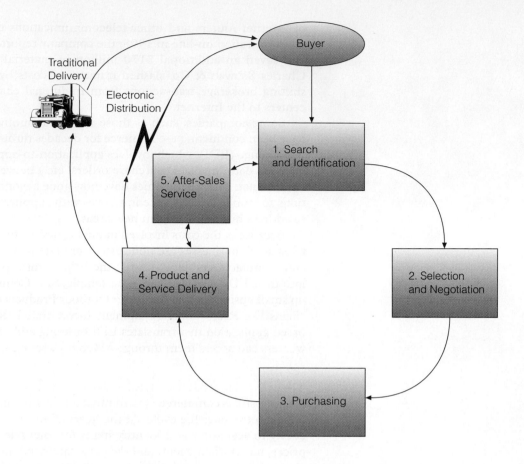

FIGURE 8.1

Multistage Model for E-Commerce (B2B and B2C)

Applied Industrial Technologies is a $2 billion distributor of industrial parts in Cleveland, Ohio, that carries parts used to maintain and repair heavy equipment. It is developing a Web-based on-line catalog containing some 300,000 parts from over 1,400 manufacturers. In addition to searching by part number and description, customers can use a "find-it" feature that lets customers locate complex parts based on engineering specifications, rather than by part number or manufacturer.[6]

Selection and Negotiation

Once the price quotations have been received from each supplier, the employee examines them and clicks on the request-for-quotation form which of the items, if any, will be ordered from a given supplier. The employee also specifies the desired delivery date. This data is used as input into the supplier's order processing TPS. In addition to price, an item's quality and the supplier's service and speed of delivery can be important in selection and negotiation.

Purchasing Products and Services Electronically

The employee completes the purchase order by sending a completed electronic form to the supplier. Complications may arise in paying for the products. Typically, a corporate buyer who makes several purchases from the supplier each year has established credit with the supplier in advance, and all purchases are billed to a corporate account. But when individuals make their first, and perhaps only, purchase from the supplier, additional safeguards and measures are required. Part of the purchase transaction can involve the customer providing a credit card number. However, computer criminals could capture this data and use the information to make their own purchases. To avoid this problem, some companies have developed security programs and procedures. For example, Secure Electronic Transactions (SET) is endorsed by IBM, Microsoft, MasterCard, and others. Another approach to paying for goods and services purchased over the

Internet is using electronic money, which can be exchanged for hard cash; CheckFree is one example. Pico-Pay, a Melbourne, Australia-based company, provides users with access to subscription content by enabling them to build up credits by viewing advertisements. A content provider may decide to charge 25 cents for access to a piece of content—a music track, photo, magazine article, or anything else that can be delivered by the Web. Pico-Pay presents the user with a list of advertisements and the amount of credit for viewing each one. Once sufficient credit has been accumulated by viewing these ads, the real content is delivered to the user. The advertiser pays for the delivery of those ads, and the publisher receives payment for the content.[7] These and many other security procedures make buying products and services over the Internet easier and safer. Read the "Ethical and Societal Issues" special-interest box to learn about some of the problems in trying to establish an effective payment system.

Product and Service Delivery

The Internet can also be used to deliver products and services, primarily software and written material. You can download software, reports on the stock market, information on individual companies, and a variety of other written reports and documents directly from the Internet. Often called *electronic distribution*, sending software, music, pictures, and written material through the Internet is faster and can be less expensive than with regular order processing. Electronic distribution can also eliminate inventory problems for manufacturers, who do not have to stock hundreds or thousands of copies of the software, reports, or documents; one copy can be downloaded to customers' computers when needed.

As more and more people use the Internet, the electronic distribution of products and services could be a major revenue source for software and publishing companies. Yet, most products cannot be delivered over the Internet, so they are delivered in a variety of ways: overnight carrier, regular mail service, truck, or rail. In some cases, the customer may elect to drive to the supplier and pick up the product.

Many manufacturers and retailers have outsourced the physical logistics of delivering merchandise to cybershoppers—the storing, packing, shipping, and tracking of products. To provide this service, DHL, Federal Express, United Parcel Service, and other delivery firms have developed software tools and interfaces that directly link customers' ordering, manufacturing, and inventory systems with their own system of highly automated warehouses, call centers, and worldwide shipping network. The goal is to make the transfer of all information and inventory—from the manufacturer to the delivery firm to the consumer—fast and simple.

For example, when a customer orders a printer at the Hewlett-Packard Web site, that order actually goes to FedEx, which stocks all the products that HP sells on-line at a dedicated e-distribution facility in Memphis, a major FedEx shipping hub. FedEx ships the order, which triggers an e-mail notification to the customer that the printer is on its way and an inventory notice to HP that the FedEx warehouse now has one less printer in stock (Figure 8.2). For product returns, HP enters return information into its own system, which is linked to FedEx. This signals a FedEx courier to pick up the unwanted item at the customer's house or business. Customers don't need to fill out shipping labels or package the item. Instead, the FedEx courier uses information transmitted over the Internet to a computer in his or her truck to print a label from a portable printer attached to his belt. FedEx has control of the return, and HP can monitor its progress from start to finish.

After-Sales Service

In addition to capturing the information to complete the order, comprehensive customer information is captured from the order and stored in the supplier's customer database. This information can include customer name, address, telephone numbers, contact person, credit history, and some order details. If, for example, the customer later telephones the supplier to complain that not all items were received, that some arrived damaged, or even that product use

PayPal Service Runs into Start-Up Problems

PayPal enables any business or consumer with an e-mail address to send and receive payments on-line. The company boasts over 13 million users around the world. Its network builds on the existing financial infrastructure of bank accounts and credit cards to create a global, real-time payment solution. PayPal members spend more than $10 million per day for approximately 200,000 transactions. The size of the PayPal network and widening acceptance of its services has helped it become a popular payment network for on-line auction Web sites.

You can create a PayPal account by providing your credit card or bank information and e-mail address. Once you have an account, you can send money by going to the PayPal site and entering the name and e-mail address of the person to be paid, plus the amount you wish to send. The payee receives the cash in his/her PayPal account at the speed of e-mail, billed to your PayPal account (which is tied electronically to your credit card, debit card, or checking account). The payee can ship you the item you bought right away, with no need to wait for the mail to arrive and for the check to clear. You can even send money to someone who does not have a PayPal account, but that person must open a PayPal account to collect it.

PayPal promises to investigate any sellers using its service who do not ship what was promised, and PayPal gives dissatisfied customers any money it can recover. Although PayPal accounts are not insured by the government the way banks are, PayPal does insure them up to $100,000 against unauthorized withdrawals.

In its haste to begin operation of what appeared to be a great business model, PayPal ran into a spate of problems at start-up. First, PayPal was called to task for failing to follow state registration procedures for transferring funds. Next, a rival on-line security company filed a patent infringement lawsuit against PayPal in early February 2002, saying that elements of the PayPal system were identical to its own. Two weeks later, PayPal was hit with a class action suit charging it with improperly administering users' accounts and poor customer service. The law firm for this class action suit gathered many of the complaints from a link posted on one of several Web sites set up by angry PayPal users. Yet another class action suit was filed in mid-March alleging that PayPal's two-tiered customer service practices violate the Electronic Funds Transfer Act, which requires companies to provide a phone number for customers to inquire about their transfers. (Under the then-existing PayPal terms-of-use agreement, nonpaying customers were required to seek help through the PayPal Web site and self-help links.)

In March 2002, a spokesman for PayPal said the company was not free to comment on these problems because PayPal was in a "post IPO (initial public stock offering) quiet period," an imprecise Securities and Exchange Commission restriction that prevents companies from speaking freely to the public while registering to sell shares. The shroud of silence also covers the month or so before the registration is filed. (The stock opened on the NASDAQ exchange on February 15, 2002.) However, the spokesman said that the lawsuits are without merit and that the company has hired 200 additional customer service representatives for its processing center in Omaha. He also pointed to PayPal's e-mail notification customer process and its no-risk guarantee for transactions between PayPal members with addresses verified by the company.

In July 2002, eBay announced that it had agreed to buy PayPal in a stock swap valued at $1.5 billion. The acquisition, which is expected to close by the end of 2002, is subject to approval by stockholders and regulatory agencies.

Discussion Questions

1. Visit the PayPal home page at http://www.paypal.com, click on "Send Money," then click on "See Demo" to follow the process of sending someone money. Write a paragraph describing your experience and reactions.
2. Why would a service such as PayPal be of great interest to a Web auction site such as eBay?

Critical Thinking Questions

3. Do you think that the problems identified in these complaints were the result of intentional actions on the part of PayPal or simply caused by the rapid growth in the number of customers?
4. Some think that records should be kept of those who welsh on Internet deals—either by not sending money to pay for items purchased or not sending the items for which others have paid. Consumer-to-consumer Internet transactions would be screened against this list of known "welshers" and the "innocent" party would be notified. Do you think that such a transaction screening system is a good idea? Why or why not?

Sources: Adapted from "About Us," PayPal Web site at http://www.paypal.com, accessed on April 9, 2002; "PayPal Partners with Discover Card to Expand Payment Network," PayPal press release, January 2, 2002, http://www. paypal.com; Daniel Greenberg, "PayPal Adds Security, Convenience to E-Payments," *USA Today*, May 14, 2001, http://www.usatoday.com; Deborah Radcliff, "Lawsuits Highlight PayPal's Growing Pains," *Computerworld*, April 1, 2002, accessed at http://www.computerworld.com; Scarlett Pruitt, "Ebay Makes Bid for PayPal," Computerworld, July 5, 2002, accessed at http://www.computerworld.com.

instructions are not clear, all customer service representatives can retrieve the order information from the database via a personal computer on their desks. Companies are adding the capability to answer many after-sales questions to their Web sites—how to maintain a piece of equipment, how to effectively use the product, how to receive repairs under warranty, and so on.

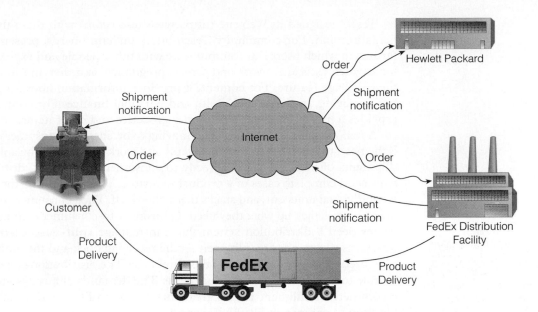

FIGURE 8.2

Product and Information Flow for HP Products Ordered over the Web

E-COMMERCE CHALLENGES

A number of challenges must be overcome for a company to convert its business processes from the traditional form to e-commerce processes. This section summarizes a few.

The first major challenge is for the company to define an effective e-commerce model and strategy. Although a number of different approaches can be used, the most successful e-commerce models include three basic components: community, content, and commerce, as shown in Figure 8.3. Message boards and chat rooms are used to build a loyal *community* of people who are interested in and enthusiastic about the company and its products and services. Providing useful, accurate, and timely *content*—such as industry and economic news and stock quotes—is a sound approach to get people to return to your Web site time and again. *Commerce* involves consumers and businesses paying to purchase physical goods, information, or services that are posted or advertised on-line.

FIGURE 8.3

Three Basic Components of a Successful E-Commerce Model

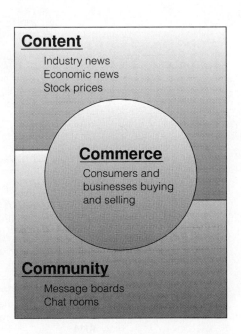

Tesco designed its Web site (http://www.tesco.com) with these three components in mind. For community, Tesco offers bulletin boards, presentations, and forums in which users can communicate with other people and experts in various topics such as beauty, work and career, pregnancy, and diet and fitness through their iVillage feature. For content, it provides information about topics such as healthy living, you and your child, and personal finance. For commerce, Tesco provides a means for shoppers to order groceries and other items.

A major challenge for companies moving to business-to-consumer e-commerce is the need to change distribution systems and work processes to be able to manage shipments of individual units directly to consumers. Traditional distribution systems send complete cases of a product to a store. The store opens the cases, takes the individual units out, and stacks them on a shelf. Then consumers walk through the aisles and pick up what they need. In business-to-consumer e-commerce, companies need a distribution system that can manage **split-case distribution**, in which cases of goods are split open on the receiving dock and the individual items are stored on shelves or in bins in the warehouse. The distribution system must also be able to ship and track individual items. The demands of business-to-consumer e-commerce fulfillment are so great that many on-line vendors outsource the function to companies like FedEx and UPS.

Another tough challenge for e-commerce is the integration of new Web-based order processing systems with traditional mainframe computer-based inventory control and production planning systems. It's one thing to allow sales reps to place orders over the Web, but another to let them find out what products are in (or soon to be in) inventory and available for sale. That means that front-end Web-enabled applications such as order taking need to be tightly integrated to traditional back-end applications such as inventory control and production planning, as shown in Figure 8.4.

THE E-COMMERCE SUPPLY CHAIN

As discussed in Chapter 2, all business organizations contain a number of value-added processes. The supply chain management process is a key value chain that, for most companies, offers tremendous business opportunities if converted

split-case distribution
a distribution system that requires cases of goods to be opened on the receiving dock and the individual items from the cases are stored in the manufacturer's warehouse

FIGURE 8.4

Web-Based Order Processing Must Be Linked to Traditional Back-End Systems

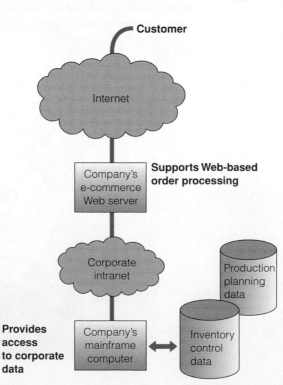

supply chain management
a key value chain composed of demand planning, supply planning, and demand fulfillment

to e-commerce. **Supply chain management** is composed of three subprocesses: demand planning to anticipate market demand, supply planning to allocate the right amount of enterprise resources to meet demand, and demand fulfillment to fulfill demand quickly and efficiently (Figure 8.5). The objective of demand planning is to understand customers' buying patterns and develop aggregate, collaborative long-term, intermediate-term, and short-term forecasts of customer demand. Supply planning includes strategic planning, inventory planning, distribution planning, procurement planning, transportation planning, and supply allocation. The goal of demand fulfillment is to provide fast, accurate, and reliable delivery for customer orders. Demand fulfillment includes order capturing, customer verification, order promising, backlog management, and order fulfillment.

Conversion to e-commerce supply chain management provides businesses an opportunity to achieve operational excellence by increasing revenues, decreasing costs, improving customer satisfaction, and reducing inventory. But to achieve this goal requires integrating all subprocesses that exchange information and move goods between suppliers and customers, including manufacturers, distributors, retailers, and any other enterprise within the extended supply chain.

Increased Revenues and Decreased Costs

By eliminating or reducing time-consuming and labor-intensive steps throughout the order and delivery process, more sales can be completed in the same time period and with increased accuracy.

Improved Customer Satisfaction

Increased and more detailed information about delivery dates and current status can increase customer loyalty.

Inventory Reduction across the Supply Chain

With increased speed and accuracy of customer order information, companies can reduce the need for inventory—from raw materials, to safety stocks, to finished goods—at all the intermediate manufacturing, storage, and transportation points. Some companies are increasing inventory levels above this minimum to make sure that they can meet changes in forecasted customer needs. Nippon Steel is using supply chain management software from i2 Technologies to cut delivery lead times for products and reduce inventories both for itself and its customers. The initial phase of the deployment provides customers with Web access to ordering, production, and quality data from the company's steel materials, coil centers, and steel plan divisions.[8]

FIGURE 8.5

Supply Chain Management

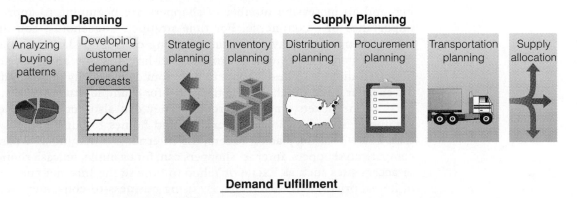

Demand Planning

| Analyzing buying patterns | Developing customer demand forecasts |

Supply Planning

| Strategic planning | Inventory planning | Distribution planning | Procurement planning | Transportation planning | Supply allocation |

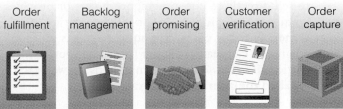

Demand Fulfillment

| Order fulfillment | Backlog management | Order promising | Customer verification | Order capture |

Through improved supply chain management, companies can set their sights not only on improving their profitability and service but also on transforming entire industries. For example, General Motors is working to be able to reach far down into its supply chain to take advantage of the design skills of its suppliers. Suppliers who currently do design work are either located on site with GM engineers or use batch communications over private networks. The automaker plans to link as many as 10,000 suppliers into a new electronic engineering and design network. Initially, 500 of GM's most-critical suppliers will be connected to the network, which is based on Electronic Data Systems' new E-vis 4.0 design-collaboration, product life-cycle management, and graphics tools. This e-commerce application will let suppliers work on-line in real time with GM designers, creating and editing three-dimensional computerized models using EDS's Unigraphics product.[9]

BUSINESS TO BUSINESS (B2B)

Although the business-to-consumer (B2C) market grabs more of the news headlines, the business-to-business (B2B) market is considerably larger and is growing much more rapidly. Business-to-business e-commerce offers enormous opportunities. It allows manufacturers to buy at a low cost worldwide, and it offers enterprises the chance to sell to a global market right from the start. Moreover, e-commerce offers great promise for developing countries, helping them to enter the prosperous global marketplace, and hence helping reduce the gap between rich and poor countries.

The rapid development of e-commerce presents great challenges to society. Even though e-commerce is creating new job opportunities, it could also cause a loss of employment in traditional job sectors. Many companies may fail in the intense competitive environment of e-commerce and find themselves out of business. Therefore, it is vital that the opportunities and implications of e-commerce be understood.

BUSINESS TO CONSUMER (B2C)

Although it is gaining acceptance, e-commerce for consumers is still in its early stages. Many shoppers are not yet convinced that it is worthwhile to connect to the Internet, search for shopping sites, wait for the images to download, try to figure out the ordering process, and then worry about whether their credit card numbers will be stolen by a hacker. But attitudes are changing, and an increasing number of shoppers are beginning to appreciate the importance of e-commerce. For time-strapped households, consumers are asking themselves, Why waste time fighting crowds in shopping malls when from the comfort of home I can shop on-line anytime and have the goods delivered directly? These shoppers have found that many goods and services are cheaper when purchased via the Web—for example, stocks, books, newspapers, airline tickets, and hotel rooms. They can also get information about automobiles, cruises, loans, insurance, and homes to cut better deals. More than a new tool for placing orders, the Internet is emerging as a paradise for comparison shoppers. Internet shoppers can, for example, unleash shopping bots or access sites such as Excite or Yahoo to browse the Internet and obtain lists of items, prices, and merchants. By using business-to-consumer e-commerce to sell directly to consumers, producers or providers of consumer services can eliminate the middlemen, or intermediaries, between them and the end consumer. In many cases, this squeezes costs and inefficiencies out of the supply chain and can lead to higher profits and lower prices for consumers. The elimination of intermediate organizations between the producer and the consumer is called **disintermediation**.

disintermediation
the elimination of intermediate organizations between the producer and the consumer

CONSUMER TO CONSUMER (C2C)

Consumer-to-consumer (C2C) e-commerce involves consumers selling directly to other consumers. Often this exchange is done through Web auction sites like eBay that enabled people to sell over $9 billion in merchandise in 2001 to other consumers by auctioning them off to the highest bidder. The growth of C2C is responsible for reducing the use of the classified pages of a newspaper to advertise and sell personal items.

GLOBAL E-COMMERCE

The use of the Internet is growing rapidly in markets throughout Europe, Asia, and Latin America. So, e-commerce sites need to broaden their focus from North American consumers. The majority of Internet users will live outside the United States by 2003, and the U.S. share of all e-commerce revenues is projected to shrink from 69 percent in 2000 to 59 percent by 2003. On-line retail sales in Europe are estimated to grow at a rate of 98 percent annually over the next five years, soaring from 2.9 billion euros in 1999 to 175 billion euros in 2005. Obviously, companies that want to succeed on the Web cannot ignore this global shift. Developing a sound global e-commerce strategy is critical for ensuring that Web sites are relevant to the consumers and businesses a company wants to reach, whether those customers are in Cleveland, Singapore, or Frankfurt.

The first step in developing a global e-commerce strategy is to determine which global markets make the most sense for selling products or services on-line. One approach is to target regions and countries in which a company already has on-line customers. Companies can track the country domains from which current users of a site are visiting, and established global companies can look to their overseas offices to help determine the languages and countries to target for their Web sites.

Once a company decides which global markets it wants to reach with its Web site, it must adapt an existing U.S.-centric site to another language and culture—a process called *localization*. Localization requires companies to have a deep understanding of the country, its people, and the market, which means either building a physical presence in the country or forming partnerships so that detailed knowledge can be gathered. Companies must take painstaking steps to ensure that e-commerce customers have a local experience even though they're shopping at the Web site of a foreign company.

Some of the steps involved in localization are the following:

- recognizing and conforming to the nuances, subtleties, and tastes of local cultures
- supporting basic trade laws such as each country's currency, payment preferences, taxes, and tariffs
- ensuring that technological capabilities match local connection speeds

Tailoring a site to another country is not easy. When Dell Computer launched an e-commerce site to sell PCs to consumers in Japan, it made the mistake of surrounding most of the site's content with black borders, a negative sign in Japanese culture. Japanese Web shoppers took one look at the site and fled. Also, support for Asian languages is difficult because Asian alphabets are more complex and not all Web development tools are capable of handling them. As a result, many companies choose to tackle Asian markets last. In addition, great care must be taken to choose icons that are relevant to a country. For example, the use of mailboxes and shopping carts may not be familiar to global consumers. Users in European countries don't take their mail from large, tubular receptacles, nor do many of them shop in stores large enough for wheeled carts.

One of the most important and most difficult decisions in a company's global Web strategy is whether Web content should be generated and updated centrally or locally. Companies that expand through international partnerships may be tempted to hand control to the new international entities to take the greatest advantage of the expertise of employees in the new markets. But turning over too much control can lead to a muddle of country-specific sites with no consistency and a scattered corporate message. A mixed model of control may be best. Decisions about corporate identity, brand representation, and the technology used for the Web sites can be made centrally to minimize Web development and support effort as well as to present a consistent corporate and brand message. But a local authority can decide on content and services best tailored for given markets.

Companies must also be aware that consumers outside the United States will access sites with different devices and modify their site design accordingly. In Europe, for example, closed-system iDTVs (interactive digital televisions) are becoming popular for accessing on-line content, with iDTVs projected to reach 80 million European households by 2005. Such devices have better resolution and more screen space than the PC monitors U.S. consumers use to access the Internet. So iDTV users expect more ambitious graphics.

A new group of software and service vendors has emerged to address Web globalization issues. The group includes companies such as Idiom, GlobalSight, and Uniscape.com. Their software can integrate with popular e-commerce and Web content management software from vendors such as Vignette, BroadVision, and Interwoven. The multilingual Web site management software can work especially well for global sites with central management.

On the Web, ultimately the only way to compete with global companies is to be a global company. Successful firms operate with a portfolio of sites designed for each target market, with shared sourcing and infrastructure to support the network of stores, and with local marketing and business development teams to take advantage of local opportunities. Service providers continue to emerge to solve the cross-border logistics, payments, and customer service needs of these pan-European retailers.

E-COMMERCE APPLICATIONS

Since B2B, B2C, C2C, and global e-commerce use is spreading, it's important to examine some of the most common current uses. E-commerce is being applied to retail and wholesale, manufacturing, marketing, investment and finance, and auctions.

RETAIL AND WHOLESALE

electronic retailing (e-tailing)
the direct sale from business to consumer through electronic storefronts, typically designed around an electronic catalog and shopping cart model

cybermall
a single Web site that offers many products and services at one Internet location

There are numerous examples of e-commerce in retail and wholesale. **Electronic retailing**, sometimes called *e-tailing*, is the direct sale from business to consumer through electronic storefronts, which are typically designed around the familiar electronic catalog and shopping cart model. Companies such as Office Depot, Wal-Mart, and many others have used the same model to sell wholesale to employees of corporations. There are tens of thousands of electronic retail Web sites—selling literally everything from soup to nuts. In addition, cybermalls are another means to support retail shopping. A **cybermall** is a single Web site that offers many products and services at one Internet location—the basic idea of a regular shopping mall. An Internet cybermall pulls together multiple buyers and sellers into one virtual place, easily reachable through a Web browser.

Giant retailer Sears, Roebuck and Co. provides an example of how e-commerce is transforming retail selling. Sears gives its shoppers the chance to order on-line and pick up items in its stores. To offer that capability, Sears had to implement technology to enable near-real-time inventory checks so customers can determine

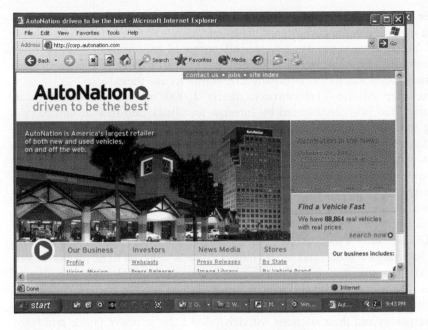

AutoNation.com is the largest auto dealership network, with 378 dealerships and 37 used-vehicle megastores.

whether an item is in stock at a given store. The item is then plucked from the shelf and sent to merchandise pickup, triggering an e-mail confirmation to the customer. Sears joins trailblazer Circuit City Stores as one of the few retailers that can perform the inventory checks necessary to enable in-store pickup.[10]

Office Depot and Amazon.com provide yet another example of how e-commerce is transforming retail selling. In September 2002, Amazon.com signed an e-commerce alliance deal with Office Depot to host an office products store on its site. More than 50,000 office products are available for sale on-line. While Office Depot has long had its own on-line sales operation, this deal greatly expands its sales market. Customers can pick up their on-line purchases in one of Office Depot's stores in the United States. Amazon.com processes the transaction for the customer while Office Depot manages inventory and product fulfillment. This deal is one of several Amazon marketing alliances with other companies, including Circuit City, Marshall Field's, Target, and Toys "R" Us.[11]

Spending on manufacturing, repair, and operations (MRO) goods and services—from simple office supplies to mission-critical equipment, such as the motors, pumps, compressors, and instruments that keep manufacturing facilities up and running smoothly—provides an example of e-commerce applied to wholesale selling. Spending on MRO goods and services often approaches 40 percent of a manufacturing company's total revenues. Despite this, spending can be haphazard, without automated controls for purchasing materials. In addition to these external purchase costs, companies face significant internal costs resulting from ineffective and cumbersome MRO management processes. The result is lost productivity and capacity. Estimates show that a high percentage of manufacturing downtime is often a result of not having the right part at the right time in the right place. E-commerce software for plant operations provides powerful comparative searching capabilities to enable identification of duplicate and functionally equivalent items, giving the insight to spot opportunities for cost savings. Comparing various suppliers, coupled with consolidating more spending with fewer suppliers, leads to decreased costs. In addition, automated workflows enable the implementation of best practices, such as requesting and approving part additions and deletions in a master record in the plant maintenance software.

MANUFACTURING

One approach taken by many manufacturers to raise profitability and improve customer service is to move their supply chain operations onto the Internet. Here they can form an **electronic exchange** to join with competitors and suppliers alike using computers and Web sites to buy and sell goods, trade market information, and run back-office operations, such as inventory control, as shown in Figure 8.6. With such an exchange, the business center is not a physical building but a network-based location where business interactions occur. This approach has greatly speeded the movement of raw materials and finished products among all members of the business community, thus reducing the amount of inventory that must be maintained. It has also led to a much more competitive marketplace

electronic exchange
an electronic forum where manufacturers, suppliers, and competitors buy and sell goods, trade market information, and run back-office operations

and lower prices. Private exchanges are owned and operated by a single company. The owner uses the exchange to trade exclusively with established business partners. Public exchanges are owned and operated by industry groups. They provide services and a common technology platform to their members and are open, usually for a fee, to any company that wants to use them.

At the turn of the 21st century, nearly 1,000 on-line marketplaces in 70 industries had been announced by Internet and brick-and-mortar companies, including the following: the Worldwide Retail Exchange (https://www.worldwide retailexchange.org) currently led by 61 retailers around the world; Covisint (http://www.covisint.com), a global exchange for the auto industry originally formed by automakers Ford, General Motors, and DaimlerChrysler; and TradeRanger in the energy and petrochemical industry (http://www.trade-ranger.com). More than 10,000 on-line exchanges are expected to sweep the business world between 2000 and 2003; however, it's anyone's guess how many will survive. To date, only a few exchanges are seeing trickles of revenue from transaction fees, software licensing, and other charges; none is believed to be profitable yet.

Several strategic and competitive issues are associated with the use of exchanges. Many companies distrust their corporate rivals and fear they may lose trade secrets through participation in such exchanges. Suppliers worry that the on-line marketplaces and their auctions will drive down the prices of goods and favor buyers. Suppliers also can spend a great deal of money in the setup to participate in multiple exchanges. For example, more than a dozen new exchanges appeared in the oil industry, and the printing industry was up to more than 20 on-line marketplaces. Until a clear winner emerges in particular industries, suppliers are more or less forced to sign on to several or all of them. Yet another issue is potential government scrutiny of exchange participants—anytime competitors get together to share information, it raises questions of collusion or antitrust behavior.

FIGURE 8.6

Model of an Electronic Exchange

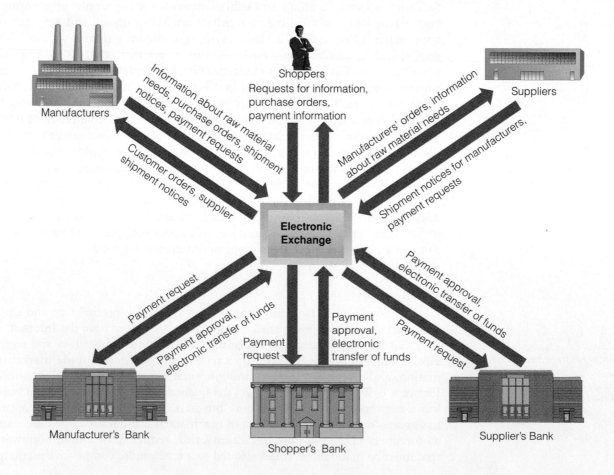

Airbus, headquartered in Toulouse, France, is a leading aircraft manufacturer. Its market share includes about half of all orders for airliners with more than 100 seats, including the double-deck A380, the world's largest commercial airliner. It spends in excess of $13 billion per year with thousands of suppliers involved with the design and development of new aircraft. While the company considered participating in public e-marketplaces, it rejected that idea because of security and competitive considerations. Instead, Airbus formed its own private B2B marketplace using software from Ariba to support its procurement strategy, establish contracts with suppliers, maintain information on suppliers, and hold auctions for commodities or general procurement. Using this system, Airbus is able to publish its procurement needs and invite suppliers from around the world to bid on them. This enables Airbus to select the best suppliers and enter into negotiations on major purchases.[12]

Many companies that already use the Internet for their private exchanges have no desire to share their on-line expertise with competitors. At Wal-Mart, the world's number one retail chain, executives turned down several invitations to join exchanges in the retail and consumer goods industries. Wal-Mart is pleased with its in-house exchange, Retail Link, which connects the company to 7,000 worldwide suppliers that sell everything from toothpaste to furniture.

In addition to formal exchanges, manufacturers are using e-commerce to improve the efficiency of the selling process. Perkins Engines, a UK-based division of Caterpillar, has implemented an on-line parts catalog for its 4,000 distributors in 160 countries that enables them to increase their sales of diesel engine parts used for maintenance and repair. When a user places an engine part in the "holding bay" (in the site's shopping cart), the software suggests related parts to purchase. For example, if users order a piston, they will need an engine gasket to finish the job. It also automatically recommends a newer part when a superseded part number is entered into the holding bay.[13]

MARKETING

The nature of the Web allows firms to gather much more information about customer behavior and preferences than they could using other marketing approaches. Marketing organizations can measure a large number of activities as customers and potential customers gather information and make their purchase decisions. Analysis of this data is complicated because of the Web's interactivity and because each visitor voluntarily provides or refuses to provide personal data such as name, address, e-mail address, telephone number, and demographic data. Internet advertisers use the data they gather to identify specific portions of their markets and target them with tailored advertising messages. This practice, called **market segmentation**, divides the pool of potential customers into segments, which are usually defined in terms of demographic characteristics such as age, gender, marital status, income level, and geographic location.

Technology-enabled relationship management has become possible when promoting and selling on the Web. **Technology-enabled relationship management** occurs when a firm obtains detailed information about a customer's behavior, preferences, needs, and buying

market segmentation
the identification of specific markets to target them with advertising messages

technology-enabled relationship management
the use of detailed information about a customer's behavior, preferences, needs, and buying patterns to set prices, negotiate terms, tailor promotions, add product features, and otherwise customize the entire relationship with that customer

On-line marketing firm Nielsen//Net-Ratings provides its clients with customized media and market research services, helping them to gain a competitive edge.

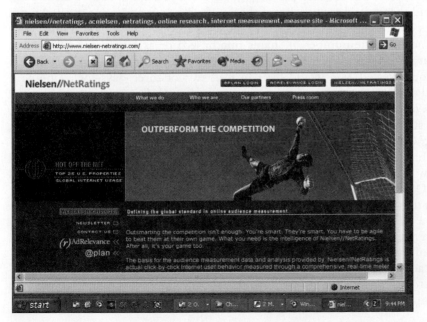

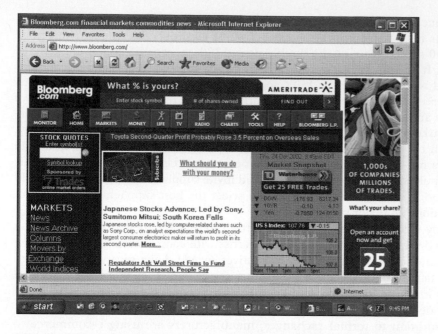

Investment and finance sites provide a multitude of services with the click of a mouse. Bloomberg.com is one of the top sites on the Web for news and financial information. Browsers can find stock market, mortgage, and other financial tracking information, as well as tips and advice on many finance subjects.

patterns and uses that information to set prices, negotiate terms, tailor promotions, add product features, and otherwise customize its entire relationship with that customer.

DoubleClick is a leading global Internet advertising company that leverages technology and media expertise to help advertisers use the power of the Web to build relationships with customers. The DoubleClick Network is its flagship product, a collection of high-traffic and well-recognized sites on the Web (AltaVista, Dilbert, US News, Macromedia, and more than 1,500 others). This network of sites is coupled with DoubleClick's proprietary DART targeting technology, which allows advertisers to target their best prospects based on the most precise profiling criteria available. DoubleClick then places a company's ad in front of those best prospects. Comprehensive on-line reporting lets advertisers know how their campaign is performing and what type of users are seeing and clicking on their ads. This high-level targeting and real-time reporting provide speed and efficiency not available in any other medium. The system is also designed to track advertising transactions, such as impressions and clicks, to summarize these transactions in the form of reports, and to compute DoubleClick Network member compensation.

INVESTMENT AND FINANCE

The Internet has revolutionized the world of investment and finance. Perhaps the changes have been so great because this industry had so many built-in inefficiencies and so much opportunity for improvement.

On-Line Stock Trading

Before the World Wide Web, if you wanted to invest in stocks, you called your broker and asked what looked promising. He'd tell you about two or three companies and then would try to sell you shares of a stock or perhaps a mutual fund. The sales commission was well over $100 for the stock (depending on the price of the stock and the number of shares purchased) or as much as an 8 percent sales charge on the mutual fund. If you wanted information about the company before you invested, you would have to wait two or three days for a one-page Standard and Poor's stock report providing summary information and a chart of the stock price for the past two years to arrive in the mail. Once you purchased or sold the stock, it would take two days to get an order confirmation in the mail, detailing what you paid or received for the stock.

The brokerage business adapted to the Internet faster than any other arm of finance, with Net brokers grabbing 45 percent of the New York Stock Exchange and NASDAQ trades by early 2000. But with the stock market decline of 2000–2001 and with day traders counting their losses, that share is down to 22 percent.[14] Still, to make a trade, all you need to do is log on to the Web site of your on-line broker and, with a few keystrokes and a few clicks of your mouse to identify the stock and number of shares involved in the transaction, you can buy and sell securities in seconds. In addition, an overwhelming amount of free information is available to on-line investors—from the latest Securities & Exchange filings to the rumors spread in chat rooms. See Table 8.1 for a short list of the more valuable sites.

Name of Site	URL	Description
411 Stocks	www.411stocks.com	One-stop location to get lots of information about a stock—price data, news, discussion groups, charts, basic data, financial statements, and delayed quotes
MarketReporter	www.marketreporter.com	Provides financial news, recommendations, upgrades, downgrades, message boards, stock market simulation game
Thomson Investors Network	www.thomsoninvest.com	Financial commentary from a number of stock market publications, including *First Watch* and *Stocks to Watch*
Elite Trader	www.elitetrader.com	Virtual gathering place for day traders with bulletin boards and chat rooms
Dayinvestor.com	www.dayinvestor.com	News and stock alerts with frequent briefs on market activity and rumors
DRIP Advisor	www.dripadvisor.com	Covers the basics of dividend reinvestment programs (DRIPs), what companies offer DRIPs, and how to start a DRIP
The Raging Bull	ragingbull.lycos.com	Contains lots of message boards; guest experts produce news, commentary, and analysis
EDGAR Online	www.edgar-online.com	Provides access to company filings with the Securities and Exchange Commission (SEC)
Federal Filings Online	www.fedfil.com	Dow Jones directory of documents filed with the federal government, including bankruptcy proceedings, initial public offering (IPO) filings, SEC reports, and court cases

TABLE 8.1

Web Sites Useful to Investors

One indispensable tool of the on-line investor is a portfolio tracker. This tool allows you to enter information about the securities you own—ticker symbol, number of shares, price paid, and date purchased—at a tracker Web site. You can then access the tracker site to see how your stocks are doing. (There is typically a 15- to 20-minute delay between the price displayed at the site and the price at which the stock is actually being sold.) In addition to reporting the current value of your portfolio, most sites provide access to news, charts, company profiles, and analyst ratings on each of your stocks. You can also program many of the trackers to watch for certain events (e.g., stock price change of more than +/– 3 percent in a single day). When one of the events you specified occurs, an "alert" symbol is posted next to the affected stock. Table 8.2 lists a number of the more popular tracker Web sites.

TABLE 8.2

Popular Stock Tracker Web Sites

Name of the Web Stock Tracker Site	URL
MSN MoneyCentral	moneycentral.msn.com/investor
Quicken.com	www.quicken.com
The Motley Fool	www.fool.com
Yahoo!	quote.yahoo.com
Morningstar	www.morningstar.com

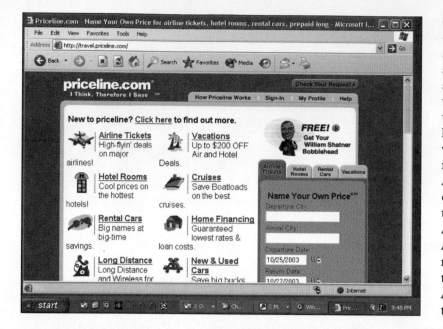

Priceline.com is a patented Internet bidding system that enables consumers to save money by naming their own price for goods and services.

electronic bill presentment
a method of billing whereby the biller posts an image of your statement on the Internet and alerts you by e-mail that your bill has arrived

On-Line Banking

On-line banking customers can check balances of their savings, checking, and loan accounts; transfer money among accounts; and pay their bills. With on-line banking, these customers think they can gain a better current knowledge of how much they have in the bank, eliminate the need to write checks in longhand, and reduce how much they spend on envelopes and stamps. All of the nation's major banks and many of the smaller banks enable their customers to pay bills on-line. In 2001, 15 million Americans paid bills on-line, and Americans took out at least $160 billion in mortgages on-line in 2001, 8 percent of the total market. The number of Americans who pay bills online is expected to reach 46 million by 2005.[15]

Here's how electronic bill payment works. You first set up a list of frequent payees, along with their addresses and a code describing the type of payment, such as "home mortgage." Then when you go on-line to pay your bills, you simply enter the code or name assigned to the check recipient, the amount of the check, and the date you want it paid. In many cases, the bank still prints and mails a check, so you have to time your on-line transactions to allow for bank processing and mail delays. But most bill-paying programs allow you to schedule recurring payments for every week, month, or quarter, which you might want to do for your auto loan or health insurance bill.

The next advance in on-line bill paying is **electronic bill presentment**, which eliminates all paper, right down to the bill itself. Under this process, the biller posts an image of your statement on the Internet and alerts you by e-mail that your bill has arrived. You then direct your bank to pay it. CheckFree (http://www.checkfree.com) offers such a service, enabling more than 5.9 million consumers to receive and pay bills over the Internet or electronically with over 430 companies. Several major banks, including Union Bank of California, J.P. Morgan Chase & Co., Wachovia, and Wells Fargo have contracted with on-line bill payment company Metavante to offer consumers the ability to view and pay bills.[16]

AUCTIONS

As discussed in Chapter 7, the Internet has created many new options for C2C, including electronic auctions, where geographically dispersed buyers and sellers can come together. A special type of auction called *bidding* allows a prospective buyer to place only one bid for an item or a service. Priceline.com is the patented Internet bidding system that enables consumers to achieve significant savings by naming their own price for goods and services. Priceline.com takes these consumer offers and then presents them to sellers, who can fill as much of that guaranteed demand as they wish at price points determined by buyers.

Now that we've examined some of the applications of e-commerce, let's look at some of the more technical issues related to information systems and technology that make it possible.

TECHNOLOGY INFRASTRUCTURE

For e-commerce to succeed, a complete technology infrastructure must be in place. These infrastructure components must be chosen carefully and integrated to be capable of supporting a large volume of transactions with customers, suppliers, and other business partners worldwide. On-line consumers complain that poor Web site performance (e.g., slow response time, poor customer support, and "lost" orders) drives them to abandon some e-commerce sites in favor of those with better, more reliable performance. This section provides a brief overview of the key technology infrastructure components (Figure 8.7).

HARDWARE

A Web server hardware platform complete with the appropriate software is a key e-commerce infrastructure ingredient. The amount of storage capacity and computing power required of the Web server depends primarily on two things—the software that must run on the server and the volume of e-commerce transactions that must be processed. Although it is possible for information systems staff to define the software to be used, sometimes much guesswork is involved in estimating how much traffic the site will generate. As a result, the most successful e-commerce solutions are designed to be highly scalable so that they can be upgraded to meet unexpected user traffic.

A key decision facing new e-commerce companies is whether to host their own Web site or to let someone else do it. Many companies decide that using a third-party Web service provider is the best way to meet initial e-commerce needs. The third-party company rents space on its computer system and provides a high-speed connection to the Internet, which minimizes the initial out-of-pocket costs for e-commerce start-up. The third party can also provide personnel trained to operate, troubleshoot, and manage the Web server. Of course, a number of companies decide to take full responsibility for acquiring, operating, and supporting the Web server hardware and software themselves, but this approach requires considerable up-front capital and a set of skilled and trained individuals. Whichever approach is taken, there must be adequate hardware backup to avoid a major business disruption in case of a failure of the primary Web server.

WEB SERVER SOFTWARE

In addition to the Web server operating system, each e-commerce Web site must have Web server software to perform a number of fundamental services, including security and identification, retrieval and sending of Web pages, Web site tracking,

FIGURE 8.7

Key Technology Infrastructure Components

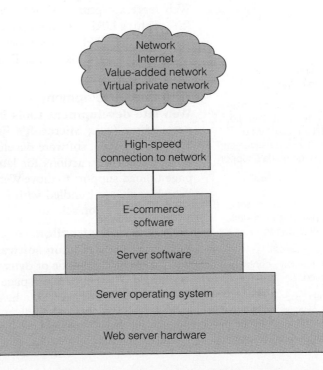

Web site development, and Web page development. The two most popular Web server software packages are Apache HTTP Server and Microsoft Internet Information Server.

Security and Identification

Security and identification services are essential for intranet Web server implementation to identify and verify employees coming into the server from the Internet. Access controls provide or deny access to files based on the user name or URL. Web servers support encryption processes for transmitting private information securely over the public Internet.

In addition to managing security and identification services, Web sites must be designed to protect against malicious attacks. A denial-of-service attack is one of the most difficult Internet threats to thwart and can be costly if it knocks an e-commerce site out of commission. During a **denial-of-service attack**, the attacker takes command of many computers on the Internet and causes them to flood the target Web site with requests for data and other small tasks that keep the target machine too busy to serve legitimate users. Many Web sites including Amazon, CNN, eBay, and Yahoo have suffered these attacks. Cervalis is an outsourcer and on-line managed hosting vendor with headquarters in Stamford, Connecticut. Because it must ensure its customers of reliable, around-the-clock network operations, it started testing Riverhead Network's Guard hardware and software device in January 2002. This product defends against a denial-of-service attack by passively monitoring network traffic for unusual activity. Once an attack is detected, all traffic is diverted to Guard and only traffic from legitimate users is passed through Guard to the user's network.[17]

Retrieving and Sending Web Pages

The fundamental purpose of a Web server is to process and respond to client requests that are sent using the HTTP protocol. In response to such a request, the Web server program locates and fetches the appropriate Web page, creates an HTTP header, and appends the HTML document to it. For dynamic pages, the server involves other programs, retrieves the results from the back-end process, formats the response, and sends the pages and other objects to the requesting client program.

Web Site Tracking

Web servers capture visitor information, including who is visiting the Web site (the visitor's URL), what search engines and key words they used to find the site, how long their Web browser viewed the site, the date and time of each visit, and which pages were displayed. This data is placed into a **Web log file** for future analysis.

Web Site Development

Web site development tools include features such as an HTML/visual Web page editor (e.g., Microsoft's FrontPage, NetStudio's NetStudio, SoftQuad's HoTMetaL Pro), software development kits that include sample code and code development instructions for languages such as Java or Visual Basic, and Web page upload support to move Web pages from a development PC to the Web site. Which tools are bundled with the Web server software depends on which Web server software you select.

Web Page Construction

Web page construction software uses Web editors and extensions to produce Web pages—either static or dynamic. **Static Web pages** always contain the same information—for example, a page that provides text about the history of the company or a photo of corporate headquarters. **Dynamic Web pages** contain variable information and are built in response to a specific Web visitor's request. For example, if a Web site visitor inquires about the availability of a certain product by entering a product identification number, the Web server will search the product

---

denial-of-service attack
an on-line attack of a Web site in which the attacker takes command of many computers on the Internet and causes them to flood the target site with requests for data and other tasks, keeping it too busy to serve legitimate users

Web log file
a file that contains information about visitors to a Web site

Web site development tools
tools used to develop a Web site, including HTML or visual Web page editor, software development kits, and Web page upload support

Web page construction software
software that uses Web editors and extensions to produce both static and dynamic Web pages

static Web pages
Web pages that always contain the same information

dynamic Web pages
Web pages containing variable information that are built in response to a specific Web visitor's request

inventory database and generate a dynamic Web page based on the current product information it found, thus fulfilling the visitor's request. This same request by another visitor later in the day may yield different results due to ongoing changes in product inventory. A server that handles dynamic content must be able to access information from a variety of databases. The use of open database connectivity enables the Web server to assemble information from different database management systems, such as SQL Server, Oracle, and Informix.

E-COMMERCE SOFTWARE

e-commerce software
software that supports catalog management, product configuration, shopping cart facilities, e-commerce transaction processing, and Web traffic data analysis

Once you have located or built a host server, including the hardware, operating system, and Web server software, you can begin to investigate and install e-commerce software. There are five core tasks that **e-commerce software** must support: catalog management, product configuration, shopping cart facilities, e-commerce transaction processing, and Web traffic data analysis.

The specific e-commerce software you choose to purchase or install depends on whether you are setting up for B2B or B2C. For example, B2B transactions do not include sales tax calculations, and software to support B2B must incorporate electronic data transfers between business partners, such as purchase orders, shipping notices, and invoices. B2C software, on the other hand, must handle the complication of accounting for sales tax based on the current laws and rules in effect in the various states.

Catalog Management

catalog management software
software that automates the process of creating a real-time interactive catalog and delivering customized content to a user's screen

Any company that offers a wide range of product offerings requires a real-time interactive catalog to deliver customized content to a user's screen. **Catalog management software** combines different product data formats into a standard format for uniform viewing, aggregating, and integrating catalog data into a central repository for easy access, retrieval, and updating of pricing and availability changes. The data required to support large catalogs is almost always stored in a database on a computer that is separate from, but accessible to, the e-commerce server machine. The effort to build and maintain online catalogs can be substantial. For example, Fastenal is a leading distributor of nuts, bolts, screws, and other products used by manufacturing and construction companies. The company has a group of more than two dozen product marketers and information systems staff who help the company assemble and update its e-commerce catalogs. Its goal is to expand its on-line catalog to the full line of over 1 million products that it distributes.[18]

Product Configuration

product configuration software
software used by buyers to build the product they need on-line

Customers need help when an item they are purchasing has many components and options. Product configuration software tools were originally developed in the 1980s to assist B2B salespeople to match their company's products to customer needs. Buyers use the new Web-based **product configuration software** to build the product they need on-line with little or no help from salespeople. For example, Dell customers use product configuration software to build the computer of their dreams. Use of such software can expand into the service arena as well, with consumer loans and financial services to help people decide what sort of loan or insurance is best for them.

Shopping Cart

electronic shopping cart
a model commonly used by many e-commerce sites to track the items selected for purchase, allowing shoppers to view what is in their cart, add new items to it, and remove items from it

Today many e-commerce sites use an **electronic shopping cart** to track the items selected for purchase, allowing shoppers to view what is in their cart, add new items to it, or remove items from it, as shown in Figure 8.8. To order an item, the shopper simply clicks that item. All the details about it—including its price, product number, and other identifying information—are stored automatically. If the shopper later decides to remove one or more items from the cart, he or she can do so by viewing the cart's contents and removing any unwanted items. When the shopper is ready to pay for the items, he or she clicks a button (usually labeled "proceed to checkout") and begins a purchase transaction. Clicking the "Checkout" button displays another screen that usually asks the shopper to fill out billing, shipping, and payment method information and to confirm the order.

E-COMMERCE TRANSACTION PROCESSING

E-commerce transaction processing software connects participants in the e-commerce economy and enables communication between trading partners, regardless of their technical infrastructure. This software fully automates transaction processes from order placement to reconciliation.

Basic transaction processing software takes data from the shopping cart and calculates volume discounts, sales tax, and shipping costs to arrive at the total cost. In some cases, the software determines shipping costs by connecting directly to shipping companies such as UPS, FedEx, and Airborne. In other cases, shipping cost may be a predetermined amount for each item ordered.

More and more e-commerce companies are outsourcing the actual inventory management and order fulfillment process to a third party. In this situation, the e-commerce transaction processing software must actually route order information to one of the shipping companies to ship from inventory under their management. For example, Hewlett-Packard has an arrangement with FedEx to manage orders for HP printers.

Web Traffic Data Analysis

It is necessary to run third-party **Web site traffic data analysis software** to make sense of all the data captured in the Web log file—to turn it into useful information to improve Web site performance. For example, when someone queries a search engine for a keyword related to your site's products or services, does your page appear in the top ten matches, or does your competitor's? If you're listed but not within the first two or three pages of results, you lose, no matter how many engines you submitted your site to.

NETWORK AND PACKET SWITCHING

Several technologies must be in place for e-commerce to work. The most obvious is that some sort of network must be in place, since e-commerce depends on the secure transmission of data over networks, whether the Internet, corporate extranets,

e-commerce transaction processing software
software that provides the basic connection between participants in the e-commerce economy, enabling communications between trading partners, regardless of their technical infrastructure

Web site traffic data analysis software
software that processes and analyzes data from the Web log file to provide useful information to improve Web site performance

FIGURE 8.8

Electronic Shopping Cart

An electronic shopping cart (or bag) allows on-line shoppers to view their selections and add or remove items.

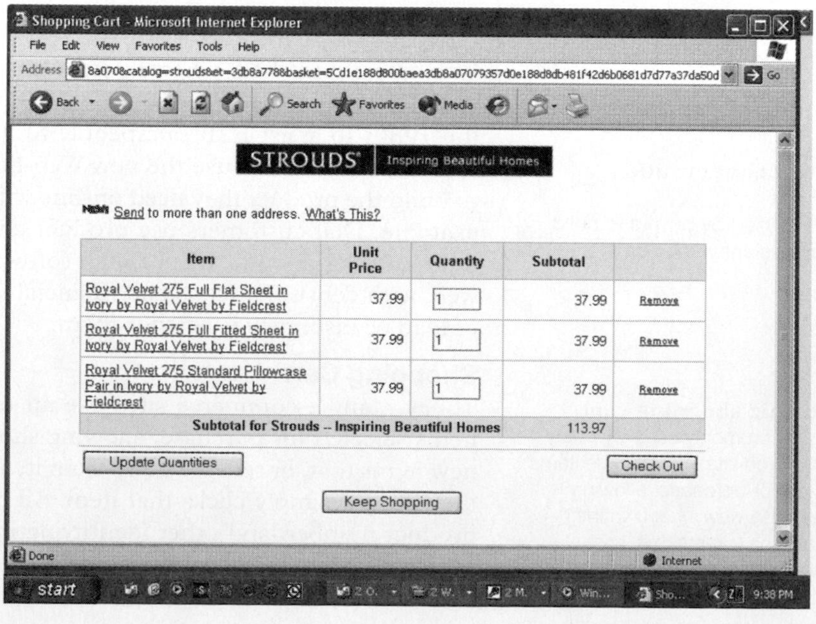

value-added networks (VANs), or virtual private networks (VPNs, discussed in Chapter 7). All these approaches rely on basic packet switching technology and the use of routers to help each packet arrive at its destination in the quickest and most economical manner. In choosing among the options, the information systems staff must weigh cost, availability, reliability, security, and redundancy issues. Cost includes the initial development cost as well as the ongoing operational and support costs. Availability concerns the hours the network is *scheduled* to be available for normal use. Reliability is the percentage of available time the network must actually be fully operational, typically 99 percent or more. Security is the ability to keep messages from being intercepted. And redundancy is the ability of the network to keep operating if key elements fail. All are important factors that must be considered.

ELECTRONIC PAYMENT SYSTEMS

digital certificate
an attachment to an e-mail message or data embedded in a Web page that verifies the identity of a sender or a Web site

certificate authority (CA)
a trusted third party that issues digital certificates

secure sockets layer (SSL)
a communications protocol used to secure sensitive data

The SSL (secure sockets layer encryption) communications protocol assures customers that information they provide to retailers, such as credit card numbers, cannot be viewed by anyone else on the Web.

Electronic payment systems are a key component of the e-commerce infrastructure. The U.S. Federal Reserve System released results of a November 2001 survey that suggested that check writing is giving way to electronic payments. The Fed found that the use of checks has declined from approximately 85 percent of noncash payments since the last study in 1979 to about 60 percent, with an estimated 50 billion checks written annually in the U.S. for a total of $48 trillion in payments.[19] Current e-commerce technology relies on user identification and encryption to safeguard business transactions. Actual payments are made in a variety of ways, including electronic cash; electronic wallets; and smart, credit, charge, and debit cards.

As discussed in Chapter 7, authentication technologies are used by organizations to confirm the identity of a user requesting access to information or assets. A **digital certificate** is an attachment to an e-mail message or data embedded in a Web site that verifies the identity of a sender or Web site. A **certificate authority (CA)** is a trusted third-party organization or company that issues digital certificates. The CA is responsible for guaranteeing that the individuals or organizations granted these unique certificates are, in fact, who they claim to be. Digital certificates thus create a trust chain throughout the transaction, verifying both purchaser and supplier identities.

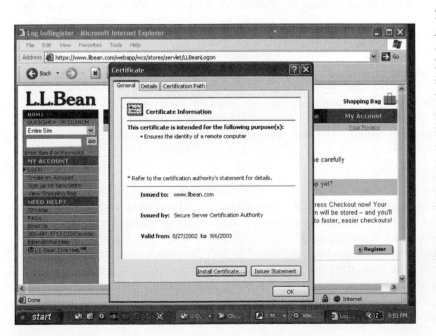

SECURE SOCKETS LAYER

All on-line shoppers fear the theft of credit card numbers and banking information. To help prevent this from happening, the **secure sockets layer (SSL)** communications protocol is used to secure sensitive data. The SSL communications protocol sits above the TCP layer of the OSI model discussed in Chapter 7, and other protocols such as Telnet and HTTP can be layered on top of it. SSL includes a handshake stage, which authenticates the server (and the client, if needed), determines the encryption and hashing algorithms to be used, and exchanges encryption keys. The handshake may use public key encryption. Following the handshake stage, data may be transferred. The data is always encrypted. This ensures that your transactions are not subject to interception or "sniffing" by a third party. For companies wishing to conduct serious e-commerce, such as receiving credit card numbers or other sensitive information, SSL is a must. Although SSL handles the encryption part of a secure e-commerce transaction, a digital certificate is necessary to provide server identification.

One tip to the security of a transaction is visible on-screen. Look at the bottom left corner of your browser before sending your credit card number to an e-commerce vendor. If you use Netscape Navigator, make sure you see a solid key in a small blue rectangle. If you use Microsoft Explorer, the words "Secure Web site" appear near a little gold lock. And, if you're worried about how secure a secure connection is, visit Netcraft at http://www.netcraft.com/security/. At this site you can type in any Web site address and determine the equipment being used for secure transactions. One more tip: to ensure security, you should always use the newest browser available; the newer the browser, the better the security.

ELECTRONIC CASH

electronic cash
an amount of money that is computerized, stored, and used as cash for e-commerce transactions

Electronic cash is an amount of money that is computerized, stored, and used as cash for e-commerce transactions. A consumer must open an account with a bank and show some identification to establish identity to obtain electronic cash. Then whenever the consumer wants to withdraw electronic cash to make a purchase, he or she accesses the bank via the Internet and presents proof of identity—typically a digital certificate issued by a certification authority. After the bank verifies the consumer's identity, it issues the consumer the requested amount of electronic cash and deducts the same amount from the consumer's account. The electronic cash is stored in the consumer's electronic wallet on his or her computer's hard drive, or on a smart card (both are discussed later).

Consumers can spend their electronic cash when they locate e-commerce sites that accept electronic cash for payment. The consumer sends electronic cash to the merchant for the specified cost of the goods or services. The merchant validates the electronic cash to be certain it is not forged and belongs to the customer. Once the goods or services are shipped to the consumer, the merchant presents the electronic cash to the issuing bank for deposit. The bank then credits the merchant's account for the transaction amount, minus a small service charge.

There are two distinct types of electronic cash: identified electronic cash and anonymous electronic cash (also known as *digital cash*). Identified electronic cash contains information revealing the identity of the person who originally withdrew the money from the bank. Also, in much the same manner as credit cards, identified electronic cash enables the bank to track the money as it moves through the economy. The bank can determine what people or organizations bought, where they bought it, when they bought it, and how much they paid. Anonymous electronic cash works just like real paper cash. Once anonymous electronic cash is withdrawn from an account, it can be spent or given away without leaving a transaction trail. A few of the companies that provide electronic cash mechanisms include VeriSign, Mondex, InternetCash, and Visa Cash.

ELECTRONIC WALLETS

electronic wallet
a computerized stored value that holds credit card information, electronic cash, owner identification, and address information

On-line shoppers quickly tire of repeatedly entering their shipment and payment information each time they make a purchase. An **electronic wallet** holds credit card information, electronic cash, owner identification, and address information. It provides this information at an e-commerce site's checkout counter. When consumers click on items to purchase, they can then click on their electronic wallet to order the item, thus making on-line shopping much faster and easier.

Household International, a leading provider of consumer finance, credit card, auto finance, and credit insurance products, in partnership with General Motors Corporation and CyberCash (acquired by VeriSign), a provider of e-commerce technologies and services for merchants, introduced the GM Card easyPay electronic wallet marketed to a large portion of GM card members. The GM Card easyPay Wallet stores a shopper's name, credit card information, shipping details, and other pertinent facts that can be called up to make an on-line purchase with a single click of a computer mouse. The wallet is available to GM card members and interested consumers at www.GMCard.com.[20]

SMART, CREDIT, CHARGE, AND DEBIT CARDS

On-line shoppers use credit and charge cards for the majority of their Internet purchases. A credit card, such as Visa or MasterCard, has a preset spending limit based on the user's credit limit, and each month the user can pay off a portion of the amount owed or the entire credit card balance. Interest is charged on the unpaid amount. A charge card, such as American Express, carries no preset spending limit, and the entire amount charged to the card is due at the end of the billing period. Charge cards do not involve lines of credit and do not accumulate interest charges.

Debit cards look like credit cards or automated teller machine (ATM) cards, but they operate like cash or a personal check. While a credit card is a way to "buy now, pay later," a debit card is a way to "buy now, pay now." Debit cards allow you to spend only what is in your bank account. It is a quick transaction between the merchant and your personal bank account. When you use a debit card, your money is quickly deducted from your checking or savings account. Credit, charge, and debit cards currently store limited information about you on a magnetic stripe. This information is read each time the card is swiped to make a purchase. All credit card customers are protected by law from paying any more than $50 for fraudulent transactions. At Visa, on-line purchases account for the highest amount of purchase fraud—24 cents for every $100 spent, compared with 6 cents for every $100 overall.[21] Indeed, the risk of bogus credit card transactions has slowed the growth of e-commerce by exposing merchants to substantial losses and making on-line shoppers nervous. Credit card fraud accounted for $1.2 billion of total on-line sales of $65 billion in 2001, with merchants forced to cover most of those losses. Banks charge merchants an average fee of 2.5 percent for on-line transactions compared with 1.5 percent for in-store purchases to offset the costs of credit card fraud.[22]

On-line payment processor VeriSign processes about one in every four on-line transactions in the United States. It recently adopted a credit card identification system developed by MasterCard to reduce credit card fraud. The system relies on a "Universal Cardholder Authentication Field" that enables merchants to verify that on-line shoppers are using credit cards that actually belong to them by entering a special password associated with the card. MasterCard is providing merchants with a powerful incentive to sign up for the new program—those merchants who verify transactions through the new system won't have to pay for losses should the transaction turn out to be illegitimate. Instead, the card issuer will be liable.[23]

smart card
a credit card–sized device with an embedded microchip to provide electronic memory and processing capability

The **smart card** is a credit card–sized device with an embedded microchip to provide electronic memory and processing capability. Smart cards can be used for a variety of purposes, including storing a user's financial facts, health insurance data, credit card numbers, and network identification codes and passwords. They can also store monetary values for spending.

Smart cards are better protected from misuse than conventional credit, charge, and debit cards because the smart card information is encrypted. Conventional credit, charge, and debit cards clearly show your account number on the face of the card. The card number, along with a forged signature, is all that a thief needs to purchase items and charge them against your card. A smart card makes credit theft practically impossible because a key to unlock the encrypted information is required, and there is no external number that a thief can identify and no physical signature a thief can forge.

Smart cards have been around for over a decade and are widely used in Europe, Australia, and Japan, but they have not caught on in the United States. Use has been limited because there are so few smart card readers to record payments, and U.S. banking regulations have slowed smart card marketing and acceptance as well. American Express launched its Blue card smart card in 1999. You can use a smart card reader that attaches to your PC monitor to make on-line purchases with your American Express card. You must visit the American Express Web site to get an electronic wallet to store your credit card information and shipping address. When you want to buy something on-line, you go to the checkout screen of a Web merchant, swipe your Blue card through the reader, type in a password, and you're done. The digital wallet automatically tells the vendor your credit card number, its expiration date, and your shipping information.[24] Read the "IS Principles in Action" special-interest box to find out more about smart cards.

THREATS TO E-COMMERCE

As with any revolutionary change, a host of issues must be dealt with to ensure that e-commerce transactions are safe and consumers are protected. Many represent significant threats to its continued growth. The following sections summarize a number of these threats and present practical ideas on how to minimize their impact.

E-COMMERCE INCIDENTS

intellectual property
music, books, inventions, paintings, and other special items protected by patents, copyrights, or trademarks

As mentioned previously, organizations use identification technologies to confirm the identity of a user requesting access to information or assets. Just as shoppers are concerned with the legitimacy of sites, e-businesses are concerned with the legitimacy of their customers. As a result, an increasing number of companies are investing in biometric technology to protect both sides. A growing number of financial services firms (e.g., Citibank, First Financial Credit Union, Huntington Bankshares, Perdue Employees Federal Credit Union) are considering the use of biometric systems that use people's unique physical or behavioral characteristics, such as fingerprints or voice patterns, to identify them. Heightened security concerns in the wake of the September 11 terrorist attacks have increased interest in the use of biometrics to develop faster and more effective passenger screening systems.

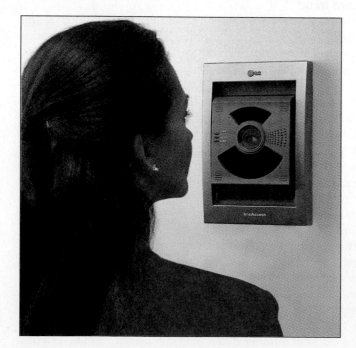

Biometric technology such as iris scanning can provide security for accessing files or obtaining clearance to enter a building.

(Source: Courtesy of Iridian Technologies, Inc.)

Biometric technology, which digitally encodes physical attributes of a person's voice, eye, face, or hand and associates them with biological attributes stored in a file, are commonly used in organizations such as the FBI to allow clearance into a building, for instance. Currently, using the technology to secure on-line transactions is rare for both cost and privacy reasons. It can be expensive to outfit every customer with a biometric scanner, and it is difficult to convince consumers to supply something as personal and distinguishing as a fingerprint.

THEFT OF INTELLECTUAL PROPERTY

Lawsuits over **intellectual property** (music, books, inventions, paintings, and other special items protected by patents, copyrights, or trademarks) have created a virtual e-commerce war zone. From music files to patented on-line coupon schemes, hardly a day goes by without some new suit being filed. Every year, American businesses lose billions of dollars from the importation

Visa Implements Smart Card Technology

Visa cards today can be used to pay for purchases at more than 42 million merchant locations in 300 countries and territories. They can also be used to obtain cash at over 700,000 ATMs in 136 countries. More than 14,000 U.S. financial institutions rely on the VisaNet processing system to facilitate over $765 billion in annual transaction volume—including more than half of all Internet payments.

Visa continues to work on behalf of its member financial institutions, merchants, and consumers to continually deliver better ways to pay. The company announced the smart Visa card in May 2001 as a means of providing a highly personalized card capable of changing as your needs change and as new services become available. The Visa smart card uses an embedded chip that is programmed to accept, store, and send data. It is also designed to minimize card duplication and forgery. An especially useful feature of the smart Visa is that it enables you to consolidate many of the store and discount cards filling your wallet onto its memory chip. Visa's vision is that smart cards will continue to add new services and levels of convenience to consumers' everyday lives by, for example, functioning as electronic keys for a home, office, or car or offering payment services through a personal computer, mobile phone, or personal digital assistant (PDA).

Smart Visa card readers are credit card–sized devices that connect to your personal computer to provide additional features. You can pay for on-line purchases by simply inserting your smart Visa card into the reader. The password-protected smart Visa card works with Verified by Visa to authenticate your identity and your card. Verified by Visa allows you to add a personal password to your existing Visa card. When you make purchases at participating on-line stores, you validate your identity by entering your password in a special Verified by Visa window. This gives you added safety and reassurance that only you can use your Visa card on-line.

As smart card technology evolves and new services become available, you will be able to use the reader to download these new features onto your smart Visa card. The smart Visa card and reader also provide secure access to your account through your financial institution's Web site. You can obtain smart Visa card readers from the financial institutions that issue the smart Visa cards or from a variety of consumer electronics retailers.

Visa is confident that smart card acceptance and usage will increase significantly in the United States. However, for this to happen, merchants must pay up to $1,000 for new point-of-sale (POS) devices that are smart-card-enabled. As a critical first step in providing smart card payments, Visa brought together key industry players, including Hypercom, Ingenico Fortronic, and VeriFone, to build smart card POS devices, networking equipment, and software. Importantly, all hardware and software had to conform to EMV (Europay Co., Master Card International, and Visa International) standards. These standards cover specifications for financial transaction systems and smart-card-based credit cards, and they must be adhered to for smart cards to be universally compatible in world markets.

Discussion Questions

1. What are the advantages of a smart card over an ordinary credit card or debit card? Are there any disadvantages?
2. Identify all the technology infrastructure components that are required to make the smart card from Visa operate effectively. Identify all of the IS technology component providers mentioned in this special-interest box and describe their role.

Critical Thinking Questions

3. Why would competitors Europay, MasterCard, and Visa work together to develop smart card standards?
4. Imagine that you are the owner of a small, local retail store with annual sales in the $2 to $5 million range. What would cause you to install three or four new POS terminals at a cost of $1,000 to accept the new smart cards?

Sources: Adapted from smart Visa Card at the Visa-USA Web site, http://www.usa.visa.com; Business Editors, "Visa U.S.A. Drives Industry Collaboration to Build Smart Card Acceptance in U.S.," *Business Wire*, May 7, 2001, http://www.businesswire.com; Lucas Mearian, "Visa Smart Card Technology Almost Ready for Prime Time," *Computerworld*, May 8, 2001, http://www.computerworld.com; Maria Trombly, "Visa Offers to Help E-Merchants Meet New Security Guidelines," *Computerworld*, March 2, 2001, http://www.computerworld.com.

and sale of counterfeit goods and the infringement of copyrights, trademarks, and patents. The entertainment and travel industries have their futures riding on the outcomes of these intellectual property battles.

Some argue that unscrupulous e-commerce companies threaten creativity by enabling people to steal the original works of innovators and artists. For example, in August 2002, the Recording Industry Association of America (RIAA) filed a lawsuit demanding that the four major Internet service providers (AT&T Broadband, Cable & Wireless USA, Sprint Corp., and WorldCom Inc.'s UUnet Technologies Inc. division) block access to a Web site that provided pirated music. The site, www.listen4ever.com, was registered in China, but the RIAA claimed that it targeted U.S. consumers because it was in English and featured free music from top-selling U.S. artists. The suit was dropped within a week when the site dropped off the Web.[25] In response to abuses, the life of patents has been increased (now lasting 20 years from date of application), copyrights now last 95 years, and more kinds of copyright infringements are now criminal offenses.

Another class of intellectual property lawsuits involves patents on business processes. For example, Overture Services filed suit against Google in April 2002 for allegedly infringing on a patent related to the company's bid-for-placement products. Overture's bid-for-placement search service allows companies to bid for search-result placement based on designated keywords. Advertisers pay Overture a premium to have their firm appear higher in the list of search items retrieved to attract traffic to their sites. Overture also sued FindWhat.com earlier in the year for allegedly infringing on the same patent.[26] Other companies have been sued because their Web site copies the "look and feel" of another company's site. Such lawsuits highlight a potentially debilitating problem: What is the extent of property rights in an information-based economy? The flurry of e-commerce patent lawsuits raises worries of overly aggressive protection of intellectual property rights and of stalling economic activity.

FRAUD

As more people use the Internet for e-commerce, they need to know that the merchant with whom they're dealing is legitimate. The Better Business Bureau's OnLine Reliability seal helps distinguish trustworthy companies among the thousands of businesses on-line, allowing consumers to easily identify them.

On-line swindlers include everyone from whiz-kid vandals to the equivalent of the old boiler-room brokers who hawk bogus offshore tax shelters. The good news is that you can enjoy the advantages of e-commerce and still keep yourself largely protected. Following are descriptions of the most common scams and some advice to help protect you.

On-Line Auction Fraud

On-line auctions brought in about $1 billion in sales in a recent year, and they represent the number-one Internet fraud, according to the National Consumers League. The majority of that fraud comes from so-called person-to-person auctions, which account for roughly half the auction sites. On these sites, it is up to the buyer and seller to resolve details of payment and delivery; the auction sites offer no guarantees. Sticking with auction sites like eBay (www.ebay.com) that ensure the delivery and quality of all the items up for bidding can help buyers avoid trouble.

Spam

spam
e-mail sent to a wide range of people and Usenet groups indiscriminately

E-mail that is sent to a wide range of people and usenet groups indiscriminately is called **spam**. Spam allows peddlers to hawk their products instantly to thousands of people at virtually no cost. And obtaining e-mail addresses to spam is now a snap, thanks to so-called *harvester programs* that snoop usenet chat groups and collect thousands of e-mail addresses in a single day. Do not respond to spam; responding will only confirm to the spammer that your e-mail address is accurate and active. Instead, simply delete spam as soon as you get it. If the spam is truly offensive or obviously fraudulent, forward the entire message to your Internet service provider, the Federal Trade Commission (www.ftc.gov), or the National Fraud Information Center (www.fraud.org) and ask that the sender be barred from sending additional messages.

In early 2001, AOL Time Warner sued the operators of a Web-based pornography company for allegedly sending millions of spam messages with sexual content to its subscribers. As you might suspect, such unsolicited messages angered subscribers and resulted in numerous complaints.[27] In an effort to combat spam, Aetna, the Hartford, Connecticut–based insurer, installed antispam software called InterScan from Trend Micro Inc. The software helps network administrators block spam and other unwanted e-mail on a network by checking incoming e-mail against a spam source and keyword list that is updated automatically over the Internet. Aetna also uses similar software from Cupertino, California–based Symantec Corp. and continues to do so because it believes that most security products fail eventually, in some way. But when they do, they don't all fail in the same way.[28]

Pyramid Schemes

Traditional pyramid schemes work by getting new investors to pay their recruiters to join the pyramid at the bottom. The new investors then (theoretically) get rich by recruiting additional new investors who will funnel money up the pyramid. Of course, eventually the people on the bottom of the pyramid have trouble finding new recruits, and it collapses. For that reason, pyramids are illegal. On-line pyramids often bundle the sale of a product or service like vitamins, credit cards, or even electricity to justify downstream recruitment fees. Usually, the product that these supposed multilevel-marketing (MLM) companies hawk is so overpriced or unwanted that the company relies mainly on recruiting fees to provide its cash flow. Be wary of any company that depends on recruitment fees to pay you. Also, avoid "opportunities" that force you to buy costly inventory. Finally, check with the Better Business Bureau (www.bbb.com) and your state attorney general's office before joining any company you're unsure about.

Investment Fraud

The North American Securities Administrators Association estimates that up to $10 billion will be lost in a year to investment fraud, primarily through the sale of bogus investments. Not all this fraud is committed on-line. Some of the more colorful scams involve eel farms, imported diamonds, and super-fast-growing trees that can reach 80 feet in three years. Call your state securities board to make sure the seller and its "security" are registered. Avoid dealing with unregistered securities and brokers.

Stock Scams

Before the Internet, most individuals had a tough time spreading rumors about a stock, but now anybody can do it. Net chat rooms like usenet's misc.invest.stocks make it cheap and easy for scammers to talk up (or down) a stock based on false information. The scammer buys the stock at a low price, spreads false rumors that help drive the stock price up, and then sells at an artificially high price before the bottom falls out. Always verify tips and information you get on-line and in chat rooms with the company directly or with reputable financial advisers. To be truly safe, avoid thinly traded stocks or penny stocks that sell for less than $1 a share, because they are the most easily manipulated.

INVASION OF CONSUMER PRIVACY

On-line consumers are more at risk today than ever before. One of the primary factors causing higher risk is *on-line profiling*—the practice of Web advertisers' recording on-line behavior for the purpose of producing targeted advertising. **Clickstream data** is the data gathered based on the Web sites you visit and what items you click on. From the marketers' perspective, the use of on-line profiling allows businesses to market to customers electronically one to one. The benefit to customers is better, more effective service; the benefit to providers is the increased business that comes from building relationships and encouraging customers to return for subsequent purchases. From the consumers' perspective, on-line profiling squeezes out anonymity, which remains crucial to privacy on the Internet. And consumers also fear that the personal information thus gathered will be shared with others without their knowledge.

clickstream data
data gathered based on the Web sites you visit and what items you click on

FIGURE 8.9

TRUSTe Seal

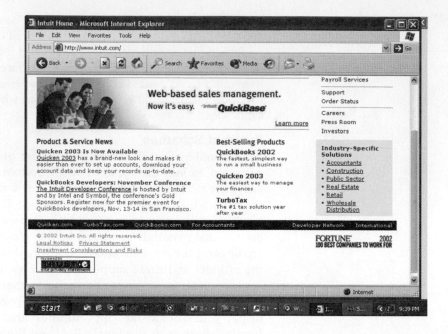

For example, a number of lawsuits have been filed against on-line marketers who intended to create massive data warehouses of consumer information by linking clickstream data with data (such as automobile registrations and product warranty registrations) from other sources.

The U.S. government has not implemented a federal privacy policy and instead relies on e-commerce self-regulation in matters of data privacy. The Better Business Bureau Online and TRUSTe are independent, nonprofit privacy initiatives dedicated to building users' trust and confidence on the Internet and supporting the accelerating growth of e-commerce. The BBB On-line Privacy seal program, which helps consumers identify on-line businesses that honor privacy protection policies, covers over 5,000 Web sites and has hundreds of applications in process. Figures 8.9 and 8.10 show the TRUSTe and BBB On-line Privacy seals, respectively. The TRUSTe program is based on a multifaceted assurance process that establishes Web site credibility, thereby making users more comfortable when making on-line purchases and providing personal information.

FIGURE 8.10

BBB On-line Privacy Seal

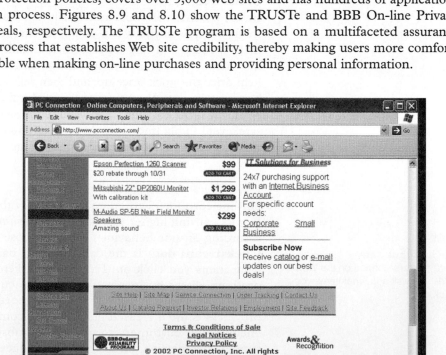

Tip	Rationale
Never give out personal information, especially your social security number.	Assume that any information you give out about yourself will appear on a database somewhere. Even something as innocuous as a warranty card may ask for income and phone numbers—don't provide such data.
Before registering with a Web site, check out its privacy policy.	Look for the BBB or TRUSTe seal. If a site doesn't explain how it plans to use your information, beware!
Be discreet in sending e-mail, posting messages to a usenet newsgroup, and "talking" in chat rooms.	Anything you send in these forums can be viewed and used by a marketer to update your personal profile.
Update your browser software.	The newest versions of Netscape Navigator and Microsoft Internet Explorer, by far the two most popular Web browsers, have the best encryption technology available. Generally speaking, if these companies release an update, someone has found a hole in their security.
Consider the purchase of a digital ID to use in place of your real name.	This permits users to retrieve Web pages anonymously.

TABLE 8.3

How to Protect Your Privacy While On-Line

safe harbor principles
a set of principles that address the e-commerce data privacy issues of notice, choice, and access

A set of **safe harbor principles** has been established that address the issues of notice, choice, and access associated with data privacy. A company following these principles must notify consumers of the purpose of data collection, allow consumers to opt out of having their data shared with third parties, and provide users with access to their personal information to review and possibly correct it. With an opt-out policy, the Web site is free to gather and sell information about you unless you specifically tell it that it cannot, which typically involves clicking on a button. With an opt-in policy, the gathering or selling of your data is forbidden unless you specifically give the Web site permission. The safe harbor principles also require organizations using personal information to "take reasonable precautions to protect it from loss, misuse, and unauthorized access, disclosure, alteration, and destruction." Table 8.3 lists several tips to help safeguard your privacy.

STRATEGIES FOR SUCCESSFUL E-COMMERCE

With all the constraints to e-commerce just covered, it is important for a company to develop an effective Web site—one that is easy to use and accomplishes the goals of the company, yet is affordable to set up and maintain. We cover several issues for a successful e-commerce site here.

DEVELOPING AN EFFECTIVE WEB PRESENCE

Most companies are still developing a Web presence—more than half of U.S. companies have in-house Web sites but less than 25 percent actually sell anything over the Web. When building a Web site, the first thing to decide is which tasks the site must accomplish. Most people agree that an effective Web site is one that creates an attractive presence and that meets the needs of its visitors, including the following:

- Obtaining general information about the organization
- Obtaining financial information for making an investment decision
- Learning the organization's position on social issues
- Learning about the products or services that the organization sells
- Buying the products or services that the company offers
- Checking the status of an order
- Getting advice or help on effective use of the products

- Registering a complaint about the organization's products
- Registering a complaint concerning the organization's position on social issues
- Providing a product testimonial or idea for a product improvement or new product
- Obtaining information about warranties or service and repair policies for products
- Obtaining contact information for a person or department in the organization

Once a company determines which objectives its site should accomplish, it can proceed to the details of actually putting up a site.

PUTTING UP A WEB SITE

Companies large and small can establish Web sites. Previously, companies had to develop their sites in-house or find and hire contractors to develop their sites. But no longer must you learn the intricacies of HTML or Java, master Web design software, or hire someone to build your site. Web site hosting services and the use of storefront brokers are two options for designing, building, operating, and maintaining a Web site. Both options offer the advantage of getting the Web site running faster and cheaper than doing it yourself, especially for a firm with few or no experienced Web developers.

Web Site Hosting Services

Web site hosting companies
companies that provide the tools and services required to set up a Web page and conduct e-commerce within a matter of days and with little up-front cost

Web site hosting companies such as Bigstep.com and freemerchant.com have made it possible to set up a Web page and conduct e-commerce within a matter of days and with little up-front cost. Such companies have packaged all the basic development tools and made them available for free. Using only a browser, you can choose a Web site template suited to your specific type of business, whatever it may be—clothing, collectibles, or sports equipment. Using the tools provided, you can write the descriptive text and add images of the products you want to sell. Within hours, your site is up on the Web and accessible to millions.

These companies can also provide free hosting for your store, but to allow visitors to pay for merchandise with credit cards, you need a merchant account with a bank. If your company doesn't already have one, then it must establish one. Typically there is a monthly charge of $10 to $50 plus a transaction fee of $.05 to $.25, and it takes at least a few days to establish a merchant account. Without such an account, you must accept payment by check or money order. In many cases, your site won't have its own distinct Web address—www.wearmyhats.com, for instance. Instead, your store will appear as a subsection of your service provider's URL.

Storefront Brokers

storefront broker
companies that act as middlemen between your Web site and on-line merchants that have the products and retail expertise

Another model for setting up a Web site is the use of a **storefront broker**, which serves as a middleman between your Web site and on-line merchants that have the actual products and retail expertise. At sites such as Bigstep (www.bigstep.com) or Vstore (www.vstore.com), you pick and choose what to sell to match the themes of your site and the products you think might interest your visitors. For example, you can set up to be a storefront operator with Vstore.com and build your own on-line store with its own unique URL. You create the product categories, choose the products, and customize the store as you see fit. You can build links to your own pages or other sites to display the products, creating, in effect, a virtual storefront stocked with merchandise that is actually handled by another on-line merchant.

The storefront broker deals with the details of the transactions, including who gets paid for what, and is responsible for bringing together merchants and reseller sites. The storefront broker is similar to a distributor in standard retail operations, but in this case no product moves—only electronic data flows back and forth. Products are ordered by a customer at your site, orders are processed through a user interface provided by the storefront broker, and product is shipped by the merchant.

Storefront brokers make their money by taking a commission from the merchant—anywhere from 5 percent to 25 percent—or by collecting a finder's fee per customer from the merchant. The storefront operator usually takes a commission, again in the range of 5 to 25 percent. Although these brokered sales represent a loss of margin for the original merchant, the increased volume can often make up for lost margins. By multiplying the number of outlets for the products—sometimes by the thousands—merchants stand to make more money.

BUILDING TRAFFIC TO YOUR WEB SITE

The Internet includes hundreds of thousands of e-commerce Web sites. With all those potential competitors, a company must take strong measures to ensure that the customers it wants to attract can find its Web site. The first step is to obtain and register a domain name. It helps if your domain name says something about your business. For instance, stuff4u might seem to be a good catchall, but it doesn't describe the nature of the business—it could be anything. If you want to sell soccer uniforms and equipment, then you'd try to get a domain name like www.soccerstuff4u.com, www.soccerequipment.com, or www.stuff4soccer.com. The more specific the Web address, the better.

meta tag
a special HTML tag, not visible on the displayed Web page, that contains keywords representing your site's content, which search engines use to build indexes pointing to your Web site

The next step to attracting customers is to make your site search-engine-friendly by including a meta tag in your store's home page. A **meta tag** is a special HTML tag, not visible on the displayed Web page, that contains keywords representing your site's content, which search engines use to build indexes pointing to your Web site. Again, the selection of keywords is critical to attracting customers, so they should be chosen carefully.

You can also use Web site traffic data analysis software to turn the data captured in the Web log file into useful information. This data can tell you the URLs from which your site is being accessed, the search engines and keywords that find your site, and other useful information. Using this data can help you identify search engines to which you need to market your Web site, allowing you to submit your Web pages to them for inclusion in the search engine's index.

MAINTAINING AND IMPROVING YOUR WEB SITE

Web site operators must constantly monitor the traffic to their site and the response times experienced by visitors. Nothing will drive potential customers away faster than if they experience unbearable delays in trying to view or order your products or services. It may be necessary over time to modify the software, databases, or hardware on which the Web site runs to ensure good response times.

Web site operators must also continually be alert to new trends and developments in the area of e-commerce and be prepared to take advantage of new opportunities. For example, recent studies show that customers more frequently visit Web sites they can customize. Users who personalize such sites also more frequently subscribe to paid sites, use on-line bill payment services, and promote products via e-mail to their friends, all of which make them an advertiser's dream. *The Wall Street Journal* spent $28 million to overhaul its WSJ.com site to attract higher-value, repeat customers by letting them more easily customize their home pages with columns, stock quotes, and other regular WSJ.com features.[29]

Personalization is another approach for increasing traffic to a Web site. Personalization is the process of tailoring Web pages to specifically target individual consumers. The goal is to meet the customer's needs more effectively and efficiently, make interactions faster and easier, and, consequently, increase customer satisfaction and the likelihood of repeat visits. Building a better understanding of customer preferences also can aid in cross-selling related products and selling more expensive products. The most basic form of personalization involves simply using the consumer's name in an e-mail campaign or in a greeting on the Web page. A more advanced form of personalization is that employed by Amazon.com, where each repeat customer is greeted by name and a list of new product recommendations based on the customer's previous purchases.

Businesses use two types of personalization techniques to capture data and build customer profiles. *Implicit personalization* techniques capture data from actual customer Web sessions—primarily, which pages were viewed and which ones weren't. *Explicit personalization* techniques capture user-provided information, such as information from warranties, surveys, user registrations, and contest entry forms filled out on-line.[30] Data can also be gathered through access to other data sources such as the Bureau of Motor Vehicles, Bureau of Vital Statistics, and marketing affiliates (firms that share marketing data). Marketing firms aggregate this information to build databases containing a huge amount of consumer behavioral data. During each customer interaction, both types of data are analyzed in real time using powerful algorithms to predict the consumer's needs and interests. This analysis makes it possible to deliver new, targeted information before the customer moves on. Because personalization depends on the gathering and use of personal user information, privacy issues are a major concern.

These tips and suggestions are only a few ideas that can help a company set up and maintain an effective e-commerce site. With technology and competition changing continually, managers should read articles in print and on the Web to keep up to date on ever-evolving issues.

SUMMARY

PRINCIPLE *E-commerce is a new way of conducting business, and as with any other new application of technology, it presents both opportunities for improvement and potential problems.*

Businesses and individuals use e-commerce to reduce transaction costs, speed the flow of goods and information, improve the level of customer service, and enable the close coordination of actions among manufacturers, suppliers, and customers. E-commerce also enables consumers and companies to gain access to worldwide markets.

Business-to-business e-commerce allows manufacturers to buy at a low cost worldwide, and it offers enterprises the chance to sell to a global market right from the start.

Although it is gaining acceptance, business-to-consumer, or e-commerce for consumers, is still in its early stages. Yet, e-commerce offers great promise for developing countries, helping them to enter the prosperous global marketplace, and hence helping reduce the gap between rich and poor countries. By using business-to-consumer e-commerce to sell directly to consumers, a producer or provider of consumer services can eliminate the middlemen, or intermediaries, between them and the end consumer. In many cases, this squeezes costs and inefficiencies out of the supply chain and can lead to higher profits and/or lower prices for consumers.

Consumer-to-consumer (C2C) e-commerce involves consumers selling directly to other consumers.

A successful e-commerce system must address the many stages consumers experience in the sales life cycle. At the heart of any e-commerce system is the ability of the user to search for and identify items for sale; select those items; negotiate prices, terms of payment, and delivery date; send an order to the vendor to purchase the items; pay for the product or service; obtain product delivery; and receive after-sales support.

The first major challenge is for the company to define an effective e-commerce strategy. Although there are a number of different approaches to e-commerce, the most successful models include three basic components: community, content, and commerce. Another major challenge for companies moving to business-to-consumer e-commerce is the need to change their distribution systems and work processes to be able to manage shipments of individual units directly to consumers and deal with split-case distribution. A third tough challenge for e-commerce is the integration of new Web-based order processing systems with traditional mainframe computer-based inventory control and production planning systems.

Supply chain management is composed of three subprocesses: demand planning to anticipate market demand, supply planning to allocate the right amount of enterprise resources to meet demand, and demand fulfillment to fulfill demand quickly and efficiently. Conversion to e-commerce supply chain management provides businesses an opportunity to achieve excellence by increasing revenues, decreasing costs, improving customer satisfaction, and reducing inventory. But to achieve this goal requires integrating all subprocesses that exchange information and move goods between suppliers and customers, including manufacturers, distributors, retailers, and any other enterprise within the extended supply chain.

The use of the Internet is growing rapidly in markets throughout Europe, Asia, and Latin America. The first step in developing a global e-commerce strategy is to determine which global markets make the most sense for selling products or services on-line. Once a company decides which global markets it wants to reach with its Web site, it must adapt an existing U.S.-centric site to another language and culture—a process called *localization*. One of the most important and most difficult decisions in a company's global Web strategy is whether Web content should be generated and updated centrally or locally. Companies must also be aware that consumers outside the United States will access sites with different devices and modify their site design accordingly.

Many manufacturers and retailers have outsourced the physical logistics of delivering merchandise to cybershoppers. To provide this service, delivery firms have developed software tools and interfaces that directly link customers' ordering, manufacturing, and inventory systems with their own system of highly automated warehouses, call centers, and worldwide shipping network. The goal is to make the transfer of all information and inventory—from the manufacturer to the delivery firm to the consumer—fast and simple.

Electronic retailing (e-tailing) is the direct sale from business-to-consumer through electronic storefronts designed around an electronic catalog and shopping cart model. A cybermall is a single Web site that offers many products and services at one Internet location. Manufacturers are joining electronic exchanges where they can join with competitors and suppliers to use computers and Web sites to buy and sell goods, trade market information, and run back-office operations such as inventory control. They are also using e-commerce to improve the efficiency of the selling process by moving customer queries about product availability and prices on-line. The nature of the Web allows firms to gather much more information about customer behavior and preferences than they could using other marketing approaches. This new technology has greatly enhanced the practice of market segmentation and enabled companies to establish closer relationships with their customers. Detailed information about a customer's behavior, preferences,

needs, and buying patterns allow companies to set prices, negotiate terms, tailor promotions, add product features, and otherwise customize a relationship with a customer. The Internet has also revolutionized the world of investment and finance, especially on-line stock trading and on-line banking. The Internet has also created many options for electronic auctions, where geographically dispersed buyers and sellers can come together.

PRINCIPLE *E-commerce requires the careful planning and integration of a number of technology infrastructure components.*

A number of infrastructure components must be chosen and integrated to support a large volume of transactions with customers, suppliers, and other business partners worldwide. These components include hardware, Web server software, e-commerce software, and network and packet switching.

Current e-commerce technology relies on the use of identification and encryption to safeguard business transactions. Web site operators must protect against a denial-of-service attack, where the attacker takes command of many computers on the Internet and causes them to flood the target Web site with requests for data and other small tasks that keep the target machine too busy to serve legitimate users. A digital certificate is an attachment to an e-mail message or data embedded in a Web page that verifies the identity of a sender or a Web site. To help prevent the theft of credit card numbers and banking information, the secure sockets layer communications protocol is used to secure all sensitive data. Actual payments are made in a variety of ways including electronic cash, electronic wallets, and smart, credit, charge, and debit cards.

PRINCIPLE *Users of the new e-commerce technology must take safeguards to protect themselves.*

As with any new revolutionary change, there are a host of issues associated with e-commerce. Among these issues are security, intellectual property rights, fraud, and privacy. E-commerce shoppers must be on constant guard to protect their rights, security, and personal privacy.

PRINCIPLE *Organizations must define and execute a strategy to be successful in e-commerce.*

Most people agree that an effective Web site is one that creates an attractive presence and meets the needs of its visitors. These are many and varied. E-commerce start-ups must choose whether they will build and operate the Web site themselves or outsource this function. Web site hosting services and storefront brokers provide options to building your own Web site. It is also critical to build traffic to your Web site by registering a domain name that is relevant to your business, making your site search-engine-friendly by including a meta tag in your home page, and using Web site traffic data analysis software to attract additional customers. Web site operators must constantly monitor the traffic and response times associated with their site and adjust

software, databases, and/or hardware to ensure that visitors have a good experience when they visit the site.

Personalization is the process of tailoring Web pages to specifically target individual consumers. The goal is to meet the customer's needs more effectively and efficiently, make interactions faster and easier, and, consequently, increase customer satisfaction and the likelihood of repeat visits.

CHAPTER 8 SELF-ASSESSMENT TEST

E-commerce is a new way of conducting business, and as with any other new application of technology, it presents both opportunities for improvement and potential problems.

1. Which of the following statements is NOT a reason for the success of UK-based on-line grocer Tesco?
   A. They have a profit margin of 8% compared with 2% for most U.S.-based grocers.
   B. They made a multimillion-dollar investment in automated distribution facilities.
   C. Workers, not automated conveyors, are used to pick items to fill customer orders.
   D. Orders are routed directly from the Web site to the customer's local Tesco store.

2. eBay is an example of what form of e-commerce?
   A. A2B
   B. B2B
   C. C2B
   D. C2C

3. On-line sales of products to consumers now exceed traditional retail sales. True False

4. Business processes that are strong candidates for conversion to e-commerce are those that are paper based and time consuming and those that can make business more convenient for customers. Thus, it comes as no surprise that the first business processes that companies converted to an e-commerce model were those related to _____ and _____.

5. E-commerce is expected to grow at a faster rate in countries outside North America than within the U.S. True False

E-commerce requires the careful planning and integration of a number of technology infrastructure components.

6. A multistage model for purchasing over the Internet includes search and identification, selection and negotiation, purchasing, product or service delivery, and _____ _____.

7. _____ is the sending of software, music, pictures, and written material through the Internet; it is faster and can be less expensive than with regular order processing.

8. The physical logistics of delivering merchandise to cybershoppers—the storing, packing, shipping, and tracking of products—is so critical to the success of an organization that few manufacturers and retailers have outsourced these activities. True False

9. Which of these is a challenge that must be overcome for a company to convert its business processes from the traditional form to business-to-consumer e-commerce processes?
   A. Define an effective e-commerce model and strategy.
   B. Change distribution systems and work processes to be able to manage shipments of individual units directly to consumers
   C. Integrate new Web-based order processing systems with traditional mainframe computer–based inventory control and production planning systems.
   D. All of the above

10. In addition to the Web server operating system, each e-commerce Web site must have Web server _____ to perform a number of fundamental services, including security and identification, retrieval and sending of Web pages, Web site tracking, Web site development, and Web page development.

11. Which of the following contain variable information and are built in response to a specific Web visitor's request?
    A. static pages
    B. artificial pages
    C. dynamic pages
    D. virtual pages

Users of the new e-commerce technology must take safeguards to protect themselves.

12. An attachment to an e-mail message or data embedded in a Web site that verifies the identity of a sender or Web site is called
    A. digital certificate
    B. computer virus
    C. encryption code
    D. none of the above

13. To help prevent the theft of credit card numbers from on-line shoppers, the _____ communications protocol is used to secure sensitive data. It sits above the TCP layer of the OSI model and other protocols such as Telnet and HTTP can be layered on top of it.

14. A smart card makes credit theft practically impossible because a key to unlock the encrypted information is required, and there is no external number that a thief can identify and no physical signature a thief can forge. True False

15. Helps consumers identify on-line businesses that honor privacy protection policies, covers over 5,000 Web sites and has hundreds of applications in process.
    A. ICRA rating system
    B. TRUSTe and BBB privacy seals
    C. safe harbor principles
    D. electronic certificate

Organizations must define and execute a strategy to be successful in e-commerce.

16. A _____ is not visible on a displayed Web page but contains keywords that search engines use to direct users to a Web site.
    A. cyberlink
    B. direct index
    C. meta tag
    D. traffic controller

Chapter 8 Self-Assessment Test Answers:

1. b; 2. d; 3. False; 4. buying, selling; 5. True; 6. after-sales support; 7. Electronic distribution; 8. False; 9. d; 10. software; 11. c; 12. a; 13. secure sockets layer (SSL); 14. True; 15. b; 16. c.

KEY TERMS

REVIEW QUESTIONS

1. Define the term *e-commerce*. Identify and briefly describe three different forms of e-commerce. Which form is the largest in terms of dollar volume?

2. What sort of business processes are good candidates for conversion to e-commerce?

3. A major challenge for companies moving to business-to-consumer e-commerce is the need to change distribution systems and work processes to be able to manage shipments of individual units directly to consumers. What sort of changes are required and why?

4. Briefly describe the three subprocesses that make up supply chain management.

5. What are some of the business opportunities presented by business-to-business e-commerce? What are some of the key issues?

6. Briefly describe the multistage model for purchasing over the Internet.

7. Distinguish between EDI and e-commerce.

8. Briefly describe how on-line banking works.

9. What is an electronic exchange? How does it work?

10. What are some of the issues associated with the use of electronic exchanges?

11. What are some of the advantages of using a service like Bigstep.com to set up a Web site? What are some of the disadvantages?

12. Why is it necessary to continue to maintain and improve an existing Web site?

13. What role do digital certificates and certificate authorities play in e-commerce?

14. What is the secure sockets layer and how does it support e-commerce?

15. Briefly explain the differences among smart, credit, charge, and debit cards.

16. What actions can you take to minimize the risk of being a victim of e-commerce fraud?

17. What actions can you take to safeguard your individual privacy as you surf the Web?

18. What is meant by localization of a global Web site?

DISCUSSION QUESTIONS

1. Why is it important in effective e-commerce for front-end Web-enabled applications like order taking to be tightly integrated to back-end applications like inventory control and production planning?

2. Why are many manufacturers and retailers outsourcing the physical logistics of delivering merchandise to shoppers? What advantages does such a strategy offer? Are there any potential issues or disadvantages?

3. Distinguish between a B2B and B2C e-commerce company.

4. Would you agree with the statement that e-commerce is dead or dying? Why or why not?

5. Wal-Mart, the world's number one retail chain, has turned down several invitations to join exchanges in the retail and consumer goods industries. Is this good or bad for the overall U.S. economy? Why?

6. What are some of the ethical and societal issues associated with integrating clickstream, demographic, and transactional data into one comprehensive data warehouse? What actions can you take to minimize these issues?

7. Briefly discuss actions e-commerce companies have taken to ensure the legitimacy of their customers. Which of these approaches do you think is best? Why?

8. Discuss the pros and cons of e-commerce companies capturing data about you as you visit their sites.

9. Do you think that companies such as Pico-Pay that provide users with access to subscription content by enabling them to build up credits by viewing advertisements will be successful? Why or why not?

10. Outline the key steps in developing a corporate global e-commerce strategy.

PROBLEM-SOLVING EXERCISES

1. As a team, develop a set of criteria you would use to evaluate various business-to-consumer Web sites on the basis of ease of use, protection of consumer data, security of payment process, etc. Develop a simple spreadsheet containing these criteria and use it to evaluate five different Web sites.

2. Do research to get current data about the growth of B2B, B2C, or C2C e-commerce—either in the United States or worldwide. Use a graphics software package to create a line graph representing this growth. Extend the growth line five years beyond the available data using two different modeling tools available with the software package. Write a paragraph discussing the issues and assumptions that affect the accuracy of your five-year projection and the likelihood that e-commerce will achieve this forecast.

TEAM ACTIVITIES

1. Split up your team and do research to identify as many organizations as possible that have as one of their goals safeguarding individual data privacy. Try to document each organization's position on on-line profiling. Next, choose a country (other than the United States) and document its position on on-line privacy laws. Report your findings back to the team for group discussion.

2. As a team, choose an idea for a Web site—products or services you would provide. Develop an implementation plan that outlines the steps you need to take and the decisions you must make to set up the Web site and make it operational.

3. Find a Web site that provides portfolio tracking. As a team, select 5 stocks to make an "imaginary" purchase and allocate $100,000 among them. Check back in one week and determine the change in value in your investment. Make a list of the various features and analyses that are available at this Web site. Compare the change in the value of your portfolio with that of other teams. Compare the various features and analyses available at your Web site with the sites used by other teams.

WEB EXERCISES

1. Access the Web sites of two package-delivery services and document the services they provide for cybershoppers in terms of product delivery and return. Which company offers the superior services? Why do you think so?

2. Identify and visit the Web sites of three public exchanges that are owned and operated by an industry group. Write a brief summary of each exchange outlining its purpose, members, and information about its operation.

3. Do research to identify three companies that provide biometric security products. Briefly summarize each firm's products and current applications.

CASES

CASE 1

DaimlerChrysler Joins Covisint Exchange

Procurement decisions affect long-term supplier relations and directly determine product costs and quality. DaimlerChrysler, the German-based company, is using e-commerce to optimize the core tasks of supply and parts procurement, including the analysis of information on suppliers and markets; preparation of procurement criteria with product development and logistics; price research procedures; and contract management. Improving the procurement process has the potential to significantly reduce processing times and expenditures, which can translate into lower costs and reduced time to market.

Recognizing the potential benefits of a streamlined procurement system, Ford and General Motors sought to establish competing automotive procurement marketplaces (Ford's AutoXchange and GM's TradeXchange). The two companies eventually dropped those plans and decided to work with Covisint to develop a worldwide portal for the automobile industry that would ensure a standard method of connecting suppliers and manufacturers on the Internet—from development through logistics. DaimlerChrysler (Germany), Renault (France), Nissan (Japan), and Peugeot Citroën (France) agreed to join Ford and GM in the development of Covisint and to use this electronic trade exchange. Automobile industry suppliers benefit from the industry standard electronic marketplace as well by eliminating time-consuming and costly separate interfaces and applications to the various manufacturers and their own suppliers.

The Covisint exchange uses Commerce One's MarketSite software to process transactions, conduct on-line auctions, and manage the content of a parts catalog. Covisint uses Oracle's enterprise resource planning applications to run its internal operations and Oracle's Exchange Marketplace to provide security, single sign-on, and registration capabilities. (Single sign-on is a session/user authentication process that permits a user to access all the applications they have been given the rights to on the server, and eliminates future authentication prompts when the user switches applications during that particular session.) The portal integrates all original equipment manufacturers and supplier applications worldwide via one common framework. Covisint's backers hope to channel more than $300 billion in annual transactions through the exchange.

Over a 12-month period, DaimlerChrysler managed 512 on-line bidding events valued at $9 billion using Covisint.

In May 2001, DaimlerChrysler staged an on-line bidding event with an order volume of $3 billion in just four days. In total, 43 percent of the total value of the parts for a future Chrysler model series was negotiated on-line with over 50 on-line bidding events in the third quarter of 2001 alone.

The business benefits from finding new suppliers and reducing the price of products are substantial; however, the biggest benefits come from reducing the time it takes to close deals. DaimlerChrysler (and the other automakers) are trying to reduce the time required to get new products to market. The use of e-commerce has reduced the time required for order placement by up to 80 percent, cutting both time and business process costs. As a result, DaimlerChrysler's management stated that the economic effects achieved with e-commerce in the first year of implementation had already covered the costs of its investment in e-commerce and that it holds great potential for the future, too.

Discussion Questions
1. Why would Ford and GM scrap plans to build their own exchanges in favor of an industrywide standard electronic marketplace?
2. Are there any disadvantages or potential problems in the use of the Covisint electronic exchange?

Critical Thinking Questions
3. If the major automobile manufacturers and parts suppliers all use the same industrywide electronic marketplace, is it possible for DaimlerChrysler to gain an advantage over the others in the important area of procurement? Why or why not?
4. What issues might there be for a small parts supplier in getting connected to Covisint?

Sources: Adapted from "e-Business at DaimlerChrysler Is Paying Off: Savings Exceed Present Investment," February 4, 2002, DaimlerChrysler Web site at http://www.daimlerchrysler.de; "DaimlerChrysler Meets Combined Global Procurement Requirements for PCs via Online Bidding Event," November 5, 2001, DaimlerChrysler Web site at http://www.daimlerchrysler.de; "DaimlerChrysler Selects Covisint to Develop New Global Supplier Portal," January 23, 2002, DaimlerChrysler Web site at http://www.daimlerchrysler.de; Lee Copeland, "Covisint's Stalled Start," *Computerworld*, December 17, 2001, http://www.computerworld.com; Lee Copeland, "Covisint Technology Partners Sign Equity Agreements," *Computerworld*, January 1, 2001, http://www.computerworld.com; Heather Harreld, "Covisint Taps MatrixOne for Collaboration Tools," *Computerworld*, September 5, 2001, http://www.computerworld.com.

CASE 2

Lowe's Fights Tooth and Nail to Be No. 1

Lowe's employs over 100,000 people and is the 14th largest U.S. retailer, with more than 750 stores in 42 states. Lowe's caters to the Do-It-Yourselfer, as well the commercial business customer. Its stores carry more than 40,000 home improvement items including plumbing and electrical products, tools, building materials, hardware, outdoor equipment, appliances, lumber, nursery and gardening products, millwork, paint, sundries, cabinets, and furniture.

Annual sales exceed $22 billion, making the company No. 2 behind industry leader The Home Depot. However, competition is heating up. Home Depot is looking over its shoulder at a rival that is coming on strong. Lowe's is in the midst of an aggressive $2 billion expansion plan, opening more than two stores every week. Its new superstores are the largest in the industry—at approximately 150,000 square feet of retail space. And Lowe's is a leader among home improvement stores in having a presence on the Web—beating The Home Depot by 5 months in offering its products on-line.

While neither company provides on-line sales figures, both acknowledge that the volume is small so far. It is clear, however, that both have identified e-commerce as a growth area and are investing in it even as they scale back in some other areas. Competition for in-store sales is getting so intense that the incremental business gained from a well-designed Web site can make a difference in the battle for No. 1.

For its part, Lowe's is continuing to build its business on the Internet by opening a new portal, Accent & Style, which offers home decorating tips. The goal is for this site to drive on-line sales and draw people into stores. Lowe's executives believe that the biggest opportunity on the Internet is to educate customers and prepare them to make buying decisions.

Most shoppers still want to see and touch what they're buying, but there is evidence that appliances—a major growth category for Lowe's and Home Depot—can sell well on the Internet. Maytag, the nation's third largest appliance maker, launched its Maytag.com Web site in January 2001 and has derived 70 percent of its sales from major appliances costing $600 or more. Lowe's understands that the Web has its limits, and so it carries only about 35 percent of its total store inventory on-line.

Lowe's has its eye on controlling the fulfillment costs of on-line orders to ensure that its e-commerce initiative becomes profitable. Lowe's customers can choose to have items delivered or pick them up at the store. Lowe's contracts with NFI Interactive, a logistics company, to make its home deliveries. Often, orders above a certain amount are delivered for no additional charge. For customer pickups, Lowe's store personnel package orders that come in from the Web.

Discussion Questions

1. Do some quick research to compare total sales for Lowe's versus The Home Depot for the past two years. Which one is number 1 in sales? Is the gap closing or increasing?
2. Visit the Web site of each company. Write a brief paragraph describing your on-line experience as you try to find the closest store and purchase some items of interest to you. Which Web site would you rate as being better? Why?

Critical Thinking Questions

3. Do you agree with the Lowe's company executives who think that the biggest opportunity on the Internet is to educate customers and prepare them to make buying decisions? Or should increasing on-line sales be the goal of the company Web site?
4. What is your opinion in terms of how important an on-line presence is for Lowe's?

Sources: Adapted from "About Lowe's," Lowe's Web site, http://www.lowes. com, accessed on April 7, 2002; Eric Young, "Home Improvement Chains Battle Online," *Computerworld*, May 14, 2001, http://www.computerworld. com; Amy Tsao, "How Home Depot and Lowe's Measure Up," *Business Week Online*, December 5, 2001, http://www.businessweek.com; Steve Ulfelder, "The Web's Last Gap," *Computerworld*, June 18, 2001, http://www. computerworld.com.

CASE 3

Port of Seattle Turns to E-Commerce

The Port of Seattle is the Northwest's largest seaport and airport, and it employs over 1,750 workers. Its Seattle-Tacoma International Airport processes some 28 million passengers a year, and its deepwater harbor is the fifth largest container port in the United States. (A container is a large, rectangular box used to transport goods on ships, trucks, or railroad cars.) Managers at the Port of Seattle developed a strategy to use e-commerce to help the airport better handle the increasing service expectations of airline customers and become more competitive with seaports across the country for the approximately $740 billion in cargo that flows through the U.S. annually.

The Port of Seattle chose IBM as a partner to help it implement this new strategy. Together they completed the first phase of the port's e-business initiative in April 2001. This phase developed the port's vision and strategy, assessed existing technology, and broadly defined infrastructure needs.

It also defined some initial e-commerce projects, including internal expense reporting, construction document management to support the $3 billion capital improvement program at Seattle-Tacoma International Airport, an on-line parking payment system at the airport, and on-line reservation systems for various port facilities and services.

The Port of Seattle commissioners initially budgeted $11 million to cover hardware, software, and implementation costs needed for these projects. The budget may need to be revised as they define the scope and requirements for each project. As part of its contract with the port, IBM is deploying consulting teams, project management staff, business analysts, and technology gurus to help the port complete these projects. Additional contracts may be negotiated as the port's e-business initiative continues.

The initial e-commerce projects will provide greater convenience and service to the port's customers. For example, the on-line reservation system for port facilities such as Bell Harbor Marina is now completed. It allows boat operators to make reservations and pay for moorage over the Internet, at any time that is convenient to them. Marina staff spend less time manually entering reservations and more time serving customers at the marina. Easy-to-follow instructions allow boaters to create vessel profiles, select and confirm reservation dates and boat slips, and make credit card payments for the first night's stay. The system can be used to make reservations for a single boat, a charter vessel, or a group of boats, day or night.

Discussion Questions

1. The Port of Seattle wisely divided the work to be done into phases with the first phase focused on developing a vision, assessing existing technology, defining infrastructure needs, and outlining future projects. What do you think is the goal of the second phase of this work?
2. What sort of benefits might the port gain from implementation of the remaining initial e-commerce projects?

Critical Thinking Questions

3. How do you think the initial five e-commerce projects were identified? Which one or two projects might be more strategic for the port?
4. Discuss the pros and cons of using an outside firm like IBM to help define and implement the port's e-commerce projects.

Sources: Adapted from "Click into a Slip at Bell Harbor Marina," Port of Seattle press release, February 25, 2002, http://www.portseattle.org; Business Editors/Hi-Tech Writers, "Port of Seattle Taps IBM as Its E-Business Technology Consultant; Agreement Will Raise the Stakes in Competition Among Ports," *BusinessWire*, August 16, 2001; Linda Rosencrance, "Port of Seattle Taps IBM as E-Business Partner," *Computerworld*, August 21, 2002, http://www.computerworld.com.

NOTES

Sources for the opening vignette on p. 319: Adapted from "Preliminary Statement of Results 52 Weeks Ending 23 February 2002," Corporate Info found at http://www.tesco.com; Christopher T. Heun, "Grocers Count on IT to Keep Cash Registers Ringing," *InformationWeek*, December 24, 2001, accessed at http://www.informationweek.com; Todd R. Weiss, "Online Grocer Webvan Crashes with a Thud," *Computerworld,* July 16, 2001, accessed at http://www.computerworld.com; Todd R. Weiss, "HomeRuns.com Latest Online Grocer to Bow Out," *Computerworld*, July 13, 2001, accessed at http://www.computerworld.com.

1. Don Blancharski, "Is E-Commerce Dead?," *ITT World.com*, October 23, 2001, accessed at http://www.itworld.com.
2. Linda Rosencrance, "Online Sales Hit $10 Billion in Q4 2001," *Computerworld*, February 20, 2002, accessed at http:// www.computer world.com.
3. Jane Black, "Online Extra: Where the Web Is Really Revolutionizing Business," *Business Week Online*, August 27, 2001, accessed at http:// www.businessweek.com.

4. Linda Rossetti, "The Big Bounce," *Computerworld*, March 12, 2001, accessed at http://www.computerworld.com.

5. Mark Hall, "The Weakest Link," *Computerworld*, December 17, 2001, accessed at http://www.computerworld.com.

6. Steve Konicki, "A Page from Amazon's Book," *Information Week*, September 17, 2001, accessed at http://www.informationweek.com.

7. Stephan Withers, "Australian Company Touts Web Payment System," *ZDNet Australia*, April 12, 2002, accessed at http://chkpt.zdnet.com.

8. Richard Karpinski, "Nippon Steel Goes Live with Big i2 Deployment," *InternetWeek*, August 30, 2002, accessed at http://www.internetweek.com.

9. Steve Konicki, "GM to Mine Supply Chain for Design Partners," *Information Week*, April 15, 2002, accessed at http://www.information week.com.

10. Carol Sliva, "Online Sales Strong, but E-Tailers Cautious on Spending," *Computerworld*, December 10, 2001, accessed at http://www.computer world.com.

11. Beth Cox, "Office Depot Scores Amazon Deal," *InternetNews*, September 6, 2002, accessed at http://www.internetnews.com.

12. Tom Smith, "Airbus Automates Sup@ir-Sized Supply Chain," *InternetWeek*, August 22, 2002, accessed at http://www.internetweek.com.

13. Mitch Betts, "Perkins Takes Smart Approach to Online Parts Catalog," *Computerworld*, December 17, 2001, accessed at http://www.computer world.com.

14. Timothy J. Mullaney and Darnell Little, "Online Finance Hits Its Stride," *Business Week Online*, April 22, 2002, accessed at http://www.businessweek.com.

15. Timothy J. Mullaney and Darnell Little, "Online Finance Hits Its Stride," *Business Week Online*, April 22, 2002, accessed at http://www.businessweek.com.

16. Linda Rosencrance, "Metavante to Acquire Assets of Rival," *Computerworld*, July 30, 2002, accessed at http://www.computerworld.com.

17. George V. Hulme, "Protect Network Traffic with Latest Security Gear," *Information Week*, May 20, 2002, accessed at http://www.information week.com.

18. Alorie Gilbert, "E-Catalogs: Long Journey to Rewards," *Information Week*, August 6, 2001, pp. 51–52.

19. Lucas Mearian, "Research Points to Sharp Rise in Number of E-Billing Users," *Computerworld*, December 3, 2001, accessed at http://www.computerworld.com.

20. About the Card accessed at http://www.gmcard.com on April 19, 2002.

21. Lucas Mearian, "Visa Pushes Online Security Software on Merchants and Banks," *Computerworld*, May 11, 2001, accessed at http://www.computerworld.com.

22. Michael Liedtke, "VeriSign Embraces MasterCard's Online Anti-Fraud System," *Information Week*, September 4, 2002, accessed at http://www.informationweek.com.

23. Michael Liedtke, "VeriSign Embraces MasterCard's Online Anti-Fraud System," *Information Week*, September 4, 2002, accessed at http://www.informationweek.com.

24. "Blue from American Express," American Express Web site at http://home4americanexpress4.com, accessed on April 20, 2002.

25. Scarlett Pruitt, "Recording Industry Drops Suit Ordering ISPs to Block Chinese Site," *Computerworld*, August 22, 2002, accessed at http://www.computerworld.com.

26. Scarlett Pruitt, "Google Finds Itself in Patent Suit," *Computerworld*, April 5, 2002, accessed at http://www.computerworld.com.

27. Todd R. Weiss, "AOL Sues Company over Alleged E-Mail Spamming," *Computerworld*, January 4, 2001, accessed at http://www.computerworld.com.

28. Jennifer Disabatino, "Bottom Line Hit Hard by Need to Fend Off Spam and Viruses," *Computerworld*, June 17, 2002, accessed at http://www.computerworld.com.

29. Julia King, "WSJ.com Completes Web Site Overhaul," *Computerworld*, February 11, 2002, accessed at http://www.computerworld.com.

30. Todd Hollowell and Gauray Verma, "Customers Want the Personal Touch," *Information Week*, June 24, 2002, accessed at http://www.informationweek.com.

Transaction Processing and Enterprise Resource Planning Systems

CHAPTER 9

PRINCIPLES	LEARNING OBJECTIVES
An organization's TPS must support the routine, day-to-day activities that occur in the normal course of business and help a company add value to its products and services.	• Identify the basic activities and business objectives common to all transaction processing systems. • Describe the inputs, processing, and outputs for the transaction processing systems associated with order processing. • Describe the inputs, processing, and outputs for the transaction processing systems associated with purchasing. • Describe the inputs, processing, and outputs for the transaction processing systems associated with accounting business processes.
TPSs help multinational corporations form business links with their business partners, customers, and subsidiaries.	• Identify the challenges the multinational corporations must face in planning, building, and operating their TPSs.
Implementation of an enterprise resource planning system enables a company to achieve numerous business benefits through the creation of a highly integrated set of systems.	• Define the term *enterprise resource planning system* and discuss the advantages and disadvantages associated with the implementation of such a system.

[Wal-Mart and Kmart]

Transaction Processing Systems Distinguish Major Retailers

Wal-Mart is the world's #1 retailer, with more than 4,500 stores, including discount stores (Wal-Mart), combination discount and grocery stores (Wal-Mart Supercenters and ASDA in the United Kingdom), and membership-only warehouse stores (Sam's Club). For over a decade, Wal-Mart has made major IS investments and business-process changes to improve its basic transaction processing systems and business operations, primarily inventory and warehouse management.

Key to Wal-Mart's success is its pricing strategy based on everyday low prices, which results in much more predictable customer demand, enabling Wal-Mart to improve its inventory control systems and processes while lowering costs. Wal-Mart also built an e-business system to regularly communicate sales and inventory data from every store to its thousands of suppliers and buyers. It deployed a private trading hub to consolidate its purchasing globally and bring suppliers on-line to bid on contracts. The success of these improvements to its basic transaction processing systems lowered Wal-Mart's costs, and much of the savings were passed on to customers in the form of lower prices.

At the start of 2002, Kmart was the #3 U.S. retailer (behind Wal-Mart and Target). It had more than 2,100 stores in all 50 states and U.S. territories, but in sharp contrast to Wal-Mart, its sales were lagging. In spite of all its efforts, a dismal holiday season and an erosion of supplier confidence forced Kmart to file for Chapter 11 bankruptcy protection in January 2002. The company has closed more than 200 stores and laid off some 22,000 of its 250,000 employees. It is currently in a battle for its corporate life.

Kmart employs a pricing model that relies on special sales (including its "blue-light specials") to attract customers. Heavy use of these promotions creates sharp peaks and valleys in demand for products. Kmart was unable to build basic shipment planning, shipment execution, and inventory control systems to deal effectively with its widely fluctuating customer demand. Also, Kmart's inability to provide routine and reliable information and to build tight links with its suppliers made it impossible for suppliers to respond quickly enough to meet the retailer's needs.

The stark contrast between the two retailers' inventory control systems is readily apparent when visiting their stores. At Kmart, it isn't unusual to find shelves empty, with products piled up in stockrooms. Wal-Mart's shelves are always full, and there are almost no storage areas because its vendor-managed inventory system makes suppliers responsible for delivering products when Wal-Mart needs them, typically just in time to be sold.

As you read this chapter, consider the following:
- ⊙ What are the day-to-day business activities that transaction processing systems must support?
- ⊙ How does a company's business strategy affect the types of systems it uses?
- ⊙ How are sophisticated integrated systems employed to achieve real business benefits?

As you can see in the opening vignette, businesses rely on information systems to integrate their daily transaction activities. The many business activities associated with supply, distribution, sales, marketing, accounting, and taxation can be performed quickly while avoiding waste and mistakes. The goal of this computerization is ultimately to satisfy a business's customers and provide a competitive advantage by reducing costs and improving service.

Transaction processing was one of the first business processes to be computerized, and without information systems, recording and processing business transactions would consume huge amounts of an organization's resources. The transaction processing system (TPS) also provides employees involved in other business processes—the management information system/decision support system (MIS/DSS) and the special-purpose information systems—with data to help them achieve their goals. A transaction processing system serves as the foundation for the other systems (Figure 9.1). Transaction processing systems perform routine operations such as sales ordering and billing, often performing the same operations daily or weekly. The amount of support for decision making that a TPS directly provides managers and workers is low.

These systems require a large amount of input data and produce a large amount of output without requiring sophisticated or complex processing. As we move from transaction processing to management information/decision support, and special-purpose information systems, we see less routine, more decision support, less input and output, and more sophisticated and complex processing and analysis. But the increase in sophistication and complexity in moving from transaction processing does not mean that it is less important to a business. In most cases, all these systems start as a result of one or more business transactions.

AN OVERVIEW OF TRANSACTION PROCESSING SYSTEMS

Every organization has manual and automated *transaction processing systems (TPSs)*, which process the detailed data necessary to update records about the fundamental business operations of the organization. These systems include order entry, inventory control, payroll, accounts payable, accounts receivable, and general ledger, to name just a few. The input to these systems includes basic business transactions such as customer orders, purchase orders, receipts, time cards, invoices, and customer payments. The result of processing business transactions is that the organization's records are updated to reflect the status of the operation at the time of the last processed transaction. Automated TPSs consist of all the components of a CBIS, including databases, telecommunications, people, procedures, software, and hardware devices used to process transactions. The processing activities include data collection, data edit, data correction, data manipulation, data storage, and document production.

For most organizations, TPSs support the routine, day-to-day activities that occur in the normal course of business that help a company add value to its products and services. Depending on the customer, value may mean lower price, better service, higher quality, or uniqueness of product. By adding a significant

FIGURE 9.1

TPS, MIS/DSS, and Special Information Systems

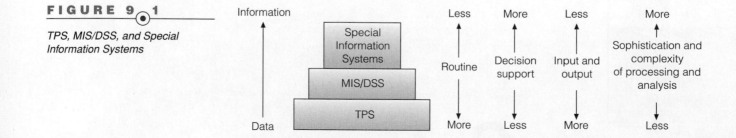

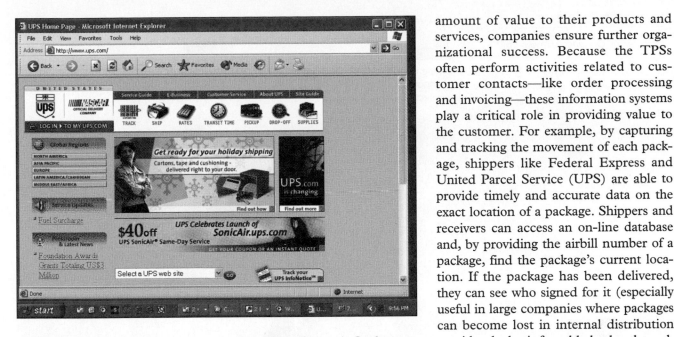

UPS adds value to its service by providing timely and accurate data on-line on the exact location of a package.

amount of value to their products and services, companies ensure further organizational success. Because the TPSs often perform activities related to customer contacts—like order processing and invoicing—these information systems play a critical role in providing value to the customer. For example, by capturing and tracking the movement of each package, shippers like Federal Express and United Parcel Service (UPS) are able to provide timely and accurate data on the exact location of a package. Shippers and receivers can access an on-line database and, by providing the airbill number of a package, find the package's current location. If the package has been delivered, they can see who signed for it (especially useful in large companies where packages can become lost in internal distribution systems and mailrooms). Such a system provides the basis for added value through improved customer service.

TRADITIONAL TRANSACTION PROCESSING METHODS AND OBJECTIVES

batch processing system
method of computerized processing in which business transactions are accumulated over a period of time and prepared for processing as a single unit or batch

When computerized transaction processing systems first evolved, only one method of processing was available. All transactions were collected in groups, called *batches*, and processed together. With **batch processing systems**, business transactions are accumulated over a period of time and prepared for processing as a single unit or batch (Figure 9.2a). The time period during which transactions are accumulated is whatever length of time is needed to meet the needs of the users of that system. For example, it may be important to process invoices and customer payments for the accounts receivable system daily. On the other hand, the payroll system may receive time cards and process them biweekly to create checks and update employee earnings records as well as to distribute labor costs. The essential characteristic of a batch processing system is that there is some delay between the occurrence of the event and the eventual processing of the related transaction to update the organization's records.

on-line transaction processing (OLTP)
computerized processing in which each transaction is processed immediately, without the delay of accumulating transactions into a batch

Today's computer technology allows another processing method, called *on-line, real-time*, or **on-line transaction processing (OLTP)**. With this form of data processing, each transaction is processed immediately, without the delay of accumulating transactions into a batch (Figure 9.2b). As soon as the input is available, a computer program performs the necessary processing and updates the records affected by that single transaction. Consequently, at any time, the data in an on-line system always reflects the current status. When you make an airline reservation, for instance, the transaction is processed and all databases, such as seat occupancy and accounts receivable, are updated immediately. This type of processing is absolutely essential for businesses that require data quickly and update it often, such as airlines, ticket agencies, and stock investment firms. Many companies have found that OLTP helps them provide faster, more efficient service—one way to add value to their activities in the eyes of the customer. Increasingly, companies are using the Internet to perform many OLTP functions. A third type of transaction processing, called *on-line entry with delayed processing*, is a compromise between batch and on-line processing. With this type of system, transactions are entered into the computer system when they occur, but they are not processed

FIGURE 9.2

Batch versus On-Line Transaction Processing

Batch processing (a) inputs and processes data in groups. In on-line processing (b), transactions are completed as they occur.

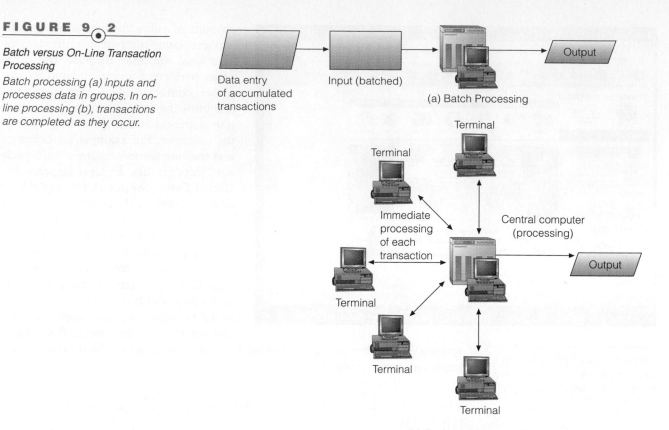

Data entry of accumulated transactions

Input (batched)

Output

(a) Batch Processing

Terminal

Terminal

Immediate processing of each transaction

Central computer (processing)

Output

Terminal

Terminal

Terminal

(b) On-Line Transaction Processing

When you order a product over the phone, the vendor may use on-line entry with delayed processing. Your order is entered into the computer at the time of the call but is not processed immediately.

(Source: © 2002 PhotoDisc.)

immediately. For example, when you call a toll-free number and order a product, your order is typically entered into the computer when you make the call. However, the order may not be processed until that evening after business hours.

Even though the technology exists to run TPS applications using on-line processing, it is not done for all applications. For many applications, batch processing is more appropriate and cost-effective. Payroll transactions and billing are typically done via batch processing. Specific goals of the organization define the method of transaction processing best suited for the various applications of the company. Figure 9.3 shows the total integration of a firm's transaction processing systems. Read the "IS Principles in Action" special-interest feature to learn more about what is involved in converting from batch processing to on-line processing.

Because of the importance of transaction processing, organizations expect their TPSs to accomplish a number of specific objectives, including the following:

Process data generated by and about transactions. The primary objective of any TPS is to capture, process, and store transactions and to produce a variety of documents related to routine business activities. These business activities can be directly or indirectly related to selling products and services to customers. Processing orders, purchasing materials, controlling inventory, billing customers, and paying suppliers and employees are all business activities that result from customer orders. These activities result in transactions that are processed by the TPS.

Utilities, telecommunications companies, financial-services organizations, and indeed many businesses are under enormous

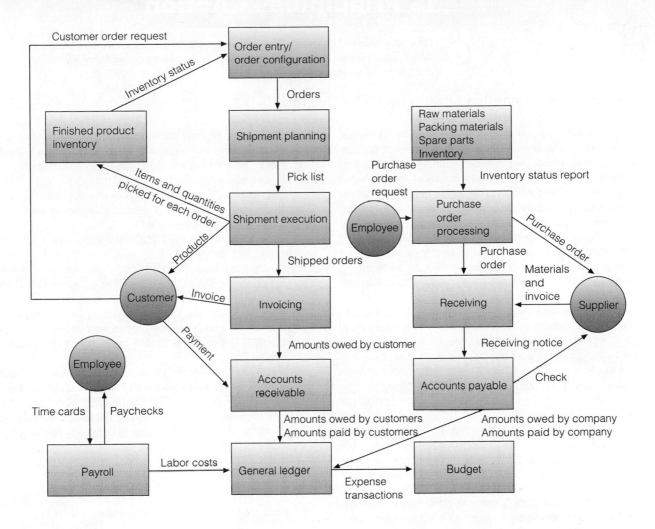

FIGURE 9.3

Integration of a Firm's TPSs

pressure to process ever-larger volumes of transactions in near-real time. Unfortunately, their information system infrastructures are struggling to handle the increased workload. But now improved data management software with data store, data analysis, data management, and transaction-processing capabilities can ease the burden. Aleri Inc. has developed data-management software that can process millions of transactions in a fraction of the time traditional database systems take. The software is in use at the Allied Irish Bank in Dublin and is being implemented at several other financial-service companies.[1]

Maintain a high degree of accuracy and integrity. One objective of any TPS is error-free data input and processing. Even before the introduction of computer technology, employees visually inspected all documents and reports introduced into or produced by the TPS. Because humans are fallible, the transactions were often inaccurate, resulting in wasted time and effort and requiring resources to correct them. An editing program, for example, should have the ability to determine that an entry that should read "40 hours" is not entered as "400 hours" or "4000 hours" because of a data entry error.

An important component of data integrity is to avoid fraudulent transactions. E-commerce companies face this problem when accepting credit or debit card information over the Internet. How can these companies make sure that the people making the purchases are who they say they are? One approach is to use a digital certificate. A digital certificate is a small computer file that serves as both an ID card and a signature. Some believe that digital certificates, which use complex mathematical codes, are almost fraud proof.

PRINCIPLE

An organization's TPS must support the routine, day-to-day activities that occur in the normal course of business and help a company add value to its products and services.

Financial Industry Considers Move to Trade Plus One Day (T+1)

Although stock trades seem to occur effortlessly and instantaneously to on-line customers, who type a stock symbol and click the mouse a few times, the behind-the-scenes processing requires several steps. First, both the seller and the buyer must register the transaction with a central clearing organization, such as the Depository Trust & Clearing Corporation (DTCC). Next, the two accounts involved in the trade must be reconciled, any errors corrected, and money has to change hands. Finally, the actual stock certificates must be exchanged. Currently, this system of clearing and settling stock trades involves overnight batch processing of transactions and must be completed within three days.

The Securities and Exchange Commission (SEC) is pushing the financial industry to clear and settle all trades within one day (trade plus one day, or T+1) to increase the stability of the global banking system. Just as someone else's bad check can have a ripple effect when you've already written other checks assuming those funds are in your account, large trading firms can have the same problem and set off ripples throughout the financial world. For example, a 1987 stock market downturn caused a few securities firms to go under. The then five-day settlement period resulted in funds shortages rippling throughout the industry, causing other firms to fail. This incident spurred the SEC to require financial institutions to move from a T+5 to a T+3 clearing and settlement process in 1995.

Aside from market stability, there are other reasons for moving to T+1. The United States must do so to maintain its leadership position in the financial markets. In Hong Kong and Singapore, financial institutions have already implemented T+1; Japan and Europe are planning to do so as well. Implementation of a T+1 process will also allow the industry to handle increased trading volume—the current batch processes are reaching the limits of their capacity. It is estimated that the switch to T+1 will save industry participants about $18 billion over five years by reducing manual processing, lowering error rates, and speeding payments. The savings will more than offset the estimated industry cost of around $8 billion.

The change to T+1 will require brokerages and other financial institutions to completely reengineer all of their trading processes, as well as their underlying information system infrastructures. This will result in substantial costs to the many affected financial firms. T+1 will require Wall Street's traditional batch processing systems to be changed to a real-time processing network that never crashes. One approach is to move to straight-through processing, which provides a nonstop flow of information from trade execution to settlement. A group called the Global Straight Through Processing Association (GSTPA) plans to build a global network for brokerages, custodians, and other firms to improve and speed cross-border trading at a cost of more than $100 million.

Perhaps the biggest challenge to implementing T+1 is to get all participants in the securities transaction chain to agree on exactly how things will work. Consistent customer databases and settlement instruction databases must be implemented to ensure an efficient process for clearing and settlement data. In addition, a master database shared by all user institutions must be created.

Clearinghouses like the DTCC must also make substantial changes. Not only must they upgrade all their internal processes, but efficient, reliable data transmission links must also be established to all the other industry players. DTCC has been working on this for more than two years. It has begun moving to a multibatch system, replacing its one overnight process with a series of batch processes that run throughout the day. DTCC expects to spend more than $100 million to prepare for T+1.

The T+1 initiative was slowed by September 11, 2001, disaster recovery considerations. The Securities Industry Association delayed the target date for the launch of T+1 to June 2005, citing the fact that many brokerage houses and banks are focusing more on business continuity planning than on shortening settlement times.

Discussion Questions

1. What is the business justification for moving to T+1? Which of these reasons do you find the most compelling? Why?
2. What are the biggest barriers to implementing T+1?

Critical Thinking Questions

3. Given the substantial benefits and need to convert to T+1, do you think the Securities Industry Association made the right decision to delay its implementation so companies could focus on business continuity planning? How might the eventual implementation of T+1 affect those business continuity plans?
4. Check out the Web site for the Global Straight Through Processing Association at http://www.gstpa.org to find out the current status of implementing T+1. Write a brief report to your instructor summarizing your findings.

Sources: Adapted from Lucas Mearian, "One-Day Stock Trade Settlements in Doubt," *Computerworld*, November 5, 2001, http://www.computerworld.com; "SEI Investments Partners with Omgeo to Speed Trade Settlements, Move Toward T+1," *PRNewswire*, March 26, 2002, accessed at http://www.findarticles.com; Lucas Mearian, "Wall St. Aims for One-Day Trading," *Computerworld*, January 14, 2002, http://www.computerworld.com; "Bloomberg to Support Siebel 7; Bloomberg to Deliver Integrated Broker Desktop Solutions with Siebel 7," *PRNewswire*, October 2, 2001, accessed at http://www.findarticles.com; Lucas Mearian, "Straight-Through Trade Processing Pressures IT," *Computerworld*, January 21, 2002, http://www.computerworld.com; Steve Crosby, SIA Annual Operations Conference Presentation, accessed at the GSTPA Web site at http://www.gstpa.org, May 9, 2002.

As the volume of data being processed and stored increases, it becomes more difficult for individuals and machines to review all input data. Doing so is critical, however, because data and information generated by the TPS are often used by other information systems in an organization. So, a company must ensure both data integrity and accuracy.

Produce timely documents and reports. Manual transaction processing systems can take days to produce routine documents. Fortunately, the use of computerized TPSs significantly reduces this response time. Improvements in information technology, especially hardware and telecommunications links, allow transactions to be processed in a matter of seconds. The ability to conduct business transactions in a timely way can be very important for the profitable operation of the organization. For instance, if bills (invoices) are sent out to customers a few days earlier than usual, payment may be received earlier. A number of transaction processing systems have built-in capabilities to monitor how timely a company is when processing transactions and producing reports and documents. Some monitoring software packages can compare actual performance with corporate goals and objectives.

GiantLoop Network is a networking supplier that must maintain high service levels for its customers. The firm built an operational support center based on application monitoring software from Dirig Software's RelyENT product to increase the reliability of its operations. The software collects data and statistics on system-, application-, and component-level computer resources; provides a snapshot of their use, performance levels, and overall health; and can even perform scripted actions in response to problems or events.[2] As a result, GiantLoop is helping its customers' transaction processing systems perform more reliably.

Timing is also crucial for related applications such as order processing, invoicing, accounts receivable, inventory control, and accounts payable. Because of electronic recording and transmission of sales information, transactions can be processed in seconds rather than overnight, thus improving companies' cash flow. Thus customers find credit card charges they made on the final day of the billing period on their current monthly bill.

Increase labor efficiency. Before computers existed, manual business processes often required rooms full of clerks and equipment to process the necessary business transactions. Today, organizations have implemented transaction processing systems to substantially reduce clerical and other labor requirements. A small minicomputer linked to a company's cash registers has replaced a room full of clerks, typewriters, and filing cabinets.

Organizations are interested in gaining even greater labor efficiency by further streamlining business processing. For example, Procter & Gamble (P&G) implemented a contract and trade management system built around a software package from I-many Inc. Now when contracts in the food services division are signed, the software automatically calculates service fees, volume discounts, and other sales incentives. It then audits each sales transaction to ensure that all terms of the contract are being met. P&G's mainframe transaction processing systems just weren't flexible enough to handle the variables involved in selling through so many distribution channels and to so many customers. The system eliminates many billing problems and greatly reduces the amount of clerical effort directed at bill resolution. As a result, the system paid for itself in the first year of operation.[3]

Help provide increased service. Without question, we are quickly becoming a service-oriented economy. Even strong manufacturing companies, including household appliance makers and automobile manufacturers, realize the importance of providing superior customer service. A transaction processing system for AllEventsTickets.com or TicketsNow.com, for example, allows concert enthusiasts to order tickets over the Internet instead of standing in line for hours, or even days. One objective of TPSs is to assist the organization in providing efficient service. Some companies' EDI systems (see Chapter 6) allow customers to place orders electronically, thus bypassing slower and more error-prone methods of written or oral communication.

Help build and maintain customer loyalty. A firm's transaction processing systems are often the means for customers to communicate. It is important that the customer interaction with these systems keeps customers satisfied and returning.

Achieve competitive advantage. A goal common to almost all organizations is to gain and maintain a competitive advantage. As discussed in Chapter 2, a competitive advantage provides a significant and long-term benefit for the organization. When a TPS is developed or modified, the personnel involved should carefully consider the significant and long-term benefits the new or modified system might provide.

Collaborative Planning, Forecasting, and Replenishment (CPFR) is an information system protocol used by retailers and suppliers when exchanging real-time data to anticipate demand and to manage and update inventory as needed. Many manufacturers are using CPFR to form tighter links with their suppliers and to gain a competitive advantage. For example, Ace Hardware uses CPFR technology to manage about 6,000 items made by 23 manufacturers. Product promotions have improved because with CPFR, suppliers are able to connect to Ace's inventory databases and see the same inventory-control screens Ace sees. As a result, Ace and participating suppliers spend less time discussing why certain items are out of stock and more time planning promotions and introducing new items. The improved planning has helped Ace reduce labor costs associated with receiving merchandise from its CPFR partners by up to 20 percent. Freight costs have also been cut, enabling Ace to reduce its retail prices.[4]

Some of the ways that companies can use transaction processing systems to achieve competitive advantage are summarized in Table 9.1.

Depending on the specific nature and goals of the organization, any of these objectives may be more important than others. By meeting these objectives, TPSs can support corporate goals such as reducing costs; increasing productivity, quality, and customer satisfaction; and running more efficient and effective operations. For example, overnight delivery companies such as FedEx expect their TPSs to increase customer service. These systems can locate a client's package at any time—from initial pickup to final delivery. This improved customer information

TABLE 9.1

Examples of Transaction Processing Systems for Competitive Advantage

Competitive Advantage	Example
Customer loyalty increased	Use of customer interaction system to monitor and track each customer interaction with the company
Superior service provided to customers	Use of tracking systems that are accessible by customers to determine shipping status
Better relationship with suppliers	Use of an Internet marketplace to allow the company to purchase products from suppliers at discounted prices
Superior information gathering	Use of order configuration system to ensure that products ordered will meet customer's objectives
Costs dramatically reduced	Use of warehouse management system employing scanners and bar-coded product to reduce labor hours and improve inventory accuracy
Inventory levels reduced	Use of collaborative planning, forecasting, and replenishment to ensure the right amount of inventory is in stores

allows companies to produce timely information and be more responsive to customer needs and queries.

Brokerage firms employ specialized TPSs that process orders, provide customer statements, and produce managerial reports. Increasingly, brokerage firms offer their clients software to monitor their own accounts on-line. These systems indicate all buy and sell orders and associated commissions, the securities held, the purchase price, the current market price, the dividends or interest per period, the dividends per year, and the anticipated yield on each security, as well as the anticipated yield of each client's portfolio of securities. The objectives are to increase customer service and provide timely reports. Obviously, maintaining a high degree of data accuracy and integrity is also important. Discount brokerage firms are even offering software that allows clients to enter their own buy and sell requests without first speaking with a broker. This increases labor efficiency but raises issues about the level of commissions that should be charged.

TRANSACTION PROCESSING ACTIVITIES

Along with having common characteristics, all transaction processing systems perform a common set of basic data processing activities. TPSs capture and process data that describes fundamental business transactions. This data is used to update databases and to produce a variety of reports people both within and outside the enterprise use (Figure 9.4). The business data goes through a **transaction processing cycle** that includes data collection, data editing, data correction, data manipulation, data storage, and document production (Figure 9.5).

Data Collection

The process of capturing and gathering all data necessary to complete transactions is called **data collection**. In some cases it can be done manually, such as by collecting handwritten sales orders or changes to inventory. In other cases, data collection is automated via special input devices like scanners, point-of-sale devices, and terminals.

Data collection begins with a transaction (e.g., taking a customer order) and results in the origination of data that is input to the transaction processing system. Data should be captured at its source, and it should be recorded accurately, in a timely fashion, with minimal manual effort, and in a form that can be directly

---

transaction processing cycle
the process of data collection, data editing, data correction, data manipulation, data storage, and document production

data collection
the process of capturing and gathering all data necessary to complete transactions

FIGURE 9.4

A Simplified Overview of a Transaction Processing System

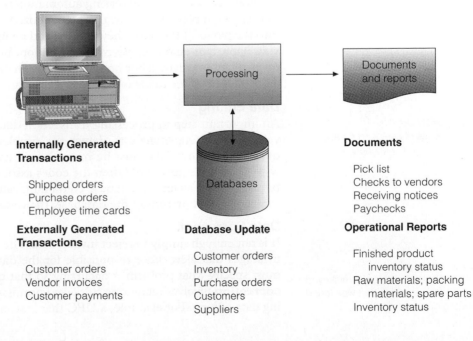

Data entry and input

Processing

Documents and reports

Internally Generated Transactions

Shipped orders
Purchase orders
Employee time cards

Externally Generated Transactions

Customer orders
Vendor invoices
Customer payments

Databases

Database Update

Customer orders
Inventory
Purchase orders
Customers
Suppliers

Documents

Pick list
Checks to vendors
Receiving notices
Paychecks

Operational Reports

Finished product
     inventory status
Raw materials; packing
     materials; spare parts
Inventory status

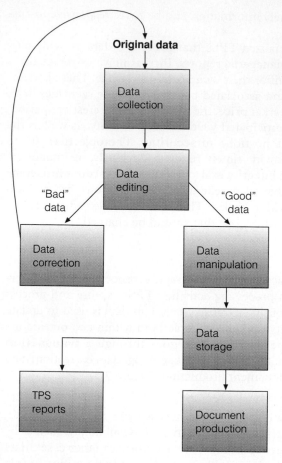

Original data

Data collection

Data editing

"Bad" data

"Good" data

Data correction

Data manipulation

Data storage

TPS reports

Document production

FIGURE 9●5

Data Processing Activities Common to Transaction Processing Systems

entered into the computer rather than keying the data from some type of document. This approach is called *source data automation.* An example of source data automation is the use of scanning devices at a retail store checkout to read the Universal Product Code (UPC) automatically. Reading the UPC bar codes is quicker and more accurate than having a cash register clerk enter codes manually. The scanner reads the bar code for each item and looks up its price in the item database. The point-of-sale TPS uses the price data to determine the customer's bill. The number of units of this item purchased, the date, the time, and the price are also used to update the store's inventory database, as well as its database of detailed purchases. The inventory database is used to generate a management report notifying the store manager to reorder items whose sales have reduced the stock below the reorder quantity. The detailed purchases database can be used by the store (or sold to market research firms or manufacturers) for detailed analysis of sales (Figure 9.6).

Procter & Gamble and other manufacturers and retailers are testing the use of radio-frequency identification devices (RFID) to replace bar codes for tracking goods through the supply chain and onto the store shelf. The small RFID chips emit radio waves that can be scanned through a carton, allowing the entire inventory to be scanned in the same time it takes to scan a single UPC bar-coded item. P&G and twenty-five other manufacturers and retailers have anted up $300,000 to join MIT's Auto-ID Center, which is working out the issues of building a global infrastructure to pick up RFID signals and a global registry of domain names to make it work.[5]

Many grocery stores combine point-of-sale scanners and coupon printers. The systems are programmed so that each time a specific product—say, a box of cereal—crosses a checkout scanner, an appropriate coupon—perhaps a milk coupon—is printed. Companies can pay to be promoted through the system, which is then reprogrammed to print those companies' coupons if the customer buys a competitive brand. These TPSs help grocery stores to increase profits by improving their repeat sales and bringing in revenue from other businesses.

Another example of increasing automation is industrial data collection devices that allow employees to scan their magnetized employee ID cards to enter data into the payroll TPS when they start or end a job. Not only do these devices provide important employee payroll information, but they also help an organization determine how many labor hours are spent on each job or project so that staffing adjustments can be made or future projects can be accurately planned.

Data Editing

data editing
the process of checking data for validity and completeness

An important step in processing transaction data is to perform **data editing** for validity and completeness to detect any problems with the data. For example, quantity and cost data must be numeric and names must be alphabetic; otherwise, the data is not valid. Often the codes associated with an individual transaction are edited against a database containing valid codes. If any code entered (or scanned) is not present in the database, the transaction is rejected.

Data Correction

data correction
the process of reentering miskeyed or misscanned data that was found during data editing

It is not enough simply to reject invalid data. The system should also provide error messages that alert those responsible for the data edit function. Error messages must specify what problem is occurring so that corrections can be made. A **data correction** involves reentering miskeyed or misscanned data that was found during data editing. For example, a UPC that is scanned must be in a master table of valid UPCs. If the code is misread or does not exist in the table, the checkout clerk is given an instruction to rescan the item or key in the information manually.

Data Manipulation

data manipulation
the process of performing calculations and other data transformations related to business transactions

Another major activity of a TPS is **data manipulation**, the process of performing calculations and other data transformations related to business transactions. Data manipulation can include classifying data, sorting data into categories, performing calculations, summarizing results, and storing data in the organization's database for further processing. In a payroll TPS, for example, data manipulation includes multiplying an employee's hours worked by the hourly pay rate. Overtime calculations, federal and state tax withholdings, and deductions are also performed.

Data Storage

data storage
the process of updating one or more databases with new transactions

Data storage involves updating one or more databases with new transactions. Once the update process is complete, this data can be further processed and manipulated by other systems so that it is available for management decision making. Thus, although transaction databases can be considered a by-product of transaction processing, they have a pronounced effect on almost all other information systems and decision-making processes in an organization.

Document Production and Reports

document production
the process of generating output records and reports

TPSs produce important business documents. **Document production** involves generating output records and reports. These documents may be hard-copy paper reports or displays on computer screens (sometimes referred to as *soft copy*). Paychecks, for example, are hard-copy documents produced by a payroll TPS, while an outstanding balance report for invoices might be a soft-copy report displayed by an accounts receivable TPS. Often, results from one TPS are passed downstream as input to other systems (as shown in Figure 9.6), where the results of updating the inventory database are used to create the stock exception report (a type of management report) of items whose inventory level is less than the reorder point.

In addition to major documents like checks and invoices, most TPSs provide other useful management information and decision support, such as reports that help managers and employees perform various activities. These reports can be

FIGURE 9.6

Point-of-Sale Transaction Processing System

Scanning items at the checkout stand updates a store's inventory database and its database of purchases.

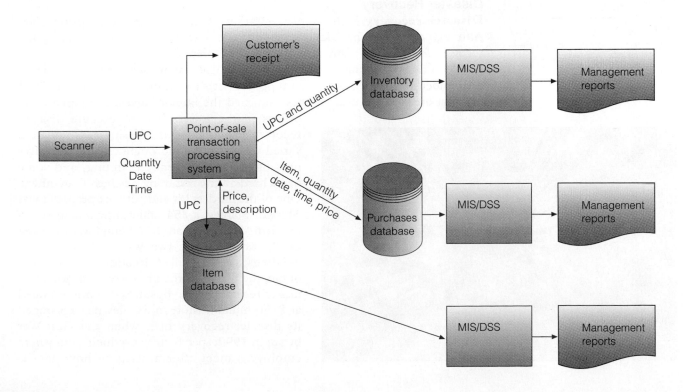

printed or displayed on a computer screen. A report showing current inventory is one example; another might be a document listing items ordered from a supplier to help a receiving clerk check the order for completeness when it arrives. A TPS can also produce reports required by local, state, and federal agencies, such as statements of tax withholding and quarterly income statements.

Throughout this chapter we will look at some ways companies have employed TPSs to help them meet organizational goals.

CONTROL AND MANAGEMENT ISSUES

Transaction processing systems are the backbone of any organization's information systems. They capture facts about the fundamental business operations of the organization—facts without which orders cannot be shipped, customers cannot be invoiced, and employees and suppliers cannot be paid. In addition, the data captured by the TPSs flow downstream to the other systems of the organization. Like any structure, an organization's information systems are only as good as the foundation on which they are built. In fact, most organizations would grind to a screeching halt if their TPSs failed.

Business Resumption Planning

business resumption planning
the process of anticipating and providing for disasters

Business resumption planning is the process of anticipating and minimizing the effects of disasters. Disasters can be natural emergencies such as a flood, a fire, or an earthquake or interruptions in business processes such as labor unrest, terrorist activity, or erasure of an important file. Business resumption planning focuses primarily on two issues: maintaining the integrity of corporate information and keeping the information system running until normal operations can be resumed.

One of the first steps of business resumption planning is to identify potential threats or problems, such as natural disasters, employee misuse of personal computers, and poor internal control procedures. Business resumption planning also involves disaster preparedness. IS managers should occasionally hold an unannounced "test disaster"—similar to a fire drill—to ensure that the disaster plan is effective.

Disaster Recovery

disaster recovery
the implementation of the business resumption plan

Disaster recovery is the implementation of the business resumption plan. Although companies have known about the importance of disaster planning and recovery for decades, many do not adequately prepare.

The value of backup and recovery for critical systems was brought home on September 11, 2001. Nearly 3,000 people perished in the September 11, 2001 terrorist attacks. But even as they mourned the human losses, companies with offices in the towers had to execute disaster recovery plans to resume business. Companies varied widely in the thoroughness and effectiveness of their contingency planning, and some had a harder time resuming business than others. One of the hardest hit and yet best prepared was Morgan Stanley, a $54 billion investment banking firm. The firm had 3,700 employees located on 25 floors in the two World Trade Center buildings; six were killed. In addition to the loss of life, the firm suffered property damages and a loss of revenue due to business downtime valued at $150 million. Morgan Stanley had revamped its disaster-recovery plan when the Gulf War began in 1990, specifying everything from where employees meet after a crisis to how data is

Companies like Iron Mountain provide a secure, off-site environment for records storage. In the event of a disaster, vital data can be recovered.

(Source © 2002 PhotoDisc.)

protected. As a result, backup data centers in Manhattan and New Jersey were able to keep all critical information systems running without interruption. There was no loss of any computer-based data, and its e-mail service was restored within 72 hours. On the other extreme of preparedness was May Davis Group, a privately held financial services firm with 13 employees on the 87th floor of 1 World Trade Center. The firm had no disaster-recovery plan at the time of the attack. One employee perished in the attack, property damage exceeded $100,000, and the loss of revenue due to business downtime and data loss exceeded $1 million.[6]

Transaction Processing System Audit

The accounting scandals of 2001–2002 spurred corporate board members, lawmakers, regulators, and stockholders to pressure corporate executives to produce accurate financial reports and to do so in a timely fashion. In July 2002, the Sarbanes-Oxley Act was enacted, setting deadlines for public companies to implement procedures that ensure their audit committees can document underlying financial data to validate earnings reports and meet demands for accuracy in numbers.[7] Various actions are being taken to improve financial reporting. Business managers are demanding that their financial systems provide real-time data feeds and updates of expenditures and sales so they can ensure their numbers are right. PeopleSoft, for instance, sells CFO Portal, a browser-based system that provides such capabilities. Companies are beginning to use standardized electronic financial reports using eXtensible Business Reporting Language that enables them to provide hyperlinks connecting numbers in financial statements to explanatory footnotes so that financial statements are easier to understand.[8]

Clearly the CIO must take a role in preventing the kind of accounting irregularities that got companies into trouble and erased investor confidence. One key step is to conduct a **transaction processing system audit** that attempts to answer four basic questions:

- Does the system meet the business need for which it was implemented?
- What procedures and controls have been established?
- Are these procedures and controls being used properly?
- Are the information systems and procedures producing accurate and honest reports?

In addition to these four basic auditing questions, other areas are typically investigated during an audit. These areas include the distribution of output documents and reports, the training and education associated with existing and new systems, and the time necessary to perform various tasks and to resolve problems and bottlenecks in the system. General areas of improvement are also investigated and reported during the audit.

Two types of audits exist. An internal audit is conducted by employees of the organization; an external audit is performed by accounting firms or companies and individuals not associated with the organization. In either case, a number of steps are performed. The auditor inspects all programs, documentation, control techniques, the disaster plan, insurance protection, fire protection, and other systems management concerns such as efficiency and effectiveness of the disk or tape library. This check is accomplished by interviewing IS personnel and performing a number of tests on the computer system. External audits are important for stockholders and others outside the company, in addition to managers and employees inside the company. A number of Internet startup companies, for example, have overstated their income, which has resulted in high stock evaluations in some cases. An external audit by a reputable auditing company can help uncover these reporting problems.

In establishing the integrity of the computer programs and software, an audit trail must be established. The **audit trail** allows the auditor to trace any

transaction processing system audit
an examination of the TPS to answer whether the system meets the business need for which it was implemented, what procedures and controls have been established, whether these procedures and controls are being used properly, and whether the information systems and procedures are producing accurate and honest reports

audit trail
documentation that allows the auditor to trace any output from the computer system back to the source documents

output from the computer system back to the source documents. With many of the real-time and time-sharing systems available today, it is extremely difficult to follow an audit trail. In many cases, no record of inputs to the system exists; thus, the audit trail is destroyed. In such cases, the auditor must investigate the actual processing in addition to the inputs and outputs of the various programs. In an attempt to safeguard the privacy of medical records, the federal Health Insurance Portability and Accountability Act (HIPAA) requires that health care organizations, health care providers, and insurance companies establish an audit trail for each patient record. As an individual's medical record moves, each application it touches must imprint it with an identifier that cites every person who handled it and for what purpose.[9]

While the CIO must take an active role in ensuring the integrity of financial reporting systems, that's not the same as guaranteeing the material accuracy of financial statements. At some of the companies under scrutiny, not even the most conscientious and informed CIO could have uncovered the financial shenanigans going on.[10]

TRADITIONAL TRANSACTION PROCESSING APPLICATIONS

In this section we present an overview of several common transaction processing systems that support the order processing, purchasing, and accounting business processes (Table 9.2).

ORDER PROCESSING SYSTEMS

order processing systems
systems that process order entry, sales configuration, shipment planning, shipment execution, inventory control, invoicing, customer relationship management, and routing and scheduling

Order processing systems include order entry, sales configuration, shipment planning, shipment execution, inventory control, invoicing, customer relationship management, and routing and scheduling. The business processes supported by these systems are so critical to the operation of an enterprise that the order processing systems are sometimes referred to as the "lifeblood of the organization." Figure 9.7 is a system-level flowchart that shows the various systems and the information that flows between them. A rectangle represents a system, a line represents the flow of information from one system to another, and a circle represents any entity outside the system—in this case, the customer.

Order Entry

order entry system
process that captures the basic data needed to process a customer order

The **order entry system** captures the basic data needed to process a customer order. Orders may come through the mail or via a telephone ordering system, be gathered by a staff of sales representatives, arrive via EDI transactions directly from a customer's computer over a wide area network, or be entered directly over the Internet by the customer using a data entry form on the firm's Web site. Figure 9.8 is a data flow diagram of a typical order entry system. The data flow diagram is more detailed than the system-level flowchart. It shows the various

TABLE 9.2

The Systems That Support Order Processing, Purchasing, and Accounting Functions

Order Processing	Purchasing	Accounting
• Order entry	• Inventory control (raw materials, packing materials, spare parts, and supplies)	• Budget
• Sales configuration		• Accounts receivable
• Shipment planning	• Purchase order processing	• Payroll
• Inventory control (finished product)	• Receiving	• Asset management
• Invoicing and billing	• Accounts payable	• General ledger
• Customer interaction		
• Routing and scheduling		

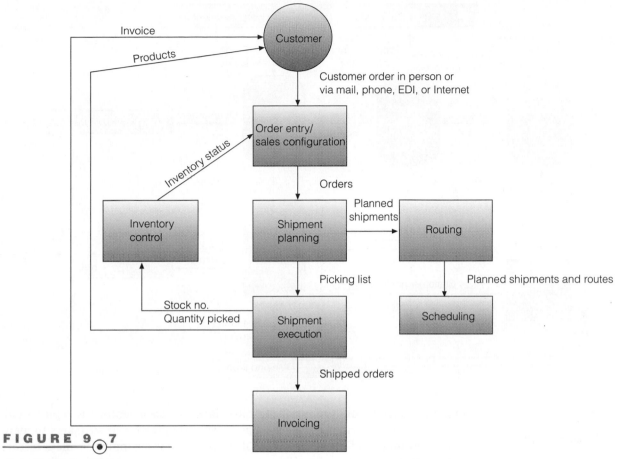

FIGURE 9.7

Order Processing Systems

Safeway, a leading grocery retailer in the UK, has a sophisticated software system that gathers detailed information about every item in a customer's grocery cart. This data can be used to tailor individual mailings, analyze product performance, or forecast shopping patterns.

business processes that are supported by a system and the flow of data between processes. A rectangle with rounded corners represents a business process.

PepsiAmericas is the second-largest independent Pepsi bottler in the United States, with a territory that encompasses a large part of the Midwest and South. It invested over $16 million to upgrade its order processing system. The change is driven by increased competition from Coca-Cola Co. as both firms strive to introduce new drinks and compete for limited shelf space. One part of the investment involves providing 1,500 of its salespeople with Pocket PCs from Symbol Technologies. The sales force uses these devices to presell orders for a growing line of Pepsi products including bottled water, energy drinks, and Mountain Dew Red, as well as the traditional Pepsi products. The sales force enters orders into their Pocket PCs and transmits the information to a PeopleSoft-based order-shipping-billing system via a Cellular Digital Packet Data system operated by Verizon Wireless.[11]

With an on-line order processing system, such as one used by direct retailers, the status of each inventory item (also called a *stock keeping unit,* or *SKU*) on the order is checked to determine whether sufficient finished product is available. If an order item cannot be filled, a substitute item may be suggested or a back

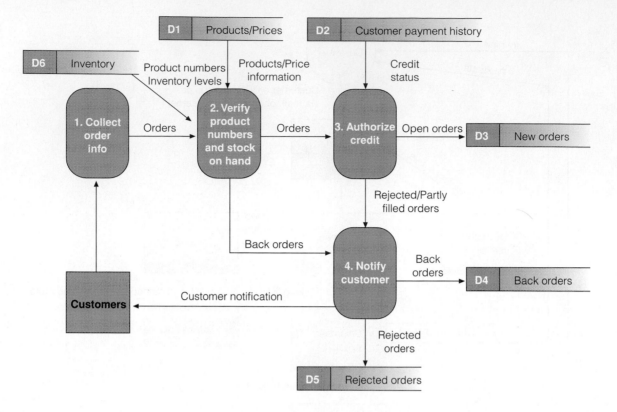

FIGURE 9 ⊙ 8

Data Flow Diagram of an Order Entry System

Orders are received by mail, phone, EDI, or the Internet from customers or sales reps and entered into the order processing system. This application affects accounting, inventory, warehousing, finance, and invoicing applications. Note that in an integrated order processing system, order entry personnel have access to back order, inventory, and customer information from separate data files or directly through the processing mechanism.

(Source: George W. Reynolds, *Information Systems for Managers*, 3rd ed., St. Paul, MN: West Publishing Co., 1995, p. 198. Reprinted with permission from Course Technology.)

order is created—the order will be filled later, when inventory is replenished. Order processing systems can also suggest related items for order takers to mention to promote add-on sales. Order takers also review customer payment history data from the accounts receivable system to determine whether credit can be extended.

Once an order is entered and accepted, it becomes an open order. Typically, a daily sales journal (which includes customer information, products ordered, quantity discounts, and prices) is generated.

Electronic data interchange (EDI) can be an important part of the order entry TPS. With EDI, a customer or client organization can place orders directly from its purchasing TPS into the order processing TPS of another organization. Or, the order processing TPS of the supplier companies and the purchasing TPS of the customers could be linked indirectly through a third-party clearinghouse. In any event, this computer-to-computer link allows efficient and effective processing of sales orders and enables an organization to lock in customers and lock out competitors through enhanced customer service. With EDI, orders can be placed anytime of the day or night, and immediate notification of order receipt and processing can be made. Today, more and more companies are using electronic data interchange to make paperless business transactions a reality.

As discussed in Chapter 8, order processing is being done through e-commerce and Internet systems to a greater extent today. However, many multinational companies use a wide variety of methods for order processing. For example, Coca-Cola de Mexico supplies 17 Mexican bottling groups with the Coke mixture. Each bottler provides periodic forecasts on how much of the mixture they need. Complicating this process is the fact that each bottler has its own level of technological sophistication. Some have state-of-the-art e-commerce or electronic data interchange systems in place, while others still rely on fax and telephone communications. As part of an upgrade to its order processing transaction system, Coca-Cola intends to establish a standardized ordering process and insist that all bottlers procure the mixture on a fixed weekly basis instead of randomly.[12]

Sales Configuration

sales configuration system
process that ensures that the products and services ordered are sufficient to accomplish the customer's objectives and will work well together

Another important aspect of order processing is sales configuration. The **sales configuration system** ensures that the products and services ordered are sufficient to accomplish the customer's objectives and will work well together. For example, using a sales configuration program, a sales representative knows that a computer printer needs a certain cable and a LAN card so that it can be connected to the LAN. Without a sales configuration program, a sales representative might sell a customer the wrong cable or forget the LAN card.

Sales configuration programs also suggest optional equipment. For example, if a customer orders a palmtop computer, the sales configuration program will suggest an AC adapter, backup software and cables, and a modem to allow the palmtop computer to connect to the Internet. If a company is buying a 747 aircraft from Boeing, a sales configuration program can help the sales representative work with the company to determine the number of seats that are needed, the most appropriate navigation systems to install, the type of landing gear that should be used, and hundreds of other available options that can be specified for the 747.

Sales configuration software can also solve customer problems and answer customer questions. For example, a sales configuration program can determine whether a factory robot made by one manufacturer can be controlled by a computer system developed by another manufacturer. Sales configuration programs can eliminate mistakes, reduce costs, and increase revenues. These advantages have led companies such as Hewlett-Packard, Boeing, and Silicon Graphics to implement these systems.

Shipment Planning

shipment planning system
system that determines which open orders will be filled and from which location they will be shipped

shipment execution system
system that coordinates the outflow of all products from the organization, with the objective of delivering quality products on time to customers

New orders received and any other orders not yet shipped (open orders) are passed from the order entry system to the shipment planning system. The **shipment planning system** determines which open orders will be filled and from which location they will be shipped. This is a trivial task for a small company with lots of inventory, only one shipping location, and a few customers concentrated in a small geographic area. But it is an extremely complicated task for a large global corporation with limited inventory (not all orders for all items can be filled), dozens of shipping locations (plants, warehouses, contract manufacturers, etc.), and tens of thousands of customers. The trick is to minimize shipping and warehousing costs while still meeting customer delivery dates.

The output of the shipment planning system is a plan that shows where each order is to be filled and a precise schedule for shipping with a specific carrier on a specific date and time. The system also prepares a picking list that is used by warehouse operators to select the ordered goods from the warehouse. These outputs may be in paper form, or they may be computer records that are transmitted electronically. The picking list document, an example of which is shown in Figure 9.9, lists the customer name, number, order number, and all items that have been ordered. A description of all items, along with the number to be shipped, is also included.

Many companies use a sophisticated warehouse management system to speed order processing time and improve inventory accuracy.

(Source: © Roger Tully/Getty Images.)

American Eagle Outfitters Inc. rolled out the Pick Ticket Management System (PkMS) application from Manhattan Associates and reduced labor costs by over $600,000 per year. The application sped up material handling by generating pick lists with products listed in optimal order to reduce order fulfillment time, employing bar codes on products, and optimizing the allocation of products shipped to its 640 stores.[13]

Shipment Execution

The **shipment execution system** coordinates the outflow of all products and goods from the organization, with the objective of delivering quality products on time to customers. The shipping department is usually given responsibility for physically packaging

FIGURE 9 9

A Picking List

This document guides warehouse employees in locating items to fill an order. Note the second and third columns of the slip, which instruct the warehouse workers where to locate the items. Also note that in this instance, the third item ordered was back-ordered three cases. Once items are picked from inventory, this data is entered into the data transaction processing system, and a packing slip and shipping notice are generated.

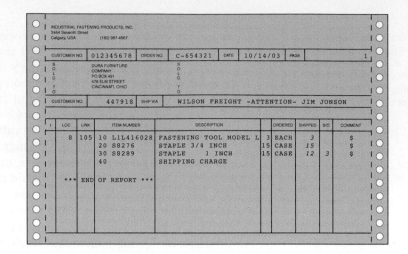

and delivering all products to customers and suppliers. This delivery can include mail services, trucking operations, and rail service. The system receives the picking list from the shipment planning system.

Sometimes orders cannot be filled exactly as specified. One reason is "out-of-stocks," meaning the warehouse does not have sufficient quantity of an item to fill a customer's order. Shortages can be caused if a production run did not produce the expected quantity of an item because of manufacturing problems. The company policy may be not to ship any of the item, ship as many units of the item as are available and create a back-order request for the remainder, or to substitute another item. Thus, as items are picked and loaded for shipment, warehouse operators must enter data about the exact items and quantity of each that are loaded for each order. When the shipment execution system processing cycle is complete, it passes the "shipped orders" business transactions "downstream" to the invoicing system. These transactions specify exactly the items that were shipped, the quantity of each, and the person or company to whom the order was shipped. This data is used to generate a customer invoice. The shipment execution system also produces packing documents, which are enclosed with the items being shipped, to tell customers what items are in the shipment, what is back-ordered, and the exact status of all items in the order. Soft-copy data—such as that provided by advanced shipment notices and shipment tracking systems—is also made available to other business functions.

Guide Corporation is an automobile headlight maker that implemented a shipment execution system to connect with GM's receiving and inventory control information system. Bar codes on the headlight assemblies are scanned at a Guide warehouse as they are loaded for shipping, and the information is stored on a computer located at Guide's plant. When the shipment arrives at GM, workers can access the stored shipment data to match it against what they receive. This process greatly speeds the receipt of goods, improves the accuracy of GM inventory, and simplifies the process of reconciling any differences between the number of items shipped and received and settling claims for damaged products.[14]

Inventory Control

inventory control system
system that updates the computerized inventory records to reflect the exact quantity on hand of each stock-keeping unit

For each item picked during the shipment execution process, a transaction providing the stock number and quantity picked is passed to the **inventory control system**. In this way, the computerized inventory records are updated to reflect the exact quantity on hand of each stock-keeping unit. Thus, when order takers check the inventory level of a product, they receive current information.

Once products have been picked out of inventory, other documents and reports are initiated by the inventory control application. For example, the inventory status

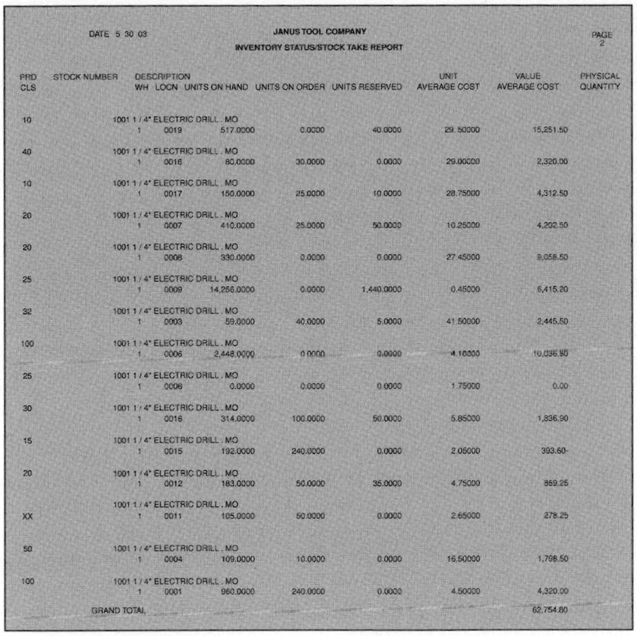

DATE 5 30 03

JANUS TOOL COMPANY

INVENTORY STATUS/STOCK TAKE REPORT

PAGE 2

PRD CLS	STOCK NUMBER	DESCRIPTION	WH	LOCN	UNITS ON HAND	UNITS ON ORDER	UNITS RESERVED	UNIT AVERAGE COST	VALUE AVERAGE COST	PHYSICAL QUANTITY
10	1001	1 / 4" ELECTRIC DRILL . MO	1	0019	517.0000	0.0000	40.0000	29. 50000	15,251.50	
40	1001	1 / 4" ELECTRIC DRILL . MO	1	0016	80.0000	30.0000	0.0000	29.00000	2,320.00	
10	1001	1 / 4" ELECTRIC DRILL . MO	1	0017	150.0000	25.0000	10.0000	28.75000	4,312.50	
20	1001	1 / 4" ELECTRIC DRILL . MO	1	0007	410.0000	25.0000	50.0000	10.25000	4,202.50	
20	1001	1 / 4" ELECTRIC DRILL . MO	1	0008	330.0000	0.0000	0.0000	27.45000	9,058.50	
25	1001	1 / 4" ELECTRIC DRILL . MO	1	0009	14,256.0000	0.0000	1,440.0000	0.45000	6,415.20	
32	1001	1 / 4" ELECTRIC DRILL . MO	1	0003	59.0000	40.0000	5.0000	41.50000	2,445.50	
100	1001	1 / 4" ELECTRIC DRILL . MO	1	0006	2,448.0000	0.0000	0.0000	4.10000	10,036.80	
25	1001	1 / 4" ELECTRIC DRILL . MO	1	0006	0.0000	0.0000	0.0000	1.75000	0.00	
30	1001	1 / 4" ELECTRIC DRILL . MO	1	0016	314.0000	100.0000	50.0000	5.85000	1,836.90	
15	1001	1 / 4" ELECTRIC DRILL . MO	1	0015	192.0000	240.0000	0.0000	2.05000	393.60-	
20	1001	1 / 4" ELECTRIC DRILL . MO	1	0012	183.0000	50.0000	35.0000	4.75000	869.25	
XX	1001	1 / 4" ELECTRIC DRILL . MO	1	0011	105.0000	50.0000	0.0000	2.65000	278.25	
50	1001	1 / 4" ELECTRIC DRILL . MO	1	0004	109.0000	10.0000	0.0000	16.50000	1,798.50	
100	1001	1 / 4" ELECTRIC DRILL . MO	1	0001	960.0000	240.0000	0.0000	4.50000	4,320.00	
	GRAND TOTAL								62,754.80	

FIGURE 9.10

An Inventory Status Report

This output from the inventory application summarizes all inventory items shipped over a specified time period.

report (Figure 9.10) summarizes all inventory items shipped over a specified time period. It can include stock numbers, descriptions, number of units on hand, number of units ordered, back-ordered units, average costs, and related information. It is used to determine when to order more inventory and how much of each item to order, and it helps minimize "stockouts" and back orders. Data from this report is used as input to other information systems to help production and operations managers analyze the production process.

For almost all companies, inventory must be tightly controlled. One objective is to minimize the amount of cash tied up in inventory by placing just the right amount of inventory on the factory or warehouse floor.

To gain a competitive advantage, many manufacturing organizations are moving to real-time inventory control systems based on bar coding the finished product, scanners and radio display terminals mounted on forklifts, and wireless LAN communications to track each time an item is moved in the warehouse. One significant advantage is that the inventory data is more accurate and current for people performing order entry, production planning, and shipment planning. In addition, warehouse operations can be streamlined by providing directions to the forklift drivers.

In addition to being useful for physical goods such as automobiles and home appliances, inventory control is essential for industries in the service sector. Such organizations as hotels, airlines, rental car agencies, and universities, which primarily provide services, can use inventory applications to help them monitor use of rooms, airline seats, car rentals, and classroom capacity. Airlines face an especially difficult inventory problem. Empty airline seats (inventory) have absolutely no value after a plane takes off. Yet overbooking can result in too many seats being sold and customer complaints. Sophisticated reservation systems allow airlines to quickly update and add seating assignments.

The Men's Wearhouse captures real-time sales data to enable it to follow sales at any store on a given day. The clothing chain carries unusual sizes, from 56 on the large end down to 22 on the small end, but doesn't order a lot of items in these sizes for each store. Allocation of items with very limited inventory has caused problems for the Men's Wearhouse in the past. So, the firm built a telecommunications network connecting cash registers in its 400 stores with its Houston headquarters to capture sales data. Linking this data to its inventory control system helps the Men's Wearhouse move inventory among its stores because it understands what's selling and where and what's not selling.[15]

Invoicing

Customer invoices are generated based on records received from the shipment execution TPS. This application encourages follow-up on existing sales activities, increases profitability, and improves customer service. Most invoicing programs automatically compute discounts, applicable taxes, and other miscellaneous charges (Figure 9.11). Because most computerized operations contain elaborate databases on customers and inventory, many invoicing applications require only information on the items ordered and the client identification number; the invoicing application does the rest. It looks up the full name and address of the customer, determines whether the customer has an adequate credit rating, automatically computes discounts, adds taxes and other charges, and prepares invoices and envelopes for mailing.

THANK YOU FOR YOUR BUSINESS, OUR
FAX NO. IS 1-555-123-4567

PINNACLE MFG. & SUPPLY CO.

LOCATION 390
CINCINNATI, OH 45200
PHONE 555-765-4321
PLEASE REMIT TO THE ABOVE ADDRESS

INVOICE NO.	
001	1020345
PLEASE REFER TO THIS INVOICE NUMBER IN ALL CORRESPONDENCE	
PAGE	-1

TERMS:
NET 30TH
1 1/2% SERVICE CHARGE WILL BE CHARGED MONTHLY ON PAST DUE ACCOUNTS

S O L D   T O
BEST DEAL FOODS   (1)   041017
1714 PLEASANT AVENUE
CINCINNATI, OH 45200

S H I P   T O
BEST DEAL FOODS
1714 PLEASANT AVENUE
CINCINNATI, OH 45200

CUSTOMER ORDER NO.	ORDER DATE	DATE SHIPPED	INVOICE DATE	JOB NAME/JOB NUMBER	SHIPPED VIA
YZ-000066	7/30/03	7/31/03	8/04/03	OLEAN RESUPPLY 1	JASON 2 DEL.

QUANTITY ORDERED	QUANTITY SHIPPED	QUANTITY BACKORDERED	ITEM NO.	DESCRIPTION	UNIT PRICE	UNIT MEAS	TOTAL
26	23	3	12165	1 X 3/4 150# T304 SS REDUCER SS4R134	6.250	EA	143.75
30	30		12162	3/4 X 1/2 150# T304 SS REDUCER SS4R3412	3.570	EA	107.10
18	13	5	12160	1/2 X 3/8 150# T304 SS REDUCER SS4R1238	2.760	EA	35.88

MAIN OFFICE: CINCINNATI, OH. **BRANCH OFFICES:** COLUMBUS, OH. LOUISVILLE, KY. • LEXINGTON, KY. • INDIANAPOLIS, IN.

BILLING DISCREPANCIES MUST BE REPORTED WITHIN 30 DAYS OF THE INVOICE

A RESTOCKING CHARGE WILL BE CHARGED ON RETURNED MERCHANDISE. NO MERCHANDISE RETURNED WITHOUT OUR WRITTEN AUTHORIZATION. PAST DUE AMOUNTS SUBJECT TO SERVICE CHARGE AT THE HIGHEST LEGAL RATE ALLOWABLE AND REASONABLE ATTORNEY'S FEES IF ACCOUNT IS PLACED FOR COLLECTION.

MATERIAL		286.73
OH SALES TAX		.00
TOTAL AMOUNT DUE		286.73

DEDUCT   .00 FROM TOTAL IF PAID BY TERMS

FIGURE 9 ⊙ 11

An output of the invoicing system, a customer invoice reflects the value of the current invoice, as well as which products the customer purchased.

customer relationship management (CRM) system
a collection of people, processes, software, and Internet capabilities that help an enterprise manage customer relationships effectively and systematically

Invoicing in a service organization can be even more complicated than invoicing in manufacturing and retail firms. The trick is to match all services rendered with a specific customer and to include all appropriate rates and charges in calculating the bill. This is especially difficult if the data needed for billing has not been accurately and completely captured in a TPS.

Customer Relationship Management (CRM)

A **customer relationship management (CRM) system** is a collection of people, processes, software, and Internet capabilities that help an enterprise manage customer relationships effectively and systematically (Figure 9.12). The goal of CRM is to understand and anticipate the needs of current and potential customers to increase customer retention and loyalty while optimizing the way products and services are sold.

CRM software automates and integrates the functions of sales, marketing, and service in an organization. The objective is to capture data about every contact a company has with a customer through every channel and store it in the CRM system to enable the company to truly understand customer actions. CRM software helps an organization build a database about its customers that describes relationships in sufficient detail so that management, salespeople, customer service providers, and even customers can access information to match customer needs with product plans and offerings, remind customers of service requirements, know what other products a customer had purchased, and any number of things.

CRM software is now the number one selling software application in the world, having surpassed enterprise resource planning applications in 2001 in terms of total license revenues. The top 15 rated CRM software packages for 2002 are listed in alphabetical order in Table 9.3.

The focus of CRM involves much more than installing new software. Moving from a culture of simply selling products to placing the customer first is essential to a successful CRM deployment. Before any software is loaded onto a computer, a company must retrain employees. Who handles customer issues and when must be clearly defined, and computer systems need to be integrated so all pertinent information is available immediately, whether a customer calls a sales representative or customer service representative.[16] Blue Cross Blue Shield of Massachusetts has implemented a CRM software package from Pegasystems to help it improve service and sell additional products to existing customers. The CRM system builds a central repository of member-contact histories to ensure that all lines of business and all channels for communication, such as e-mail, fax, phone, and Web site, provide a consistent view of the customer. Blue Cross Blue Shield customers can go on-line and enter address and ID-card changes. The application also lets members retrieve specific information about their policies and benefits, including information on copayment requirements and covered services. The CRM software provides scripting capabilities that let Blue Cross Blue Shield's managers customize service scripts that prompt a customer-service agent to offer specific information to a customer when requested.[17]

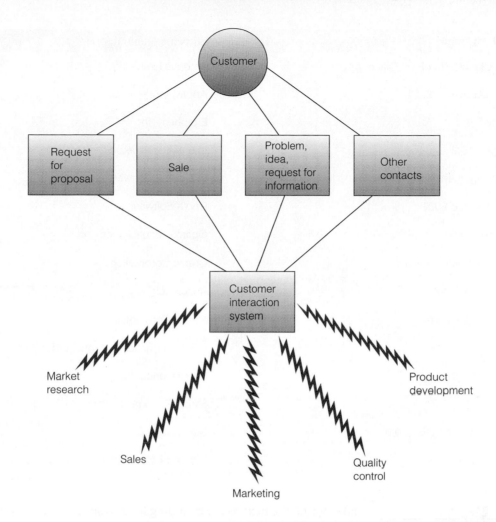

FIGURE 9.12

*Customer Relationship
Management System*

E.piphany has designed its CRM system to make it easier for sales staff to access customer data, receive relevant sales information, and track customer inquiries. Its new ActivePath technology pulls all of the relevant information from various TPS applications and presents it to the salesperson—including all relevant product and sales data and customer interaction history. Sales associates who open the calendar within the sales application see an icon next to each customer's name. When highlighted with a mouse, the icon provides specific suggestions for the salesperson, depending on the status of the account. If the customer hasn't bought a product yet, the system may suggest ways to prepare for the sale. The system can also pull external data, such as driving directions to the customer site from MapQuest or company information from Hoover's Online. Users can also view marketing campaigns and evaluate suggested cross-selling opportunities.[18]

In addition to using stationary computers, some CRM systems can be accessed via wireless devices. Firstwave Technologies offers eCRM Suite 7.0, a package of Web-based CRM applications for small to medium-sized businesses. The new suite includes eSales, which lets sales professionals access customer data via mobile devices and the Internet.[19]

Routing and Scheduling

Many computer manufacturers and software firms have developed specialized transaction processing systems for companies in the distribution industry. Some distribution applications are for wholesale operations; others are for retail or specialized applications. Trucking firms, beverage distributors, electrical distributors, and oil and natural gas distribution companies are only a few examples.

Like airlines, distribution companies must also determine the best use of their resources. For example, a motor freight company might have 100 deliveries to

Software Package	Software Manufacturer
Amdocs ClarifyCRM eFrontOffice v. 10.2	Amdocs Limited
Applix iEnterprise v. 8.3	Applix, Inc.
E.piphany E.5	E.piphany, Inc.
SalesLogix v. 5.2	Interact Commerce Corporation
Client Management Software 5.0	Oncontact Software Corporation
ONYX Enterprise 2001	ONYX Software
ExSellence v. 4.0	Optima Technologies, Inc.
Oracle CRM Suite 11i v. 5.5	Oracle Corporation
PeopleSoft 8.1 CRM	PeopleSoft, Inc.
Pivotal eRelationship	Pivotal Corporation
e-point 5.2	Point Information Systems, Ltd.
mySAP CRM 3.0	SAP America, Inc.
Siebel 7	Siebel Systems, Inc.
Staffware Process Suite v. 9.0	Staffware PLC
Worldtrak v. 5.3	Worldtrak Corporation

T A B L E   9 . 3

The 15 Top-Rated CRM Software Packages

Source: Excerpted from "ISM Unveils the 2002 Top 30 CRM Software Packages," February 13, 2002, accessed at http://www.ismguide.com. Used with permission.

routing system
system that determines the best way to get products from one location to another

scheduling system
system that determines the best time to deliver goods and services

purchasing transaction processing systems
systems that include inventory control, purchase order processing, receiving, and accounts payable

make during the next week, including loads from Miami to Boston and Seattle to Salt Lake City. A **routing system** helps determine the best way to get products from one location to another.

The **scheduling system** determines the best time to pick up or deliver goods and services. For example, trucks can be scheduled to deliver automobile transmission systems from California to Michigan during the second week of September, when oil and gas prices are low. Other objectives are to carry a profitable load on the return trip and to minimize total distance traveled, which can result in lower fuel, driver, and truck maintenance costs. For these reasons, many distribution companies have designed TPSs to help determine which routes will allow for efficient service, while making cost-effective use of drivers and trucks. For firms such as these, scheduling and routing programs are connected to the organization's order and inventory transaction processing system. FedEx uses an automated vehicle routing and scheduling system that uses the customer information to determine which packages should go on which vehicles as well as the delivery routes drivers should take and their sequence of stops. FedEx also uses a geographical information system to generate computerized maps and turn-by-turn directions for each driver. The technology ensures that drivers cover the fewest miles in the shortest time thus saving time and reducing costs.

PURCHASING SYSTEMS

The **purchasing transaction processing systems** include inventory control, purchase order processing, receiving, and accounts payable (Figure 9.13).

Inventory Control

A manufacturing firm has several kinds of inventory, such as raw materials, packing materials, finished goods, and maintenance parts. Each day manufacturers must

FIGURE 9 13

*Purchasing Transaction
Processing System*

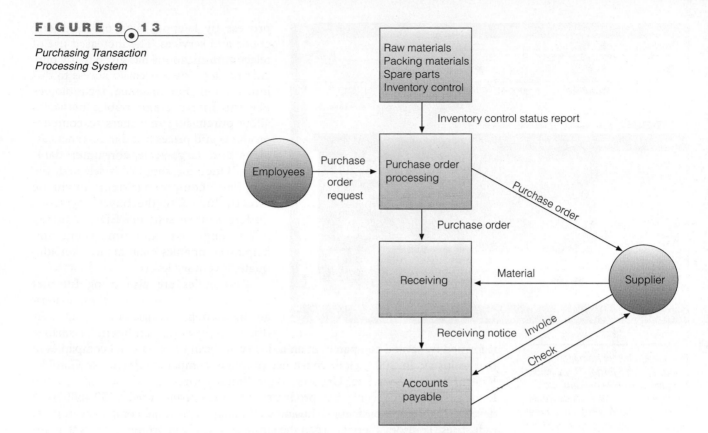

determine how much of these items to buy to make and ship the products customers have ordered. If they buy too much, they will tie up too much cash in idle
inventory, but if they don't keep enough of these items on hand, they will be
unable to meet demand and lose sales revenue. A mistake in tracking current raw
material inventory can cost millions of dollars, causing the manufacturer to miss
profitability targets. We have already discussed the use of an inventory control
system for finished product inventory. A similar transaction processing system
can be used to manage the inventory of these items.

Sun Microsystems uses an accurate inventory control system and demand
forecast data that factors in market volatility, the price of components, their rate
of depreciation, and their availability to determine how many of each server component to order. The system performs a complex set of statistical and probability
calculations using the type of data just mentioned to make purchasing recommendations for the most critical components on the production line.[20]

Purchase Order Processing

An organization's purchasing department typically has a number of employees
who are responsible for all purchasing activities for the organization. Whenever
materials or high-cost items are purchased, the purchasing department is
involved. The **purchase order processing system** helps purchasing departments complete their transactions quickly and efficiently. Every organization has
its own policies, practices, and procedures for purchasing supplies and equipment. DaimlerChrysler AG's Mercedes Benz auto production operation in Spain
has reduced procurements costs by $18 million by employing purchasing soft-
ware. Use of the software plus refinement of work processes has reduced the
number of workers required to check invoices and process orders, enabled consolidation of contracts with vendors, identified opportunities to negotiate better
contracts with vendors, and cut the paperwork required for orders.[21]

Companies are increasingly purchasing needed supplies through the Internet
or an Internet exchange. The purchasing department can facilitate the buying

**purchase order processing
system**
system that helps purchasing
departments complete their transactions quickly and efficiently

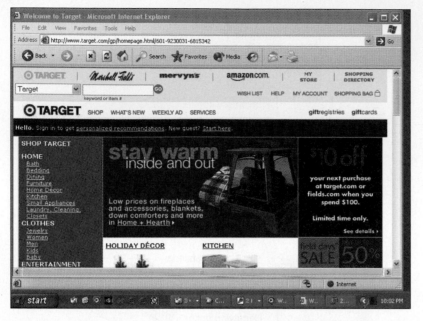

Big name retailers from around the world, such as Target, Royal Ahold, Kingfisher, and Dairy Farm, are members of The WorldWide Retail Exchange (the WWRE), an Internet-based business-to-business (B2B) exchange in the retail e-market-place. The WWRE enables retailers and suppliers in the food, general merchandise, textile/home, and drugstore sectors to substantially reduce costs across product development, e-procurement, and supply chain processes.

process by keeping data on suppliers' goods and services. The increased use of telecommunications has given many purchasing departments easier access to this information. For instance, technologies like the Internet and public networks allow purchasing managers to compare products and prices listed in Internet catalogs and large-scale consumer databases. Once the supplier is selected, the suppliers' computer systems might be directly linked to the buyer's systems. Orders can be sent via EDI, reducing purchasing costs and time spent and helping companies maintain low, yet adequate, inventory levels.

Companies are also using Internet exchanges to help them purchase materials and supplies at discounted prices. As discussed previously, an Internet exchange is formed by several companies in an industry and can be open to all companies in that industry. In 2000, giant consumer products companies Procter & Gamble, Kraft, Nabisco, Sara Lee, Unilever, Pepsi Bottling Group, and forty-three other food, beverage, and consumer products companies, committed $250 million to develop Transora, a business-to-business exchange open to all companies in these industries. Transora's services span the entire supply chain, from suppliers to manufacturers to retailers, and provide procurement, vendor and product catalogs, online order management, supply chain collaboration, and financial services. Suppliers benefit by gaining access to a larger customer base and are able to reduce customer acquisition costs. Participating manufacturers gain from improved customer service with retailers and wholesalers. Retailers and wholesalers are able to simplify their ordering process and improve order accuracy.

Volkswagen developed an exchange to help it purchase factory parts, tools, office equipment, and other products. In its first 15 months of operation, Volkswagen AG spent $5.2 billion buying parts, PCs, and raw materials over the exchange and estimates that it has cut procurement costs by 40 to 50 percent while reducing contract negotiations from as long as three months to one day.[22] Companies can use software to help them take advantage of purchasing materials and suppliers using multiple Internet exchanges.

Instead of searching for the lowest prices from a list of suppliers, many companies form strategic partnerships with one or two major suppliers for important parts and materials. The partners are chosen based on prices and their ability to deliver quality products on time consistently. For example, the automotive companies request that their suppliers have plants or offices close to their operations in Michigan. The ability to conduct electronic commerce following EDI standards is also a key factor.

Receiving

Like centralized purchasing, many organizations have a centralized receiving department responsible for taking possession of all incoming items, inspecting them, and routing them to the people or departments that ordered them. In addition, the receiving department notifies the purchasing department when items have been received. This notification may be done using a paper form called a *receiving report* or electronically through a business transaction created by entering data into the receiving TPS.

An important function of many receiving departments is quality control by inspection. Inspection procedures and practices are set up to monitor the quality of incoming items. Any items that fail inspection are sent back to the supplier, or adjustments are made to compensate for faulty or defective products.

receiving system
system that creates a record of expected receipts

Many suppliers now send their customers advance shipment notices. This business transaction is input to the customer's **receiving system** to create a record of expected receipts. In addition, items are shipped with a bar code identification on the container. At the receiving dock, the worker scans the bar code on each container, and a transaction is sent to the receiving system, where the bar code identification number is matched against the file of expected receipt records. This additional check improves the accuracy of the receiving process, eliminates the need to perform manual data entry, and reduces the manual effort required.

Accounts Payable

accounts payable system
system that increases an organization's control over purchasing, improves cash flow, increases profitability, and provides more effective management of current liabilities

The **accounts payable system** attempts to increase an organization's control over purchasing, improve cash flow, increase profitability, and provide more effective management of current liabilities. Checks to suppliers for materials and services are the major outputs. Most accounts payable applications strive to manage cash flow and minimize manual data entry. Input from the purchase order processing system provides an electronic record to the accounts payable application that updates the accounts payable database to create a liability record showing that the firm has made a commitment to purchase a specific good or service. Once the accounts payable department receives a bill from a supplier, the bill is verified and checked for accuracy. Upon receiving notice that the goods and services have been delivered in a satisfactory manner, the data is entered into the accounts payable application. A typical check from an accounts payable application is shown in Figure 9.14. In addition to containing standard information found on any check, most accounts payable checks include the items ordered, invoice date, invoice numbers, amount of each item, any discounts, and the total amount of the check. This information allows the company to consolidate several invoices and bills into a single payment. In addition to checks, companies can also pay their suppliers electronically using EDI, the Internet, or other electronic payment systems.

In spite of the ability to use e-commerce to buy goods and services, the reconciliation of purchase orders, supplier invoices, and shipping documents with one another so bills can be paid is still a time-intensive manual process. General Electric plans to launch an application that will automate the reconciliation process through its EDI subsidiary, GE Global eXchange Services Inc. (GXS). GXS will translate EDI, Web-based, and spreadsheet documents into an XML data format and then make them available on-line to suppliers and GE's buyers. Data translation, workflow, and the storing of the actual database will all be

FIGURE 9.14

A Check Generated by an Accounts Payable Application

The check stub details items ordered, invoice dates, invoice numbers, cost of each item, discounts, and the total amount of the check.

MEM	INVOICE	INVOICE NUMBER	AMOUN	DISCOUN	NET
LAWN ORNAM	07 01 03	1701	166.67	1.67	165.00
LAWN STATU	07 05 03	2211	110.00	1.10	108.90
	DETACH BEFORE		276.67	2.77	273.90

handled by GXS. The goal is to speed the purchasing process, enabling companies to achieve the quick-payment discounts and lower inventory costs that were expected from on-line procurement projects.[23]

A common report produced by the accounts payable application is the purchases journal. As shown in Figure 9.15, this report summarizes all the organization's bill-paying activities for a particular period. Financial managers use this report to analyze bills that have been paid by the organization. This information is also used to help analyze current and future cash flow needs. Data is summarized for each supplier or parts manufacturer. Invoice number, description, amounts, discounts, and total checking activity are included. In addition, many purchasing reports include the total amount of checks generated by the accounts payable application on a daily, weekly, or monthly basis.

The accounts payable application ties into other information systems, including cash flow analysis, which helps an organization ensure that sufficient funds are available for the accounts payable application and can show the best sources of funds for payments that must be made via the accounts payable application.

ACCOUNTING SYSTEMS

accounting systems
systems that include budget, accounts receivable, payroll, asset management, and general ledger

The primary **accounting systems** include the budget, accounts receivable, payroll, asset management, and general ledger (Figure 9.16).

Budget

budget transaction processing system
system that automates many of the tasks required to amass budget data, distribute it to users, and consolidate the prepared budgets

In an organization, a budget can be considered a financial plan that identifies items and dollar amounts that the organization estimates it will spend. In some organizations, budgeting can be an expensive and time-consuming process of manually distributing and consolidating information. The **budget transaction processing system** automates many of the tasks required to amass budget data, distribute it to users, and consolidate the prepared budgets. Automating the budget process allows financial analysts more time to manage it to meet organizational goals by setting enterprisewide budgeting targets, ensuring a consistent budget model and assumptions across the organization, and monitoring the status of each department's spending.

Mission St. Joseph's Health System of Asheville, North Carolina, cut the time required to prepare its annual budget in half by using Hyperion's Planning software. The new system replaced a process that relied on Excel spreadsheets, paper

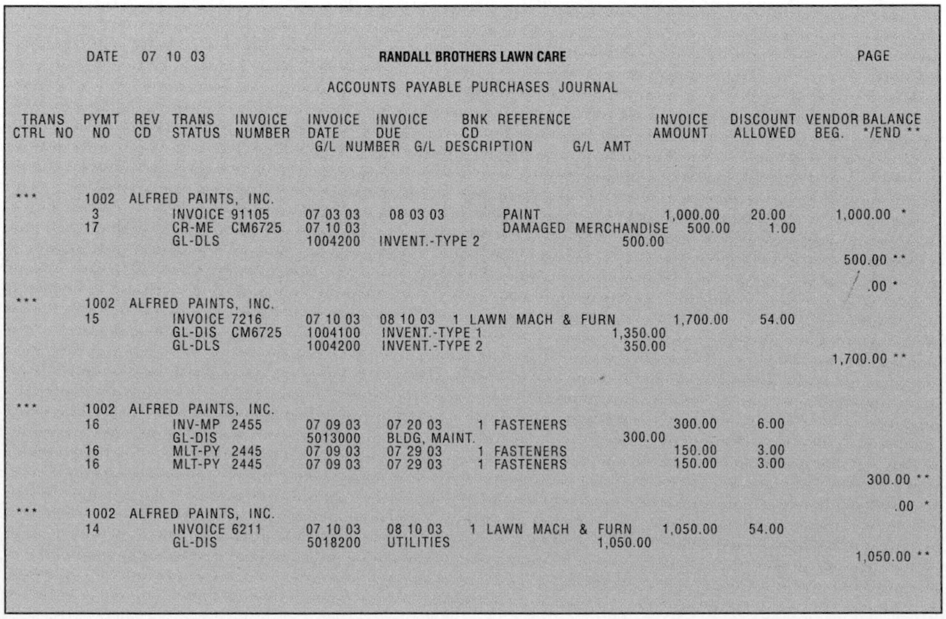

```
      DATE   07 10 03                    RANDALL BROTHERS LAWN CARE                            PAGE

                                       ACCOUNTS PAYABLE PURCHASES JOURNAL

 TRANS   PYMT  REV  TRANS   INVOICE   INVOICE  INVOICE     BNK REFERENCE           INVOICE   DISCOUNT  VENDOR BALANCE
CTRL NO   NO   CD   STATUS  NUMBER    DATE     DUE         CD                      AMOUNT    ALLOWED   BEG.   */END **
                                       G/L NUMBER  G/L DESCRIPTION      G/L AMT

***      1002  ALFRED PAINTS, INC.
          3         INVOICE 91105     07 03 03   08 03 03        PAINT                1,000.00    20.00        1,000.00 *
         17         CR-ME   CM6725    07 10 03                   DAMAGED MERCHANDISE   500.00      1.00
                    GL-DLS            1004200   INVENT.-TYPE 2                500.00
                                                                                                                500.00 **
                                                                                                              /   .00 *
***      1002  ALFRED PAINTS, INC.
         15         INVOICE 7216      07 10 03  08 10 03  1 LAWN MACH & FURN          1,700.00    54.00
                    GL-DIS  CM6725    1004100   INVENT.-TYPE 1              1,350.00
                    GL-DLS            1004200   INVENT.-TYPE 2                350.00
                                                                                                              1,700.00 **

***      1002  ALFRED PAINTS, INC.
         16         INV-MP  2455      07 09 03   07 20 03     1 FASTENERS              300.00     6.00
                    GL-DIS            5013000    BLDG, MAINT.                 300.00
         16         MLT-PY  2445      07 09 03   07 29 03     1 FASTENERS              150.00     3.00
         16         MLT-PY  2445      07 09 03   07 29 03     1 FASTENERS              150.00     3.00
                                                                                                                 300.00 **
                                                                                                                    .00 *
***      1002  ALFRED PAINTS, INC.
         14         INVOICE 6211      07 10 03   08 10 03    1 LAWN MACH & FURN    1,050.00    54.00
                    GL-DIS            5018200    UTILITIES                 1,050.00
                                                                                                               1,050.00 **
```

FIGURE 9.15

An Accounts Payable Purchases Journal

Generated by the accounts payable application, this report summarizes an organization's bill-paying activities for a particular period.

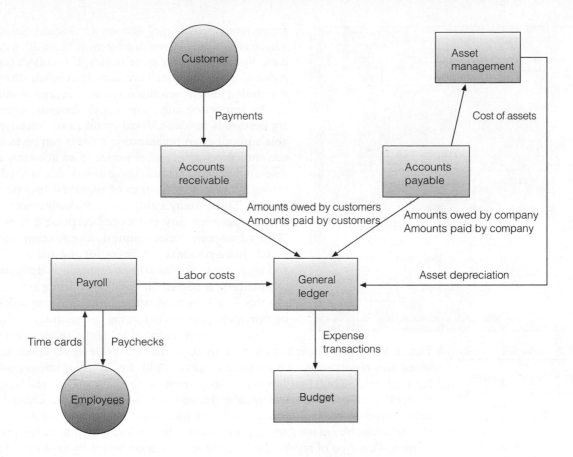

FIGURE 9.16

Financial Systems

accounts receivable system
system that manages the cash flow
of the company by keeping track
of the money owed the company
on charges for goods sold and
services performed

worksheets, and custom applications. While the decision to change to a new
budget system was initially greeted with skepticism, the more than 100 people
involved in departmental budget preparation soon embraced the software
because of its ease of use. Training for each user took about 30 minutes.[24]

Accounts Receivable

The **accounts receivable system** manages the cash flow of the company by keeping track of the money owed the company on charges for goods sold and services
performed. When goods are shipped to a customer, the customer's accounts payable
system receives a business transaction from the invoicing system, and the customer's
account is updated in the accounts receivable system of the supplier. A statement
reflecting the balance due is sent to active customers. Upon receipt of payment, the
amount due from that customer is reduced by the amount of payment.

The major output of the accounts receivable application is monthly bills or
statements sent to customers. As you can see in Figure 9.17, a bill sent to a customer can include the date items are purchased, descriptions, reference numbers, and amounts. In addition, bills can include amounts for various periods,
totals, and allowances for discounts. The accounts receivable application should
monitor sales activity, improve cash flow by reducing the time between a customer's receipt of items ordered and payment of bills for those items, and ensure
that customers continue to contribute to profitability. Most systems can handle
payment in a variety of ways, including standard bank checks, credit cards,
money-wiring services, and electronic funds transfer via EDI. Increasingly, companies are using the Internet for their accounts receivable application. Using
these Internet systems, customers can pay their bills while connected to the
Internet on their home PC, at a retail store, or using a handheld computer with a
wireless connection to the Internet.

The accounts receivable system is vital to managing the cash flow of the firm.
One major way to increase cash flow is by identifying overdue accounts. Reports

STATEMENT OF ACCOUNT

ABBEY WINDOW CLEANING CO. 163 BERKSHIRE AVE. ANY CITY, ANYWHERE 12345	STATEMENT DATE	ACCOUNT NUMBER	
	05/27/03	1001001	

Amount Paid $ _____

PLEASE DETACH AT PERFORATION AND RETURN TOP PORTION WITH YOUR CHECK

STATEMENT DATE	ACCOUNT NO.		PREVIOUS BALANCE:
05-27-03	1001001		267.27

DATE	DESCRIPTION	REF. NO.	REFERENCE ONLY	DEBITS	CREDITS
05-01-03	INVOICE	1700		65.95	
05-15-03	INVOICE	3064		203.32	
05-01-03		3067		60.25	
05-15-03	INVOICE	3069		18.33	
05-01-03		3072		254.82	
05-15-03	INVOICE	3076		1,222.51	
05-15-03		3079		7.57	

UNPAID BALANCE:
2,402.27

SUM OF PERIODS - CREDITS = UNPAID BALANCE				
PERIOD 1	PERIOD 2	PERIOD 3	PERIOD 4	PERIOD 5
1,832.75	56.40	17.56	106.36	74.95

FIGURE 9.17

An Accounts Receivable Statement

Generated by the accounts receivable application, a bill is sent to a customer (usually monthly) and details items purchased, dates of purchase, and amounts due.

are generated that "age" accounts to identify customers whose accounts are overdue by more than 30, 60, or 90 days. Special action may be initiated to collect funds or reduce the customer's credit limit with the firm, depending on the amount owed and degree of lateness.

An important function of the accounts receivable application is to identify bad credit risks. Because a sizable amount of an organization's assets can be tied up in accounts receivable, one objective of an accounts receivable application is to minimize losses due to bad debts through early identification of potential bad-debt customers. Thus, many companies routinely check a customer's payment history before accepting a new order. With advances in telecommunications, companies can search huge national databases for the names of firms and individuals who have been reported as delinquent on payments or as bad credit risks. When using external data like this in a TPS application, however, companies must be extremely cautious regarding the accuracy of the data.

The accounts receivable aging report, shown in Figure 9.18, is a valuable aspect of an accounts receivable application. In most cases, this report sorts all outstanding debts or bills by date. For unpaid bills that have remained outstanding for a predetermined amount of time, "reminder notices" can also be automatically generated. The accounts receivable aging report gives managers an immediate look into large bills that are long overdue so that they can be followed up to prevent more orders being sent to delinquent customers. This type of report can be produced on a customer-by-customer basis or in a summary format.

The two primary outputs of the payroll system are the payroll check and stub, which are distributed to the employees, and the payroll register, which is a summary report of all payroll transactions. In addition, the payroll system prepares W-2 statements at the end of the year for tax purposes. In a manufacturing firm, hours worked and labor costs may be captured by job so that this information can be passed on to the manufacturing costs system.

FIGURE 9.18

An Accounts Receivable Aging Report

The output from an accounts receivable application tells managers what bills are overdue, either customer by customer or in a summary format.

6588-DIAGE	TOOL DISTRIBUTORS INC.	0001 ACCOUNTS RECEIVABLE AGING ANALYSIS AS OF SEP 25 2003		SEP 25 2003	PAGE 1	

TYPE	OPEN-ITEM NUMBER	ITM-DATE	REFERENCE NUMBER	INVOICE/PAYMENT AMOUNT	CURRENT	31-60 DAYS	61-90 DAYS	91-120 DAYS	OVER 120 DAYS

CLASS 05	CUST 94367	NAME TOOLS OF AMERICA	3225 N PARKWAY	TEL 513-2664731	CONT B. BROWN			

INV	95361	03/17/03	23,058.37					23,058.37
OM	96853	03/20/03	5,589.42					1,589.42
INV	105395	07/04/03	2,923.45		2,923.45			
OM	116594	07/06/03	198.32		198.32			
INV	123984	07/15/03	23,087.28					
DSC	123984	08/17/03	1,204.36					
CA	968351	08/19/03	21,882.92					
INV	147296	07/23/03	19,709.57					
CA	83495	07/31/03	18,725.21					
CM	995473	08/19/03	984.59					
INV	149384	08/29/03	23,831.37	21,831.59				
INV	158439	09/10/03	30,086.68	30,086.68				
INV	161236	09/23/03	25,520.37	25,520.37				

TOTL RECEIVABLE FOR CUST 94367	107,208.20	79,438.64	3,121.77	0.00	0.00	24,647.79
CA	3,121.77		3,121.77			
CA	22,640.01	22,640.01				
CM	279.84	279.84				

| UNAPPLIED CREDITS | 26,041.62 | 22,919.85 | 3,121.77 | 0.00 | 0.00 | 0.00 |

| NET RECEIVABLE FOR CUST 94367 | 81,166.58 | 56,518.79 | 0.00 | 0.00 | 0.00 | 24,647.79 |

TOTAL FOR CLASS	05	42	PRINTED		TOTAL RECEIVABLE	UNAPPLIED CREDITS	NET RECEIVABLE
				CURRENT	161,506.12	67,832.57	93,673.55
				31-60 DAYS	31,494.14	18,984.68	12,509.46
				61-90 DAYS	11,569.32	4,267.91	7,301.59
				91-120 DAYS	27,764.18	0.60	27,746.18
				TOTAL	238,823.97	93,232.49	145,581.08

Responsibility for running this TPS application can be outsourced to a service company. Rather than write their own payroll application, many firms rely on a purchased software application for payroll processing. Some of these packages are tailored for a specific industry; others can accommodate a wide range of uses. In most cases, the number of hours worked by each employee is collected using a variety of data entry devices, including time clocks, time cards, and industrial data-collection devices. Once collected, payroll data is used to prepare weekly, biweekly, or monthly employee paychecks (Figure 9.19). Payroll systems can handle overtime, vacation pay, variable and multirate salary structures, incentive programs, and commissions. Most payroll applications automatically generate both federal and state tax forms related to payroll and process deductions, including tax-deferred annuities, savings plans, and U.S. government savings bonds. Often payroll applications have EDI arrangements with employees' banks to make direct deposits into employees' accounts.

payroll journal
a report that contains employees' names, the area where employees worked during the week, hours worked, the pay rate, a premium factor for overtime pay, earnings, earnings type, various deductions, and net pay calculations

As you can see in Figure 9.19, the payroll program has produced a weekly paycheck, which includes the employee's hourly rate, total hours worked, regular pay, premium pay, federal and state tax withholdings, and other deductions. In addition to paychecks, most payroll programs produce a **payroll journal**, shown in Figure 9.20. A typical payroll journal contains employees' names, the areas where employees worked during the week, hours worked, the pay rate, a premium factor for overtime pay, earnings, the earnings type, various deductions, and net pay calculations. Financial managers at the operational level use the payroll journal to monitor and control pay to individual employees. As you can see in Figure 9.20, the payroll journal also includes totals for hours worked, earnings, deductions, and net pay.

Payroll TPS applications also provide input into various weekly, quarterly, and yearly reports. Most of these reports are used by financial managers to help control payroll costs and cash flows. Payroll applications also provide necessary audit trails through the use of payroll master files and documents used by internal accountants and external auditors to make sure that the application is functioning as intended.

Like many other transaction processing applications, the payroll application interfaces with other applications. All payroll entries are entered to the general ledger systems. Furthermore, there can be a direct link between payroll activities and production/inventory control operations. Direct links are often used for manufacturing operations or job-shop systems because data collected on hours worked from the payroll application helps determine the total cost of completing various jobs. For example, if an employee who earns $15 per hour spends 20 hours completing a particular job, the labor costs for that job are $300. This type of information from the payroll application is useful in determining the cost to produce a product or render a service and thus in determining its profitability.

Asset Management

Capital assets represent major investments for the organization whose value appears on the balance sheet under fixed assets. These assets have a useful life of several years or more, over which their value is depreciated, resulting in a tax reduction.

FIGURE 9.19

A typical paycheck stub details the employee's hours worked for the period, salary, vacation pay, federal and state taxes withheld, and other deductions.

TENDER CARE DAY CARE CENTER, INC. CINCINNATI, OH 45200				4207
Vacation Taken This Check		0.000	Sick Taken This Check	0.000
Vacation Available		0.667	Sick Available	0.667
Earnings			Deductions	
Description	Hours	Amount	Description	Amount
Regular Pay	36.320	199.76	FICA Withheld	15.28
Overtime	0.000	0.00	Fed. Tax W/H	15.00
SICK TIME	0.000	0.00	State Tax W/H	1.67
PERSONAL	0.000	0.00	Other W/H #1	4.19
VACATION	0.000	0.00	Other W/H #2	0.00
HOLIDAY PAY	0.000	0.00	Other W/H #3	0.00
Gross Pay#4	0.000	0.00	Other W/H #4	0.00
Total		199.76	Total	36.14
YEAR TO DATE				
Total Earnings		397.76	FICA	30.43

FIGURE 9.20

A Payroll Journal

Generated by the payroll application, this report helps managers monitor total payroll costs for an organization and the impact of those costs on cash flow.

```
DATE  12 25 03                    RANDALL BROTHERS LAWN CARE                              PAGE  003

                                        PAYROLL JOURNAL
                                             PR020

   EMPLOYEE NAME
ERRORS            EMPLOYEE   COST           PREM                EARN                  DED            PAYMENT
WARNINGS           NUMBER   CENTER  HOURS  RATE  FACTOR EARNINGS TYPE  DEDUCTIONS    TYPE  NET PAY  LOCATION

   DANIEL JACKSON
                     112      30    40.00  6.00         240.00   001                         44.19   CIN
                     112      30    10.00  6.00   1.5    90.00   030
                     112                                               5.00         220
                     112                                              15.81         801
                     112                                              44.19         802

   HOWARD SIMPSON
                     126      30    32.00  5.25        168.00    001                         12.82   CIN
                     126                                               5.00         220
                     126                                               2.00         223
                     126                                              10.16         801
                     126                                              12.82         802
```

asset management transaction processing system
system that controls investments in capital equipment and manages depreciation for maximum tax benefits

The **asset management transaction processing system** controls investments in capital equipment and manages depreciation for maximum tax benefits. Key features of this application include efficient handling of a wide range of depreciation methods, country-specific tax reporting and depreciation structures for the various countries in which the firm does business, and workflow-managed processes to easily add, transfer, and retire assets.

General Ledger

Every monetary transaction that occurs within an organization must be properly recorded. Payment of a supplier's invoice, receipt of payment from a customer, and payment to an employee are examples of monetary transactions. A computerized **general ledger system** is designed to allow automated financial reporting and data entry. The general ledger application produces a detailed list of all business transactions and activities. Reports, including profit and loss (P&L) statements, balance sheets, and general ledger statements, can be generated (Figure 9.21). Furthermore, historical data can be kept and used to generate trend analyses and reports for various accounts and groups of accounts used in the general ledger package. Various income and expense accounts can be generated for the current period, year to date, and month to date as required. The reports generated by the general ledger application are used by accounting and financial managers to monitor the profitability of the organization and to control cash flows.

general ledger system
system designed to automate financial reporting and data entry

Financial reports that summarize sales by customer and inventory items can also be produced. These reports are used by marketing and financial managers to determine which customers are contributing to sales and inventory items that are selling as expected.

A key to the proper recording and reporting of financial transactions is the corporation's chart of accounts (Table 9.4). This chart provides codes for each type of expense or revenue. By entering transactions consistent with the chart of accounts, financial data can be reported in a simple and consistent fashion across all organizations of the enterprise, even if it is a multinational corporation.

INTERNATIONAL ISSUES

Businesses are increasingly operating across country borders or around the globe. Numerous complications arise that multinational corporations must address in planning, building, and operating their TPSs. Different languages and cultures, disparities in information system infrastructure, varying laws and customs rules, and multiple currencies are among the challenges of linking all the business partners, customers, and subsidiaries of a multinational company.

MOORE CHEMICALS INC.
INCOME AND EXPENSE STATEMENT
FOR 3RD QUARTER ENDED
SEPTEMBER 30, 2003
(UNAUDITED)

	CURRENT PERIOD	% OF BASE	YEAR-TO-DATE	% OF BASE
GROSS SALES	$ 590,471.96	100.0	$ 1,654,493.31	100.00
COST OF GOODS SOLD	395,029.72	65.2	1,059,279.54	64.00
GROSS PROFIT	205,442.24	34.8	596,213.77	36.0
OPERATING EXPENSES	$ 10,244.0			
DEPRECIATION	78,799.			
SALARIES AND COMPENSATION	13,026.1			
OTHER OPERATING EXPENSES				
OFFICERS COMPENSATION	$ 24,152.8			
SALARIES	4,999.8			
BONUSES	6,700.0			
RESEARCH				
TOTAL OPERATING EXPENSES	137,912.8			
INCOME - BEFORE TAXES	67,529.4			
INCOME TAX EXPENSE	37,141.1			
NET INCOME (LOSS)	$ 30,386.2			
EARNINGS PER SHARE	1.5			

MOORE CHEMICALS INC.
BALANCE SHEET
AT
SEPTEMBER 30, 2003
(UNAUDITED)

ASSETS		TH...
CURRENT ASSETS		
CASH	$ 75,	
ACCOUNTS PAYABLE	263,	
PREPAID INSURANCE	10,	
INVENTORY	216,	
MARKETABLE SECURITIES	2,	
TOTAL CURRENT ASSETS	568,	
FIXED ASSETS		
LAND AND IMPROVEMENTS		
PLANT AND EQUIPMENT		
ACCUMULATED DEPRCIATION	(10?	
TOTAL FIXED ASSETS		
OTHER ASSETS-SEE SCHEDULE		
TOTAL ASSETS	$	

MOORE CHEMICALS INC.
BALANCE SHEET
AT
SEPTEMBER 30, 2003
(UNAUDITED)

LIABILITIES	THIS YEAR	LAST YEAR
CURRENT LIABILITIES		
ACCOUNTS PAYABLE	$ 66,702.68	$ 72,668.57
NOTES PAYABLE-SEE SCHED.	31,520.00	43,200.00
ACCRUED EXPENSES	127,200.62	145,447.99
ACCRUED TAXES	92,087.49	53,042.39
TOTAL CURRENT LIABILITIES	317,510.79	314,358.95
LONG-TERM LIABILITIES		
NOTES PAYABLE LONG-TERM (NOTE 2)	73,120.00	148,000.00
DEFERRED U.S. INCOME TAX (NOTE 3)	49,210.00	58,713.00
%6 DEBENTURES DUE MAY '93	66,840.00	85,700.00
TOTAL LONG-TERM LIAB	189,170.00	292,413.00
TOTAL LIABILITIES	506,680.79	606,771.93
STOCKHOLDER'S QUITY		
PREFERRED STOCK PAR VALUE - $100 (NOTE 4)	40,950.00	40,950.00
COMMAN STOCK NO PAR VALUE (NOTE 5)	53,990.00	42,210.00
RETAINED EARNINGS	328,137.91	241,086.61
TOTAL STOCKHOLDER'S EQUITY	423,077.91	324,246.61
TOTAL LIABILITIES AND STOCKHOLDERS' EQUITY	$ 929,758.70	$ 931,018.56

DATE 02 28 03   MOORE CHEMICALS INC.
GENERAL LEDGER
BATCH NUMBER: 1   OPERATION CODE: JLG

ACCOUNT NUMBER	TITLE	DATE - PER	NO	SJC	REFER	DE...		
001001.400	CASH IN BANK 3	05 28 02 03	231	GL	JE	AN...		
001001.100	CASH ON HAND	05 28 02 03	231	GL	JE JNL	CA...		
005034.000	CASH ON HAND	05 28 02 03	233	GL	JE	PA...		
005008.000	EQUIPMENT REPAIRS	05 28 02 03	234	GL	JE	REPAIR OF ADDING MACHINE	99.95	
001001.000	CONTRIBUTIONS	05 28 02 03	235	GL	JE	SCHOOL CONTRIBUTION	300.25	
002001.000	ACCOUNTS PAYABLE	05 28 02 03	236	GL	JE	ABC LEASE CO.		406.00
005010.000	EQUIPMENT RENTALS	05 28 02 03	237	GL	JE	OFFICE EQUIPMENT	406.00	
009999.900	CATCHALL CLEARING ACCT	05 28 02 03	238	GL	JE	CATCHALL ACCOUNT		15,000.00
005016.000	PROFESSIONAL FEES	05 28 02 03	239	GL	JE	CONSULT FEE FOR COMPUTER	15,000.00	

BATCH HEADER-REC. COUNT	9	TOTAL DEBITS	16,500.00
MACHINE RECORD COUNT	9	TOTAL CREDITS	16,500.00
BATCH HEADER TOT. DR/CR	16.5000.00	DIFFERENCE	0.00

FIGURE 9-21

Outputs from a General Ledger System

An income statement (top) details sales, costs, and operating expenses to produce a statement of income for an organization. A balance sheet (center left and right) breaks down assets and liabilities so that managers can see at a glance whether income is covering expenses. A general ledger statement (bottom) tracks credits and debits.

Different Languages and Cultures

It is difficult to get people from several diverse countries who speak different languages and who were raised in different cultures to agree on a single work process. In some cultures, people are not used to working as teams in a networked environment. Despite these complications, many multinational companies are able to establish close connections with their business partners and roll out standard IS applications for all to use. However, those standard applications often don't account for all the differences among business partners and employees operating in other parts of the world. So, sometimes they require extensive and costly customization. For example, even though English has become a standard business language among executives and senior managers, many people within organizations do not speak English. As a result, software may need to be designed with local language interfaces to ensure the successful implementation of a new IS.

Major Account Name	Type of Expense	Subaccount Code Used to Identify Transaction
Wages and Benefits	Management salaries and benefits	MSALSB
	Nonmanagement salaries and benefits	NMALSB
	Overtime	OVT
Travel and Training	Travel-related expenses	TRAVEL
	Tuition for training classes	TUITION
Professional Services	Fees paid to consultants, contractors, trainers, and other professionals	PROFSV
Maintenance Expense	Maintenance labor	MAINTL
	Maintenance parts	MAINTP
	Maintenance supplies	MAINTS

TABLE 9.4

Sample Partial Chart of Accounts

Customization may also be needed to handle date fields correctly (e.g., the U.S. data format is month/day/year, the European format is day/month/year, and Japan uses year/month/day). Even so, users may also have to implement manual processes and overrides to enable systems to function correctly.

One company facing these challenges is Hunt Corporation, a U.S.-based maker and distributor of office supplies and graphics products. Hunt currently uses faxes and telephone calls to process transactions with its suppliers in Asia, a process that slows the movement of goods through the company's supply chain. Hunt is considering establishing business relationships with new suppliers in Latin America. If it does so, Hunt will install data entry terminals in its facilities to create a direct interface to its TPSs and reduce the amount of manual processing required. But Hunt will have to convince business partners that it is the worth the cost in time and money to convert to the new system. Many have already participated in unsuccessful software projects, so they are wary of investing in them. Success will also require a large amount of training to ensure that the business partners' workers follow the correct procedures and use the proper documentation for business transactions.[25]

Disparities in Information System Infrastructure

The lack of a robust or even a common information infrastructure can also create problems. The U.S. telecommunications industry is highly competitive, with many options for high-quality service at relatively low rates. Many other countries' telecommunications services are controlled by a government bureaucracy or a monopoly, with no motivation to provide fast and inexpensive customer service. For example, much of Latin America lags other parts of the world in Internet usage, and on-line marketplaces are almost nonexistent there. This gap makes it difficult for multinational companies to get on-line with their Latin American business partners. Even something as mundane as the power plug on a piece of equipment built in one country may not fit into the power socket of another country.

Varying Laws and Customs Rules

Numerous laws can affect the collection and dissemination of data. For one example, labor laws in some countries prohibit the recording of worker performance data. Also, some countries have passed laws limiting the transborder flow of data linked to individuals. Specifically, European Community Directive 95/96/EC of 1998 requires that any company doing business within the borders of the 15 Western European Union member nations put in place privacy directives on the fair and appropriate use of information. It bars the export of data to countries that do not have data-protection standards comparable to the EU's. Initially, the EU countries were concerned that the U.S.'s largely voluntary system of data privacy did not meet the EU directive's stringent standards. Eventually, the U.S. Department of Commerce worked out an agreement to allow American companies to import and export data. Failure to gain this compromise would have severely limited the exchange of information about employees and consumers.

Trade custom rules between nations are international laws that describe a practice followed by two or more nations when dealing with each other. They cover imports and exports and the systems and procedures dealing with quotas, visas, entry documents, commercial invoices, foreign trade zones, the payment of duty and taxes, and many other related issues. For example, the North American Free Trade Agreement (NAFTA) of 1994 is a set of trade custom rules that address the flow of goods throughout the North American continent. The great number of these custom rules and their changes over time create nightmares for people who must keep existing TPSs consistent with the rules.

Multiple Currencies

The TPSs of multinational companies must conduct transactions in multiple currencies. To do so, a set of exchange rates is defined, and the ISs apply these rates to translate from one currency to another. The systems must be current with foreign currency exchange rates, handle reporting and other transactions such as cash receipts, issue vendor payments and customer statements, record retail store payments, and generate financial reports in the currency of choice.

ENTERPRISE RESOURCE PLANNING

Flexibility and quick response are hallmarks of business competitiveness. Access to information at the earliest possible time can help businesses serve customers better, raise quality standards, and assess market conditions. Enterprise resource planning (ERP) is a key factor in instant access. Although some think that ERP systems are only for extremely large companies, this is not the case. Medium-sized companies can also benefit from the ERP approach. A few leading vendors of ERP systems are listed in Table 9.5.

AN OVERVIEW OF ENTERPRISE RESOURCE PLANNING

The key to ERP is real-time monitoring of business functions, which permits timely analysis of key issues such as quality, availability, customer satisfaction, performance, and profitability. Financial and planning systems receive "triggered" information from manufacturing and distribution. When something happens on

TABLE 9.5

Some ERP Software Vendors

Software Vendor	Name of Software
Oracle	Oracle Manufacturing
SAP America	SAP R/3
Baan	Triton
PeopleSoft	PeopleSoft
J. D. Edwards	WorldSoftware and One World
Ross Systems	iRenaissance
QAD	MFG/Pro

the manufacturing line that affects a business situation—for example, packing material inventory drops to a certain level, which affects the ability to deliver an order to a customer—a message is triggered for the appropriate person in purchasing.

In addition to manufacturing and finance, ERP systems can also support human resources, sales, and distribution. This sort of integration breaks through traditional corporate boundaries. For example, in August 2001, Chick-fil-A upgraded to Oracle's ERP software by installing the human resource, financial, and payroll applications. Later it added receivables, cash management, and order management software. Chick-fil-A's 1,000 restaurants in 34 states are connected to its data center's core ERP system running on Hewlett-Packard Unix servers. ERP helped Chick-fil-A automate its accounting system and enabled the company to open new stores without having to add a commensurate number of information system employees to support them.[26]

Successful implementation of a comprehensive ERP system can have a dramatic impact across the entire organization. For a manufacturing organization, the planning process begins with the preparation of a long-term demand forecast. This is prepared weekly for up to 18 months in advance and attempts to predict the amount of each product to be purchased over this time period. As finished products are withdrawn from inventory in response to customer demand, additional new, finished products need to be produced. The ERP production planning module uses the demand forecast and finished product inventory data to determine the week-by-week production schedule. This plan may reveal interesting insights, such as the need to build additional manufacturing capacity, hire additional workers, or develop new suppliers to provide sufficient raw materials. These new requirements can be input to the purchasing system and human resource modules of the ERP system so that managers in those areas can develop future plans. All this data can be fed into the financial module of the ERP system to prepare a profit and loss forecast statement to assess the firm's future profitability. This profit forecast in turn can be used to help establish new budget limits for the upcoming year.

ERP systems accommodate the different ways each company runs its business by either providing vastly more functions than one business could ever need or including customization tools that allow firms to fine-tune what should already be a close match. SAP R/3 is the undisputed king of the first approach. R/3 is easily the broadest and most feature-rich ERP system on the market. Thus, rather than compete on size, most competitors focus on customizability. ERP systems have the ability to configure and reconfigure all aspects of the IS environment to support whatever way your company runs its business.

ADVANTAGES AND DISADVANTAGES OF ERP

Increased global competition, new needs of executives for control over the total cost and product flow through their enterprises, and ever-more-numerous customer interactions are driving the demand for enterprisewide access to real-time information. ERP offers integrated software from a single vendor to help meet those needs. The primary benefits of implementing ERP include elimination of inefficient systems, easing adoption of improved work processes, improving access to data for operational decision making, and technology standardization. ERP vendors have also developed specialized systems for specific applications and market segments. Most ERP vendors have also developed a customer relationship management (CRM) package for their ERP system. Osram Sylvania, a lighting manufacturer, plans on using CRM to allow lighting buyers to place orders directly over the Internet. Of course, developing custom packages for every market need and segment would be a huge undertaking for ERP vendors. As a result, major ERP vendors are increasingly seeking help from other software vendors to develop specialized programs to tie directly into their ERP systems.

Even with the benefits of ERP, most companies have found it surprisingly difficult to justify implementation of an ERP system based strictly on cost savings.

Elimination of Costly, Inflexible Legacy Systems

Adoption of an ERP system enables an organization to eliminate dozens or even hundreds of separate systems and replace them with a single integrated set of applications for the entire enterprise. In many cases, these systems are decades old, the original developers are long gone, and the systems are poorly documented. As a result, the systems are extremely difficult to fix when they break, and adapting them to meet new business needs takes too long. They become an anchor around the organization that keeps it from moving ahead and remaining competitive. An ERP system helps match the capabilities of an organization's information systems to its business needs—even as these needs evolve.

Improvement of Work Processes

Competition requires companies to structure their business processes to be as effective and customer-oriented as possible. ERP vendors do considerable research to define the best business processes. They gather requirements of leading companies within the same industry and combine them with research findings from research institutions and consultants. The individual application modules included in the ERP system are then designed to support these **best practices**, the most efficient and effective ways to complete a business process. Thus, implementation of an ERP system ensures good work processes based on best practices. For example, for managing customer payments, the ERP system's finance module can be configured to reflect the most efficient practices of leading companies in an industry. This increased efficiency ensures that everyday business operations follow the optimal chain of activities, with all users supplied the information and tools they need to complete each step.

best practices
the most efficient and effective ways to complete a business process

Increase in Access to Data for Operational Decision Making

ERP systems operate via an integrated database and use essentially one set of data to support all business functions. So, decisions on optimal sourcing or cost accounting, for instance, can be run across the enterprise from the start, rather than looking at separate operating units and then trying to coordinate that information manually or reconciling data with another application. The result is an organization that looks seamless, not only to the outside world but also to the decision makers who are deploying resources within the organization.

The data is integrated to provide excellent support for operational decision making and allows companies to provide greater customer service and support, strengthen customer and supplier relationships, and generate new business opportunities. For example, once a salesperson makes a new sale, the business data captured during the sale is distributed to related transactions for the financial, sales, distribution, and manufacturing business functions in other departments.

Upgrade of Technology Infrastructure

An ERP system provides an organization with the opportunity to upgrade and simplify the information technology it employs. In implementing ERP, a company must determine which hardware, operating systems, and databases it wants to use. Centralizing and formalizing these decisions enables the organization to eliminate the hodgepodge of multiple hardware platforms, operating systems, and databases it is currently using—most likely from a variety of vendors. Standardization on fewer technologies and vendors reduces ongoing maintenance and support costs as well as the training load for those who must support the infrastructure. Remy Corporation, a $22 million Denver-based professional services firm, needed its front-office applications (those that interact directly with customers) to more easily integrate with its back-office systems. It decided to eliminate its collection of systems from a variety of software manufacturers and move to a PeopleSoft ERP system to avoid potential system integration issues.[27]

Expense and Time in Implementation

Getting the full benefits of ERP is not simple or automatic. Although ERP offers many strategic advantages by streamlining a company's transaction processing system, ERP is time-consuming, difficult, and expensive to implement. Some companies have spent years and tens of millions of dollars implementing ERP systems. And when there are problems with an ERP implementation, it can be expensive. GM Locomotive Group, a $2 billion subsidiary of the auto manufacturer that makes locomotives, diesel engines, and armored vehicles such as tanks, installed an ERP system to improve its financial reporting and its ability to forecast spare parts needs. It encountered such severe problems during system rollout that its spare parts business virtually ground to a halt and caused GM to launch an emergency turnaround effort. The software wasn't configured well enough to match internal business processes, and mainframe data wasn't properly formatted for the new system.[28] Teddy bear maker Russ Barrie and Company also failed with its first attempt to implement an ERP system. The company had previously updated its systems to avoid problems in its distribution, financial, and customer service systems. But when it then attempted to implement a packaged ERP application, it encountered problems severe enough that many of the new applications had to be taken off-line.[29]

Difficulty Implementing Change

In some cases, a company has to make radical changes in how it operates to conform with the work processes (best practices) supported by the ERP. These changes can be so drastic to long-time employees that they retire or quit rather than go through the change. This exodus can leave a firm short of experienced workers.

Difficulty Integrating with Other Systems

Most companies have other systems that must be integrated with the ERP. These systems can include financial analysis programs, Internet operations, and other applications. Many companies have experienced difficulties making these other systems operate with their ERP system. Other companies employ additional software to create these links. General Mills uses Tidal Software's Enterprise Scheduler to link the systems used to manage customer orders and its SAP R/3 ERP system. The Enterprise Scheduler converts 2.5 million orders sent annually via EDI transactions to SAP's format and imports them to R/3. It then routes orders within the company and alerts General Mills's personnel to any problems associated with the orders, such as difficulties in scheduling production or meeting customer desired delivery dates.[30]

Risks in Using One Vendor

The high cost to switch to another vendor's ERP system makes it extremely unlikely that a firm will do so. So, once a company has adopted an ERP system, the vendor knows it has a "captive audience" and has less incentive to listen and respond to customer issues. The high cost to switch also creates a high level of risk—in the event the ERP vendor allows its product to become outdated or goes out of business. Picking an ERP system involves not just choosing the best software product but also choosing the right long-term business partner.

Implementing an ERP system is extremely challenging and requires tremendous amounts of resources, the best IS people, and plenty of management support. Many companies have failed with their initial attempts, causing major business disruptions. Public companies, facing quarterly financial pressures, have become more willing to hold software suppliers publicly responsible for problems tied to the use of their products. The negative impact on the software supplier can be severe. Read the "Ethical and Societal Issues" special-interest box for an example of what can go wrong.

Nike Stumbles Implementing ERP

Nike is the world's leading shoe company and sells its products throughout the United States and 140 other countries. Nike also sells Cole Haan dress and casual shoes and a line of athletic apparel and equipment. In addition, Nike operates Niketown shoe and sportswear stores and is opening Nike Goddess stores catering to women.

The firm issued an earnings warning in February 2001, blaming its $400 million project to roll out a new demand- and inventory-management system—as well as lower shoe sales—for an expected earnings shortfall of $100 million. Nike said that after implementing a new ERP system in the summer of 2000, orders for some shoes were placed twice—once by the new system and once by its existing order-management system. Also, orders for many new shoe styles were lost and never processed. The heavily customized ERP system included modules from Dallas-based i2 Technologies Inc. i2 provides software that helps manufacturers plan and schedule production and related operations such as raw materials procurement and product delivery.

In its defense, i2 said the company's software modules represented only about 10 percent of the $400 million ERP project. The installation was also large and complex, requiring a high degree of customization on the I2 applications that were then linked with other ERP and back-end systems. In addition, the wide range of apparel products sold in a multitude of sizes and styles led to further difficulties in tailoring the i2 software to match to Nike's internal business processes. i2 said that Nike failed to follow i2's recommendation to minimize customization, to adopt i2's best practices for the footwear and apparel business, and to deploy the system gradually and in stages. Instead, Nike heavily customized the software and brought the system to thousands of suppliers and distributors at once.

The Nike announcement created a serious public relations problem for i2, and potential customers began to question the viability of the company's software. Shares of i2 stock dropped from just over $25 in March 2001 to under $4 in May 2002.

(The stock had been as high as $104 in March 2000.) The Nike incident was not the sole or even primary cause, but it was certainly a contributing factor.

Interestingly, brokerage firm Wells Fargo Van Kasper lowered its rating of Nike in December 2001, in part based on the perceived problem with Nike's implementation of SAP software. In addition to the difficulties Nike has already encountered, investors were concerned that the installation of a new SAP software system in the company's core U.S. business segment would impede the flow of Nike's spring 2002 merchandise line.

Discussion Questions

1. What business benefits would Nike likely gain from the successful implementation of an ERP system?
2. Do research on the Web on i2 Technologies and write a paragraph summarizing its current business state.

Critical Thinking Questions

3. Supply-chain software vendor i2 says Nike pushed the $400 million system into production too quickly, insisted on too much customization, and went live with too many suppliers and distributors at once. That could be—but for $400 million, did Nike have a right to set some high expectations of the software vendors?
4. This isn't the first time a company has attributed lowered earnings to new software mishaps. How might a financial auditor or IS consultant pinpoint the amount of the earnings shortfall due to weak sales and the amount due to IS problems? Would knowing this information make any difference?

Sources: Adapted from Bob Evans, "Listening Post," *InformationWeek*, June 4, 2001, http://www.informationweek.com; Marc Songini, "Nike Says Profit Woes IT Based," *Computerworld*, March 5, 2001, http://www.computerworld.com; John Soat, "IT Confidential," *InformationWeek*, March 5, 2001, http://www.informationweek.com; Steve Konicki, "Nike Just Didn't Do It Right," *InformationWeek*, March 5, 2001, http://www.informationweek.com; Aaron Ricadela, "The State of Software Quality," *InformationWeek*, May 21, 2001, http://www.informationweek.com; John Soat, "IT Confidential," *InformationWeek*, December 17, 2001, http://www.informationweek.com.

EXAMPLE OF AN ERP SYSTEM

SAP R/3 has been called one of the most complex packages ever written for use in corporations. However, it is also the most widely used ERP solution in the world. In response to criticisms about its complexity, SAP has worked hard to try to simplify SAP R/3 and to develop streamlined versions of the system. For example, SAP Business One is a reduced set of ERP applications that's designed for use by small and midsize companies. SAP is also developing eleven industry-specific application packages for midsize companies that have more sophisticated transaction needs.[31] Hershey Foods made business headlines in 1999 when it had problems in a failed $112 million attempt to deploy SAP AG's R/3 software and other business applications. But the candy maker was successful with a 2002 upgrade to the Web-enabled version 4.6 of R/3. The second attempt was completed 20 percent under budget and without any of the order processing and product-shipment disruptions that marred the initial attempt. Hershey credited enhancements in the software for reducing costs and simplifying the implementation.[32]

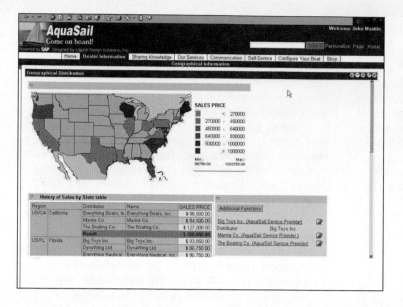

The following sections provide a brief description of the fundamental design and architecture of the SAP R/3 ERP system.

The SAP ERP system was developed from the perspective of a corporation as a whole, rather than any specific business department. All data is entered only once in the system, and all SAP programs use the same database with little data redundancy. Each data item is clearly documented in a data dictionary. The software is flexible enough to be configured to meet the customer's business requirements. It is based on a three-level client/server architecture consisting of clients, application servers, and database servers (Figure 9.22). R/3 will run on a wide variety of hardware, from a small Windows NT server up to massively parallel systems.

Clients in the SAP System

The R/3 system typically supports hundreds or even thousands of clients. Clients are usually desktop computers with fast processors and at least 32 MB of RAM. Users of the clients request services from the application servers.

Application Servers in the SAP System

There are many application servers in a typical R/3 system. The servers are powerful midrange or even mainframe computers. The job of the server is to reply to

FIGURE 9.22

SAP Three-Tier Client/Server Architecture

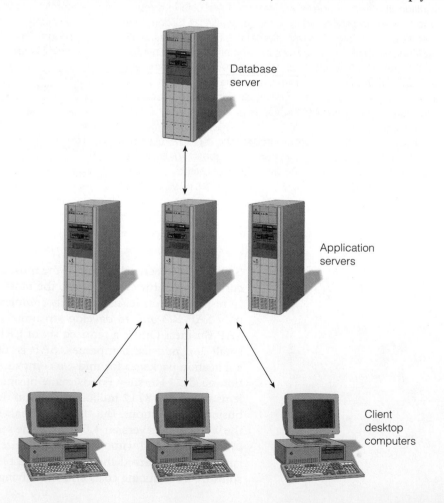

Database server

Application servers

Client desktop computers

all requests made of it, including requests for data, communication of messages, and update of master files. The request from the client travels along the network to the application server. A dispatcher program running on the application server manages the queues of user requests to determine which should be executed next. Application servers are grouped by the SAP R/3 system administrator into classes depending on the applications they run. One class might, for example, run the financial modules while another class might run the sales and distribution modules. The application server can contain either third-party or user-developed software, as long as it is written in ABAP/4, SAP's fourth-generation programming language.

Business Application Programming Interfaces (BAPIs)

Business application programming interfaces, or BAPIs, are public interfaces. These interfaces were developed with SAP customers, software development organizations, and standards organizations to enable SAP customers to develop their own applications to interface with SAP. SAP then has the flexibility to change the underlying software, as long as the interface itself is not changed (Figure 9.23). Thus, new SAP software versions can be introduced without invalidating existing systems. An example of a BAPI is "customer order" to allow checking the status of the customer order.

Database Server in the SAP System

The database server in the R/3 system holds the data and is accessed and updated constantly. Depending on the hardware selected, the database may be distributed among multiple machines or reside on a single computer. The SAP update process is designed to accommodate hundreds or even thousands of users on a single database server and still provide satisfactory response times.

Objects in the SAP System

object
a collection of data and programs

Like many popular databases, SAP has adopted objects as one of its key implementation concepts. An SAP **object** is a collection of data and programs. "Purchase order" and "customer" are examples of SAP business objects used in business processes. Attributes contain the details of an object, like name, date of employment, and address of an employee.

Repository

All development objects of the ABAP/4 Development Workbench are stored in the ABAP/4 repository. These objects include ABAP/4 programs, screens, documentation, and other tools that a system developer would need. The SAP R/3 database makes constant use of the repository. The repository sits between the database and the application modules and provides the logical mappings of data

FIGURE 9.23

Business Application Programming Interface (BAPI)

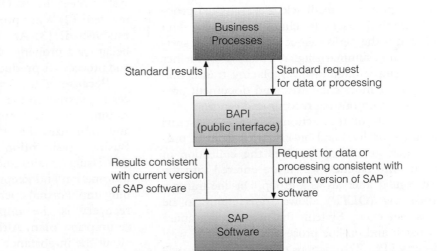

relationships and conditions. The repository serves primarily as a tool to enter, manage, and evaluate information about a company's data. It is an active data dictionary, so new or changed data in the repository is immediately available to all system components. As a result, application programs and screens are always supplied with up-to-date information.

Tables

There are three major types of tables: system configuration tables, control tables, and application data tables. All are defined in the repository. System configuration tables are maintained primarily by SAP and define the structure of the system; clients do not change these tables. To customize the system, the client's ERP project team uses both control tables and application data tables. Control tables define functions that guide the user in his or her activities. For example, a control table might be set up to require a customer service representative to enter a line item to reference data about the product from the material master table before a purchase order is accepted. Application data tables are divided into two types—transactions and master data tables. Transaction tables are the largest, since they contain the daily operations data, such as orders, payments received, invoices, and shipments. Master data files describe sets of basic business entities such as customers, vendors, products, materials, and the like.

This section has focused on the key elements of the SAP R/3 ERP system; however, the systems from PeopleSoft, Oracle, and other vendors are based on similar basic design elements—the use of client/server architecture, standard business application interfaces, object-oriented programming, database, and data tables.

SUMMARY

PRINCIPLE   *An organization's TPS must support the routine, day-to-day activities that occur in the normal course of business and help a company add value to its products and services.*

Transaction processing systems (TPSs) are at the heart of most information systems in businesses today. TPSs consist of all the components of a CBIS, including databases, telecommunications, people, procedures, software, and hardware devices to process transactions. All TPSs perform the following basic activities: data collection involves the capture of source data to complete a set of transactions; data edit checks for data validity and completeness; data correction involves providing feedback of a potential problem and enabling users to change the data; data manipulation is the performance of calculations, sorting, categorizing, summarizing, and storing for further processing; data storage involves placing transaction data into one or more databases; and document production involves outputting records and reports.

The methods of transaction processing systems include batch, on-line, and on-line with delayed processing. Batch processing involves the collection of transactions into batches, which are entered into the system at regular intervals as a group. On-line transaction processing (OLTP) allows transactions to be entered as they occur. Systems that use a compromise between batch and on-line processing are on-line with delayed-entry TPSs. Transactions may be entered as they occur, but processing is not performed immediately.

Organizations expect TPSs to accomplish a number of specific objectives, including processing data generated by and about transactions, maintaining a high degree of accuracy, ensuring data and information integrity, compiling accurate and timely reports and documents, increasing labor efficiency, helping provide increased and enhanced service, and building and maintaining customer loyalty.

The CIO must take a role in preventing the kind of accounting irregularities that can get companies into trouble and erase investor confidence. One key step is to conduct a transaction processing system audit that attempts to answer four basic questions: (1) Does the system meet the business need for which it was implemented? (2) What procedures and controls have been established? (3) Are these procedures and controls being used properly? (4) Are the information systems and procedures producing accurate and honest reports?

Because of the importance of a transaction processing system to the ongoing operation of an organization, a business resumption plan that anticipates and minimizes the effects of disasters is mandatory. Business resumption planning focuses primarily on two issues: maintaining the integrity of corporate information and keeping the information system running until normal operations can be resumed. Disaster recovery is the implementation of the business resumption plan. Although companies have known about the importance of disaster planning and recovery for decades, many do not adequately prepare.

TPS applications are seen throughout an organization. The order processing systems include order entry, sales configuration, shipment planning, shipment execution, inventory control, invoicing, customer relationship management, and routing and scheduling. Order entry captures the basic data needed to process a customer order. Once an order is entered and accepted, it becomes an open order. Sales configuration ensures that the products and services offered are sufficient for customer needs. Shipment planning determines which open orders will be filled and from which location they will be shipped. The system prepares an order confirmation notice that is sent to the customer and a picking list used by warehouse operators to fill the order. The shipment execution system is used by the warehouse operators to enter data on what was actually shipped to the customer. It passes shipped order transactions downstream to the invoicing system. For each item picked during the shipment execution process, a transaction providing stock number and quantity is passed to the finished product inventory control system. The invoicing system generates customer invoices based on the records received from the shipment execution system. The customer relationship management system monitors and tracks each customer interaction to ensure first-quality service and maximum profits. Routing and scheduling systems are used in distribution functions to determine the best use of a company's resources.

The purchasing information systems include inventory control, purchase order processing, accounts payable, and receiving. The inventory control system tracks the level of all packing materials and raw materials. It provides information to users on when to order additional materials. The purchase order processing system supports the policies, practices, and procedures of the purchasing department. The accounts payable system monitors and controls the outflow of funds to an organization's suppliers. The receiving system captures data about receipts of specific materials from suppliers so that approval for payment can be granted or refused.

The accounting systems include the budget, accounts receivable, payroll, asset management, and general ledger. The budget system automates many of the tasks required to amass budget data, distribute it to users, and consolidate the prepared budgets. The accounts receivable system manages the cash flow of ⊙ the company by keeping track of the money owed the company. The payroll processing application processes employee paychecks and performs numerous calculations relating to time worked, deductions, commissions, and taxes. The outputs are used to help control payroll costs and cash flows and develop reports to the federal government. The asset management system controls investments in capital equipment and manages depreciation for maximum tax benefits. The general ledger system records every monetary transaction and enables production of automated financial reporting.

PRINCIPLE   *TPSs help multinational corporations form business links with their business partners, customers, and subsidiaries.*

Numerous complications arise that multinational corporations must address in planning, building, and operating their TPSs. These include dealing with different languages and cultures, disparities in information system infrastructure, and varying laws and customs.

PRINCIPLE   *Implementation of an enterprise resource planning system enables a company to achieve numerous business benefits through the creation of a highly integrated set of systems.*

Enterprise resource planning (ERP) software is a set of integrated programs that manage a company's vital business operations for an entire multisite, global organization. It must be able to support multiple legal entities, multiple languages, and multiple currencies. Although the scope of an ERP system may vary from vendor to vendor, most ERP systems provide integrated software to support manufacturing and finance. In addition to these core business processes, some ERP systems are capable of supporting additional business functions such as human resources, sales, and distribution.

Implementation of an ERP system can provide many advantages, including elimination of costly, inflexible legacy systems; providing improved work processes; providing access to data for operational decision making; and creating the opportunity to upgrade technology infrastructure. Some of the disadvantages associated with an ERP system are that they are time-consuming, difficult, and expensive to implement.

CHAPTER 9 SELF-ASSESSMENT TEST

An organization's TPS must support the routine, day-to-day activities that occur in the normal course of business and help a company add value to its products and services.

1. Identify three possible methods of processing associated with transaction processing systems.

A. order processing, customer processing, and supplier processing
B. batch, on-line, and on-line entry with delayed processing
C. data collection, data edit, and data correction
D. data manipulation, data storage, document production

2. The primary objective of any TPS is to capture, process, and store transactions and to produce a variety of documents related to routine business activities. True False

3. All transaction processing systems perform a common set of basic data processing activities. Which of the following is NOT one of the basic data processing activities?
   A. data collection and data editing
   B. data correction and data manipulation
   C. data storage and document production
   D. data duplication and data elimination

4. The process of capturing and gathering all data necessary to complete transactions is called _____ _____.

5. Data should be captured at its source, and it should be recorded accurately, in a timely fashion, with minimal manual effort, and in a form that can be directly entered into the computer rather than keying the data from some type of document. True False

6. _____ _____ involves reentering miskeyed or misscanned data that was found during data editing.

7. Which of the following statements is TRUE?
   A. Disaster recovery is the implementation of the business resumption plan.
   B. Business resumption planning is the process of anticipating and minimizing the effects of disasters.
   C. A hot site is often a compatible mainframe system that is operational and ready to use. If the primary mainframe has problems, the hot site can be used immediately as a backup.
   D. All of the above.

8. The _____ _____ captures the basic data needed to process a customer order.

9. The systems associated with order processing should be tightly integrated, with output from one becoming input to another. True False

10. Inventory control, purchase order processing, receiving, and accounts payable systems make up a set of systems called the _____ systems.

11. The _____ transaction processing system automates many of the tasks required to amass budget data, distribute it to users, and consolidate the prepared budgets.

12. Which of the following systems is NOT considered to be part of a firm's accounting systems?

A. accounts receivable
B. inventory control
C. payroll
D. general ledger

13. A computerized _____ _____ system is designed to allow automated financial reporting and data entry and produces a detailed list of all business transactions and activities.

TPSs help multinational corporations form business links with their business partners, customers, and subsidiaries.

14. Many multinational companies roll out standard IS applications for all to use. However, those standard applications often don't account for all the differences among business partners and employees operating in other parts of the world. Which of the following is a frequent modification that is needed to standard software?
   A. Software may need to be designed with local language interfaces to ensure the successful implementation of a new IS.
   B. Customization may be needed to handle date fields correctly.
   C. Users may also have to implement manual processes and overrides to enable systems to function correctly.
   D. All of the above.

15. Fortunately, the telecommunications industry worldwide is highly competitive, with many options for high-quality service at relatively low rates. True False

Implementation of an enterprise resource planning system enables a company to achieve numerous business benefits through the creation of a highly integrated set of systems.

16. Which of the following is a primary benefit of implementing an ERP system?
   A. elimination of inefficient systems
   B. easing adoption of improved work processes
   C. improving access to data for operational decision making
   D. all of the above

17. The individual application modules included in an ERP system are designed to support the _____ _____, the most efficient and effective ways to complete a business process.

18. Because it is so critical to the operation of an organization, most companies are able to implement an ERP system without major difficulty. True False

19. ERP systems accommodate the different ways each company runs its business by either providing

vastly more functions than one business could ever need or including customization tools that allow firms to fine-tune what should already be a close match. Which vendor is the undisputed king of the first approach?

A. Oracle

B. SAP

C. PeopleSoft

D. J. D. Edwards

Chapter 9 Self-Assessment Test Answers

1. b; 2. True; 3. d; 4. data collection; 5. True; 6. Data correction; 7. d; 8. order entry system; 9. True; 10. purchasing; 11. budget; 12. b; 13. general ledger; 14. d; 15. False; 16. d; 17. best practices; 18. False; 19. b.

KEY TERMS

REVIEW QUESTIONS

1. What specific objectives do organizations hope to accomplish through the use of transaction processing systems?

2. List several characteristics of transaction processing systems.

3. What is the common set of basic transaction processing activities performed by all transaction processing systems?

4. What is an enterprise resource planning system?

5. What is the purpose of a customer relationship management system?

6. Identify four complications that multinational corporations must address in planning, building, and operating their TPSs.

7. What is the difference between batch processing, on-line processing, and on-line entry with delayed processing systems?

8. A business resumption plan focuses on what two issues?

9. What is the difference between business resumption planning and disaster recovery?

10. What questions are answered through successful completion of a transaction processing audit?

11. Describe an order entry system for taking orders over the Internet.

12. What systems are included in the purchasing family of systems?

13. Identify the various subsystems included as part of the order processing system.

14. What systems are included in the accounting family of systems?

15. Why is the general ledger application key to the generation of accounting information and reports?

16. Give an example of how transaction processing systems can be used to gain competitive advantage.

DISCUSSION QUESTIONS

1. Assume that you are the owner of a small dry cleaning business. Describe the day-to-day transaction processing activities that you would encounter.

2. Your company is a medium-sized service company with revenue of $500 million per year. You've decided that the organization will implement a

CRM system to capture and report information about all customer interactions. What are some of the key questions that must be answered to further define the scope of this effort?

3. Imagine that you are the new IS manager for a Fortune 1000 company. Your internal information

systems audit has revealed that your firm's systems are lacking disaster recovery plans and backup procedures. How would you decide whether your firm should invest in implementation of an ERP system?

4. What is the role of the C10 in ensuring the accuracy and integrity of a firm's financial reports?

5. What is the advantage of implementing ERP as an integrated solution to link multiple business processes? What are some of the issues and potential problems?

6. You are the key user of the firm's accounts receivable system and have been asked to perform an information system audit of this system. Outline the steps you would take to complete the audit.

7. You are building your firm's first-ever customer relationship management system. Discuss the features you would design into the system. How might you include suggestions from your customers into your design? Should the system be built using Internet technology?

8. You are in charge of a complete overhaul of your firm's purchasing systems. How would you define the requirements for this collection of systems? What features would you want to include?

9. How would you develop a business resumption plan and prepare for a potential disaster recovery? What steps would you use to plan for a potential disaster?

PROBLEM-SOLVING EXERCISES

1. Assume that you are starting a video store that will do advertising via fliers and creation of its own Web site. Using a graphics program, draw a diagram that shows the different ways you will interact with your customers. Use a word processing program to develop a list of key facts you would like to capture about each customer and about each contact with a customer.

2. The rental (order processing) application in your video store has three databases: rental item,

title, and customer. The video database contains information about every tape or DVD available for rental. The title database contains information about each specific movie title; any one tape or DVD title (for example, *Casablanca*) may have multiple videos or DVDs associated with it because more than one copy may be available for rental. The customer database contains each customer's ID number and address.

Specific fields in each database are listed in the accompanying table. Key fields are indicated with an asterisk.

Title
Title ID ★
Tape or DVD Indicator ★
Category
Rating
Rental rate
Rental Amount Y-T-D

Rental Item
Title ID ★
Tape or DVD Indicator ★
Unique ID Number ★
Status

Customer
Customer number ★
Customer name
Address
Phone number
Rental charges Y-T-D

Build a simple transaction processing system to support the video store operation. Enter the complete database definitions into your database management software and create a data-entry screen to allow store personnel to efficiently enter customer rentals and returns. This screen must include basic information such as DVD or tape name and number, customer name and number, date out, date returned, and charge for rental per day. The data from the screen updates all appropriate data in each database.

A. Enter several of your favorite movies to create at least ten entries for the title database.

B. For each title entry, create one to five entries in the video database.

C. Make up at least ten entries for the customer database.

D. Enter the data necessary to handle the checkout of at least six specific videos by different customers. Can your simple transaction processing system handle a situation in which one customer checks out more than one video at a time? What happens if a customer wants to check out a video but there are no copies remaining?

E. Check to see whether the Y-T-D fields in the title and customer databases are updated correctly.

TEAM ACTIVITIES

1. Assume that your team has formed a consulting firm to evaluate the business resumption plans of companies. Develop a list of at least 10 questions you would ask as you audit a firm's plan. Visit a company and perform the audit based on these questions.

2. Your team should interview a business owner about the company's order processing transaction processing system. Develop a report that describes this company's order processing transaction processing system.

WEB EXERCISES

1. A number of companies sell customer relationship management software to gather useful information about a company's customers. Search the Internet to get more information about CRM or one of the companies that makes and sells this powerful software. Develop a one-page report or send an e-mail message to your instructor about what you found.

2. Using the Internet, identify several companies that provide back-up-site processing locations. Choose one and write a paragraph describing its services.

CASES

CASE 1

MetLife Implements CRM

MetLife is a leading provider of insurance and other financial services to individual and institutional customers. It serves 10 million individual U.S. households and 64,000 companies and institutions with 33 million employees and members. MetLife is one of the largest U.S. insurers, offering life and property/casualty insurance (including home and auto coverage), as well as savings, retirement, and other financial services for groups and individuals. It also has international insurance operations in 13 countries. MetLife demutualized and sold about a third of the company to the public in 2000.

The business environment for insurance companies has changed dramatically in recent years. Consumers have many more purchasing options thanks to the 1999 Gramm-Leach-Bliley Act, which allowed banks to merge with securities and insurance companies. The act enabled insurance companies and financial institutions to sell a broader array of products and, in turn, created a highly competitive environment. Insurance consumers' buying habits have changed—instead of agents pushing products, consumers are now seeking out information, often from both insurance companies and banks that are developing hybrid insurance and securities products.

Insurance companies are aggressively pursuing new business strategies that will help them keep their existing customers and win new customers from new and old competitors. Many of them are turning to customer relationship management systems to market their services. MetLife, in particular, is concentrating its efforts on implementing a CRM system and customer-centric service strategy to help it retain consumers and, in turn, boost sales.

MetLife has been working with software vendor DWL Inc. since 2000 to develop and deploy DWL Customer, a real-time transactional application that consolidates customer data. This application creates a single master record for each customer by pulling information from over 30 transaction processing systems. The goal is to ensure that every business unit and every MetLife employee has a consistent and current view of a customer's data. Doing so enables the sales department to better target customers for cross-selling opportunities. For example, a service rep could sell a life-insurance policy to someone who holds a health-insurance policy with the company.

Creating a master record for each customer will also help MetLife keep records up to date and identify any data accuracy problems. For example, if a customer has a life-insurance policy that states his age is 32, but he later opens a mutual fund and gives his age to the agent as 52, the system will alert the agent of the problem.

Successful implementation of the CRM system inevitably will change the way MetLife employees do their jobs. Sales and service representatives, for example, will be expected to deal with all aspects of their customers' financial needs, not just the one or two product lines they've traditionally handled. MetLife's management believes that customer service is imperative not only to MetLife's ability to grow but also to its ability to survive. Changes in work processes and roles coupled with successful implementation of the CRM system will enable the company to connect with customers in a way that provides intrinsic value and growth for the future.

Discussion Questions

1. What challenges is MetLife facing that is driving it to invest in a CRM system?
2. What benefits does MetLife expect to achieve through successful implementation of CRM?

Critical Thinking Questions

3. Gaining the desired business benefits from the CRM system requires people to change they way they operate. What sort of changes must be made? What can MetLife management do to help ensure that employees are willing to make these changes?
4. Imagine that you are a MetLife service agent with 15 years of experience. Make a list of all the pros and cons you can imagine that such an individual would associate with moving to the new way of doing business.

Sources: Adapted from Jennifer Maselli, "Data Central," *InformationWeek*, January 21, 2002, pp. 45–46; Jennifer Maselli, "Insurers Look to CRM for Profits," *InformationWeek*, May 6, 2002, http://www.informationweek.com;

CASE 2

Verizon Restores Telecommunications Service in Wake of 9/11 Disaster

Verizon was formed in 2000 when Bell Atlantic bought GTE. It is the number one phone company in the U.S. as well as the number one U.S. wireless provider. Its 140 West Street location is directly across the street from the World Trade Center complex and served as a super telecommunications switching center routing voice and data calls from Wall Street to the rest of the world.

When 7 World Trade Center collapsed a few hours after the Twin Towers fell on September 11, its debris crashed into Verizon's switching facility. Damage to the building was so great that the switching center completely failed. The collapse of the towers sent a massive steel beam slicing through a bundle of fiber-optic cables buried eight feet below ground, destroying high-speed access lines and rupturing water pipes. The basement of the building was filled with more than 10 million gallons of water. Two Verizon employees lost their lives at the scene; another was killed in the Pentagon attack.

The circuits that served the financial district were knocked out, including data lines that served the New York Stock Exchange. The communication disruption was so complete that it threatened a lengthy trading outage completely destroying the exchange's ability to run its trade execution transaction processing systems and damaging our country's economic stability. The Bush administration recognized the severity of the situation and, once emergency response and rescue efforts were given the support they required, the White House made restoring Wall Street's telecommunications connections the highest priority. In the days following the attack, perhaps no event came to symbolize America's resilience more than the reopening of the New York Stock Exchange.

Verizon played a key role in getting the world's largest stock exchange rewired so that stock trades could be passed to stock trading systems. Restoring communications required a virtual army of federal and industry technicians. Setting their grief aside, Verizon workers returned to Ground Zero the next day to restore service by any means possible. Fiber-optic cables had

"MetLife Launches New Company Web Site," *BusinessWire*, April 30, 2002, accessed at http://www.news.moneycentral.msn.com; and "About Us," MetLife Web site at http://www.metlife.com, accessed May 12, 2002.

to be dragged from street level up through the building's fifth- and eighth-floor windows to tie them into the main switching fabric because it was impossible to run the cables through the flooded basement. With the massive effort, the firm was able to reroute traffic around buildings and connect new circuits in just a few days.

Verizon's rival carriers even helped by sharing network capacity and choking off nonessential traffic to the area when call volume got too high. Lucent Technologies, one of Verizon's main systems providers, rushed a 100,000-line switch to the scene to replace a massive switch that had been destroyed.

As the first workers began filtering into the Wall Street area on Monday, September 17, Verizon quietly turned on the last piece of the data network for the securities industry. And then it continued restoring service to other customers served by the West Street office.

The work Verizon did to restore service following the destruction of the World Trade Center certainly represents a triumph and provides confidence to all those businesses so dependent on telecommunications. But the disaster also underscores the dependence of key transaction processing systems on modern telecommunications networks and the need for businesses to consider the use of multiple telecommunications carriers in defining their business resumption plans.

Discussion Questions

1. What elements of a business resumption plan did Verizon have in place to be able to restore critical services so quickly?
2. Why was restoring communications to the New York Stock Exchange given such a high priority?

Critical Thinking Questions

3. What business resumption planning lessons do you think companies learned as a result of the 9/11 disaster?
4. With the benefit of 20-20 hindsight, is there anything Verizon could have done to be better prepared for a disaster?

Sources: Adapted from Dan Verton, "Digital Destruction Was Worst Imaginable," *Computerworld*, March 4, 2002, http://www.computerworld.com; Linda Rosencrance, "AT&T Working to Connect Customers Affected by NY Attack," *Computerworld*, September 19, 2001, http://www.computerworld.com; John Rendleman, "Back Online," *InformationWeek*, October 29, 2001, http://www.informationweek.com; Stephanie N. Mehta, "Telco on the Frontline," *Fortune*, October 15, 2001, pp.139–142.

CASE 3

Upgrade Required for Sabre Transaction Processing Systems

Sabre Holdings Company is a travel-service provider and operator of the Sabre on-line transaction processing airline reservations system for more than 50 airlines. It recently acquired Travelocity.com, a leading on-line B2C travel site, and owns GetThere, a leading provider of Web-based B2B travel reservation systems. Headquartered in Dallas–Fort Worth, it employs 5,500 workers in 45 countries, and recent revenues exceeded $2 billion.

In a multiyear deal costing over $100 million, Sabre is replacing its aging transaction processing systems that run on mainframe computers with state-of-the-art applications that run

on NonStop Himalaya servers and modern database software from Hewlett-Packard. The goal is to upgrade its air shopping applications by the end of 2005. Customers and travel agents use the air shopping information systems to search for available seats and check airfares. Plans to upgrade Sabre's booking, ticketing, and check-in applications are pending.

IBM implemented the initial Sabre system in the late 1950s based on technology called Transaction Processing Facility (TPF). Surprisingly, TPF still constitutes the technology underlying almost all airline reservation systems. The code is old, and making changes to it is much more difficult than it is in a modern system. Continued reliance on TPF-based technology will place Sabre at a severe competitive disadvantage. For example, in the wake of the September 11 terrorist attacks, the U.S. Department of Transportation identified ways to improve airline security. One idea was to require airline reservation systems to match booking passengers' names to a database of known terrorists and flag any names for follow-up. While such capability is relatively easy to add to a modern relational database, it is extremely difficult in the TPF system. According to industry experts, adding the name-screening capability could also cause bottlenecks that would grind the airlines' systems to a screeching halt.

The new Sabre reservation system must be more flexible—to allow changes in business needs such as those required to provide improved security. It must also provide faster response times to enable Sabre to meet the increasing demands of travelers who book on the Internet—whose number is expected to grow as much as 70 percent yearly.

Access to fare information is through a huge database that includes over 45 million fares with some 300,000 changes on a typical day. The new system will be able to process transactions even as new fares are being loaded into the system. In the old mainframe environment, requests for information on seat availability had priority over requests for information on a departing flight, a meal, or maintenance information. Add to that the millions of Internet users trying to book flights, and the result was lengthy delays in system response time.

Discussion Questions

1. What advantages will the new Sabre system provide?
2. Visit the Travelocity.com Web site and identify the lowest fares to your favorite destination. Write a brief paragraph summarizing your experience related to system performance.

Critical Thinking Questions

3. Why has Sabre waited until now to upgrade the underlying technology of its airline reservation system?
4. What sort of improvements would you like to see in airline booking, ticketing, and check-in systems?

Sources: Adapted from "Sabre and Compaq Join Forces to Advance State-of-the-Art Travel Technology," Sabre Holdings Corp. News Release, August 28, 2001, accessed at http://www.sabre.com; Jennifer Disabatino, "Sabre Sheds Its Mainframe Legacy," *Computerworld*, September 3, 2001, http://www.computerworld.com; Rick Whiting, "Sabre Pours Millions into Technology Upgrade," *InformationWeek*, September 3, 2001, http://www.informationweek.com; Jennifer DiSabatino, "Air Security May Require IT Overhaul," *Computerworld*, October 22, 2001, http://www.computerworld.com.

NOTES

Sources for the opening vignette on p. 365: Adapted from "About Kmart," accessed at Kmart Web site at http://www.kmartcorp.com/corp/story/index.stm, accessed May 14, 2002; Carol Sliwa, "Beyond IT: Business Strategy Was a Problem, Too," *Computerworld*, January 25, 2002, http://www.computerworld.com; Carol Sliwa, "IT Difficulties Help Take Kmart Down," *Computerworld*, January 28, 2002, http://www.computerworld.com; Steve Konicki, "Now In Bankruptcy, Kmart Struggled with Supply Chain," *InformationWeek*, January 28, 2002, http://www.informationweek.com; Frank Hayes, "Lessons from Kmart," *Computerworld*, January 28, 2002, accessed at http://www.computerworld.com; Julia King, "McKinsey: Stand-Alone IT Investments Are a Strategic Mistake," *Computerworld*, December 3, 2001, accessed at http://www.computerworld.com.

1. Rick Whiting, "Lightning Fast Data Processing," *InformationWeek*, April 22, 2002, accessed at http://www.informationweek.com.
2. Amy Helen Johnson, "Monitoring Tools Keep Web Apps on Track," *Computerworld*, November 12, 2001, accessed at http://www.computerworld.com.
3. Marc L. Songini, "P&G Unit Aims IT at Contract Monitoring," *Computerworld*, January 28, 2002, accessed at http://www.computerworld.com.
4. Carol Sliwa, "Ace Has a Place for CPFR," *Computerworld*, February 11, 2002, accessed at http://www.computerworld.com.
5. Cheryl Rosen, "RFID-Driven Supply Chain Tested," *InformationWeek*, June 20, 2001, accessed at http://www.informationweek.com.
6. Deirdre Lanning, "The I.T. Toll," *Business 2.0*, December 2001, accessed at http://www.business2.com.
7. Eileen Colkin and Rick Whiting, "Inadequate IT," *InformationWeek*, September 9, 2002, accessed at http://www.informationweek.com.
8. Eileen Colkin, Diane Rezendes Khirallah, and Sandra Swanson, "Where Was IT?" *InformationWeek*, July 1, 2002, accessed at http://www.informationweek.com.
9. Antone Gonsalves, "Forcing the Applications Integration Issue," *InformationWeek*, February 25, 2002, accessed at http://www.informationweek.com.
10. Eileen Colkin, Diane Rezendes Khirallah, and Sandra Swanson, "Where Was IT?" *InformationWeek*, July 1, 2002, accessed at http://www.informationweek.com.
11. Bob Brewin, "PepsiAmericas Pours Out Pocket PCs to Top Salespeople, Drivers," *Computerworld*, June 22, 2002, accessed at http://www.computerworld.com.
12. Marc L. Songini, "Global Supply Chains Rife with Challenges," *Computerworld*, March 12, 2001, accessed at http://www.computerworld.com.
13. Marc L. Songini, "Chain Reactions," *Computerworld*, April 29, 2002, accessed at http://www.computerworld.com.
14. Primm Fox, "Auto Supplier Cuts IT Costs by $12 Million," *Computerworld*, October 9, 2001, accessed at http://www.computerworld.com.
15. Christopher T. Heun, "Pumping Up Retail," *InformationWeek*, October 29, 2001, accessed at http://www.informationweek.com.
16. Antone Gonsalves, "Measuring the Benefits of CRM," *InformationWeek*, March 1, 2002, accessed at http://www.informationweek.com.
17. Jennifer Maselli, "Insurers Look to CRM for Profits," *InformationWeek*, May 6, 2002, accessed at http://www.informationweek.com.
18. Jennifer Maselli, "'How May I Help You?' Could Mean So Much More," *InformationWeek*, March 18, 2002, accessed at http://www.informationweek.com.
19. Jennifer Maselli, "Vendor Wants Smaller Businesses to Catch the CRM Wave," *InformationWeek*, March 11, 2002, accessed at http://www.informationweek.com.
20. Alorie Gilbert, "Rapt Lets Sun Maintain Inventory Control," *InformationWeek*, January 1, 2001, accessed at http://www.informationweek.com.
21. Steve Konicki, "E-Procurement Software Drives Savings at Mercedes," *InformationWeek*, December 24–31, 2001, p. 19.
22. Lee Copeland, "Security Shoves Auctions In-House," *Computerworld*, July 23, 2001, accessed at http://www.computerworld.com.
23. Michael Meehan, "GE Looks to Automate E-Procurement Billing," *Computerworld*, January 28, 2002, accessed at http://www.computerworld.com.
24. Eileen Colkin and Rick Whiting, "Inadequate IT," *InformationWeek*, September 9, 2002, accessed at http://www.informationweek.com.
25. Marc L. Songini, "Global Supply Chains Rife with Challenges," *Computerworld*, March 12, 2001, accessed at http://www.computerworld.com.
26. Marc Songini, "Burger King Upgrades to mySAP.com," *Computerworld*, July 15, 2002, accessed at http://www.computerworld.com.
27. Steve Konicki, "With Applications, Less Is More," *InformationWeek*, February 5, 2002, p. 45.
28. Marc L. Songini, "GM Locomotive Unit Puts ERP Rollout Back on Track," *Computerworld*, February 11, 2002, accessed at http://www.computerworld.com.
29. Marc L. Songini, "Teddy Bear Maker Prepares for Second Attempt at ERP Rollout," *Computerworld*, February 2, 2002, accessed at http://www.computerworld.com.
30. Steve Konicki, "Job Scheduling Puts General Mills on Top of Orders," *InformationWeek*, February 4, 2002, p. 45.
31. Marc L. Songini, "SAP to Simplify, Streamline Apps," *Computerworld*, June 10, 2002, accessed at http://www.computerworld.com.
32. Todd R. Weiss, "Hershey Upgrades R/3 ERP System without Hitches," *Computerworld*, September 9, 2002, accessed at http://www.computerworld.com.

Information and Decision Support Systems

PRINCIPLES

LEARNING OBJECTIVES

PRINCIPLES	LEARNING OBJECTIVES
Good decision-making and problem-solving skills are the key to developing effective information and decision support systems.	• Define the stages of decision making. • Discuss the importance of implementation and monitoring in problem solving.
The management information system (MIS) must provide the right information to the right person in the right fashion at the right time.	• Define the term *MIS* and clearly distinguish the difference between a TPS and an MIS. • Discuss information systems in the functional areas of business organizations.
Decision support systems (DSSs) are used when the problems are more unstructured.	• List and discuss important characteristics of DSSs that give them the potential to be effective management support tools. • Identify and describe the basic components of a DSS.
Specialized support systems, such as group decision support systems (GDSSs) and executive support systems (ESSs), use the overall approach of a DSS in situations such as group and executive decision making.	• State the goals of a GDSS and identify the characteristics that distinguish it from a DSS. • Identify the fundamental uses of an ESS and list the characteristics of such a system.

[Shearman & Sterling]

Using Information Systems to Harness the Knowledge of the Firm

The global law firm Shearman & Sterling has a reputation for successfully managing big business transactions. The firm represented Viacom in its $36 billion acquisition of CBS, the largest broadcast deal in history. It also facilitated the $189 billion merger of SmithKline Beecham and Glaxo Wellcome, creating the world's largest pharmaceutical company. The firm's portfolio includes more than the depth of big clients; it also contains considerable breadth, with more than 1,000 lawyers located in all of the world's financial capitals—making it one of the few truly global law firms.

Like many other successful law firms, Shearman & Sterling has a high level of technology diffusion. In designing its information system, the firm strove to save its lawyers from wasting time "reinventing the wheel" for each new case they handle. Recognizing the value of its collective brainpower, Shearman & Sterling began capturing, cataloging, and indexing its rapidly accumulating pool of information several years ago. Today, its lawyers are able to search through thousands of case histories on their company notebook computers from anywhere in the world in a fraction of the time that it once took to comb through a law library.

In addition to knowledge management, Shearman & Sterling provide their lawyers with a number of other decision support tools. Over the years, the firm has built a powerful system of collaborative work spaces using Lotus Notes, Lotus Domino, Lotus Sametime, and Lotus QuickPlace, enabling its attorneys to share documents and other information on client cases. Legal teams can share schedules, depositions, witness lists, and related cases. They can also use their notebook computers for virtual group meetings to brainstorm over the case at hand. A lawyer in need of advice can consult with other firm members on-line or can seek out and consult with a specialist in a particular legal area.

By cataloging and indexing the firm's knowledge and case histories, Shearman & Sterling have been able to break the traditional time-and-materials billing system and offer their clients fixed-rate billing for many of their services. Their information system allows them to query their data, producing reports that indicate the average bill for each type of case. Since the system is tied to the firm's financial information system, lawyers can record their hours and submit their bills from remote locations.

Using IBM's Websphere software platform, attorneys at Shearman & Sterling can access all of the firm's information systems from their Web browser. They can categorize their content by topic area on tabbed pages, which resemble physical file folders, and organize their files, e-mails, and other forms of information in areas called "personal places." "Community places" provide areas for team collaboration.

Shearman & Sterling has outsourced the majority of its systems development to IBM and Lotus. However, the firm's own systems programmers develop Web service applications using Lotus software tools that work in conjunction with the outsourced software. This capability allows the firm's own systems programmers to develop additional ad hoc decision support systems as needed.

Shearman & Sterling have proved that even in the most traditional and time-honored professions, information systems can assist us in being more productive. The laws may not have changed significantly since the days Perry Mason graced our black-and-white television sets, but practicing law certainly has.

As you read this chapter, consider the following:
- What types of decisions do lawyers and other professionals need to make? How can an information system assist them with their decision making?
- What challenges does Shearman & Sterling face in its attempt to catalog and index the knowledge of its lawyers? How might a database of this knowledge be organized? What types of database queries and reports might be useful?

As seen in the opening vignette, information and decision support systems are the lifeblood of today's organizations. Thanks to information and decision support systems, managers and employees can obtain useful information in real time. As we saw in Chapter 9, the TPS captures a wealth of data. When this data is filtered and manipulated, it can provide powerful support for managers and employees. The ultimate goal of management information and decision support systems is to help managers and executives at all levels make better decisions and solve important problems. The result can be increased revenues, reduced costs, and the realization of corporate goals. We begin by investigating decision making and problem solving.

DECISION MAKING AND PROBLEM SOLVING

Every organization needs effective decision making to reach its objectives and goals. In most cases, strategic planning and the overall goals of the organization set the stage for value-added processes and the decision making required to make them work. Often, information systems assist with strategic planning and problem solving. Good decision analysis, for example, contributed about a billion dollars to Eastman Kodak's profits during the 1990s.[1]

DECISION MAKING AS A COMPONENT OF PROBLEM SOLVING

In business, one of the highest compliments you can get is to be recognized by your colleagues and peers as a "real problem solver." Problem solving is a critical activity for any business organization. Once a problem has been identified, the problem-solving process begins with decision making. A well-known model developed by Herbert Simon divides the **decision-making phase** of the problem-solving process into three stages: intelligence, design, and choice. This model was later incorporated by George Huber into an expanded model of the entire problem-solving process (Figure 10.1).

decision-making phase
the first part of problem solving, including three stages: intelligence, design, and choice

FIGURE 10.1

How Decision Making Relates to Problem Solving

The three stages of decision making—intelligence, design, and choice—are augmented by implementation and monitoring to result in problem solving.

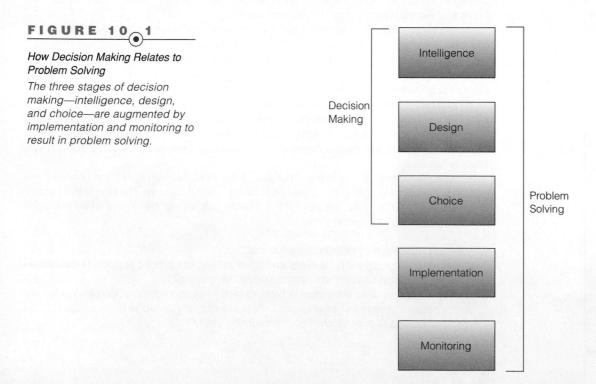

412

intelligence stage
the first stage of decision making, in which potential problems or opportunities are identified and defined

design stage
the second stage of decision making, in which alternative solutions to the problem are developed

choice stage
the third stage of decision making, which requires selecting a course of action

problem solving
a process that goes beyond decision making to include the implementation stage

implementation stage
a stage of problem solving in which a solution is put into effect

monitoring stage
final stage of the problem-solving process, in which decision makers evaluate the implementation

programmed decisions
decisions made using a rule, procedure, or quantitative method

Ordering more inventory when inventory levels drop to specified levels is an example of a programmed decision.

(Source: Courtesy of Symbol Technologies)

The first stage in the problem-solving process is the **intelligence stage**. During this stage, potential problems or opportunities are identified and defined. Information is gathered that relates to the cause and scope of the problem. During the intelligence stage, resource and environmental constraints are investigated. For example, exploring the possibilities of shipping tropical fruit from a farm in Hawaii to stores in Michigan would be done during the intelligence stage. The perishability of the fruit and the maximum price consumers in Michigan are willing to pay for the fruit are problem constraints. Aspects of the problem environment that must be considered in this case include federal and state regulations regarding the shipment of food products.

In the **design stage**, alternative solutions to the problem are developed. In addition, the feasibility of these alternatives is evaluated. In our tropical fruit example, the alternative methods of shipment, including the transportation times and costs associated with each, would be considered. During this stage the problem solver might determine that shipment by freighter to California and then by truck to Michigan is not feasible because the fruit would spoil.

The last stage of the decision-making phase, the **choice stage**, requires selecting a course of action. In our tropical fruit example, the Hawaiian farm might select the method of shipping by air to Michigan as its solution. The choice stage would then conclude with selection of the actual air carrier. As we will see later, various factors influence choice; the apparently easy act of choosing is not as simple as it might first appear.

Problem solving includes and goes beyond decision making. It also includes the **implementation stage,** when the solution is put into effect. For example, if the Hawaiian farmer's decision is to ship the tropical fruit to Michigan as air freight using a specific air freight company, implementation involves informing the farming staff of the new activity, getting the fruit to the airport, and actually shipping the product to Michigan.

The final stage of the problem-solving process is the **monitoring stage**. In this stage, decision makers evaluate the implementation to determine whether the anticipated results were achieved and to modify the process in light of new information. Monitoring can involve feedback and adjustment. For example, after the first shipment of fruit, the Hawaiian farmer might learn that the flight of the chosen air freight firm routinely makes a stopover in Phoenix, Arizona, where the plane sits on the runway for a number of hours while loading additional cargo. If this unforeseen fluctuation in temperature and humidity adversely affects the fruit, the farmer might have to readjust his solution to include a new air freight firm that does not make such a stopover, or perhaps he would consider a change in fruit packaging.

PROGRAMMED VERSUS NONPROGRAMMED DECISIONS

In the choice stage, various factors influence the decision maker's selection of a solution. One such factor is whether the decision can be programmed. **Programmed decisions** are made using a rule, procedure, or quantitative method. For example, to say that inventory should be ordered when inventory levels drop to 100 units is to adhere to a rule. Programmed decisions are easy to computerize using traditional information systems. It is simple, for example, to program a computer to order more inventory when inventory levels for a certain item reach 100 units or fewer. Most of the processes automated through transaction processing systems share this characteristic: the relationships between system elements are fixed by rules, procedures, or numerical relationships. Management information systems are also used to solve programmed decisions by providing reports on problems that are routine and in which the relationships are well defined (structured problems).

nonprogrammed decisions
decisions that deal with unusual or exceptional situations

Nonprogrammed decisions, however, deal with unusual or exceptional situations. In many cases, these decisions are difficult to quantify. Determining the appropriate training program for a new employee, deciding whether to start a new type of product line, and weighing the benefits and drawbacks of installing a new pollution control system are examples. Each of these decisions contains many unique characteristics for which the application of rules or procedures is not so obvious. Today, decision support systems are used to solve a variety of nonprogrammed decisions, in which the problem is not routine and rules and relationships are not well defined (unstructured or ill-structured problems).

OPTIMIZATION, SATISFICING, AND HEURISTIC APPROACHES

In general, computerized decision support systems can either optimize or satisfice. An **optimization model** will find the best solution, usually the one that will best help the organization meet its goals. For example, an optimization model can find the appropriate number of products an organization should produce to meet a profit goal, given certain conditions and assumptions. Optimization models utilize problem constraints. A limit on the number of available work hours in a manufacturing facility is an example of a problem constraint. Some spreadsheet programs, such as Excel, have optimizing features (Figure 10.2). CTI, an appliance manufacturer, used an optimization program called *optimization-based decision support system (OBDSS)* to help it reduce the time and cost of manufacturing appliances.[2] The company estimated that profits increased more than $3 million as a result of the OBDSS.

optimization model
a process to find the best solution, usually the one that will best help the organization meet its goals

A **satisficing model** is one that will find a good—but not necessarily the best—problem solution. Satisficing is usually used because modeling the problem properly to get an optimal decision would be too difficult, complex, or costly. Satisficing normally does not look at all possible solutions but only at those likely to give good results. Consider a decision to select a location for a new plant. To find the optimal (best) location, you would have to consider all cities in the United States or the world. A satisficing approach would be to consider only five or ten cities that might satisfy the company's requirements. Limiting the options may not result in the best decision, but it will likely result in a good decision, without spending the time and effort to investigate all cities. Satisficing is a good alternative modeling method because it is sometimes too expensive to analyze every alternative to get the best solution.

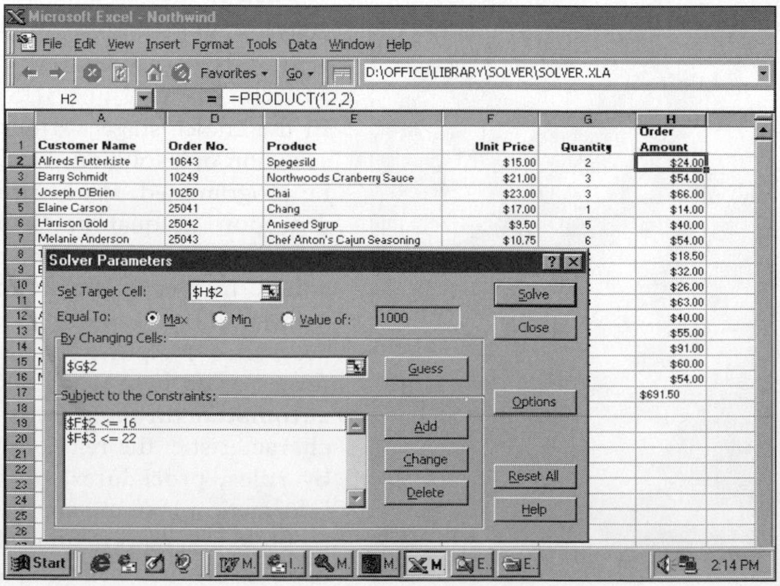

satisficing model
a model that will find a good—but not necessarily the best—problem solution

FIGURE 10.2

Some spreadsheet programs, such as Excel, have optimizing routines. This figure shows Solver, which can find an optimal solution given certain constraints.

heuristics
commonly accepted guidelines or procedures that usually find a good solution

Heuristics, often referred to as "rules of thumb"—commonly accepted guidelines or procedures that usually find a good solution—are very often used in decision making. A heuristic that baseball team managers use is to place batters most likely to get on base at the top of the lineup, followed by the power hitters who'll drive them in to score. An example of a heuristic used in business is to order four months' supply of inventory for a particular item when the inventory level drops to 20 units or fewer; even though this heuristic may not minimize total inventory costs, it may be a very good rule of thumb to avoid stockouts without too much excess inventory. Chunghwa Telecom in Taiwan, for example, used a heuristic called *SDHTOOL* to cut network costs by more than 15 percent.[3] Chunghwa, Taiwan's only full-service telecommunications carrier, used SDHTOOL to help make Taiwan an Asian-Pacific telecommunications center and improve Internet service to the nation. One way to achieve better decision making and problem solving is through a management information system.

AN OVERVIEW OF MANAGEMENT INFORMATION SYSTEMS

Management information systems (MISs) can often give companies a competitive advantage by providing the right information to the right people in the right format and at the right time. In many cases, companies and individuals are willing to pay companies for this type of information.

MANAGEMENT INFORMATION SYSTEMS IN PERSPECTIVE

The primary purpose of an MIS is to help an organization achieve its goals by providing managers with insight into the regular operations of the organization so that they can control, organize, and plan more effectively and efficiently. One important role of the MIS is to provide the right information to the right person in the right fashion at the right time. In short, an MIS provides managers with information, typically in reports, that support effective decision making and provides feedback on daily operations. Figure 10.3 shows the role of MISs within the flow of an organization's information. Note that business transactions can enter the organization through traditional methods or via the Internet or an extranet connecting customers and suppliers to the firm's transaction processing systems. The use of management information systems spans all levels of management. That is, they provide support to and are used by employees throughout the organization.

INPUTS TO A MANAGEMENT INFORMATION SYSTEM

Data that enters an MIS originates from both internal and external sources (Figure 10.3). The most significant internal source of data for an MIS is the organization's various TPSs and ERP systems and related databases. One of the major activities of a TPS is to capture and store the data resulting from ongoing business transactions. With every business transaction, various TPS applications make changes to and update the organization's databases. For example, the billing application helps keep the accounts receivable database up to date so that managers know who owes the company money. These updated databases are a primary internal source of data for the management information system. In companies that have implemented an ERP system, the collection of databases associated with this system are an important source of internal data for the MIS. As discussed in Chapter 5, companies also use data warehouses and data marts to store valuable business information. Other internal data comes from specific functional areas throughout the firm. According to Robert Anderson, a research director for Gartner, "Now more than ever, enterprises need to leverage the massive stores of operational data found within [enterprise resource planning] systems and begin exploiting it for competitive advantage."[4]

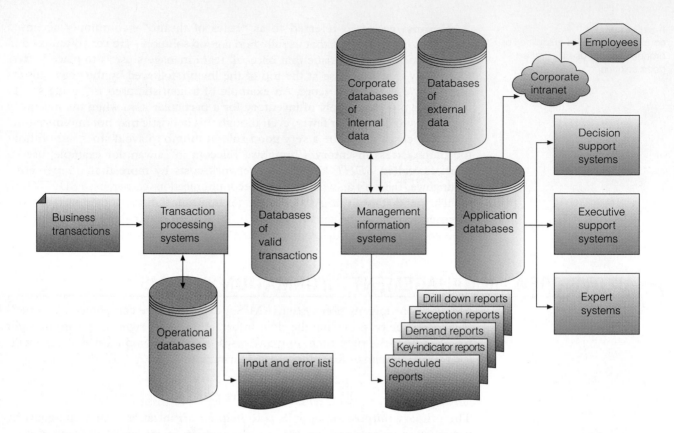

External sources of data can include customers, suppliers, competitors, and stockholders, whose data is not already captured by the TPS, as well as other sources, such as the Internet. In addition, many companies have implemented extranets to link them to these entities and allow for the exchange of data and information.

The MIS uses the data obtained from these sources and processes it into information more usable to managers, primarily in the form of predetermined reports. For example, rather than simply obtaining a chronological list of sales activity over the past week, a national sales manager might obtain her organization's weekly sales data in a format that allows her to see sales activity by region, by local sales representative, by product, and even in comparison with last year's sales.

OUTPUTS OF A MANAGEMENT INFORMATION SYSTEM

The output of most management information systems is a collection of reports that are distributed to managers. These reports can come from the company's various databases through a data mining process, first introduced in Chapter 5. Data mining allows a company to filter through a vast amount of data stored in databases, data warehouses, and data marts to produce a variety of reports, including scheduled reports, key-indicator reports, demand reports, exception reports, and drill down reports (Figure 10.4).

Scheduled Reports

scheduled reports
reports produced periodically, or
on a schedule, such as daily,
weekly, or monthly

Scheduled reports are produced periodically, or on a schedule, such as daily, weekly, or monthly. For example, a production manager could use a weekly summary report that lists total payroll costs to monitor and control labor and job costs. A manufacturing report generated once a day to monitor the production of a new item is another example of a scheduled report. Other scheduled reports can help managers control customer credit, the performance of sales representatives, inventory levels, and more. Trimac, a Canadian bulk hauling company, uses a huge database to produce a variety of scheduled reports for its managers.[5]

FIGURE 10.4

Reports Generated by an MIS

The five types of reports are (a) scheduled, (b) key-indicator, (c) demand, (d) exception, and (e–h) drill down.

(Source: George W. Reynolds, *Information Systems for Managers,* 3rd ed. St. Paul, MN: West Publishing Co., 1995. Reprinted with permission from Course Technology.)

(a) Scheduled Report

Daily Sales Detail Report

Prepared: 08/10/XX

Order #	Customer ID	Salesperson ID	Planned Ship Date	Quantity	Item #	Amount
P12453	C89321	CAR	08/12/01	144	P1234	$3,214
P12453	C89321	CAR	08/12/01	288	P3214	$5,660
P12454	C03214	GWA	08/13/01	12	P4902	$1,224
P12455	C52313	SAK	08/12/01	24	P4012	$2,448
P12456	C34123	JMW	08/13/01	144	P3214	$ 720
....	....	....	....	....	....	....

(b) Key-Indicator Report

Daily Sales Key Indicator Report

	This Month	Last Month	Last Year
Total Orders Month to Date	$1,808	$1,694	$1,914
Forecasted Sales for the Month	$2,406	$2,224	$2,608

(c) Demand Report

Daily Sales by Salesperson Summary Report

Prepared: 08/10/XX

Salesperson ID	Amount
CAR	$42,345
GWA	$38,950
SAK	$22,100
JWN	$12,350
....	....

(d) Exception Report

Daily Sales Exception Report—Orders Over $10,000

Prepared: 08/10/XX

Order #	Customer ID	Salesperson ID	Planned Ship Date	Quantity	Item #	Amount
P12345	C89321	GWA	08/12/01	576	P1234	$12,856
P22153	C00453	CAR	08/12/01	288	P2314	$28,800
P23023	C32832	JMN	08/11/01	144	P2323	$14,400
....	....	....	....	....	....	....
....	....	....	....	....	....	....

According to Len More, a Trimac project manager, "This initiated with our Business Intelligence Project. One of the tasks was to deliver reports."

A **key-indicator report** summarizes the previous day's critical activities and is typically available at the beginning of each workday. These reports can summarize inventory levels, production activity, sales volume, and the like. Key-indicator reports are used by managers and executives to take quick, corrective action on significant aspects of the business.

Demand Reports

Demand reports are developed to give certain information upon request. In other words, these reports are produced on demand. For example, an executive may want to know the production status of a particular item—a demand report can be generated to give the requested information. Suppliers and customers can also use demand reports. FedEx, for example, provides demand reports on its Web site to allow its customers to track packages from their source to their final destination.[6] On average, the bar code of a typical FedEx package is scanned a mind-boggling

key-indicator report
summary of the previous day's critical activities; typically available at the beginning of each workday

demand reports
reports developed to give certain information at a manager's request

23 times as it travels through the FedEx system. Penske Logistics uses wireless terminals to provide customers with critical delivery information on demand.[7] Other examples of demand reports include reports requested by executives to show the hours worked by a particular employee, total sales to date for a product, and so on.

Exception Reports

Exception reports are reports that are automatically produced when a situation is unusual or requires management action. For example, a manager might set a parameter that generates a report of all inventory items with fewer than the equivalent of 5 days of sales on hand. This unusual situation requires prompt action to avoid running out of stock on the item. The exception report generated by this parameter would contain only items with fewer than 5 days of sales in inventory. As with key-indicator reports, exception reports are most often used to monitor aspects important to an organization's success. In general, when an exception report is produced, a manager or executive takes action. Parameters, or *trigger points*, for an exception report should be set carefully. Trigger points that are set too low may result in an abundance of exception reports; trigger points that are too high could mean that problems requiring action are overlooked. For example, if a manager wants a report that contains all projects over budget by $100 or more, he may find that almost every company project exceeds its budget by at least this amount. The $100 trigger point is probably too low. A trigger point of $10,000 might be more appropriate.

exception reports
reports automatically produced when a situation is unusual or requires management action

FIGURE 10.4 *(cont.)*

Reports Generated by an MIS

The five types of reports are (a) scheduled, (b) key-indicator, (c) demand, (d) exception, and (e–h) drill down.

(Source: George W. Reynolds, *Information Systems for Managers,* 3rd ed. St. Paul, MN: West Publishing Co., 1995. Reprinted with permission from Course Technology.)

(e) First-Level Drill Down Report

Earnings by Quarter (Millions)			
	Actual	**Forecast**	**Variance**
2nd Qtr. 2002	$12.6	$11.8	6.8%
1st Qtr. 2002	$10.8	$10.7	0.9%
4th Qtr. 2001	$14.3	$14.5	-1.4%
3rd Qtr. 2001	$12.8	$13.3	-3.8%

(f) Second-Level Drill Down Report

Sales and Expenses (Millions)			
Qtr: 2nd Qtr. 2002	**Actual**	**Forecast**	**Variance**
Gross Sales	$110.9	$108.3	2.4%
Expenses	$ 98.3 $	$ 96.5	1.9%
Profit	12.6	$ 11.8	6.8%

(g) Third-Level Drill Down Report

Sales by Division (Millions)			
Qtr: 2nd Qtr. 2002	**Actual**	**Forecast**	**Variance**
Beauty Care	$ 34.5	$ 33.9	1.8%
Health Care	$ 30.0	$ 28.0	7.1%
Soap	$ 22.8	$ 23.0	-0.9%
Snacks	$ 12.1	$ 12.5	-3.2%
Electronics	$ 11.5	$ 10.9	5.5%
Total	$110.9	$108.3	2.4%

(h) Fourth-Level Drill Down Report

Sales by Product Category (Millions)			
Qtr: 2nd Qtr. 2002 **Division: Health Care**	**Actual**	**Forecast**	**Variance**
Toothpaste	$12.4	$10.5	18.1%
Mouthwash	$ 8.6	$ 8.8	-2.3%
Over-the-Counter Drugs	$ 5.8	$ 5.3	9.4%
Skin Care Products	$ 3.2	$ 3.4	-5.9%
Total	$30.0	$28.0	7.1%

drill down reports
reports providing increasingly
detailed data about a situation

Drill Down Reports

Drill down reports provide increasingly detailed data about a situation. Through the use of drill down reports, analysts are able to see data at a high level first (similar to a bag of cookies), then at a more detailed level (say, an Oreo), and then a very detailed level (an Oreo double-filling cookie's components).

Developing Effective Reports

Management information system reports can help managers develop better plans, make better decisions, and obtain greater control over the operations of the firm. It is important to recognize that various types of reports can overlap. For example, a manager can demand an exception report or set trigger points for items contained in a key-indicator report. In addition, some software packages, such as ReportSaveV3 by Ascend Software, can be used to produce, gather, and distribute reports from different computer systems.[8] Certain guidelines should be followed in designing and developing reports to yield the best results. Table 10.1 explains these guidelines.

CHARACTERISTICS OF A MANAGEMENT INFORMATION SYSTEM

Scheduled, key-indicator, demand, exception, and drill down reports have all helped managers and executives make better, more timely decisions. When the guidelines for developing effective reports are followed, higher revenues and lower costs can be realized. Read the "Ethical and Societal Issues" box to see how one energy company capitalized on the strength of its MIS. In general, management information systems perform the following functions:

- *Provide reports with fixed and standard formats.* For example, scheduled reports for inventory control may contain the same types of information placed in the same locations on the reports. Different managers may use the same report for different purposes.
- *Produce hard-copy and soft-copy reports.* Some MIS reports are printed on paper and are considered hard-copy reports. Most output soft copy, using

TABLE 10.1

Guidelines for Developing MIS Reports

Guidelines	Reason
Tailor each report to user needs.	The unique needs of the manager or executive should be considered, requiring user involvement and input.
Spend time and effort producing only reports that are useful.	Once instituted, many reports continue to be generated even though no one uses them anymore.
Pay attention to report content and layout.	Prominently display the information that is most desired. Do not clutter the report with unnecessary data. Use commonly accepted words and phrases. Managers can work more efficiently if they can easily find desired information.
Use management by exception reporting.	Some reports should be produced only when there is a problem to be solved or an action that should be taken.
Set parameters carefully.	Low parameters may result in too many reports; high parameters mean valuable information could be overlooked.
Produce all reports in a timely fashion.	Outdated reports are of little or no value.
Periodically review reports.	Review reports at least once a year to make sure all reports are still needed. Review report content and layout. Determine whether additional reports are needed.

The Energy-Conscious MIS

The millennium arrived to find the U.S. West Coast struggling with an energy crisis of unprecedented proportions. Rolling blackouts left residents and businesses suddenly—and unexpectedly—in the dark as energy producers were unable to meet their customer's demand. Puget Sound Energy (PSE), Washington State's leading energy utility, was in a unique position to offer its customers solutions.

In 1997, anticipating a highly competitive energy market due to deregulation, PSE invested heavily in a new information system to gain a competitive edge. It deployed an automated meter-reading (AMR) system—at a cost of $45 million—and connected it to its customer MIS. The goal was to capture, analyze, and share real-time data about each customer's power consumption. The new system did away with meter readers and collected data automatically from wireless devices installed at each customer's residence or business. PSE counted on the new system to greatly reduce its operational costs and provide a return on its investment within 10 years. What it didn't realize was that it would provide much more.

By spring 2000, PSE's customer service needs shifted significantly. As wholesale gas and electric energy prices spiked, the company took the customer information from its AMR and billing systems and made it available to customers through an Internet portal. The 450,000 residential gas and electric customers who signed up for the company's portal-based Personal Energy Management (PEM) program were able to see their energy usage for any given day, month, or year and compare it with past usage statistics.

Regulators gave PSE permission to launch a pilot time-of-day billing program for 300,000 residential customers, who would be billed rates based on the times they consumed energy rather than the traditional flat rates. Rates were higher during peak demand in the day and lower later in the evening. This new unique billing option, combined with customer's ability to view real-time energy consumption data, empowered customers to make better-informed energy consumption decisions. Customer response to the program was overwhelmingly positive. Of the customers who signed up, 89 percent, or 267,000, had shifted some of their energy usage from peak to off-peak hours, allowing PSE overall to switch 5 to 6 percent of its energy production to the more economical off-peak rates. In addition, not only were customers switching consumption to off-peak hours, but they were using less energy overall.

In 2001, PSE extended its pilot to 20,000 commercial customers, and in 2002, it received permission to expand the residential pilot to 800,000 customers. Eventually, the utility wants to offer consumers a variety of rate packages similar to those offered by cable companies, says Brian Pollom, director of metering network services at PSE.

With automated meter-reading already in place when the energy crunch hit, PSE was able to quickly react to their customer's needs. By analyzing detailed energy consumption information provided by MIS reports, PSE was able to pinpoint the times of day when residential energy consumption was excessive. Based on this information, it created a billing system to entice customers to change their consumption habits. Accessing the same information system from the Web, customers were able to study their own consumption patterns and change their usage to save money.

The success of PSE's system earned it the 2001 Edison Award from the Washington-based Edison Electric Institute, which recognizes energy companies for outstanding contributions to the energy industry. Many utility companies have followed PSE's lead and are implementing similar systems and services for their customers. Time-of-day billing, and the technology that makes it possible, is likely to become the carrot that motivates consumers to consume energy more responsibly.

Discussion Questions

1. What forms of input are used for PSE's customer MIS described here?
2. What types of information and reports should this MIS generate to assist PSE managers? How about for PSE customers?

Critical Thinking Questions

3. PSE expected that the automated meter-reading (AMR) system would greatly reduce its operational costs. How could it do so?
4. Compare PSE's service of delivering real-time data to its customers with another business or industry with which you're familiar. How are the two systems similar and different?

Sources: Melissa Solomon, "Powering Down," *Computerworld*, July 22, 2002, http://www.computerworld.com; Puget Sound Energy Web site, http://www.pse.com, accessed July 24, 2002; SchumbergerSema Corp., Real-Time Energy Management Web site, http://www.slb.com/utilities/tech_sol/rt_energy_mgmt.html, accessed July 24, 2002.

visual displays on computer screens. Soft-copy output is typically formatted in a reportlike fashion. In other words, a manager might be able to call an MIS report up directly on the computer screen, but the report would still appear in the standard hard-copy format. Hard copy is still the most often used form of the MIS report.

- *Use internal data stored in the computer system.* MIS reports use primarily internal sources of data that are contained in computerized databases. Some MISs use external sources of data about competitors, the marketplace, and so on. The Internet and extranets are frequently used sources for external data.

- *Allow end users to develop their own custom reports.* Although analysts and pro- grammers may be involved in developing and implementing complex MIS reports that require data from many sources, end users are increasingly devel- oping their own simple programs to query a database and produce basic reports. This capability, however, can result in several end users developing the same or similar reports, which can result in more total time expended and additional storage requirements, compared with having an analyst develop one report for all users.
- *Require user requests for reports developed by systems personnel.* When informa- tion systems personnel develop and implement MIS reports, a formal request to the information systems department may be required. If a man- ager, for example, wants a production report to be used by several people in his or her department, a formal request for the report is often required. End user–developed reports require much less formality.

FUNCTIONAL ASPECTS OF THE MIS

Most organizations are structured along functional lines or areas. This functional structure is usually apparent from an organization chart, which typically shows vice presidents under the president. Some of the traditional functional areas are accounting, finance, marketing, personnel, research and development (R&D), legal services, operations/production management, and information systems. The MIS can be divided along those functional lines to produce reports tailored to individual functions (Figure 10.5).

FINANCIAL MANAGEMENT INFORMATION SYSTEMS

financial management information system
an information system that pro- vides financial information to all financial managers within an organization and a broader set of people who need to make better decisions

A **financial management information system** provides financial information not only for executives but also for a broader set of people who need to make bet- ter decisions on a daily basis. Finding opportunities and quickly identifying prob- lems can mean the difference between a business's success and failure. Specifically, the financial MIS performs the following functions:

- Integrates financial and operational information from multiple sources, including the Internet, into a single MIS
- Provides easy access to data for both financial and nonfinancial users, often through use of the corporate intranet to access corporate Web pages of finan- cial data and information
- Makes financial data available on a timely basis to shorten analysis turnaround time
- Enables analysis of financial data along multiple dimensions—time, geography, product, plant, customer
- Analyzes historical and current financial activity
- Monitors and controls the use of funds over time

Figure 10.6 shows typical inputs, function-specific subsystems, and outputs of a financial MIS.

In addition to providing information for internal control and management, financial MISs often are required to provide information to outside individu- als and groups, including stockholders and federal agencies. Public companies are required to disclose their financial results to stockholders and the public. The federal government also requires financial statements and information systems. As a result of antiterrorism legislation signed into law by President Bush, financial service firms must now implement new information systems designed to make it easier for law enforcement agencies to find and freeze assets owned by suspected terrorists.[9] The legislation also attempts to uncover money laundering.

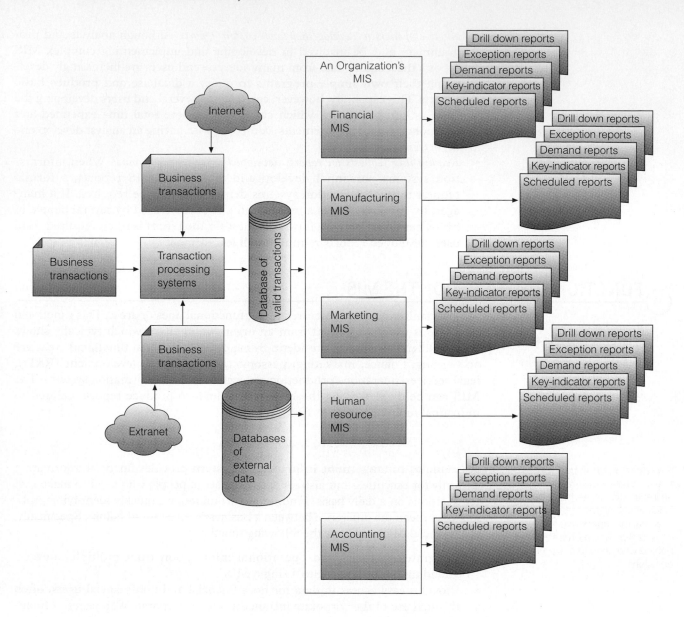

FIGURE 10.5

The MIS is an integrated collection of functional information systems, each supporting particular functional areas.

Depending on the organization and its needs, the financial MIS can include both internal and external systems that assist in acquiring, using, and controlling cash, funds, and other financial resources. These subsystems of the financial MIS have a unique role in adding value to a company's business processes. For example, a real estate development company might use a financial MIS subsystem to help it use and manage funds. Suppose the firm takes $10,000 deposits on condominiums in a new development. Until construction begins, the company will be able to invest these surplus funds. By using reports produced by the financial MIS, finance staff can analyze investment alternatives. The company might invest in new equipment or purchase global stocks and bonds. The profits generated from the investment can be passed along to customers in different ways. The company can pay stockholders dividends, buy higher quality materials, or sell the condominiums at a lower cost.

Other important financial subsystems include profit/loss and cost accounting, and auditing. Each subsystem interacts with the TPS in a specialized way and has information outputs that assist financial managers in making better decisions. These outputs include profit/loss and cost accounting reports, internal and external auditing reports, and uses and management of funds reports.

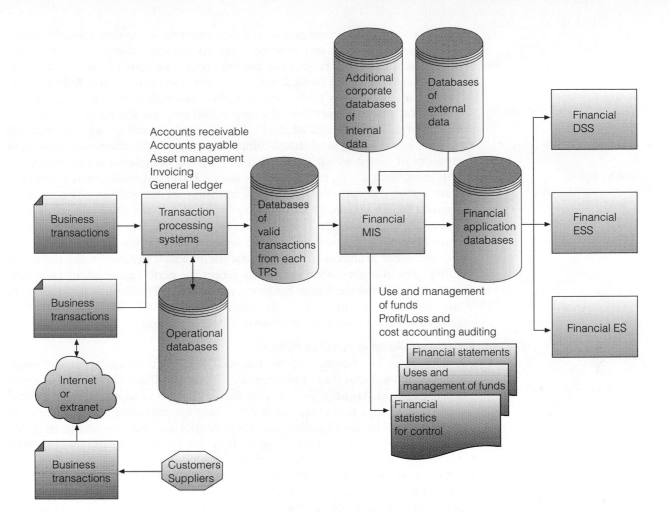

FIGURE 10.6

Overview of a Financial MIS

profit centers
departments within an organization that track total expenses and net profits

revenue centers
divisions within a company that track sales or revenues

cost centers
divisions within a company that do not directly generate revenue

auditing
analyzing the financial condition of an organization and determining whether financial statements and reports produced by the financial MIS are accurate

Profit/Loss and Cost Systems

Two specialized financial functional systems are profit/loss and cost systems, which organize revenue and cost data for the company. Revenue and expense data for various departments is captured by the TPS and becomes a primary internal source of financial information for the MIS.

Many departments within an organization are **profit centers**, which means they track total expenses and net profits. An investment division of a large insurance or credit card company is an example of a profit center. Other departments may be **revenue centers**, which are divisions within the company that primarily track sales or revenues, such as a marketing or sales department. Still other departments may be **cost centers**, which are divisions within a company that do not directly generate revenue, such as manufacturing or research and development. These units incur costs with little or no direct revenues. Vertex Pharmaceuticals, for example, constructed a supercomputer with 112 processors to help it accelerate drug research and development.[10] The company hopes that the estimated $500,000 cost center will result in new or improved drugs. Data on profit, revenue, and cost centers is gathered (mostly through the TPS but sometimes through other channels as well), summarized, and reported by the profit/loss and cost subsystems of the financial MIS.

Auditing

Auditing involves analyzing the financial condition of an organization and determining whether financial statements and reports produced by the financial MIS are accurate. Because financial statements, such as income statements and balance

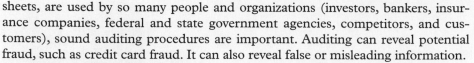

sheets, are used by so many people and organizations (investors, bankers, insurance companies, federal and state government agencies, competitors, and customers), sound auditing procedures are important. Auditing can reveal potential fraud, such as credit card fraud. It can also reveal false or misleading information.

internal auditing
auditing performed by individuals within the organization

Internal auditing is performed by individuals within the organization. For example, the finance department of a corporation may use a team of employees to perform an audit. Typically, an internal audit is conducted to see how well the organization is meeting established company goals and objectives—no more than five weeks of inventory on hand, all travel reports completed within one week of returning from a trip, and similar measures. **External auditing** is performed by an outside group, such as an accounting or consulting firm such as PricewaterhouseCoopers, Deloitte & Touche, or one of the other major, international accounting firms. The purpose of an external audit is to provide an unbiased picture of the financial condition of an organization. Auditing can also uncover fraud and other problems. In some cases, the financial picture from an external auditing firm may not always completely reflect the performance of the company. Some critics point to the Enron bankruptcy fiasco of 2001–2002, which resulted in many employees and investors losing huge sums of money, as an example of an external audit not showing the true picture of the company.[11]

external auditing
auditing performed by an outside group

Uses and Management of Funds

Another important function of the financial MIS is funds usage and management. Companies that do not manage and use funds effectively often have lower profits or face bankruptcy. To help with the funds usage and management, some banks are backing a new computerized payment system called Straight-Through Processing.[12] The new system has the potential to clear payments in a day instead of several days or more. Outputs from the funds usage and management subsystem, when combined with other subsystems of the financial MIS, can locate serious cash flow problems and help the organization increase profits.

Internal uses of funds include additional inventory, new or updated plants and equipment, additional labor, the acquisition of other companies, new computer systems, marketing and advertising, raw materials, land, investments in new products, and research and development. External uses of funds are typically investment related. On occasion, a company might have excess cash from sales that is placed into an external investment. External uses of funds often include bank accounts, stocks, bonds, bills, notes, futures, options, and foreign currency.

Departments such as manufacturing or research and development are cost centers since they do not directly generate revenue.

(Source: © 2000 Photodisc)

MANUFACTURING MANAGEMENT INFORMATION SYSTEMS

More than any other functional area, manufacturing has been revolutionized by advances in technology. As a result, many manufacturing operations have been dramatically improved over the last decade. Also, with the emphasis on greater quality and productivity, having an efficient and effective manufacturing process is becoming even more critical. The use of computerized systems is emphasized at all levels of manufacturing— from the shop floor to the executive suite. The use of the Internet has also streamlined all aspects of manufacturing. Figure 10.7 gives an overview of some of the manufacturing MIS inputs, subsystems, and outputs.

The subsystems and outputs of the manufacturing MIS monitor and control the flow of materials, products, and services through the organization. The objective of the manufacturing MIS is to produce products that meet

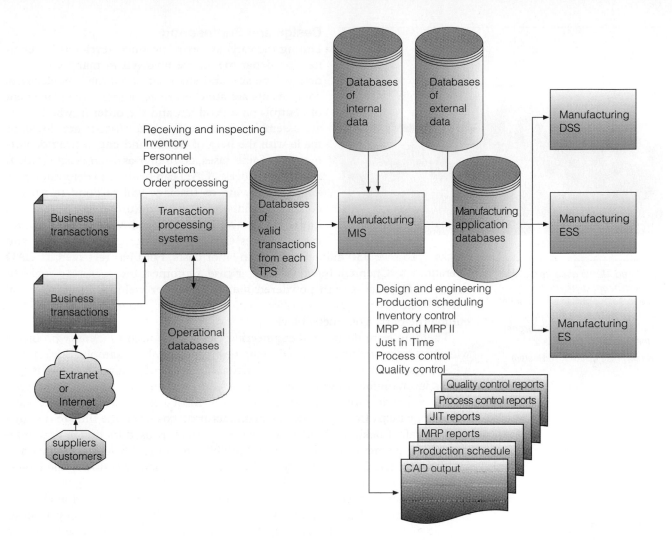

Receiving and inspecting
Inventory
Personnel
Production
Order processing

Design and engineering
Production scheduling
Inventory control
MRP and MRP II
Just in Time
Process control
Quality control

Quality control reports
Process control reports
JIT reports
MRP reports
Production schedule
CAD output

FIGURE 10.7

Overview of a Manufacturing MIS

customer needs—from the raw materials provided by suppliers to finished goods and services delivered to customers—at the lowest possible cost. Cunningham Motor Co., for example, is attempting to manufacture a pricey sports coupe with a 600 horsepower engine that sells for about $250,000.[13] Started by Robert Lutz, a former vice president of Chrysler, the company will not put one penny into manufacturing facilities. Instead, the new car will be manufactured entirely by others. According to Lutz, "Cunningham will be the world's most virtually integrated car company." Virtual organizations were discussed in Chapter 2.

As raw materials are converted to finished goods, the manufacturing MIS monitors the process at almost every stage. New bar codes called *smart labels* could make this process easier.[14] The smart labels, made of chips and tiny radio transmitters, allow materials and products to be monitored through the entire manufacturing process. Procter & Gamble, Gillette, Wal-Mart, and Target have helped to fund research into this new manufacturing MIS. Car manufacturers, which convert raw steel, plastic, and other materials into a finished automobile, also monitor the manufacturing process. Auto manufacturers add thousands of dollars of value to the raw materials they use in assembling a car. If the manufacturing MIS also lets them provide customized paint colors on any of their models, it has further added value (although less tangible) by ensuring a direct customer fit. In doing so, the MIS helps provide the company the edge that can differentiate it from competitors. The success of an organization can depend on the manufacturing function. Some common information subsystems and outputs used in manufacturing are discussed next.

Lockheed Martin used its design and engineering information systems to help it obtain a contract to build Joint Strike Fighter military jets. The contract was the largest defense contract in history.

(Source: Photo by Lockheed Martin)

Design and Engineering

During the early stages of product development, engineering departments are involved in many aspects of design. The size and shape of parts, the way electrical components are attached to equipment, the placement of controls on a product, and the order in which parts are assembled into the finished product are decisions made with the help of design and engineering departments. In some cases, computer-assisted design (CAD) assists this process. CAD can be used to determine how an airplane wing or fuselage will respond to various conditions and stresses while in use. CAD is also used in the automotive industry. Ford Motor Company, for example, uses CAD to help design its new cars and trucks.[15] To save $30 million or more in labor costs, Ford has relocated its CAD operations to Chennai, India. In another case, Optomec, Inc. is using a sophisticated CAD process with powdered metals and laser welding to make metal parts.[16] The unique process makes metal parts from powdered metal instead of cutting the parts from metal blocks.

The data from design and engineering can also be used to identify problems with existing products and help develop new products. For example, Boeing uses a CAD system to develop a complete digital blueprint of an aircraft before it ever begins its manufacturing process. As mock-ups are built and tested, the digital blueprint is constantly revised to reflect the most current design. Using such technology helps Boeing reduce its manufacturing costs and the time to design a new aircraft. Lockheed Martin, a defense contractor, used its design and engineering departments to help obtain a $200 billion contract from the Pentagon to build fighter jets for the military.[17] The Joint Strike Fighter contract, the largest defense contract in history, uses information systems to help design, engineer, and manufacture these sophisticated military jets. According to an analyst for AMR Research, "When you firm up a design, that's when you firm up your costs and the manufacturer's process capability and ability to produce." The Lockheed Martin contract will produce 3,000 X-35 fighter jets, starting in 2009 and will continue for over 20 years.

Master Production Scheduling and Inventory Control

Scheduling production and controlling inventory are critical for any manufacturing company. The overall objective of master production scheduling is to provide detailed plans for both short-term and long-range scheduling of manufacturing facilities (Figure 10.8). Master production scheduling software packages can

FIGURE 10.8

A master production schedule for computer disks and CDs indicates the quantity of each to be produced each week in thousands.

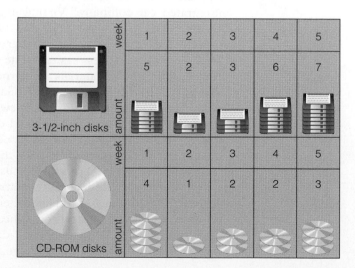

	week	1	2	3	4	5
3-1/2-inch disks	amount	5	2	3	6	7
	week	1	2	3	4	5
CD-ROM disks	amount	4	1	2	2	3

include forecasting techniques that attempt to determine current and future demand for products and services. After current demand has been determined and future demand has been estimated, the master production scheduling package can determine the best way to use the manufacturing facility and all its related equipment. The result of the process is a detailed plan that reveals a schedule for every item that will be manufactured.

An important key to the manufacturing process is inventory control. Great strides have been made in developing cost-effective inventory control programs and software packages that allow automatic reordering, forecasting, generation of shop documents and reports, determination of manufacturing costs, analysis of budgeted costs versus actual costs, and the development of master manufacturing schedules, resource requirements, and plans. A furniture company, for example, uses an approach, called "simple, quick, and affordable (SQA)" to keep inventory levels and costs low. Once an order is received, it is broken down into the inventory parts that are needed to successfully complete the order on time. An SQA Web site is used to make sure that the needed inventory is available to complete the order. Procter & Gamble, which produces consumer products that range from Pampers to Pepto-Bismol, uses quick-response inventory control systems to speed products to market.[18] According to a company spokesman, "A key benefit for consumers is that the products are fresher." In another case, Ford Motor Company decided to use UPS Logistics to help the company speed the delivery of parts to factories and finished cars to dealerships.[19] The new inventory control system has reduced by four days the time it typically takes to ship a finished vehicle to a dealership. But more importantly, the new system has also reduced vehicle inventory by about $1 billion, saving the company $125 million in annual inventory carrying costs, which dramatically improves Ford's profitability. Many inventory control techniques like Ford's attempt to minimize inventory related costs.

There are a number of inventory control techniques.[20] Most determine how much and when to order. One method of determining how much inventory to order is called the **economic order quantity (EOQ)**. This quantity is determined in such a way as to minimize the total inventory costs. The "When to order?" question is based on inventory usage over time. Typically, the question is answered in terms of a **reorder point (ROP)**, which is a critical inventory quantity level. When the inventory level for a particular item falls to the reorder point, or critical level, a report might be output so that an order is immediately placed for the EOQ of the product. Another inventory technique used when the demand for one item is dependent on the demand for another is called **material requirements planning (MRP)**. The basic goal of MRP is to determine when finished products, like automobiles or airplanes, are needed and then to work backward to determine deadlines and resources needed, such as engines and tires, to complete the final product on schedule. Krone Communications in Peenya, India, uses MRP to plan manufacturing with fluctuating, seasonal demand for communications products.[21]

Manufacturing resource planning (MRPII) refers to an integrated, companywide system based on network scheduling that enables people to run their business with a high level of customer service and productivity, while lowering costs and inventories. MRPII is broader in scope than MRP; thus, the latter has been dubbed "little MRP." MRPII places a heavy emphasis on planning. This helps companies ensure that the right product is in the right place at the right time.

Just-in-time (JIT) inventory and manufacturing is an approach that maintains inventory at the lowest levels without sacrificing the availability of finished products. With this approach, inventory and materials are delivered just before they are used in a product. A JIT inventory system would arrange for a car windshield to be delivered to the assembly line only a few moments before it is secured to the automobile, rather than having it sitting around the manufacturing facility

---

economic order quantity (EOQ)
the quantity that should be reordered to minimize total inventory costs

reorder point (ROP)
a critical inventory quantity level

material requirements planning (MRP)
a set of inventory control techniques that help coordinate thousands of inventory items when the demand of one item is dependent on the demand for another that determines when to order more inventory

manufacturing resource planning (MRPII)
an integrated, companywide system based on network scheduling that enables people to run their business with a high level of customer service and productivity

just-in-time (JIT) inventory approach
a philosophy of inventory management in which inventory and materials are delivered just before they are used in manufacturing a product

Computer-assisted manufacturing systems control complex processes on the assembly line and provide users with instant access to information.

(Source: © Lester Lefkowitz CORBIS)

computer-assisted manufacturing (CAM)
a system that directly controls manufacturing equipment

computer-integrated manufacturing (CIM)
using computers to link the components of the production process into an effective system

flexible manufacturing system (FMS)
an approach that allows manufacturing facilities to rapidly and efficiently change from making one product to another

quality control
a process that ensures that the finished product meets the customers' need

while the car's other components are being assembled. Although JIT has many advantages, it also renders firms more vulnerable to process disruptions. Ford Motor Company, for example, altered its JIT program by stockpiling some inventory to avoid production shutdowns that could result from possible parts shortages.[22] Other automotive companies are doing the same. According to a Deloitte Consulting expert, "We really have regressed to the stage where we're asking plants to carry more buffer inventory."[23]

Process Control

Managers can use a number of technologies to control and streamline the manufacturing process. For example, computers can be used to directly control manufacturing equipment, using systems called **computer-assisted manufacturing (CAM)**. CAM systems can control drilling machines, assembly lines, and more. Some of them operate quietly, are easy to program, and have self-diagnostic routines to test for difficulties with the computer system or the manufacturing equipment.

Computer-integrated manufacturing (CIM) uses computers to link the components of the production process into an effective system. CIM's goal is to tie together all aspects of production, including order processing, product design, manufacturing, inspection and quality control, and shipping. CIM systems also increase efficiency by coordinating the actions of various production units. In some areas, CIM is used for even broader functions. For example, it can be used to integrate all organizational subsystems, not just the production systems. In automobile manufacturing, design engineers can have their ideas evaluated by financial managers before new components are built to see whether they are economically viable, saving not only time but also money.

A **flexible manufacturing system (FMS)** is an approach that allows manufacturing facilities to rapidly and efficiently change from making one product to another. In the middle of a production run, for example, the production process can be changed to make a different product or to switch manufacturing materials. Often a computer is used to direct and implement the changes. By using an FMS, the time and cost to change manufacturing jobs can be substantially reduced, and companies can react quickly to market needs and competition.

FMS is normally implemented using computer systems, robotics, and other automated manufacturing equipment. New product specifications are fed into the computer system, and the computer then makes the necessary changes. Although few companies have a fully implemented FMS, recently use of the overall FMS approach has increased. DaimlerChrysler, for example, is going to use its new FMS to build the 2004 Pacifica minivan on the same manufacturing line as other leading minivans.24

Quality Control and Testing

With increased pressure from consumers and a general concern for productivity and high quality, today's manufacturing organizations are placing more emphasis on **quality control**, a process that ensures that the finished product meets the customers' needs. For a continuous process, control charts are used to measure weight, volume, temperature, or similar attributes (Figure 10.9). Then, upper and lower control chart limits are established. If these limits are exceeded, the manufacturing equipment is inspected for possible defects or potential problems.

When the manufacturing operation is not continuous, sampling can allow the producer or consumer to review and accept or reject one or more products. Acceptance sampling is used for items as simple as nuts and bolts or as complex as airplanes. The development of the control chart limits and the specific acceptance sampling plans can be fairly complex. So, quality-control software programs have been used to generate them.

FIGURE 10.9

Industrial Control Chart

This chart is used to monitor product quality in an industrial chemical application. Product variances exceeding certain tolerances cause those products to be rejected.

marketing MIS

information system that supports managerial activities in product development, distribution, pricing decisions, and promotional effectiveness

Marketing research data yields valuable information for the development and marketing of new products.

(Source: Grandpix/Index Stock)

Whether the manufacturing operation is continuous or discrete, the results from quality control are analyzed closely to identify opportunities for improvements. Teams using the total quality management (TQM) or continuous improvement process (see Chapter 2) often analyze this data to increase the quality of the product or eliminate problems in the manufacturing process. The result can be a cost reduction or increase in sales.

Information generated from quality-control programs can help workers locate problems in manufacturing equipment. Quality-control reports can also be used to design better products. With the increased emphasis on quality, workers should continue to rely on the reports and outputs from this important application.

MARKETING MANAGEMENT INFORMATION SYSTEMS

A **marketing MIS** supports managerial activities in product development, distribution, pricing decisions, promotional effectiveness, and sales forecasting. Marketing functions are increasingly being performed on the Internet. A number of companies are developing Internet marketplaces to advertise and sell products. Customer relationship management (CRM) programs, available from some ERP vendors, help a company manage all aspects of customer encounters. CRM software can help a company collect customer data, contact customers, educate customers on new products, and sell products to customers through an Internet site. Alaska Airlines used a CRM system to notify customers about flight changes after the September 11th tragedy.[25] According to Karen Wells-Fletcher, manager of network operations at Alaska Airlines, "I don't know if we could have made it through September 11th and all the rescheduling without this system. We just don't have the manpower to call everyone personally." Other airlines have also benefited from CRM.[26] Delta Airlines passengers can now go to the Delta Web site (www.delta.com) to look up possible wait times.

Crane Engineering, an industrial equipment distributor in Kimberly, Wisconsin, uses CRM to help manage customer interactions.[27] According to one company representative, "Salespeople want to know what's in it for them; it's not enough to tell them they have to do it. But give them a panoramic view of what their customer is doing in call centers and on the company Web site, such as buying other products or complaints. That's a very powerful motivator—they respond to revenue potential and growing their customer base." Yet, not all CRM systems and marketing sites on the Internet are successful. According to Meta Group, 55 to 75 percent of CRM projects fail to meet their objectives. Customization and ongoing maintenance of a CRM system can be expensive.[28] In addition, some companies have trouble successfully selling products on their Internet sites. Figure 10.10 shows the inputs, subsystems, and outputs of a typical marketing MIS.

Subsystems for the marketing MIS include marketing research, product development, promotion and advertising, and product pricing. These subsystems and their outputs help marketing managers and executives increase sales, reduce marketing expenses, and develop plans for future products and services to meet the changing needs of customers.

Marketing Research

Surveys, questionnaires, pilot studies, and interviews are popular marketing research tools. The purpose of marketing research is to conduct a formal study of the market and customer preferences. Marketing research can identify prospects (potential future customers) as well as the features that current customers really want in a good or service (such as green ketchup or vanilla-flavored cola). Such attributes as

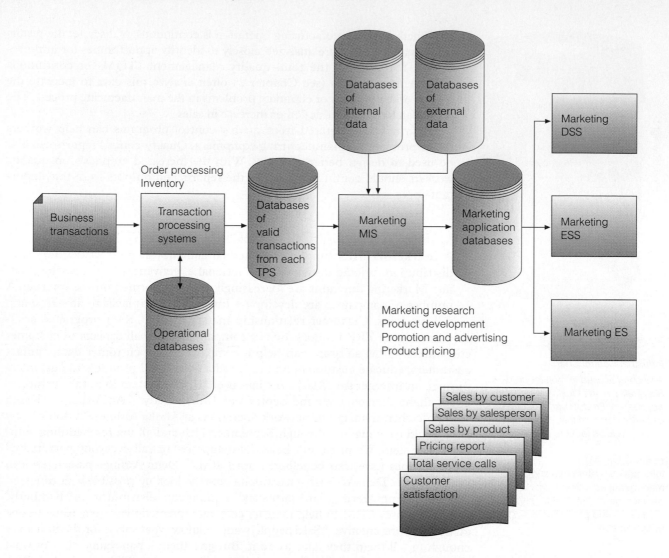

FIGURE 10 10

Overview of a Marketing MIS

style, color, size, appearance, and general fit can be investigated through marketing research. Pricing, distribution channels, guarantees and warranties, and customer service can also be determined. Once entered into the marketing MIS, data collected from marketing research projects is manipulated to generate reports on key indicators like customer satisfaction and total service calls. Reports generated by the marketing MIS help marketing managers be better informed to help the organization meet its performance goals.

Forecasting demand can be an important result of marketing research and sophisticated software.[29] The parts division of Hyundai Motor America, for example, uses marketing research and software to predict the demand for car parts. The software from Demand Management helped the company reduce delivery time for key auto parts by 20 percent. Other companies, including Colgate-Palmolive and Unilever, also use sophisticated software and marketing research data to forecast demand for their products. Demand forecasts for products and services are also critical to make sure raw materials and supplies are properly managed.[30] Doug Thomas, an assistant professor at Pennsylvania State University, believes that this type of forecasting can help prevent "suppliers from building up huge stockpiles of inventory when a product is not selling."

The Internet is changing the way many companies think about marketing research. Conventional methods of collecting data often cost millions of dollars. For a fraction of these costs, companies can put up Internet information servers and launch discussion groups on topics that their customers care about. These information sites must be well designed, or they won't be visited, but a frequently visited site can provide feedback worth a fortune. Companies that are viewed as credible, not just clever, will win enormous advantages. Presence and intelligent interaction, not just advertising, are the keys that will unlock commercial opportunities on-line. Some people, however, consider Internet marketing research to be a nuisance or even harmful. Some companies gather information on customers using cookies, which collect data on people's Internet surfing habits, and sell it to others. To protect customer privacy and keep the valuable marketing research data to themselves, some companies, including General Motors, Ford, and Procter & Gamble, are starting to block Internet ad servers from getting the data. According to one General Motors spokesperson, "We've never given away anything."

Product Development

Product development involves the conversion of raw materials into finished goods and services and focuses primarily on the physical attributes of the product. Many factors, including plant capacity, labor skills, engineering factors, and materials are important in product development decisions. In many cases, a computer program is used to analyze these various factors and to select the appropriate mix of labor, materials, plant and equipment, and engineering designs. Make-or-buy decisions can also be made with the assistance of computer programs. Faucet maker Moen decided to carry a variety of products with different colors and styles.[31] It concluded that it was not in the business of selling hardware but instead should be selling fashion and jewelry for bathrooms and kitchens. Using the Internet and product-development software, the 50 engineers responsible for new product development were able to reduce the time from design to placement on store shelves from 24 months to only 16 months.

Promotion and Advertising

One of the most important functions of any marketing effort is promotion and advertising. Product success is a direct function of the types of advertising and sales promotion done. Dole Food Company, for example, promoted its products by putting a $10 electronic incentive on 30 million packages of its Fruit Bowl.[32] The $10 in electronic currency could be spent at seven participating on-line retailers, including Art.com, Cooking.com, Kbkids.com, and SunglassHut.com. The size of the promotion budget and the allocation of this budget among various campaigns are important factors in planning the campaigns that will be launched—everything from placing ads during the Super Bowl to offering coupons in a grocery store. Television coverage, newspaper ads, promotional brochures and literature, and training programs for salespeople are all components of these campaigns. Because of the time and scheduling savings they offer, computer programs are used to set up the original budget and to monitor expenditures and the overall effectiveness of various promotional campaigns.

Product Pricing

Product pricing is another important and complex marketing function. Retail price, wholesale price, and price discounts must be set. A major factor in determining pricing policy is an analysis of the demand curve, which attempts to determine the relationship between price and sales. Most companies try to develop pricing policies that will maximize total sales revenues—usually a function of price elasticity. If the product is very price sensitive, a reduction in price can generate a substantial increase in sales, which can result in higher revenues. A product that is relatively insensitive to price can have its price substantially increased without a large reduction in demand. Computer programs can help determine price elasticity and various pricing policies, such as supply and demand curves for pricing analysis (Figure 10.11). Typically, a marketing executive has the ability to make alterations in price on the computer system, which analyzes price changes and their impact on total revenues. The rapid feedback now obtainable through computer communications networks enables managers to determine the results of pricing decisions much more quickly than in the past. This ability facilitates more

Typical Supply and Demand Curve for Pricing Analysis

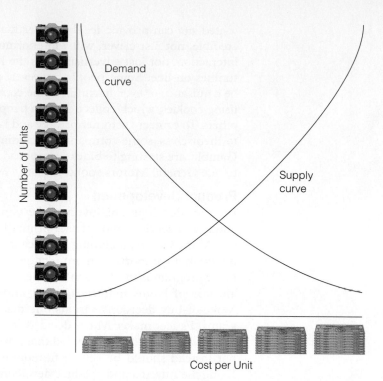

aggressive pricing strategies, which can be quickly adjusted to meet market needs. One critical pricing decision is when to mark down product prices.[33] Using sophisticated software, ShopKo has been able to reduce the number and amount of price cuts, which has helped increase profitability.

Sales analysis is also important to identify products, sales personnel, and customers that contribute to profits and those that do not. Several reports can be generated to help marketing managers make good sales decisions (Figure 10.12). The sales-by-product report lists all major products and their sales for a period of time, such as a month. This report shows which products are doing well and which ones need improvement or should be discarded altogether. The sales-by-salesperson report lists total sales for each salesperson for each week or month. This report can also be subdivided by product to show which products are being sold by each salesperson. The sales-by-customer report is a tool to use to identify high- and low-volume customers.

HUMAN RESOURCE MANAGEMENT INFORMATION SYSTEMS

human resource MIS
an information system that is concerned with activities related to employees and potential employees of an organization, also called a personnel MIS

A **human resource MIS**, also called the *personnel MIS*, is concerned with activities related to employees and potential employees of the organization. Because the personnel function relates to all other functional areas in the business, the human resource MIS plays a valuable role in ensuring organizational success. Some of the activities performed by this important MIS include workforce analysis and planning; hiring; training; job and task assignment; and many other personnel-related issues. Personnel issues can include offering new hires attractive stock option and incentive programs. One company, for example, offered new engineers a two-year lease on a sporty BMW roadster as a signing bonus. An effective human resource MIS will allow a company to keep personnel costs at a minimum while serving the required business processes needed to achieve corporate goals. Figure 10.13 shows some of the inputs, subsystems, and outputs of the human resource MIS.

Human resource subsystems and outputs range from the determination of human resource needs and hiring through retirement and outplacement. Most medium and large organizations have computer systems to assist with human

FIGURE 10.12

Reports Generated to Help Marketing Managers Make Good Decisions

(a) This sales-by-product report lists all major products and their sales for the period from August to December. (b) This sales-by-salesperson report lists total sales for each salesperson for the same time period. (c) This sales-by-customer report lists sales for each customer for the period. Like all MIS reports, totals are provided automatically by the system to show managers at a glance the information they need to make good decisions.

(a) Sales by Product

Product	August	September	October	November	December	Total
Product 1	34	32	32	21	33	152
Product 2	156	162	177	163	122	780
Product 3	202	145	122	98	66	633
Product 4	345	365	352	341	288	1,691

(b) Sales by Salesperson

Salesperson	August	September	October	November	December	Total
Jones	24	42	42	11	43	162
Kline	166	155	156	122	133	732
Lane	166	155	104	99	106	630
Miller	245	225	305	291	301	1,367

(c) Sales by Customer

Customer	August	September	October	November	December	Total
Ang	234	334	432	411	301	1,712
Braswell	56	62	77	61	21	277
Celec	1,202	1,445	1,322	998	667	5,634
Jung	45	65	55	34	88	287

resource planning, hiring, training and skills inventory, and wage and salary administration. Outputs of the human resource MIS include reports such as human resource planning reports, job application review profiles, skills inventory reports, and salary surveys.

Human Resource Planning

One of the first aspects of any human resource MIS is determining personnel and human needs. The overall purpose of this MIS subsystem is to put the right number and kinds of employees in the right jobs when they are needed. Effective human resource planning requires defining the future number of employees needed and anticipating the future supply of people for these jobs. For companies involved with large projects, such as military contractors and large builders, human resource plans can be generated directly from data on current and future projects.

Suppose a construction company obtains a contract from a group of investors to build a 250-unit apartment complex. Forecasting programs and project management software packages can be used to develop reports that describe what people are needed and when they are needed during the entire construction project. A typical output would be a human resource needs and planning report, which might specify that ten employees will be needed in August to pour concrete slabs, and eight carpenters and four painters will be needed in October. Alternatively, a factory might use this report to list total number of employees needed, broken down according to skill level, such as highly technical, technical, semitechnical, and so on.

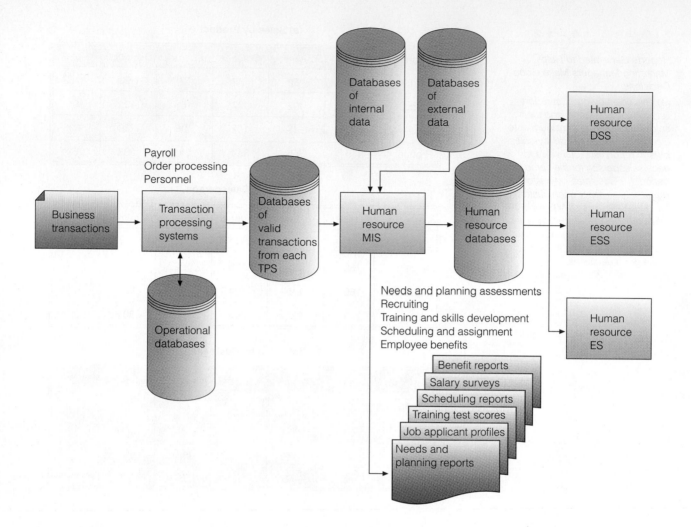

FIGURE 10.13

*Overview of a Human
Resource MIS*

Personnel Selection and Recruiting

If the human resource plan reveals that additional personnel are required, the next logical step is recruiting and selecting personnel. This subsystem performs one of the most important and critical functions of any organization, especially in service organizations, where employees can define the company's success. Companies seeking new employees often use computers to schedule recruiting efforts and trips and to test potential employees' skills. Some software companies, for example, use computerized testing to determine a person's programming skills and abilities. Management information systems can be used to help rank and select potential employees. For every applicant, the results of interviews, tests, and company visits can be analyzed by the system and printed. This report, called a *job applicant review profile*, can assist corporate recruiting teams in final selection. Some software programs can even analyze this data to help identify job applicants most likely to stay with the company and perform according to corporate standards.

Many companies now use the Internet to screen for job applicants. Applicants use a template to load their resume onto the Internet site. HR managers can then access these resumes and identify applicants they are interested in interviewing.

Training and Skills Inventory

Some jobs, such as programming, equipment repair, and tax preparation, require very specific training. Other jobs may require general training about the organizational culture, orientation, dress standards, and expectations of the organization. Today, many organizations conduct their own training, with the assistance of information systems and technology. Self-paced training can involve computerized tutorials, video programs, and CD-ROM books and materials. Distance learning, where training and classes are conducted over the Internet, is also becoming a viable alternative to more traditional training and learning approaches. This text and supporting material, for example, can be used in a distance-learning environment.

When training is complete, employees may be required to take computer-scored tests to reveal their mastery of skills and new material. The results of these tests are usually given to the employee's supervisor in the form of training or skills inventory reports. In some cases, skills inventory reports are used for job placement. For instance, if a particular position in the company needs to be filled, managers might wish to hire internally before they recruit. The skills inventory report would help them evaluate current employees to determine their potential for the position. They can also be part of employee evaluations and determine raises or bonuses. These types of tests, however, must be valid and reliable to avoid mistakes in job placement and bonuses.

Technology can also be used to reduce training needs and costs. Buy.com, for example, used Finali and Net Sage to help employees answer customer questions.[34] The Finali and Net Sage products helped Buy.com reduce training and customer service costs by 40 percent.

Scheduling and Job Placement

Scheduling people and jobs can be relatively straightforward or extremely complex. For some small service companies, scheduling and job placements are based on which customers walk through the door. Determining the best schedule for flights and airline pilots, placing military recruits into jobs, and determining what truck drivers and equipment should be used to transport materials across the country normally require sophisticated computer programs. In most cases, various schedules and job placement reports are generated. Employee schedules are developed for each employee, showing their job assignments over the next week or month. Job placements are often determined based on skills inventory reports, which show which employee might be best suited to a particular job.

Wage and Salary Administration

The last of the major human resource MIS subsystems involves determining wages, salaries, and benefits, including medical payments, savings plans, and retirement accounts. Wage data, such as industry averages for positions, can be taken from the corporate database and manipulated by the human resource MIS to provide wage information and reports to higher levels of management. These reports, called *salary surveys*, can be used to compare salaries with budget plans, the cost of salaries versus sales, and the wages required for any one department or office. The reports also help show backup of key positions in the company. Wage and salary administration also entails designing retirement programs for employees. Some companies use computerized retirement programs to help employees gain the most from their retirement accounts and options.

OTHER MANAGEMENT INFORMATION SYSTEMS

In addition to finance, manufacturing, marketing, and human resource MISs, some companies have other functional management information systems. For example, most successful companies have well-developed accounting functions and a supporting accounting MIS. Also, many companies make use of geographic information systems for presenting data in a useful form.

Accounting MISs

accounting MIS
information system that provides aggregate information on accounts payable, accounts receivable, payroll, and many other applications

In some cases, accounting works closely with financial management. An **accounting MIS** performs a number of important activities, providing aggregate information on accounts payable, accounts receivable, payroll, and many other applications. The organization's TPS captures accounting data, which is also used by most other functional information systems.

Some smaller companies hire outside accounting firms to assist them with their accounting functions. These outside companies produce reports for the firm using raw accounting data. In addition, many excellent integrated accounting programs, such as QuickBooks, are available for personal computers in small companies. Depending on the needs of the small organization and its personnel's computer experience, using these computerized accounting systems can be a very cost-effective approach to managing information.

Geographic Information Systems

Increasingly, managers want to see data presented in graphical form. A **geographic information system (GIS)** is a computer system capable of assembling, storing, manipulating, and displaying geographically referenced information, that is, data identified according to their locations. A GIS enables users to pair maps or map outlines with tabular data to describe aspects of a particular geographic region. For example, sales managers may want to plot total sales for each county in the states they serve. Using a GIS, they can specify that each county be shaded to indicate the relative amount of sales—no shading or light shading represents no or little sales and deeper shading represents more sales. As seen in the "IS Principles in Action" box, a GIS can be invaluable in helping to eradicate a forest pest, prevent forest damage, and save millions of dollars.

Because the GIS works with any data represented in tabular form, graphical capability is finding its way into spreadsheets. For example, Excel and Lotus include a mapping tool that lets you plot spreadsheet data as a demographic map. Such applications show up frequently in scientific investigations, resource management, and real-estate development planning. Retail, government, and utility organizations are frequent users of GISs. Retail chains, for example, need spatial analysis to determine where potential customers are located and where their competition is.

We saw earlier in this chapter that management information systems (MISs) provide useful summary reports to help solve structured and semistructured business problems. Decision support systems (DSSs) offer the potential to assist in solving both semistructured and unstructured problems.

geographic information system (GIS)
a computer system capable of assembling, storing, manipulating, and displaying geographic information, i.e., data identified according to their locations

AN OVERVIEW OF DECISION SUPPORT SYSTEMS

A DSS is an organized collection of people, procedures, software, databases, and devices used to support problem-specific decision making and problem solving. The focus of a DSS is on decision-making effectiveness when faced with unstructured or semistructured business problems. Decision support systems offer the potential to generate higher profits, lower costs, and better products and services. For example, healthcare organizations use DSSs to track and reduce costs. As with a TPS and an MIS, a DSS should be designed, developed, and used to help an organization achieve its goals and objectives.

Decision support systems, although skewed somewhat toward the top levels of management, are used at all levels. To some extent, today's managers at all levels are faced with less structured, nonroutine problems, but the quantity and magnitude of these decisions increase as a manager rises higher in an organization. Many organizations contain a tangled web of complex rules, procedures, and decisions. DSSs are used to bring more structure to these problems to aid the decision-making process. In addition, because of the inherent flexibility of decision support systems, managers at all levels are able to use DSSs to assist in some relatively routine, programmable decisions in lieu of more formalized management information systems.

The management information system (MIS) must provide the right information to the right person in the right fashion at the right time.

Geographic Information System Saves Trees

A war is being waged in the northeastern United States. It began in 1869 when an invading force arrived near Boston, and it has raged ever since. The enemy has spread to all or part of 17 states and the District of Columbia. Numerous battles are being waged to contain the enemy—from the Upper Peninsula of Michigan, in eastern Wisconsin, across northern Illinois, central Ohio, and West Virginia to the coast of North Carolina. The enemy? Gypsy moths. The cost of war: yearly defoliation reaching into the millions of acres, with costs running into tens of millions of dollars. Our valiant defenders: the U.S. Department of Agriculture's Forest Service.

The gypsy moth, *Lymantria dispar*, is one of North America's most devastating forest pests. It was accidentally introduced near Boston by E. Leopold Trouvelot. About 10 years after this introduction, the first outbreaks began in Trouvelot's neighborhood and in 1890 the state and federal government began attempts to eradicate the gypsy moth. These attempts ultimately failed, and since that time, the gypsy moth has continued to spread. The gypsy moth is known to feed on the foliage of hundreds of species of plants in North America, but its most common hosts are oaks and aspen. When densities reach very high levels, trees may become completely defoliated. Several successive years of defoliation, along with contributions by other stresses, may ultimately result in death of a tree.

In 1999 the USDA Forest Service, along with state and federal participants, implemented the National Gypsy Moth Slow the Spread (STS) Project across the 1,200 mile gypsy moth frontier from North Carolina through the Upper Peninsula of Michigan. The goal of the project is to use novel integrated pest management (IPM) strategies to reduce the rate of gypsy moth spread into uninfested areas. The STS project is expected to:

- decrease the new territory invaded by the gypsy moth each year from 15,600 square miles to 6,000 square miles
- protect forests, forest-based industries, urban and rural parks, and private property
- avoid at least $22 million per year in damage and management costs.

While traditional approaches to gypsy moth management address the high concentrations occurring in infested areas, the STS project focuses on low-level populations in transition zones on the margins.

What does all this have to do with information systems? At the heart of the STS is a Web-based geographic information system (GIS) that supplies participants with valuable data gathered from the transition zone. Using this information, rangers can easily determine where new infestation is taking place and treat the area while the infestation can be controlled. In effect, the GIS compiles and stores data from the entire project.

Gypsy moth traps are placed along the uninfested side of a current boundary line every two kilometers. The traps are checked at the conclusion of the moth's flight season, and the numbers of moths found in each trap counted and recorded in a central database that feeds the GIS.

The GIS uses the Web to display a picture of a map of the northeastern United States, which when clicked on zooms in on a specific region. Through the use of color coding, graphic icons, geometric shapes, and other markings, researchers can easily spot areas on the map with higher densities of moths and order these regions to be studied more closely. Traps are laid out every 500 meters in these areas to determine the specific location of the swarm. Trapped moths are counted, and the results stored in the database and accessed via the GIS. As swarm locations are pinpointed, they are eradicated to slow the spread.

The GIS is used to:

- locate traps and list the moth count for each trap
- identify areas for treatment
- identify the vegetation in any given area (shrubs, deciduous forest, evergreen forest, wetlands, urban forest)
- identify the elevation of a given location
- track the history of treatment of a given area
- examine the results of the treatment

The USDA Forest Service refers to the GIS as its decision support system, since it assists the rangers in deciding what areas need closer examination, and ultimately which areas need treatment. By using a GIS, they are able to see the patterns of new infestation and decide on an informed defensive action. The GIS brings data, numbers, locations, and quantities to life and effectively supports strategic planning. Whether an organization is tracking moths or customer sales in retail outlet stores, GISs assist managers in deciding what strategies are working and where to focus their attention.

Discussion Questions

1. How would the management of the STS project differ if the Internet and this GIS were not available?
2. The STS project GIS is a Web-based data access tool. It allows system users to view data but not to enter data. Would it be even more helpful if it could be used to enter data into the system? How so?

Critical Thinking Questions

3. How might a GIS be adapted to serve the needs of a retail marketing firm? In a marketing GIS, what might replace the six bulleted items in the article? For example, rather than locating traps and moth counts for each trap, what if the GIS is adapted to display franchise locations and sales at each franchise?
4. What types of problems and scenarios are best suited for a GIS solution? Why?

Sources: "Slow the Spread" Web site, http://www.ento.vt.edu/STS/, accessed May 4, 2002; "Gypsy Moth in North America," the National Forest Service Web site, http://www.fs.fed.us/ne/morgantown/4557/gmoth/, accessed May 4, 2002.

CHARACTERISTICS OF A DECISION SUPPORT SYSTEM

Decision support systems have a number of characteristics that allow them to be effective management support tools. Of course, not all DSSs work the same—some are small in scope and offer only some of these characteristics. In general, a decision support system can perform the following functions:

Handle large amounts of data from different sources. For instance, advanced database management systems and data warehouses have allowed decision makers to search for information with a DSS even when some data resides in different databases on different computer systems or networks. Other sources of data may be accessed via the Internet or over a corporate intranet. Using the Internet, oil giant BP was able to use a decision support system to save an estimated $300 million in one year.[35] The Internet allowed BP to coordinate a large amount of drilling and exploration data from around the globe. According to John Leggate, vice president of digital business for BP, "The Net allows us to stop being a conglomerate and become a single, smooth global corporation."

Provide report and presentation flexibility. Managers can get the information they want, presented in a format that suits their needs. Furthermore, output can be displayed on computer screens or printed, depending on the needs and desires of the problem solvers.

Offer both textual and graphical orientation. Today's decision support systems can produce text, tables, line drawings, pie charts, trend lines, and more. By using their preferred orientation, managers can use a DSS to get a better understanding of a situation and to convey this understanding to others.

Support drill down analysis. A manager can get more levels of detail when needed by drilling down through data. For example, a manager can get more detailed information for a project—viewing the overall project cost or drilling down and seeing the cost for each phase, activity, and task.

Perform complex, sophisticated analysis and comparisons using advanced software packages. Marketing research surveys, for example, can be analyzed in a variety of ways using programs that are part of a DSS. Many of the analytical programs associated with a DSS are actually stand-alone programs, and the DSS brings them together.

Support optimization, satisficing, and heuristic approaches. By supporting all types of decision-making approaches, a DSS gives the decision maker a great deal of flexibility in computer support for decision making. For example, **what-if analysis**, the process of making hypothetical changes to problem data and observing the impact on the results, can be used to control inventory. Given the demand for products, such as automobiles, the computer can determine the necessary parts and components, including engines, transmissions, windows, and so on. With "what-if" analysis, a manager can make changes to problem data (the number of automobiles needed for next month) and immediately see the impact on the parts requirements.

Multicriteria decision making allows managers to take a number of important goals into account. A car manufacturer, for example, may want to maximize profits while keeping all its plants open for the next few months and avoiding a labor strike. Multicriteria decision-making approaches allow managers to seek several goals at the same time.

Goal-seeking analysis is the process of determining the problem data required for a given result. For example, a financial manager may be considering an investment with a certain monthly net income, and the manager might have a goal to earn a return of 9 percent on the investment. Goal seeking allows the manager to determine what monthly net income (problem data) is needed to have a return of 9 percent (problem result). Some spreadsheets can be used to perform goal-seeking analysis (see Figure 10.14).

Simulation is the ability of the DSS to duplicate the features of a real system. In most cases, probability or uncertainty are involved. For example, the mean

what-if analysis
the process of making hypothetical changes to problem data and observing the impact on the results

goal-seeking analysis
the process of determining the problem data required for a given result

simulation
the ability of the DSS to duplicate the features of a real system

FIGURE 10.14

With a spreadsheet program, a manager can enter a goal, and the spreadsheet will determine the needed input to achieve the goal.

(Source: Courtesy of Lotus Development Corporation.)

FIGURE 10.14

With a spreadsheet program, a manager can enter a goal, and the spreadsheet will determine the needed input to achieve the goal.

(Source: Courtesy of Lotus Development Corporation.)

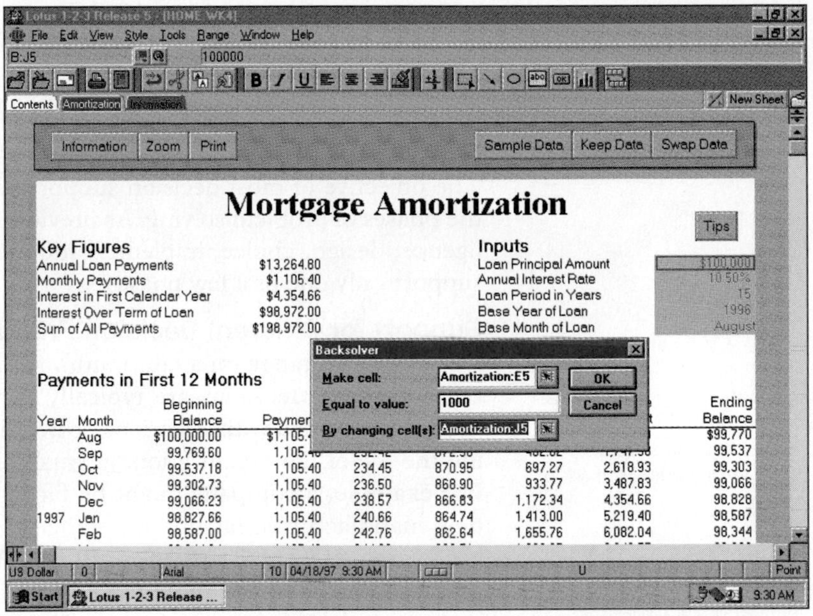

time between failure and the mean time to repair key components of a manufacturing line can be calculated to determine the impact on the number of products that can be produced each shift. Engineers can use this data to determine which components need to be reengineered to increase the mean time between failures and which components need to have an ample supply of spare parts to reduce the mean time to repair. Drug companies are using simulated trials to reduce the need for human participants and reduce the time and costs of bringing a new drug to market.[36] Drug companies are hoping that this use of simulation will help them identify successful drugs earlier in development.

CAPABILITIES OF A DECISION SUPPORT SYSTEM

Developers of decision support systems strive to make them more flexible than management information systems and to give them the potential to assist decision makers in a variety of situations. Table 10.2 lists a few DSS applications. DSSs can assist with all or most problem-solving phases, decision frequencies,

TABLE 10.2

Selected DSS Applications

Company or Application	Description
Cinergy Corporation	The electric utility developed a DSS to reduce lead time and effort required to make decisions in purchasing coal.
RCA	The company developed a DSS called Industrial Relation Information System (IRIS) to help solve personnel problems and issues.
U.S. Army	It developed a DSS to help recruit, train, and educate enlisted forces. The DSS uses simulation that incorporates what-if features.
National Audubon Society	It developed a DSS called Energy Plan (EPLAN) to analyze the impact of U.S. energy policy on the environment.
Hewlett-Packard	The computer company developed a DSS called Quality Decision Management to help improve the quality of its products and services.
Virginia	The state of Virginia developed the Transportation Evacuation Decision Support System (TEDSS) to determine the best way to evacuate people in case of a nuclear disaster at its nuclear power plants.

and different degrees of problem structure. DSS approaches can also help at all levels of the decision-making process. In this section we investigate these DSS capabilities. (An actual DSS may provide only a few of these capabilities, depending on the uses and scope of the DSS.)

Support for Problem-Solving Phases

The objective of most decision support systems is to assist decision makers with the phases of problem solving. As previously discussed, these phases include intelligence, design, choice, implementation, and monitoring. A specific DSS might support only one or a few phases.

Support for Different Decision Frequencies

Decisions can range on a continuum from one-of-a-kind to repetitive decisions. One-of-a-kind decisions are typically handled by an **ad hoc DSS**. An ad hoc DSS is concerned with situations or decisions that come up only a few times during the life of the organization; in small businesses, they may happen only once. For example, a company might be faced with a decision on whether to build a new manufacturing facility in another area of the country. Repetitive decisions are addressed by an **institutional DSS**. An institutional DSS handles situations or decisions that occur more than once, usually several times a year or more. An institutional DSS is used repeatedly and refined over the years. Examples of institutional DSSs include systems that support portfolio and investment decisions and production scheduling. These decisions may require decision support numerous times during the year. Between these two extremes are decisions managers make several times, but not regularly or routinely.

Support for Different Problem Structures

As discussed previously, decisions can range from highly structured and programmed to unstructured and nonprogrammed. **Highly structured problems** are straightforward, requiring known facts and relationships. **Semistructured or unstructured problems**, on the other hand, are more complex. The relationships among the data are not always clear, the data may be in a variety of formats, and the data is often difficult to manipulate or obtain. In addition, the decision maker may not know the information requirements of the decision in advance.

Support for Various Decision-Making Levels

Decision support systems can offer help for managers at different levels within the organization. Operational-level managers can get assistance with daily and routine decision making. Tactical-level decision makers can be supported with analysis tools to ensure proper planning and control. At the strategic level, DSSs can help managers by providing analysis for long-term decisions requiring both internal and external information (Figure 10.15).

A COMPARISON OF DSS AND MIS

A DSS differs from an MIS in numerous ways, including the type of problems solved; the support given to users; the decision emphasis and approach; and the type, speed, output, and development of the system used. Table 10.3 lists brief descriptions of these differences.

ad hoc DSS
a DSS concerned with situations or decisions that come up only a few times during the life of the organization

institutional DSS
a DSS that handles situations or decisions that occur more than once, usually several times a year or more. An institutional DSS is used repeatedly and refined over the years.

highly structured problems
problems that are straightforward and require known facts and relationships

semistructured or unstructured problems
more complex problems in which the relationships among the data are not always clear, the data may be in a variety of formats, and the data is often difficult to manipulate or obtain

FIGURE 10.15

Decision-Making Level
Strategic-level managers are involved with long-term decisions, which are often made infrequently. Operational-level managers are involved with decisions that are made more frequently.

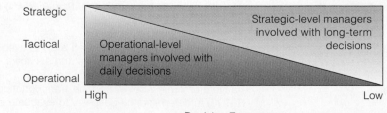

Factor	DSS	MIS
Problem Type	A DSS is good at handling unstructured problems that cannot be easily programmed.	An MIS is normally used only with more structured problems.
Users	A DSS supports individuals, small groups, and the entire organization. In the short run, users typically have more control over a DSS.	An MIS supports primarily the organization. In the short run, users have less control over an MIS.
Support	A DSS supports all aspects and phases of decision making; it does not replace the decision maker—people still make the decisions.	This is not true of all MIS systems—some make automatic decisions and replace the decision maker.
Emphasis	A DSS emphasizes actual decisions and decision-making styles.	An MIS usually emphasizes information only.
Approach	A DSS is a direct support system that provides interactive reports on computer screens.	An MIS is typically an indirect support system that uses regularly produced reports.
System	The computer equipment that provides decision support is usually on-line (directly connected to the computer system) and related to real time (providing immediate results). Computer terminals and display screens are examples—these devices can provide immediate information and answers to questions.	An MIS, using printed reports that may be delivered to managers once a week, may not provide immediate results.
Speed	Because a DSS is flexible and can be implemented by users, it usually takes less time to develop and is better able to respond to user requests.	An MIS's response time is usually longer.
Output	DSS reports are usually screen oriented, with the ability to generate reports on a printer.	An MIS, however, typically is oriented toward printed reports and documents.
Development	DSS users are usually more directly involved in its development. User involvement usually means better systems that provide superior support. For all systems, user involvement is the most important factor for the development of a successful system.	An MIS is frequently several years old and often was developed for people who are no longer performing the work supported by the MIS.

TABLE 10.3

Comparison of DSSs and MISs

COMPONENTS OF A DECISION SUPPORT SYSTEM

dialogue manager
user interface that allows decision makers to easily access and manipulate the DSS and to use common business terms and phrases

At the core of a DSS are a database and a model base. In addition, a typical DSS contains a **dialogue manager**, which allows decision makers to easily access and manipulate the DSS and to use common business terms and phrases. Finally, access to the Internet, networks, and other computer-based systems permits the DSS to tie into other powerful systems, including the TPS or function-specific subsystems. Internet software agents, for example, can be used in creating powerful decision support systems. Figure 10.16 shows a conceptual model of a DSS.

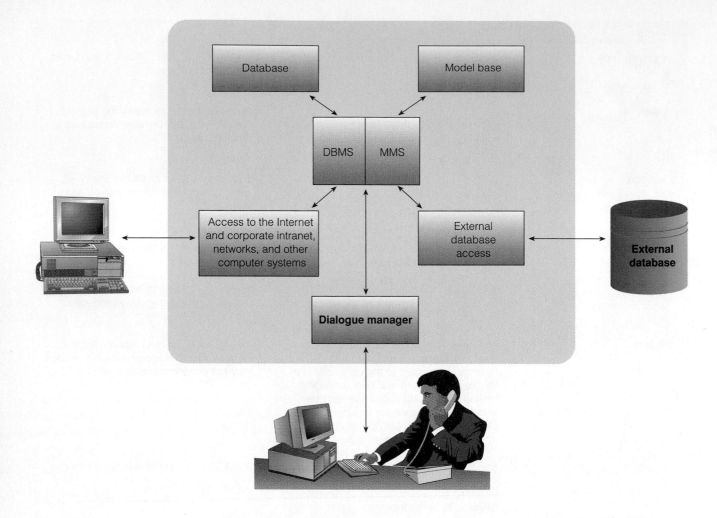

FIGURE 10.16

Conceptual Model of a DSS

DSS components include a model base; database; external database access; access to the Internet and corporate intranet, networks, and other computer systems; and a dialogue manager.

THE DATABASE

The database management system allows managers and decision makers to perform *qualitative analysis* on the company's vast stores of data in databases, data warehouses, and data marts, discussed in Chapter 5. A *data-driven DSS* primarily performs qualitative analysis based on the company's databases. Data-driven DSSs tap into vast stores of information contained in the corporate database, retrieving information on inventory, sales, personnel, production, finance, accounting, and other areas. Data mining, a process introduced in Chapter 5, is often used in a data-driven DSS. TWA, for example, used a data-driven DSS to help it identify customers for round trip flights between Los Angeles and San Juan, Puerto Rico.[37] Using a system from NCR, TWA searched its data warehouse and contacted 30,000 customers who might be interested in the flight. An astounding 25 percent either booked the flight or booked another Caribbean flight. Foxwoods Resort Casino uses a data-driven DSS to search a 200-GB database to get detailed information on its patrons.[38] It can tell how many kids a customer has, how much he or she spends a day on gambling, and more. According to Brian Charette, director of gaming systems, "We know who these people are and cater to them. We make sure they have flowers in the room, a drink in the hand, and reservations at the restaurant."

A database management system can also connect to external databases to give managers and decision makers even more information and decision support. External databases can include the Internet, libraries, government databases, and more. The combination of internal and external database access can give key decision makers a better understanding of the company and its environment.

THE MODEL BASE

model base
part of a DSS that provides decision makers access to a variety of models and assists them in decision making

The **model base** allows managers and decision makers to perform *quantitative analysis* on both internal and external data. A *model-driven DSS* primarily performs mathematical or quantitative analysis. The model base gives decision makers access to a variety of models and assists them in the decision-making process. **Model management software (MMS)** is often used to coordinate the use of models in a DSS, including financial, statistical analysis, graphical, and project management models. Depending on the needs of the decision maker, one or more of these models can be used (Table 10.4).

model management software (MMS)
software that coordinates the use of models in a DSS

THE DIALOGUE MANAGER

The dialogue manager allows users to interact with the DSS to obtain information. It assists with all aspects of communications between the user and the hardware and software that constitute the DSS. In a practical sense, to most DSS users, the dialogue manager is the DSS. Upper-level decision makers are often less interested in where the information came from or how it was gathered than that the information is both understandable and accessible.

GROUP DECISION SUPPORT SYSTEMS

The DSS approach has resulted in better decision making for all levels of individual users. However, many DSS approaches and techniques are not suitable for a group decision-making environment. Although not all workers and managers are involved in committee meetings and group decision-making sessions, some tactical and strategic-level managers can spend more than half their decision-making time in a group setting. Such managers need assistance with group decision making. A **group decision support system (GDSS)**, also called *group support system* and *computerized collaborative work system*, consists of most of the elements in a DSS, plus GDSS software needed to provide effective support in group decision-making settings (Figure 10.17).[39]

group decision support system (GDSS)
software application that consists of most elements in a DSS, plus software needed to provide effective support in group decision making; also called *group support system* or *computerized collaborative work system*

Group decision support systems are used in most industries. Architects are increasingly using GDSS to help them collaborate with other architects and builders to help them develop the best plans and to compete for contracts.[40] Caterpillar, the $20 billion producer of mining and construction equipment, used a collaborative system to link raw material suppliers to the company.[41] "When you choose to collaborate, you cannot escape the fact that you've now changed the business process. Different people have to be notified and different

TABLE 10.4

DSSs often use financial, statistical, graphical, and project management models.

Model Type	Description	Software That Can Be Used
Financial	Provides cash flow, internal rate of return, and other investment analysis	Spreadsheet, such as Excel
Statistical	Provides summary statistics, trend projections, hypothesis testing, and more	Statistical program, such SPSS or SAS
Graphical	Assists decision makers in designing, developing, and using graphic displays of data and information	Graphics programs, such as PowerPoint
Project Management	Handles and coordinates large projects; also used to identify critical activities and tasks that could delay or jeopardize an entire project if they are not completed in a timely and cost-effective fashion	Project management software, such as Project

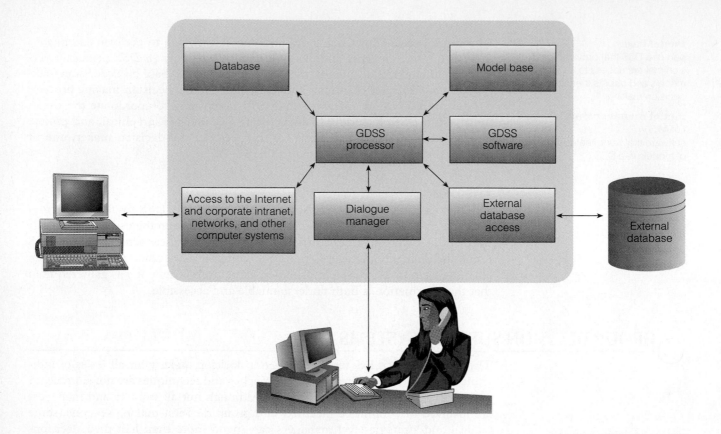

FIGURE 10 ● 17

Configuration of a GDSS

A GDSS contains most of the elements found in a DSS, plus software to facilitate group member communications.

actions triggered," says Bruce Anderson of IBM's industrial consulting sector, which developed the program for Caterpillar. Caterpillar's collaborative system links suppliers to its network to help reduce costs and improve the process of getting raw materials to its plants.

Pharmative Corporation, a producer of vitamin products, is using a new system called collaborative planning, forecasting, and replenishment to improve inventory planning and customer service.[42] Industry Directions, a marketing research firm, estimates that almost 70 percent of manufacturers, distributors, and retail firms surveyed planned to implement the system in the near future.

CHARACTERISTICS OF A GDSS THAT ENHANCE DECISION MAKING

It is often said that two heads are better than one. When it comes to decision making, a GDSS's unique characteristics have the potential to result in better decisions. Developers of these systems try to build on the advantages of individual support systems while realizing that new and additional approaches are needed in a group decision-making environment. For example, some GDSSs can allow the exchange of information and expertise among people without meetings or direct face-to-face interaction. Following are some characteristics that can improve and enhance decision making.

Special Design

The GDSS approach acknowledges that special procedures, devices, and approaches are needed in group decision-making settings. These procedures must foster creative thinking, effective communications, and good group decision-making techniques.

Ease of Use

Like an individual DSS, a GDSS must be easy to learn and use. Systems that are complex and hard to operate will seldom be used.[43] Many groups have less tolerance than do individual decision makers for poorly developed systems.

Flexibility

Two or more decision makers working on the same problem may have different decision-making styles and preferences. Each manager makes decisions in a unique way, in part because of different experiences and cognitive styles. An effective GDSS not only has to support the different approaches that managers use to make decisions but also must find a means to integrate their different perspectives into a common view of the task at hand.

Decision-Making Support

A GDSS can support different decision-making approaches, including the **delphi approach**, in which group decision makers are geographically dispersed throughout the country or the world. This approach encourages diversity among group members and fosters creativity and original thinking in decision making. Another approach, called **brainstorming**, which often consists of members offering ideas "off the top of their heads," fosters creativity and free thinking. The **group consensus approach** forces members in the group to reach a unanimous decision. With the **nominal group technique**, each decision maker can participate; this technique encourages feedback from individual group members, and the final decision is made by voting, similar to a system for electing public officials.

Anonymous Input

Many GDSSs allow anonymous input, where the person giving the input is not known to other group members. For example, some organizations use a GDSS to help rank the performance of managers. Anonymous input allows the group decision makers to concentrate on the merits of the input without considering who gave it. In other words, input given by a top-level manager is given the same consideration as input from employees or other members of the group. Some studies have shown that groups using anonymous input can make better decisions and have superior results compared with groups that do not use anonymous input. Anonymous input, however, can result in flaming, where an unknown team member posts insults or even obscenities on the GDSS system.

Reduction of Negative Group Behavior

One key characteristic of any GDSS is the ability to suppress or eliminate group behavior that is counterproductive or harmful to effective decision making. In some group settings, dominant individuals can take over the discussion, which can prevent other members of the group from presenting creative alternatives. In other cases, one or two group members can sidetrack or subvert the group into areas that are nonproductive and do not help solve the problem at hand. Other times, members of a group may assume they have made the right decision without examining alternatives—a phenomenon called *groupthink*. If group sessions are poorly planned and executed, the result can be a tremendous waste of time. Today, many GDSS designers are developing software and hardware systems to reduce these types of problems. Procedures for effectively planning and managing group meetings can be incorporated into the GDSS approach. A trained meeting facilitator is often employed to help lead the group decision-making process and to avoid groupthink.

Parallel Communication

With traditional group meetings, people must take turns addressing various issues. One person normally talks at a time. With a GDSS, it is possible for every group member to address issues or make comments at the same time by entering them into a PC or workstation. These comments and issues are displayed on every group member's PC or workstation immediately. Parallel communication can speed meeting times and result in better decisions.

Automated Record Keeping

Most GDSSs can keep detailed records of a meeting automatically. Each comment that is entered into a group member's PC or workstation can be anonymously

delphi approach
a decision-making approach in which group decision makers are geographically dispersed; this approach encourages diversity among group members and fosters creativity and original thinking in decision making

brainstorming
decision-making approach which often consists of members offering ideas "off the top of their heads"

group consensus approach
decision-making approach that forces members in the group to reach a unanimous decision

nominal group technique
decision-making approach that encourages feedback from individual group members, and the final decision is made by voting, similar to the way public officials are elected

GDSS software allows work teams to collaborate and reach better decisions—even if they work across town, in another region, or on the other side of the globe.

(Source: © Mark Richards/PhotoEdit)

recorded. In some cases, literally hundreds of comments can be stored for future review and analysis. In addition, most GDSS packages have automatic voting and ranking features. After group members vote, the GDSS records each vote and makes the appropriate rankings.

GDSS SOFTWARE

GDSS software, often called *groupware* or *workgroup software*, helps with joint work group scheduling, communication, and management. One popular package, Lotus Notes, can capture, store, manipulate, and distribute memos and communications that are developed during group projects. It can also incorporate knowledge management, discussed in Chapter 5, into the Lotus Notes Package. Some companies, like ExxonMobil Corporation, standardize on messaging and collaboration software, such as Lotus Notes.[44] Microsoft's NetMeeting product supports application sharing in multiparty calls. Exchange from Microsoft is another example of groupware. This software allows users to set up electronic bulletin boards, schedule group meetings, and use e-mail in a group setting. NetDocuments Enterprise was a *PC Magazine* Editor's Choice for providing Web collaboration.[45] The groupware is intended for legal, accounting, and real-estate businesses. A Breakout Session feature allows two people to take a copy of a document to a shared folder or director for joint revision and work. The software also permits digital signatures and the ability to download and work on shared documents on handheld computers. Other GDSS software packages include Collabra Share, OpenMind, and TeamWare. All of these tools can aid in group decision making.

In addition to stand-alone products, GDSS software is increasingly being incorporated into existing software packages. Today, some transaction processing and enterprise resource planning packages include collaboration software.[46] SAP, a popular ERP package discussed in Chapter 9, has developed mySAP Technology to facilitate collaboration and to allow SAP users to integrate applications from other vendors into the SAP system of programs.

GDSS ALTERNATIVES

Group decision support systems can take on a number of network configurations, depending on the needs of the group, the decision to be supported, and the geographic location of group members. The frequency of GDSS use and the location of the decision makers are two important factors (Figure 10.18).

FIGURE 10.18

GDSS Alternatives

The decision room may be the best alternative for group members who are located physically close together and who need to make infrequent decisions as a group. By the same token, group members who are situated at distant locations and who frequently make decisions together may require a wide area decision network to accomplish their goals.

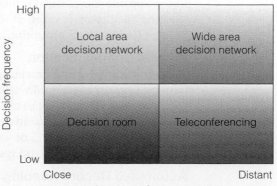

The Decision Room

decision room
a room that supports decision making, with the decision makers in the same building, combining face-to-face verbal interaction with technology to make the meeting more effective and efficient

The **decision room** is ideal for situations in which decision makers are located in the same building or geographic area and the decision makers are occasional users of the GDSS approach. In these cases, one or more decision rooms or facilities can be set up to accommodate the GDSS approach. Groups, such as marketing research teams, production management groups, financial control teams, or quality-control committees, can use the decision rooms when needed. The decision room alternative combines face-to-face verbal interaction with technology-aided formalization to make the meeting more effective and efficient.[47] A typical decision room is shown in Figure 10.19.

The Local Area Decision Network

The local area decision network can be used when group members are located in the same building or geographic area and under conditions in which group decision making is frequent. In these cases, the technology and equipment of the GDSS approach is placed directly into the offices of the group members. Usually this is accomplished via a local area network (LAN).

The Teleconferencing Alternative

Teleconferencing is used for situations in which the decision frequency is low and the location of group members is distant. These distant and occasional group meetings can tie together multiple GDSS decision-making rooms across the country or around the world. Using long-distance communications technology, these decision rooms are electronically connected in teleconferences and videoconferences. This alternative can offer a high degree of flexibility. The GDSS decision rooms can be used locally in a group setting or globally when decision makers are located throughout the world. GDSS decision rooms are often connected through the Internet.

The Wide Area Decision Network

The wide area decision network is used when the decision frequency is high and the location of group members is distant. In this case, the decision makers require frequent or constant use of the GDSS approach. Decision makers located throughout the country or the world must be linked electronically through a wide

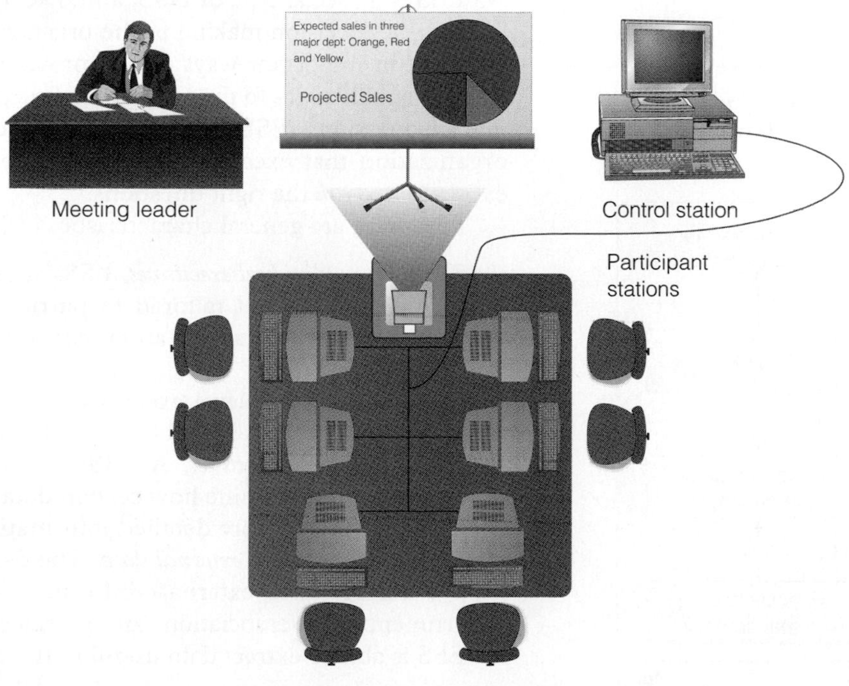
Expected sales in three major dept: Orange, Red and Yellow
Projected Sales
Meeting leader
Control station
Participant stations

FIGURE 10.19

The GDSS Decision Room

For group members who are in the same location, the decision room is an optimal GDSS alternative. This approach can use both face-to-face and computer-mediated communication. By using networked computers and computer devices, such as project screens and printers, the meeting leader can pose questions to the group, instantly collect their feedback, and, with the help of the governing software loaded on the control station, process this feedback into meaningful information to aid in the decision-making process.

virtual workgroups
teams of people located around the world working on common problems

area network (WAN). The group facilitator and all group members are geographically dispersed. In some cases, the model base and database are also geographically dispersed. This GDSS alternative allows people to work in **virtual workgroups**, where teams of people located around the world can work on common problems.

The Internet is increasingly being used to support wide area decision networks. As discussed in Chapters 7 and 8, a number of technologies, including video conferencing, instant messaging, chat rooms, and telecommuting, can be used to assist the GDSS process. In addition, many specialized wide area decision networks make use of the Internet for group decision making and problem solving.

EXECUTIVE SUPPORT SYSTEMS

executive support system (ESS), or executive information system (EIS)
specialized DSS that includes all hardware, software, data, procedures, and people used to assist senior-level executives within the organization

Because top-level executives often require specialized support when making strategic decisions, many companies have developed systems to assist executive decision making. This type of system, called an **executive support system (ESS)**, is a specialized DSS that includes all hardware, software, data, procedures, and people used to assist senior-level executives within the organization. In some cases, an ESS, also called an **executive information system (EIS)**, supports the actions of members of the board of directors, who are responsible to stockholders. These top-level decision-making strata are shown in Figure 10.20.

An ESS can also be used by individuals farther down in the organizational structure. Once targeted at the top-level executive decision makers, ESSs are now marketed to—and used by—employees at other levels in the organization. In the traditional view, ESSs give top executives a means of tracking critical success factors. Today, all levels of the organization share information from the same databases. However, for our discussion, we will assume ESSs remain in the upper management levels, where they indicate important corporate issues, indicate new directions the company may take, and help executives monitor the company's progress.

EXECUTIVE SUPPORT SYSTEMS IN PERSPECTIVE

An ESS is a special type of DSS, and, like a DSS, an ESS is designed to support higher-level decision making in the organization. The two systems are, however, different in important ways. DSSs provide a variety of modeling and analysis tools to enable users to thoroughly analyze problems—that is, they allow users to *answer* questions. ESSs present structured information about aspects of the organization that executives consider important—in other words, they allow executives to *ask* the right questions.

Following are general characteristics of ESSs:

- *Tailored to individual executives.* ESSs are typically tailored to individual executives; DSSs are not tailored to particular users. An ESS is an interactive, hands-on tool that allows an executive to focus, filter, and organize data and information.
- *Easy to use.* A top-level executive's most critical resource can be his or her time. Thus, an ESS must be easy to learn and use and not overly complex.
- *Have drill down abilities.* An ESS allows executives to drill down into the company to determine how certain data was produced. Drill down allows an executive to get more detailed information if needed.
- *Support the need for external data.* The data needed to make effective top-level decisions is often external—information from competitors, the federal government, trade associations and journals, consultants, and so on. An effective ESS is able to extract data useful to the decision maker from a wide variety of

FIGURE 10.20

The Layers of Executive Decision Making

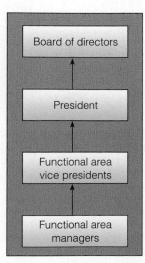

sources including the Internet and other electronic publishing sources such as LexisNexis.

- *Can help with situations that have a high degree of uncertainty.* There is a high degree of uncertainty with most executive decisions. Handling these unknown situations using modeling and other ESS procedures helps top-level managers measure the amount of risk in a decision.
- *Have a future orientation.* Executive decisions are future oriented, meaning that decisions will have a broad impact for years or decades. The information sources to support future-oriented decision making are usually informal—from golf partners to members of social clubs or civic organizations.
- *Are linked with value-added business processes.* Like other information systems, executive support systems are linked with executive decision making about value-added business processes. For instance, executive support systems can be used by car-rental companies to analyze trends.

CAPABILITIES OF EXECUTIVE SUPPORT SYSTEMS

The responsibility given to top-level executives and decision makers brings unique problems and pressures to their jobs. The following is a discussion of some of the characteristics of executive decision making that are supported through the ESS approach. As you will note, most of these are related to an organization's overall profitability and direction. An effective ESS should have the capability to support executive decisions with many of these capabilities, such as strategic planning and organizing, crisis management, and more.

Support for Defining an Overall Vision

One of the key roles of senior executives is to provide a broad vision for the entire organization. This vision includes the organization's major product lines and services, the types of businesses it supports today and in the future, and its overriding goals.

Support for Strategic Planning

ESSs also support strategic planning. **Strategic planning** involves determining long-term objectives by analyzing the strengths and weaknesses of the organization, predicting future trends, and projecting the development of new product lines. It also involves planning the acquisition of new equipment, analyzing merger possibilities, and making difficult decisions concerning downsizing and the sale of assets if required by unfavorable economic conditions.

strategic planning
determining long-term objectives by analyzing the strengths and weaknesses of the organization, predicting future trends, and projecting the development of new product lines

Support for Strategic Organizing and Staffing

Top-level executives are concerned with organization structure. For example, decisions concerning the creation of new departments or downsizing the labor force are made by top-level managers. Overall direction for staffing decisions and effective communication with labor unions are also major decision areas for top-level executives. ESSs can be employed to help analyze the impact of staffing decisions, potential pay raises, changes in employee benefits, and new work rules.

Support for Strategic Control

Another type of executive decision relates to strategic control, which involves monitoring and managing the overall operation of the organization. Goal seeking can be done for each major area to determine what performance these areas need to achieve to reach corporate expectations. Effective ESS approaches can help top-level managers make the most of their existing resources and control all aspects of the organization.

Support for Crisis Management

Even with careful strategic planning, a crisis can occur. Major disasters, including hurricanes, tornadoes, floods, earthquakes, fires, and terrorist activities such as

9/11, can totally shut down major parts of the organization. Handling these emergencies is another responsibility for top-level executives. In many cases, strategic emergency plans can be put into place with the help of an ESS. These contingency plans help organizations recover quickly if an emergency or crisis occurs.

Decision making is a vital part of managing businesses strategically. IS systems such as information and decision support, group decision support, and executive support systems help managers and employees by tapping existing databases and providing them with current, accurate information. The increasing integration of all business information systems—from TPSs to MISs to DSSs—can help organizations monitor their competitive environment and make better-informed decisions. Organizations can also use specialized business information systems, discussed in the next chapter, to achieve their goals.

SUMMARY

PRINCIPLE *Good decision-making and problem-solving skills are the key to developing effective information and decision support systems.*

Every organization needs effective decision making and problem solving to reach its objectives and goals. Problem solving begins with decision making. A well-known model developed by Herbert Simon divides the decision-making phase of the problem-solving process into three stages: intelligence, design, and choice. During the intelligence stage, potential problems or opportunities are identified and defined. Information is gathered that relates to the cause and scope of the problem. Constraints on the possible solution and the problem environment are investigated. In the design stage, alternative solutions to the problem are developed and explored. In addition, the feasibility and implications of these alternatives are evaluated. Finally, the choice stage involves selecting the best course of action. In this stage, the decision makers evaluate the implementation of the solution to determine whether the anticipated results were achieved and to modify the process in light of new information learned during the implementation stage.

Decision making is a component of problem solving. In addition to the intelligence, design, and choice steps of decision making, problem solving also includes implementation and monitoring. Implementation places the solution into effect. After a decision has been implemented, it is monitored and modified if needed.

Decisions can be programmed or nonprogrammed. Programmed decisions are made using a rule, procedure, or a quantitative method. Ordering more inventory when the level drops to 100 units or less is an example of a programmed decision. A nonprogrammed decision deals with unusual or exceptional situations. Determining the best training program for a new employee is an example of a nonprogrammed decision.

Decisions can use optimization, satisficing, or heuristic approaches. Optimization finds the best solution. Optimization problems often have an objective like maximizing profits given production and material constraints. When a problem is too complex for optimization, satisficing is often used. Satisficing finds a good, but not necessarily the best, decision. Finally, a heuristic is a "rule of thumb" or commonly used guideline or procedure used to find a good decision.

PRINCIPLE *The management information system (MIS) must provide the right information to the right person in the right fashion at the right time.*

A management information system is an integrated collection of people, procedures, databases, and devices that provide managers and decision makers with information to help achieve organizational goals. An MIS can help an organization achieve its goals by providing managers with insight into the regular operations of the organization so that they can control, organize, and plan more effectively and efficiently. The primary difference between the reports generated by the TPS and those generated by the MIS is that MIS reports support managerial decision making at the higher levels of management.

Data that enters the MIS originates from both internal and external sources. The most significant internal source of data for the MIS is the organization's various TPSs and ERP systems. Data warehouses and data marts also provide important input data for the MIS. External sources of data for the MIS include extranets, customers, suppliers, competitors, and stockholders.

The output of most management information systems is a collection of reports that are distributed to managers. These reports include scheduled reports, key-indicator reports, demand reports, exception reports, and drill down reports. Scheduled reports are produced periodically, or on a schedule, such as daily, weekly, or monthly. A key-indicator report is a special type of scheduled report. Demand reports are developed to give certain information at a manager's request. Exception reports are automatically produced when a situation is unusual or requires management action. Drill down reports provide increasingly detailed data about situations.

Management information systems have a number of common characteristics, including producing scheduled, demand, exception, and drill down reports; producing reports with fixed and standard formats; producing hard-copy and soft-copy reports; using internal data stored in organizational computerized databases; and having reports developed and implemented by IS personnel or end users.

Most MISs are organized along the functional lines of an organization. Typical functional management information systems include accounting, manufacturing, marketing, and human resources. Each system is composed of inputs, processing subsystems, and outputs. The primary sources of input to functional MISs include the corporate strategic plan, data from the TPS, information from other functional areas, and external sources including the Internet. The primary output of these functional MISs are summary reports that assist in managerial decision making.

A financial management information system provides financial information to all financial managers within an organization, including the chief financial officer (CFO). Subsystems are financial forecasting, profit/loss and cost systems, use and management of funds, and auditing.

A manufacturing MIS accepts inputs from the strategic plan, the TPS, and external sources. The TPSs involved support the business processes associated with the receiving and inspecting of raw materials and supplies; inventory tracking of raw materials, work in process, and finished goods; labor and personnel management; management of assembly lines, equipment and machinery, inspection, and maintenance; and order processing. The subsystems involved are design and engineering, master production scheduling, inventory control, process control, and quality control and testing.

A marketing MIS supports managerial activities in the areas of product development, distribution, pricing decisions, promotional effectiveness, and sales forecasting. Subsystems include marketing research, product development and reporting, promotion and advertising, and product pricing.

A human resource MIS is concerned with activities related to employees of the organization. Subsystems include human resource planning, personnel selection and recruiting, training and skills inventories, scheduling and job placement, and wage and salary administration.

An accounting MIS performs a number of important activities, providing aggregate information on accounts payable, accounts receivable, payroll, and many other applications. The organization's TPS captures accounting data, which is also used by most other functional information systems. Geographic information systems provide regional data in graphical form.

PRINCIPLE *Decision support systems (DSSs) are used when the problems are more unstructured.*

A decision support system (DSS) is an organized collection of people, procedures, software, databases, and devices working to support managerial decision making. DSS characteristics include the ability to handle large amounts of data; obtain and process data from different sources; provide report and presentation flexibility; perform complex statistical analysis; offer textual and graphical orientations; support optimization, satisficing, and heuristic approaches; and perform "what-if" simulation and goal-seeking analysis.

DSSs provide support assistance through all phases of the problem-solving process. Different decision frequencies also require DSS support. An ad hoc DSS addresses unique, infrequent decision situations; an institutional DSS handles routine decisions. Highly structured problems, semistructured problems, and unstructured problems can be supported with a DSS. A DSS can also support different managerial levels, including strategic, tactical, and operational-level managers. A common database is often the link that ties together a company's TPS, MIS, and DSS.

The components of a DSS are the database, model base, dialogue manager, and a link to external databases, the Internet, the corporate intranet, extranets, networks, and other systems. The database can use data warehouses and data marts. A data-driven DSS primarily performs qualitative analysis based on the company's databases. Data-driven DSSs tap into vast stores of information contained in the corporate database, retrieving information on inventory, sales, personnel, production, finance, accounting, and other areas. Data mining is often used in a data-driven DSS. The model base contains the models used by the decision maker, such as financial, statistical, graphical, and project management models. A model-driven DSS primarily performs mathematical or quantitative analysis. Model management software (MMS) is often used to coordinate the use of models in a DSS. The dialogue manager provides a dialogue management facility to assist in communications between the system and the user. Access to other computer-based systems permits the DSS to tie into other powerful systems, including the TPS or function-specific subsystems.

PRINCIPLE *Specialized support systems, such as group decision support systems (GDSSs) and executive support systems (ESSs), use the overall approach of a DSS in situations such as group and executive decision making.*

A group decision support system (GDSS), also called a *computerized collaborative work system*, consists of most of the elements in a DSS, plus software needed to provide effective support in group decision-making settings. GDSSs are typically easy to learn and use and can offer specific or general decision-making support. GDSS software, also called *groupware*, is specially designed to help generate lists of decision alternatives and perform data analysis. These packages let people work on joint documents and files over a network.

The frequency of GDSS use and the location of the decision makers will influence the GDSS alternative chosen. The decision room alternative supports users in a single location that meet infrequently. Local area networks can be used when group members are located in the same geographic area and users meet regularly.

Teleconferencing is used when decision frequency is low and the location of group members is distant. A wide area network is used when the decision frequency is high and the location of group members is distant.

Executive support systems (ESSs) are specialized decision support systems designed to meet the needs of senior management. They serve to indicate issues of importance to the organization, indicate new directions the company may take, and help executives monitor the company's progress. ESSs are typically easy to use, offer a wide range of computer resources, and handle a variety of internal and external data. In addition, the ESS performs sophisticated data analysis, offers a high degree of specialization, and provides flexibility and comprehensive communications abilities. An ESS also supports individual decision-making styles. Some of the major decision-making areas that can be supported through an ESS are providing an overall vision, strategic planning and organizing, staffing and labor relations, crisis management, and strategic control.

CHAPTER 10 SELF-ASSESSMENT TEST

Good decision-making and problem-solving skills are the key to developing effective information and decision support systems.

1. The first stage of the decision-making process is:
   A. Initiation phase
   B. Intelligence phase
   C. Design phase
   D. Choice phase

2. Problem solving is one of the phases of decision making. True False

3. _____ is the final stage of problem solving.

4. A decision that inventory should be ordered when inventory levels drop to 500 units is an example of
   A. Synchronous decision
   B. Asynchronous decision
   C. Nonprogrammed decision
   D. Programmed decision

5. A(n) _____ model will find the best solution, usually the one that will best help the organization meet its goals.

6. A heuristic is often referred to as a "rule of thumb." It is a commonly accepted guideline or procedure that usually find a good solution. True False

The management information system (MIS) must provide the right information to the right person in the right fashion at the right time.

7. What summarizes the previous day's critical activities and is typically available at the beginning of each workday?
   A. Key-indicator report
   B. Demand report
   C. Exception report
   D. Database report

8. Inventory control is a subsystem of
   A. The marketing MIS
   B. The financial MIS
   C. The manufacturing MIS
   D. The auditing MIS

9. Another name for the _____ MIS is the personnel MIS because it is concerned with activities related to employees and potential employees of the organization.

Decision support systems (DSSs) are used when the problems are more unstructured.

10. The focus of a decision support system is on decision-making effectiveness when faced with unstructured or semistructured business problems. True False

11. _____ analysis is the process of determining the problem data required for a given result.

12. What component of a decision support system allows decision makers to easily access and manipulate the DSS and to use common business terms and phrases?
    A. The knowledge base
    B. The model base
    C. The dialogue manager
    D. The expert system

Specialized support systems, such as group decision support systems (GDSSs) and executive support systems (ESSs), use the overall approach of a DSS in situations such as group and executive decision making.

13. In a GDSS, what approach or technique uses voting to make the final decision in a group setting?
    A. Group consensus
    B. Group think
    C. Nominal group technique
    D. Delphi

14. _____ helps with joint work group scheduling, communication, and management.

15. The local area decision network is the ideal GDSS alternative for situations in which decision makers are located in the same building or geographic area and the decision makers are occasional users of the GDSS approach. True False

16. A(n) _____ supports the actions of members of the board of directors, who are responsible to stockholders.

Chapter 10 Self-Assessment Test Answers

1. B, 2. False, 3. Monitoring, 4. D, 5. optimization, 6. True, 7. A, 8. C, 9. human resource, 10. True, 11. Goal seeking, 12. C, 13. C, 14. Groupware or workgroup software, 15. False, 16. Executive support system (ESS).

KEY TERMS

REVIEW QUESTIONS

1. Define the term *management information system (MIS)*.
2. What are the basic kinds of reports produced by an MIS?
3. What guidelines should be followed in developing reports for management information systems?
4. Identify the functions performed by all MISs.
5. What are the functions performed by a financial MIS?
6. Describe the functions of a manufacturing MIS.
7. What is a human resource MIS? What are its outputs?
8. List and describe some other types of MISs.
9. What are the stages of problem solving?
10. What is the difference between decision making and problem solving?
11. What is a geographic information system?
12. Describe the difference between a structured and an unstructured problem and give an example of each.

13. Define *decision support system.* What are its characteristics?
14. Describe the difference between a data-driven and a model-driven DSS.
15. What is the difference between "what-if" analysis and goal-seeking analysis?
16. What are the components of a decision support system?
17. Describe four models used in decision support systems.
18. State the objective of a group decision support system (GDSS) and identify three characteristics that distinguish it from a DSS.
19. Identify three group decision-making approaches often supported by a GDSS.
20. What is an executive support system? Identify three fundamental uses for such a system.

DISCUSSION QUESTIONS

1. What is the relationship between an organization's transaction processing systems and its management information systems? What is the primary role of management information systems?
2. How can data warehouses and data marts be used in a company's MIS?
3. How can management information systems be used to support the objectives of the business organization?
4. Describe a financial MIS for a *Fortune* 1000 manufacturer of food products. What are the primary inputs and outputs? What are the subsystems?
5. How can a strong financial MIS provide strategic benefits to a firm?
6. Why is auditing so important in a financial MIS? Give an example of an audit that failed to disclose the true nature of the financial position of a firm. What was the result?
7. What is the difference in roles played by an internal auditing group versus an external auditing group?
8. You have been hired to develop a management information system and a decision support system for a manufacturing company. Describe what information you would include in printed reports and what information you would provide using a screen-based decision support system.

9. You have been hired to develop a DSS for a car company like Ford or GM. Describe how you would use both data-driven and model-driven DSSs.
10. Imagine that you are the CFO for a services organization. You are concerned with the integrity of the firm's financial data. What steps might you take to ascertain the extent of problems?
11. What functions do decision support systems support in business organizations? How does a DSS differ from a TPS and a MIS?
12. How is decision making in a group environment different from individual decision making, and why are information systems that assist in the group environment different? What are the advantages and disadvantages of making decisions as a group?
13. You have been hired to develop group support software. Describe the features you would include in your new GDSS software.
14. The use of ESSs should not be limited to the executives of the company. Do you agree or disagree? Why?
15. Imagine that you are the vice president of manufacturing for a *Fortune* 1000 manufacturing company. Describe the features and capabilities of your ideal ESS.

PROBLEM-SOLVING EXERCISES

 1. You have been asked to select GDSS software to help your company make better decisions to market a new product. Using the Internet, find and investigate three types of software that your company could use for collaborative decision making. Use your word processor to describe what you found and the advantages and disadvantages of each GDSS software package.

2. Review the summarized consolidated statement of income for the manufacturing company shown here.

Operating Results (in millions)

Operating Revenues	$2,924,177
Operating Expenses (including taxes)	2,483,687
Operating Income	440,490
Other Income and Expenses	13,497
Income before Interest and Other Charges	453,987
Interest and Other Charges	262,845
Net Income	191,142
Average Common Shares Outstanding	147,426
Earnings per Share	$1.30

Use graphics software to prepare a set of bar charts that shows data for this year compared with the data for last year.

A. Operating revenues increase by 3.5% while operating expenses increase 2.5%.

B. Other income and expenses decrease to $13,000.

C. Interest and other charges increase to $265,000.

If you were a financial analyst tracking this company, what detailed data might you need to perform a more complete analysis? Write a brief memo summarizing your data needs.

3. As the head buyer for a major supermarket chain, John is constantly being asked by manufacturers and distributors to stock their new products. Over 50 new items are introduced each week. Many times these products are launched with national advertising campaigns and special promotional allowances to both retailers, such as John's firm, and consumers. The store has only a limited amount of shelf and floor space to stock items. Thus, to add new products, the amount of shelf space allocated to existing products must be reduced, or items must be eliminated altogether. Develop a spreadsheet that John can use to estimate the change in profits from adding or deleting an item from inventory. The spreadsheet will include input such as estimated weekly sales in units, shelf space allocated to stock an item (measured in units), total cost per unit, sales price per unit, and promotional allowance earned per unit. The spreadsheet should calculate total profit by item and then sort the rows in descending order based on total profit. Because of the limited amount of shelf space, the spreadsheet should also calculate the accumulated shelf space based on the items stocked.

TEAM ACTIVITIES

1. Divide the class into teams of three or four classmates and use the following role-playing scenario: Your consulting team is designing a management information system for a small manufacturing organization that produces ten different models of high-performance bicycles. These cycles are sold to distributors and bicycle shops throughout North America. One of the major problems this company faces is poor inventory control. It always seems that too many bikes are in stock and yet the "hottest selling" model is out of stock. The owner has suggested

that your team look at the feasibility of implementing a manufacturing MIS to help deal with this problem. Prepare a brief memo that describes the subsystems and outputs associated with the typical manufacturing MIS. Draw a systems level flowchart and outline how these subsystems need to be integrated and discuss where the inventory control system fits. What are some of the issues that will need to be addressed to develop a single integrated MIS to meet the needs of this organization? Are there some prerequisites that must be completed before the inventory control system can be built? If so, what are they? What additional benefits, besides better inventory control, can be expected?

2. Have your team work together in making a group decision, such as to develop a list of the 10 best companies to start a career in marketing. Appoint one or two members of the team to disrupt the meeting with negative group behavior. After the meeting, have your team describe how to prevent this negative group behavior. What GDSS software

features would you suggest to prevent the negative group behavior your team observed?

3. Imagine that you and your team have decided to develop an ESS software product to support senior executives in the music recording industry. What are some of the key decisions these executives must make? Make a list of the capabilities that such a system must provide to be useful. Identify at least six sources of external information that will be useful to its users.

WEB EXERCISES

1. Most companies typically have a number of functional MISs, such as finance. Find the sites of two finance companies, such as a bank or a brokerage company. Compare these sites. Which one do you prefer? How could these sites be improved? (Hint: If you are having trouble, try Yahoo. It should have a listing for "Business and Economy" on its home page. From there you can go to "companies" and then "finance." There will be several menu choices from there.) You may be asked to develop a report or send an e-mail message to your instructor about what you found.

2. This chapter mentioned a number of companies that use a DSS. Using the Internet, investigate

one company to determine what products they produce. Write a report describing additional DSSs that the company could develop to help them increase profits or reduce costs.

3. Software, such as the Excel spreadsheet, is often used to find an optimal solution to maximize profits or minimize costs. Search the Internet using Yahoo, Google, or another search engine to find other software packages that offer optimization features. Write a report describing one or two of the optimization software packages. What are some of the features of the package?

CASES

CASE 1

Marketing MIS Gives Entertainment UK the Upper Hand

When residents of the United Kingdom purchase music and video CDs and DVDs, there is a one in four chance that the product they purchased was supplied by Entertainment UK. Entertainment UK links the recording industry and the public through their two divisions:

1. *Entertainment UK*—Wholesale supplier to retailers like Tesco, Woolworth's, and Safeway

2. *Entertainment UK Direct*—Retail supplier to the public through the Internet, kiosks, catalogues, digital TV, or special in-store orders
   Entertainment UK invests substantially in technology, reflecting a firm commitment to growth through quality and efficiency. Since the entertainment business is governed by frequently changing trends and overnight shifts in popular opinion, companies that are most in tune with popular opinion will earn a significant market share of the industry. Entertainment UK has recognized this fact and employed Microstrategy business

intelligence software tools to help keep its finger on the pulse of the market.

Entertainment UK uses MicroStrategy software to enhance several areas of its business, including:

- Managing each category—CD, DVD, VHS
- Understanding the business of each retailer
- Offering the most suitable products and promotions
- Monitoring the effects of marketing and promotional campaigns on the performance of each product and individual retail outlet

Entertainment UK collects transaction data from all of its customers and retailers and combines it into one powerful information system. The Microstrategy software allows Entertainment UK marketing specialists to query the transaction data using Web-based tools, to see which items are selling, which promotions are working, and what price is optimal for each product. It provides a library of over 150 analytical functions that provide insight into problem areas and opportunities. Marketing specialists can use the software to create visually pleasing presentations and reports to assist their retail suppliers in choosing the products best suited for their client base.

Entertainment UK combines its sophisticated marketing MIS with a supply chain management system, sophisticated call-center systems, and an automated distribution system to provide its customers with every benefit technology has to offer.

Ian McKee, information systems manager at Entertainment UK, states, "Entertainment UK provides just in time stock control on a sale or return basis, so it is critical that we have accurate market information to make good decisions quickly. Our role as a distributor is to be the hub of the supply chain, sharing this market knowledge with our suppliers and customers to maximize sales by ensuring the right products are on the right shelves at the right time."

Discussion Questions

1. How can a marketing MIS be used to support the four bulleted services listed in the article? Specifically, what data can be queried to yield helpful statistics?
2. A just-in-time inventory approach is difficult to manage for items whose consumption rate is not easily predictable. In the entertainment industry where the market fluctuates daily, it becomes nearly impossible to maintain. How can suppliers like Entertainment UK predict the market to know what to deliver just in time?

Critical Thinking Questions

3. Acting as middleman between the record/video producers and the public places Entertainment UK in a position to offer both sides special services. We've heard about the services Entertainment UK provides for its clients, the retailers and consumers. What services can Entertainment UK provide for the record/video producers? Is Entertainment UK a valuable middleman, or would consumers and the music/video producers be better off doing away with the middleman and dealing directly with each other? Why?
4. Sometimes it's difficult to tell whether the consumer is leading the market or market promotions are leading the consumer. This is particularly true in fad industries such as clothing and entertainment. How will a marketing MIS differ for these fad industries, from a traditional marketing MIS for long-standing successful products?

Sources: "Entertainment UK Charts Performance with MicroStrategy Software," from Microstrategy's Success Stories Web site, http://www.microstrategy.com, accessed May 4, 2002; Entertainment UK Web site, http://www.entuk.com/index2.html, accessed May 4, 2002; "Even Small Call Centers Can Benefit from Specialist Products," *Business Info Magazine*, February 2, 2002, http://www.entuk.com/coverage.php?coverageID=10

CASE 2

CampusShip Provides a Centralized Shipping System for Distributed Organizations

Businesses and organizations that are spread over large geographic areas such as various locations in a large building, multiple buildings around a large campus, or multiple locations around a city or state face unique information management challenges. One such challenge is to provide the semi-isolated office locations with the software tools they need to excel at their jobs, while still allowing central decision makers control and visibility of the information flow. When it comes to corporate shipping systems, UPS has provided a solution for customers who face such challenges: CampusShip.

UPS has been successful in the very competitive shipping market by being early adopters of information system technology and using information systems as an integral part of their value-added process. Superior problem solving and decision making have allowed UPS to bring products and services to their customers ahead of the competition. The development of its most recent product offers an ideal example of the problem-solving process at its best.

Through ongoing intelligence gathering, UPS discovered that some of its customers, those with widely dispersed facilities, were having a difficult time managing their shipping information. They needed a way to manage their shipping from a central location. Moving from the intelligence stage to the design stage of the problem-solving process, UPS needed to determine what, if anything, could be done for its customers.

In this case, identifying the problem was key to UPS's good fortune. Once identified, it was clear that if UPS could deliver a solution, it would earn the appreciation of its customers—along with a competitive advantage. After examining several proposals from its systems analysts, UPS executives chose and implemented the system that was to be named CampusShip.

CampusShip is free to UPS customers and assists mailroom administrators in managing the shipping transactions of businesses with widely scattered facilities. Tim Geiken, vice president of e-commerce marketing at UPS, said the company is targeting customers in the financial, insurance, real estate, legal, automotive, and retail industries, as well as in higher education and government, where employees are located on corporate campuses. Companies using CampusShip include The Home Depot, DaimlerChrysler, and Best Buy.

CampusShip uses UPS's OnLine Tools, which allow shippers to track packages, print labels, manage an address book, and handle e-mail shipping notification to a recipient right from their desktops or laptops. Users need only Internet access and their companies' permission to use the new tools.

At the same time, the system offers transportation and mailroom decision makers centralized control over shipping procedures and costs. It's highly customizable, letting those decision makers specify reference fields that control shipping activity. And it helps simplify billing by generating the detailed shipping histories that professional service firms require for client invoicing.

The complete list of CampusShip features include:
- Web-based shipping accessible from any computer connected to the Internet
- Access to global and personal address books
- Customization features for a business's needs
- Ability to import address books, employee data, shipper account numbers, and reference codes
- E-mail notification to package recipients
- Shipping reports to help businesses efficiently allocate costs to either internal departments or specific clients

Jim Hay, general transportation manager of small package solutions at Eden Prairie, Minnesota–based Best Buy, claims that CampusShip "saves 15 minutes per user per package. The biggest area of savings is in time and [in fewer] lost and misrouted packages due to poor handwriting or bad addresses," he said.

Ramala Ravi, an analyst at IDC in Framingham, Massachusetts, said CampusShip is a smart way for UPS to reach out to its existing customers. "This is a great way to generate more business from its installed customer base," she said.

UPS continues to monitor the implementation of this and its other information systems and services, refining them as needed and watching for other opportunities as they present themselves.

Discussion Questions

1. What advantages does CampusShip offer over an in-house developed shipping information system? What are the disadvantages, dangers, and limitations of allowing UPS to store your shipping data? Do the advantages outweigh the disadvantages?

2. If you were a player in the shipping industry, what methods would you employ to continuously improve service to your clients (high-tech and low-tech)?

Critical Thinking Questions

3. Is UPS in the shipping business, the information systems development business, or both? Why do you think UPS has been successful in seeing a return on its sizable technology investment, while others have taken a more cautious approach to technology investment?

4. When it comes to shipping, how valuable is service and convenience? Do you think Jim Hay of Best Buy (quoted in the case) would be willing to pay more to UPS with CampusShip than to a competitor who doesn't supply a CampusShip-like service? Is CampusShip more valuable to UPS as a free service? Why?

Sources: Linda Rosencrance, "UPS Launches CampusShip Shipping," *Computerworld*, March 7, 2002 , http://www.computerworld.com; "New Solution Tailors UPS Shipping Tools for Multi-Location and Corporate Campus Customers," *Business Wire*, March 5, 2002; UPS Website, http://www.ups.com/bussol/solutions/campusship.html, accessed May 5, 2002.

CASE 3

Storebrand Uses WebService to Automate Data Entry

Storebrand ASA is Norway's largest provider of pension plans, insurance, and other financial services. Storebrand supplies 390,000 employees of 6,500 companies with pension plans. Records for all these Storebrand members are kept in an IBM DB2 Universal Database.

Until recently, updating member pension plan data was an onerous task. Whenever a member's salary or personal information changed, the employer would need to revise its records and send the updates to Storebrand through FTP, mail, or fax. Storebrand customer service reps would then enter the changes into the DB2 database.

The whole update process was time-intensive and error-prone. It dawned on Karsten Heslien, Storebrand's chief technology officer, that the updated information the company needed already existed in the employers' human resource MIS. Human resource employees were duplicating their efforts by entering updated information in their own database and then sending the same information to Storebrand. If it were possible to connect the two information systems, Storebrand's and the client's, then the update to Storebrand's database could take place automatically without human intervention, and the possibility of human error.

The development of Web Services technology arrived just in time to solve Storebrand's problem. Web Services are applications designed to communicate with each other over the Internet using accepted Web standards. Working with IBM, Storebrand was able to develop Web Services applications to run on its client's human resource MISs to provide input to the Storebrand MIS. Now, when clients run their payroll applications, they can choose to add members to—or delete members from—their Storebrand pension plan, update members' records, or reconcile their local pension plan records with Storebrand records.

Since the implementation of the new system, Storebrand has enjoyed significant savings in customer support costs, along with improved customer satisfaction. Also, Storebrand

plans to leverage its investment in this project to deliver additional Web Services to its partners. Storebrand can publish its Web Services on the Internet for independent insurance distributors to access and incorporate into their applications. Storebrand can also extend the architecture it created for the pension plan update application to serve its other lines of business.

Storebrand's solution to this problem demonstrates a trend in information system development accelerated by the birth of Web services. Through tightly controlled interfaces, the output of one information system can be used as the input for another. As Web Services become increasingly popular, the boundaries between information systems become less defined. The information system itself takes on more responsibility for acquiring data, while the human users benefit from more reliable and accurate information, provided the moment the data comes into existence.

Discussion Questions

1. What complications do you think Storebrand encountered when designing Web service applications to run on its clients' payroll applications? How will Web service technology affect the relationships between software companies? Would circumstances be better if all involved used software created by the same manufacturer?

2. Consider some of the tedious data entry processes that you have seen on more than one occasion: registration forms, applications, multiple-choice exams, tax forms. Which of these tasks might be automated, and which will necessarily always need to be handled manually?

Critical Thinking Questions

3. The automation of tedious activities such as data entry and transfer are designed to reduce labor but, in doing so, will most likely leave some employees without work. Rather than laying off employees, shrinking the workforce, and making the existing employees nervous about job security, what can employers do to build a healthy and loyal workforce while elevating the general feeling of job fulfillment?

4. Storebrand found a way to do away with redundant data entry. Consider the number of times you have needed to

supply companies with your own personal information: name, phone, address, social security number, etc. Will there come a day when we each manage our own personal information systems? What types of transactions might we be able to simplify using our own IS and Web services? Do such services exist today?

Sources: Success Stories on the IBM Web site, http://www4.ibm.com, accessed May 5, 2002; Storebrand Web site, http://www.storebrand.com, accessed May 5, 2002.

NOTES

Sources for the opening vignette on p. 411: "Shearman & Sterling Shares Knowledge to Think Big and Think Fast," IBM Software Success Stories Web site, http://www3.ibm.com; accessed May 1, 2002; Jennifer Disabatino, "Lotus Ties Notes, Domino to Java for Web Services," *Computerworld*, May 1, 2002, http://www.computerworld.com; Shearman & Sterling Web site, http://www.shearmanandsterling.com, accessed May 1, 2002.

1. Clemen, Robert, and Kwit, Robert, "The Value of Decision Analysis at Eastman Kodak Company, 1990–1999," *Interfaces*, September–October 2001, p. 74.

2. Gazmuri, Pedro, et al., "Developing and Implementing a Production Planning DSS for CTI Using Structured Modeling," *Interfaces*, July–August 2001, p. 22.

3. Shyur, Ching-Chir et al., "SDHTOOL: Planning Survivable and Cost-Effectiveness SDH Networks at Chunghwa," *Interfaces*, July–August 2001, p. 87.

4. Solomon, Melissa, "Bean Counting," *Computerworld*, November 5, 2001, p. 46.

5. Rosencrance, Linda, "Data Warehouse Gives Trimac Information for the Long Haul," *Computerworld*, July 2, 2001, p. 47.

6. Brewin, Bob, and Rosencrance, Linda, "Follow That Package," *Computerworld*, March 19, 2001, p. 58.

7. Brewin, Bob, "Penske Outfits Fleet with Wireless Terminals," *Computerworld*, June 11, 2001, p. 10.

8. Pacchiano, Ronald, "Centralizing Reporting," *PC Magazine*, September 4, 2001, p. 40.

9. Thibodeau, P. and Mearian, L. "Terrorism Taxes IT Planning," *Computerworld*, November 5, 2001, p. 1.

10. Weiss, Todd, "Linux Supercomputer to Be Used for Drug Research," *Computerworld*, August 27, 2001, p. 14.

11. London, Simon, "Accounting: New U.S. Rules," *Financial Times*, March 5, 2002, pg. 14.

12. Mearian, Lucas, "Bank's Project to Speed Payment Process," *Computerworld*, January 7, 2002, p. 6.

13. Dolan, K., and Meredith, R., "Ghost Cars, Ghost Brands," *Forbes*, April 30, 2001.

14. Crissey, Mike, "Miniature 'Smart Labels' to Track Goods from the Plant to the Pantry," *The Rocky Mountain News*, February 25, 2002, p. 6B.

15. Gladwin, Lee, "Ford Opens IT Hub in India to Save Millions," *Computerworld*, March 19, 2001, p. 7.

16. Brunt, Charles, "N.M.-Based Firm Gets License to Market Metal Parts Making Process," *Albuquerque Journal*, April 17, 2001.

17. Verton, Dan, "Fighter Jet Contract a Proving Ground for IT," *Computerworld*, November 12, 2001, p. 10.

18. Songini, Marc, "Quick Response," *Computerworld*, February 19, 2001, p. 47.

19. Nash, Kim, "Ford's Vehicle Delivery Project Ahead of Plan," *Computerworld*, October 1, 2001, p. 30.

20. Raghunathan, S. and Yeh, A., "Continuous Replenishment Program (CRP) Between Manufacturer and Its Retailer," *Information Systems Research*, December 2001, p. 406.

21. Claus, David, "QAD Helps Krone Communications, India," *Business Wire*, October 1, 2001.

22. Konicki, Steve, "Ford Starts Stockpiling," *InformationWeek Online*, September 9, 2001.

23. Pereira, et al., "Chrysler Averts a Parts Crisis," *The Wall Street Journal*, September 24, 2001, p. B1.

24. "Segment-Busting 2004 Chrysler Pacifica to Be Built at Windsor Assembly Line," *PR Newswire*, February 13, 2002.

25. Songini, Marc, "Airlines Spurred to Focus on CRM Software after Attacks," *Computerworld*, January 28, 2002, p. 14.

26. Disabatino, Jennifer, "Delta Aims CRM Tool at Holiday Travelers," *Computerworld*, November 26, 2001, p. 14.

27. McDonnell, Sharon, "Putting CRM to Work," *Comptuerworld*, March 12, 2001, p. 48.

28. Songini, Marc, "Customizing a CRM Application Is Risky," *Computerworld*, November 5, 2001, p. 48.

29. Schoenberger, Chana, "The Weakest Link," *Forbes*, October 1, 2001, p. 114.

30. Songini, Marc, "As Economy Slows, Companies Turn to Supply Forecasting," *Computerworld*, May 21, 2001, p. 8.

31. Keenan, Faith, "Opening the Spigot," *BusinessWeek*, June 4, 2001, p. EB 17.

32. Rosen, Cheryl, "Dole Uses Electronic Currency in Online Promotion," *InformationWeek Online*, May 25, 2001.

33. Merrick, Amy, "Retailers Try to Get Leg Up on Markdowns with New Software," *The Wall Street Journal*, August 7, 2001, p. A1.

34. Forgrieve, Janet, "Poor Response to E-Mails Gold Mine for Company," *The Rocky Mountain News*, January 9, 2002, p. 5B.

35. Echikson, William, "When Oil Gets Well Connected," *Business Week*, December 3, 2001, p. EB 28.

36. Greenemeier, Larry, "Sim Software High on Biotech's Wish List," *InformationWeek Online*, December 17, 2001.

37. Klebnikov, Paul, "The Resurrection of NCR," *Forbes*, July 9, 2001, p. 70.

38. Nash, Kim, "Casino Hit Jackpot with Customer Data," *Computerworld*, July 2, 2001, p. 16.

39. Khalifa et al., "GSS Facilitation Restrictiveness in Collaborative Learning," *34th Annual Hawaii International Conference on Systems Science*, 2001, p. 29.

40. Bowen, Smalley, "Building Collaboration," *Computerworld*, November 5, 2001, p. 39.

41. Sliwa, Carol, "Caterpillar to Link Suppliers," *Computerworld*, September 17, 2001, p. 1.

42. Songini, Marc, "Pharmative to Launch Collaborative System with K Mart," *Computerworld*, February 5, 2001, p. 7.

43. Ragu, et al., "Collaborative Decision Making: A Connectionist Paradigm for Dialectical Support," *Information Systems Research*, December 2001, p. 363.

44. Disabatino, Jennifer, "While Lotus Pushes KM, Users Focus on R5 Migration," *Computerworld*, January 22, 2001, p. 10.

45. Alwang, Greg, "Web Collaboration," *PC Magazine*, October 30, 2001, p. 32.

46. Kontzer, Tony, "New SAP Infrastructure Stresses Interoperability, Collaboration," *InformationWeek Online*, November 7, 2001.

47. Ocker, R. "The Effects of Face-to-Face and Computer-Mediated Communication," *34th Annual Hawaii International Conference on Systems Science*, 2001, p. 16.

Specialized Business Information Systems: Artificial Intelligence, Expert Systems, Virtual Reality, and Other Specialized Systems

CHAPTER 11

PRINCIPLES	LEARNING OBJECTIVES
Artificial intelligence systems form a broad and diverse set of systems that can replicate human decision making for certain types of well-defined problems.	• Define the term *artificial intelligence* and state the objective of developing artificial intelligence systems. • List the characteristics of intelligent behavior and compare the performance of natural and artificial intelligence systems for each of these characteristics. • Identify the major components of the artificial intelligence field and provide one example of each type of system.
Expert systems can enable a novice to perform at the level of an expert but must be developed and maintained very carefully.	• List the characteristics and basic components of expert systems. • Identify at least three factors to consider in evaluating the development of an expert system. • Outline and briefly explain the steps for developing an expert system. • Identify the benefits associated with the use of expert systems.
Virtual reality systems have the potential to reshape the interface between people and information technology by offering new ways to communicate information, visualize processes, and express ideas creatively.	• Define the term *virtual reality* and provide three examples of virtual reality applications.
Special-purpose systems can help organizations and individuals achieve their goals.	• Discuss examples of special-purpose systems for organizational and individual use.

[Artificial Life, Inc.]

Intelligent Bots: The New Citizens of Cyberspace

Meet the newest form of customer service: intelligent robots, or "bots," for the Internet. Artificial Life, Inc., develops, markets, and supports the bots for business use. Its creations "live" on Web sites and are anxious to assist the next visitor. Each appears as an animated three-dimensional person (commonly referred to as an *avatar*) with his or her own personality. There's Lucy McBot, a redhead in a form-fitting green polo shirt who is rather arrogant and vain, but good humored. Roy is described as a very nice guy, cheerful, romantic, and a little naïve but always willing to help. Then there's Andrew, who comes with a tongue-in-cheek disclaimer: This bot has a tendency to exhibit behavior not suitable for children under the age of 17.

These and the other Artificial Life bots are created to serve as electronic representatives of the businesses and organizations that purchase them. Artificial Life's smart bots are custom-designed expert systems capable of engaging a customer in a real conversation. Through a keyboard-based discussion, a customer can be guided, as well as consulted. The smart bot can display Web pages, answer questions, and even make recommendations. It engages customers in the same conversational manner as a human sales or support person would. But unlike its human counterpart, it is always available, does not grow weary of answering the same question repeatedly, and is under no pressure to quickly conclude a conversation to attend to another waiting customer.

Each bot is tailored to meet the needs and taste of the customer. Bots are also matched to the industry and environment in which they will serve. A bot representing a bank might look clean-cut and act conservatively, while one used to sell the latest pop music would act and dress more casually. Once a bot is created and given a personality, a business expert uses software tools to teach the bot everything it needs to know to answer customer questions intelligently. For example, the bank bot would be programmed with information about interest rates and customer accounts, while the pop music bot would become familiar with the latest music releases.

If the bot is unable to assist a customer with a particular question or need, it can connect the customer to its human counterpart, who can then take over the conversation. This system serves both the business and the customer: It saves valuable employee time by automating routine questions and requests. Customer service representatives spend less time answering frequently asked questions and more time engaged in complex problem solving. The customers benefit by getting faster results—not having to wait for a customer service representative or traverse a labyrinth of Web page links—and smart bots are more helpful and engaging than traditional search engines.

Artificial Life's smart bots come with a sophisticated suite of software that allows a business to program and supervise the bot. The SmartBot Suite also provides utilities to generate reports and analyze statistics on the bot's conversations with users. The data feedback allows businesses to add to the bot's knowledge to most effectively serve customers.

Bots can also field e-mail questions and interact with customers through instant messaging. Once added to a customer's instant messaging buddy list, a bot provides one-click access to company information and services. Imagine making a hair appointment by clicking on Eduardo in your buddy list—the bot from the corner salon.

In the future, we will be interacting with increasing numbers of computers that act on behalf of a person or organization. At times they may frustrate us, but other times they will assist or amuse. The hope is that they will simplify our lives. One thing is for certain: as computing power increases and research in artificial intelligence progresses, it will become increasingly difficult to tell the bots from the humans—at least on-line.

As you read through this chapter, consider the following:
- What does it take to design and build an effective artificial intelligence system?
- What business needs can be addressed by an expert system?
- What are some practical applications for virtual reality?

Science fiction movies give us a glimpse of the future, but many practical applications of artificial intelligence exist today, among them medical diagnostics, mechanical diagnostics, and development of computer systems.

(Source: The Kobal Collection/Amblin/ Dreamworks/Stanley Kubrick/WB/ James, David.)

At a Dartmouth College conference in 1956, John McCarthy proposed the use of the term artificial intelligence (AI) to describe computers with the ability to mimic or duplicate the functions of the human brain. Many AI pioneers attended this first conference; a few predicted that computers would be as "smart" as people by the 1960s. The prediction has not yet been realized, but the benefits of artificial intelligence in business and research can be seen today, and research continues.

⦿ AN OVERVIEW OF ARTIFICIAL INTELLIGENCE

Science fiction novels and popular movies have featured scenarios of computer systems and intelligent machines taking over the world. Stephen Hawking, who is the Lucasian Professor of Mathematics at Cambridge University (a position once held by Isaac Newton) and author of *A Brief History of Time,* said, "In contrast with our intellect, computers double their performance every 18 months. So the danger is real that they could develop intelligence and take over the world."[1] Computer systems such as Hal in the classic movie *2001: A Space Odyssey* and the movie *A.I.* are futuristic glimpses of what might be. These accounts are fictional, but we see the real application of many computer systems that use the notion of AI. These systems help to make medical diagnoses, explore for natural resources, determine what is wrong with mechanical devices, and assist in designing and developing other computer systems. In this chapter we explore the exciting applications of artificial intelligence, expert systems, virtual reality, and some specialized systems to see what the future really might hold.

ARTIFICIAL INTELLIGENCE IN PERSPECTIVE

artificial intelligence systems
people, procedures, hardware, software, data, and knowledge needed to develop computer systems and machines that demonstrate the characteristics of intelligence

Artificial intelligence systems include the people, procedures, hardware, software, data, and knowledge needed to develop computer systems and machines that demonstrate characteristics of intelligence. Researchers, scientists, and experts on how humans think are often involved in developing these systems. The objective in developing contemporary AI systems is not to replace human decision making completely but to replicate it for certain types of well-defined problems. As with other information systems, the overall purpose of artificial intelligence applications in business is to help an organization achieve its goals.

THE NATURE OF INTELLIGENCE

intelligent behavior
the ability to learn from experiences and apply knowledge acquired from experience, handle complex situations, solve problems when important information is missing, determine what is important, react quickly and correctly to a new situation, understand visual images, process and manipulate symbols, be creative and imaginative, and use heuristics

From the early AI pioneering stage, the research emphasis has been on developing machines with **intelligent behavior**. Some of the specific characteristics of intelligent behavior include the ability to do the following:

Learn from experience and apply the knowledge acquired from experience. Being able to learn from past situations and events is a key component of intelligent behavior and is a natural ability of humans, who learn by trial and error. However, learning from experience is not natural for computer systems. This ability must be carefully programmed into the system. Today, researchers are developing systems that have this ability. For instance, computerized AI chess software can learn to improve while playing human competitors.

Computers like Deep Blue attempt to learn from past chess moves. The powerful supercomputer's logic system was able to calculate the ramifications of up to 100 billion chess maneuvers within the allotted time for each move.

(Source: Courtesy of the Association for Computing Machinery.)

In addition to learning from experience, people apply what they have learned to new settings and circumstances. In a number of cases, individuals have taken what they have learned and succeeded with in one endeavor and applied it to another. For example, a company that developed a dishwashing product effective for cleaning greasy dishes developed a variation of the product for use in cleaning up messy highway spills. Although humans have the ability to apply what they have learned to new settings, this characteristic is not automatic with computer systems. Developing computer programs to allow computers to apply what they have learned can be difficult.

Handle complex situations. Humans are involved in complex situations. World leaders face difficult political decisions regarding terrorism, conflict, global economic conditions, hunger, and poverty. In a business setting, top-level managers and executives are faced with a complex market, challenging competitors, intricate government regulations, and a demanding workforce. Even human experts make mistakes in dealing with these situations. Developing computer systems that can handle perplexing situations requires careful planning and elaborate computer programming.

Solve problems when important information is missing. The essence of decision making is dealing with uncertainty. Quite often, decisions must be made even when we lack information or have inaccurate information, because obtaining complete information is too costly or impossible. You have probably seen movies in which computers have responded to human commands with statements like "Does not compute" and "Insufficient information." Today, AI systems can make important calculations, comparisons, and decisions even when missing information.

Determine what is important. Knowing what is truly important is the mark of a good decision maker. Every day we are bombarded with facts and must process large amounts of data, filtering out what is unnecessary. Determining which items are crucial can make the difference between good decisions and those that ultimately lead to problems or failures. Computers, on the other hand, do not have this natural ability. Developing programs and approaches to allow computer systems and machines to identify important information is not a simple task.

React quickly and correctly to a new situation. A small child, for example, can look over a ledge or a drop-off and know not to venture too close. The child reacts quickly and correctly to a new situation. Computers, on the other hand, do not have this ability without complex programming.

Understand visual images. Interpreting visual images can be extremely difficult, even for sophisticated computers. People and animals can look at objects interacting in our environment and understand exactly what is going on. For instance, we can see a man sitting at a table and know that he has legs and feet that we cannot see. Being able to understand and correctly interpret visual images is an extremely complex process for computer systems. Moving through a room of chairs, tables, and other objects can be trivial for people but extremely complex for machines, robots, and computers. Such machines require an extension of understanding visual images, called a **perceptive system**. Having a perceptive system allows a machine to approximate the way a human sees, hears, and feels objects.

perceptive system
A system that approximates the way a human sees, hears, and feels objects

Process and manipulate symbols. People see, manipulate, and process symbols every day. Visual images provide a constant stream of information to our brains. By contrast, computers have difficulty handling symbolic processing and reasoning. Although computers excel at numerical calculations, they aren't as good at dealing with symbols and three-dimensional objects. Recent developments in machine-vision hardware and software, however, allow some computers to process and manipulate symbols on a limited basis.

Be creative and imaginative. Throughout history, some people have turned difficult situations into advantages by being creative and imaginative. For instance, when shipped a lot of defective mints with holes in the middle, an enterprising entrepreneur decided to market these new mints as Life Savers instead of returning them to the manufacturer. Ice cream cones were invented at the St. Louis World's Fair when an imaginative store owner decided to wrap ice cream with a waffle from his grill for portability. Developing new and exciting products and services from an existing (perhaps negative) situation is a human characteristic. Few computers have the ability to be truly imaginative or creative in this way, although software has been developed to enable a computer to write short stories.

Use heuristics. With some decisions, people use heuristics (rules of thumb arising from experience) or even guesses. In searching for a job, we may decide to rank companies we are considering according to profits per employee. Companies making more profits might pay their employees more. In a manufacturing setting, a corporate president may decide to look at only certain locations for a new plant. We make these types of decisions using general rules of thumb, without completely searching all alternatives and possibilities. Today, some computer systems also have this ability. They can, given the right programs, obtain good solutions that use approximations instead of trying to search for an optimal solution, which would be technically difficult or too time-consuming.

This list of traits only partially defines intelligence. Unlike virtually every other field of information systems research in which the objectives can be clearly defined, the term *intelligence* is a formidable stumbling block. One of the problems in artificial intelligence is arriving at a working definition of real intelligence against which to compare the performance of an artificial intelligence system.

THE DIFFERENCE BETWEEN NATURAL AND ARTIFICIAL INTELLIGENCE

Since the term *artificial intelligence* was defined in the 1950s, experts have disagreed about the difference between natural and artificial intelligence. For instance, is there a difference between carbon life (human or animal life) and silicon life (a computer chip) in terms of behavior? Can computers be programmed to have "common sense"? Profound differences exist, but they are declining in number (Table 11.1). One of the driving forces behind AI research is an attempt to understand how humans actually reason and think. It is believed that the ability to create machines that can reason will be possible only once we truly understand our own processes for doing so. Read the accompanying "Ethical and Societal Issues" box to see one attempt to create computers with common sense.

THE MAJOR BRANCHES OF ARTIFICIAL INTELLIGENCE

AI is a broad field that includes several specialty areas, such as expert systems, robotics, vision systems, natural language processing, learning systems, and neural networks (Figure 11.1). Many of these areas are related; advances in one can occur simultaneously with or result in advances in others.

Expert Systems

An expert system consists of hardware and software that stores knowledge and makes inferences, similar to a human expert. Because of their many business

Attributes	Natural Intelligence (Human)	Artificial Intelligence (Machine)
The ability to use sensors (eyes, ears, touch, smell)	High	Low
The ability to be creative and imaginative	High	Low
The ability to learn from experience	High	Low
The ability to be adaptive	High	Low
The ability to afford the cost of acquiring intelligence	High	Low
The ability to use a variety of information sources	High	High
The ability to acquire a large amount of external information	High	High
The ability to make complex calculations	Low	High
The ability to transfer information	Low	High
The ability to make a series of calculations rapidly and accurately	Low	High

TABLE 11 • 1

A Comparison of Natural and Artificial Intelligence

robotics
Mechanical or computer devices that perform tasks requiring a high degree of precision or that are tedious or hazardous for humans

applications, expert systems are discussed in more detail in the next several sections of the chapter.

Robotics

Robotics involves developing mechanical or computer devices that can paint cars, make precision welds, and perform other tasks that require a high degree of precision or are tedious or hazardous for humans. Contemporary robotics combines both high-precision machine capabilities with sophisticated controlling software. The controlling software in robots is what is most important in terms of AI. The processor in an advanced industrial robot today works at about 10 million instructions per second (MIPS)—no smarter than an insect. To achieve anything

FIGURE 11 • 1

A Conceptual Model of Artificial Intelligence

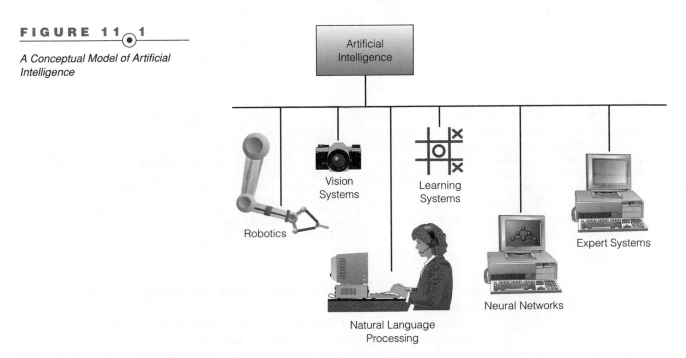

Artificial Intelligence

Robotics

Vision Systems

Learning Systems

Natural Language Processing

Neural Networks

Expert Systems

Cyc: The System That Is an Expert at Common Sense

In 1984 AI pioneer Doug Lenat began formalizing the rules for human common sense and entering them into a computer program he named Cyc (short for encyclopedia). Lenat's goal was to develop a rational computer program that could make independent assertions. He has labored years to codify facts such as "Once people die, they stop buying things." He uses a form of symbolic logic called *predicate calculus* to classify and show the properties of information in a standard way. Now, 19 years later, with over 600 person-years and $60 million invested, the Cyc knowledge base contains over 3 million rules of thumb that the average person knows about the world, plus about 300,000 terms or concepts. And Lenat's intelligent child is ready to begin earning its keep.

What service can Cyc provide to businesses? Possibly many. "I see this more as a power source rather than a single application," Lenat states. "[For any given application], you need common-sense knowledge and [content] knowledge. We are building in the common-sense knowledge." For example, Cyc could serve as a smart Web search engine that finds what you want because it understands content, instead of just matching key words. Cyc could automatically cleanse files, databases, and spreadsheets of errors and inconsistencies. Cyc also might be the final piece both in developing software that doesn't trip over the many ambiguities of the English language and in giving computers that ability. Current applications for Cyc that have been released or are currently in development include the following:

- CycSecure: An intelligent network security system that seeks out and identifies network security vulnerabilities by thinking like a hacker or disgruntled employee.
- Semantic data mining: Data mining capabilities that can understand relationships or the context of data. For example, a typical personnel database may know that "Fred Utz" is a "professor" who works for "Harvard University," but it does not know what a person, professor, or university is. Cyc, however, does know.
- Guided Integration of Structured Terminology (GIST): Vocabulary function that allows users to import and simultaneously manage and integrate multiple industry-specific terms and synonyms.
- "Smart" interfaces: A system that responds to user commands more intuitively. Rather than clicking an option in a menu, you could type in what you wish to happen, and Cyc would empower the computer to understand.
- Intelligent character simulation: Capability that makes games behave more like people than machines.
- Enhanced virtual reality environments: More lifelike interactions that would be responsive to humans, such as conversational dialog.

Cyc is now available to the public through the recent release of OpenCyc, a slightly pared-down version of the original (free download from http://www.opencyc.org). "Up until now, the only people adding knowledge were a small priesthood of logicians," says Lenat. "Now, suddenly, millions of people can add their knowledge to Cyc. Because of the acceleration, we'll be at 10 million assertions a year from now." "A typical person knows about 100 million things about the world. I see us crossing that point in five years. It's difficult to predict the course thereafter."

Lenat is not concerned with creating computers that think like humans. "Absolutely none of my work is based on a desire to understand how human cognition works," says Lenat. "I don't understand, and I don't care to understand. It doesn't matter to me how people think; the important thing is what we know, not how do we know it." This echoes the view of most researchers in the field. The focus of artificial intelligence today is no longer on psychology but on goals shared by the rest of computer science: the development of systems to augment human abilities. Let people do what they do best and machines do what they do best. In short, it doesn't make sense to try to make a machine into a person, since we already have a successful method for creating people.

Discussion Questions
1. Besides the applications listed, what other business applications can you think of for a computer that has common sense?
2. How will the introduction of common sense into computer systems affect human–computer relations? Will users react more positively or negatively to a PC that exhibits common sense?

Critical Thinking Questions
3. When Cyc knows as much about the world as a typical person, could it be considered a conscious being?
4. How can Cyc improve systems such as medical expert systems used to diagnose a patient's symptoms?

Sources: Gary H. Anthes, "Computerizing Common Sense," *Computerworld*, April 8, 2002, http://www.computerworld.com/news/2002/story/0,11280, 69881,00.html; Mitchell Leslie, "Wise Up, Dumb Machine," *Stanford Magazine*, March/April 2002, http://www.stanfordalumni.org/news/magazine/2002/marapr/departments/brightideas.html; Michael Hiltzik, "A.I. Reboots," *MIT Technology Review*, March 2002, http://www.technologyreview.com/articles/print_version/hiltzik0302.asp; Cycorp Web site, http://www.cyc.com, accessed June 2002.

even approaching human intelligence, the robot processor must achieve 100 trillion operations per second.

Many applications of robotics exist, and research into these unique devices continues.[2] Manufacturers use robots to assemble and paint products. Welding robots have enabled firms to manufacture top-quality products and reduce

Robots can be used in situations that are hazardous or difficult in other ways, such as disabling a bomb.

(Source: AP/Wide World Photos.)

labor costs while shortening delivery time to their customers. Honda Motor Co. developed a humanoid robot called Asimo.[3] The experimental robot can shake hands, dance, and even reply to simple questions. The technology used in the robot's legs may offer improvements to people with disabilities. The project was started as an exercise in Honda's motorcycle division. According to one of the engineers, the staff wanted "to accomplish something that could be helpful somehow to human life." In Albuquerque, New Mexico, a remote-controlled robot was used to lay fiber-optic cable.[4] Funny-looking and -acting robots are even being used to entertain disabled children while they receive needed therapy and exercise.[5]

The military uses of robots are moving beyond movies, such as *Star Wars Episode II—Attack of the Clones,* to become real weapons. The U.S. Navy is developing a robot that has the appearance of a large lobster with eight legs.[6] The underwater robot has sensors that can hear and smell. The Navy hopes the robot will be operational by 2010 to locate mines and other underwater objects. The Air Force is developing a smart robotic jet fighter.[7] Often called *unmanned combat air vehicles (UCAVs),* these robotic war machines, like the X-45A, will be able to identify and destroy targets without human pilots. UCAVs can send pictures and information to a central command center and can be directed to strike military targets. These new machines will be able to extend the current Predator and Global Hawk technologies the military used in Afghanistan after the September 11th terrorist attacks.

Although robots are essential components of today's automated manufacturing and military systems, future robots will find wider applications in banks, restaurants, homes, doctors' offices, and hazardous working environments such as nuclear stations. The Da Vinci Surgical System, for example, allows doctors to operate using a robotic arm.[8] Sitting at a console with familiar controls, the surgeon can replace a heart valve or remove a tumor. The robotic arm can be accurately controlled and only requires a small incision in the patient, making surgery more precise and the recovery easier. Microrobotics, also called *micro-electro-mechanical systems (MEMS)* that are the size of a grain of salt, are also being developed.[9] MEMS can be used in air bags, a person's blood to monitor the body, cell phones, refrigerators, and more. According to Paul McWhorter, cofounder of MEMX, "Micromachines will be unseen parts of virtually every new product, but since they'll be embedded even more deeply than microprocessors, only their effect will matter—they'll be totally transparent."

A robot must not only execute tasks programmed by the user but must also be able to interact with its environment through its sensors and actuators, sense and avoid unforeseen obstacles, and perform its duties much the same way humans do. According to Dr. Brian Scasselate, the head of the Cog Robot Project at MIT, "In the 20-year range, we'll see robotic systems that are social. Robotic systems will become part of daily life and part of ourselves."[10]

Vision Systems

vision systems
The hardware and software that permit computers to capture, store, and manipulate visual images and pictures

Another area of AI involves vision systems. **Vision systems** include hardware and software that permit computers to capture, store, and manipulate visual images and pictures. The U.S. Justice Department uses vision systems to perform fingerprint analysis, with almost the same level of precision as human experts. The speed with which the system can search through a huge database of fingerprints has brought quick resolution to many long-standing mysteries. Vision systems are also effective at identifying people based on facial features. Canesta, a start-up firm in California, uses infrared light and photographic chips to give computer devices three-dimensional images of objects.[11] Light is beamed at the object and reflected back to the computer device. Small differences in the time it takes for the light to return gives the device a 3-D image of the object.

Vision systems can be used in conjunction with robots to give these machines "sight." Robots such as those used in factory automation typically perform mechanical tasks with little or no visual stimuli. Robotic vision extends the capability of these systems, allowing the robot to make decisions based on visual input. Generally, robots with vision systems can recognize black and white and some gray shades but do not have good color or three-dimensional vision. Other systems concentrate on only a few key features in an image, ignoring the rest. It may take years before a robot or other computer system can "see" in full color and draw conclusions from what it sees, the way humans do.

Natural Language Processing

natural language processing
Processing that allows the computer to understand and react to statements and commands made in a "natural" language, such as English

As discussed in Chapter 4, **natural language processing** allows a computer to understand and react to statements and commands made in a "natural" language, such as English.[12] There are three levels of voice recognition: command (recognizes dozens to hundreds of words), discrete (recognizes dictated speech with pauses between words), and continuous (recognizes natural speech). For example, a natural language processing system can be used to retrieve important information without typing in commands or searching for key words. With natural language processing, it is possible to speak into a microphone connected to a computer and have the computer convert the electrical impulses generated from the voice into text files or program commands. With some simple natural language processors, you say a word into a microphone and type the same word on the keyboard. The computer then matches the sound with the typed word. With more advanced natural language processors, recording and typing words is not necessary. Upstart Natural Machine is making its Java code for verbal AI available to others. The company hopes that everyone will benefit by making the computer code open and available.

Brokerage services are a perfect fit for voice-recognition technology to replace the existing "press 1 to buy or sell a stock" touch-pad telephone menu system. People buying and selling stock use a vocabulary too varied for easy access through menus and touch pads but still small enough for software to process in real time. Several brokerages—including Charles Schwab & Co., Fidelity Investments, DLJdirect, and TD Waterhouse Group—offer voice recognition services. Schwab is using a natural language search engine to help customers navigate its Web site.[13] According to Debbie Naganuma, director of electronic brokerage product development at Schwab, "We have a good site, but [users said] it's hard to traverse." The new natural language search engine should help customers navigate the Schwab site. T. Rowe Price, another brokerage firm, uses natural language voice recognition to let customers access retirement accounts, check balances, and get stock quotes by using simple voice commands over the phone.[14] Eventually, T. Rowe Price hopes the technology will allow transactions to be made using voice commands over the phone. TD Waterhouse uses a natural-language-processing search engine that allows customers to get their questions answered through the brokerage firm's call center.[15] One of the big advantages of voice recognition is that the number of calls routed to the customer service department drops considerably once new voice features are added. That is desirable to brokerages because it helps them staff their call centers correctly—even in volatile markets.

Other companies are also profitably using voice-recognition technology.[16] Hewlett-Packard has implemented a voice-recognition system in its call centers to provide assistance and reduce costs. Hewlett-Packard uses a voice-recognition program called SpeachWorks. It took H-P five months of intensive programming to build the necessary software. While a typical person uses a vocabulary of about 20,000 words or fewer, Hewlett-Packard's voice-recognition software has a built-in

Dragon Systems' Naturally Speaking 6 Essentials uses continuous voice recognition or natural speech, allowing the user to speak to the computer at a normal pace without pausing between words. The spoken words are transcribed immediately onto the computer screen.

(Source: Courtesy of ScanSoft, Inc.)

vocabulary of 85,000 words. Other SpeachWorks customers include FedEx and Continental Airlines. SpeachWorks claims that its software is so good that some customers forget they are talking to a computer and start discussing the weather or sports scores. Some companies are starting to share their natural language experience and computer code with others.[17]

Learning Systems

learning systems
A combination of software and hardware that allows the computer to change how it functions or reacts to situations based on feedback it receives

Another part of AI deals with **learning systems**, a combination of software and hardware that allows a computer to change how it functions or reacts to situations based on feedback it receives. For example, some computerized games have learning abilities. If the computer does not win a game, it remembers not to make the same moves under the same conditions again. Learning systems software requires feedback on the results of actions or decisions. At a minimum, the feedback needs to indicate whether the results are desirable (winning a game) or undesirable (losing a game). The feedback is then used to alter what the system will do in the future.

Neural Networks

neural network
A computer system that can simulate the functioning of a human brain

An increasingly important aspect of AI involves neural networks. A **neural network** is a computer system that can act like or simulate the functioning of a human brain. The systems use massively parallel processors in an architecture that is based on the human brain's own meshlike structure. In addition, neural network software can be used to simulate a neural network using standard computers. Neural networks can process many pieces of data at once and learn to recognize patterns. Chevron Phillips Chemical Company, for example, uses neural network software to analyze a large amount of data to control chemical reactors.[18] Some of the specific features of neural networks include the following:

- The ability to retrieve information even if some of the neural nodes fail
- Fast modification of stored data as a result of new information
- The ability to discover relationships and trends in large databases
- The ability to solve complex problems for which all the information is not present

Neural networks excel at pattern recognition. For example, neural network computers can be used to read bank check bar codes despite smears or poor-quality printing. Many retail stores use Falcon Fraud Manager, a neural network system, to detect possible credit card fraud.[19] Falcon Fraud Manager is used to protect more than 450 million credit card accounts. Some hospitals use neural networks to determine a patient's likelihood of contracting cancer or other diseases. The speed of genomic research can be increased with software that includes neural network features, such as NAG data mining software.[20]

Neural nets work particularly well when it comes to analyzing detailed trends. Large amusement parks and banks use neural networks to figure out staffing needs based on customer traffic—a task that requires precise analysis, down to the half-hour. Increasingly, businesses are firing up neural nets to help them navigate ever-thicker forests of data and make sense of myriad customer traits and buying habits. Computer Associates has developed Neugents, neural intelligence agents that "learn" patterns and behaviors and predict what will happen next. For example, Neugents can be used to track the habits of insurance customers and predict which ones will not renew, say, an automobile policy. They can then suggest to an insurance agent what changes might be made in the policy to get the consumer to renew it. The technology also can be employed to track individual users and their on-line preferences so that users at e-commerce sites don't have to input the same information each time they log on—their purchasing history and other data will be recalled each time they access a Web site.

AI Trilogy is a neural network software program that can run on a standard PC.[21] The software can make predictions with NeuroShell Predictor and make classifications with NeuroShell Classifier. The software package also contains GeneHunter, which uses a genetic algorithm to get the best result from the neural network system.

Searchspace is an example of pattern-recognition software that uses neural networks.[22] The software has the ability to analyze hundreds of millions of bank, brokerage, and insurance accounts involving a trillion dollars or more. The software can uncover money laundering and other suspicious money transfers. Today, partially as a result of the September 11th attacks, all transfers of $10,000 or more across U.S. borders must be reported to the federal government. Software such as Searchspace can help in uncovering illegal transfers.

AN OVERVIEW OF EXPERT SYSTEMS

As mentioned earlier, an expert system behaves similarly to a human expert in a particular field. Computerized expert systems have been developed to diagnose problems, predict future events, and solve energy problems. They have also been used to design new products and systems, determine the best use of lumber, and increase the quality of healthcare. Like human experts, computerized expert systems use heuristics, or rules of thumb, to arrive at conclusions or make suggestions. Expert systems have also been used to determine credit limits for credit cards. The research conducted in AI during the past two decades is resulting in expert systems that explore new business possibilities, increase overall profitability, reduce costs, and provide superior service to customers and clients.

CHARACTERISTICS OF AN EXPERT SYSTEM

Expert systems have a number of characteristics and capabilities, including the following:

Can explain their reasoning or suggested decisions. A valuable characteristic of an expert system is the capability to explain how and why a decision or solution was reached. For example, an expert system can explain the reasoning behind the conclusion to approve a particular loan application. The ability to explain its reasoning processes can be the most valuable feature of a computerized expert system. The user of the expert system thus gains access to the reasoning behind the conclusion.

Can display "intelligent" behavior. Considering a collection of data, an expert system can propose new ideas or approaches to problem solving. A few of the applications of expert systems are an imaginative medical diagnosis based on a patient's condition, a suggestion to explore for natural gas at a particular location, and providing job counseling for workers.

Can draw conclusions from complex relationships. Expert systems can evaluate complex relationships to reach conclusions and solve problems. For example, one proposed expert system would work with a flexible manufacturing system to determine the best use of tools. Another expert system can suggest ways to improve quality-control procedures.

Can provide portable knowledge. One unique capability of expert systems is that they can be used to capture human expertise that might otherwise be lost. A classic example of this is the expert system called DELTA (Diesel Electric Locomotive Troubleshooting Aid), which was developed

Credit card companies often use expert systems to determine credit limits for credit cards.

(Source: © David Young Wolff/Getty Images.)

to preserve the expertise of the retiring David Smith, the only engineer competent to handle many highly technical repairs of such machines.

Can deal with uncertainty. One of an expert system's most important features is its ability to deal with knowledge that is incomplete or not completely accurate. The system deals with this problem through the use of probability, statistics, and heuristics.

Even though these characteristics of expert systems are impressive, other characteristics limit their current usefulness. Some of these characteristics are as follows:

Not widely used or tested. Even though successes occur, expert systems are not used in a large number of organizations. In other words, they have not been widely tested in corporate settings.

Difficult to use. Some expert systems are difficult to control and use. In some cases, the assistance of computer personnel or individuals trained in the use of expert systems is required to help the user get the most from these systems. Today's challenge is to make expert systems easier to use by decision makers who have limited computer programming experience.

Limited to relatively narrow problems. Whereas some expert systems can perform complex data analysis, others are limited to simple problems. Also, many problems solved by expert systems are not that beneficial in business settings. An expert system designed to provide advice on how to repair a machine, for example, is unable to assist in decisions about when or whether to repair it. In general, the narrower the scope of the problem, the easier it is to implement an expert system to solve it.

Cannot readily deal with "mixed" knowledge. Expert systems cannot easily handle knowledge that has a mixed representation. Knowledge can be represented through defined rules, through comparison with similar cases, and in various other ways. An expert system in one application might not be able to deal with knowledge that combined both rules and cases.

Possibility of error. Although some expert systems have limited abilities to learn from experience, the primary source of knowledge is a human expert. If this knowledge is incorrect or incomplete, it will affect the system negatively. Other development errors involve poor programming practices. Because expert systems are more complex than other information systems, the potential for such errors is greater.

Cannot refine its own knowledge. Expert systems are not capable of acquiring knowledge directly. A programmer must provide instructions to the system that determine how the system is to learn from experience. Also, some expert systems cannot refine their own knowledge—such as eliminating redundant or contradictory rules.

Difficult to maintain. Related to the preceding point is the fact that expert systems can be difficult to update. Some are not responsive or adaptive to changing conditions. Adding new knowledge and changing complex relationships may require sophisticated programming skills. In some cases, a spreadsheet used in conjunction with an expert system shell can be used to modify the system. In others, upgrading an expert system can be too difficult for the typical manager or executive. Future expert systems are likely to be easier to maintain and update.

expert system shells
A collection of software packages and tools used to develop expert systems

May have high development costs. Expert systems can be expensive to develop when using traditional programming languages and approaches. Development costs can be greatly reduced through the use of software for expert system development. **Expert system shells**, a collection of software packages and tools used to develop expert systems, can be implemented on most popular PC platforms to reduce development time and costs.

Raise legal and ethical concerns. People who make decisions and take action are legally and ethically responsible for their behavior. A person, for example, can be taken to court and punished for a crime. When expert systems are used to make decisions or help in the decision-making process, who is legally and ethically

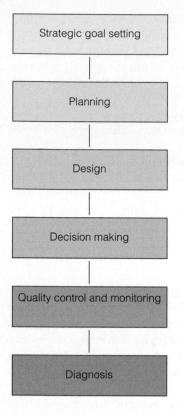

FIGURE 11.2

Solutions Offered by Expert Systems

responsible? The human experts used to develop the knowledge on which the system relies, the expert system developer, the user, or someone else? For example, if a doctor uses an expert system to make a diagnosis and the diagnosis is wrong, who is responsible? These legal and ethical issues have not been completely resolved.

CAPABILITIES OF EXPERT SYSTEMS

Compared with other types of information systems, expert systems offer a number of powerful capabilities and benefits. For example, one expert system, called XCON, is often used in designing computer system configurations because it consistently does a better job than human beings.

Expert systems can be used to solve problems in every field and discipline and can assist in all stages of the problem-solving process. Past successes have shown that expert systems are good at strategic goal setting, planning, design, decision making, quality control and monitoring, and diagnosis (Figure 11.2).

Strategic Goal Setting

Developing strategic goals for an organization is one of the most important functions of top-level decision makers. Strategic goals provide a framework for all other activities throughout the organization. Expert systems can suggest strategic goals and explore the impact of adopting them. Strategic goals can include identifying opportunities in the marketplace, analyzing the strengths of the existing organization, determining the power and position of competitors, and understanding the existing labor force. For example, say a California wine maker is currently perceived as a low-cost/low-quality producer. An expert system can help the company's top-level management determine the costs and benefits involved in producing higher quality wines and changing its image in the marketplace.

Planning

Expert systems have been employed to assist in the planning process. The ability to reach overall corporate objectives, the impact of plans on organizational resources, and the ways specific plans will help an organization compete in the marketplace can be investigated via expert systems. A manufacturing company, for example, might be exploring the possibility of building a new plant. An expert system can assist with this planning process by suggesting factors that should be considered in making the final decision, based on facts supplied by management.

Design

Designing new products and services requires experience, judgment, and an understanding of the marketplace. Some expert systems have been developed to assist in designing a variety of products, such as computer chips and systems. These types of expert systems use general design principles, an understanding of manufacturing procedures, and a collection of design rules.

Decision Making

Wouldn't it be nice to have an expert help us make our day-to-day decisions? Expert systems have provided this type of support for many individuals and organizations. Acting as advisors or counselors, these systems can suggest possible alternatives, ways of looking at problems, and logical approaches to the decision-making process. In addition, expert systems can improve the learning process for those who are not as experienced in decision making.

Quality Control and Monitoring

Measuring the quality of products and services, determining whether an existing computer system is operating as intended, analyzing the efficiency of a manufacturing plant, and determining the overall effectiveness of a hospital or nursing home are some of the abilities of monitoring systems. Computerized expert systems can assist in monitoring various systems and proposing solutions

A U.S. government agency developed an expert system that analyzes weather patterns and crop data. After it reviews all relevant information, it advises farmers about irrigation, fertilization, and the optimal time to harvest to maximize crop yields.

(Source: Scott Bauer/USDA.)

to system problems. Expert systems can also be used to monitor product quality. When machines are malfunctioning, the expert system can assist in determining possible causes.

Diagnosis

Monitoring and diagnosis go hand in hand. Monitoring determines the current state of a system; diagnosis looks at the causes and proposes solutions. In medicine, expert systems have been employed to diagnose difficult patient conditions. An expert system can analyze test results and patient symptoms. Some systems put probability estimates on potential diseases, given the data and analysis performed. An expert system can provide the doctor with the probable cause of the medical problem and propose treatments or interventions. In a business setting, an expert system can diagnose potential problems of, for example, a chemical distillation facility that is not operating as expected or desired.

WHEN TO USE EXPERT SYSTEMS

Sophisticated expert systems can be difficult, expensive, and time-consuming to develop. This is especially true for large expert systems implemented on mainframes. The following is a list of factors that normally make expert systems worth the expenditure of time and money:

* Provide a high potential payoff or significantly reduce downside risk
* Can capture and preserve irreplaceable human expertise
* Can develop a system more consistent than human experts
* Can provide expertise needed at a number of locations at the same time or in a hostile environment that is dangerous to human health
* Can provide expertise that is expensive or rare
* Can develop a solution faster than human experts can
* Can provide expertise needed for training and development to share the wisdom and experience of human experts with a large number of people

COMPONENTS OF EXPERT SYSTEMS

An expert system consists of a collection of integrated and related components, including a knowledge base, an inference engine, an explanation facility, a knowledge base acquisition facility, and a user interface. A diagram of a typical expert system is shown in Figure 11.3. In this figure, the user interacts with the user interface, which interacts with the inference engine. The inference engine interacts with the other expert system components. These components must work together in providing expertise.

FIGURE 11.3

Components of an Expert System

The Knowledge Base

The knowledge base stores all relevant information, data, rules, cases, and relationships used by the expert system. As seen in Figure 11.4, a knowledge base is a natural extension of a database (presented in Chapter 5) and an information and decision support system (presented in Chapter 10). As discussed in Chapter 5, raw facts can be used to perform basic business transactions but are seldom used without manipulation in decision making. As we move to information and decision support, data is filtered and manipulated to produce a variety of reports to help managers make better decisions. With a knowledge base, we try to understand patterns and relationships in data as a human expert does in making intelligent decisions.

FIGURE 11.4

The Relationships among Data, Information, and Knowledge

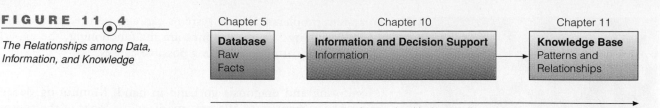

A knowledge base must be developed for each unique application. For example, a medical expert system will contain facts about diseases and symptoms. The knowledge base can include generic knowledge from general theories that have been established over time and specific knowledge that comes from more recent experiences and rules of thumb. Knowledge bases, however, go far beyond simple facts, also storing relationships, rules or frames, and cases. For example, certain telecommunications network problems may be related or linked; one problem may cause another. In other cases, rules suggest certain conclusions, based on a set of given facts. In many instances, these rules are stored as **if-then statements**, such as "If a certain set of network conditions exists, then a certain network problem diagnosis is appropriate." Cases can also be used. This technique involves finding instances, or cases, that are similar to the current problem and modifying the solutions to these cases to account for any differences between the previously solved cases stored in the computer and the current situation or problem.

if-then statements
Rules that suggest certain conclusions

Assembling Human Experts   One challenge in developing a knowledge base is to assemble the knowledge of multiple human experts. Typically, the objective in building a knowledge base is to integrate the knowledge of individuals with similar expertise (e.g., many doctors may contribute to a medical diagnostics knowledge base). A knowledge base that contains information from numerous experts can be extremely powerful and accurate in terms of its predictions and suggestions. Unfortunately, human experts can disagree on important relationships and interpretations of data, presenting a dilemma for designers and developers of knowledge bases and expert systems in general. Some human experts are more expert than others; their knowledge, experience, and information are better developed and more accurately represent reality. When human experts disagree on important points, it can be difficult for expert systems developers to determine which rules and relationships to place in the knowledge base.

The Use of Fuzzy Logic   Another challenge for expert system designers and developers is capturing knowledge and relationships that are not precise or exact.[23] Computers typically work with numerical certainty; certain input values will always result in the same output. In the real world, as you know from experience, this is not always the case. To handle this dilemma, a specialty research area in computer science, called **fuzzy logic**, has been developed. Research into fuzzy logic has been going on for decades, but its application to expert systems is just beginning to show results in a variety of areas.

fuzzy logic
A special research area in computer science that allows shades of gray and does not require everything to be simple black or white, yes/no, or true/false

Instead of the usual black and white, yes/no, or true/false conditions of typical computer decisions, fuzzy logic allows shades of gray, or what are known as "fuzzy sets." The criteria on whether a subject or situation fits into a set are given in percentages or probabilities. For example, a weather forecaster might state, "if it is very hot with high humidity, the likelihood of rain is 75 percent." The imprecise terms of "very hot" and "high humidity" are what fuzzy logic must determine to formulate the chance of rain. Fuzzy logic rules help computers evaluate the imperfect or imprecise conditions they encounter and make "educated guesses" based on the likelihood or probability of correctness of the decision. This ability to estimate whether a condition fits a situation more closely resembles the judgment a person makes when evaluating situations.

Fuzzy logic is used in embedded computer technology—for example, autofocus cameras, medical equipment that monitors patients' vital signs and makes automatic corrections, and temperature sensors attached to furnace controls. MediS Diagnostics, a British company, has developed a computer process based on fuzzy logic to detect lung cancer earlier, giving the potential to dramatically improve survival rates from the deadly disease.[24]

rule
A conditional statement that links given conditions to actions or outcomes

The Use of Rules A **rule** is a conditional statement that links given conditions to actions or outcomes. As we saw earlier, a rule is constructed using if-then constructs. If certain conditions exist, then specific actions are taken or certain conclusions are reached. In an expert system for a weather forecasting operation, for example, the rules could state that if certain temperature patterns exist with a given barometric pressure and certain previous weather patterns over the last 24 hours, then a specific forecast will be made, including temperatures, cloud coverage, and the wind-chill factor. Rules are often combined with probabilities, such as if the weather has a particular pattern of trends, then there is a 65 percent probability that it will rain tomorrow. Likewise, rules relating data to conclusions can be developed for any knowledge base. Most expert systems prevent users from entering contradictory rules. Figure 11.5 shows the use of expert system rules in helping to determine whether a person should receive a mortgage loan from a bank. In general, as the number of rules an expert system knows increases, the precision of the expert system increases.

The Use of Cases As mentioned previously, an expert system can use cases in developing a solution to a current problem or situation. This process involves (1) finding cases stored in the knowledge base that are similar to the problem or situation at hand and (2) modifying the solutions to the cases to fit or accommodate the current problem or situation. Cases stored in the knowledge base can be identified and selected by comparing the parameters of the new problem with the cases stored in the computer system. For example, a company may be using an

FIGURE 11.5

Rules for a Credit Application

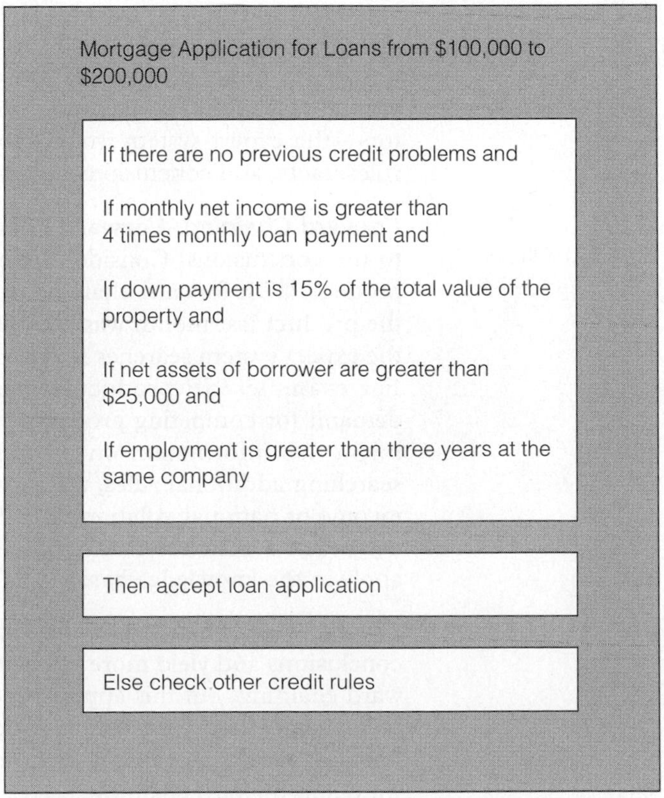

Mortgage Application for Loans from $100,000 to $200,000

If there are no previous credit problems and

If monthly net income is greater than 4 times monthly loan payment and

If down payment is 15% of the total value of the property and

If net assets of borrower are greater than $25,000 and

If employment is greater than three years at the same company

Then accept loan application

Else check other credit rules

expert system to determine the best location of a new service facility in the state of New Mexico. Labor and transportation costs may be the most important factors. The expert system may identify two previous cases involving the location of a service facility where labor and transportation costs were also important—one in the state of Colorado and the other in the state of Nevada. The expert system will modify the solution to these two cases to determine the best location for a new facility in New Mexico. The result might be to locate the new service facility in the city of Santa Fe.

The Inference Engine

The overall purpose of an **inference engine** is to seek information and relationships from the knowledge base and to provide answers, predictions, and suggestions the way a human expert would. In other words, the inference engine is the component that delivers the expert advice.

The process of retrieving relevant information and relationships from the knowledge base is not simple. As you have seen, the knowledge base is a collection of facts, interpretations, and rules. The inference engine must find the right facts, interpretations, and rules and assemble them correctly. In other words, the inference engine must make logical sense out of the information contained in the knowledge base, the way the human mind does when sorting out a complex situation. The inference engine has a number of ways of accomplishing its tasks, including backward and forward chaining.

inference engine
Part of the expert system that seeks information and relationships from the knowledge base and provides answers, predictions, and suggestions the way a human expert would

Backward Chaining **Backward chaining** is the process of starting with conclusions and working backward to the supporting facts. If the facts do not support the conclusion, another conclusion is selected and tested. This process is continued until the correct conclusion is identified. Consider an expert system that forecasts product sales for next month. With backward chaining, we start with a conclusion, such as "Sales next month will be 25,000 units." Given this conclusion, the expert system searches for rules in the knowledge base that support the conclusion, such as "IF sales last month were 21,000 units and sales for competing products were 12,000 units, THEN sales next month should be 25,000 units or greater." The expert system verifies the rule by checking sales last month for the company and its competitors. If the facts are not true—in this case, if last month's sales were not 21,000 units or 12,000 units for competitors—the expert system would start with another conclusion and proceed until rules, facts, and conclusions match.

backward chaining
The process of starting with conclusions and working backward to the supporting facts

Forward Chaining **Forward chaining** starts with the facts and works forward to the conclusions. Consider the expert system that forecasts future sales for a product. With forward chaining, we start with a fact, such as "The demand for the product last month was 20,000 units." With the forward-chaining approach, the expert system searches for rules that contain a reference to product demand. For example, "IF product demand is over 15,000 units, THEN check the demand for competing products." As a result of this process, the expert system might use information on the demand for competitive products. Next, after searching additional rules, the expert system might use information on personal income or national inflation rates. This process continues until the expert system can reach a conclusion using the data supplied by the user and the rules that apply in the knowledge base.

forward chaining
The process of starting with the facts and working forward to the conclusions

Comparison of Backward and Forward Chaining Forward chaining can reach conclusions and yield more information with fewer queries to the user than backward chaining, but this approach requires more processing and a greater degree of sophistication. Forward chaining is often used by more expensive expert systems. Some systems also use mixed chaining, which is a combination of backward and forward chaining.

The Explanation Facility

explanation facility
Component of an expert system that allows a user or decision maker to understand how the expert system arrived at certain conclusions or results

An important part of an expert system is the **explanation facility**, which allows a user or decision maker to understand how the expert system arrived at certain conclusions or results. A medical expert system, for example, may have reached the conclusion that a patient has a defective heart valve given certain symptoms and the results of tests on the patient. The explanation facility allows a doctor to find out the logic or rationale of the diagnosis made by the expert system. The expert system, using the explanation facility, can indicate all the facts and rules that were used in reaching the conclusion. This facility allows doctors to determine whether the expert system is processing the data and information correctly and logically.

The Knowledge Acquisition Facility

A difficult task in developing an expert system is the process of creating and updating the knowledge base. In the past, when more traditional programming languages were used, developing a knowledge base was tedious and time-consuming. Each fact, relationship, and rule had to be programmed into the knowledge base. In most cases, an experienced programmer had to create and update the knowledge base.

knowledge acquisition facility
Part of the expert system that provides convenient and efficient means of capturing and storing all the components of the knowledge base

Today, specialized software allows users and decision makers to create and modify their own knowledge bases through the knowledge acquisition facility (Figure 11.6). The overall purpose of the **knowledge acquisition facility** is to provide a convenient and efficient means for capturing and storing all components of the knowledge base. Knowledge acquisition software can present users and decision makers with easy-to-use menus. After filling in the appropriate attributes, the knowledge acquisition facility correctly stores information and relationships in the knowledge base, making the knowledge base easier and less expensive to set up and maintain. Knowledge acquisition can be a manual process or a mixture of manual and automated procedures. Regardless of how the knowledge is acquired, it is important to validate and update the knowledge base frequently to make sure it is still accurate.

The User Interface

Specialized user interface software is employed for designing, creating, updating, and using expert systems. The main purpose of the user interface is to make the development and use of an expert system easier for users and decision makers. At one time, skilled computer personnel created and operated most expert systems; today, the user interface permits decision makers to develop and use their own expert systems. Because expert systems place more emphasis on directing user activities than do other types of systems, text-oriented user interfaces (using menus, forms, and scripts) may be more common in expert systems than the graphical interfaces often used with DSSs.

EXPERT SYSTEMS DEVELOPMENT

Like other computer systems, expert systems require a systematic development approach for best results (Figure 11.7). This approach includes determining the requirements for the expert system, identifying one or more experts in the area or discipline under investigation, constructing the components of the expert system, implementing the results, and maintaining and reviewing the complete system.

FIGURE 11.6

The knowledge acquisition facility acts as an interface between experts and the knowledge base.

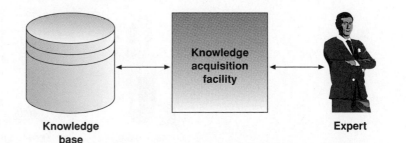

Knowledge base

Knowledge acquisition facility

Expert

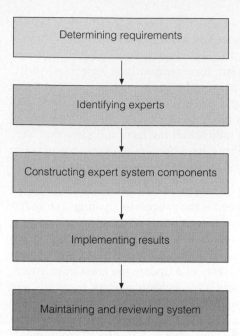

FIGURE 11.7

Steps in the Expert System Development Process

domain expert
The individual or group who has the expertise or knowledge one is trying to capture in the expert system

knowledge engineer
An individual who has training or experience in the design, development, implementation, and maintenance of an expert system

FIGURE 11.8

Participants in Expert Systems Development and Use

The Development Process

Specifying the requirements for an expert system begins with identifying the system's objectives and its potential use. Identifying experts can be difficult. In some cases, a company will have human experts on hand; in other cases, experts outside the organization will be required. Developing the expert system components requires special skills. Implementing the expert system involves placing it into action and making sure it operates as intended. Like other computer systems, expert systems should be periodically reviewed and maintained to make sure they are delivering the best support to decision makers and users.

Many companies are only now beginning to use and develop expert systems. Expert system development is a team effort, but experienced personnel and users may be in high demand within an organization. Because development can take months or years, the cost of bringing in consultants for development can be high. It is critical, therefore, to find and assemble the right people to assist with development.

Participants in Developing and Using Expert Systems

Typically, several people are involved in developing and using an expert system (Figure 11.8).

The Domain Expert Because of the time and effort involved in the task, an expert system is developed to address only a specific area of knowledge. This area of knowledge is called the domain. The **domain expert** is the individual or group who has the expertise or knowledge one is trying to capture in the expert system. In most cases, the domain expert is a group of human experts. The domain expert (individual or group) usually has the ability to do the following:

- Recognize the real problem
- Develop a general framework for problem solving
- Formulate theories about the situation
- Develop and use general rules to solve a problem
- Know when to break the rules or general principles
- Solve problems quickly and efficiently
- Learn from past experience
- Know what is and is not important in solving a problem
- Explain the situation and solutions of problems to others

The Knowledge Engineer and Knowledge Users A **knowledge engineer** is an individual who has training or experience in the design, development, implementation, and maintenance of an expert system, including training or experience with

Domain expert   Knowledge engineer   Knowledge user

knowledge user
The individual or group who uses
and benefits from the expert system

expert system shells. The **knowledge user** is the individual or group who uses
and benefits from the expert system. Knowledge users do not need any previous
training in computers or expert systems.

Expert Systems Development Tools and Techniques

Theoretically, expert systems can be developed from any programming language.
Since the introduction of computer systems, programming languages have become
easier to use, more powerful, and increasingly able to handle specialized requirements. In the early days of expert systems development, traditional high-level languages, including Pascal, FORTRAN, and COBOL, were used (Figure 11.9).
LISP was one of the first special languages developed and used for artificial intelligence applications. PROLOG, a more recent language, was also developed for AI
applications. Since the 1990s, however, other expert system products (such as
shells) are available that remove the burden of programming, allowing nonpro-
grammers to develop and benefit from the use of expert systems.

Expert System Shells and Products As discussed, an expert system shell is a
collection of software packages and tools used to design, develop, implement,
and maintain expert systems. Expert system shells exist for both personal computers and mainframe systems. Some shells are inexpensive, costing less than
$500. In addition, off-the-shelf expert system shells are available that are com-
plete and ready to run. The user enters the appropriate data or parameters, and
the expert system provides output to the problem or situation.

There are a number of expert system shells and products. AiroPeek, for
example, is an expert system shell that analyzes LAN networks.[25] Aircuity, Inc.,
has developed an expert system shell to monitor air quality in commercial buildings.[26] RigInsight System is an expert system shell that analyzes oil and drilling
operations.[27] Cycorp, Inc., is planning to offer an open-source version of its Cyc
knowledge base to the general public.[28] The knowledge base was assembled over
almost 20 years by industry experts. Anyone using the Cyc knowledge base, however, is asked to share the general knowledge they added with others. Other
expert system shells are summarized in Table 11.2.

FIGURE 11.9

*Software for expert systems devel-
opment has evolved greatly since
1980, from traditional programming
languages to expert system shells.*

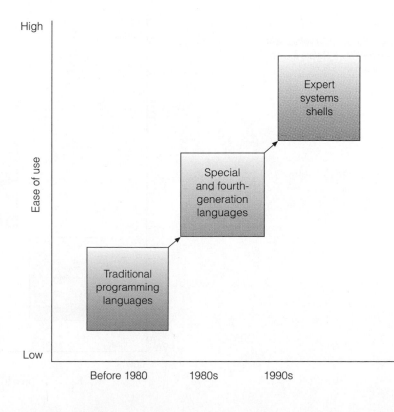

Name of Shell	Application and Capabilities
Financial Advisor	Analyzes financial investments in new equipment, facilities, and the like; requests the appropriate data and performs a complete financial analysis.
G2[29]	Assists in oil and gas operations. Transco, a British company, uses it to help in transport of gas to more than 20 million commercial and domestic customers.
RAMPART[30]	Analyzes risk. The U.S. General Services Administration uses it to analyze risk to the approximately 8,000 federal buildings it manages.
HazMat Loader[31]	Analyzes hazardous materials in truck shipments. According to Pennsylvania State Senator Roger Madigan, "September 11 has brought to the consciousness of the American public the risks and hazards that are part of our national transportation system."
MindWizard	Enables development of compact expert systems ranging from simple models that incorporate their business decision rules to highly sophisticated models; PC based and inexpensive.
LSI Indicator[32]	Helps determine property values; developed by one of the largest residential title and closing companies.

TABLE 11 2

Popular Expert System Shells

Expert Systems Development Alternatives

Expert systems can be developed from scratch by using an expert system shell or by purchasing an existing expert system package. A graph of the general cost and time of development alternatives is shown in Figure 11.10. It is usually faster and less expensive to develop an expert system using an existing package or an expert system shell. Note that there will be an additional cost of developing an existing package or acquiring an expert system shell if the organization does not already have this type of software.

In-House Development: Develop from Scratch Developing an expert system from scratch is usually more costly than the other alternatives, but an organization has more control over the features and components of the system. Such

FIGURE 11 10

Some Expert Systems Development Alternatives and Their Relative Cost and Time Values

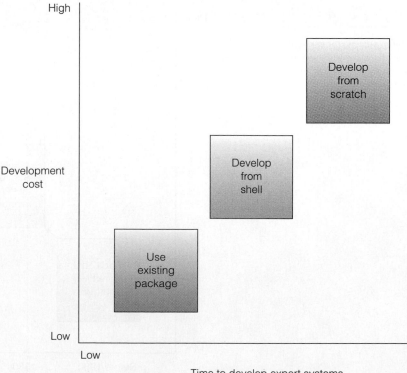

customization also has a downside; it can result in a more complex system, with higher maintenance and updating costs.

In-House Development: Develop from a Shell As you have seen, an expert system shell consists of one or more software products that assist in the development of an expert system. In some instances, the same shell can be used to develop many expert systems. Developing an expert system from a shell can be less complex and easier to maintain than developing one from scratch. However, the resulting expert system may need to be modified to tailor it to specific applications. In addition, the capabilities and features of the expert system can be more difficult to control.

Off-the-Shelf Purchase: Use Existing Packages Using an existing expert system package is the least expensive and fastest approach, in most cases. An existing expert system package is one that has been developed by a software or consulting company for a specific field or area, such as the design of a new computer chip or a weather forecasting and prediction system. The advantages of using an existing package can go beyond development time and cost. These systems can also be easy to maintain and update over time. A disadvantage of using an off-the-shelf package is that it may not be able to satisfy the unique needs or requirements of an organization.

APPLICATIONS OF EXPERT SYSTEMS AND ARTIFICIAL INTELLIGENCE

Expert systems and artificial intelligence are being used in a variety of ways. Some of the applications of these systems are summarized next.

Credit granting and loan analysis. Many banks employ expert systems to review an individual's credit application and credit history data from credit bureaus to make a decision on whether to grant a loan or approve a transaction. KPMG Peat Marwick uses an expert system called Loan Probe to review its reserves to determine whether sufficient funds have been set aside to cover the risk of some uncollectible loans.

Stock Picking. Some expert systems are used to help investment professionals pick stocks and other investments. Read the "IS Principles in Action" box to see how JJX Capital uses a variety of artificial intelligence and expert systems tools to analyze over 700 indicators for stocks and bonds.

Catching cheats and terrorists. Some gambling casinos use an expert system called NORA (Non-Obvious Relationship Awareness) from Systems Research & Development (SRD) to catch gambling cheaters, as well as employees involved in collusion.[33] NORA finds connections between people and organizations that are both obvious and nonobvious. For example, an arrested slot cheater may be found to have lived at the same address as the slot shift supervisor. In-Q-Tel, the venture capital arm of the Central Intelligence Agency, invested in SRD because it determined the NORA technology had potential national security benefits.

Budgeting. Ford Motor Company uses Prototype Optimization Model (POM) to help it budget, plan, and coordinate its prototype testing program.[34] The expert system should save the company more than $250 million in prototype vehicle costs annually.

Games. Some expert systems are used for entertainment. For example, Proverb is an expert system designed to solve standard American crossword puzzles given the grid and clues.

Writing. Expert systems, such as E-rater by Educational Testing Service, can evaluate or rate writing.[35] The Graduate Management Admissions Test requires two essays. In the past the essays were read and evaluated by a human reader. Today, E-rater and a human reader evaluate the essays.

NORA (Non-Obvious Relationship Awareness) from Systems Research & Development is a program that can ingest 100 million rows of data and produce an alpha page to a beeper, warning security personnel of connections between people and organizations that are both obvious and nonobvious.

(Source: Courtesy of Systems Research & Development.)

PRINCIPLE

⊙ *Expert systems can enable a novice to perform at the level of an expert but must be developed and maintained very carefully.*

Beating Wall Street with an Intelligent Supercomputer

John Fitzpatrick has purchased the most powerful supercomputer ever built for commercial use. It has one purpose: to beat Wall Street. Fitzpatrick is president of Cayman Islands–based JJX Capital. He and his associates believe their machine will help them predict, with unprecedented speed and accuracy, the future price movements of every stock, bond, and commodity traded in the United States.

Professional money managers who invest on behalf of banks, insurance companies, and investment brokerages have different systems for choosing what investments to make and when to enter and exit the market to increase profit or decrease loss. Each professional uses varying amounts of computer power to help make these decisions on a daily basis.

Although each professional has a different system, all draw from the same set of 700+ technical indicators to analyze the markets and make investment decisions. Some managers choose 5 of the 700; some choose 10; some choose 60 or 70. Each manager is limited in his or her selections only by the amount of computer power the company uses.

JJX Capital's supercomputer will be able to use all 700+ indicators all the time, reducing investment risk by reducing unknowns in decision making. The system currently allows JJX to exceed the technical ability of every fund manager in the world.

"What we're talking about here is doing investment strategy optimizations that would take a matter of hours or days on a PC, being able to have them done in a matter of seconds, and do that not just for one stock but the whole market," says Steve Ward, CEO of Frederick, Maryland–based Ward Systems, which is providing artificial intelligence software for the project. The software incorporates a combination of artificial intelligence technologies. Drawing from the knowledge of financial experts, it incorporates fuzzy logic, neural networks, and genetic algorithms to help predict the performance of an investment.

The supercomputer on which the AI system runs consists of 2,048 parallel 1-GHz processors, linked to a massive shared memory warehouse. With a capacity to conduct 2 trillion floating point operations per second (2 teraflops), its creators say the machine will race through calculations almost three times faster than the next-fastest commercially dedicated supercomputer, which is owned by San Francisco–based brokerage Charles Schwab.

Skeptics argue that reliably predicting the movements of even a single stock is beyond the capabilities of any computer or software today. Dr. David J. Leinweber, a visiting professor at the California Institute of Technology and former partner and managing director of investment firm First Quadrant, based in Pasadena, California, states, "On a daily basis, things like artificial intelligence and supercomputers have some incremental value. But the market is the reflection of not only underlying economic laws but also of human behavior; and until someone can explain all aspects of human behavior, you can't accurately predict the market."

John Fitzpatrick is gambling that human behavior, at least in the stock market, can be predicted. If he's right, the future of investing will become a battle of supercomputers and AI systems with the spoils of victory awarded to the investor with the biggest and smartest machine.

Discussion Questions

1. If Fitzpatrick is right, and his supercomputer beats the stock market, how will investment strategies be affected? Will life change for the thousands of money managers making their living predicting the market? If so, how?
2. The more complex the models become in financial and statistical analysis applications, the longer they take to run on conventional computer systems. What types of business activities would profit from the use of supercomputers?

Critical Thinking Questions

3. As we increasingly depend on expert systems to handle complex processing and direct our decision making, we reap benefits balanced with risks. What risks do we face when we trust an expert system to guide our investment strategies? What can be done to minimize those risks?
4. Do you agree with Leinweber and other skeptics who say that reliably predicting the movements of even a single stock is beyond the capabilities of any computer or software today? On what do you base your opinion?

Sources: Joe Ashbrook Nickell, "Crunching for Dollars: AI Takes Aim at Wall Street," *Technology Review*, April 19, 2002, http://www.techreview.com/articles/wo_nickell041902.asp; JJX Capital Web site, http://www.jjxcapital.com, Ward Systems Web site, http://www.wardsystems.com, and NeuroShell Trader Software Web site, http://www.neuroshell.com, all accessed June 2002.

Information management and retrieval. The explosive growth of information available to decision makers has created a demand for devices to help manage the information. Expert systems can aid this process through the use of bots. Businesses might use a bot to retrieve information from large distributed databases or a vast network like the Internet. Expert system agents help managers find the right data and information while filtering out irrelevant facts that might impede timely decision making.

AI and expert systems embedded in products. The antilock braking system on modern automobiles is an example of a rudimentary expert system. A processor senses when the tires are beginning to skid and releases the brakes for a fraction of a second to prevent the skid. AI researchers are also finding ways to use neural networks and robotics in everyday devices, such as toasters, alarm clocks, and televisions.

Plant layout. FLEXPERT is an expert system that uses fuzzy logic to perform plant layout. The software helps companies determine the best placement for equipment and manufacturing facilities.

Hospitals and medical facilities. Some hospitals use expert systems to determine a patient's likelihood of contracting cancer or other diseases. Hospitals, pharmacies, and other healthcare providers can use CaseAlert by MEDecision to determine possible high-risk or high-cost patients.[36] MYCIN is an expert system developed at Stanford University to analyze blood infections. UpToDate is another expert system used to diagnose patients.[37] According to one physician, "I could go on the Web and click around the medical literature and try to find answers, but UpToDate is so current and well-written, I can just whip out my laptop and find what I need." A medical expert system used by the Harvard Community Health Plan allows members of the HMO to get medical diagnoses via home personal computers. For minor problems, the system gives uncomplicated treatments; for more serious conditions, the system schedules appointments. The system is highly accurate, diagnosing 97 percent of the patients correctly (compared with the doctors' 78 percent accuracy rating).

In order to help doctors in the diagnosis of thoracic pain MatheMEDics has developed THORASK, a straightforward, easy-to-use program, requiring only the input of carefully obtained clinical information. The program helps the less experienced to distinguish the three principal categories of chest pain from each other. It does what a true medical expert system should do without the need for complicated user input. You answer basic questions about the patient's history and directed physical findings, and the program immediately displays a list of diagnoses. The diagnoses are presented in decreasing order of likelihood, together with their estimated probabilities. The program also provides concise descriptions of relevant clinical conditions and their presentations, as well as brief suggestions for diagnostic approaches. For purposes of record keeping, documentation, and data analysis, there are options for saving and printing cases.

Help desks and assistance. Customer service help desks use expert systems to provide timely and accurate assistance. Kaiser Permanente, a large HMO, uses an expert system and voice response to automate its help desk function. The automated help desk frees up staff to handle more complex needs, while still providing more timely assistance for routine calls.

Employee performance evaluation. An expert system developed by Austin-Hayne, called Employee Appraiser, provides managers with expert advice for use in employee performance reviews and career development. Expert systems software can also analyze an employee's strengths by analyzing the e-mail messages he or she sends.[38]

Virus detection. IBM is using neural network technology to help create more advanced software for eradicating computer viruses, a major problem in American businesses. IBM's neural network software deals with "boot sector" viruses, the most prevalent type, using a form of artificial intelligence that mimics the human brain and generalizes by looking at examples. It requires a vast number of training samples, which in the case of antivirus software are three-byte virus fragments.

Repair and maintenance. ACE is an expert system used by AT&T to analyze the maintenance of telephone networks. IET-Intelligent Electronics uses an expert system to diagnose maintenance problems related to aerospace equipment. General Electric Aircraft Engine Group uses an expert system to enhance maintenance performance levels at all sites and improve diagnostic accuracy.

Shipping. CARGEX-Cargo Expert System is used by Lufthansa, a German airline, to help determine the best shipping routes.

Marketing. CoverStory is an expert system that extracts marketing information from a database and automatically writes marketing reports.

Warehouse optimization. United Distillers uses an expert system to determine the best combinations of liquor stocks to produce its blends of Scottish whiskey. This information is then supplemented with information about location of the casks for each blend. The system optimizes the selection of required casks, keeping to a minimum the number of "doors" (warehouse sections) from which the casks must be taken and the number of casks that need to be moved to clear the way. Other constraints must be satisfied, such as the current working capacity of each warehouse and the maintenance and restocking work that may be in progress.

INTEGRATING EXPERT SYSTEMS

As with the other information systems, an expert system can be integrated with other systems in an organization through a common database. An expert system that identifies late-paying customers who should not receive additional credit may draw data from the same database as an invoicing MIS that produces weekly reports on overdue bills. The same database—a by-product of the invoicing transaction processing system—might also be used by a decision support system to perform "what-if" analysis to determine the impact of late payments on cash flows, revenues, and overall profit levels.

In many organizations, these systems overlap. A TPS might be expanded to provide management information, which in turn may provide some DSS functions, and so on. In each progressive phase of this overlap, the information system assists with the decision-making process to a greater extent. Of all these information systems, expert systems provide the most support, proposing decisions based on specific problem data and a knowledge base. Understanding the capabilities and characteristics of expert systems is the first step in applying these systems to support managerial decision making and organizational goals.

⦿ VIRTUAL REALITY

The term *virtual reality* was initially coined by Jason Lanier, founder of VPL Research, in 1989. Originally, the term referred to *immersive virtual reality* in which the user becomes fully immersed in an artificial, three-dimensional world that is completely generated by a computer. Immersive virtual reality may represent any three-dimensional setting, real or abstract, such as a building, an archaeological excavation site, the human anatomy, a sculpture, or a crime scene reconstruction. Through immersion, the user can gain a deeper understanding of the virtual world's behavior and functionality.

virtual reality system
A system that enables one or more users to move and react in a computer-simulated environment

A **virtual reality system** enables one or more users to move and react in a computer-simulated environment. Virtual reality simulations require special interface devices that transmit the sights, sounds, and sensations of the simulated world to the user. These devices can also record and send the speech and movements of the participants to the simulation program, enabling users to sense and manipulate virtual objects much as they would real objects. This natural style of interaction gives the participants the feeling that they are immersed in the simulated world.

INTERFACE DEVICES

To see in a virtual world, often the user will wear a head-mounted display (HMD) with screens directed at each eye. The HMD also contains a position tracker to monitor the location of the user's head and the direction in which the user is looking. Using this information, a computer generates images of the virtual world—a slightly different view for each eye—to match the direction the user is looking and displays

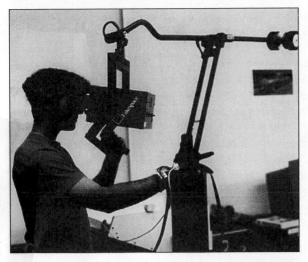

The BOOM, a head-coupled display device.

(Source: Courtesy of University of Michigan Virtual Reality Laboratory.)

these images on the HMD. With current technology, virtual-world scenes must be kept relatively simple so that the computer can update the visual imagery quickly enough (at least ten times a second) to prevent the user's view from appearing jerky and from lagging behind the user's movements.

Alternative concepts—BOOM and CAVE—were developed for immersive viewing of virtual environments to overcome the often uncomfortable intrusiveness of a head-mounted display. The BOOM (Binocular Omni-Orientation Monitor) from Fakespace Labs is a head-coupled stereoscopic display device. Screens and optical systems are housed in a box that is attached to a multilink arm. The user looks into the box through two holes, sees the virtual world, and can guide the box to any position within the virtual environment. Head tracking is accomplished via sensors in the links of the arm that holds the box.

The Electronic Visualization Laboratory at the University of Illinois at Chicago introduced a room constructed of large screens on which the graphics are projected onto the three walls and the floor. The CAVE, as this room is called, provides the illusion of immersion by projecting stereo images on the walls and floor of a room-sized cube. Several persons wearing lightweight stereo glasses can enter and walk freely inside the CAVE. A head-tracking system continuously adjusts the stereo projection to the current position of the leading viewer.

Users hear sounds in the virtual world through earphones. The information reported by the position tracker is also used to update audio signals. When a sound source in virtual space is not directly in front of or behind the user, the computer transmits sounds to arrive at one ear a little earlier or later than at the other and to be a little louder or softer and slightly different in pitch.

The *haptic* interface, which relays the sense of touch and other physical sensations in the virtual world, is the least developed and perhaps the most challenging to create. Currently, with the use of a glove and position tracker, the computer locates the user's hand and measures finger movements. The user can reach into the virtual world and handle objects; however, it is difficult to generate the sensations that are felt when a person taps a hard surface, picks up an object, or runs a finger across a textured surface. Touch sensations also have to be synchronized with the sights and sounds users experienced.

FORMS OF VIRTUAL REALITY

Viewing the Detroit Midfield Terminal in an immersive CAVE system.

(Source: Courtesy of University of Michigan Virtual Reality Laboratory.)

Aside from immersive virtual reality, which we just discussed, virtual reality can also refer to applications that are not fully immersive, such as mouse-controlled navigation through a three-dimensional environment on a graphics monitor, stereo viewing from the monitor via stereo glasses, stereo projection systems, and others. Actuality Systems makes what some think is the world's highest-resolution 3-D images.[39] The system uses a clear crystal that is about 20 inches in diameter. Images are displayed inside the crystal, which you can walk around to get different views.

On Web sites, businesses, such as software and communications companies, are using *avatars,* or virtual people, to bring their Web sites to life. As we saw in the opening vignette, avatars can be useful for customer service, allowing companies to personalize their sites and provide assistance whenever it is needed.

Ryan Patterson, a 17-year-old high school student in Colorado, developed a virtual reality glove that translates sign language into text on a thin, portable computer screen.[40] The young inventor has received hundreds of thousands of dollars and numerous scholarships for his inventions.

Some virtual reality applications allow views of real environments with super-imposed virtual objects. Motion trackers monitor the movements of dancers or athletes for subsequent studies in immersive virtual reality. Telepresence systems (e.g., telemedicine, telerobotics) immerse a viewer in a real world that is captured by video cameras at a distant location and allow for the remote manipulation of real objects via robot arms and manipulators. Many believe that virtual reality will reshape the interface between people and information technology by offering new ways to communicate information, visualize processes, and express ideas creatively.

USEFUL APPLICATIONS

There are literally hundreds of applications of virtual reality, with more being developed as the cost of hardware and sofware declines and people's imaginations are opened to the potential of virtual reality. Here is a summary of some of the more interesting applications.

Medicine

Surgeons in France performed the first successful closed-chest coronary bypass operation. Instead of cutting open the patient's chest and breaking his breast-bone, as is usually done, surgeons used a virtual reality system that enabled them to operate through three tiny half-inch holes between the patient's ribs. They inserted thin tubes to tunnel to the operating area and protect the other body tissue. Then three arms were inserted into the tubes. One was for a 3-D camera; the other two held tiny artificial wrists to which a variety of tools—scalpels, scissors, needle—were attached. The virtual reality system mimicked the movements of the surgeon's shoulders, elbows, and wrists. The surgeon sits at a computer work-station several feet from the operating table and moves instruments that control the ones inside the patient. The instruments inside the patient mimic the motion of the surgeon's hands so accurately that it is possible to sew up a coronary artery as thin as a thread. The surgeon watches his progress on a screen that enlarges the artery in 3-D to the size of a garden hose.

In yet another medical application of virtual reality, researchers are trying to translate biomechanical measurements so that they can be stored in a computer database. Data from actual patients will be gathered from sensors mounted on the fingertips of virtual-reality gloves to create a database that will be made available to examining physicians. Upon examination of a new patient, a feedback system hooked up to the virtual reality gloves would capture sensations felt during the examination and compare them with data in the database. This approach could allow surgeons to go right into the operating room without having to obtain a CAT scan, thus saving time—and with many injuries, that's absolutely critical.

Virtual reality technology can also be used to link stroke patients to physical therapists.[41] Patients put on special gloves and other virtual reality devices at home that are linked to the physical therapist's office. The physical therapist can see whether the patient is performing the correct exercises without having to travel to the patient's home or hospital room. Using virtual reality can cut costs and motivate patients to exercise regularly. According to Grigore Burdea, Director of Rutgers University's Center for Advanced Information Processing, "There's tremendous potential here to have the therapy monitored at a distance and to have it done at home."

Education and Training

Virtual environments are used in education to bring exciting new resources into the classroom. Students can stroll among digital bookshelves, learn anatomy on a simulated cadaver, or participate in historical events, all in virtual reality.

Third-grade students at John Cotton Tayloe School in Washington, North Carolina, are able to take an exciting virtual trip down the Nile for an integrated-curriculum lesson on ancient Egypt. This interactive virtual reality computer lesson integrates social studies, geography, music, art, science, math, and language arts. The software used to design the lesson was 3-D Website Builder (created by the Virtus Corporation), which allowed the designers to create a desert landscape complete with an oasis, camels, and a pyramid. Students could view the scenes from all angles, including front and top views, and could even enter a pyramid and view the sarcophagus holding the mummy in the middle of a room containing different Egyptian items. On one wall was a hieroglyphic message that was part of the lesson. On another wall was artwork depicting life in ancient Egypt, and against another were pieces of Egyptian furniture and a harp.

Virtual technology has also been applied by the military. To help with aircraft maintenance, a virtual reality system has been developed to simulate an aircraft and give a user a sense of touch, while computer graphics give the senses of sight and sound. The user sees, touches, and manipulates the various parts of the virtual aircraft during training. The Virtual Aircraft Maintenance System simulates real-world maintenance tasks that are routinely performed on the AV8B vertical takeoff and landing aircraft used by the U.S. Marines. The Pentagon is using a virtual reality training lab to prepare for a military crisis.[42] The virtual reality system, developed by media giant Viacom for Hollywood productions, helps with war scenarios. Hollywood screen writers and computer scientists are also helping through the Institute of Creative Technologies, which was launched with a $45 million grant from the Army, to develop immersive virtual reality simulations for the military.

Real Estate Marketing and Tourism

Virtual reality has been used to increase real estate sales in several powerful ways. From Web publishing to laptop display to a potential buyer, virtual reality provides excellent exposure for properties and attracts potential clients. Clients can take a virtual walk through properties and eliminate wrong choices without wasting valuable time. Virtual walk-throughs can be mailed on diskettes or posted on the Web as a convenience for nonlocal clients. A CD-ROM containing all virtual reality homes can also be sent to clients and other agents. Realatrends Real Estate Service offering homes for sale in Orange County, California (http://www.realatrends.com/virtual_tours.htm), is just one of many real estate firms offering this service. In another Web application, the U.S. government created a virtual tour of the White House while the facility was closed for security concerns.[43] The virtual tour allowed people to see a 360-degree view of rooms on the Internet. According to Ari Fleisher, White House spokesman, "It's pretty remarkable what technology allows on the computer screen."

Computer-generated image technology is used widely in sports simulations. Shown here is Microsoft's Links 2003.

(Source: Courtesy of Microsoft Corporation.)

Entertainment

Computer-generated image technology, or CGI, has been around since the 1970s. A number of movies used this technology to bring realism to the silver screen, including *Star Wars Episode II—Attack of the Clones*. In other settings, a team of artists rendered the roiling seas and crashing waves of *Perfect Storm* almost entirely on computers using weather reports, scientific formulas, and their imagination. There was also *Dinosaur* with its realistic talking reptiles, *Titan A.E.*'s beautiful 3-D space-scapes, and the casts of computer-generated crowds and battles in *Gladiator* and *The Patriot*. CGI can also be used for sports simulation to enhance the viewers' knowledge and enjoyment of a game.

The rock group Aerosmith plans to update its Web site with 3-D images of members of the band.[44] These 3-D avatars on the Web will allow people to hear guitar riffs and see bands perform as if they were at a rock concert.

OTHER SPECIALIZED SYSTEMS

In addition to artificial intelligence, expert systems, and virtual reality, a number of other interesting and exciting specialized systems have appeared. In 2001, for example, the Segway personal transporter received international news coverage.[45] After much speculation about its purpose and practicality—the system was kept top secret during development—the electric scooter made its debut. The technology used to keep the scooter upright is impressive. Using sophisticated software, sensors, and gyro motors, the device can transport standing people through warehouses, offices, and downtown sidewalks. It is being tested for postal delivery and could be used for short commutes. The inventor, Dean Kamen, has other inventions to his credit—he developed the first wearable kidney dialysis pump for home use and the iBOT, a computer-controlled wheelchair.

Another specialized system claims to use brain waves to control a computer.[46] It has been reported that wearing a red plastic cap embedded with electrodes, Cathal O'Philbin was able to concentrate and use willpower to enter three words into a computer screen: "Arsenal football club." Some believe that the technology, called *adaptive brain interface technology,* might eventually assist many disabled people. After the experiment, O'Philbin said, "I was totally exhausted at the end of it." It took about three and a half hours for Mr. O'Philbin to use his willpower to enter the three words into the computer.

Accenture Technology Labs is developing a small, wearable device called a *personal awareness assistant (PAA).*[47] The device can be used to continuously record interactions and store them for future recall. The PAA uses defined phrases, such as "nice to meet you," to trigger its recording system for conversations or its camera for a few pictures of an individual. With the PAA, you could record your entire life, store it electronically, and recall it at any time. Some critics, among them Chris Hoffnagle, legislative counsel for the Electronic Privacy Information Center, question whether such a device would be an invasion of privacy. Currently, the prototype PAA is stored in a fanny pack with a 400-MHz Pentium processor, 1-GB hard disk, and 256 MB of memory. Eventually, Accenture hopes the device can be made as small as a cell phone.

A number of special-purpose telecommunications systems are also being developed. Apparent Technologies, a company formed by Eastman Kodak, is developing small radio transceivers that can be placed in other products, like cell phones.[48] The radio transceivers will allow cell phones and other devices to connect to the Internet, cellular phone service, and other devices that use the technology. Associated Food Stores, for example, is using this technology to place tags on its produce.[49] The new system could save up to $100,000 per year by reducing the amount of spoiled produce. The system uses $50 radio frequency tags that are placed on trailers with adhesives or mechanical fasteners. The cost of the radio tags should be paid for in about a year.

Microsoft and Accenture are developing automotive telematics software and services to allow cars and trucks to be connected to the Internet and corporate networks.[50] The new telematics software will be able to track driver speed and location, allow gas stations to remotely charge for fuel and related services, and more. According to Umar Riaz of Accenture, "Everyone talks about consumer applications, but we believe the potential for telematics is just limitless. The

Segway, a human transport device, uses sophisticated software, sensors, and gyro motors to transport people in an upright position.

(Source: © Reuters NewMedia Inc./CORBIS.)

insurance industry is very interested in being able to monitor driving behavior and usage, which would enable them to price their premiums more accurately."

Special-purpose bar codes are also being introduced in a variety of settings.[51] In Accenture's Office of the Future, each consultant and office space has a bar code. Instead of having permanent offices, Accenture assigns consultants to offices as needed. The bar codes help to make sure that a consultant's work, mail, and other materials are delivered to the right Accenture office space when they are needed. Accenture estimates that it saves about $8 million a year in rent by not having permanent office spaces sitting idle for its consultants. "A can of peas has its own bar code, why not us?" says Doug Picker of Symbol Technologies, a company that makes mobile bar code systems.

Three-dimensional images have been in existence for decades. Older 3-D systems, however, typically require clumsy glasses. New display screens, by companies such as Dimension Technologies, are improving.[52] The new flat-panel displays beam two different images simultaneously. Instead of displaying the images side-by-side, the two images are shown in interleaving vertical strips. The results are promising. One moment you are looking at a normal display screen. The next moment, you are looking at objects in three dimensions that appear to be coming through the screen and into thin air. The new screens are impressive, but the image quality is slightly fuzzy and the viewing distance from the screen and viewing angle are important to get the best results. Experts believe that this newer type of screen will have important uses in industrial design, medicine, architecture, and games and simulations.

SUMMARY

PRINCIPLE *Artificial intelligence systems form a broad and diverse set of systems that can replicate human decision making for certain types of well-defined problems.*

The term *artificial intelligence* is used to describe computers with the ability to mimic or duplicate the functions of the human brain. The objective of building AI systems is not to replace human decision making completely but to replicate it for certain types of well-defined problems.

Intelligent behavior encompasses several characteristics including the abilities to learn from experience and apply this knowledge to new experiences; handle complex situations and solve problems for which pieces of information may be missing; determine relevant information in a given situation, think in a logical and rational manner, and give a quick and correct response; and understand visual images and processing symbols. Computers are better than humans at transferring information, making a series of calculations rapidly and accurately, and making complex calculations, but humans are better than computers at all other attributes of intelligence.

Artificial intelligence is a broad field that includes several key components, such as expert systems, robotics, vision systems, natural language processing, learning systems, and neural networks. An expert system consists of the hardware and software to produce systems that behave as a human expert would in a specialized field or area (e.g., credit analysis). Robotics uses mechanical or computer devices to perform tasks that require a high degree of precision or are tedious or hazardous for humans (e.g., stacking cartons on a pallet). Vision systems include hardware and software that permit computers to capture, store, and manipulate images and pictures (e.g., face-recognition software). Natural language processing allows the computer to understand and react to statements and commands made in a "natural" language, such as English. Learning systems use a combination of software and hardware to allow a computer to change how it functions or reacts to situations based on feedback it receives (e.g., a computerized chess game). A neural network is a computer system that can simulate the functioning of a human brain (e.g., disease diagnostics system).

PRINCIPLE *Expert systems can enable a novice to perform at the level of an expert but must be developed and maintained very carefully.*

Expert systems can explain their reasoning or suggested decisions, display intelligent behavior, manipulate symbolic information and draw conclusions from complex relationships, provide portable knowledge, and deal with uncertainty. They are not yet widely used; some are difficult to use, are limited to relatively narrow problems, cannot readily deal with mixed knowledge, present the possibility for error, cannot refine their own knowledge base, are difficult to maintain, and may have high development costs. Their use also raises legal and ethical concerns. The capabilities of an expert system include strategic goal setting, planning, design, decision making, quality control and monitoring, and diagnosis.

An expert system consists of a collection of integrated and related components, including a knowledge base, an inference engine, an explanation facility, a knowledge acquisition facility, and a user interface. The knowledge base is an extension of a database discussed in Chapter 5 and an information and decision support system discussed in Chapter 10. It contains all the relevant data, rules, and relationships used in the expert system. The rules are often composed of if-then statements, which are used for drawing conclusions. Fuzzy logic allows expert systems to incorporate facts and relationships into expert system knowledge bases that may be imprecise or unknown.

The inference engine processes the rules, data, and relationships stored in the knowledge base to provide answers, predictions, and suggestions the way a human expert would. Two common methods for processing include backward and forward chaining. Backward chaining starts with a conclusion, then searches for facts to support it; forward chaining starts with a fact, then searches for a conclusion to support it. Mixed chaining is a combination of backward and forward chaining.

The explanation facility of an expert system allows the user to understand what rules were used in arriving at a decision. The knowledge acquisition facility helps the user add or update knowledge in the knowledge base. The user interface makes it easier to develop and use the expert system.

The individuals involved in the development of an expert system include the domain expert, the knowledge engineer, and the knowledge users. The domain expert is the individual or group who has the expertise or knowledge being captured for the system. The knowledge engineer is the developer whose job is the extraction of the expertise from the domain expert. The knowledge user is the individual who benefits from the use of the developed system.

Following is a list of factors that normally make expert systems worth the expenditure of time and money: a high potential payoff or significantly reduced downside risk, the ability to capture and preserve irreplaceable human expertise, the ability to develop a system more consistent than human experts, expertise needed at a number of locations at the same time, and expertise needed in a hostile environment that is dangerous to human health. The expert system solution can be developed faster than the solution from human experts. An ES also provides expertise needed for training and development to share the wisdom and experience of human experts with a large number of people.

The steps involved in the development of an expert system include determining requirements, identifying experts, constructing expert system components, implementing results, and maintaining and reviewing the system.

Expert systems can be implemented in several ways. Previously, traditional high-level languages, including Pascal, FORTRAN, and COBOL, were used. LISP and PROLOG are two languages specifically developed for creating expert systems from scratch. A faster and less expensive way to acquire an expert system is to purchase an expert system shell or existing package. The shell program is a collection of software packages and tools used to design, develop, implement, and maintain expert systems. Advantages of expert system shells include ease of development and modification, use of satisficing, use of heuristics, and development by knowledge engineers and end users. The approach selected depends on the benefits compared with cost, control, and complexity considerations.

The benefits of using an expert system go beyond the typical reasons for using a computerized processing solution. Expert systems display "intelligent" behavior, manipulate symbolic information and draw conclusions, provide portable knowledge, and can deal with uncertainty. Expert systems can be used to solve problems in many fields or disciplines and can assist in all stages of the problem-solving process. Past successes have shown that expert systems are good at strategic goal setting, planning, design, decision making, quality control and monitoring, and diagnosis.

There are number of applications of expert systems and artificial intelligence, including credit granting and loan analysis, catching cheats and terrorists, budgeting, games, writing, information management and retrieval, AI and expert systems embedded in products, plant layout, hospitals and medical facilities, help desks and assistance, employee performance evaluation, virus detection, repair and maintenance, shipping, and warehouse optimization.

PRINCIPLE *Virtual reality systems have the potential to reshape the interface between people and information technology by offering new ways to communicate information, visualize processes, and express ideas creatively.*

A virtual reality system enables one or more users to move and react in a computer-simulated environment. Virtual reality simulations require special interface devices that transmit the sights, sounds, and sensations of the simulated world to the user. These devices can also record and send the speech and movements of the participants to the simulation program. Thus, users are able to sense and manipulate virtual objects much as they would real objects. This natural style of interaction gives the participants the feeling that they are immersed in the simulated world.

Virtual reality can also refer to applications that are not fully immersive, such as mouse-controlled navigation through a three-dimensional environment on a graphics monitor, stereo viewing from the monitor via stereo glasses, stereo projection systems, and others. Some virtual reality applications allow views of real environments with superimposed virtual objects. Virtual reality applications are found in medicine, education and training, real estate and tourism, and entertainment.

PRINCIPLE *Special-purpose systems can help organizations and individuals achieve their goals.*

A number of special-purpose systems have recently appeared to assist organizations and individuals in new and exciting ways. Segway, for example, is an electric scooter that uses sophisticated software, sensors, and gyro motors to transport people through warehouses, offices, downtown sidewalks, and through other spaces. Another specialized system uses brain waves to control a computer. Wearing a red plastic cap embedded with electrodes, one person was able to concentrate and use willpower to enter three words into a computer screen. A small, wearable device called a *personal awareness assistant (PAA)* can continuously record interactions and store them for future use. There are also a number of special-purpose telecommunications systems that can be placed in products for varied uses.

CHAPTER 11 SELF-ASSESSMENT TEST

Artificial intelligence systems form a broad and diverse set of systems that can replicate human decision making for certain types of well-defined problems.

1. The objective in developing contemporary AI systems is to completely replace human decision making. True False

2. _____ are rules of thumb arising from experience or even guesses.

3. What is an important attribute for artificial intelligence?
   A. The ability to use sensors
   B. The ability to learn from experience
   C. The ability to be creative
   D. The ability to acquire a large amount of external information

4. _____ involves mechanical or computer devices that can paint cars, make precision welds, and perform other tasks that require a high degree of precision or are tedious or hazardous for humans.

5. What branch of artificial intelligence involves a computer system that can simulate the functioning of a human brain?
   A. Expert systems
   B. Neural networks
   C. Natural language processing
   D. Vision systems

6. A(n) _____ is a combination of software and hardware that allows the computer to change how it functions or reacts to situations based on feedback it receives.

Expert systems can enable a novice to perform at the level of an expert but must be developed and maintained very carefully.

7. What is a disadvantage of an expert system?
   A. The inability to solve complex problems.
   B. The inability to deal with uncertainty
   C. The limitation of dealing with relatively narrow problems
   D. The inability to draw conclusions from complex relationships

8. A(n) _____ is a collection of software packages and tools used to develop expert systems that can be implemented on most popular PC platforms to reduce development time and costs.

9. The ability to perform strategic goal setting is one of the capabilities of an expert system. True False

10. What stores all relevant information, data, rules, cases, and relationships used by the expert system?
    A. The knowledge base
    B. The data interface
    C. The database
    D. The acquisition facility

11. A disadvantage of an expert system is the inability to provide expertise needed at a number of locations at the same time or in a hostile environment that is dangerous to human health. True False

12. What is NOT used in the development and use of expert systems?
    A. Fuzzy logic
    B. The use of rules
    C. The use of cases
    D. The use of natural language processing

13. An important part of an expert system is the _____, which allows a user or decision maker to understand how the expert system arrived at certain conclusions or results.

14. In an expert system, the domain expert is the individual or group who has the expertise or knowledge one is trying to capture in the expert system. True False

Virtual reality systems have the potential to reshape the interface between people and information technology by offering new ways to communicate information, visualize processes, and express ideas creatively.

15. A(n) _____ enables one or more users to move and react in a computer-simulated environment.

16. What type of virtual reality is used to make humans feel as though they are in a three-dimensional setting, such as a building, an archaeological excavation site, the human anatomy, a sculpture, or a crime scene reconstruction?
    A. Chaining
    B. Relative
    C. Immersive
    D. Visual

Special-purpose systems can help organizations and individuals achieve their goals.

17. A _____ is a small, wearable special-purpose device developed by Accenture Technology used to continuously record interactions and store them for future recall.

Chapter 11 Self-Assessment Test Answers

1. False, 2. Heuristics, 3. D, 4. Robotics, 5. B, 6. learning system, 7. C, 8. expert system shell, 9. True, 10. A, 11. False, 12. D, 13. explanation facility, 14. True, 15. virtual reality system, 16. C, 17. personal awareness assistant (PAA).

KEY TERMS

artificial intelligence systems, 462	if-then statements, 474	natural language processing, 468
backward chaining, 476	inference engine, 476	neural network, 469
domain expert, 478	intelligent behavior, 462	perceptive system, 463
expert system shells, 471	knowledge acquisition facility, 477	robotics, 465
explanation facility, 477	knowledge engineer, 478	rule, 475
forward chaining, 476	knowledge user, 479	virtual reality system, 484
fuzzy logic, 474	learning systems, 469	vision systems, 467

REVIEW QUESTIONS

1. Define the term *artificial intelligence*.
2. What is a vision system? Discuss two applications of such a system.
3. What is natural language processing? What are the three levels of voice recognition?
4. Describe three examples of the use of robotics. How can a microrobot be used?
5. What is a learning system? Give a practical example of such a system.
6. What is a neural network? Describe two applications of neural networks.
7. What is meant when it is said that neural networks learn to program themselves?
8. What are the capabilities of an expert system?
9. Under what conditions is the development of an expert system likely to be worth the effort?
10. Identify the basic components of an expert system and describe the role of each.
11. What are fuzzy sets and fuzzy logic?
12. How are rules used in expert systems?
13. Expert systems can be built based on rules or cases. What is the difference between the two?
14. Describe the roles of the domain expert, the knowledge engineer, and the knowledge user in expert systems.
15. What are the primary benefits derived from the use of expert systems?

16. Identify three approaches for developing an expert system.
17. Describe three applications of expert systems or artificial intelligence.
18. Identify three special interface devices developed for use with virtual reality systems.
19. Identify and briefly describe three specific virtual reality applications.
20. Give three examples of other special-purpose systems.

DISCUSSION QUESTIONS

1. What are the requirements for a computer to exhibit human-level intelligence? How long will it be before we have the technology to design such computers? Do you think we should push to try to accelerate such a development? Why or why not?
2. What are some of the tasks at which robots excel? Which human tasks are difficult for them to master? What fields of AI are required to develop a truly perceptive robot?
3. You have been hired to capture the knowledge of a brilliant attorney who has an outstanding track record for selecting jury members favorable to her clients during the pretrial jury selection process. This knowledge will be used as the basis for an expert system to enable other attorneys to have similar success. Is this system a good candidate for an expert system? Why or why not?
4. Briefly explain why human decision making often does not lead to optimal solutions to problems.
5. What is the purpose of a knowledge base? How is one developed?
6. What is the relationship between a database and a knowledge base?
7. Imagine that you are developing the rules for an expert system to select the strongest candidates for a medical school. What rules or heuristics would you include?
8. What skills does it take to be a good knowledge engineer? Would knowledge of the domain help or hinder the knowledge engineer in capturing knowledge from the domain expert?
9. Which interface is the least developed and most challenging to create in a virtual reality system? Why do you think this is so?
10. What application of virtual reality has the most potential to generate increased profits in the future?

PROBLEM-SOLVING EXERCISES

1. You are a senior vice-president of a company that manufacturers kitchen appliances. You are considering using robots to replace up to 10 of your skilled workers on the factory floor. Using a spreadsheet, analyze the costs of acquiring several robots to paint and assemble some of your products versus the cost savings in labor. How many years would it take to pay for the robots from the savings in fewer employees? Assume that the skilled workers make $20 per hour, including benefits.

2. Assume you live in an area where there is a wide variation in weather from day to day. Develop a simple expert system to provide advice on the type of clothes to wear based on the weather. The system needs to help you decide which clothes and accessories (umbrella, boots, etc.) to wear for sunny, snowy, rainy, hot, mild, or cold days. Key inputs to the system include last night's weather forecast, your observation of the morning temperature and clouds, yesterday's weather, and the activities you have planned for the day. Using your word processing program, create seven or more rules that could be used in such an expert system. Create five cases and use the rules you developed to determine the best course of action.

3. Using a graphics program, diagram the components of a virtual reality system that could be used to market real estate. Carefully draw and label each component. Use the same graphics program to make a one-page outline of a presentation to a real estate company interested in your virtual reality system.

TEAM ACTIVITIES

1. With two or three of your classmates, do research to identify three real examples of robotics in use. Discuss the problems solved by each of these systems. Which has the greatest potential for cost savings? What are the other advantages of each robotic system?
2. Form a team to debate other teams from your class on the following topic: "Are Expert Systems Superior to Humans When It Comes to Making Objective Decisions?" Develop several points supporting either side of the debate.
3. With members of your team, think of an idea for a virtual reality system for a new, exciting game. What are the main features of the game that make it unique and highly marketable?

WEB EXERCISES

1. Use the Internet to get information about fuzzy logic. Describe three examples of how this technology is used.
2. This chapter discussed several examples of expert systems. Search the Internet for two examples of the use of expert systems. Which one has the greatest potential to increase profits for the firm? Explain your choice.
3. Use the Internet to get more information about one of the special-purpose systems discussed at the end of chapter. Write a report about what you found. Give an example of a new special-purpose system that has great promise in the future.

CASES

CASE 1

PixAround.com Brings 3-D Worlds to Your Desktop

Been working hard? How about a quick 10-minute break at a Mediterranean seaside resort? PixAround.com can bring that beach to your PC or Palm computer with such clarity that you'll almost be able to feel the spray on your face. Singapore-based PixAround is the world's leading provider of 360-degree interactive digital imaging. The company has created software that will transform digital images from your camera or camcorder into a virtual environment for your computer. While immersive technologies may bring a sense of reality to virtual reality, the hardware to run them is not yet affordable enough to become standard equipment in today's households. The virtual reality applications getting the most use now do not require expensive hardware accessories. PixAround realizes these current limitations, so its systems use a standard mouse and simple navigation buttons to allow users to walk through the virtual environments it provides.

PixAround targets its software to service providers such as on-line marketing professionals, Web developers, and virtual tour companies. The software is primarily used for two purposes: to create virtual tours and to present a 360-degree view of a product.

PixAround's software is a popular tool in the real estate, travel and tourism, and hospitality businesses. By clicking a mouse and dragging across an on-screen image, users can view an entire room as if they were actually standing in the middle of the room and turning a full circle. Using the zoom buttons, users can zoom in to look at the details of a sculpture resting on a shelf in the corner. Travel agents use PixAround's product to give a prospective customer a feel for what it's like to be in the middle of a Las Vegas casino or resting on a beach in Bermuda.

Car dealers use PixAround's software to allow customers to rotate and examine a vehicle on the screen. By clicking and dragging across an image, users can rotate the car and scrutinize it from every angle—even the undercarriage and roof. The software is also valuable for customers who prefer to shop on-line rather than visiting the local mall. Being able to examine products from all angles reassures customers that they are purchasing a high-quality product.

The most recent version of PixAround's PixMaker software includes a utility that allows a company to place product logos, pictures, and hyperlinks within the virtual environment. Imagine touring a virtual model home, discovering an interesting blender on the kitchen counter, which when clicked on opens a window with a 3-D view of the blender or the manufacturer's Web site. Or perhaps visitors of the virtual home will find a business card on the kitchen table at the end of the tour, which when clicked opens an e-mail form.

PixAround's virtual 3-D worlds can run on Web sites using Java, be packaged in an executable file and sent via e-mail as a postcard, or be inserted into a PowerPoint presentation.

PixAround has also recently released a version of its product for PDAs. We'll no doubt be seeing much more of this technology in the coming years.

Discussion Questions

1. Compare and contrast the impact of marketing using 360-degree interactive digital imaging compared with traditional Web marketing for real estate and other products.
2. List five products, other than those already mentioned in the case, that could profit from using 360-degree interactive digital imaging. What types of products are best suited for this technology? Why?

Critical Thinking Questions

3. What features could PixAround add to its product to improve it? What features might make this virtual experience even more lifelike?
4. What are the limitations of supplying virtual and immersive technologies over the Internet? How might this list of limitations shrink in the coming years?

Sources:"PixMakerPro—the 360 Degree Digital Imaging Solution for Professionals; Customizable Digital Marketing Solutions That Allow Marketers to Do It Their Way," *PR Newswire*, May 28, 2002; PixAround.com Web site, http://pixaround.com, accessed June 2002.

CASE 2

Transco Uses Gensym's Expert System to Control Natural Gas Flow

Complex, volatile systems, such as those used in manufacturing and production, telecommunications, supply chain management, and distribution, typically require technicians to monitor them continuously to safeguard against unexpected problems. Failure to catch telltale signs of trouble in some cases could lead to disaster. Take for example Transco, the company responsible for delivering natural gas to over 20 million industrial, commercial, and domestic customers in the United Kingdom.

Transco maintains over 275,000 km of natural gas pipeline, comprising high-pressure national and regional transmission systems and lower-pressure distribution systems. Gas is pumped through the network by 24 compressor stations located around the country. Each compressor station is staffed with a team of technicians who monitor the pressure within the system, watching for increases in pressure that could lead to explosions or decreases in pressure that could indicate a leak of the poisonous gas.

Such work is tedious and tiring. The stream of data varies continuously and requires compensating adjustments for each fluctuation. Operators can't afford a lapse in concentration, since failure in the system would be disastrous. This scenario is ripe for automation. Enter Gensym.

Gensym is a leading provider of expert system software products that monitor, diagnose, control, and optimize complex operations in real time. Gensym's flagship product, G2, is programmed to emulate the reasoning of human experts as they assess, diagnose, and respond to unusual operating situations or as they seek to optimize operations—a perfect solution to Transco's problem.

Gensym consultants worked with Transco experts and other suppliers to develop a system that could monitor and respond to the continuously varying flow of data just as a human expert would. With Gensym's G2 software as part of its new control system, Transco is able to deploy advanced systems, utilizing expert knowledge, to enhance technicians' ability to maximize the performance of its pipeline network.

Gensym has provided expert systems to many other large corporations around the world. For example, *Fortune* 1000 manufacturers such as ExxonMobil, DuPont, LaFarge, Eli Lilly, and Seagate all use G2 to help operators detect problems early and provide advice that avoids production errors and unexpected shutdowns. Manufacturers and government agencies use G2 to optimize their supply chain and logistics operations. And communications companies such as AT&T, Ericsson Wireless, and Nokia use G2 to troubleshoot network faults so that service levels are maximized.

As the complexity of business operations increases, it is increasingly difficult to interpret and respond to time-critical data. Gensym's G2 is designed to bring complexity under control by:

1. transforming complex real-time data into useful information through knowledge-based reasoning and analysis
2. monitoring for potential problems before they can adversely impact operations
3. diagnosing root causes of problems to speed their resolution
4. recommending or taking corrective actions to help ensure successful recovery
5. coordinating activities and information to optimize operational processes

The system enables businesses to capture the knowledge of the best operations experts and to intelligently combine that knowledge with real-time data, archived information, and even business policies. Powerful reasoning engines analyze all of these inputs in real time to develop the best possible operating decisions, either as recommendations or as automated actions. The bottom line: saving businesses time and money.

Discussion Questions

1. List the benefits and drawbacks of automating tedious, dangerous jobs such as Transco's.
2. List five industries that manage hazardous, complex systems. How could an expert system assist each of these industries in managing the system?

Critical Thinking Questions

3. List some jobs that could not be automated with an expert system. What characteristics make these jobs unprogrammable?
4. Is there a danger in becoming dependent on expert systems? If so, at what point does it become dangerous? If not, why not?

Sources: "UK's Transco Selects Gensym's G2 Software; Large Gas Pipeline Company to Deploy Knowledge-Based Operator-Support Systems," *Business Wire*, February 6, 2002; Gensym Web site, http://www.gensym.com, and Transco Web site, http://www.transco.uk.com, both accessed June 2002.

IBM's eLiza: Computer Systems That Care for Themselves

A recent report stated that system administration can account for 75 percent of the total cost of maintaining an enterprisewide information system. These large special-purpose systems depend on mainframe servers to provide access to the central corporate database. When such a system goes down, a business can grind to a halt. Corporations invest considerable amounts in support personnel and redundant systems to safeguard against system failure. Even with significant investment and safeguards, there are still high rates of system downtime.

With this in mind, IBM has launched project eLiza to automate many system administrator duties and save its customers big bucks. The effort aims to create servers that respond to unexpected capacity demands and system glitches without human intervention. The goals? New highs in reliability, availability, and serviceability and new lows in downtime and cost of ownership.

IBM has classified a system administrator's duties into four areas: system configuration, maintenance, security, and efficiency. By analyzing the details of each area, IBM has been able to automate many tasks to create servers that are smart enough to care for themselves. The goal is to create servers that are:

1. Self-configuring: the ability for servers to define themselves on the fly. This self-management capability means that new features, software, and servers can be added to the infrastructure with no disruption of services.
2. Self-healing: the ability to recover from a failing component by first detecting and isolating it, taking it off-line, fixing or isolating the component, and introducing a new or repaired component into service without any disruption.
3. Self-protecting: the ability to define and manage access from users to all the resources within the enterprise, protect against unauthorized access, detect intrusions and report them as they occur, and provide backup and recovery capabilities that are as secure as the original resource management systems.
4. Self-optimizing: the ability to maximize resource use to meet end-user needs with no human intervention.

To manage capacity demands more efficiently, IBM has created Enterprise Workload Manager, software that governs not just single servers but groups, monitoring the machines and shifting work among them. But IBM isn't the only company aggressively researching ways to get groups of servers to work together without human intervention. Sun Microsystems revealed its N1 software, which treats groups of computers as if they were a single pool of processing and storage power. And Hewlett-Packard plans to develop a utility data center to simplify management of equipment-filled data centers.

Many components of project eLiza have already been released either as part of IBM server operating systems or as add-on tools. Many more components are in development and soon to be released. IBM is also developing tools to automate many of the tasks involved in database administration. IBM's SMART (Self Managing and Resource Tuning) database initiative offers administrators simplified recovery features, along with a range of diagnostic and self-managing capabilities designed to minimize database outages.

Intelligent technology that allows IBM servers to care for themselves is also being used by many other technology companies. Smart networks are monitoring and correcting network problems and smart security applications are watching for security breaches. Soon we will have smart PCs that won't crash or freeze when applications misbehave. Could it be that our days of technological frustration are numbered?

Discussion Questions

1. What types of problems have you experienced with PCs, networks, and the Internet that might have been identified and eliminated with self-healing technology?
2. What dangers are involved in transferring system administration tasks from people to the machines? Could users become disadvantaged by this technology? Do the benefits outweigh the liabilities?

Critical Thinking Questions

3. Self-healing devices use technology to maintain technology. This might strike some as a chicken and egg scenario: Who or what will maintain the technology that is maintaining the technology? What role will human system administrators play in this new maintenance system?
4. This new technology is intended to reduce the cost of ownership of a system, so it may possibly replace a significant number of system administrators. Yet, often when technology reduces the needs of human experts in one area, it creates a need for human experts in another. What new careers will this technology create?

Sources: Stephen Shankland, "IBM Empowers Self-Healing Devices," *ZDNet News*, May 2, 2002, http://www.zdnet.com/2100-1103-897212.html; Ed Scannell, "IBM Brings Self-Healing to Databases." *InfoWorld*, June 3, 2002, http://www.idg.net/go.cgi?id=693800; John Longwell, "Checkpoint 'Smart Defenses' System," Global News Wire—Asia Africa Intelligence Wire, May 13, 2002; IBM Web site, http://www.ibm.com.

NOTES

Sources for the opening vignette on p. 461: "Artificial Life Announces Full Integration of Instant Messaging and eCRM Technology," *Business Wire*, March 19, 2002; "Artificial Life Wins New Clients and Contracts," *PR Newswire*, January 10, 2002; Artificial Life's Web site, http://www.artificial-life.com, accessed June 2002.

1. Ewait, David, "Stephen Hawking Warns That Machines Could Take Over the World," *InformationWeek Online*, September 5, 2001.

2. Fahey, Jonathan, "The Science of Small," *Forbes*, February 5, 2001, p. 124.
3. Zaun, Todd, "What's 4 Feet Tall, Cost Millions of Dollars, and Does the Waltz?" *The Wall Street Journal*, September 4, 2002, p. A1.
4. Cope, James, "Robot Lays Fiber-Optic Net in Albuquerque," *Computerworld*, February 26, 2001, p. 10.
5. Ridgeway, Nicole, "Robo-Therapy," *Forbes*, May 14, 2001, p. 216.
6. Squedo, Anni Marie, "Meet the Newest Recruits: Robots," *The Wall Street Journal*, December 13, 2001, p. B1.

7. Holmes, Stanley, "Planes That Know What to Bomb," *Business Week*, November 12, 2001, p. 91.

8. Huber, Peter, "The Dexterous Robot," *Forbes*, February 18, 2002, p. 88.

9. Pfeiffer, Eric, "Micromachines: They're Huge," *Forbes ASAP*, April 2, 2001, p. 41.

10. MKF, "A.I. in the Real World," *PC Magazine*, September 4, 2001, p. 30.

11. Arnst, Catherine, "The Vision Thing Gets Closer for Computers," *Business Week*, April 15, 2002, p. 79.

12. Alwang, Greg, "Voice Recognition: Getting Better," *PC Magazine*, February 26, 2002, p. 29.

13. Mearian, Lucas, "Schwab Taps Natural-Language Search Engine," *Computerworld*, July 16, 2001, p. 8.

14. Copeland, Ron, "Heeding Customer Requests," *InformationWeek Online*, July 9, 2001.

15. Mearian, Lucas, "Brokerage Launches Search Engine to Aid Call Center," *Computerworld*, March 18, 2002, p. 10.

16. Holly, Susan, "Speak," *PC Magazine*, August 2001, p. IBIZ 5.

17. Brock, Kelly, "Does the Future of AI Belong to Open Source Java Programmers," *PR Newswire*, March 5, 2002.

18. George, Amy, "Chevron Phillips Chemical Company LP Renews Pavilion Agreement," *Business Wire*, February 25, 2002.

19. Roth, Susan, "Federated Implements HNC Falcon Fraud Manager," *Business Wire*, February 4, 2002.

20. "NAG Data Mining Components Speed Application Development," *Worldwide Database*, March 2002.

21. Roush, W.B. "AI Trilogy," *OR/MS Today*, February 2001, p. 64.

22. Kellner, Tomas, "Cybersleuth," *Forbes*, November 26, 2001, p. 202.

23. "Fuzzy Logic Software Corporation Announces Closing the Share Purchase Agreement," *Business Wire*, February 25, 2002.

24. Harvey, Fiona, "Test May Improve Lung Cancer Survival Rates," *Financial Times*, September 26, 2001, p. 10.

25. "WildPackets and Atheors Partner to Provide Wireless LAN Management," *Business Wire*, February 27, 2002.

26. Hague, Steve, "Aircuity, Inc, The Emerging Leader of Intelligent Air Quality Monitoring for Commercial Buildings," *Business Wire*, December 19, 2001.

27. Carter, Steven, "Transocean Sedco Forex Subscribes to RihInsight System," *Business Wire*, February 26, 2002.

28. Kontzer, Tony, "Artificial Intelligence to Hit the Mainstream," *InformationWeek Online*, August 31, 2001.

29. Siegel, David, "Transco Selects Gensym's G2 Software," *Business Wire*, February 6, 2002.

30. Ray, Mike, "NeoSafety Award Licensing Agreement from Sandia," *PR Newswire*, February 6, 2002.

31. Shuey, Craig, "PA State Senator Distributes RegScan Compliance System," *PR Newswire*, January 18, 2002.

32. Rozmus, Paul, "LSI and ATSI Provide Link for Collateral Assessment Services," *Business Wire*, December 21, 2001.

33. Disabatino, Jennifer, "CIA-Backed Analysis Tool Eyed for Passenger Checks," *Computerworld*, January 1, 2002, p. 12.

34. Chelst, et al., "Rightsizing and Management of Prototype Vehicle Testing at Ford Motor Company," *Interfaces*, January-February 2001, p. 91.

35. Seligman, Dan, "The Computers Rate the Writers," *Forbes*, October 29, 2001, p. 122.

36. Kohler, Tracey, "MEDecision's CaseAlert Software to Be Used in Innovative Medicaid Disease Management Program," *PR Wire*, February 21, 2002.

37. Landro, Laura, "New Medical Software Gives Physicians Clues When They're Stumped," *The Wall Street Journal*, June 29, 2001, p. B1.

38. Murphy, Victoria, "You've Got Expertise," *Forbes*, February 5, 2001, p. 132.

39. Kay, Russell, "3-D Vision Speaks Volumes," *Computerworld*, April 1, 2002, p. 44.

40. Crecentre, Brian, "Young Scientist's Creation a Labor of Glove," *Rocky Mountain News*, March 13, 2002, p. 27A.

41. Blough, Kay, "Virtual Reality to Aid Stroke Therapy," *InformationWeek Online*, 02-19-02.

42. Lippman, John, "As Hollywood Casts About for a War Role, Virtual Reality Is Star," *The Wall Street Journal*, November 9, 2001, p. A1.

43. Ewalt, David, "White House Launches Virtual Holiday Tour," *InformationWeek Online*, December 12, 2001.

44. Tran, Murphy, "Avatars Widen Realm of Virtual Reality on the Internet," *The Wall Street Journal*, January 24, 2001, p. B1.

45. Machrone, Bill, "Segue — From Ginger to Segway," *PC Magazine*, January 29, 2002, p. 53.

46. Mitchener, Brandon, "Disabled Could Be Liberated by Brain-Wave Technology," *The Wall Street Journal*, March 14, 2001, p. B1.

47. "Recording Your Life," *InformationWeek Online*, December 19, 2001.

48. "Kodak Forms Company to Rival Bluetooth," *InformationWeek Online*, January 1, 2001.

49. Songini, Marc, "Radio Tagging of In-Transit Material Speeds Data Gathering in the Supply Chain," *Computerworld*, April 8, 2002, p. 46.

50. "Microsoft, Accenture to Sell Software to Automakers," *InformationWeek Online*, January 11, 2001.

51. Hwang, Suein, "In Office of the Future, We'll All Be Scanned Like a Can of Peas," *The Wall Street Journal*, April 10, 2002, p. B1.

52. Kay, Russell, "True 3-D Without Glasses," *Computerworld*, April 30, 2001, p. 53.

WORLD VIEWS CASES

This interesting World Views Case explores the use of the Internet to provide outstanding customer service in Beijing, China. The company uses an effective management information system to transfer Internet orders to its local sites around Beijing. The management information system also helps to measure performance in terms of customer service, quality, and delivery speed. Because of its success, the company is investigating expansion in Beijing and starting service in other cities.

A Happy Valentine's Day in Virtual China

Bin Xie
Tsinghua University, Beijing People's Republic of China

Xiaoyang Wang
Tsinghua University, Beijing, People's Republic of China

John Paynter
University of Auckland, New Zealand

On February 14, 2001, Zhang Yongqing, the CEO of eGuo.com (the name translates to eCountry in English), had been busy. eGuo is an on-line gift company that currently sells about 6,000 items, including 1,000 food items, and Valentine's Day is a big gift-giving holiday for the Chinese. The enthusiasm of Beijing youth for this first Valentine's Day of the 21st century kept eGuo busy filling orders. Now it was 10:00 P.M., and Zhang Yongqing could relax for a while and look at the results of the day's order fulfillment. The total number of fulfilled orders was about 1,200, including 350 for flowers and 50 for perfume. Mr. Zhang was happy with the performance. Since the NASDAQ stock index had dropped in April 2000, the efficacy of the B2C model had been questioned, and venture capital investors had lost enthusiasm for the dot-coms. Under such difficult circumstances, the investors would be pleased with eGuo's Valentine's Day efforts. However, the large number of orders challenged eGuo's logistics system to keep its promise of one-hour delivery in Beijing. That was just the opportunity Zhang was thinking of when he set up eGuo.com in October 1999. At that time, the biggest problems Chinese e-commerce faced were logistics and credit payments. He wanted to use the one-hour delivery promise to solve that problem. Zhang stated his e-commerce strategy this way: "The customer needs convenience, quick and cheap service; otherwise, we would lose the meaning of 'e.' It would be best if you can get what you want as soon as you click (like sending an e-mail). The customer expects instantaneous service. From the traditional viewpoint, it usually takes one hour to go to the closest market to buy something. Psychologically, one hour is a kind of limit separating 'short' time from 'long' time. So one hour can be considered instantaneous service. In addition, one-hour delivery can solve the problem of payment; eGuo can get money back when it sends products."

E-commerce got its start in China in 2000 during an e-commerce symposium. During the symposium, there were purchasing tests on B2C Web sites in order to demonstrate e-commerce. eGuo was the first to deliver the orders. So eGuo became a lab for Beijingers to test e-commerce.

On April 15, 2000, eGuo launched free delivery service within the Fourth Ring Road (within a 20-kilometer radius of the center of Beijing), including Zhongguanchun and the Asian Games Village in Beijing. eGuo used a special promotion of "Cool One Hour—Buy Ten Get One" that included Pepsi. That is, for every 10 RMB (1 RMB = U.S. $0.12) of products purchased from eGuo, customers would get a can of Pepsi. It was very hot in the summer of 2000—up to 40 degrees centigrade in Beijing—so the promotion was well timed. People who ordered from eGuo could stay at home enjoying "Cool" Pepsi. The eGuo one-hour delivery service exceeded the expectations of customers, especially when compared with the four- to five-day delivery time for typical on-line shops. eGuo received a positive response from its promotion and great media attention. For example, it also provided tickets to cultural shows and to public attractions such as the Beijing Aquatic Animal Show.

The eGuo distribution network operates on a hub and spoke model, with one centralized hub and 10 spoke stations located in Beijing to achieve the one-hour delivery standard. eGuo also opened three spoke stations as "eGuo Marts" so that customers could buy products directly. Communication is the key to fulfilling the orders between the headquarters and warehouses. Initially, the company only used the telephone to relay orders, but this resulted in many errors. A management information system (MIS) was developed to exchange order and inventory information using the Internet, with the hub computers at each location connected via the Internet to the eGuo server at headquarters using dial-up telephone lines. In this way, the inventory at each spoke station is known, and the hub can coordinate inventory sharing. The spoke stations can submit replenishment plans to the hub based on the order history and stock-taking records.

eGuo has a relatively loose work environment and a culture based on teamwork. It provides opportunities for staff to develop within the company. eGuo also provides comprehensive staff training, such as customer service training, for call center and delivery staff, as well as more specialized training in logistics for the warehouse staff. Staff performance is measured by calling customers to monitor the quality of service. For example, are the correct goods delivered at the correct price and received on time from the delivery service? The organizational model attracts staff to eGuo from other companies. For instance, call center staff have come from pager-service companies and spoke station supervisors from traditional retail organizations such as Carrefour (www.carrefour.com), one of the world's largest retailers, with more than 300,000 employees. Because its employees receive a relatively high income and have sound job prospects in a growth business, they are proud of their work at eGuo.

Zhang was now considering four options for the future development of his company: building one of the largest shopping malls in Beijing within three years, opening 100 eGuo Marts in Beijing within the next two years, duplicating the Beijing B2C model in Shanghai and Guangzhou, and acquiring eToys in the United States to set up a B2B model between Chinese toy manufacturers and the American retailer.

Note: The eGuo site is located at www.eGuo.com, but currently no English translation version is available.

Discussion Questions
1. Describe the history and development of the business model for eGuo.
2. How does eGuo integrate virtual and physical operations?

Critical Thinking Questions
3. Which option would you choose to develop eGuo if you were in Zhang's position? For example, if you were Zhang, what technologies would you investigate to keep your business competitive?
4. What are the threats to eGuo?

Sources: Bin Xie and Xiaoyang Wang, "Case Study: e-Guo.com," *Proceedings of the Twelfth Annual Conference of the Production and Operations Management Society*, POM-2001, March 20–April 2, 2001, Orlando, Florida; Yanbin QI, "The Largest Online Toy Market is Going to Lay Off All Its Employees, eGuo's Acquisition of eToys Will Have a Result Next Month," *Beijing Youth News*, February 11, 2001, p. 13; QI Quan, "eGuo Wants to Open Offline Markets," *Economics Daily*, March 1, 2001, p. 13.

Scheduling workers for jobs has long been a difficult task for many industries, including airlines, banks, and hospitals. This World Views Case reveals how a computer-based solution was found for the Hospital Authority of Hong Kong, which used an optimization technique called constraint programming. *This award-winning application shows how it is possible to use a sophisticated information system to reduce costs and keep managers and employees happy.*

Staff Rostering System at the Hospital Authority of Hong Kong

Hakman A. Wan
Open University of Hong Kong

The Hospital Authority of Hong Kong (http://www.ha.org.hk) is the statutory body that manages all public hospitals in Hong Kong, a special administration region in Southern China. The authority oversees more than 28,000 hospital beds and employs some 50,000 staff in over 1,000 wards.

Staff rostering had once been a big problem for ward managers. Based on their own experiences and operational needs, they manually scheduled the rosters of individual nurses, student nurses, and clinical supporting staff according to their ranks, shift duties, and special regulations of the wards and hospitals concerned. The task was time-consuming, and sometimes complaints arose when staff found that they were not given an equal number of days off on weekends.

With the help of the City University of Hong Kong, the Hospital Authority started to implement a Staff Rostering System (SRS) in 1998. It was first piloted in 1999, and new enhancements have been added periodically since then. The current SRS, version 2.8, is now available in 11 large hospitals.

Andy Chun led a systems development group of more than 20 analysts and programmers during the peak development period. Chun opted for a technique in artificial intelligence called *constraint programming*. While letting the IT department of the Hospital Authority take care of the database and user interface, Chun and his group built a model identifying relevant variables and their constraints—for example, who should work on a particular day and on what shift. What remained was to use the system to determine the optimal nurse roster so that it met the personnel demands on each shift in each hospital, while offering nurses the fairest possible roster, such as the same number of days off on weekends per year.

Nurse rostering is an example of a scheduling problem that is notorious for having a huge number of variables and constraints. In real life, the multiple effect of these variables generates an enormous search space (i.e., possibilities from which the problem solver needs to look for an optimal solution). However, Chun tackled the problem with the constraint programming technique, which reduced the number of possibilities tremendously and limited the search space to a manageable size.

The final SRS was developed by using Microsoft Visual Studio and C++ class libraries from RogueWave (http://www.roguewave.com). The front-end graphical user interface was written in Visual Basic.

Hospital ward managers are generally happy with the system. Some are impressed by the comprehensive functionality of SRS in keeping leaves and compensatory time-off records; some report that they now spend less time generating a roster for their staff. The SRS is considered a strategic success. It received an "Innovative Applications of AI Award" by the American Association for Artificial Intelligence (AAAI) in 2000. The Hospital Authority has identified two directions for the future development of the SRS: to integrate the system with the human resources and payroll system and with the executive information system of the organization.

Discussion Questions

1. Designing the right timetable for courses offered at your university each semester is usually the job of the registrar's office. If you were to produce the timetable for the next semester, what are the variables that you would consider? What are the constraints that a timetable must satisfy? Compare the complexity of nurse rostering and course offerings in a university.

2. Mathematical methods are generally more reliable and efficient in solving engineering problems. Why do you think the Hospital Authority of Hong Kong did not consider algorithmic solutions but instead used artificial intelligence in developing its nurse rosters?

Critical Thinking Questions

3. What would you think are the advantages and disadvantages of using the SRS from the ward managers' perspective?

4. It is possible that the SRS does not generate the best possible roster all the time. What do you think a ward manager should do when he/she finds that a small manual adjustment on a roster produced by the SRS can make more nurses happier?

Sources: Chun, AHW et. al., "Nurse Rostering at the Hospital Authority of Hong Kong," *Proceedings of the 17th National Conference on Artificial Intelligence and 12th Conference on Innovative Applications of Artificial Intelligence*, July 30–August 3, 2000, Austin, Texas, AAAI Press/The MIT Press, pp. 951–956; also available at *http://www.e-optimization.com/resources/uploads/ACFAA.pdf*; "Staff Rostering System, Hong Kong Hospital Authority," accessed at http://www.ha.org.hk, September 2002; BonVision's Intelligent Solution—Hospital Authority Rolls Out New Release of Nurse Rostering Engine, 2001, accessed at http://www.aotl.com/universal/news_event/news.asp?news_id=7.

SYSTEMS
DEVELOPMENT

Systems Investigation and Analysis

CHAPTER 12

<table>
<tr><th>PRINCIPLES</th><th>LEARNING OBJECTIVES</th></tr>
</table>

PRINCIPLES	LEARNING OBJECTIVES
Effective systems development requires a team effort of stakeholders, users, managers, systems development specialists, and various support personnel, and it starts with careful planning.	• Identify the key participants in the systems development process and discuss their roles. • Define the term *information systems planning* and list several reasons for initiating a systems project. • Identify important system performance requirements for applications that run on the Internet or a corporate intranet or extranet. • Discuss three trends that illustrate the impact of enterprise resource planning software packages on systems development.
Systems development often uses tools to select, implement, and monitor projects, including net present value (NPV), prototyping, rapid application development, CASE tools, and object-oriented development.	• Discuss the key features, advantages, and disadvantages of the traditional, prototyping, rapid application development, and end-user systems development life cycles. • Identify several factors that influence the success or failure of a systems development project. • Discuss the use of CASE tools and the object-oriented approach to systems development.
Systems development starts with investigation and analysis of existing systems.	• State the purpose of systems investigation. • Discuss the importance of performance and cost objectives. • State the purpose of systems analysis and discuss some of the tools and techniques used in this phase of systems development.

[Comcast Corporation]

The Challenges of Providing Broadband Internet Access

Comcast, the third largest cable provider in the United States, found itself in a tough spot when its partner Excite@Home declared bankruptcy. Excite@Home had been responsible for supplying 950,000 Comcast customers with high-speed Internet access via Comcast's cable.

Comcast partnered with Excite@Home to expand its cable operations to deliver high-speed Internet service. Excite@Home provided Internet services such as e-mail and Web access over Comcast cable. After establishing a healthy broadband business for Comcast, at the peak of its apparent success, Excite@Home filed for bankruptcy while being sued by investors for not being up-front in their financial reports. Providing broadband Internet access is costly, and Excite@Home was burning through its cash at a substantially higher rate than indicated in its filings with the Securities and Exchange Commission.

Comcast found itself holding the bag. How could it continue supplying service to its 950,000 Internet customers? Should it scramble to find another partner? After evaluating several alternatives and with full knowledge of the complications that brought down its former partner, Comcast decided to develop its own high-speed Internet network and e-mail system. By deciding to develop its own system, Comcast eliminated the danger of being let down by another partner, but exchanged that risk for the possibility of failing on its own and discrediting the organization. It was a big gamble, and Comcast realized the amount of effort it would require to succeed was considerable. Comcast paid $160 million to keep the Excite service running through the end of February 2002, at which point it hoped its new system would be in place.

The development of the new system was an 18-month process that was not without its share of problems. Dave Watson, vice president for marketing at Comcast, said that some customers experienced a "small" delay in receiving e-mail. But some customers said the delays were more substantial than the company indicated. They claimed they were unable to send or receive e-mail for days. "It is a total nightmare," said Michael Gunther, a Comcast customer in Summit, New Jersey, noting that it had become a regular habit for him to call Comcast customer service: "I had to put them on speed dial because I talk to them more than I talk to my wife." Most frustrating of all, some said, was that their calls to customer service were not answered, were met with busy signals, or were answered by technicians who lacked the expertise to solve their problems.

Why did Comcast customers put up with this ordeal? Some customers said that they considered switching to another Internet provider but found few competitors in their markets offering high-speed access. Comcast gambled that its monopoly in the market would sustain its customer base through the hardships of the transition, a gamble that paid off.

After months of gradually transferring customers to its own network, service returned to normal for Comcast customers; the overwhelming number of calls to customer service dwindled. Comcast survived the ordeal. In fact, Comcast did more than survive—it prospered! By transforming itself into an independent broadband Internet service provider, Comcast made itself more attractive to other players in the industry and profited from the resulting mergers, acquisitions, and partnerships.

As you read through this chapter, consider the following:
- How do systems development projects affect customers and employees during and after the process?
- How can companies prepare for the effects of system development process prior to implementation?

When an organization needs to accomplish a new task or change a work process, how does it do it? It develops a new system or modifies an existing one. Systems development is the activity of creating or modifying existing business systems. It refers to all aspects of the process—from identifying problems to be solved or opportunities to be exploited to the implementation and refinement of the chosen solution. As seen in the opening vignette, the bankruptcy of a partner can even launch a systems development project.

The results of systems development can mean the success or failure of an entire organization. Successful systems development has resulted in huge increases in revenues and profits. Companies that don't innovate with new systems development initiatives or fail to successfully complete a systems development effort can lose millions. Understanding and being able to apply a systems development life cycle, tools, and techniques, which are discussed in this chapter and the next, will help ensure the success of the development projects in which you participate. It can also help your career and the financial success of your company.

AN OVERVIEW OF SYSTEMS DEVELOPMENT

In today's businesses, managers and employees in all functional areas work together and use business information systems. As a result, users of all types are helping with development and, in many cases, leading the way. This chapter and the next will provide you with a deeper appreciation of the systems development process and help you avoid costly failures.

PARTICIPANTS IN SYSTEMS DEVELOPMENT

Effective systems development requires a team effort. The team usually consists of stakeholders, users, managers, systems development specialists, and various support personnel. This team, called the *development team*, is responsible for determining the objectives of the information system and delivering a system that meets these objectives to the organization. Many development teams use a project manager to head the systems development effort and the project management approach to help coordinate the systems development process. A *project* is a planned collection of activities that achieves a goal, such as constructing a new manufacturing plant or developing a new decision-support system. All projects have a defined starting point and ending point, normally expressed as dates such as August 4th and November 11th. Most have a budget, such as $150,000. A *project manager* is the individual responsible for coordinating all people and resources needed to complete a project on time. In systems development, the project manager can be an information systems person inside the organization or an external consultant hired to complete the project. Project managers need technical, business, and people skills. In addition to completing the project on time and within the specified budget, the project manager is usually responsible for controlling project quality, training, communications, risks, and the acquisition of any necessary equipment, including office supplies and sophisticated computer systems.

stakeholders
individuals who, either themselves or through the organization they represent, ultimately benefit from the system development project

users
individuals who will interact with the system regularly

In the context of systems development, **stakeholders** are individuals who, either themselves or through the area of the organization they represent, ultimately benefit from the systems development project.[1] One systems development methodology, called *agile modeling*, calls for very active participation of customers and other stakeholders in the systems development process.[2] **Users** are individuals who will interact with the system regularly. They can be employees, managers, or suppliers. For large-scale systems development projects, where the investment in and value of a system can be quite high, it is common to have senior-level managers, including the company president and functional vice presidents (of finance, marketing, and so on), be part of the development team.

The IS competence of managers can have a big impact on the systems development effort.[3] A study of the IS competence of business managers revealed that those managers with more knowledge and skill in computer technology were more willing to form partnerships with IS people and to lead and participate in systems development projects. Users who do not agree with a systems development project, however, can be hostile to the project and may even try to disrupt it.[4]

Depending on the nature of the systems project, the development team might include systems analysts and programmers, among others. A **systems analyst** is a professional who specializes in analyzing and designing business systems. Systems analysts play various roles while interacting with the stakeholders and users, management, vendors and suppliers, external companies, software programmers, and other IS support personnel (Figure 12.1). Like an architect developing blueprints for a new building, a systems analyst develops detailed plans for the new or modified system. The **programmer** is responsible for modifying or developing programs to satisfy user requirements. Like a contractor constructing a new building or renovating an existing one, the programmer takes the plans from the systems analyst and builds or modifies the necessary software.

The other support personnel on the development team are mostly technical specialists, including database and telecommunications experts, hardware engineers, and supplier representatives. One or more of these roles may be outsourced to outside experts or consultants. Depending on the magnitude of the systems development project and the number of IS systems development specialists on the team, the team may also include one or more IS managers. The composition of a development team may vary over time and from project to project. For small businesses, the development team may consist of a systems analyst and the business owner as the primary stakeholder. For larger organizations, formal IS staff can include hundreds of people involved in a variety of activities, including systems development. Every development team should have a team leader. This individual can be from the IS department, a manager from the company, or a consultant from outside the company. The team leader needs both technical and people skills.

systems analyst
professional who specializes in analyzing and designing business systems

programmer
specialist responsible for modifying or developing programs to satisfy user requirements

FIGURE 12.1

The systems analyst plays an important role in the development team and is often the only person who sees the system in its totality. The one-way arrows in this figure do not mean that there is no direct communication between other team members. Instead, these arrows just indicate the pivotal role of the systems analyst—an individual who is often called on to be a facilitator, moderator, negotiator, and interpreter for development activities.

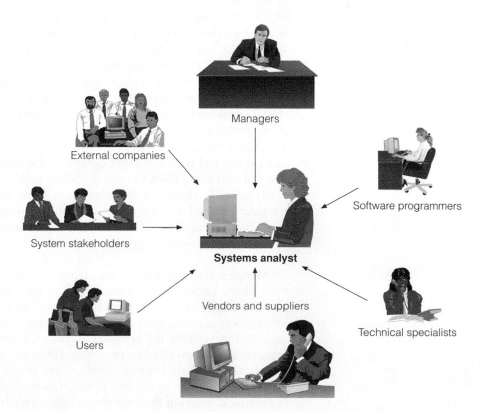

Managers

External companies

Software programmers

System stakeholders

Systems analyst

Users

Vendors and suppliers

Technical specialists

Development projects place great demands on staff who may already have day-to-day responsibilities. To keep IS personnel motivated and reduce stress, some companies have adopted a consultancy approach to systems development.[5] Harrah's Entertainment, for example, used this approach to help cut costs, improve productivity, and reduce burnout of its IS staff. The large casino and hotel operator let its IS staff select upcoming projects and gave systems development teams the resources they needed to complete projects on time. According to one company executive, "The new structure consists of a 'floating pool' of 150 developers and three resource managers. This model also affords employees more flexibility in scheduling."

At some point in your career, you will likely be a participant in systems development. You could be involved in a systems development team—as a user, as a manager of a business area or project team, as a member of the IS department, or maybe even as a CIO. With the increasing power and ease of software suites and Web authoring tools, you could also develop some of your own systems or modify existing ones. You might develop a spreadsheet to analyze financial alternatives or a personal Web site, for example.

Regardless of the specific nature of a project, systems development creates or modifies systems, which ultimately means change. Managing this change effectively requires development team members to communicate well. It is important that you learn communication skills, because you probably will participate in systems development during your career. You may even be the individual who initiates systems development.

INITIATING SYSTEMS DEVELOPMENT

Systems development begins when an individual or group capable of initiating organizational change perceives a need for a new or modified system. Such individuals have a stake in the development of the system. Executives at Delta Airlines, for example, initiated a systems development project when they decided to expand the company's Web site.[6] The new Web site allows employees to log on to the site to give them deeply discounted flight and travel opportunities. FleetBoston Financial Corporation initiated a systems development effort to introduce a new customer relationship management (CRM) system to hundreds of its banks.[7] According to Maura Fairbanks, manager of the retail distribution for the bank, "It allows the branch managers to do customization to fit their [local] markets." FleetBoston is one of the first financial institutions to use data from the CRM system to help employees and workers analyze market needs and provide customized solutions. In another case, TWA decided to start a systems development project to change its reservation system.[8] The new reservation system, Sabre Holding Corporation, will allow TWA to check its passengers against local, state, and federal agency databases for suspected terrorists. The new system will also allow TWA to reduce staffing needs and costs.

Systems development initiatives arise from all levels of an organization and are both planned and unplanned. Solid planning and managerial involvement helps ensure that these initiatives support broader organizational goals. Systems development projects may be initiated for a number of reasons, as shown in Figure 12.2.

Mergers and acquisitions can trigger many systems development projects. Because the information systems for companies are usually different, a large systems development effort is typically required to unify systems. Even with similar information systems, the procedures, culture, training, and management of the information systems are typically different, requiring a realignment of the IS departments. When chicken processor Tyson acquired IBP, a meat-packing conglomerate, a massive systems development project was initiated to integrate the information systems of the two firms.[9] The multimillion-dollar project had some IS personnel working 80-hour weeks to integrate the two information systems.

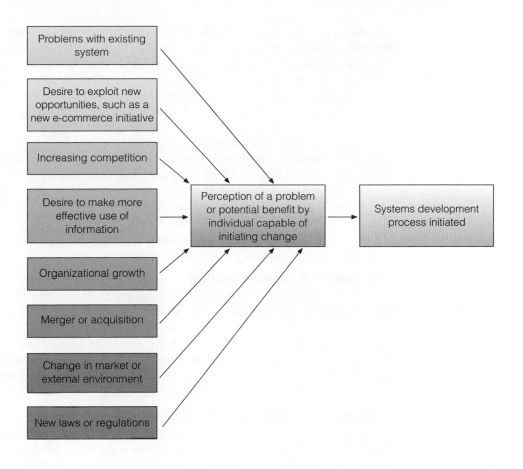

FIGURE 12.2

*Typical Reasons to Initiate a
Systems Development Project*

The external environment is another cause for new systems development. The Centers for Disease Control (CDC) is developing a new system to help it deal with the anthrax threat and other possible crises.[10] Some people in the department compared the old system to a "pony express" system that used paper-based reports and phone calls. The systems development effort will result in a new Internet system to speed information flow and help CDC executives manage health problems. According to one CDC manager, "Early detection and response is critical, and it all hinges on communications and information technology."

The federal government also fosters new systems development projects in the private sector.[11] The CIA, for example, has funded a venture-capital firm called In-Q-Tel with $28.5 million. In-Q-Tel seeks help from corporations and their systems development departments to build new spy tools for security. "We have succeeded at luring companies to come in and show us stuff that even we are shocked by," says Christopher Tucker, chief strategist at In-Q-Tel. The "Q" in In-Q-Tel is from James Bond creator Ian Fleming's fictional inventor and spy gadget creator.

Corporate litigation, which appears to be on the rise, also has initiated systems development projects.[12] New systems are often needed to protect a company from lawsuits and help them in court if one is brought against the company. According to Tom DeMarco, a consultant for Cutter Consortium, "Big systems integrators often have 50 concurrent lawsuits pending." Some have estimated that from 10 to 15 percent of the information systems budget for a typical company is spent on legal costs and the systems that are built to fight litigation.

INFORMATION SYSTEMS PLANNING AND ALIGNING CORPORATE AND IS GOALS

Because an organization's strategic plan contains both organizational goals and a broad outline of steps required to reach them, the strategic plan affects the type of system an organization needs. For example, a strategic plan may identify a doubling of sales revenue within five years, a 20 percent reduction of administrative expenses over three years, acquisition of at least two competing companies within a year, or market leadership in a given product category as organizational goals. Organizational commitments to policies such as continuous improvement are also reflected in the strategic plan. Such goals and commitments set broad outlines of system performance.

Often, a section of the strategic plan lists guidelines for meeting specific goals that relate to units or departments. Examples of these guidelines might be improving customer service for luxury car buyers, expanding international distribution by purchasing existing distributors, and using a specific amount of money to buy back company stock. The strategic plan also provides general direction to the functional areas within an organization, including marketing, production, finance, accounting, and human resources. For the IS department, these directions are encompassed in the information systems plan.

Information Systems Planning

information systems planning
the translation of strategic and organizational goals into systems development initiatives

The term **information systems planning** refers to the translation of strategic and organizational goals into systems development initiatives (Figure 12.3). The Marriott hotel chain, for example, invites its chief information officer to board meetings and other top-level management meetings. Proper IS planning ensures that specific systems development objectives support organizational goals. See the "IS Principles in Action" box to learn how British Telecommunications was able to plan and implement a new Internet billing system.

Aligning Corporate and IS Goals

Aligning organizational goals and IS goals is critical for any successful systems development effort.[13] Since ISs support other business activities, both IS staff and people in other departments need to understand each other's responsibilities and tasks. Aetna hired a new CIO to educate its managers about the potential of information systems in general.[14] According to Aetna's CIO, "Historically, IT as a profession has not succeeded in educating business leaders on what the IT discipline is all about. So IT investments, by and large, have not been optimized." Determining whether organizational and IS goals are aligned can be difficult, so researchers have increasingly tackled the problem. Jerry Luftman, executive director and distinguished service professor at the Stevens Institute of Technology in New Jersey, developed a measure of alignment that uses five levels ranging from ad hoc processes (Level 1 Alignment) to optimized processes (Level 5 Alignment).[15] The five-level measure can be used by companies to determine the extent IS projects are aligned and compatible with organizational goals. A study by the management-consulting firm McKinsey & Co. also confirmed the importance of aligning corporate and IS goals. According to Mike Nevens, an analyst with the company, "Two things are surprising to us from this research. The first is how large the benefit is if companies get all of the business factors aligned with IT. The other big surprise is how few companies are actually able to do it."

One of the primary benefits of IS planning and alignment of goals is a long-range view of information technology's use in the organization. Specific systems development initiatives may spring from the IS plan, but the IS plan must also provide a broad framework for future success. The IS plan should guide development of the IS infrastructure over time. Another benefit of IS planning is that it ensures better use of IS resources—including funds, personnel, and time for scheduling specific projects. The steps of IS planning are shown in Figure 12.4.

Overall IS objectives are usually distilled from the relevant aspects of the organization's strategic plan. IS projects can be identified either directly from the objectives determined in the first step or may be identified by others, such as managers within the various functional areas. Setting priorities and selecting

FIGURE 12.3

Information systems planning transforms organizational goals outlined in the strategic plan into specific system development activities.

British Telecommunications Adopts On-Line Account Management and Billing System

British Telecommunications (BT) PLC employed a traditional paper-based billing system that was costly in both postage and labor. So, staff decided to investigate the costs and benefits of a new electronic billing system that would allow customers to use the Internet to manage their accounts and pay their bills. They began their research by examining U.S. statistics, where many companies were already offering on-line billing services.

Initial research revealed that 30 percent of on-line adults in the United States view their credit card bills on-line, and 20 percent view their telecommunications and cable bills over the Web. About 6 percent view their utility and insurance bills on-line. Encouraged by these facts, BT examined the methods by which U.S. citizens were paying their bills on-line. It discovered that although U.S. banks offer bill-paying services, consumers prefer to open accounts with individual companies. These findings indicated that BT could expect anywhere between 10 and 20 percent of its on-line customers to use an on-line payment service.

BT also conducted research into its own operations to determine how much it would save by converting customers from traditional billing to on-line payments. Although the company would not disclose the results of this research, Gartner Inc. analyst Avivah Litan estimates the savings is substantial—up to 45 cents per bill. For a company the size of BT, those savings could add up to millions of dollars, she said. If BT could get 90 percent of its more than 21 million customers signed up and paying their bills on-line, it could save close to $110 million annually. Litan said another advantage of on-line billing is that companies can see a return on investment as soon as they sign up about 9 percent of their customers. For a company like BT, she said, that benchmark should take about a year to reach.

Preliminary research clearly indicated that BT should move forward with the development of an on-line payment system. After studying current systems and discussing the matter with its systems development staff, BT decided to go with a product from edocs Inc. rather than developing its own system. Edocs was able to provide an on-line payment solution that didn't require significant changes to BT's existing systems.

Since edocs is in the business of developing on-line account management and billing systems, its product is more robust than anything BT could have affordably developed and includes many features to entice BT customers to use the product. The software allows customers to analyze phone bills—to check who is called the most and when. The system also allows customers to replace phone numbers with names. So, instead of getting a list of phone numbers on the bill, the customer will see a list of names of people and businesses called. It also allows them to build an address book. This last feature is important for keeping customers since they are not likely to switch phone companies if they have spent time creating their address book.

BT invested a lot of time to assure the success of its new system. It determined that the system would benefit the stakeholders, would be attractive to the users, and would most likely provide a quick return on investment. So, it was worth the time and investment. Looking down the road, BT expects usage of its on-line payment system to increase as access to broadband connections grows—a service and new system that BT is also developing for its customers.

Discussion Questions

1. Is it possible that BT's new on-line payment system might hurt the company? What are the dangers involved in this project? Can there ever be a new system proposal that is 100 percent guaranteed to produce positive results? Why or why not?

2. What considerations contribute to a company's decision to develop information systems in-house rather than go with an outside provider?

Critical Thinking Questions

3. What other services might BT offer to its customers on-line that are traditionally offered by phoning customer service?

4. BT took advantage of U.S. statistics. How might it have predicted public response to on-line payments if it had been the first to offer the service?

Sources: Brian Sullivan, "BT Pushing Online Customer Bill-Paying," *Computerworld*, June 17, 2002, http://www.computerworld.com; BT Together Web site, http://www.bt.com/together/index.jsp, accessed July 2002; edocs Inc. Web site, http://www.edocs.com, accessed July 2002.

projects typically requires the involvement and approval of senior management. When objectives are set, planners consider the resources necessary to complete the projects including employees (systems analysts, programmers, and others); equipment (computers, network servers, printers, and other devices); expert advice (specialists and other consultants); and software, among others.

Developing a Competitive Advantage

In today's business environment, many companies seek systems development projects that will provide them with a competitive advantage. Thinking competitively usually requires creative and critical analysis. For example, a company may

FIGURE 12.4

The Steps of IS Planning
Some projects are identified through overall IS objectives, whereas additional projects, called unplanned projects, are identified from other sources. All identified projects are then evaluated in terms of their organizational priority.

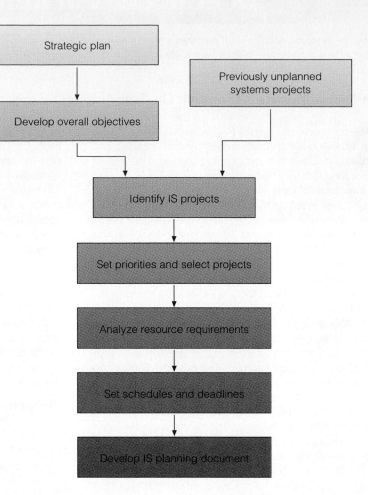

creative analysis
the investigation of new approaches to existing problems

critical analysis
the unbiased and careful questioning of whether system elements are related in the most effective or efficient ways

want to achieve a competitive advantage by improving its customer-supplier relationship. Linking customers and suppliers can result in superior products and services.

Creative analysis involves the investigation of new approaches to existing problems. By looking at problems in new or different ways and by introducing innovative methods to solve them, many firms have gained a competitive advantage. Typically, these new solutions are inspired by people and events not directly related to the problem.

Critical analysis requires unbiased and careful questioning of whether system elements are related in the most effective or efficient ways. It involves considering the establishment of new or different relationships among system elements and perhaps introducing new elements into the system. Critical analysis in systems development involves the following actions:

*   *Going beyond automating manual systems.* Many organizations use systems development simply to automate existing manual systems, which may result in relatively faster, more efficient systems. However, if the underlying manual system is flawed, automating it might just magnify its impact. In addition, automating existing manual systems might miss many opportunities by only considering doing things in the same old way. For example, rather than automate current customer service systems, many companies are implementing customer service systems for the Internet and Web technology so that customers can provide a high degree of self-service. Critical analysis in systems development involves asking why things are done a certain way and considering alternative approaches.
*   *Questioning statements and assumptions.* Questioning users about their needs and clarifying their initial responses can result in better systems and more

accurate predictions. Too often, stakeholders and users specify certain system requirements because they assume their needs can only be met that way. Often, an alternative approach would be better. For example, a stakeholder may be concerned because there is always too much of some items in stock and not enough of other items. So, the stakeholder might request a totally new and improved inventory control system. An alternative approach is to identify the root cause for poor inventory management. This latter approach might determine that sales forecasting is inaccurate and needs improvement or that production is not capable of meeting the set production schedule. All too often solutions are selected before a complete understanding of the nature of the problem itself is obtained.

- *Identifying and resolving objectives and orientations that conflict.* Different departments in an organization can have different objectives and orientations. The buying department may want to minimize the cost of spare parts by always buying from the lowest-cost supplier, but engineering might want to buy more expensive, higher-quality spare parts to reduce the frequency of replacement. These differences must be identified and resolved before a new purchasing system is developed or an existing one modified.

ESTABLISHING OBJECTIVES FOR SYSTEMS DEVELOPMENT

The overall objective of systems development is to achieve business goals, not technical goals, by delivering the right information to the right person, in the right format, at the right time. The impact a particular system has on an organization's ability to meet its goals determines the true value of that system to the organization. Although all systems should support business goals, some systems are more pivotal in continued operations and goal attainment than others. These systems are called **mission-critical systems**. An order-processing system, for example, is usually considered mission critical. Without it, few organizations could continue daily activities, and they clearly would not meet set goals.

The goals defined for an organization will in turn define the objectives that are set for a system. A manufacturing plant, for example, might determine that minimizing the total cost of owning and operating its equipment is critical to meet production and profit goals. **Critical success factors (CSFs)** are factors that are essential to the success of certain functional areas of an organization. The CSF for manufacturing—minimizing equipment maintenance and operating costs—would be converted into specific objectives for a proposed system. One specific objective might be to alert maintenance planners when a piece of equipment is due for routine preventive maintenance (e.g., cleaning and lubrication). Another objective might be to alert the maintenance planners when the necessary cleaning materials, lubrication oils, or spare parts inventory levels are below specified limits. These objectives could be accomplished either through automatic stock replenishment via electronic data interchange or through the use of exception reports. One study found that different CSFs might be important during different phases of a systems development project.[16]

Regardless of the particular systems development effort, the development process should define a system with specific performance and cost objectives. The success or failure of the systems development effort will be measured against these objectives.

Performance Objectives

The extent to which a system performs as desired can be measured through its performance objectives. System performance is usually determined by such factors as the following:

- *The quality or usefulness of the output.* Is the system generating the right information for a value-added business process or by a goal-oriented decision maker?

---

mission-critical systems
systems that play a pivotal role in an organization's continued operations and goal attainment

critical success factors (CSFs)
factors that are essential to the success of a functional area of an organization

- *The accuracy of the output.* Is the output accurate and does it reflect the true situation? As a result of the accounting scandals of 2002, where some companies overstated revenues or understated expenses, accuracy is becoming more important and top corporate officers are being held responsible for the accuracy of all corporate reports.
- *The quality or usefulness of the format of the output.* Is the output generated in a form that is usable and easily understood? For example, objectives often concern the legibility of screen displays, the appearance of documents, and the adherence to certain naming conventions.
- *The speed at which output is generated.* Is the system generating output in time to meet organizational goals and operational objectives? Objectives such as customer response time, the time to determine product availability, and throughput time are examples.

In some cases, the achievement of performance objectives can be easily measured (e.g., by tracking the time it takes to determine product availability). The achievement of performance objectives is sometimes more difficult to ascertain in the short term. For example, it may be difficult to determine how many customers are lost because of slow responses to customer inquiries regarding product availability. These outcomes, however, are often closely associated with corporate goals and are vital to the long-term success of the organization.

Cost Objectives

The benefits of achieving performance goals should be balanced with all costs associated with the system, including the following:

- *Development costs.* All costs required to get the system up and running should be included.
- *Costs related to the uniqueness of the system application.* A system's uniqueness has a profound effect on its cost. An expensive but reusable system may be preferable to a less costly system with limited use.
- *Fixed investments in hardware and related equipment.* Developers should consider costs of such items as computers, network-related equipment, and environmentally controlled data centers in which to operate the equipment.
- *Ongoing operating costs of the system.* Operating costs include costs for personnel, software, supplies, and such resources as the electricity required to run the system.

CDNow.com is a company that uses the Internet as its only sales channel.

Balancing performance and cost objectives within the overall framework of organizational goals can be challenging. Systems development objectives are important, however, because they allow an organization to allocate resources effectively and efficiently and measure the success of a systems development effort.

SYSTEMS DEVELOPMENT AND E-COMMERCE

As seen in the many examples and cases in this and other chapters, companies are increasingly converting at least some portion of their business to run over the Internet, intranets, or extranets. Applications that are being moved to the Internet include those that support selling products to customers, placing orders with suppliers, and letting customers access information about production, inventory, orders, or accounts receivable. In addition, a number of companies sell their products and services only on the Internet. Internet technology enables companies to extend their information systems beyond their boundaries to reach their customers, suppliers, and partners, allowing them to conduct business much faster, to interact with more people, and to keep one step ahead of the competition. Some companies have been willing to sustain losses from their e-commerce initiatives to gain an advantage for the future.

Building a static Web site to display simple text and graphics is fairly straightforward. However, implementing a dynamic core business application that runs over the Web is much more complicated. Such applications must meet special business needs. They must be able to scale up to support highly variable transactions from potentially thousands of users. Ideally, they can scale up instantly when needed. They must be reliable and fault tolerant, providing continuous availability while processing all transactions accurately. They must also integrate with existing infrastructure, including customer and order databases, existing applications, and enterprise resource planning systems. Development and maintenance must be quick and easy, as business needs may require changing applications on the fly.

Many tools are available for building and running Web applications. The best tools provide components to support applications on an enterprise scale while speeding development. Several vendors provide what is known as an applications server to provide remote access to databases via a corporate intranet, including NetDynamics, SilverStream, WebLogic, Novera Software, Netscape Communications, Microsoft, and IBM. Thomson Financial Services used an application server to build two applications—one tracks job candidates and the second monitors consultants' work hours.

TRENDS IN SYSTEMS DEVELOPMENT AND ENTERPRISE RESOURCE PLANNING

Enterprise resource planning software has reached beyond business processes and is increasingly affecting systems development. Not only are planners considering different types of systems that include ERP software, but the ERP software that is already in place is driving planners to explore and develop different types of systems. Other ERP users are moving from just using the software to run the business to trying to use it to make business decisions.

An important trend in systems development and the use of ERP systems is that companies wish to stay with their primary ERP vendor (SAP, Oracle, PeopleSoft, etc.) instead of looking elsewhere for answers to their data warehousing and production planning needs or developing in-house solutions. Thus, they look to their original ERP vendor to provide these solutions.

A second trend is that many software vendors are building software that integrates with the ERP vendor's package. For example, Aspect Development, a leading supplier of electronic catalogs for high-tech components, has now entered the market for on-line catalogs of maintenance, repair, and operations (MRO) supplies. The company has a product called Morocco that is an MRO supplies catalog for users of SAP's ERP software. Morocco works in conjunction with the

PeopleSoft is a provider of enterprise applications that enable organizations to reduce costs and increase productivity by directly connecting customers, suppliers, partners, and employees to business processes on-line, in real time.

purchasing module of SAP and directs buyers to preferred suppliers specified by the purchasing department. Again, there is less in-house development and more dependence on ERP vendors and their strategic partners to provide enhancements and add-ons to the original ERP package.

A third interesting trend is the increase in the number of companies that, once they have successfully implemented their own company's ERP project, are branching out to provide consulting to other companies.

SYSTEMS DEVELOPMENT LIFE CYCLES

The systems development process is also called a *systems development life cycle* (*SDLC*) because the activities associated with it are ongoing. As each system is being built, the project has timelines and deadlines, until at last the system is installed and accepted. The life of the system continues as it is maintained and reviewed. If the system needs significant improvement beyond the scope of maintenance, if it needs to be replaced because of a new generation of technology, or if the IS needs of the organization change significantly, a new project will be initiated and the cycle will start over.

A key fact of systems development is that the later in the SDLC an error is detected, the more expensive it is to correct (Figure 12.5). Barry Boehm documents this analysis in a classic work.[17] One reason for the mounting costs is that if an error is found in a later phase of the SDLC, the previous phases must be reworked to some extent. Another reason is that the errors found late in the SDLC have an impact on more people. For example, an error found after a system is installed may require retraining users once a "workaround" to the problem has been found. Thus, experienced system developers prefer an approach that will catch errors early in the project life cycle.

Four common systems development life cycles exist: traditional, prototyping, rapid application development (RAD), and end-user development. In addition, companies can outsource the systems development process. With some companies, these approaches are formalized and documented so that system developers have a well-defined process to follow; in other companies, less formalized approaches are used. Keep Figure 12.5 in mind as you are introduced to alternative SDLCs in the next section.

FIGURE 12.5

The later that system changes are made in the SDLC, the more expensive these changes become.

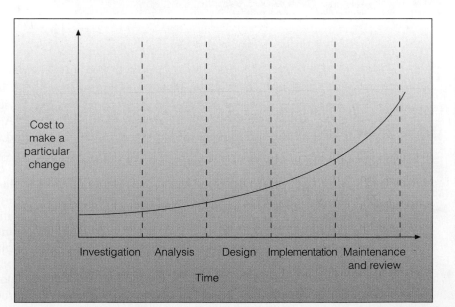

THE TRADITIONAL SYSTEMS DEVELOPMENT LIFE CYCLE

Traditional systems development efforts can range from a small project, such as purchasing an inexpensive computer program, to a major undertaking. The steps of traditional systems development may vary from one company to the next, but most approaches have five common phases: investigation, analysis, design, implementation, and maintenance and review (Figure 12.6).

In the systems investigation phase, potential problems and opportunities are identified and considered in light of the goals of the business. Systems investigation attempts to answer the question "What is the problem, and is it worth solving?" The primary result of this phase is a defined information system project for which business problems or opportunity statements have been created, to which some organizational resources have been committed and for which systems analysis is recommended. Systems analysis attempts to answer the question "What must the information system do to solve the problem?" This phase involves the study of existing systems and work processes to identify strengths, weaknesses, and opportunities for improvement. The major outcome of systems analysis is a list of requirements and priorities. **Systems design** seeks to answer the question "How will the information system do what it must do to obtain the problem solution?" The primary result of this phase is a design that either describes the new system or describes how existing systems will be modified. The system design details system outputs, inputs, and user interfaces; specifies hardware, software, database, telecommunications, personnel, and procedure components; and shows how these components are related. Systems implementation involves creating or acquiring the various system components detailed in the systems design, assembling them, and placing the new or modified system into operation. An important task during this phase is to train the users. Systems implementation results in an installed, operational information system that meets the business needs for which it was developed. The purpose of systems maintenance and review is to ensure the system operates and to modify the system so that it continues to meet changing business needs. As shown in Figure 12.6, a system under development moves from one phase of the traditional SDLC to the next.

systems design
the systems development phase that defines how the information system will do what it must do to obtain the problem solution

FIGURE 12.6

The Traditional Systems Development Life Cycle

Sometimes, information learned in a particular phase requires cycling back to a previous phase.

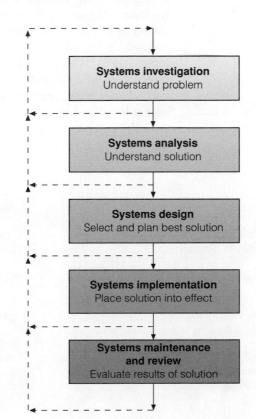

The traditional SDLC allows for a large degree of management control. At the end of each phase, a formal review is performed and a decision is made whether to continue with the project, terminate the project, or perhaps repeat some of the tasks of the current phase. Use of the traditional SDLC also creates much documentation, such as entity-relationship diagrams. This documentation, if kept current, can be useful when it is time to modify the system. The traditional SDLC also ensures that every system requirement can be related to a business need. In addition, resulting products can be reviewed to verify that they satisfy the system requirements and conform to organizational standards.

A number of companies use a standard SDLC. Some are very similar to the one shown in Figure 12.6. The consulting company Recho, for example, uses an SDLC that includes information collection, user requirements, detailed system analysis, design, programming, testing, implementation, and maintenance.[18] Paragon Development Systems uses an SDLC that includes planning, procurement, deployment, management, support, and retirement.[19]

A major problem with the traditional SDLC is that the user does not have access to the solution until the system is nearly complete. Quite often, users get a system that does not meet their real needs because its development was based on the development team's understanding of the needs. The traditional approach is also inflexible. Changes in user requirements cannot be accommodated during development. In spite of its limitations, however, the traditional SDLC is still used for large, complex systems that affect entire businesses, such as TPS and MIS systems. It is also frequently employed on government projects because of the strengths mentioned previously. Table 12.1 lists advantages and disadvantages of the traditional SDLC.

PROTOTYPING

prototyping
an iterative approach to the systems development process

Prototyping takes an iterative approach to the systems development process. During each iteration, requirements and alternative solutions to the problem are identified and analyzed, new solutions are designed, and a portion of the system is implemented. Users are then encouraged to try the prototype and provide feedback (Figure 12.7). Prototyping begins with the creation of a preliminary model of a major subsystem or a scaled-down version of the entire system. For example, a prototype might be developed to show sample report formats and input screens. Once developed and refined, the prototypical reports and input screens are used as models for the actual system, which may be developed using an end-user programming language such as Visual Basic. The first preliminary model is refined to form the second- and third-generation models, and so on until the complete system is developed (Figure 12.8). A number of tools can be used to help implement prototyping.[20] One expert, for example, recommends

TABLE 12.1

Advantages and Disadvantages of Traditional SDLC

Advantages	Disadvantages
Formal review at the end of each phase allows maximum management control.	Users get a system that meets the needs as understood by the developers; this may not be what was really needed.
This approach creates considerable system documentation.	Documentation is expensive and time-consuming to create. It is also difficult to keep current.
Formal documentation ensures that system requirements can be traced back to stated business needs.	Often, user needs go unstated or are misunderstood.
It produces many intermediate products that can be reviewed to see whether they meet the users' needs and conform to standards.	Users cannot easily review intermediate products and evaluate whether a particular product (e.g., data flow diagram) meets their business requirements.

FIGURE 12.7

Prototyping Is an Iterative Approach to Systems Development

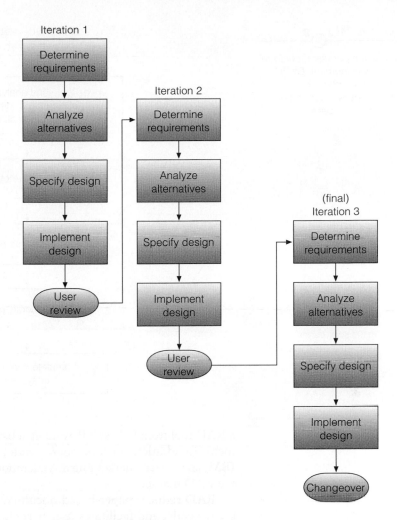

operational prototype

a functioning prototype that accesses real data files, edits input data, makes necessary computations and comparisons, and produces real output

nonoperational prototype

a mockup, or model, that includes output and input specifications and formats

rapid application development (RAD)

a systems development approach that employs tools, techniques, and methodologies designed to speed application development

using virtual reality to help users visualize the completed system for manufacturing companies. This approach can reduce the risks of developing prototypes that don't meet user expectations.

Types of Prototypes

Prototypes can be classified as operational or nonoperational. An **operational prototype** is a prototype that works—accesses real data files, edits input data, makes necessary computations and comparisons, and produces real output. Fully developed financial reports are examples. The operational prototype may access real files but perhaps does no editing of input. A **nonoperational prototype** is a mockup, or model. It typically includes output and input specifications and formats. The outputs include printed reports to managers and the screen layout of reports displayed on personal computers or terminals. The inputs reveal how data is captured, what commands users must enter, and how the system accesses other data files. The primary advantage of a nonoperational prototype is that it can be developed much faster than an operational prototype. Nonoperational prototypes can be discarded, and a fully operational system can be built based on what was learned from the prototypes. The advantages and disadvantages of prototyping are summarized in Table 12.2.

RAPID APPLICATION DEVELOPMENT, AGILE DEVELOPMENT, AND JOINT APPLICATION DEVELOPMENT

Rapid application development (RAD) employs tools, techniques, and methodologies designed to speed application development.[21] For example, PowerBuilder,

FIGURE 12.8

Prototyping is a popular technique in systems development. Each generation of prototype is a refinement of the previous generation based on user feedback.

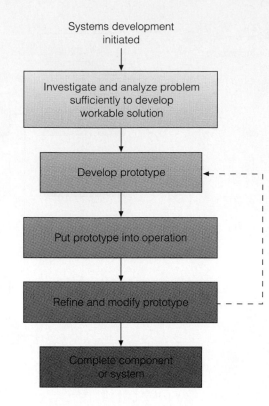

TABLE 12.2

Advantages and Disadvantages of Prototyping

a RAD tool from Sybase's Powersoft subsidiary, is popular with the federal government. In addition, such database vendors as Computer Associates International, IBM, and Oracle market fourth-generation languages and other products targeting the RAD market.

RAD reduces paper-based documentation, automatically generates program source code, and facilitates user participation in design and development activities. With RAD, entire systems are developed in less than six months. The ultimate goal is to accelerate the process so that applications can go into production much sooner than when using other approaches.[22] Prudential Real Estate, for example, used a RAD systems development methodology to reduce the time for it to update its password-protected Internet site.[23] Using RAD, the Internet site is now updated four times a year.

RAD makes adapting to changing system requirements easier. Often called *agile* or *extreme programming* (*EP*), these approaches allow the systems to change as they are being developed.[24] Agile development often requires frequent face-to-face

Advantages	Disadvantages
Users can try the system and provide constructive feedback during development.	Each iteration builds on the previous one. The final solution may be only incrementally better than the initial solution.
An operational prototype can be produced in weeks.	Formal end-of-phase reviews may not occur. Thus, it is very difficult to contain the scope of the prototype, and the project never seems to end.
As solutions emerge, users become more positive about the process and the results.	System documentation is often absent or incomplete, since the primary focus is on development of the prototype.
Prototyping enables early detection of errors and omissions.	System backup and recovery, performance, and security issues can be overlooked in the haste to develop a prototype.

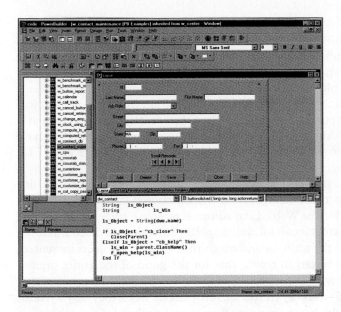

PowerBuilder, a RAD tool from Sybase, is used in both the public and private sectors.

(Source: Courtesy of Sybase, Inc.)

joint application development (JAD)
process for data collection and requirements analysis

TABLE 12.3

Advantages and Disadvantages of RAD

meetings with the systems developers and users. Some predict that agile programming will eventually be used by most IT departments.[25] While the development process can be fluid and flexible, it can become complex and time consuming with larger projects. As a result, agile development is most appropriate for smaller projects.

RAD makes extensive use of the **joint application development (JAD)** process for data collection and requirements analysis. Originally developed by IBM Canada in the 1970s, JAD involves group meetings in which users, stakeholders, and IS professionals work together to analyze existing systems, propose possible solutions, and define the requirements of a new or modified system. JAD groups consist of both problem holders and solution providers. A group normally requires one or more top-level executives who initiate the JAD process, a group leader for the meetings, potential users, and one or more individuals who act as secretaries and clerks to record what is accomplished and to provide general support for the sessions. Many companies have found that groups can develop better requirements than individuals working independently and have assessed JAD as a very successful development technique.

Throughout the RAD project, users and developers work together as one team. This teamwork promotes healthy risk taking and team-based decision making, resulting in better systems with shorter delivery dates. If the entire system is too large to be completed in less than six months, it is broken into subsystems and delivered subsystem by subsystem. The first subsystem may be delivered in three to four months, with no delivery date more than six months after the last one. This subdividing leads to less waste, because even if there is a serious error in the system, only one subsystem has to be rebuilt.

RAD should not be used on every software development project. In general, it is best suited for decision support and management information systems and less well suited for transaction processing applications. During a RAD project, the level of participation of stakeholders and users is much higher than in other approaches. They become working members of the team and can be expected to spend more than 50 percent of their time producing project outcomes. This time commitment can be a problem if the users are also needed to perform their normal business role. For this reason, RAD team participants are often taken off their normal assignments and put full-time on the RAD project. Because of the full-time commitment and intense schedule deadlines, RAD is a high-pressure development approach that can easily result in employee burnout. Table 12.3 lists advantages and disadvantages of RAD.

Advantages	Disadvantages
For appropriate projects, this approach puts an application into production sooner than any other approach.	This intense SDLC can burn out systems developers and other project participants.
Documentation is produced as a by-product of completing project tasks.	This approach requires systems analysts and users to be skilled in RAD system development tools and RAD techniques.
RAD forces teamwork and lots of interaction between users and stakeholders.	RAD requires a larger percentage of stakeholders' and users' time than other approaches.

THE END-USER SYSTEMS DEVELOPMENT LIFE CYCLE

Systems development initiatives arise from a wide variety of individuals and organizational areas, including users. The proliferation of general-purpose information technology and the flexibility of many packaged software programs have allowed non-IS employees to independently develop information systems that meet their needs. Such employees have believed that, by bypassing the formal requisitioning of resources from the IS department, they can develop systems more quickly. In addition, these individuals often believe that they have better insight into their own needs and can develop systems better suited for their purposes.

End-user-developed systems range from the very small (e.g., a software routine to merge form letters) to those of significant organizational value (such as customer contact databases for the Web). Like all projects, some end-user-developed systems fail, and others are successful. Initially, IS professionals discounted the value of these projects. As the number and magnitude of these projects increased, however, IS professionals began to realize that for the good of the entire organization, their involvement with these projects needed to increase.

end-user systems development
any systems development project in which the primary effort is undertaken by a combination of business managers and users

Today, the term **end-user systems development** describes any systems development project in which the primary effort is undertaken by a combination of business managers and users. Rather than ignoring these initiatives, astute IS professionals encourage them by offering guidance and support. Technical assistance, communication of standards, and the sharing of "best practices" throughout the organization are just some of the ways IS professionals work with motivated managers and employees undertaking their own systems development. In this way, end-user-developed systems can be structured as complementary to, rather than in conflict with, existing and emerging information systems. In addition, this open communication among IS professionals, managers of the affected business area, and users allows the IS professionals to identify specific initiatives so that additional organizational resources, beyond those available to business managers or users, are provided for its development.

There are disadvantages of end-user systems development. Some end users don't have the training to effectively develop and test a system. Multimillion dollar mistakes, for example, can be made using faulty spreadsheets that were never tested. Some end-user systems are poorly documented. When they are updated, errors can be introduced that make the system error prone. In addition, some end users spend time and corporate resources developing systems that were already available.

Many end users are already demonstrating their systems development capability by designing and implementing their own PC-based systems.

(Source: © Eyewire/Getty Images.)

OUTSOURCING

Many companies hire an outside consulting firm that specializes in systems development to take over some or all of its development activities. American Express, for example, hired IBM for about $4 billion over 7 years to manage its Web site, network servers, data storage, and help-desk operations.[26] The deal also moved about 2,000 American Express employees to IBM's Global Services division. This approach, called *outsourcing*, is gaining in popularity. In another outsourcing deal, AT&T has agreed to pay Accenture, a computer consulting company, about $2.6 billion to provide guidance on cutting costs and improving efficiency for AT&T's long-distance division. Table 12.4 describes the circumstances in which outsourcing is a good idea.

Reducing costs, obtaining state-of-the-art technology, eliminating staffing and personnel problems, and increasing technological flexibility are reasons that companies have used the outsourcing approach.

Reason	Example
When a company believes it can cut costs.	PacifiCare outsourced its IS operations to IBM and Keane, Inc.[27] PacifiCare hopes the outsourcing will save it about $400 million over ten years.
When there is limited opportunity for the firm to distinguish itself competitively through a particular information system operation or application.	Kodak outsourced its IS operations, including mainframe processing, telecommunications, and personal computer support, because there was limited opportunity to distinguish the company through these IS operations. Kodak kept application development and support inhouse because it thought that these activities had competitive value.
When uninterrupted information system service is not crucial.	Airline reservations or catalog shopping systems are "mission critical" and should not be trusted outside the firm.
When outsourcing does not strip the company of technical know-how required for future IS innovation.	Firms must ensure that their IS staffs remain technically up-to-date and have the expertise to develop future applications.
When the firm's existing information system capabilities are limited, ineffective, or technically inferior.	A company might use outsourcing to help it make the transition from a centralized mainframe environment to a distributed client/server environment.
When a firm is downsizing. The decision to outsource systems development is often a response to downsizing, which reduces the number of employees or managers, equipment and systems, and even functions and departments. Outsourcing allows companies to downsize the IS department and alleviate difficult financial situations by reducing payroll and other expenses.	First Fidelity, a major bank, used outsourcing as part of a program to reduce the number of employees by 1,600 and slash expenses by $85 million.

TABLE 12.4

When to Use Outsourcing for Systems Development

Reducing costs is a primary reason for outsourcing. Ingersoll-Rand, an industrial products company, used outsourcing for its network security system to avoid the costs of hiring network security specialists.[28] According to Dave Malicoat, manager of Internet Services, "We don't have the people necessary to do the required security tasks around the clock." Shutterfly, Inc., used outsourcing to handle e-mail questions from customers because of the high cost of IS personnel and real estate in Silicon Valley.[29] Shutterfly is using a 65,000-square-foot facility in Bangalore, India, to respond to customer e-mails. Some companies also realize that it can be hard to find talented IS people. "There's not enough smart security people on the planet for all companies to have their own network defense centers," says Don Walker, CEO of Vertict.[30] To improve the quality and consistency of its 25 Web sites, Gillette Company outsourced them all to two companies.[31]

There are disadvantages to the outsourcing approach, however. Companies using outsourcing may be asked to sign complicated and restrictive legal contracts that may be difficult to change. Internal expertise and loyalty can suffer under an outsourcing arrangement. When a company outsources, key information systems personnel with expertise in technical and business functions are no longer needed. Once these IS employees leave, their experience with the organization and expertise in information systems is lost. In addition, outsourcing large and complex projects can be very expensive in the long run.

A number of companies offer outsourcing services—from general systems development to specialized services. IBM's Global Services, for example, is one of the largest full-service outsourcing and consulting services available.[32] IBM has consultants located in offices around the world. Electronic Data Systems (EDS) is another large company that specializes in consulting and outsourcing.[33]

EDS has approximately 140,000 employees in almost 60 countries and more than 9,000 clients worldwide. In 2001, the company signed $31.4 billion in new contracts for consulting and outsourcing. Accenture, which was once part of Arthur Andersen, is another company that specializes in consulting and outsourcing.[34] The company has more than 75,000 employees in 47 countries.

FACTORS AFFECTING SYSTEMS DEVELOPMENT SUCCESS

Successful systems development means delivering a system that meets user and organizational needs—on time and within budget.[35] As seen in the "Ethical and Societal Issues" box, companies can spend substantial sums on getting different systems to work together. Systems development leaders have identified factors that can contribute to successful systems development efforts—at a reasonable cost. These factors are discussed next.

DEGREE OF CHANGE

A major factor that affects the quality of systems development is the degree of change associated with the project. The scope can vary from implementing minor enhancements to an existing system to major reengineering. The project team needs to recognize where they are on this spectrum of change.

Continuous Improvement versus Reengineering

As discussed in Chapter 2, continuous improvement projects do not require significant business process or information system changes nor retraining of individuals; thus, they have a high degree of success. Typically, because continuous improvements involve minor improvements, they also have relatively modest benefits. On the other hand, reengineering involves fundamental changes in how the organization conducts business and completes tasks. The factors associated with successful reengineering are similar to those of any development effort, including top management support, clearly defined corporate goals and systems development objectives, and careful management of change. Major reengineering projects tend to have a high degree of risk but also a high potential for major business benefits (Figure 12.9).

Managing Change

The ability to manage change is critical to the success of systems development. New systems inevitably cause change. For example, the work environment and habits of users are invariably affected by the development of a new information system. Unfortunately, not everyone adapts easily. Managing change requires the ability to recognize existing or potential problems (particularly the concerns of

FIGURE 12.9

Degree of change can greatly affect the probability of a project's success.

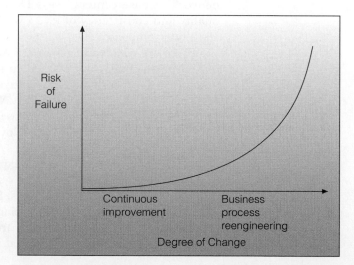

Information Systems Running Amok?

Doug Elix, a top IBM executive, has made a plea to the computer industry to work to reduce the complexity of information systems. Elix is an Australian who heads IBM's $35 billion global services division, which accounts for 40 percent of the company's revenue and gross profits. In his address to industry leaders at the World Congress on IT in Adelaide, he stated that technology has become more and more complex to manage, despite incredible advances. "It's staggering to think how much we are spending on integration of complex technology, rather than spending time and money on things that will improve how our organizations perform, or improve the way we live our lives."

According to research company IDC, U.S. companies spent $2.6 trillion on information systems during the 1990s, and companies in the Asia-Pacific region spent more than $1.3 trillion.

A study by the Gartner Group found that integrating incompatible programs and systems accounts for 40 percent of all IS spending in most organizations. Application integration ranks as one of the top strategic priorities for chief information officers around the world. Elix blames the complexity of ISs on the large number of competing companies supplying corporate customers. "Years ago, companies would have purchased all their IT requirements from a single vendor. Now there are multiple suppliers and their equipment has to be interconnected."

At the current pace, 200 million more IS workers will be needed to support the billion people, millions of businesses, and one trillion devices that will be connected by the end of this decade. In short, Elix says, "The technology is outstripping the human capacity to manage it."

Elix recommends industry collaboration and standards-based systems as a solution. He called on the industry to support common Web and networking standards, along with nonproprietary operating systems such as Linux. "I propose that we as an industry shed our last vestiges of building based on proprietary architectures." Proprietary architecture refers to systems that require unique software and are unable to communicate with systems created by another manufacturer without the use of middleware, software written to mediate between two separate systems.

Most of the major players recognize the value of collaboration in the IS industry. One effort at standardization is the Web Services Interoperability Organization (WS-I), composed of dozens of industry leaders—many aggressive competitors—that are working together to promote information system interoperability across platforms, operating systems, and programming languages.

Besides agreeing on industry standards, Elix also recommends more automation in IS management, for example, a new system that relies less on human intervention and acts more like the human body, which is the "perfect autonomic computing machine." Elix's vision of the future includes secure, self-maintaining, self-diagnosing, and self-healing computing systems that are linked by grids combining the massive processing power of several organizations. He sees grid computing turning the Net into "a living, interactive, interconnected virtual supercomputer."

Through cooperation among industry leaders, replacing proprietary architectures with standardized systems, and developing automated systems that require less human intervention, Elix maintains that we can regain control of IS complexity. Information system investment can then satisfy the needs of the users and assist in achieving the goals of the organization more fully and economically.

Discussion Questions

1. How do you think the complexity of information systems has affected organizations' willingness to expand their technological infrastructure? What types of systems development projects are more likely to be given the go-ahead?

2. What considerations are necessary during the systems investigation stage of development to ensure that a proposed change does not overcomplicate the existing system?

Critical Thinking Questions

3. How can businesses and organizations persuade tech companies to work together to reduce the cost of technology integration? What type of leverage is available?

4. Cooperation between competing companies and the establishment of industry standards often work to level the playing field and wrest monopoly power away from those that wield it. Compare and contrast a technological future where one company has monopoly control of information system technology, and a standards-based future. Which is best for technological development? Why?

Sources: Ian Grayson, "It's Running Amok, IBM Chief Warns," *Australian IT*, March 1, 2002, http://australianit.news.com.au; Adrienne Perry, "IBM: Solve Hi-Tech Complexity," *Infotech Weekly*, March 11, 2002, p.5.

users) and deal with them before they become a serious threat to the success of the new or modified system. Here are several of the most common problems:

- Fear that the employee will lose his or her job, power, or influence within the organization
- Belief that the proposed system will create more work than it eliminates

- Reluctance to work with "computer people"
- Anxiety that the proposed system will negatively alter the structure of the organization
- Belief that other problems are more pressing than those solved by the proposed system or that the system is being developed by people unfamiliar with "the way things need to get done"
- Unwillingness to learn new procedures or approaches

Preventing or dealing with these types of problems requires a coordinated effort from stakeholders and users, managers, and information systems personnel. One remedy is simply to talk with all people concerned and learn what their biggest concerns are. Management can then deal with those concerns and try to eliminate them. Once immediate concerns are addressed, people can become part of the project team.

QUALITY AND STANDARDS

Another key success factor is the quality of project planning.[36] The bigger the project, the more likely that poor planning will lead to significant problems. A federal jury, for example, found the maker of navigational software partly responsible for an airplane crash near Cali, Colombia. Poor systems development planning can be deadly.

Many companies find that large systems projects fall behind schedule, go over budget, and do not meet expectations. Although proper planning cannot guarantee that these types of problems will be avoided, it can minimize the likelihood of their occurrence. Good systems development is not automatic. Certain factors contribute to the failure of systems development projects. These factors and countermeasures to eliminate or alleviate the problem are summarized in Table 12.5.

TABLE 12.5

Project Planning Issues Frequently Contributing to Project Failure

Factor	Countermeasure
Solving the wrong problem	Establish a clear connection between the project and organizational goals.
Poor problem definition and analysis	Follow a standard systems development approach.
Poor communication	Communicate, communicate, communicate.
Project is too ambitious	Narrow the project focus to address only the most important business opportunities.
Lack of top management support	Identify the senior manager who has most to gain from the success of the project, and recruit this individual to champion the project.
Lack of management and user involvement	Identify and recruit key stakeholders to be active participants in the project.
Inadequate or improper system design	Follow a standard systems development approach.
Lack of standards	Implement a standards system, such as ISO 9001.
Poor testing and implementation	Plan sufficient time for this activity.
Users are unable to use the system effectively	Develop a rigorous user-training program and budget sufficient time in the schedule to execute it.
Lack of concern for maintenance	Include an estimate of people effort and costs for maintenance in the original project justification.

ISO 9001 are international quality standards used by IS and other organizations to ensure quality of products and services.

The development of information systems requires a constant trade-off of schedule and cost versus quality. Historically, the development of application software has put an overemphasis on schedule and cost to the detriment of quality. Techniques, such as use of the ISO 9001 standards, have been developed to improve the quality of information systems. ISO 9001 are international quality standards originally developed in Europe in 1987. The most recent version was published in 2000. These standards address customer satisfaction and are the only standards in the ISO 9000 family where third-party certification can be achieved. Adherence to ISO 9001 is a requirement in many international markets.[37] Many companies in the United States and around the world strive to achieve ISO 9001 certification.[38]

Many IS organizations have incorporated ISO 9001, total quality management, and statistical process control principles into the way they produce software. Often, to ensure the quality of the systems development process and finished product, an IS organization will form its own quality assurance groups to work with project teams and encourage them to follow established standards.

In addition to quality standards, the project management process itself profoundly affects the success of the resulting system. One study investigated the impact of project management on e-commerce applications, which were estimated to be worth $1.3 trillion by 2003.[39] The study concluded that project management needs to incorporate balance and leadership to make sure that customer and stakeholder demands are met or exceeded. Project management can be especially difficult for global projects, which often require project team members to speak different languages, adapt to different cultures and customs, gain input from local managers and stakeholders, and analyze ethical and legal issues throughout the project.[40]

THE CAPABILITY MATURITY MODEL (CMM)

Organizational experience with the systems development process is also a key factor for systems development success.[41] The Capability Maturity Model (CMM) is one way to measure this experience. It is based on research done at Carnegie Mellon University and work by the Software Engineering Institute (SEI). CMM is a measure of the maturity of the software development process in an organization. CMM grades an organization's systems development maturity using five levels from initial to optimizing (Figure 12.10). A brief description of each level follows.

1. *Initial.* This level is typical of organizations inexperienced with software and systems development. This level often has an ad hoc or even a chaotic development process.
2. *Repeatable.* The second level tracks development costs, schedules, and functionality. The discipline to repeat previous systems development success is in place.
3. *Defined.* With the third level, organizations use documented and defined procedures. All projects done by the organization use these standardized approaches to develop software and systems. Programming standards are often used at this level.

FIGURE 12.10

Systems Development Maturity Based on the Capability Maturity Model (CMM)

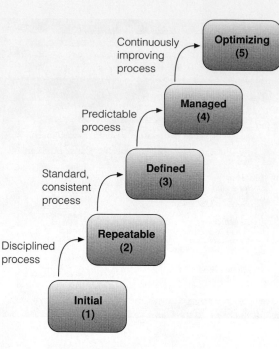

4. *Managed.* At this level, organizations use detailed measures of the systems development process to help manage the process and improve software and systems quality.

5. *Optimized.* This is the highest level of experience and maturity. Continuous improvement is used to strengthen all aspects of the systems development process. Organizations at this level often initiate innovative projects. The goal is to optimize all aspects of the systems development effort.

The CMM model has been popular in the U.S. and around the world, and SEI certifies organizations as being at one of the five levels. Any organization can seek certification, and many computer-consulting companies attempt to be certified at the highest level (optimization). Wipro GE Medical, for example, received Level 5 certification. The company develops advanced medical software for computerized tomography (CT) scanners, magnetic resonance imaging (MRI) devices, and other medical equipment. Sasken Communications Company is a telecommunications and consulting company that has also achieved Level 5 certification. According to the lead assessor for SEI, "Sasken is one of the few organizations which has achieved excellence by providing operational flexibility in projects in conjunction with consistency and standardization across all business units. The company has been able to reach Level 5 maturity spanning its product offerings as well as its services; that makes it a unique achievement."[42]

USE OF PROJECT MANAGEMENT TOOLS

Project management involves planning, scheduling, directing, and controlling human, financial, and technological resources for a defined task whose result is achievement of specific goals and objectives. A **project schedule** is a detailed description of what is to be done. Each project activity, the use of personnel and other resources, and expected completion dates are described. A **project milestone** is a critical date for the completion of a major part of the project. The completion of program design, coding, testing, and release are examples of milestones for a programming project. The **project deadline** is the date the entire project is to be completed and operational—when the organization can expect to begin to reap the benefits of the project. One company offers a 20 percent refund if it doesn't meet a client's project deadline. In addition, any additional work done after the project deadline is performed free of charge.

project schedule
detailed description of what is to be done

project milestone
a critical date for the completion of a major part of the project

project deadline
date the entire project is to be completed and operational

critical path
activities that, if delayed, would delay the entire project

In systems development, each activity has an earliest start time, earliest finish time, and slack time, which is the amount of time an activity can be delayed without delaying the entire project. The **critical path** consists of all activities that, if delayed, would delay the entire project. These activities have zero slack time. Any problems with critical-path activities will cause problems for the entire project. To ensure that critical-path activities are completed in a timely fashion, formalized project management approaches have been developed. A number of tools, such as Project by Microsoft, are available to help compute these critical project attributes.

Program Evaluation and Review Technique (PERT)
a formalized approach for developing a project schedule

Gantt chart
a graphical tool used for planning, monitoring, and coordinating projects

Although the steps of systems development seem straightforward, larger projects can become complex, requiring literally hundreds or thousands of separate activities. For these systems development efforts, formal project management methods and tools become essential. A formalized approach called **Program Evaluation and Review Technique (PERT)** creates three time estimates for an activity: shortest possible time, most likely time, and longest possible time. A formula is then applied to come up with a single PERT time estimate. A **Gantt chart** is a graphical tool used for planning, monitoring, and coordinating projects; it is essentially a grid that lists activities and deadlines. Each time a task is completed, a darkened line is placed in the proper grid cell to indicate the completion of a task (Figure 12.11).

FIGURE 12.11

Sample Gantt Chart

A Gantt chart shows progress through systems development activities by putting a bar through appropriate cells.

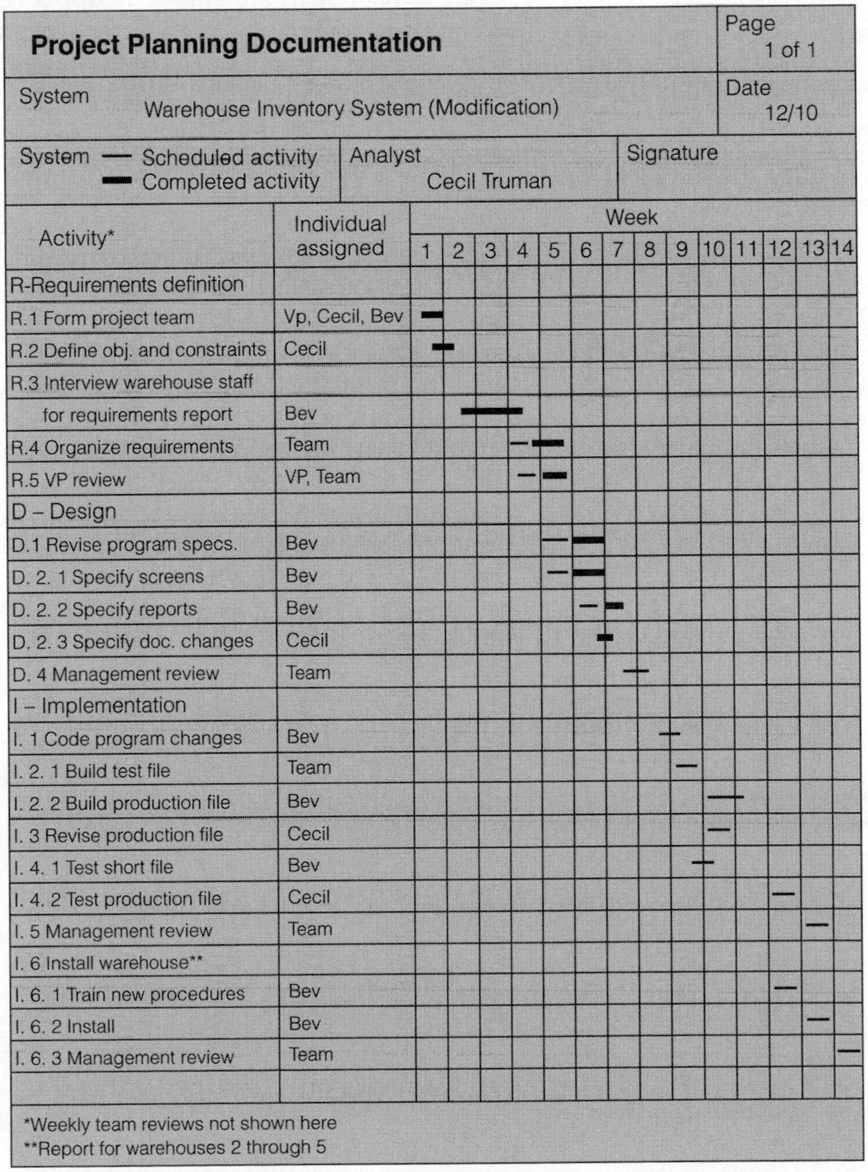

Project Planning Documentation
Page 1 of 1
System Warehouse Inventory System (Modification)
Date 12/10
System — Scheduled activity
▬ Completed activity
Analyst Cecil Truman
Signature
Activity*
Individual assigned
Week
1 2 3 4 5 6 7 8 9 10 11 12 13 14
R-Requirements definition
R.1 Form project team — Vp, Cecil, Bev
R.2 Define obj. and constraints — Cecil
R.3 Interview warehouse staff for requirements report — Bev
R.4 Organize requirements — Team
R.5 VP review — VP, Team
D – Design
D.1 Revise program specs. — Bev
D. 2. 1 Specify screens — Bev
D. 2. 2 Specify reports — Bev
D. 2. 3 Specify doc. changes — Cecil
D. 4 Management review — Team
I – Implementation
I. 1 Code program changes — Bev
I. 2. 1 Build test file — Team
I. 2. 2 Build production file — Bev
I. 3 Revise production file — Cecil
I. 4. 1 Test short file — Bev
I. 4. 2 Test production file — Cecil
I. 5 Management review — Team
I. 6 Install warehouse**
I. 6. 1 Train new procedures — Bev
I. 6. 2 Install — Bev
I. 6. 3 Management review — Team
*Weekly team reviews not shown here
**Report for warehouses 2 through 5

Both PERT and Gantt techniques can be automated using project management software. This software monitors all project activities and determines whether activities and the entire project are on time and within budget. Project management software also has workgroup capabilities to handle multiple projects and to allow a team of people to interact with the same software. Project management software helps managers determine the best way to reduce project completion time at the least cost. Many project managers, however, fear that the quality of a systems development project will suffer with shortened deadlines and that slack time should be added back to the schedule as a result.[43] One research study conducted at the Harvard Business School, however, revealed that adding slack time does not necessarily improve software quality. Several project management software packages are identified in Table 12.6.

USE OF COMPUTER-AIDED SOFTWARE ENGINEERING (CASE) TOOLS

computer-aided software engineering (CASE)
tools that automate many of the tasks required in a systems development effort and enforce adherence to the SDLC

upper-CASE tools
tools that focus on activities associated with the early stages of systems development

lower-CASE tools
tools that focus on the later implementation stage of systems development

integrated-CASE tools (I-CASE)
tools that provide links between upper- and lower-CASE packages, thus allowing lower-CASE packages to generate program code from upper-CASE package designs

Computer-aided software engineering (CASE) tools automate many of the tasks required in a systems development effort and enforce adherence to the SDLC, thus instilling a high degree of rigor and standardization to the entire systems development process.

CASE packages that focus on activities associated with the early stages of systems development are known as **upper-CASE tools**. These packages provide automated tools to assist with systems investigation, analysis, and design activities. Other CASE packages, called **lower-CASE tools**, focus on the later stages of systems development and are capable of automatically generating structured program code. Prover Technology, for example, has developed a lower-CASE tool that searches for programming bugs.[44] The CASE tool searches for all possible design scenarios to make sure that the program is error free. Some CASE tools provide links between upper- and lower-CASE packages, thus allowing lower-CASE packages to generate program code from upper-CASE package designs. These tools are called **integrated-CASE tools (I-CASE)**. Companies that produce CASE tools include Accenture, Microsoft, Oracle, and others.

As with any team, coordinating the efforts of members of a systems development team can be a problem. So, many CASE tools allow more than one person to work on the same system at the same time via a multiuser interface, which coordinates and integrates the work performed by all members of the design team. With this facility, a person working on one aspect of systems development can automatically share his or her results with someone working on another aspect of the system. Oracle Designer and Developer CASE tools, for example, can help systems analysts automate and simplify the development process for database systems. Advantages and disadvantages of CASE tools are listed in Table 12.7.

TABLE 12.6

Selected Project Management Software Packages

Software	Vendor
BeachBox '98	NetSQL Partners
Job Order	Management Software
OpenPlan	Welcom
Project	Microsoft
Project Scheduler	Scitor
Super Project	Computer Associates

TABLE 12.7

Advantages and Disadvantages of CASE Tools

Advantages	Disadvantages
Produce systems with a longer effective operational life	Produce initial systems that are more expensive to build and maintain
Produce systems that more closely meet user needs and requirements	Require more extensive and accurate definition of user needs and requirements
Produce systems with excellent documentation	May be difficult to customize
Produce systems that need less systems support	Require more training of maintenance staff
Produce more flexible systems	May be difficult to use with existing systems

OBJECT-ORIENTED SYSTEMS DEVELOPMENT

The success of a systems development effort can depend on the specific programming tools and approaches used. As mentioned in Chapter 4, object-oriented (OO) programming languages allow the interaction of programming objects—consisting of both data and the actions that can be performed on the data. So, an object could be data about an employee and all the operations (such as payroll, benefits, and tax calculations) that might be performed on the data.

Developing programs and applications using OO programming languages involves constructing modules and parts that can be reused in other programming projects. Chapter 4 discussed a number of programming languages that use the object-oriented approach, including Visual Basic, C++, and Java. These languages allow systems developers to take advantage of the OO approach, making program development faster and more efficient, resulting in lower costs. Modules can be developed internally or obtained from an external source. Once a company has the programming modules, programmers and systems analysts can modify them and integrate them with other modules to form new programs.

object-oriented systems development (OOSD)
approach to systems development that combines the logic of the systems development life cycle with the power of object-oriented modeling and programming

Object-oriented systems development (OOSD) combines the logic of the systems development life cycle with the power of object-oriented modeling and programming. OOSD follows a defined systems development life cycle, much like the SDLC. The life cycle phases can be, and usually are, completed with many iterations. Object-oriented systems development typically involves the following:

1. *Identifying potential problems and opportunities within the organization that would be appropriate for the OO approach.* This process is similar to traditional systems investigation. Ideally, these problems or opportunities should lend themselves to the development of programs that can be built by modifying existing programming modules.

2. *Defining what kind of system users require.* This analysis means defining all the objects that are part of the user's work environment (object-oriented analysis). The OO team must study the business and build a model of the objects that are part of the business (such as a customer, an order, or a payment). Many of the CASE tools discussed in the previous section can be used, starting with this step of OOSD.

3. *Designing the system.* This process defines all the objects in the system and the ways they interact (object-oriented design). Design involves developing logical and physical models of the new system by adding details to the object model started in analysis.

4. *Programming or modifying modules.* This implementation step takes the object model begun during analysis and completed during design and turns it into a set of interacting objects in a system. Object-oriented programming languages are designed to allow the programmer to create classes of objects in the computer system that correspond to the objects in the actual business process. Objects such as customer, order, and payment are redefined as computer system objects—a customer screen, an order-entry menu, or a dollar sign icon. Programmers then write new modules or modify existing ones to produce the desired programs.

5. *Evaluation by users.* The initial implementation is evaluated by users and improved. Additional scenarios and objects are added, and the cycle repeats. Finally, a complete, tested, and approved system is available for use.

6. *Periodic review and modification.* The completed and operational system is reviewed at regular intervals and modified as necessary.

With the object-oriented systems development (OOSD) approach, a project can be broken down into a group of objects that interact. Instead of requiring thousands or millions of lines of detailed computer instructions or code, the systems development project might require a few dozen or maybe a hundred objects. So, developers look for existing objects that may be reused when beginning a new systems development project. This reuse feature of OOSD is one of its big advantages. It greatly simplifies the project, reduces the time needed to complete the project, and can significantly lessen the possibility of errors or bugs. Although it is still necessary to go through systems investigation and analysis to determine business objects and system requirements, systems design and implementation are often easier and cost less because some of the programming modules have already been tested and proven. Also, maintenance can be simplified if programming objects can be reused to upgrade existing software and systems. Finally, because system objects are self-contained units, they can be changed or replaced with less disruption to the rest of the system. These benefits are the main reasons for the interest in object-oriented systems development.

SYSTEMS INVESTIGATION

As discussed earlier in the chapter, systems investigation is the first phase in the traditional SDLC of a new or modified business information system. The purpose is to identify potential problems and opportunities and consider them in light of the goals of the company. In general, systems investigation attempts to uncover answers to the following questions:

1. What primary problems might a new or enhanced system solve?
2. What opportunities might a new or enhanced system provide?
3. What new hardware, software, databases, telecommunications, personnel, or procedures will improve an existing system or are required in a new system?
4. What are the potential costs (variable and fixed)?
5. What are the associated risks?

INITIATING SYSTEMS INVESTIGATION

Because systems development requests can require considerable time and effort to implement, many organizations have adopted a formal procedure for initiating systems development, beginning with systems investigation. The **systems request form** is a document that is filled out by someone who wants the IS department to initiate systems investigation. This form typically includes the following information:

systems request form
document filled out by someone who wants the IS department to initiate systems investigation

- Problems in or opportunities for the system
- Objectives of systems investigation

- Overview of the proposed system
- Expected costs and benefits of the proposed system

The information in the systems request form helps to rationalize and prioritize the activities of the IS department. Based on the overall IS plan, the organization's needs and goals, and the estimated value and priority of the proposed projects, managers make decisions regarding the initiation of each systems investigation for such projects.

PARTICIPANTS IN SYSTEMS INVESTIGATION

Once a decision has been made to initiate systems investigation, the first step is to determine what members of the development team should participate in the investigation phase of the project. Members of the development team change from phase to phase (Figure 12.12).

Ideally, functional managers are heavily involved during the investigation phase. Other members could include users or stakeholders outside management, such as an employee who helped initiate systems development. The technical and financial expertise of others participating in the investigation would help the team determine whether the problem is worth solving. The members of the development team who participate in the investigation are then responsible for gathering and analyzing data, preparing a report justifying systems development, and presenting the results to top-level managers.

FEASIBILITY ANALYSIS

A key step of the systems investigation phase is **feasibility analysis**, which assesses technical, economic, legal, operational, and schedule feasibility (Figure 12.13). **Technical feasibility** is concerned with whether the hardware, software, and other system components can be acquired or developed to solve the problem. Quicken Loans, Inc., for example, is investigating the technical feasibility of using electronic signatures for its business-to-customer transactions.[45] According to a company representative, "We are trying to make applying for mortgages as easy as applying for a credit card."

Economic feasibility determines whether the project makes financial sense and whether predicted benefits offset the cost and time needed to obtain them. Delta Technology, Inc., a division of Delta Airlines, for example, investigates the economic feasibility for most of its projects.[46] According to Chief Technology Officer Curtis Robb, "Now, we're reviewing all projects we do in infrastructure to make sure they're absolutely necessary in terms of either lowering our operating costs or supporting new business functions." Putnam Lovell Securities investigated the economic feasibility of sending research reports electronically instead of through the mail.[47] Economic analysis revealed that the new approach could

feasibility analysis
assessment of the technical, operational, schedule, economic, and legal feasibility of a project

technical feasibility
assessment of whether the hardware, software, and other system components can be acquired or developed to solve the problem

economic feasibility
determination of whether the project makes financial sense and whether predicted benefits offset the cost and time needed to obtain them

FIGURE 12.12

The Systems Investigation Team
The team is made up of upper- and middle-level managers, a project manager, IS personnel, users, and stakeholders.

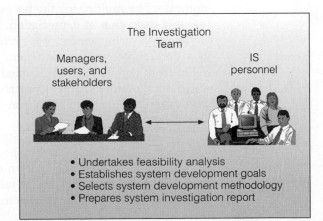

The Investigation Team

Managers, users, and stakeholders

IS personnel

- Undertakes feasibility analysis
- Establishes system development goals
- Selects system development methodology
- Prepares system investigation report

save the company up to $500,000 a year. Economic feasibility can involve cashflow analysis such as that done in net present value or internal rate of return (IRR) calculations.

Legal feasibility determines whether laws or regulations may prevent or limit a systems development project. For example, an Internet site that allowed users to share music without paying musicians or music producers was sued. Legal feasibility involves an analysis of existing and future laws to determine the likelihood of legal action against the systems development project and the possible consequences.

Operational feasibility is a measure of whether the project can be put into action or operation. It can include logistical and motivational (acceptance of change) considerations. Motivational considerations are very important because new systems affect people and data flows and may have unintended consequences. As a result, power and politics may come into play, and some people may resist the new system. Because of deadly hospital errors, a health care consortium called the Leapfrog Group is looking into the operational feasibility of developing a new computerized physician-order entry system.[48] The new system would require that all prescriptions and every order a doctor gives to staff be entered into the computer. The computer then checks for drug allergies and interactions between drugs. If operationally feasible, the new system could save lives and lawsuits.

Schedule feasibility determines whether the project can be completed in a reasonable amount of time—a process that involves balancing the time and resource requirements of the project with other projects.

Net present value is the preferred approach for ranking competing projects and for determining economic feasibility. The net present value represents the net amount by which project savings exceed project expenses, after allowing for the cost of capital and the passage of time. The cost of capital is the average cost of funds used to finance the operations of the business. It represents the minimum desired rate of return on an investment; thus, it is also called the *hurdle rate*. Net present value takes into account that a dollar returned at a later date is not worth as much as one received today, since the dollar in hand can be invested to earn profits or interest in the interim. Spreadsheet programs, such as Excel, have built-in functions to compute the net present value and internal rate of return. The net present value is the sum of the expected cash flows from each year of the project and can be expressed as follows:

$$\text{Net present value} = \sum_{t=1}^{n} (\text{CF}_t)/(1+k)^t$$

where CF_t is the expected cash flow in period t and k is the project's cost of capital.

Since income tax payments represent a disbursement of cash, all cash flows affecting taxable income are credited to the project on an after-tax basis (i.e., computed by multiplying the tax complement by the pretax cash flow), with the notable exception of depreciation. Since depreciation expense does not involve actual cash disbursement, it is excluded from cash flows. However, depreciation is a deductible expense for federal tax purposes. Therefore, the amount of tax relief from depreciation is included in cash flows. It is computed by multiplying the federal tax rate by the depreciation expense. Table 12.8 illustrates these concepts for a sample project in a firm with a tax rate of 36 percent and a cost of capital of 20 percent.

The key to making accurate estimates of the cash flows associated with a project is to involve the business managers responsible for the business functions served. Another key resource is an organization's financial manager, who should be very familiar with net present value analysis. If a systems development project is determined to be feasible, systems investigation will formally begin.

legal feasibility
determination of whether laws or regulations may prevent or limit a systems development project

operational feasibility
measure of whether the project can be put into action or operation

schedule feasibility
determination of whether the project can be completed in a reasonable amount of time

net present value
the preferred approach for ranking competing projects and determining economic feasibility

FIGURE 12.13

Technical, Economic, Legal, Operational, and Schedule Feasibility

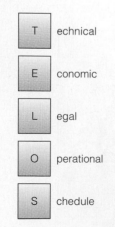

T	echnical
E	conomic
L	egal
O	perational
S	chedule

Cash Flow (in thousands of dollars)	Year 1	Year 2	Year 3	Year 4
1. Cash inflow (gross savings)	25	105	125	200
2. Cash outflow (expenses)	−135	−25	−30	−35
3. Pretax cash flow (line 1 + 2)	−110	80	95	165
4. After-tax cash flow [line 3 × (1 − tax rate)]	−70	51	61	106
5. Depreciation	60	50	40	30
6. Tax relief from depreciation (line 5 × tax rate)	22	18	14	11
7. Net after-tax cash flow (line 4 + 6)	−48	69	75	117
8. Discounted cash flow [line 7 ÷ $(1 + \text{cost of capital})^{\text{Year}}$]	−40	48	43	56
9. Net present value (sum of amounts in row 8)	107			

TABLE 12.8

Sample Net Present Value Calculation

OBJECT-ORIENTED SYSTEMS INVESTIGATION

The object-oriented approach can be used during all phases of systems development, from investigation to maintenance and review. In addition to identifying key participants and performing basic feasibility analysis, key objects can be identified during systems investigation. Consider a kayak rental business in Maui, Hawaii, where the owner wants to computerize its operations. There are many system objects for this business, including the kayak rental clerk, renting kayaks to customers, and adding new kayaks into the rental program. These objects can be diagrammed in a use case diagram (Figure 12.14). As you can see, the kayak rental clerk rents kayaks to customers and adds new kayaks to the current inventory of kayaks available for rent. The stick figure is an example of an *actor*, and the ovals each represent an event, called a *use case*. In our example, the actor (the kayak rental clerk) interacts with two use cases (Rent kayaks to customers and Add new kayaks to inventory). The use case diagram is part of the Unified Modeling Language that is used in object-oriented systems development.

systems investigation report
summary of the results of the systems investigation and the process of feasibility analysis and recommendation of a course of action

steering committee
an advisory group consisting of senior management and users from the IS department and other functional areas

THE SYSTEMS INVESTIGATION REPORT

The primary outcome of systems investigation is a **systems investigation report**. This report summarizes the results of systems investigation and the process of feasibility analysis and recommends a course of action: continue on into systems analysis, modify the project in some manner, or drop it. A typical table of contents for the systems investigation report is shown in Figure 12.15.

The systems investigation report is reviewed by senior management, often organized as an advisory committee, or **steering committee**, consisting of

FIGURE 12.14

Use Case Diagram for a Kayak Rental Application

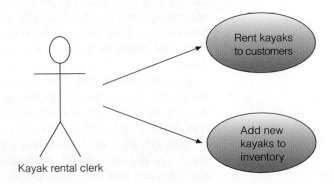

FIGURE 12.15

*A Typical Table of Contents for a
Systems Investigation Report*

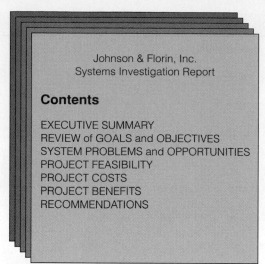

Johnson & Florin, Inc.
Systems Investigation Report

Contents

EXECUTIVE SUMMARY
REVIEW of GOALS and OBJECTIVES
SYSTEM PROBLEMS and OPPORTUNITIES
PROJECT FEASIBILITY
PROJECT COSTS
PROJECT BENEFITS
RECOMMENDATIONS

senior management and users from the IS department and other functional areas. These individuals help IS personnel with their decisions about the use of information systems in the business and give authorization to pursue further systems development activities. After review, the steering committee might agree with the recommendation of the systems development team to proceed with systems analysis or suggest a change in project focus to concentrate more directly on meeting a specific company objective. Another alternative is that everyone may decide that the project is not feasible for one reason or another and cancel the project.

SYSTEMS ANALYSIS

After a project has been approved for further study, the next step is to answer the question "What must the information system do to solve the problem?" The process needs to go beyond mere computerization of existing systems. The entire system, and the business process with which it is associated, should be evaluated.[49] Often, a firm can make great gains if it restructures both business activities and the related information system simultaneously. The overall emphasis of analysis is gathering data on the existing system, determining the requirements for the new system, considering alternatives within these constraints, and investigating the feasibility of the solutions. The primary outcome of systems analysis is a prioritized list of systems requirements.

GENERAL CONSIDERATIONS

Systems analysis starts by clarifying the overall goals of the organization and determining how the existing or proposed information system helps meet them. A manufacturing company, for example, might want to reduce the number of equipment breakdowns. This goal can be translated into one or more informational needs. One need might be to create and maintain an accurate list of each piece of equipment and a schedule for preventive maintenance. Another need might be a list of equipment failures and their causes. The U.S. National Security Agency's goal of developing a more secure operating system called for help from programmers and developers.[50] "It fits in exactly with what NSA's role is—to protect U.S. information systems and oversee encryption of sensitive data," says one Linux expert.

Analysis of a small company's information system can be fairly straightforward. On the other hand, evaluating an existing information system for a large company can be a long, tedious process. As a result, large organizations evaluating a major information system normally follow a formalized analysis procedure, involving these steps:

1. Assembling the participants for systems analysis
2. Collecting appropriate data and requirements
3. Analyzing the data and requirements
4. Preparing a report on the existing system, new system requirements, and project priorities

PARTICIPANTS IN SYSTEMS ANALYSIS

The first step in formal analysis is to assemble a team to study the existing system. This group includes members of the original development team—from users and stakeholders to IS personnel and management. Most organizations usually allow key members of the development team not only to analyze the condition of the existing system but also to perform other aspects of systems development, such as design and implementation.

Once the participants in systems analysis are assembled, this group develops a list of specific objectives and activities. A schedule for meeting the objectives and completing the specific activities is also developed, along with deadlines for each stage and a statement of the resources required at each stage, such as clerical personnel, supplies, and so forth. Major milestones are normally established to help the team monitor progress and determine whether problems or delays occur in performing systems analysis.

DATA COLLECTION

The purpose of data collection is to seek additional information about the problems or needs identified in the systems investigation report. During this process, the strengths and weaknesses of the existing system are emphasized.

Identifying Sources of Data

Data collection begins by identifying and locating the various sources of data, including both internal and external sources (Figure 12.16).

FIGURE 12.16

Internal and External Sources of Data for Systems Analysis

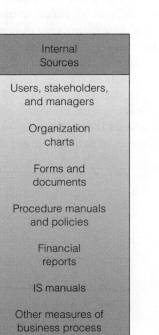

Internal Sources
Users, stakeholders, and managers
Organization charts
Forms and documents
Procedure manuals and policies
Financial reports
IS manuals
Other measures of business process

External Sources
Customers
Suppliers
Stockholders
Government agencies
Competitors
Outside groups
Journals, etc.
Consultants

Direct observation is a method of data collection. One or more members of the analysis team directly observe the existing system in action.

(Source: © 2002 PhotoDisc.)

structured interview
an interview where the questions are written in advance

unstructured interview
an interview where the questions are not written in advance

direct observation
watching the existing system in action by one or more members of the analysis team

questionnaires
a method of gathering data when the data sources are spread over a wide geographic area

statistical sampling
selection of a random sample of data and applying the characteristics of the sample to the whole group

Collecting Data

Once data sources have been identified, data collection begins. Figure 12.17 shows the steps involved. Data collection may require a number of tools and techniques, such as interviews, direct observation, and questionnaires.

Interviews may either be structured or unstructured. In a **structured interview**, the questions are written in advance. In an **unstructured interview**, the questions are not written in advance; the interviewer relies on experience in asking the best questions to uncover the inherent problems of the existing system. An advantage of the unstructured interview is that it allows the interviewer to ask follow-up or clarifying questions immediately.

With **direct observation**, one or more members of the analysis team directly observe the existing system in action. One of the best ways to understand how the existing system functions is to work with the users to discover how data flows in certain business tasks. Determining the data flow entails direct observation of users' work procedures, their reports, current screens (if automated already), and so on. From this observation, members of the analysis team determine which forms and procedures are adequate and which are inadequate and need improvement. Direct observation requires a certain amount of skill. The observer must be able to see what is really happening and not be influenced by his or her own attitudes or feelings. This approach can reveal important problems and opportunities that would be difficult to obtain using other data collection methods. An example would be observing the work procedures, reports, and computer screens associated with an accounts payable system being considered for replacement.

When many data sources are spread over a wide geographic area, **questionnaires** may be the best approach. Like interviews, questionnaires can be either structured or unstructured. In most cases, a pilot study is conducted to fine-tune the questionnaire. A follow-up questionnaire can also capture the opinions of those who do not respond to the original questionnaire.

A number of other data collection techniques can be employed. In some cases, telephone calls are an excellent method. In other cases, activities may be simulated to see how the existing system reacts. Thus, fake sales orders, stockouts, customer complaints, and data-flow bottlenecks may be created to see how the existing system responds to these situations. **Statistical sampling**, which involves taking a random sample of data, is another technique. For example, suppose we want to collect data that describes 10,000 sales orders received over the last few years. Because it is too time-consuming to analyze each of the 10,000 sales orders, a random sample of 100 to 200 sales orders from the entire batch can be collected. The characteristics of this sample are then assumed to apply to the 10,000 orders.

FIGURE 12.17

The Steps in Data Collection

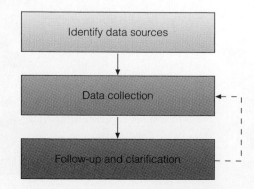

DATA ANALYSIS

The data collected in its raw form is usually not adequate to determine the effectiveness and efficiency of the existing system or the requirements for the new system. The next step is to manipulate the collected data so that it is usable for the members of the development team who are participating in systems analysis. This manipulation is called **data analysis**. Data and activity modeling, using data-flow diagrams and entity-relationship diagrams, are useful during data analysis to show data flows and the relationships among various objects, associations, and activities. Other common tools and techniques for data analysis include application flowcharts, grid charts, and CASE tools.

data analysis
manipulation of the collected data so that it is usable for the development team members who are participating in systems analysis

Data Modeling

Data modeling, first introduced in Chapter 5, is a commonly accepted approach to modeling organizational objects and associations that employ both text and graphics. The exact way data modeling is employed, however, is governed by the specific systems development methodology.

Data modeling is most often accomplished through the use of entity-relationship (ER) diagrams. Recall from Chapter 5 that an entity is a generalized representation of an object type—such as a class of people (employee), events (sales), things (desks), or places (Philadelphia)—and that entities possess certain attributes. Objects can be related to other objects in numerous ways. An entity-relationship diagram, such as the one shown in Figure 12.18a, describes a number of objects and the ways they are associated. An ER diagram is not capable in and of itself of fully describing a business problem or solution, because it lacks descriptions of the related activities. It is, however, a good place to start, since it describes object types and attributes about which data may need to be collected for processing.

Activity Modeling

To fully describe a business problem or solution, it is necessary to describe the related objects, associations, and activities. Activities in this sense are events or items that are necessary to fulfill the business relationship or that can be associated with the business relationship in a meaningful way.

Activity modeling is often accomplished through the use of data-flow diagrams. A **data-flow diagram (DFD)** models objects, associations, and activities by describing how data can flow between and around various objects. DFDs work on the premise that for every activity there is some communication, transference, or flow that can be described as a data element. DFDs describe what activities are occurring to fulfill a business relationship or accomplish a business task, not how these activities are to be performed. That is, DFDs show the logical sequence of associations and activities, not the physical processes. A system modeled with a DFD could operate manually or could be computer based; if computer based, the system could operate with a variety of technologies.

data-flow diagram (DFD)
a model of objects, associations, and activities by describing how data can flow between and around various objects

DFDs are easy to develop and easily understood by nontechnical people. Data-flow diagrams use four primary symbols, as illustrated in Figure 12.18b.

- *Data flow.* The **data-flow line** includes arrows that show the direction of data element movement.
- *Process symbol.* The **process symbol** reveals a function that is performed. Computing gross pay, entering a sales order, delivering merchandise, and printing a report are examples of functions that can be represented with a process symbol.
- *Entity symbol.* The **entity symbol** shows either the source or destination of the data element. An entity can be, for example, a customer who initiates a sales order, an employee who receives a paycheck, or a manager who gets a financial report.
- *Data store.* A **data store** reveals a storage location for data. A data store is any computerized or manual data storage location, including magnetic tape, disks, a filing cabinet, or a desk.

data-flow line
arrows that show the direction of data element movement

process symbol
representation of a function that is performed

entity symbol
representation of either a source or destination of a data element

data store
representation of a storage location for data

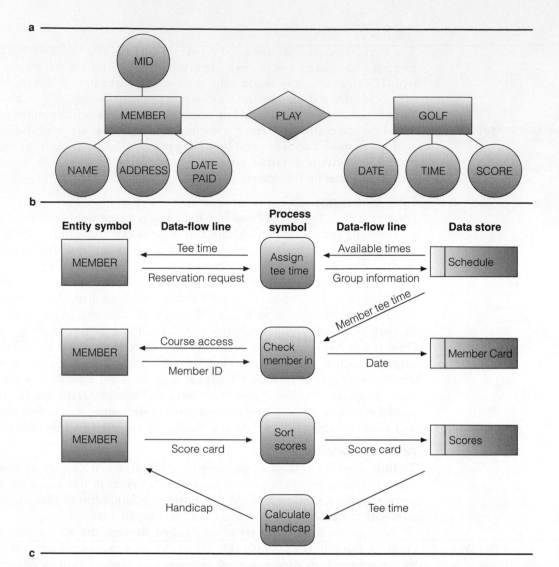

To play golf at the course, you must first pay a fee to become a member of the golf club. Members are issued member cards and are assigned member ID numbers. To reserve a tee time (a time to play golf), a member calls the club house at the golf course and arranges an available time slot with the reception clerk. The reception clerk reserves the tee time by writing the member's name and number of players in the group on the course schedule. When a member arrives at the course, he or she checks in at the reception desk where the reception clerk checks the course schedule and notes the date on the member's card. After a round of golf has been completed, the members leave their score card with the reception clerk. Member scores are tracked and member handicaps are updated on a monthly basis.

FIGURE 12.18

Data and Activity Modeling

(a) An entity-relationship diagram. (b) A data-flow diagram. (c) A semantic description of the business process.

Source: G. Lawrence Sanders, *Data Modeling* (Danvers, MA: Boyd & Fraser Publishing, 1995). Reprinted with permission from Course Technology.

application flowcharts
diagrams that show relationships among applications or systems

Comparing entity-relationship diagrams with data-flow diagrams provides insight into the concept of top-down design. Figure 12.18a and b show an entity-relationship diagram and a data-flow diagram for the same business relationship—namely, a member of a golf club playing golf. Figure 12.18c provides a brief description of the business relationship for clarification.

Application Flowcharts

Application flowcharts show the relationships among applications or systems. Assume that a small business has collected data about its order processing, inventory control, invoicing, and marketing analysis applications. Management is thinking of modifying the inventory control application. The raw facts collected, however, do not help in determining how the applications are related to each other and the databases required for each. These relationships are established through data analysis with an application flowchart (Figure 12.19). Using this tool for data analysis makes clear the relationships among the order processing, inventory control, invoicing, and marketing analysis applications.

In the simplified application flowchart in Figure 12.19, you can see that the sales ordering application provides important data to the inventory control and marketing analysis applications. The inventory control application provides data to the invoicing application. Any changes made to any one of these applications must take into account the other applications, which may provide data to or receive data from the others.

Grid Charts

grid chart
table that shows relationships among the various aspects of a systems development effort

A **grid chart** is a table that shows relationships among various aspects of a systems development effort. For example, a grid chart can be used to reveal the databases used by the various applications (Figure 12.20).

The simplified grid chart in Figure 12.20 shows that the customer database is used by the order processing, marketing analysis, and invoicing applications. The inventory database is used by the order processing, inventory control, and marketing analysis applications. The supplier database is used by the inventory control application, and the accounts receivable database is used by the invoicing application. This grid chart shows which applications use common databases and reveals that, for example, any changes to the inventory control application must investigate the inventory and supplier databases.

CASE Tools

CASE repository
a database of system descriptions, parameters, and objectives

As discussed earlier, many systems development projects use upper-CASE tools to complete analysis tasks. Most computer-aided software engineering tools have generalized graphics programs that can generate a variety of diagrams and figures. Entity-relationship diagrams, data-flow diagrams, application flowcharts, grid charts, and other diagrams can be developed using CASE graphics programs to help describe the existing system. During the analysis phase, a **CASE repository**—a database of system descriptions, parameters, and objectives—will begin to be developed.

REQUIREMENTS ANALYSIS

requirements analysis
determination of user, stakeholder, and organizational needs

The overall purpose of **requirements analysis** is to determine user, stakeholder, and organizational needs. For an accounts payable application, the stakeholders could include suppliers and members of the purchasing department. Questions that should be asked during requirements analysis include the following:

* Are these stakeholders satisfied with the current accounts payable application?
* What improvements could be made to satisfy suppliers and help the purchasing department?

For example, an analysis of Boeing's shop floor operations revealed that plant interruptions could occur as a result of supplier problems.[51] Requirements analysis revealed that the airplane manufacturer needed a system that could notify its managers of a possible supplier parts shortage that could stop production on its factory floor. The requirements analysis further revealed that the company's

FIGURE 12.19

An Application Flowchart
The flowchart shows the relationships among various applications.

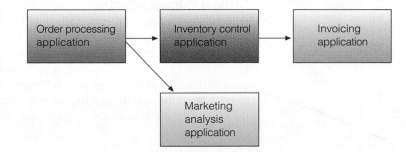

FIGURE 12.20

A Grid Chart

The chart shows the relationships among applications and databases.

Databases → Applications ↓	Customer database	Inventory database	Supplier database	Accounts receivable database
Order processing application	X	X		
Inventory control application		X	X	
Marketing analysis application	X	X		
Invoicing application	X			X

existing material requirements planning (MRP) system was inadequate in predicting plant disruptions. As a result of completing requirements analysis, Boeing was able to acquire a new software system, called iCollaboration, to help it predict and handle supplier problems and parts shortages. The Milan, Italy, clothes designer Prada performed requirements analysis and determined that its salespeople needed to do a better job in greeting customers and knowing their needs.[52] The resulting system used a handheld computer that could access customer preferences and shopping habits. The device could scan sales tags of Prada merchandise and access the Internet to get information on the styles available, fabrics that were used, and the accessories that could be purchased for the scanned item.

One of the most difficult procedures in systems analysis is confirming user or systems requirements. In some cases, communications problems can interfere with the determination of these requirements. For example, an accounts payable manager may want a better procedure for tracking the amount owed by customers. Specifically, the manager would like to have a weekly report that shows all customers who owe more than $1,000 and are more than 90 days past due on their account. A financial manager might need a report that summarizes total amount owed by customers to look at the need to loosen or tighten credit limits. A sales manager might want to review the amount owed by a key customer relative to sales to that same customer. The purpose of requirements analysis is to capture these requests in detail. Numerous tools and techniques can be used to capture systems requirements. Often, various techniques are used in the context of a JAD session.

Asking Directly

One the most basic techniques used in requirements analysis is asking directly. **Asking directly** is an approach that asks users, stakeholders, and other managers about what they want and expect from the new or modified system. This approach works best for stable systems in which stakeholders and users clearly understand the system's functions. The role of the systems analyst during the analysis phase is to critically and creatively evaluate needs and define them clearly so that the systems can best meet them.

asking directly
an approach to gather data that asks users, stakeholders, and other managers about what they want and expect from the new or modified system

Critical Success Factors

Another approach uses critical success factors (CSFs). As discussed earlier, managers and decision makers are asked to list only the factors that are critical to the success of their area of the organization. A CSF for a production manager might be adequate raw materials from suppliers; a CSF for a sales representative could be a list of customers currently buying a certain type of product. Starting from these CSFs, the system inputs, outputs, performance, and other specific requirements can be determined.

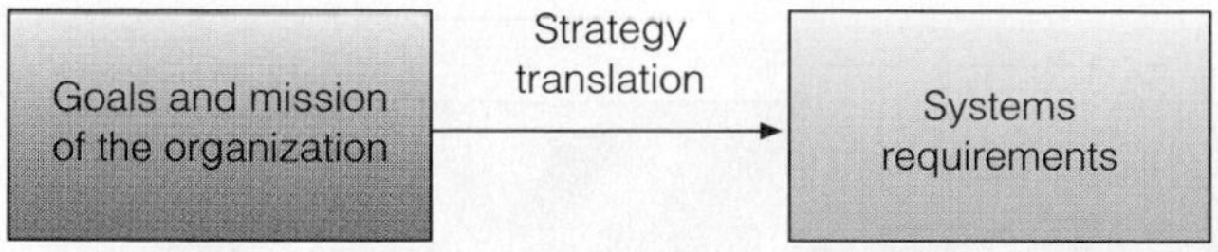

FIGURE 12.21

Converting Organizational Goals into Systems Requirements

screen layout
a technique that allows a designer to quickly and efficiently design the features, layout, and format of a display screen

report layout
technique that allows designers to diagram and format printed reports

The IS Plan

As we have seen, the IS plan translates strategic and organizational goals into systems development initiatives. The IS planning process often generates strategic planning documents that can be used to define system requirements. Working from these documents ensures that requirements analysis will address the goals set by top-level managers and decision makers (Figure 12.21). There are unique benefits to applying the IS plan to define systems requirements. Because the IS plan takes a long-range approach to using information technology within the organization, the requirements for a system analyzed in terms of the IS plan are more likely to be compatible with future systems development initiatives.

Screen and Report Layout

Developing formats for printed reports and screens to capture data and display information are some of the common tasks associated with developing systems. Screens and reports relating to systems output are specified first to verify that the desired solution is being delivered. Manual or computerized screen and report layout facilities are used to capture both output and input requirements.

Screen layout is a technique that allows a designer to quickly and efficiently design the features, layout, and format of a display screen. In general, users who interact with the screen frequently can be presented with more data and less descriptive information; infrequent users should have more descriptive information presented to explain the data that they are viewing (Figure 12.22).

Report layout allows designers to diagram and format printed reports. Reports can contain data, graphs, or both. Graphic presentations allow managers and executives to quickly view trends and take appropriate action, if necessary.

Screen layout diagrams can document the screens users desire for the new or modified application. Report layout charts reveal the format and content of various reports that the application will prepare. Other diagrams and charts can be developed to reveal the relationship between the application and outputs from the application.

Requirements Analysis Tools

A number of tools can be used to document requirements analysis. Again, CASE tools are often employed. As requirements are developed and agreed on, entity-relationship diagrams, data-flow diagrams, screen and report layout forms, and other types of documentation will be stored in the CASE repository. These requirements might also be used later as a reference during the rest of systems development or for a different systems development project.

OBJECT-ORIENTED SYSTEMS ANALYSIS

The object-oriented approach can also be used during systems analysis. Like traditional analysis, problems or potential opportunities are identified during object-oriented analysis. Identifying key participants and collecting data is still performed. But instead of analyzing the existing system using data-flow diagrams and flowcharts, an object-oriented approach is used.

In the section "Object-Oriented Systems Investigation," we introduced a kayak rental example. A more detailed analysis of that business reveals that there are two classes of kayaks: single kayaks for one person and tandem kayaks that can accommodate two people. With the OO approach, a class is used to describe different types of objects, such as single and tandem kayaks. The classes of kayaks can be shown in a generalization/specialization hierarchy diagram (Figure 12.23). KayakItem is an object that will store the kayak identification number (ID) and the date the kayak was purchased (datePurchased).

FIGURE 12.22

Screen Layouts

(a) A screen layout chart for frequent users who require little descriptive information.
(b) A screen layout chart for infrequent users who require more descriptive information.

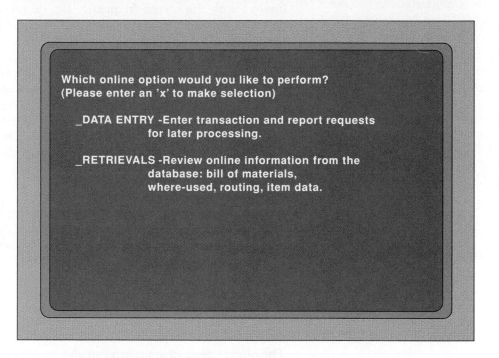

Of course, there could be subclasses of customers, life vests, paddles, and other items in the system. For example, price discounts for kayak rentals could be given to seniors (people over 65 years) and students. Thus, the Customer class could be divided into regular, senior, and student customer subclasses.

THE SYSTEMS ANALYSIS REPORT

Systems analysis concludes with a formal systems analysis report. It should cover the following elements:

1. The strengths and weaknesses of the existing system from a stakeholder's perspective
2. The user/stakeholder requirements for the new system (also called the *functional requirements*)

FIGURE 12.23

Generalization/Specialization Hierarchy Diagram for Single and Tandem Kayak Classes

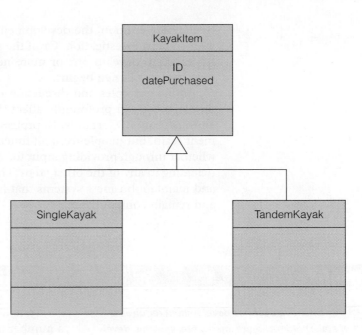

3. The organizational requirements for the new system
4. A description of what the new information system should do to solve the problem

Suppose analysis reveals that a marketing manager thinks a weakness of the existing system is its inability to provide accurate reports on product availability. These requirements and a preliminary list of the corporate objectives for the new system will be in the systems analysis report. Particular attention is placed on areas of the existing system that could be improved to meet user requirements. The table of contents for a typical report is shown in Figure 12.24.

The systems analysis report gives managers a good understanding of the problems and strengths of the existing system. If the existing system is operating better than expected or the necessary changes are too expensive relative to the benefits of a new or modified system, the systems development process can be stopped at this stage. If the report shows that changes to another part of the system might

FIGURE 12.24

A Typical Table of Contents for a Report on an Existing System

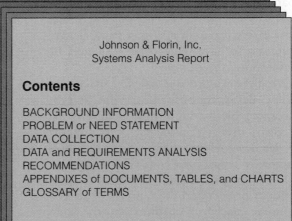

Johnson & Florin, Inc.
Systems Analysis Report

Contents

BACKGROUND INFORMATION
PROBLEM or NEED STATEMENT
DATA COLLECTION
DATA and REQUIREMENTS ANALYSIS
RECOMMENDATIONS
APPENDIXES of DOCUMENTS, TABLES, and CHARTS
GLOSSARY of TERMS

be the best solution, the development process might start over, beginning again with systems investigation. Or, if the systems analysis report shows that it will be beneficial to develop one or more new systems or to make changes to existing ones, systems design begins.

As the examples and discussion of this chapter have demonstrated, systems development can profoundly affect the current and future business activities of an organization. Certainly, IS professionals are critical members of the development team, but people from all functional areas of an organization are involved, whether through providing input for data collection, testing of prototypes, or participating in any of the other steps. The goal of development activities is to create and maintain business systems that help an organization meet its strategic goals and remain competitive.

SUMMARY

PRINCIPLE *Effective systems development requires a team effort of stakeholders, users, managers, systems development specialists, and various support personnel, and it starts with careful planning.*

The systems development team consists of stakeholders, users, managers, systems development specialists, and various support personnel. The development team is responsible for determining the objectives of the information system and delivering to the organization a system that meets its objectives.

Stakeholders are individuals who, either themselves or through the area of the organization they represent, ultimately benefit from the systems development project. Users are individuals who will interact with the system regularly. They can be employees, managers, customers, or suppliers. Managers on development teams are typically representative of stakeholders or may be stakeholders themselves. In addition, managers are most capable of initiating and maintaining change. For large-scale systems development projects, where the investment in and value of a system can be quite high, it is common to have senior-level managers be part of the development team.

A systems analyst is a professional who specializes in analyzing and designing business systems. A programmer is responsible for modifying or developing programs to satisfy user requirements. Other support personnel on the development team include technical specialists, either IS department employees or outside consultants. Depending on the magnitude of the systems development project and the number of IS systems development specialists on the team, the team may also include one or more IS managers. At some point in your career, you will likely be a participant in systems development. You could be involved in a systems development team—as a user, as a manager of a business area or project team, as a member of the IS department, or maybe even as a CIO.

Systems development projects may be initiated for a number of reasons, including the need to solve problems with an existing system, to exploit opportunities to gain competitive advantage, to increase competition, to make use of effective information, to create organizational growth, to settle a merger or corporate acquisition, and to address a change in the market or external environment. External pressures, such as potential lawsuits or terrorist attacks, can also initiate systems development.

Information systems planning refers to the translation of strategic and organizational goals into systems development initiatives. Benefits of IS planning include a long-range view of information technology use and better use of information systems resources. Planning requires developing overall IS objectives; identifying IS projects; setting priorities and selecting projects; analyzing resource requirements; setting schedules, milestones, and deadlines; and developing the IS planning document. IS planning can result in a competitive advantage through creative and critical analysis.

Establishing objectives for systems development is a key aspect of any successful development project. Critical success factors (CSFs) can be used to identify important objectives. Systems development objectives can include performance goals (quality and usefulness of the output and the speed at which output is generated) and cost objectives (development costs, fixed costs, and ongoing operating costs).

Applications must be designed to meet special business needs. Transaction processing applications for the Internet must be able to scale up to support highly variable transactions from potentially thousands of users. Ideally, they can scale up dynamically when needed. They must be reliable and fault-tolerant, providing 24 hours a day/7 days a week availability with extremely high transaction integrity. They must be able to integrate with existing infrastructure, including customer and order databases, legacy applications, and enterprise resource planning systems. Development and maintenance must be quick and easy, as business needs may require you to change applications on the fly.

The increasing use of enterprise resource planning software has affected systems development. The first trend is that companies wish to stay with their primary ERP vendor instead of looking elsewhere for answers to their data warehousing and production planning needs or developing in-house solutions. Thus, they look to their original ERP vendor to provide these solutions. A second trend is that there is less in-house development and more dependence on ERP vendors and their strategic partners to provide enhancements and add-ons to the original ERP package. A third interesting trend is the increase in the number of companies who, once they have successfully implemented their own company's ERP project, are branching out to provide consulting to other companies.

PRINCIPLE *Systems development often uses tools to select, implement, and monitor projects, including net present value (NPV), prototyping, rapid application development, CASE tools, and object-oriented development.*

The five phases of the traditional SDLC are investigation, analysis, design, implementation, and maintenance and review. Systems investigation identifies potential problems and opportunities and considers them in light of organizational goals. Systems analysis seeks a general understanding of the solution required to solve the problem; the existing system is studied in detail and weaknesses are identified. Systems design creates new or modifies existing system. Systems implementation encompasses programming, testing, training, conversion, and operation of the system. Systems maintenance and review entails monitoring the system and performing enhancements or repairs.

Advantages of the traditional SDLC include the following: it provides for maximum management control, creates considerable system documentation, ensures that system requirements can be traced back to stated business needs, and produces many intermediate products for review. Its disadvantages include the following: users may get a system that meets the needs as understood by the developers; the documentation is expensive and difficult to maintain; users' needs go unstated or may not be met; and users cannot easily review the many intermediate products produced.

Prototyping is an iterative approach that involves defining the problem, building the initial version, having users utilize and evaluate the initial version, providing feedback, and incorporating suggestions into the second version. Prototypes can be fully operational

or nonoperational, depending on how critical the system under development is and how much time and money the organization has to spend on prototyping.

Advantages of the prototyping approach include the following: users get an opportunity to try the system before it is completed; useful prototypes can be produced in weeks; users become positive about the evolving system; and errors and omissions can be detected early. Disadvantages include the following: the approach makes it difficult to start over if the initial solution misses the mark widely; it is difficult to contain the scope of the project; system documentation is often absent; and key operational considerations are often overlooked.

Rapid application development (RAD) uses tools and techniques designed to speed systems development. Its use reduces paper-based documentation, automates program source code generation, and facilitates user participation in development activities. RAD can use newer programming techniques, such as agile or extreme programming. RAD makes extensive use of the joint application development (JAD) process to gather data and perform requirements analysis. JAD involves group meetings in which users, stakeholders, and IS professionals work together to analyze existing systems, propose possible solutions, and define the requirements for a new or modified system.

RAD has the following advantages: it puts an application into production quickly; documentation is produced as a by-product; and it forces good teamwork among users, stakeholders, and developers. Its disadvantages are the following: it is an intense process that can burn out the participants; it requires participants to be skilled in advanced tools and techniques; and a large amount of time is required from stakeholders and users.

The term *end-user systems development* describes any systems development project in which the primary effort is undertaken by a combination of business managers and users.

Many companies hire an outside consulting firm that specializes in systems development to take over some or all of its information systems development activities. This approach is called *outsourcing*. Reasons for outsourcing include companies' belief that they can cut costs, their wish to achieve a competitive advantage without having the necessary IS personnel in house, their desire to obtain state-of-the-art technology, their focus on increasing technological flexibility, and their need to proceed with development despite downsizing. A number of companies offer outsourcing services, including computer vendors and specialized consulting companies.

A number of factors affect systems development success. The degree of change introduced by the project, continuous improvement and reengineering, the use of quality programs and standards, organizational experience with systems development, the use of project management tools, and the use of CASE tools and the objected-oriented approach are all factors that affect the success of a project. The greater the amount of change, the greater the degree of risk and also frequently the amount of reward. Continuous improvement projects do not require significant business process or information system changes, while reengineering involves fundamental changes in how the organization conducts business and completes tasks. Quality systems development projects often involve such factors as support from top management, strong user involvement, use of a proven methodology, clear project goals and objectives, concentration on key problems and straightforward designs, staying on schedule and within budget, good user training, and solid review and maintenance programs. Quality standards, such as ISO 9001, can also be used during the systems development process. The Capability Maturity Model (CMM) can measure an organization's experience with systems development using five levels from initial to optimized. The use of automated project management tools enables detailed development, tracking, and control of the project schedule. Effective use of a quality assurance process enables the project manager to deliver a quality system and to make intelligent trade-offs among cost, schedule, and quality. CASE tools automate many of the systems development tasks, thus reducing the time and effort required to complete them while ensuring good documentation. Object-oriented systems development can also be an important success factor. With the object-oriented systems development (OOSD) approach, a project can be broken down into a group of objects that interact. Instead of requiring thousands or millions lines of detailed computer instructions or code, the systems development project might require a few dozen or maybe a hundred objects.

PRINCIPLE   *Systems development starts with investigation and analysis of existing systems.*

In most organizations, a systems request form initiates the investigation process. Participants in systems investigation can include stakeholders, users, managers, employees, analysts, and programmers. Systems investigation is designed to assess the feasibility of implementing solutions for business problems, including technical, economic, legal, operations, and schedule feasibility. Net present value analysis is often used to help determine a project's economic feasibility. An investigation team follows up on the request and performs a feasibility analysis that addresses technical, economic, legal, operational, and schedule feasibility.

If the project under investigation is feasible, major goals are set for the system's development, including performance, cost, managerial goals, and procedural goals. Many companies choose a popular methodology so that new IS employees, outside specialists, and vendors will be familiar with the systems development tasks set forth in the approach. A systems development methodology must be selected. Objected-oriented systems investigation is being used to a greater extent. The use case diagram is part of the Unified Modeling Language that is used to document object-oriented systems development. As a final step in the investigation process, a systems investigation report should be prepared to document relevant findings.

Systems analysis is the examination of existing systems, which begins once approval for further study is received from management. Additional study of a selected system allows those involved to further understand the systems' weaknesses and potential improvement areas. An analysis team is assembled to collect and analyze data on the existing system.

Data collection methods include observation, interviews, questionnaires, and statistical sampling. Data analysis manipulates the collected data to provide information. The analysis includes grid charts, application flowcharts, and CASE tools. The overall purpose of requirements analysis is to determine user and organizational needs.

Data analysis and modeling are used to model organizational objects and associations using text and graphical diagrams. It is most often accomplished through the use of entity-relationship (ER) diagrams. Activity modeling is often accomplished through the use of data-flow diagrams (DFDs), which model objects, associations, and activities by describing how data can flow between and around various objects. DFDs use symbols for data flows, processing, entities, and data stores. Application flowcharts, grid charts, and CASE tools are also used during systems analysis.

Requirements analysis determines the needs of users, stakeholders, and the organization in general. Asking directly, using critical success factors, and determining requirements from the IS plan can be used. Often screen and report layout charts are used to document requirements during systems analysis.

Like traditional analysis, problems or potential opportunities are identified during object-oriented analysis. Object-oriented systems analysis can involve using diagramming techniques, such as a generalization/specialization hierarchy diagram.

CHAPTER 12 SELF-ASSESSMENT TEST

Effective systems development requires a team effort of stakeholders, users, managers, systems development specialists, and various support personnel, and it starts with careful planning.

1. _____ is the activity of creating or modifying existing business systems. It refers to all aspects of the process—from identifying problems to be solved or opportunities to be exploited to the implementation and refinement of the chosen solution.

2. Which of the following individuals ultimately benefit from a systems development project?
   A. Computer programmers
   B. Systems analysts
   C. Stakeholders
   D. Senior-level managers

3. A(n) _____ is a professional who specializes in analyzing and designing business systems. This individual plays various roles while interacting with the users, management, vendors and suppliers, external companies, software programmers, and other IS support personnel.

4. Like a contractor constructing a new building or renovating an existing one, the programmer takes the plans from the systems analyst and builds or modifies the necessary software. True False

5. The term _____ refers to the translation of strategic and organizational goals into systems development initiatives.

6. What factors are essential to the success of certain functional areas of an organization?
   A. Critical success factors
   B. Systems analysis factors
   C. Creative goal factors
   D. Systems development factors

Systems development often uses tools to select, implement, and monitor projects, including net present value (NPV), prototyping, rapid application development, CASE tools, and object-oriented development.

7. What is the third level in the Capability Maturity Model (CMM)?
   A. Initial
   B. Optimizing
   C. Managed
   D. Defined

8. System performance is usually determined by such factors as fixed investments in hardware and related equipment. True False

9. _____ takes an iterative approach to the systems development process. During each iteration, requirements and alternative solutions to the problem are identified and analyzed, new solutions are designed, and a portion of the system is implemented.

10. ISO 9001 can be used to increase project quality. True False

11. What consists of all activities that, if delayed, would delay the entire project?
    A. Deadline activities
    B. Slack activities
    C. RAD tasks
    D. The critical path

Systems development starts with investigation and analysis of existing systems.

12. The systems request form is a document that is filled out during systems analysis. True False

13. Feasibility analysis is typically done during which systems development stage?
    A. Investigation
    B. Analysis
    C. Design
    D. Implementation

14. Data modeling is most often accomplished through the use of _____, while activity modeling is often accomplished through the use of _____.

15. The overall purpose of requirements analysis is to determine user, stakeholder, and organizational needs. True False

Chapter 12 Self-Assessment Test Answers

1. Systems development, 2. C, 3. systems analyst., 4. True, 5. information systems planning, 6. A, 7. D, 8. False, 9. Prototyping, 10. True, 11. D, 12. False, 13. A, 14. entity-relationship (ER) diagrams, data-flow diagrams, 15. True.

KEY TERMS

application flowcharts, 540
asking directly, 542
CASE repository, 541
computer-aided software engineering
    (CASE), 530
creative analysis, 512
critical analysis, 512
critical path, 529
critical success factors (CSFs), 513
data analysis, 539
data store, 539
data-flow diagram (DFD), 539
data-flow line, 539
direct observation, 538
economic feasibility, 533
end-user systems development, 522
entity symbol, 539
feasibility analysis, 533
Gantt chart, 529
grid chart, 541

information systems planning, 510
integrated-CASE (I-CASE) tools, 530
joint application development
    (JAD), 5125
legal feasibility, 534
lower-CASE tools, 530
mission-critical systems, 513
net present value (NPV), 534
nonoperational prototype, 519
object-oriented systems
    development, 531
operational feasibility, 534
operational prototype, 519
process symbol, 539
Program Evaluation and Review
    Technique (PERT), 529
programmer, 507
project deadline, 528
project milestone, 528
project schedule, 528

prototyping, 518
questionnaires, 538
rapid application development
    (RAD), 519
report layout, 543
requirements analysis, 541
schedule feasibility, 534
screen layout, 543
stakeholders, 506
statistical sampling, 538
steering committee, 535
structured interview, 538
systems analyst, 507
systems design, 517
systems investigation report, 535
systems request form, 532
technical feasibility, 534
unstructured interview, 538
upper-CASE tools, 530
users, 506

REVIEW QUESTIONS

1. What is an information system stakeholder?
2. What is the goal of information systems planning? What steps are involved in IS planning?
3. What are the typical reasons to initiate systems development?
4. What actions can be used during critical analysis?
5. Describe the five levels of the Capability Maturity Model.
6. How is a Gantt chart developed? How is it used?
7. What are the steps of the systems development life cycle?
8. Why is it important to identify and remove errors early in the system development life cycle?

9. Identify four reasons a systems development project may be initiated.
10. List factors that have a strong influence on project success.
11. What is the purpose of systems investigation?
12. What are the steps of object-oriented systems development?
13. Define the different types of feasibility that system development must consider.
14. What is the net present value of a project?
15. What is the purpose of systems analysis?
16. How does the JAD technique support the RAD systems development life cycle?

DISCUSSION QUESTIONS

1. Why is it important for business managers to have a basic understanding of the systems development process?
2. Briefly describe the role of a system user in the systems investigation and systems analysis stages of a project.
3. Assume that you are investigating a new sales marketing program for a clothing company. Use critical

analysis to help you determine the major requirements for the new system.
4. Briefly describe when you would use the object-oriented approach to systems development instead of the traditional systems development life cycle.
5. During the systems investigation phase, how important is it to think creatively? What are some approaches to increase creativity?

6. For what types of systems development projects might prototyping be especially useful? What are the characteristics of a system developed with a prototyping technique?

7. Imagine that your firm has never developed an information systems plan. What sort of issues between the business functions and IS organization might exist?

8. Assume that you are responsible for a new payroll program. What steps would you take to ensure a high-quality payroll system?

9. How important are communications skills to IS personnel? Consider this statement: "IS personnel need a combination of skills—one-third technical skills, one-third business skills, and one-third communications skills." Do you think this is true? How would this affect the training of IS personnel?

10. Discuss three reasons why aligning overall business goals with IS goals is important.

11. Imagine that you are a highly paid consultant who has been retained to evaluate an organization's systems development processes. With whom would you meet? How would you make your assessment?

12. You are a senior manager of a functional area in which a mission-critical system is being developed. How can you safeguard this project from mushrooming out of control?

PROBLEM-SOLVING EXERCISES

1. Develop a spreadsheet program to determine net present value using the form outlined below.

Cash Flow (in thousands of dollars)

	Year 1	Year 2	Year 3	Year 4
1. Cash inflow (gross savings)				
2. Cash outflow (expenses)				
3. Pretax cash flow (line 1 + 2)				
4. After-tax cash flow [line 3 x (1 – tax rate)]				
5. Depreciation				
6. Tax relief from depreciation (line 5 x tax rate)				
7. Net after-tax cash flow (line 4 + 6)				
8. Discounted cash flow [line 7 ÷ (1 + cost of capital)^{Year}]				
9. Net present value (sum of amounts in row 8)				

Use the spreadsheet to select between two projects with the following cash flows. Project one has a gross savings of $100,000 per year and expenses of $25,000 per year, plus $15,000 per year in depreciation. Project two has a gross savings of $75,000 the first year with $125,000 per year thereafter. Expenses are $35,000 per year plus $18,000 per year in depreciation. Assume a capital cost rate of 15 percent and tax rate of 35 percent.

2. You are developing a new information system for The Fitness Center, a company that has five fitness centers in your metropolitan area, with about 650 members and 30 employees in each location. This system will be used by both members and fitness consultants to keep track of participation in various fitness activities, such as free weights, volleyball, swimming, stair climbers, and aerobic and yoga classes. One of the performance objectives of the system is that it helps members plan a fitness program to meet their particular needs. The primary purpose of this system, as envisioned by the director of marketing, is to assist The Fitness Center in obtaining a competitive advantage over other fitness clubs.

Use word processing software to prepare a brief memo to the required participants in the development team for this systems development project. Be sure to specify what roles these individuals will play and what types of information you hope to obtain from them. Assume that the relational database model will be the basis for building this system. Use a database management system to define the various tables that will make up the database.

3. You are going to start a video rental business. Using the object-oriented approach, develop a use case and a generalization/specialization hierarchy diagram using a graphics program.

TEAM ACTIVITIES

1. Systems development is more of an art and less of a science, with a wide variety of approaches in how companies perform this activity. You and the members of your team are to interview members of an information systems organization's development group. List the steps that the IS group uses in developing a new system or modifying an existing one. How does the organization's approach compare with the techniques discussed in the chapter? Consider both the traditional systems development life cycle and the object-oriented approach. Prepare a short report on your findings.

2. Your team has been hired to determine the requirements and layout of the Web pages for a company that sells fishing equipment over the Internet. Using the approaches discussed in this chapter, develop a rough sketch of at least five Web pages that you would recommend. Make sure to show the important features and the hyperlinks for each page.

WEB EXERCISES

1. A number of companies were discussed in this chapter. Locate the Web site of one of these companies. What are the goods and services that this company produces? After visiting the company Web site, describe how systems development could be used to improve the goods or services it produces. You may be asked to develop a report or send an e-mail message to your instructor about what you found.

2. Locate several companies on the Internet that have received CMM certification. What level was achieved? To what extent do these companies stress their CMM certification? What major outsourcing or IS consulting companies do not advertise achieving a CMM certification on the Web? In your opinion, how important is CMM certification for the success of an outsourcing or IS consulting company? Write a report describing what you found.

CASES

CASE 1

General Motors Reduces Dealer Costs with Procurement System

It is estimated that GM's 7,500 dealers spend around $1 billion on various materials and supplies, such as office supplies, computers, and tools. These expenses are passed on to the consumer in the form of higher sticker prices, which hurts GM by reducing sales and market share. GM's sales of cars and light trucks in the United States fell 1.1 percent to 4.86 million in 2001 and declined 2.9 percent to 1.94 million in the first half of 2002. Concerned GM managers met to discuss ways to reduce the spending.

GM management realized that its dealers were individually unable to reduce their spending, but if they combined their purchasing power, they could wield more influence with suppliers to reduce their prices. In short, GM's solution lay in the power of bulk purchasing. The challenge then was to develop a system that all 7,500 dealers could access to purchase supplies from vendors who partnered with GM to offer reduced prices.

GM's systems analysts worked with two outside companies to develop the system: Covisint LLC, an automotive purchasing exchange, and Reynolds & Reynolds Co., the developer of *GM DealerWorld*, the Internet portal all GM dealers currently used.

GM created partnerships with vendors who provided products their dealers typically need to purchase. These vendors agreed to offer their products to GM dealers at discounted prices in exchange for the larger volume of sales they were likely to incur. For instance, dealers will be able to get 5 cents off each gallon of gas they buy from a partner.

Once partnerships were created, Covisint developed the transaction processing system that would allow GM dealers to purchase from the partners, and Reynolds & Reynolds developed the Web interface within *GM DealerWorld* so that dealers could conveniently shop for the products they require.

The resulting product was named *GM Dealer Supply Advantage.* Now when a dealer needs to purchase business materials, rather than going to the yellow pages, the dealer can call up this private marketplace on-line, which sells only items that an automotive dealer might be interested in—at a discounted price. Automotive dealer nirvana! Dealers pay $360 per year to access *GM Dealer Supply Advantage* and are expected to reduce their annual procurement costs by 15 percent. The automaker estimated that about one-third of its 7,500 U.S. dealers would initially sign on to the program.

GM expects the dealer purchasing site to help boost the retailers' profits, allowing them to invest more in their outlets and sell more cars and trucks. The day that the new system was publicly announced, GM shares rose $1.28 on the New York Stock Exchange.

Discussion Questions

1. What type of information did GM need to collect from dealers to determine the need for *GM Dealer Supply Advantage*? Once the need was established, what type of information did GM need to collect from dealers to develop an effective system?
2. Why do you think GM is charging their dealers for this service?

Critical Thinking Questions

3. Why would only one-third of GM dealers adopt this system? What conditions may exist to deter dealers from jumping on board? Although this system appears to benefit GM dealers and partners, what players in the market place will be financially hurt by its implementation?
4. Besides the automotive industry, what other industries might benefit from a system such as this?

Sources: Ed Garsten, "GM Creates Web Site Designed to Drive Down Costs at Dealerships," The Associated Press State & Local Wire, June 17, 2002; Alison Fitzgerald, "GM Dealers to Get Online Purchasing: Lower Prices Promised," *National Post* (f/k/a *The Financial Post*), June 18, 2002, p. FP14; "GM Unveils Dealer Buying Site," *Journal of Commerce—JoC Online*, June 18, 2002, http://www.joc.com; Linda Rosencrance, "GM Set to Launch Procurement Web Site for Dealers," *Computerworld*, June 17, 2002, http://www.idg.net/ic_876602_4914_1-2787.html; Margaret Kane, "GM Site Offers Dealers Plenty of Bargains," *ZDNet News*, June 17, 2002; GM Web site, http://www.media.gm.com/news/releases/020617_dealerships.html, accessed July 2002.

CASE 2

Staples Provides Convenience with In-Store Kiosks

Staples, Inc., self-proclaimed inventor of the office supply superstore, installed kiosks in its retail outlet stores to provide in-store customers with access to the 50,000 products available through Staples.com; that's roughly 42,000 additional products beyond what is offered in its stores. A kiosk is a small computer station that usually incorporates a touch-sensitive display, which allows users to access information. The original installation of the kiosks allowed shoppers to peruse the on-line catalog and make credit card purchases of Staples products, which would then be shipped to their home.

While this service alone provided convenience to the customers, Michael J. Ragunas, chief technology officer at Staples.com, believed that the kiosks could offer more. Customers were inconvenienced by the need to pull out their credit card at the kiosk and then again at the checkout counter when they paid for their in-store purchases. Ragunas met with his systems development team and launched the In-Store Access Point project. The project had two primary objectives: to allow customers to pay for kiosk purchases along with in-store purchases in one transaction with cash, check, or credit card and to provide a tool for customers to design the computer system they wished to purchase. This second goal was added to address the problem of customers who wished to purchase computers with different configurations than the ones on display.

The second goal was met with the help of an outside software vendor. Staples used a configuration tool from Calico Commerce in San Jose, which allowed customers to choose the specifications of their PC. Staples built an XML interface to enable the Calico tool to interact with the Staples.com ordering system. Completed orders are sent to the manufacturers via electronic data interchange from Staples' back-office systems.

The first goal took a bit more effort. Staples IS personnel had to tweak the system so it could detect whether customers are using in-store kiosks and hold their orders until payment is received at the cash register. Customers ordering merchandise at in-store kiosks print out a ticket with a bar code, which is scanned at the cash register. The cash register pulls the kiosk order information from the back-end order management system. The register then sends the payment information to the back-office system so that the order and payment can be joined.

"Integration is hard work," says Ragunas. "There are some technologies out there that can help, but it's still hard because most of what you're working with is legacy, and you have to figure out ways to make what you already have talk to each other."

The project took a little more than a year to complete and on all accounts is considered a success. "The fact that you can consolidate your purchases and choose multiple ways to make purchases in the store puts them a level ahead at this point," said Geri Spieler, an analyst at Gartner Inc.

Staples has logged close to $4 million in sales per week on the kiosks and eliminated its inventory of PCs in more than 200 stores, Ragunas says. With the customized configuration system, customers get exactly what they want and Staples has reduced its inventory costs and freed space for other products, he adds. Customers using Staples.com in stores are more likely also to order merchandise through the company's Web site from their homes or offices. "We know that customers who shop with us in multiple channels spend more with us overall—2.5 times if two channels, 4.5 times if three," says Ragunas.

The In-Store Access Point project earned Staples, Ragunas, and his team industrywide recognition when they were awarded *Electronic Commerce World* magazine's 2nd annual EC World-Class Performance award, which recognizes companies for successful implementations of e-commerce systems.

Discussion Questions

1. What conditions and systems should exist for a store to benefit from an in-store kiosk?
2. What type of analysis do you think Staples did to determine that kiosks would bring in a return on investment, prior to installing them?

Critical Thinking Questions

3. After traveling to a store, why would a person then shop on-line on a kiosk? Couldn't that shopping be done from home? Compare the benefits and problems involved in both in-store shopping and on-line shopping. How does a kiosk bridge these worlds?

4. How do Staples in-store kiosks affect its selection of products to display in the store? What types of products would benefit sales by being in the store? Which products are better provided to customers on-line?

Sources: Carol Sliwa, "Staples Inc.," *Computerworld*, March 11, 2002, http://www.computerworld.com/softwaretopics/crm/story/0,10801,68866,00.html; "EC World Names Staples, Inc. among E-Business Award Winners," *PR Newswire*, February 21, 2002; Staples Web site, http://www.staples.com, accessed July 2002.

CASE 3

Wesco Distribution Adds Links to the Supply Chain

Wesco Distribution, Inc., a $3.9 billion Pittsburgh-based company, is one of the country's largest distributors of electrical products and other maintenance, repair, and operating supplies to large companies. The company describes itself as being "procurement specialists, helping customers lower supply chain costs and raise the efficiency of entire operations through a combination of advanced, innovative distribution capabilities."

Wesco stocks more than 140,000 items from hundreds of manufacturers. In addition, it also supplies its customers with more than 900,000 other maintenance, repair, and operating items that the company doesn't stock. For these items, Wesco representatives place orders with the manufacturers on their client's behalf. However, such transactions can quickly become complicated. When a customer requests an item it doesn't stock, the Wesco sales representative either calls the manufacturer directly or checks its Web site for pricing and availability. The Wesco rep then phones the customer back with the information, confirms the order, hangs up, gets back on-line with the manufacturer, and places the order—if the item is still in stock. Although orders for unstocked items account for only 20 percent of Wesco's business, gathering information on those purchases for customers took 40 percent of the sales force's time, says Russ Lambert, Wesco's director of e-commerce.

To solve this problem, Wesco needed to develop a system that would let its sales reps access information about unstocked items as easily as they accessed information about their in-stock items. Since this information existed on the manufacturer's information systems, Wesco needed to develop a system to link its information system with those of its manufacturers. The challenge, however, was that all of the manufacturers used different types of systems, and Wesco's own system was a 20-year-old, proprietary, green-screen dinosaur. Coming up with a way to link these heterogeneous systems was the most difficult part of the project, Lambert says.

The company built an Internet gateway with an interface for inbound and outbound pathways to the legacy system. XML code was written both to pull data into Wesco's legacy system and to integrate supplier systems with the gateway.

Wesco used Austin, Texas–based Vignette Corp.'s content management software and San Jose–based BEA Systems Inc.'s WebLogic application server as its core technologies for the project.

Since the e-procurement system went live, Wesco's 1,000 salespeople in 400 locations have been able to access the finished-goods inventory systems of major suppliers directly. Now, while a customer requesting unstocked items is still on the line, a salesperson can send a query over the Web to the supplier's system with the push of a button, receive an answer in about 10 seconds, and communicate that to the customer.

Wesco's new system has cut phone costs by reducing the duration of each call by at least six minutes. It has also increased sales of unstocked items and saved an enormous amount of time for salespeople, Lambert says. He estimates that the company could save as much as $12 million annually if the new system saves 1,000 salespeople just three hours per week. Considering that it has cost about $400,000 to date to implement the system, demonstrating return on investment has been a "slam dunk," says Lambert. The system is also bound to have a positive impact on customer perception, he says. Because of Wesco's system, customers know when they call that they're making a real-time order against in-stock supplies, Lambert says. "It has been a great proof-of-concept about the power of direct linkage in the supply chain."

Discussion Questions

1. How, would you guess, did Wesco discover the problem described in the case in its system?

2. Why wouldn't Wesco take this opportunity to replace its old antiquated system with a completely new state-of-the-art system?

Critical Thinking Questions

3. How would the solution to this problem differ if all involved were using standards-based systems rather than proprietary systems?

4. How could the order processing system be even further automated?

Sources: Jaikumar Vijayan, "Wesco Distribution Inc." March 11, 2002, *Computerworld*, http://www.computerworld.com/softwaretopics/erp/story/0,10801,68868,00.html; Wesco International Web site, http://www.wescodist.com, accessed July 2002.

NOTES

Sources for the opening vignette on p. 505: Matt Richtel, "Comcast Says Its Transition to Internet Network Is Done," *The New York Times*, March 2, 2002, http://www.nytimes.com; Matt Richtel, "Comcast Copes with Internet Problems," *The New York Times*, January 4, 2002, www.nytimes.com; "Law Firm Lovell & Stewart Announces Securities Fraud Class Action Lawsuit against AT&T Corp. Alleging Misstatements and Omissions Regarding Excite@Home," *Business Wire*, March 5, 2002; Richard J. Martin, "Seeing the AT&T-Comcast Deal in a Different Light," *Business Week*, Readers Report, Number 3785, p. 19; Comcast Web site, http://www.comcast.com, accessed July 2002.

1. Friedman, Andrew, et al., "Developing Stakeholder Theory," *The Journal of Management Studies*, January 2002, p. 1.
2. Ambler, Scott, "Know the User Before Implementing a System," *Computing Canada*, February 1, 2002, p. 13.
3. Bassellier, et al., "Information Technology Competence of Business Managers," *Journal of Management Information Systems*, Spring 2001, p. 159.
4. Yeh, Quey-Jen et al., "Two Conflict Potentials During IS Development," *Information & Management*, December 2001, p. 135.
5. Dash, Julekha, "Harrah's Bets on New Staff Structure," *Computerworld*, July 2, 2001, p. 34.
6. Meehan, Michael, "Delta Air Lines Web Site Goes Corporate," *Computerworld*, February 5, 2001, p. 6.
7. Songini, Marc, "Fleet Expands CRM Tool to Hundreds of Banks," *Computerworld*, March 12, 2001, p. 22.
8. Disabatino, Jennifer, "TWA to Shed IT Staff," *Computerworld*, October 29, 2001, p. 8.
9. Cope, James, "Tyson IT Staff Faces Meaty Integration Job," *Computerworld*, January 14, 2002, p. 10.
10. Brewin, Bob, "Anthrax Threat Exposes IT Ills," *Computerworld*, October 22, 2001, p. 1.
11. Anthes, Gary, "CIA-Funded Capital In-Q-Tel Is Investing," *Computerworld*, February 5, 2001, p. 59.
12. Buxbaum, Peter, "See You in Court," *Computerworld*, March 26, 2001, p. 42.
13. Sabherwal, Rajiv, et al., "Alignment Between Business and IS Strategies," *Information Systems Research*, March 2001, p. 11.
14. Mearian, Lucas, "Aetna Pins Recovery on IT/Business Union," *Computerworld*, November 12, 2001, p. 6.
15. Buxbaum, Peter, "Measuring Alignment," *Computerworld*, May 7, 2001, p. 46.
16. Ang, James et al., "A Multiple-Case Design Methodology for Studying MRP Success and CSFs," *Information & Management*, January 2002, p. 271.
17. Barry W. Boehm, *Software Engineering Economics* (Englewood Cliffs, N.J.: Prentice-Hall, 1981).
18. Recho Web site at http://www.recho.com, accessed on March 23, 2002.
19. PDS Web site at http://pdspc.com, accessed on March 23, 2002.
20. Choe, S. H. et al., "Modeling and Optimisation of Rapid Prototyping," *Computers in Industry*, January 2002, p. 39.
21. Eva, Malcolm, "Requirements Acquisition for Rapid Applications Development," *Information & Management*, December 2001, p. 101.
22. Eva, Malcolm, "Requirements Acquisition for Rapid Applications Development," *Information & Management*, December 2001, p. 101.
23. Harding, Elizabeth, "RAD Approach Gives Real Estate IT Team Quick Reflexes," *Software Magazine*, January 2001, p. 7.
24. Yourdon, Ed, "Can XP Projects Grow?" *Computerworld*, July 23, 2001, p. 28.

25. Sliwa, Carol, "Users Warm Up to Agile Programming," *Computerworld*, March 18, 2002, p. 8.
26. Greenemeier, Larry, "American Express, IBM Sign $4B Deal," *InformationWeek Online*, February 26, 2002.
27. Brewin, Bob, "PacifiCare Outsources IT Operations to IBM, Keane," *Computerworld*, January 7, 2002, p. 7.
28. Verton, Dan, "Experts, Users Strategize on Security at Crime Summit," *Computerworld*, March 12, 2001, p. 20.
29. Dash, Julekha, "Customer Support Moves Overseas," *Computerworld*, March 19, 2001, p. 10.
30. Verton, Dan, "Experts, Users Strategize on Security at Crime Summit," *Computerworld*, March 12, 2001, p. 20.
31. Dash, Julekha, "Gillette Outsources All Sites," *Computerworld*, January 21, 2001, p. 14.
32. IBM Web page at http://www-ibm.com.services/strategies, accessed on March 23, 2002.
33. EDS Web page at http://www.eds.com, accessed on March 23, 2002.
34. Accenture Web page at http://www.accenture.com, accessed on March 23, 2002.
35. Rai, Arun et al., "Assessing the Validity of IS Success Models," *Information Systems Research*, March 2002, p. 50.
36. Seghezzi, Hans, "Business Excellence," *Total Quality Management*, December 2001, p. 861.
37. Welcome to ISO Easy Web page at http://www.isoeasy.com, accessed on March 23, 2002.
38. "Smith & Associates Wins ISO Certification," *EBN*, March 4, 2002, p. 34.
39. Weiss, J. "Project Management Process in Early Stage E-Business," *Proceedings of the 34th Annual Hawaii International Conference on Systems Science*, January 2001, p. 231.
40. Melynuka, Kathleen, "Projects Across the Pond," *Computerworld*, October 8, 2001, p. 32.
41. Capability Maturity Model for Software home page at http://www.sei.cmu.edu.
42. "Sasken Achieves SEI CMM Level 5," *Businessline*, March 15, 2002.
43. Austin, Robert, "The Effects of Time Pressure on Quality in Software Development," *Information Systems Research*, June 2001, p. 195.
44. Classen, Monika, "Prover Technology Launches Tempo," *PR Newswire*, June 5, 2001.
45. Vijayan, Jaikumar, "Mortgage Vendor Signs on to E-Signatures," *Computerworld*, January 21, 2002, p. 8.
46. King, Julia, "Make It Bigger, Better, Faster," *Computerworld*, January 1, 2002, p. 20.
47. Kerstetter, Jim, "The Web at Your Service," *Business Week*, March 18, 2002, p. EB13.
48. Landro, Laura, "Deadly Hospital Errors Prompt Group to Push for Technological Help," *The Wall Street Journal*, March 15, 2002, p. B1.
49. Stallinger, Friedrich et al., "System Dynamics and Simulation of Collaborative Requirements Engineering," *The Journal of Systems and Software*, December 15, 2001, p. 311.
50. Weiss, Todd, "Feds Seek Developer's Help Making Linux More Secure," *Computerworld*, January 15, 2001, p. 8.
51. Cope, James, "App Helps Boeing Link Factory Floor to Suppliers," *Computerworld*, March 19, 2001, p. 12.
52. Brown, Jeanette, "Prada Gets Personal," *Business Week*, March 18, 2002, p. EB8.

Systems Design, Implementation, Maintenance, and Review

CHAPTER 13

PRINCIPLES	LEARNING OBJECTIVES
Designing new systems or modifying existing ones should always be aimed at helping an organization achieve its goals.	• State the purpose of systems design and discuss the differences between logical and physical systems design. • Outline key steps taken during the design phase. • Describe some considerations and diagrams used during object-oriented design. • Define the term *RFP* and discuss how this document is used to drive the acquisition of hardware and software. • Describe the techniques used to make systems selection evaluations.
The primary emphasis of systems implementation is to make sure that the right information is delivered to the right person in the right format at the right time.	• State the purpose of systems implementation and discuss the various activities associated with this phase of systems development. • List the advantages and disadvantages of purchasing versus developing software. • Discuss the software development process and some of the tools used in this process, including object-oriented program development tools.
Maintenance and review add to the useful life of a system but can consume large amounts of resources. These activities can benefit from the same rigorous methods and project management techniques applied to systems development.	• State the importance of systems and software maintenance and discuss the activities involved. • Describe the systems review process.

Simplifying International Shipping

In the late 1990s, FedEx CIO Robert B. Carter discovered that many customers were frustrated and intimidated by the complexities of shipping their products overseas. Global trade regulations, import duties and taxes, import or export forms, product restrictions, embargoes, and special licensing requirements can all make international trade and overseas shipping an ordeal. So, to provide more customer assistance, Carter assembled a team to develop a new system for overseas shipping.

The team members began by defining precisely what services and information the system should provide. Through internal research they were able to create a list of the countries to which they most frequently shipped and the legal documentation required for those countries. They developed a database to store and organize all the collected information and designed procedures for updating that information.

FedEx systems analysts decided that the most convenient method for making this information available to their customers was through the company's Web site. They designed a Web site with forms and wizard assistance that customers could use to walk through procedures, enter the necessary information, and determine the documentation needed for the overseas delivery. The designers and programmers also linked the Web forms with the database and provided a suite of applications to access pertinent database information and create a Web page to display it.

After three years of programming, testing, and debugging, the system was launched on time and on budget in August 2000. The resulting product was named FedEx Global Trade Manager, a free Web-based guide to international shipping for small and midsize businesses. The application helps shippers understand global trade regulations and prepare the appropriate import or export forms based on the items being shipped and the countries of origin and destination. It also alerts users to restrictions on shipping certain items, lets them know if a country is under a trade embargo, and provides information on special licensing requirements.

After the system launch, further investigation turned up an important additional service this already valuable tool could provide. System designers went back to the drawing board to create a component for determining government charges and fees for international shipments, including import duty, value-added tax, and excise duty. They named the resulting tool the Duty and Tax Estimator and included it on the FedEx Global Trade Manager Web site as a premier service, available at a cost.

This new service is a valuable business tool for companies large and small that need to provide up-front duty and tax information to customers before shipping. Imperial Graphics of Grand Rapids, Michigan, said estimates of such costs help streamline the shipping process. "International trade is complex, so it is critical that our 200 monthly international shipments don't get delayed in customs," said Dan Polkowski, distribution and logistics manager for Imperial Graphics. "We can estimate duties and taxes, but the process is very manual and time-consuming. The new FedEx duty and tax estimator speeds up the process. We can input a few facts then print out the calculation-results screen to send to our customers in advance of shipping." Polkowski also said the FedEx estimator eliminates one invoice in his company's billing process, saving Imperial Graphics "tens of thousands of dollars annually in time, paper and postage."

Donald Broughton, a transportation analyst at A.G. Edwards & Sons in St. Louis, says that although FedEx was the first company to offer such a service, other companies, including Atlanta-based United Parcel Service, now have similar tools.

FedEx continues to measure the project's success by tracking the increase in the number of customers who use Global Trade Manager, as well as the growth of international shipments. Currently, 70,000 registered customers use the services, and that number is growing 300 percent a year. Although Carter won't release the project's financial impact on shareholders, he says it is "among the most profitable at FedEx."

Carter believes that technology projects that focus on customer needs will win approval from the executive board at FedEx. Because this project increased international capabilities and offered payback to customers, it was well received throughout the company from the start.

As you read through this chapter, consider the following:
- ◉ What design steps were likely to have taken place to develop FedEx's Global Trade Manager?
- ◉ What maintenance is required for FedEx's Global Trade Manager now that it is operating and successful?

The way an information system is designed, implemented, and maintained profoundly affects the daily functioning of an organization. Like investigation and analysis covered in the previous chapter, design, implementation, maintenance, and review covered in this chapter strive to achieve organizational goals, such as reduced costs, increased profits, or improved customer service. The goal is to develop a new or modified system to deliver the right information to the right person at the right time.

Information systems must be continually updated or replaced as the needs of a business change, including hardware, software, and the infrastructure components. The opening vignette, for example, shows how FedEx was able to implement a shipping system that improved customer interaction in the United States and around the world. Once created, the new or updated systems must be flexible to meet evolving needs. Otherwise, companies are forced to repeat a costly cycle of building and replacing information systems.

Businesses in many industries develop or modify systems for a variety of uses. And new systems are designed to improve both internal and external processes. Table 13.1 lists some of the companies that were rated the best in their industry by *Computerworld* in its Premier 100 issue.[1]

TABLE 13.1

Top Uses for New Systems in Various Industries

(Source: Data from Maryfran Johnson, "Best in Class," *Computerworld,* March 11, 2002, pp. 1–21.)

Company	Web Site	Business	Application
Staples	www.staples.com	Office products and services	Developed new system that allows customers to consolidate Web and store purchases and pay using cash, check, or credit card.
Burlington Coat Factory	www.coat.com	Discount clothing	Used new Web-based applications and the Linux operating system in more than 250 stores to develop new applications and use existing ones.
State Street Corp.	www.statestreet.com	Financial services	Developed an accounting and settlement system for financial transfers and transactions.
HON Industries	www.honindustries.com	Office furniture	Developed a new system to more accurately measure warehouse capacity and usage.
GFInet, Inc.	www.gfinet.com	Online trading	Developed a virtual private network for currency trading.
FedEx	www.fedex.com	Shipping and logistics	Developed Global Trade Manager, a free Internet shipping system for small and midsize businesses. See the opening vignette for details.
Wesco Distribution, Inc.	www.wescodist.com	Electrical products distribution	Developed a standard system to get information from its suppliers using the Internet.

After an organization investigates and analyzes the need for a new system and approves its development, the project moves to the later stages of development: design, implementation, and maintenance and review. This chapter presents the basics of systems design, implementation, and maintenance and review. Both users and IS personnel need to be aware of these stages so that they can participate in good systems development no matter what field their organizations occupy.

SYSTEMS DESIGN

The purpose of systems design is to answer the question "How will the information system solve a problem?" The primary result of the systems design phase is a technical design that details system outputs, inputs, and user interfaces; specifies hardware, software, databases, telecommunications, personnel, and procedures; and shows how these components are related. The new system should overcome shortcomings of the existing system and help the organization achieve its goals. Of course, the system must also meet certain guidelines, including user and stakeholder requirements and the objectives defined during previous development phases.

Design can range in scope from individual to multicorporate projects (Table 13.2). Nearly all companies are continually involved in designing systems for individuals, workgroups, and the enterprise. Increasingly, companies undertake multicorporate design, where two or more companies form a partnership or an alliance to design a new system.

Systems design is typically accomplished using the tools and techniques discussed in the previous chapter. Depending on the specific application, these methods can be used to support and document all aspects of systems design. Two key aspects of systems design are logical and physical design.

LOGICAL AND PHYSICAL DESIGN

As we discussed earlier in the database chapter, design has two dimensions: logical and physical. The **logical design** refers to what the system will do. The **physical design** refers to how the tasks are accomplished, including how the components work together and what each component does.

logical design
description of the functional requirements of a system

physical design
specification of the characteristics of the system components necessary to put the logical design into action

Logical Design

Logical design describes the functional requirements of a system. That is, it conceptualizes what the system will do to solve the problems identified through earlier analysis. Without this step, the technical details of the system (such as which hardware devices should be acquired) often obscure the best solution. Logical design involves planning the purpose of each system element, independent of hardware and software considerations. The logical design specifications that are determined and documented include the following:

Output design Output design describes all outputs from the system and includes the types, format, content, and frequency of outputs. For example, a requirement that all company invoices reference the customer's original invoice number is a logical design specification. Screen and layout tools can be used during output design to capture the output requirements for the system. Merrill Lynch,

TABLE 13.2

The Scope of Design

Design can range in scope from multicorporate to individual projects.

Multicorporate design projects	Increasingly
Enterprise design projects	Narrow
Workgroup design projects	Scope
Individual design projects	

for example, designed its output for a new order entry system to be consistent with its other systems and to provide a common output format.[2] According to an official from the firm, "It's not necessarily always an upgrade but more about trying to get common systems, common platforms, and common interfaces." To achieve this consistency in output design, Merrill Lynch decided to standardize the order entry system for all of its operations.

Input design   Once output design has been completed, input design can begin. Input design specifies the types, format, content, and frequency of input data. For example, the requirement that the system capture the customer's phone number from his or her incoming call and use that to automatically look up the customer's account information is a logical design specification. A variety of diagrams and screen and report layouts can be used to reveal the type, format, and content of input data.

Process design   The types of calculations, comparisons, and general data manipulations required of the system are determined during process design. For example, a payroll program requires gross and net pay computations, state and federal tax withholding, and various deductions and savings plans.

File and database design   Most information systems require files and database systems. The capabilities of these systems are specified during the logical design phase. For example, the ability to obtain instant updates of customer records is a logical design specification. In many cases, a database administrator is involved with this aspect of logical design. Data-flow and entity-relationship diagrams are typically used during file and database design.

The design of an organization's file and database systems should include the cost of file and database storage and the cost to manage the organization's storage. According to a study conducted by Gartner Group, for every dollar spent on physical storage, a company can spend $7 on data and storage management.[3] Because of these high costs, some companies are considering a charge-back policy, where individual departments pay for their storage and management costs. UPMC Health Systems in Pittsburgh, for example, is considering a charge-back policy for its file and database systems. The approach would charge its departments for the cost of file and database storage and management over UPMC's network storage system. According to Karen Malik, manager of network servers for UPMC, "It would be a huge process for us to implement that, but it would likely be worth it because of the return on investment."

Telecommunications design   During logical design, the network and telecommunications systems need to be specified. For example, a hotel might specify a client/server system with a certain number of workstations that are linked to a server. From these requirements, a hybrid topology might be chosen. Graphics programs, CASE tools, and the object-oriented approach can be used to facilitate logical network design.

Procedures design   All information systems require procedures to run applications and handle problems if they occur. These important policies are captured during procedures design. Once designed, procedures can be described by using text and word processing programs. For example, the steps to add a new customer account may involve a series of both manual and computerized tasks. Written procedures would be developed to provide an efficient process for all to follow.

Controls and security design   Another important part of logical design is to determine the required frequency and characteristics of backup systems. In general, everything should have a backup, including all hardware, software, data, personnel, supplies, and facilities. Planning how to avoid or recover from a computer-related disaster should also be considered in this stage of the logical design phase. Ron Stephenson, vice president of information systems research at Downey Savings and

Loan Association, discussed the importance of planning for a potential computer-related disaster.[4] "We back up everything—everything. We have processes to get data from one place to another, but if you can't get the people over there, the data and software doesn't do any good. It's the nontechnical things that I think need to be completely revisited."

Personnel and job design Some systems require additional employees; others may need modification of the tasks associated with one or more existing IS positions. The job titles and descriptions of these positions are specified during personnel and job design. Organization charts are useful during personnel design to diagram various positions and job titles. Word processing programs are also used to describe job duties and responsibilities.

Physical Design

Physical design specifies the characteristics of the system components necessary to put the logical design into action. In this phase, the characteristics of each of the following components must be specified:

Hardware design All computer equipment, including input, processing, and output devices, must be specified by performance characteristics. For example, if the logical design specified that the database must hold large amounts of historical data, then the system storage devices must have large capacity.

Software design All software must be specified by capabilities. For example, if dozens of users must have the ability to update the database concurrently, as specified in the logical design, then the physical design must specify a database management system to allow this to occur. In some cases, software can be developed internally; in other situations it will be purchased from an IS vendor. This is often called the "make or buy" decision. Logical design specifications for program outputs, data inputs, and processing requirements are also considered during the physical design of the software. For example, the ability to access data stored on certain disk files that the program will use is specified.

Database design The type, structure, and function of the databases must be specified. The relationships between data elements established in the logical design must be mirrored in the physical design as well. These relationships include such things as access paths and file structure organization. Fortunately, many excellent database management systems exist to assist with this activity.

Telecommunications design The characteristics of the communications software, media, and devices must be specified. For example, if the logical design specifies that all members of a department must be able to share data and run common software, then the local area network configuration and the communications software that are specified in the physical design must possess this capability.

Personnel design This step involves specifying the background and experience of individuals most likely to use the system, including the job descriptions specified in the logical design. In the past, companies hired a large number of people to answer questions from customers and potential customers. The costs of these employees and their training could be staggering. As a result, many companies are starting to automate this function to reduce personnel costs and provide more accurate information. Boeing Commercial Airplanes, for example, used a knowledge management system to streamline the process.[5] According to a Boeing spokesperson, "We are trying to change support from being a liability—and the cost of answering hundreds of questions at thousands of dollars a query—to a thing of value."

Procedures and control design How each application is to run, as well as what is to be done to minimize the potential for crime and fraud, must be specified. These specifications include auditing, backup, and output distribution methods.

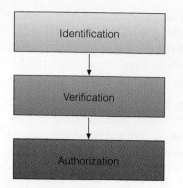

FIGURE 13.1

The Levels of the Sign-On Procedure

sign-on procedure
identification numbers, passwords, and other safeguards needed for an individual to gain access to computer resources

menu-driven system
system in which users simply pick what they want to do from a list of alternatives

help facility
a program that provides assistance when a user is having difficulty understanding what is happening or what type of response is expected

FIGURE 13.2

A menu-driven system allows you to choose what you want from a list of alternatives.

SPECIAL SYSTEM DESIGN CONSIDERATIONS

A number of special system characteristics should be considered during both logical and physical design. These characteristics include sign-on procedures, interactive processing, interactive dialogue, error prevention and detection, and emergency alternate procedures.

Procedures for Signing On

System control methods are established during systems design. With almost any system, control problems may exist, such as criminal hackers breaking into the system or an employee mistakenly accessing confidential data. Sign-on procedures are the first line of defense against these problems. A **sign-on procedure** consists of identification numbers, passwords, and other safeguards needed for an individual to gain access to computer resources. The sign-on, also called a "logon," can identify, verify, and authorize access and usage (Figure 13.1). Identification means that the computer identifies the user as valid. If you must enter an identification number and password when logging on to a mainframe, you have gone through the identification process. For systems and applications that are more sensitive or secure, verification is used. Verification involves entering an additional code before access is given. Finally, authorization restricts the user's access to certain parts of a system or application. Consider a credit-checking application for a major credit card company. To grant credit, a clerk may be given only basic credit information on the screen. A credit manager, however, will have an authorization code to get additional credit information about a client.

Interactive Processing

Today, most computer systems allow interactive processing. With this type of system, people directly interact with the processing component of the system through terminals or networked PCs. The system and the user respond to each other in real time, which means within a matter of seconds. Interactive real-time processing requires special design features for ease of use such as menu-driven systems, help commands, table lookup facilities, and restart procedures. With a **menu-driven system** (Figure 13.2), users simply pick what they want to do from a list of alternatives. Most people can easily operate these types of systems. They select their choice or respond to questions (or prompts) from the system, and the system does the rest.

Many designers incorporate a **help facility** into the system or applications program. When a user is having difficulty understanding what is happening or what type of response is expected, he or she can activate the help facility. The help

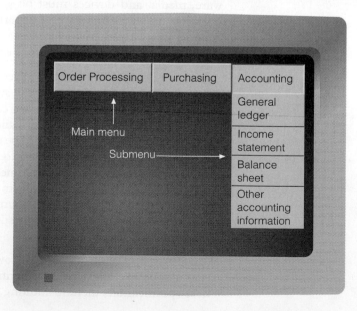

screen relates directly to the problem the user is having with the software. The program responds with information on the program status, what possible commands or selections the user can give, and what is expected in terms of data entry.

Incorporating tables within an application is another very useful design technique. **Lookup tables** can be developed and used by computer programs to simplify and shorten data entry. For example, if you are entering a sales order for a company, you simply type in its abbreviation, such as ABCO. The program will then go to the customer table, normally stored on a disk, and look up all the information pertaining to the company abbreviated ABCO that is required to complete the sales order. This information is then displayed on a screen for confirmation. The use of these tables can prevent wasting a tremendous amount of time entering the same data over and over again into the system.

If a problem occurs in the middle of an application—such as a temporary interruption of power or a printer running out of paper—the application currently being run is typically shut down. As a result, easy-to-use **restart procedures** are developed and incorporated into the design phase. With a restart procedure, it is very simple for an individual to restart an application where it left off.

Designing Good Interactive Dialogue

Dialogue refers to the series of messages and prompts communicated between the system and the user. From a user's point of view, good interactive dialogue from the computer system is essential, making data entry faster, easier, and more accurate. Poor dialogue from the computer can confuse the user and result in the wrong information being entered. If the computer system prompts the user to enter ACCOUNT, does it mean account number, type of account, or something else? The following list covers some elements that create good interactive dialogue. These elements should be considered during systems design.

Clarity The computer system should ask for information using easily understood language. Whenever possible, the users themselves should help select the words and phrases used for dialogue with the computer system.

Response time Ideally, responses from the computer system should approximate a normal response time from a human being carrying on the same sort of dialogue.

Consistency The system should use the same commands, phrases, words, and function keys for all applications. After a user learns one application, all others will then be easier to use.

Format The system should use an attractive format and layout for all screens. The use of color, highlighting, and the position of information on the screen should be considered carefully and consistently.

Jargon All dialogue should be written in easy-to-understand terms. Avoid jargon known only to IS specialists.

Respect All dialogue should be developed professionally and with respect. Dialogue should not talk down to or insult the user. Avoid statements such as "You have made a fatal error."

Preventing, Detecting, and Correcting Errors

The best and least expensive time to deal with potential errors is early in the design phase.[6] During installation or after the system is operating, it is much more expensive and time-consuming to handle errors and related problems. Good systems design attempts to prevent errors before they occur, which involves recognizing what can happen and developing steps and procedures that can prevent, detect, and correct errors. This process includes developing a good backup system that can recover from an error. Table 13.3 lists the major causes of system errors.

---

lookup tables
tables containing data that are developed and used by computer programs to simplify and shorten data entry

restart procedures
simplified process to access an application from where it left off

TABLE 13.3

Major Causes of System Errors

Human	Natural	Technical
IS Personnel	Wind	Hardware
Authorized users	Fire	Application software
Nonauthorized users	Earthquakes	Systems software
	Extreme temperatures	Database
	Floods	Communications
	Hurricanes	Electricity

OBJECT-ORIENTED DESIGN

Logical and physical design can be accomplished using either the traditional structured approach or the object-oriented (OO) approach to systems development. Many organizations today are turning to OO development because of its increased flexibility. So, we outline a few OO design considerations and diagrams here.

Using the OO approach, we design key objects and classes of objects in the new or updated system. This process includes consideration of the problem domain, the operating environment, and the user interface. The problem domain involves the classes of objects related to solving a problem or realizing an opportunity. In our Maui, Hawaii, kayak rental shop example first introduced in Chapter 12 and referring back to the generalization/specialization hierarchy showing classes we presented there, KayakItem in Figure 12.23 is an example of a problem domain object that we will use to store information on kayaks in the rental program. The operating environment for the rental shop's system includes objects that interact with printers, system software, and other software and hardware devices. The user interface for the system includes objects that users interact with, such as buttons and scroll bars in a Windows program.

During the design phase we also need to consider the sequence of events that must happen for the system to function correctly. For example, we might want to design the sequence of events that are needed to add a new kayak to the rental program. A sequence of events is often called a *scenario*, which can be diagrammed in a sequence diagram (Figure 13.3).

A sequence diagram is read starting from the top and moving down.

1. The Create arrow at the top is a message from the Kayak rental clerk to the kayakItem object to create information on a new kayak to be placed into the rental program.
2. The kayakItem object knows that it needs the ID for the kayak and sends a message to the clerk requesting the information. See the getID arrow.
3. The clerk then types the ID into the computer. This is shown with the ID arrow. The data is stored in the kayakItem object.
4. Next, kayakItem requests the purchase date. This is shown in the getDatePurchased arrow.
5. Finally, the clerk types the purchase date into the computer. The data is also transferred to KayakItem object. This is shown in the datePurchased arrow at the bottom of Figure 13.3.

This scenario is only one example of a sequence of events. Other scenarios might include entering information about life jackets, paddles, suntan lotion, and other accessories. The same types of use case and generalization/specialization hierarchy diagrams discussed in Chapter 12 can be created for each, and additional sequence diagrams will also be needed.

FIGURE 13.3

A Sequence Diagram to Add a New KayakItem Scenario

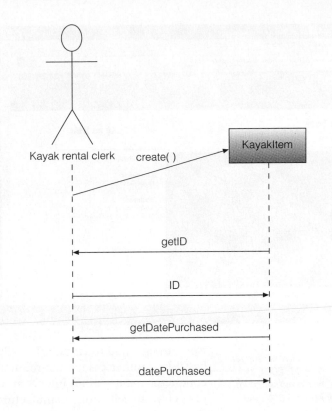

EMERGENCY ALTERNATE PROCEDURES AND DISASTER RECOVERY

As part of ongoing information systems assessment, organizations need to identify key information systems that control cash flow (e.g., invoicing, accounts receivable, payroll) and support other key business operations (e.g., inventory control, shipping customer products). Organizations typically develop a set of emergency procedures and recovery plans for these systems.

If a key information system becomes unusable, the end users need a set of emergency alternate procedures to follow to carry on with business. End users should work with IS personnel to develop these procedures. They may be manual procedures to work around the unavailable automated work process. For example, if the order entry transaction processing system is unavailable, users could resort to special preprinted forms to capture basic order data for later entry when the system is available again. In some cases, the emergency alternate procedures involve accessing a remote computer for computing resources.

Disaster planning is the process of anticipating and providing for disasters. A disaster can be an act of nature (a flood, fire, or earthquake) or a human act, terrorism, error, labor unrest, or erasure of an important file. An earthquake in Seattle, for example, caused Boeing to lose its telecommunications capabilities for days, which forced the airplane manufacturing company to shut down some of its operations.[7] In another case, Morgan Stanley planned a backup trading floor in case of a potential disaster.[8] The company was lucky to avoid losses or trading disruptions through two disasters at the World Trade Center. The backup trading operation is located about 35 miles from its current trading floor in midtown Manhattan. Merrill Lynch is planning to move its main data center from Manhattan to Staten Island as a result of the September 11th terrorist attacks.[9] The new location is also on a different power grid, which is an attractive feature of the move for Merrill Lynch. As seen in these examples, disaster planning often focuses primarily on two issues: maintaining the integrity of corporate information and keeping the information system running until normal operations can be resumed.

To help its New York clients recover from the September 11, 2001, terrorist attacks, Dell Computer promised delivery of about 5,000 new computers in 24 hours.

In some cases, a remote disaster that has no impact on the company's computer system can stop systems development projects in midstream because of the impact the disaster has on the company's business. Royal Caribbean Cruises, for example, was in the middle of a $180 million systems development project, called Leapfrog, to substantially improve its reservation and supply-chain systems. Then came September 11th. As a result of the terrorist attack and reduced demand for cruises, the company shut down the systems development project indefinitely and fired almost half of its IS employees. These actions were very painful but necessary to keep the company operating. According to the CIO, "To take the whole thing apart so quickly when we were executing on all cylinders was probably the worst thing I had to endure. It really tested every ounce of leadership skills that I had." In another case, a ten-minute fire in an Albuquerque, New Mexico, semiconductor plant had an impact around the globe.[10] The Albuquerque plant supplies chips to cell phone manufacturers Nokia in Finland and Ericsson in Sweden. Both companies had to scramble to get back into operation. Ericsson may have lost $400 million in revenues because of the lack of chips to make cell phones and related products.

One of the first steps of disaster planning is to identify potential threats or problems, such as natural disasters, employee errors, and poor internal control procedures. Disaster planning also involves disaster preparedness. IS managers should occasionally hold an unannounced "test disaster"—similar to a fire drill—to ensure that the disaster plan is effective.

Disaster recovery is defined as the implementation of the disaster plan. Although companies have known about the importance of disaster planning and recovery for decades, many do not adequately prepare. The primary tools used in disaster planning and recovery are hardware, software and database, telecommunications, and personnel backup.

Hardware Backup

Companies commonly form arrangements with their hardware vendor or a disaster recovery company to provide access to a compatible computer hardware system or additional hardware in the event of a disaster. Dell Computer, for example, called all of its 85 New York clients directly affected by the September 11th attacks to see how it could help them.[11] As a result of the calls, Dell promised to deliver about 5,000 new computers in 24 hours.

hot site
a duplicate, operational hardware system or immediate access to one through a specialized vendor

cold site
a computer environment that includes rooms, electrical service, telecommunications links, data storage devices, and the like; also called a shell

A duplicate, operational hardware system (or immediate access to one through a specialized vendor) is an example of a **hot site**. A hot site is a compatible computer system that is operational and ready to use. If the primary computer has problems, the hot site can be used immediately as a backup. Another approach is to use a **cold site**, also called a *shell*, which is a computer environment that includes rooms, electrical service, telecommunications links, data storage devices, and the like. If a problem occurs with the primary computer, backup computer hardware is brought into the cold site, and the complete system is made operational. For both hot and cold sites, telecommunications media and devices are used to provide fast and efficient transfer of processing jobs to the disaster facility.

A number of firms offer disaster recovery services. Sun Microsystems, for example, has developed Americas Command Center for its customers.[12] The new

4,400-square-foot facility will monitor about 5,000 Sun clients to prevent potential problems. According to one Sun representative, "The kind of customer we work with now can't go down, they can't go red. We have to solve most of our customers' problems before that happens." Business Recovery Management offers a Disaster Recovery Center in Pittsburgh that provides the facilities, computer systems, and other equipment for recovery from unplanned business interruptions. To provide an effective, fully equipped workplace in time of disaster or special need, the center contains technology to ensure that a business stays connected both locally and globally, including fully equipped computer rooms for local and remote processing; wiring and public branch exchange facilities to support voice, data, facsimile, and video telecommunications; complete offices and services with conference rooms, private offices, and workstation areas to accommodate more than 250 people; and operational and technical support to assist in recovery.[13] Guardian Computer Support is an international computer support company providing a variety of computer services including on-site hardware and software maintenance, contract staffing, disaster recovery, and outsourcing services.[14]

Software and Database Backup

Making duplicate copies of all programs, files, and data backs up software and databases. At least two backup copies should be made. One copy can be kept in the IS department in case of accidental destruction of the software; the other should be kept offsite in a safe, secure, fireproof, and temperature- and humidity-controlled environment. A number of service companies provide this type of backup environment. Oracle, for example, offers its customers disaster recovery products for its database systems. A key objective is to help companies keep their e-commerce applications running. Oracle Parallel Fail Safe, one of Oracle's disaster recovery products, can switch from a failed Web site to a parallel Web site and then bring the failed Web site back online within 30 seconds of its failure.

In addition to application software, system software, such as operating systems, should also be backed up.[15] One approach is to run multiple operating systems on the same computer system or server. Some experts estimate that this approach can save as much as 70 percent of traditional backup costs. The primary disadvantage is a slight reduction in speed and performance. Companies, like VMWare in California, offer software that allow computers and servers to run up to 20 different operating systems at the same time, including Windows 3.1, Windows XP, Unix, and others.

Backup is also essential for programs and data, including data warehouses and data marts. According to one software industry reporter, "The single most effective and powerful approach for avoiding catastrophic failure of a data warehouse is a profoundly distributed architecture."[16] Some companies, like GiantLoop, provide database backup services by distributing critical data to remote data storage centers.[17] The centers are linked with fiber-optic lines to allow the rapid transfer of data to needed computer systems in case of a disaster. GiantLoop also uses sophisticated software to help companies manage their distributed data centers. The advent of more distributed systems, like client/server systems, means that many users now have important, and perhaps mission-critical, data and applications on their computers.

Utility packages inexpensively provide backup features for desktop computers by copying data onto CD-ROM or CD-RW disks, magnetic disk, or tape. But software and database backup can be very difficult if an organization has a large amount of data. For some companies, making a backup of the entire database could take hours. A tight budget may also prohibit backing up significant quantities of data. As a result, some companies use **selective backup**, which involves creating backup copies of only certain files. For example, only critical files might be copied every night for backup purposes.

selective backup
creating backup copies of only certain files

Another backup approach is to make a copy of all files changed during the last few days or the last week, a technique called **incremental backup**. This approach to backup uses an **image log**, which is a separate file that contains only changes to applications. Whenever an application is run, an image log is created that contains all changes made to all files. If a problem with a database occurs, an old database with the last full backup of the data, along with the image log, can be used to re-create the current database.

Good software can also provide additional security for people, hardware, and other computer system components. Cantor Fitzgerald, a bond-trading firm formerly located in the World Trade Center, lost its offices and more than 700 employees on September 11th.[18] But two days after the attack, the firm was back in operation. How was it able to accomplish this amazing task? Its sophisticated trading software, called eSpeed, is part of the reason. Cantor's remaining employees in other locations in the United States and in London were able to use the software to keep the company operational. According to Matt Claus, Cantor's new chief technology officer, "The reason it was possible was because of our previous strategy to build a concurrent computer center that mirrored all the services in eSpeed's World Trade Facilities."

Telecommunications Backup

Most disaster recovery plans call for the backup of vital Internet and telecommunications systems. This is especially true for companies that have substantial e-commerce operations. According to a *PC Magazine* article, "Every company needs a disaster recovery plan, but e-businesses have some special needs to guarantee they're running around the clock."[19] Some plans might call for recovering whole networks. In other plans, the most critical nodes on the network are backed up by duplicate components. Using such fault-tolerant networks, which will not break down when one node or part of the network malfunctions, can be a more cost-effective approach to telecommunications backup. IBM has developed a new information system to help monitor computer networks and the Internet.[20] The new system is part of IBM's eLize project, which will allow computer systems to monitor themselves and make changes that are required as they or their environment change.

Personnel Backup

Information systems personnel must also have backup, which can be accomplished in a number of ways. One of the best approaches is to provide cross training for IS and other personnel so that each individual can perform an alternate job if required. For example, a company might train employees in accounting, finance, or other IS departments to operate the system if a disaster strikes. The company could also make an agreement with another IS department or an outsourcing company to supply IS personnel if necessary. Weyerhaeuser, for example, used outsourcing for many of its IS personnel.[21] The wood-products company hired EDS and transferred many of its employees to EDS. Blake Self makes the same commute to the same office to do basically the same work he has done for the past 15 years. But instead of working for Weyerhaeuser as he has for more than a decade, he is now employed by EDS.

Security, Fraud, and the Invasion of Privacy

Security lapses, fraud, and the invasion of privacy can present disastrous problems. For example, because of an inadequate security and control system, a futures and options trader for

Personnel backup is critical in disaster recovery. Shown is the SunGard Data Systems crisis management center in Philadelphia on September 13, 2001. SunGard team members helped New York businesses with data-recovery and accounting services.

(Source: AP/WideWorld Photos.)

a British bank lost about $1 billion. A simple system might have prevented a problem that caused the 200-year-old bank to collapse. In addition, from time to time, IRS employees have been caught looking at the returns of celebrities and others. Preventing and detecting these problems is an important part of systems design. Prevention includes the following:

- Determining potential problems
- Ranking the importance of these problems
- Planning the best place and approach to prevent problems
- Deciding the best way to handle problems if they occur

Every effort should be made to prevent problems, but companies must establish procedures to handle problems if they occur.

SYSTEMS CONTROLS

systems controls
rules and procedures to maintain data security

Most IS departments establish tight **systems controls** to maintain data security. Systems controls can help prevent computer misuse, crime, and fraud by managers, employees and others. Accounting scandals reported in 2001 and 2002, including WorldCom and Enron, have caused many IS departments to develop systems controls to make it more difficult for executives to mislead investors and employees. Some of these scandals involved billions of dollars.

closed shops
IS department in which only authorized operators can run the computers

open shops
IS department in which other people, such as programmers and systems analysts, are also authorized to run the computers

Most IS departments have a set of general operating rules that help protect the system. Some information systems departments are **closed shops**, in which only authorized operators can run the computers. Other IS departments are **open shops**, in which other people, such as programmers and systems analysts, are also authorized to run the computers. Other rules specify the conduct of the IS department.

deterrence controls
rules and procedures to prevent problems before they occur

These rules are examples of **deterrence controls**, which involve preventing problems before they occur. Making a computer more secure and less vulnerable to a break-in is another example. Good control techniques should help an organization contain and recover from problems. The objective of containment control is to minimize the impact of a problem while it is occurring, and recovery control involves responding to a problem that has already occurred.

Many types of system controls may be developed, documented, implemented, and reviewed. These controls touch all aspects of the organization, including the following:

Input controls. Input controls maintain input integrity and security. Some input controls involve the people who use the system; others relate to the data. The overall purpose is to reduce errors while protecting the computer system against improper or fraudulent input. Input controls range from using standardized input forms to eliminating data entry errors and using tight password and identification controls. For example, based on their logon identification, users are provided with access to a subset of the system and its capabilities. Some users can only view data; other users can update and view data. In addition, input controls can involve more sophisticated hardware and software that can use voice, fingerprints, and related techniques to identify and permit access to sensitive computer systems.

Processing controls. Processing controls deal with all aspects of processing and storage. In many cases, hardware and software are duplicated to provide procedures that ensure processing is as error-free as possible. In addition, storage controls prevent users from gaining access to or accidentally destroying data. The use of passwords and identification numbers, backup copies of data, and storage rooms that have tight security systems are examples of storage controls.

Output controls. Output controls are developed to ensure that output is handled correctly. In many cases, output generated from the computer system is recorded in a file that indicates the reports and documents generated, the time they were generated, and their final destinations.

Many companies use ID badges to prevent unauthorized access to sensitive areas in the information systems facility.

(Source: Courtesy of Sensomatic Electronics Corporation.)

Database controls. Database controls deal with ensuring an efficient and effective database system. These controls include the use of identification numbers and passwords, without which a user is denied access to certain data and information. Many of these controls are provided by database management systems.

Telecommunications controls. Telecommunications controls are designed to provide accurate and reliable data and information transfer among systems. Some telecommunications controls include hardware and software and other devices developed to ensure correct communications while eliminating the potential for fraud and crime. Examples are encryption devices and expert systems that can be used to protect a network from unauthorized access.

Personnel controls. Various personnel controls can be developed and implemented to make sure only authorized personnel have access to certain systems to help prevent computer-related mistakes and crime. Personnel controls can involve the use of identification numbers and passwords that allow only certain people access to certain data and information. ID badges and other security devices (such as "smart cards") can prevent unauthorized people from entering strategic areas in the information systems facility.

Once controls are developed, they should be documented in various standards manuals that indicate how the controls are to be implemented. They should then be implemented and frequently reviewed. It is common practice to measure the extent to which control techniques are used and to take action if the controls have not been implemented.

THE IMPORTANCE OF VENDOR SUPPORT

Whether an individual is purchasing a personal computer or an experienced company is acquiring an expensive mainframe computer, the system is often obtained from one or more vendors. In some cases, the vendor simply provides hardware or software. In other cases, the vendor provides additional services.[22] Some of the factors to consider in selecting a vendor are the following:

- The vendor's reliability and financial stability
- The type of service offered after the sale
- The goods and services the vendor offers and keeps in stock
- The vendor's willingness to demonstrate its products
- The vendor's ability to repair hardware
- The vendor's ability to modify its software
- The availability of vendor-offered training of IS personnel and system users
- Evaluations of the vendor by independent organizations

GENERATING SYSTEMS DESIGN ALTERNATIVES

When additional hardware and software are not required, alternative designs are often generated without input from vendors. If the new system is complex, the original development team may want to involve other personnel in generating alternative designs. If new hardware and software are to be acquired from an outside vendor, a formal request for proposal (RFP) should be made.

Request for Proposal

request for proposal (RFP)
a document that specifies in detail required resources such as hardware and software

The **request for proposal (RFP)** is one of the most important documents generated during systems development. It often results in a formal bid that is used to determine who gets a contract for new or modified systems. The RFP specifies in detail the required resources such as hardware and software. While it can take time and money to develop a high-quality RFP, it can save a company in the long run.[23] According to an individual who frequently prepares RFPs, "If I renegotiate a contract without going through the RFP process, I'll save 10 percent compared to what I currently pay. But if I issue an RFP, I'll save 40 percent." Companies that frequently generate RFPs can automate the process. Wachovia Bank, for example, purchased a software package, called The RFP Machine from Pragmatech Software, to improve the quality of its RFPs and to reduce the time it takes to produce them.[24] The RFP Machine stores important data needed to generate RFPs and automates the process of producing RFP documents.

In some cases, separate RFPs are developed for different needs. For example, a company might develop separate RFPs for hardware, software, and database systems. The RFP also communicates these needs to one or more vendors, and it provides a way to evaluate whether the vendor has delivered what was expected. In some cases, the RFP is made part of the vendor contract. The table of contents for a typical RFP is shown in Figure 13.4.

Financial Options

When it comes to acquiring computer systems, three choices are available: purchase, lease, or rent. Cost objectives and constraints set for the system play a significant role in the alternative chosen, as do the advantages and disadvantages of each. Table 13.4 summarizes the advantages and disadvantages of these financial options.

Determining which option is best for a particular company in a given situation can be difficult. Financial considerations, tax laws, the organization's policies, its sales and transaction growth, marketplace dynamics, and the organization's financial resources are all important factors. In some cases, lease or rental fees can amount to more than the original purchase price after a few years. As a result, some companies prefer to purchase their equipment.

On the other hand, constant advances in technology can make purchasing risky. A company would not want to purchase a new multimillion-dollar computer only to have newer and more powerful computers available a few months later at a lower price. Some companies employ several people to determine the best option based on all the factors. This staff can also help negotiate purchase, lease, or rental contracts.

EVALUATING AND SELECTING A SYSTEM DESIGN

The final step in systems design is to evaluate the various alternatives and select the one that will offer the best solution for organizational goals. Depending on

FIGURE 13.4

A Typical Table of Contents for a Request for Proposal

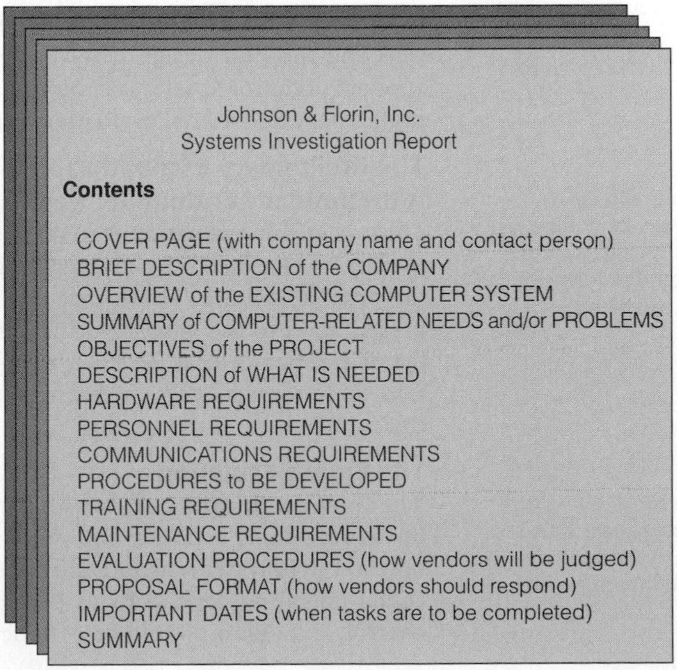

Johnson & Florin, Inc.
Systems Investigation Report

Contents

COVER PAGE (with company name and contact person)
BRIEF DESCRIPTION of the COMPANY
OVERVIEW of the EXISTING COMPUTER SYSTEM
SUMMARY of COMPUTER-RELATED NEEDS and/or PROBLEMS
OBJECTIVES of the PROJECT
DESCRIPTION of WHAT IS NEEDED
HARDWARE REQUIREMENTS
PERSONNEL REQUIREMENTS
COMMUNICATIONS REQUIREMENTS
PROCEDURES to BE DEVELOPED
TRAINING REQUIREMENTS
MAINTENANCE REQUIREMENTS
EVALUATION PROCEDURES (how vendors will be judged)
PROPOSAL FORMAT (how vendors should respond)
IMPORTANT DATES (when tasks are to be completed)
SUMMARY

TABLE 13.4

Advantages and Disadvantages of Acquisition Options

Renting (Short-Term Option)

Advantages	Disadvantages
No risk of obsolescence	No ownership of equipment
No long-term financial investment	High monthly costs
No initial investment of funds	Restrictive rental agreements
Maintenance usually included	

Leasing (Longer-Term Option)

Advantages	Disadvantages
No risk of obsolescence	High cost of canceling lease
No long-term financial investment	Longer time commitment than renting
No initial investment of funds	No ownership of equipment
Less expensive than renting	

Purchasing

Advantages	Disadvantages
Total control over equipment	High initial investment
Can sell equipment at any time	Additional cost of maintenance
Can depreciate equipment	Possibility of obsolescence
Low cost if owned for a number of years	Other expenses, including taxes and insurance

their weight, any one of these objectives may result in the selection of one design over another. For example, financial concerns might make a company choose rental over equipment purchase. Specific performance objectives—say, that the new system must perform on-line data processing—may result in a complex network design for which control procedures must be established. Evaluating and selecting the best design involves a balance of system objectives that will best support organizational goals. Normally, evaluation and selection involves both a preliminary and a final evaluation before a design is selected.

The Preliminary Evaluation

preliminary evaluation
an initial assessment whose purpose is to dismiss the unwanted proposals; begins after all proposals have been submitted

A **preliminary evaluation** begins after all proposals have been submitted. The purpose of this evaluation is to dismiss unwanted proposals. Several vendors can usually be eliminated by investigating their proposals and comparing them with the original criteria. Those that compare favorably are asked to make a formal presentation to the analysis team. The vendors should also be asked to supply a list of companies that use their equipment for a similar purpose. The organization then contacts these references and asks them to evaluate their hardware, their software, and the vendor.

The Final Evaluation

final evaluation
a detailed investigation of the proposals offered by the vendors remaining after the preliminary evaluation

The **final evaluation** begins with a detailed investigation of the proposals offered by the remaining vendors. The vendors should be asked to make a final presentation and to fully demonstrate the system. The demonstration should be as close to actual operating conditions as possible. Such applications as payroll, inventory control, and billing should be conducted using a large amount of test data.

After the final presentations and demonstrations have been given, the organization makes the final evaluation and selection. Cost comparisons, hardware performance, delivery dates, price, flexibility, backup facilities, availability of software training, and maintenance factors are considered. Although it is good to compare computer speeds, storage capacities, and other similar characteristics, it is also necessary to carefully analyze whether the characteristics of the proposed systems meet the company's objectives. In most cases, the RFP captures these objectives and goals. Figure 13.5 illustrates the evaluation process.

Note that the number of possible alternatives decreases as the firm gets closer to making a final decision.

EVALUATION TECHNIQUES

The exact procedure used to make the final evaluation and selection varies from one organization to the next. Some were first introduced in Chapter 2, including return on investment (ROI), earnings growth, market share, customer satisfaction, and total cost of ownership (TCO). In addition, there are four other approaches commonly used: group consensus, cost/benefit analysis, benchmark tests, and point evaluation.

Group Consensus

group consensus
decision making by a group that is appointed and given the responsibility of making the final evaluation and selection

In **group consensus**, a decision-making group is appointed and given the responsibility of making the final evaluation and selection. Usually, this group includes the members of the development team who participated in either systems analysis or systems design. This approach might be used to evaluate which of several screen layouts or reports formats is best.

Cost/Benefit Analysis

cost/benefit analysis
an approach that lists the costs and benefits of each proposed system. Once expressed in monetary terms, all the costs are compared with all the benefits

Cost/benefit analysis is an approach that lists the costs and benefits of each proposed system.[25] Once expressed in monetary terms, all the costs are compared with all the benefits. Table 13.5 lists some of the typical costs and benefits associated with the evaluation and selection procedure. This approach is used to evaluate options whose costs can be quantified, such as which hardware or software vendor to select.[26]

Benchmark Tests

benchmark test
an examination that compares computer systems operating under the same conditions

A **benchmark test** is an examination that compares computer systems operating under the same conditions.[27] Most computer companies publish their own benchmark tests, but some forbid disclosure of benchmark tests without prior

FIGURE 13.5

The Stages in Preliminary and Final Evaluations

Note that the number of possible alternatives decreases as the firm gets closer to making a final decision.

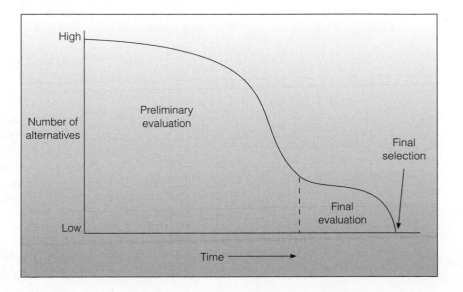

TABLE 13.5

Cost/Benefit Analysis Table

Costs	Benefits
Development costs	**Reduced costs**
Personnel	Fewer personnel
Computer resources	Reduced manufacturing costs
	Reduced inventory costs
	More efficient use of equipment
	Faster response time
	Reduced down- or crash time
	Less spoilage
Fixed costs	**Increased revenues**
Computer equipment	New products and services
Software	New customers
One-time license fees for software and maintenance	More business from existing customers
	Higher price as a result of better products and services
Operating costs	**Intangible benefits**
Equipment lease and/or rental fees	Better public image for the organization
Computer personnel (including salaries, benefits, etc.)	Higher employee morale Better service for new and existing customers
Electric and other utilities	The ability to recruit better employees
Computer paper, tape, and disks	Position as a leader in the industry
Other computer supplies	System easier for programmers and users
Maintenance costs	
Insurance	

written approval.[28] Thus, one of the best approaches is for an organization to develop its own tests, then use them to compare the equipment it is considering. Several independent companies also rate computer systems. *Computerworld*, *Datamation*, and *PC Week*, for example, not only summarize various systems but also evaluate and compare computer systems and manufacturers according to a number of criteria. This approach might be used to compare the end user system response time on two similar systems.

Point Evaluation

One of the disadvantages of cost/benefit analysis is the difficulty of determining the monetary values for all the benefits. An approach that does not employ

point evaluation system
an evaluation process in which each evaluation factor is assigned a weight, in percentage points, based on importance. Then each proposed information system is evaluated in terms of this factor and given a score ranging from 0 to 100. The scores are totaled, and the system with the greatest total score is selected

monetary values is a **point evaluation system**. Each evaluation factor is assigned a weight, in percentage points, based on importance. Then each proposed information system is evaluated in terms of this factor and given a score that might range from 0 to 100, where 0 means the alternative does not address the feature at all and 100 means the alternative addresses that feature perfectly. The scores are totaled, and the system with the greatest total score is selected. When using point evaluation, literally hundreds of factors can be listed and evaluated. Figure 13.6 shows a simplified version of this process. This approach is used when there are many factors on which options are to be evaluated, such as which software best matches a particular business's needs.

Because many elements must be considered before making a final selection, point evaluation can include a large number of factors. Performance concerns might include speed, storage capacity, and processing capabilities. Costs might include the deposit required on contract signing, payment schedules, lease and rental arrangements, maintenance costs, and availability of leasing companies. Complexity factors could include compatibility and ease of use, while control might include considerations such as training and maintenance offered by vendors, as well as system reliability and backup. When all these factors are added to the point evaluation system, a very large grid can result. The rows of the grid list the various factors important to the client company, and the columns of the grid represent the various vendors that responded to the request for proposal. Even if weights are not used, this type of chart can be very helpful. Some companies just use check marks to indicate which vendors have satisfied certain factors.

FREEZING DESIGN SPECIFICATIONS

Near the end of the design stage, an organization prohibits further changes in the design of the system. The design specifications are then said to be frozen. Freezing systems design specifications means that the user agrees in writing that the design is acceptable. Most system consulting companies insist on this step to avoid cost overruns and missed user expectations (Figure 13.7).

A problem that often arises during the implementation of any major project is that of "scope creep." As users more clearly understand how the system will work and what their needs are, they begin to request changes to the original design. Each change may be relatively minor, so the project team is strongly tempted to expand the scope of the project and incorporate the requested changes. However, the aggregate impact of many small changes can be significant; implementation of all minor changes can delay the project and/or increase the cost significantly. A common example is the request for a new report that was not included in the original design. If not managed carefully, granting the request for the first new report can lead to the request for "just one more report" and then another and another.

Prior to implementation, experienced project managers place formal controls on the project scope. A key component of the process is to assess the cost and schedule impact of each requested change, no matter how small, and to decide whether to include the change. Often the users and the project team decide to

FIGURE 13.6

An Illustration of the Point Evaluation System

In this example, software has been given the most weight (40 percent), compared with hardware (35 percent) and vendor support (25 percent). When system A is evaluated, the total of the three factors amounts to 82.5 percent. System B's rating, on the other hand, totals 86.75 percent, which is closer to 100 percent. Therefore, the firm chooses system B.

		System A			System B		
Factor's importance		*Evaluation*		*Weighted evaluation*	*Evaluation*		*Weighted evaluation*
Hardware	35%	95	35%	33.25	75	35%	26.25
Software	40%	70	40%	28.00	95	40%	38.00
Vendor support	25%	85	25%	21.25	90	25%	22.50
Totals	100%			82.5			86.75

FIGURE 13.7

Freezing Design Specifications

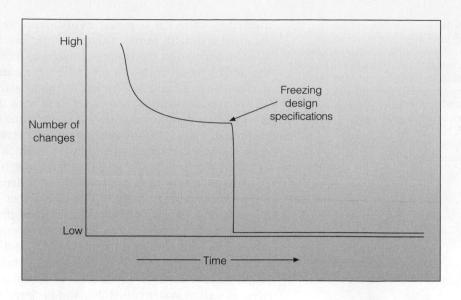

hold all changes until the original effort is completed and then prioritize the entire set of requested changes.

THE CONTRACT

One of the most important steps in systems design is to develop a good contract if new computer facilities are being acquired.[29] Finding the best terms where everyone makes a profit can be difficult.[30] According to Bart Perkins, who used to be the CIO at Dole Foods, "We negotiate so hard with suppliers that they don't make any money, and then we wonder why service is so poor." Most computer vendors provide standard contracts; however, such contracts are designed to protect the vendor, not necessarily the organization buying the computer equipment.

More and more organizations are using outside consultants and legal firms to help them develop their own contracts. Such contracts stipulate exactly what they expect from the system vendor and what interaction will occur between the vendor and the organization. All equipment specifications, software, training, installation, maintenance, and so on are clearly stated. Also, the contract stipulates deadlines for the various stages or milestones of installation and implementation, as well as actions the vendor will take in case of delays or problems. Some organizations include penalty clauses in the contract, in case the vendor is unable to meet its obligation by the specified date. Typically, the request for proposal becomes part of the contract. This saves a considerable amount of time in developing the contract, because the RFP specifies in detail what is expected from the vendors.

THE DESIGN REPORT

design report
the primary result of systems design, reflecting the decisions made for system design and preparing the way for systems implementation

System specifications are the final results of systems design. They include a technical description that details system outputs, inputs, and user interfaces, as well as all hardware, software, databases, telecommunications, personnel, and procedure components and the way these components are related. The specifications are contained in a **design report**, which is the primary result of systems design. The design report reflects the decisions made for system design and prepares the way for systems implementation. The contents of the design report are summarized in Figure 13.8.

FIGURE 13.8

A Typical Table of Contents for a System Design Report

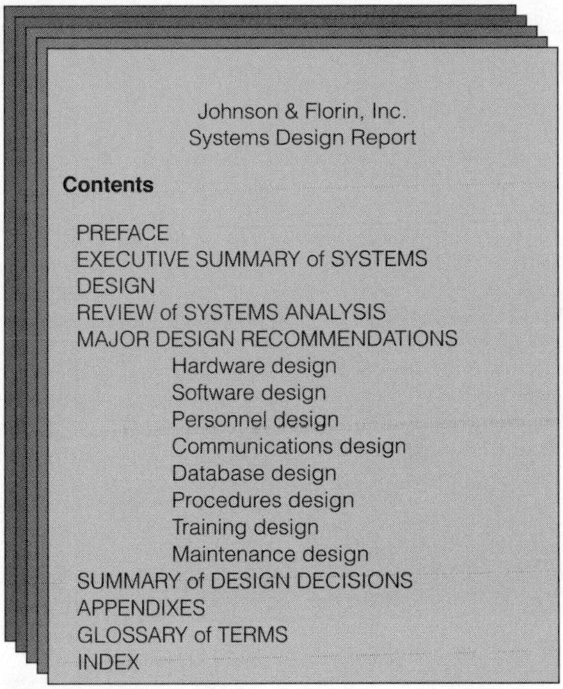

Johnson & Florin, Inc.
Systems Design Report

Contents

PREFACE
EXECUTIVE SUMMARY of SYSTEMS DESIGN
REVIEW of SYSTEMS ANALYSIS
MAJOR DESIGN RECOMMENDATIONS
- Hardware design
- Software design
- Personnel design
- Communications design
- Database design
- Procedures design
- Training design
- Maintenance design

SUMMARY of DESIGN DECISIONS
APPENDIXES
GLOSSARY of TERMS
INDEX

When developing a system, it is important to understand and thoroughly complete the systems development activities covered in this chapter. These phases provide the blueprints and groundwork for the rest of systems development. The activities of the next phases will be easier, faster, and more accurate and will result in a more efficient, effective system if the design is complete and well thought out.

SYSTEMS IMPLEMENTATION

After the information system has been designed, a number of tasks must be completed before the system is installed and ready to operate. This process, called systems implementation, includes hardware acquisition, software acquisition or development, user preparation, hiring and training of personnel, site and data preparation, installation, testing, start-up, and user acceptance. The typical sequence of these activities is shown in Figure 13.9.

ACQUIRING HARDWARE FROM AN INFORMATION SYSTEMS VENDOR

To obtain the components for an information system, organizations can purchase, lease, or rent computer hardware and other resources from an IS vendor. An IS vendor is a company that offers hardware, software, telecommunications systems, databases, IS personnel, and/or other computer-related resources. Types of information systems vendors include general computer manufacturers (e.g., IBM and Hewlett-Packard), small computer manufacturers (e.g., Dell and Gateway), peripheral equipment manufacturers (e.g., Epson and Canon), computer dealers and distributors (e.g., Radio Shack and CompUSA), and leasing companies (e.g., National Computer Leasing and Paramount Computer Rentals, plc).

It is also possible to purchase used computer equipment.[31] This option is especially attractive to firms that are experiencing an economic slowdown. Traditional

FIGURE 13.9

Typical Steps in Systems Implementation

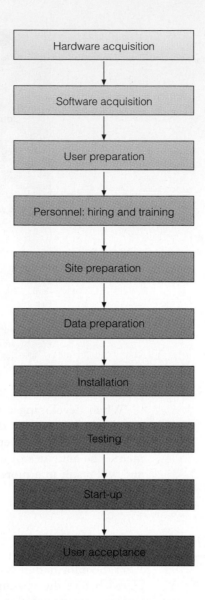

Internet auctions are often used by companies to locate used or refurbished equipment. Todd Lutwak, director for eBay's information systems marketplace, expects the company will sell more than $1 billion of computer-related equipment annually. According to the technology officer at E-Loan, Inc., "When we crunched the numbers for 2001, we averaged [paying] 21 cents on the dollar for the stuff we bought on eBay." While the price savings can be huge, buying used equipment has its disadvantages. "If you rely too much on auctions, your costs aren't predictable, and your supply definitely isn't predictable," says Bill Flanagan, an analyst for the Burton Group.

In addition, companies are increasingly turning to service providers to implement some or all of a systems development effort. As discussed in Chapter 4, an application service provider (ASP) can help a company implement software and other systems. The ASP can provide both end user support and the computers on which to run the software. ASPs often focus on high-end applications, such as database systems and enterprise resource planning packages. Salesforce.com is an ASP that monitors a company's sales data on the Web.[32] In 2001, its revenues almost doubled. According to Marc Benioff, the company's 36-year-old founder, "It took Microsoft and Oracle nine years to

reach $50 million in sales. We'll do it in three. We're going to be the most profitable software company in the world."

As mentioned in Chapter 7, an Internet service provider (ISP) assists a company in gaining access to the Internet. ISPs can also assist a company in setting up an Internet site. Some service providers specialize in specific systems or areas, such as marketing, finance, or manufacturing.

ACQUIRING SOFTWARE: MAKE OR BUY?

As with hardware, application software can be acquired several ways. As previously mentioned, it can be purchased from external developers or developed in-house. This decision is often called the **make-or-buy decision**. In some cases, companies use a blend of external and internal software development. That is, off-the-shelf or proprietary software programs are modified or customized by in-house personnel. The advantages and disadvantages of these approaches were discussed in Chapter 4.

System software, such as operating systems or utilities, is typically purchased from a software company. Increasingly, however, companies are obtaining open-source systems software, such as the Linux operating system, which can be obtained free or for a low cost. See the "Ethical and Societal Issues" box for examples of companies that are using this approach.

Computer dealers, such as CompUSA, manufacture build-to-order computer systems and sell computers and supplies from other vendors.

(Source: Courtesy of CompUSA, Inc.)

make-or-buy decision
the decision regarding whether to obtain the necessary software from internal or external sources

Newmarket International, which invented hospitality sales and event management software, created Delphi, a forecasting and tracking tool that gives salespeople instant, updated on-screen room information including transient commitments, target rates, and room type availability.

(Source: Courtesy of Newmarket International, Inc.)

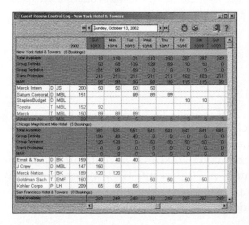

EXTERNALLY DEVELOPED SOFTWARE

Some of the reasons a company might purchase or lease externally developed software include lower costs, less risk regarding the features and performance of the package, and ease of installation. The cost of the software package is known, and there is little doubt that it will meet the company's needs. The amount of development effort is also less when software is purchased, compared with in-house development.

For example, a company may decide to purchase a general ledger software package developed by a major international consulting firm that is used widely throughout the industry. If the company were to decide to build the software, it would take many effort months (or even years), and there is a high degree of risk that when the system is implemented it may not meet the business needs as well as the software package.

If a company chooses off-the-shelf or contract software in its new systems, it must take the following steps:

Review needs, requirements, and costs. It is important to analyze the program's ability to satisfy user and organizational needs. In some cases, purchased software doesn't satisfy all of an organization's requirements or might have features that are not needed. Software costs are also important. In most cases, software can be purchased for unlimited use or leased. The specific terms of the lease are important.

Open Source vs. Proprietary Software

Increasing numbers of information systems developers are turning to Linux to power their systems. Linux (usually pronounced LIH-nuhks with a short "i") is a Unix-like operating system that is designed to provide personal computer users a free or very low-cost operating system compared with traditional and usually more expensive Unix systems. Since versions of Linux exist for all the major microprocessor platforms, it competes with all popular proprietary operating systems including Microsoft Windows and Mac OS. Unlike Windows and other proprietary systems, besides being free, Linux is publicly open and can be modified by contributors. Users can view, edit, and customize the Linux source code to have it do whatever is needed for their applications. Linux is distributed using the Free Software Foundation's *copyleft* stipulations, meaning that any modified version that is redistributed must in turn be freely available.

In Europe, oil company BP and Banca Commerciale Italiana are among the big companies that have moved to Linux. According to IBM, as many as 15 banks in central London are running Linux clusters. Korean Air, which now does all its ticketing on Linux, and U.S. motorhome manufacturer Winnebago are other high-profile examples. Most Linux adopters cite software licensing, hardware upgrade costs, and customizability as the primary reasons for the switch.

In the United States, the Walt Disney Company's animation division recently adopted Linux as its operating system of choice. In doing so they joined other major studios and special-effects houses including DreamWorks SKG, Pixar Animation Studios, Industrial Light and Magic, and Digital Domain. "For us, it's a move to less-expensive commodity technology systems," said John Carey, vice president for Walt Disney Feature Animation.

The advance of Linux into Hollywood is a sign that a technology once viewed as part of the counterculture of computing is moving steadily into the mainstream. "Hollywood is at the leading edge of computing, and it shows what Linux can do," said Martin Fink, general manager for Hewlett-Packard's Linux systems division.

Is it ethical for a software company to give away its product? Is this a fair competitive practice? The Free Software Foundation, associated with the GNU project that wrote Linux, argues the contrary: it claims that it is unethical to copyright and *own* software. It maintains that software should be freely shared and developed in a communal environment. Freely sharing the source code of operating systems like Linux and other software, they believe, promotes innovation in the development process.

Those in the commercial software industry believe that a traditional business model is needed to promote innovation in software development. It takes a significant quantity of money to employ top-notch software developers and programmers. Software manufacturers can invest hundreds of thousands of dollars in the development process. The businesses that provide us with this high-quality software could not remain afloat if they gave away their products.

While the Free Software Foundation's philosophy has not yet been taken seriously by the market, the Linux operating system certainly has. As Linux has gradually improved, through communal development, it has gained ground to the point where it is replacing proprietary Unix systems and becoming a serious competitive threat to the biggest software proprietor in the business, Microsoft.

Discussion Questions

1. Under the Free Software Foundation's philosophy, how would it be possible for software developers to earn a decent living?
2. What benefits do vendors like Microsoft offer their customers that might not be possible with uncopyrighted, free software?

Critical Thinking Questions

3. Would it be wise for Microsoft and other proprietary owners of software to make its source code available to the public for development purposes?
4. The Free Software Foundation argues that copyrights are appropriate for printed material but inappropriate for digital media such as software. Argue both sides of this statement.

Sources: Matt Loney, "More Foreign Banks Switching to Linux," *ZDNet* (UK), April 22, 2002, http://zdnet.com.com/2100-1103-887961.html; Steve Lohr, "Disney Shifting to Linux for Film Animation," *The New York Times*, June 18, 2002, http://www.nytimes.com/2002/06/18/technology/18LINU.html?todays headlines; *Search390.com*, "Linux Defined," http://search390.techtarget.com/sDefinition/0,,sid10_gci212482,00.html; Free Software Foundation's Web site, http://www.fsf.org, accessed July 2002.

Microsoft, for example, changed its purchase and lease arrangements for its popular operating systems and Office products.[33] Under the new deal, some companies will pay as much as 80 percent less, but companies that infrequently upgrade could see leasing costs increase by as much as 70 percent. How a company uses software and how often the software is upgraded can make a big difference in the cost of acquiring software from an outside vendor.

Acquire software. Many of the approaches discussed in previous sections, including the development of requests for proposals, performing financial analysis, and negotiating the software contract, should be undertaken.

Modify or customize software. Externally developed software seldom does everything the organization requires. So, it is likely that externally developed software will have to be modified to satisfy user and organizational needs. Some software vendors will assist with the modification, but others may not allow their software to be modified at all.

Acquire software interfaces. Usually, proprietary software requires a **software interface**, which consists of programs or program modifications that allow proprietary software to work with other software used in the organization. For example, if an organization purchases a proprietary inventory software package, software interfaces must allow the new software to work with other programs, such as sales ordering and billing programs.

Test and accept the software. Externally developed software should be completely tested by users in the environment in which it is to run before it is accepted.

Monitor and maintain the software and make necessary modifications. With many software applications, changes will likely have to be made over time. This aspect should be considered in advance because, as mentioned before, some software vendors do not allow their software to be modified.

When the software is not meeting organizational goals or expectations, the software can be abandoned instead of modified—often a difficult decision. The second-largest Canadian supermarket chain, for example, acquired and implemented an ERP package.[34] It cost the company almost $90 million for the software installation. After the supermarket chain experienced a five-day shutdown, which had an impact on corporate operations for almost a month, it decided to dump the software and look for another software package that could be quickly implemented and better able to meet company needs.

IN-HOUSE-DEVELOPED SOFTWARE

Another option is to make or develop software internally. This requires the company's IS personnel to be responsible for all aspects of software development. Software can be developed using the object-oriented approach or more traditional approaches. The object-oriented approach is attractive if the organization can reuse existing objects from other software packages and projects.

Some advantages inherent with in-house-developed software include meeting user and organizational requirements and having more features and increased flexibility in terms of customization and changes. Software programs developed within a company also have greater potential for providing a competitive advantage because competitors cannot easily duplicate them in the short term.

It is possible to reuse software from other development efforts to reduce the time it takes to deliver in-house software. BankAmerica, for example, reuses previously developed software to deliver new software in 90 days or less. National City Corp. was asked to reduce the average project time by 50 percent by reusing software. Initial reaction was shock: "My jaw dropped to my knees. Lunatics go for 20 percent. How can you even think about 50 percent," said Tony Hai, senior vice president and director of project services for the Cleveland financial service company. But by reusing software and employing other timesaving techniques, the company was able to achieve a 45 percent reduction in average project times.

In some cases, a company that develops its own software internally then decides to sell it to other companies. General Electric, for example, developed a better system for order processing. Now it is looking to sell the software to other companies with similar problems. If successful, this new business for General

software interface
programs or program modifications that allow proprietary software to work with other software used in the organization

Electric will help pay for the software development process. It may even turn into a profitable business.

Chief Programmer Teams

chief programmer team
a group of skilled IS professionals with the task of designing and implementing a set of programs. This team has total responsibility for building the best software possible

For software programming projects, the emphasis is on results—the finished package of computer programs. To get a smooth and efficient set of programs operating, the programming team must strive for the same overall objective. The **chief programmer team** is a group of skilled IS professionals with the task of designing and implementing a set of programs. This team has total responsibility for building the best software possible. Although the makeup of the chief programmer team varies with the size and complexity of the computer programs to be developed, a number of functions are common for all teams. A typical team has a chief programmer, a backup programmer, one or more other programmers, a librarian, and one or more clerks or secretaries (Figure 13.10).

Traditionally, programmer teams consisted of employees hired by the company. Increasingly, companies are looking to other companies or even other countries to provide the important task of programming. Russian rocket scientists, like Anatoly Gaverdovsky, have turned their skills from nuclear weapons to software development.[35] Wipro Technologies, a software company headquartered in Bangalore, India, has developed software for companies including General Electric, Nokia, and Home Depot.[36] Revenues for the Indian software company are close to $400 million. General Motors also went to Bangalore for some of its software development needs.[37] According to one IS analyst at Giga Information Group, "When done properly, offshore development can achieve cost-cutting goals, but it's not a silver bullet. It's like a marriage: If you rush in with haste, it won't work."

The Programming Life Cycle

programming life cycle
a series of steps and planned activities developed to maximize the likelihood of developing good software

Developing in-house software requires a substantial amount of detailed planning. A series of steps and planned activities can maximize the likelihood of developing good software. These phases make up the **programming life cycle**, as illustrated in Figure 13.11 and described next.

Investigation, analysis, and design activities have already been completed. So the programmer already has a detailed set of documents that describe what the system should do and how it should operate. An experienced programmer will begin with a thorough review of these documents before any code is written.

Language and software selection involves determining the best programming language for the application. Important characteristics to be considered are (1) the difficulty of the problem, (2) the type of processing to be used (batch or online), (3) the ease with which the program can be changed later, (4) the type of problem, such as business or scientific, and (5) the cost of the language. Often, a trade-off must be made between the ease of use of a language and the efficiency with which programs execute. Older programming languages are more efficient to run but more difficult and time-consuming to develop.

FIGURE 13.10

A Hierarchy Chart Showing the Typical Structure of a Chief Programmer Team

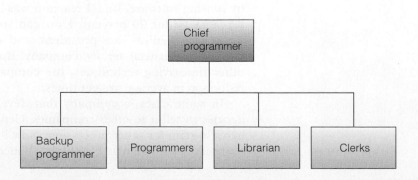

FIGURE 13.11

Steps in the Programming Life Cycle

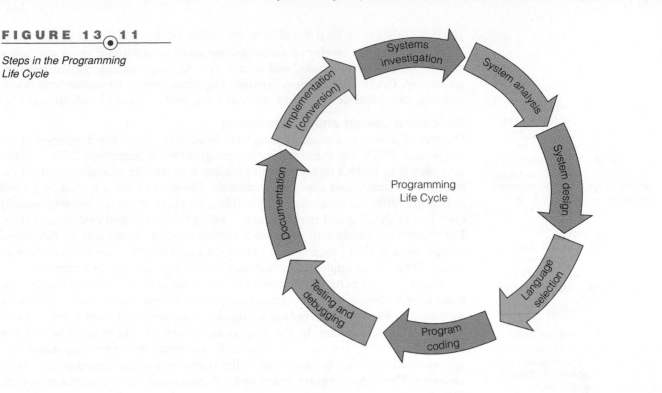

Programming Life Cycle

(Systems investigation → System analysis → System design → Language selection → Program coding → Testing and debugging → Documentation → Implementation (conversion))

Program coding is the process of writing instructions in the language selected to solve the problem. Like a contractor building a house, the computer programmer follows the plans and documents developed in the previous steps. This careful attention to detail ensures that the software actually accomplishes the desired result.

Testing and debugging are vital steps in developing computer programs.[38] In general, testing is the process of making sure the program performs as intended; debugging is the process of locating and eliminating errors.

Documentation is the next step, and it can include technical and user documentation. **Technical documentation** is used by computer operators to execute the program and by analysts and programmers in case there are problems with the program or the program needs modification. In technical documentation, the purpose of every major piece of computer code is written out and explained. Key variables are also described. **User documentation** is developed for the individuals who use the program. This type of documentation shows users, in easy-to-understand terms, how the program can and should be used. Incorporating a description of the benefits of the new application into user documentation may help stakeholders understand the reasons for the program and speed user acceptance. The software vendor often provides such documentation, or it may be obtained from a technical publishing firm. For example, Microsoft provides user documentation for its spreadsheet package Excel, but there are literally hundreds of books and manuals for this software from other sources.

Implementation, or *conversion*, is the last step in developing new computer software. It involves installing the software and making it operational. Several approaches are discussed later in the chapter when we discuss installation.

The same basic steps can be followed for programming in both fourth-generation languages (4GLs) and traditional high-level programming languages. 4GLs, however, may be easier and faster to use. They are appropriate for iterative and prototyping development techniques, because prototypes can be developed quickly. The ease of coding with fourth-generation languages also allows more emphasis on creating programs to meet user and organizational needs.

technical documentation
written details used by computer operators to execute the program and by analysts and programmers in case there are problems with the program or the program needs modification

user documentation
written description developed for individuals who use a program, showing users, in easy-to-understand terms, how the program can and should be used

TOOLS AND TECHNIQUES FOR SOFTWARE DEVELOPMENT

If software will be developed in-house, the chief programmer team can use a number of tools, techniques, and approaches. Options include structured programming, CASE tools, object-oriented implementation, cross-platform development, integrated development environments, and structured walkthroughs.

Structured Design and Programming

Structured design and programming techniques were originally developed in the 1970s and 1980s for third-generation programming languages.[39] The overall approach is to break a large, difficult problem into smaller modules, each simple enough to manage and solve independently. These modules can then be reused in new and different programs. This building block, or modular, software usually costs less to develop and maintain and is easier to modify and update over time. For example, a module to generate a certain type of report can be developed independently, then plugged into numerous programs that require this type of report. One way to implement structured design is structured programming.

The basic idea behind structured programming is to improve the logical program flow by breaking the program into groups of statements, called *structures*. As shown in Figure 13.12, only three types of structures are allowed when using structured programming. In the **sequence structure**, there must be definite starting and ending points. After starting the sequence, programming statements are executed one after another until all the statements in the sequence have been executed. Then the program either ends or continues on to another structure. The **decision structure** allows the computer to branch, depending on certain conditions. Normally, there are only two possible branches. The final structure is the **loop structure**. Actually, there are two commonly used structures for loops. One is the do-until structure, and the other is the do-while structure. Both accomplish the same thing. In the do-until structure, the loop is done until a certain condition is met. For the do-while structure, the loop is done while a certain condition exists. Structured program code development can be the key to developing good program code. Some of the characteristics of structured programming are shown in Table 13.6.

In general, a good approach to writing a large program is to start with the main module and work down to the other modules. This is called the **top-down approach** to programming and is used in structured program design and

sequence structure
a programming structure in which, after starting the sequence, programming statements are executed one after another until all the statements in the sequence have been executed. Then the program either ends or continues on to another sequence

decision structure
a programming structure that allows the computer to branch, depending on certain conditions. Normally, there are only two possible branches

loop structure
a programming structure with two commonly used structures for loops: do-until and do-while. In the do-until structure, the loop is done until a certain condition is met. For the do-while structure, the loop is done while a certain condition exists

top-down approach
a good general approach to writing a large program, starting with the main module and working down to the other modules

FIGURE 13.12

The Three Structures Used in Structured Programming

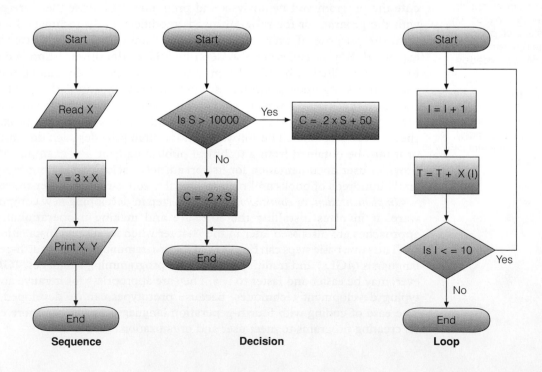

| Sequence | Decision | Loop |

TABLE 13.6

Characteristics of Structured Programming

Program code is broken into modules.
Each module has one and only one function. Such modules are said to have tight internal cohesion.
There is one and only one logic path into each module and one logic exit from each module.
The modules are loosely coupled.
GOTO statements are not allowed.

programming. Although the concept of top-down programming is simple, its use is beneficial in untangling or avoiding coding and debugging problems. The process begins by writing the main module. Then the modules at the next level are written. This procedure continues until all the modules have been written. Figure 13.13 can be used to visualize the top-down approach. In addition to program coding, the top-down approach should be used in testing and debugging. Thus, after the first, or main, module is written, it is tested and debugged. But the main module sends the computer to modules at the second level, which have not been written yet. So, simple modules at the second level are written to send the computer back to the main module so it can be fully and completely tested. If errors are found in the main module, they are corrected immediately.

CASE Tools

CASE tools are often used during software development to automate some of the techniques. For example, source code can be automatically generated using CASE tools. Of the types of CASE tools previously discussed, lower-CASE tools are most likely to be used for software programming. Lower-CASE tools can provide a graphical programming environment and include compilers, syntax checkers, and software modules that generate the actual program code. CASE tools may also have interfaces to the code generators of other vendors' CASE tools, which allows a programmer to mix and match the program code generated. Using CASE tools can help increase programmer accuracy and productivity, particularly in terms of time spent on maintenance.

Object-Oriented Implementation

Companies can also use the object-oriented approach to develop programs. With this approach, a collection of existing modules of code, or objects, can be used across a number of applications. In most cases, minimal coding changes are

FIGURE 13.13

The Top-Down Approach to Writing, Testing, and Debugging a Modular Program

Level 1 (The main module)

a. Write the main module.
b. Write any necessary dummy modules at the second level.
c. Test the main module.
d. Debug the main module.

Level 2 (This procedure is done for each module one at a time.)

a. Write the module.
b. Write any necessary dummy modules for next lower level.
c. Test the module (this will automatically test all the modules that are above this one in the structure chart).
d. Debug the module.

Level N (This procedure is repeated for all levels.)

required to mesh with the developed objects or modules of code. Even though object-oriented software development does not require the use of object-oriented languages, most developers use them for the structure and ease they provide. These languages include Java, Visual Basic, and C++ and make implementation easier and more straightforward.

Being able to reuse previously developed objects speeds software development and can improve quality. For example, the owner of the kayak rental shop, discussed in an earlier example and in Chapter 12, might also have a bicycle rental shop. If the bicycle rental shop needs objects to enter new bicycles into the computer system, they can be modified from the kayak rental system and reused in the bicycle rental business.

Cross-Platform Development

cross-platform development
development technique that allows programmers to develop programs that can run on computer systems having different hardware and operating systems, or platforms

In the past, most applications were developed and implemented using mainframe computers. Today, many applications are developed on personal computers by users of the system. In response to the growth of end-user development, software vendors now offer more tools and techniques to PC users. One software development technique, called **cross-platform development**, allows programmers to develop programs that can run on computer systems having different hardware and operating systems, or platforms. Web Service tools, such as .NET by Microsoft introduced in Chapter 7, are examples.[40] With cross-platform development, the same program might be able to run on both a personal computer and a mainframe, or on two different types of PCs. Cross-platform development can be done on all sizes of computers, including PCs. One benefit of cross-platform development is that programs can run on both small and large systems, and users can develop software on their PCs.

Integrated Development Environments

integrated development environments (IDEs)
development approach that combines the tools needed for programming with a programming language into one integrated package

Software vendors also offer integrated development environments to assist with programming on personal computers. **Integrated development environments (IDEs)** combine the tools needed for programming with a programming language into one integrated package.[41] IDE allows programmers to use simple screens, customized pull-down menus, and graphical user interfaces. Some even use different color text to allow a programmer to quickly locate sections, verbs, or errors in program code. In general, IDEs can make programming software more intuitive. Combining these tools with the language itself makes it easier for programmers to develop sophisticated programs on personal computers, making them more productive.

Eclipse Workbench is example of a product that supports IDEs that can be used with the C and C++ programming languages.[42] Eclipse Workbench includes a debugger and a compiler, along with other tools. Builder 6 Enterprise by Borland is another example of a product that supports the advantages of the IDE approach.[43] Builder 6 includes tools to help programmers write, debug, compile, and run Java programs. The overall goal of Builder 6, like other IDE tools, is to increase programmer productivity and reduce software development costs.

Structured Walkthroughs

structured walkthrough
a planned and preannounced review of the progress of a program module, a structure chart, or a human procedure

Regardless of the tools or techniques used, companies should review software throughout the development process. Companies often use a structured walkthrough technique, which is typically performed by chief programmer teams. As shown in Figure 13.14, a **structured walkthrough** is a planned and preannounced review of the progress of a program module, a structure chart, or a human procedure. The walkthrough helps team members review and evaluate the progress of components of a structured project. The structured walkthrough approach is also useful for programming projects that do not use the structured design approach.

Walkthrough Planning and Preparation

FIGURE 13.14

A structured walkthrough is a planned, preannounced review of the progress of a particular project objective.

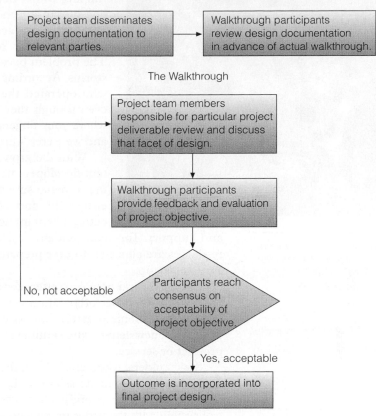

ACQUIRING DATABASE AND TELECOMMUNICATIONS SYSTEMS

Acquiring or upgrading database systems can be one of the most important steps of a systems development effort. While most companies use a relational database, some are starting to use object-oriented database systems, such as Object Store from Excelon Corporation and Objectivity from Objectivity, Inc.

Because databases are a blend of hardware and software, many of the approaches discussed earlier for acquiring hardware and software also apply to database systems. For example, an upgraded inventory control system may require database capabilities, including more hard disk storage or a new DBMS. If so, additional storage hardware will have to be acquired from an IS vendor. New or upgraded software might also be purchased or developed in-house.

With the increased use of e-commerce, the Internet, intranets, and extranets, telecommunications is one of the fastest-growing applications for today's businesses and individuals. Like database systems, telecommunications systems require a blend of hardware and software. For personal computer systems, the primary piece of hardware is a modem. For client/server and mainframe systems, the hardware can include multiplexers, concentrators, communications processors, and a variety of network equipment. Communications software will also have to be acquired from a software company or developed in-house. Again, the earlier discussion on acquiring hardware and software also applies to the acquisition of telecommunications hardware and software.

USER PREPARATION

user preparation
the process of readying managers, decision makers, employees, other users, and stakeholders for the new systems

User preparation is the process of readying managers, decision makers, employees, other users, and stakeholders for the new systems. This activity is an important but often ignored area of systems implementation. A small airline headquartered in Ft. Lauderdale, for example, didn't do adequate employee

Providing users with proper training can help ensure that the information system is used correctly, efficiently, and effectively.

(Source: Eyewire.)

training with a new software package.[44] As a result, the airline had to ground most of its flights and scramble to find hotel rooms to accommodate unhappy travelers, who were stranded in New York, Detroit, Palm Beach, and other cities. The problem became worse as a result of unexpected winter storms. According to the CIO of the company, "The people who operated the software were no longer as fast with it, even though they were trained. It's one of those situations where your fingers aren't connected to your brain anymore, and we weren't prepared for that."

With the growing trend to employee empowerment, system developers need to provide users with the proper training to make sure they use the information system correctly, efficiently, and effectively. User preparation can include active participation, marketing, training, documentation, and support. Top-management support in ensuring that sufficient time and resources are allocated to user preparation is absolutely essential to a successful system startup.

Informing and preparing users for new or modified systems can be done in a variety of ways. User preparation actually begins with user participation in system analysis. Some organizations also actively market new systems to users via brochures, newsletters, and seminars to promote them the way they would a new product or service.

Stakeholders, who benefit from the system but do not directly use or interact with it, need to be made aware of the results of the systems analysis and design effort. For example, suppose a toy manufacturer integrates a new inventory control system with the order processing and production planning systems so that it can quickly respond to changes in product demand. If a successful product promotion dramatically increases customer demand, the new application would alert managers to schedule larger or additional production runs to meet customer demand. Informed stakeholders—in this instance, customers—will likely shop at that chain, knowing that they can rely on finding stuffed toys in stock. Some companies even advertise their use of information systems to add product and service value.

Without question, training users is an essential part of user preparation, whether they are trained by internal personnel or by external training firms. In some cases, companies that provide software also train users at no charge or at a reasonable price. The cost of training can be negotiated during the selection of new software. Other companies conduct user training throughout the systems development process. Concerns and apprehensions about the new system must be eliminated through these training programs. Employees should be acquainted with the system's capabilities and limitations by the time they are ready to use it.

Continuing support provides assistance to users after a new or modified application has been installed. The overall purpose of support is to make sure users understand and benefit from the new or modified system. This support can be additional hardware, software, and service. Most vendors provide continuing support for a fee. Seminars, training programs, and consulting personnel are also popular. The preparation and distribution of user documentation is another important aspect of continuing support.

IS PERSONNEL: HIRING AND TRAINING

Depending on the size of the new system, an organization may have to hire and, in some cases, train new IS personnel. An information systems manager, systems analysts, computer programmers, data entry operators, and similar personnel may be needed for the new system.

As with users, the eventual success of any system depends on how it is used by the personnel within the organization. Training programs should be conducted for the IS personnel who will be using the computer system. These programs are similar to those for the users, although they may be more detailed in the technical aspects of the systems. Effective training will help IS personnel use the new system to perform their jobs and support other users in the organization.

SITE PREPARATION

site preparation
preparation of the location of the new system

The location of the new system needs to be prepared in a process called **site preparation**. For a small system, site preparation can be as simple as rearranging the furniture in an office to make room for a computer. With a larger system, this process is not so easy because it may require special wiring and air-conditioning. One or two rooms may have to be completely renovated, and additional furniture may have to be purchased. A special floor may have to be built, under which the cables connecting the various computer components are placed, and a new security system may be needed to protect the equipment. For larger systems, additional power circuits may also be required.

DATA PREPARATION

data preparation, or data conversion
conversion of manual files into computer files

If the organization is computerizing its work processes, all manual files must be converted to computer files in a process called **data preparation**, or **data conversion**. All permanent data must be placed on a permanent storage device, such as magnetic tape or disk. Usually the organization hires temporary, part-time data-entry operators or a service company to convert the manual data. Once the data has been converted, the temporary workers are no longer needed. A computerized database system or other software will then be used to maintain and update the computer files.

INSTALLATION

installation
the process of physically placing the computer equipment on the site and making it operational

Installation is the process of physically placing the computer equipment on the site and making it operational. Although normally the manufacturer is responsible for installing computer equipment, someone from the organization (usually the IS manager) should oversee the process, making sure that all equipment specified in the contract is installed at the proper location. After the system is installed, the manufacturer performs several tests to ensure that the equipment is operating as it should.

TESTING

Good testing procedures are essential to make sure that the new or modified information system operates as intended. Inadequate testing can result in mistakes and problems. A popular tax preparation company, for example, implemented a Web-based tax preparation system, but people could see one another's tax returns. The president of the tax preparation company called it "our worst-case scenario." Better testing can prevent these types of problems.

unit testing
testing of individual programs

system testing
testing the entire system of programs

volume testing
testing the application with a large amount of data

integration testing
testing all related systems together

acceptance testing
conducting any tests required by the user

Several forms of testing should be used, including testing each of the individual programs (**unit testing**), testing the entire system of programs (**system testing**), testing the application with a large amount of data (**volume testing**), and testing all related systems together (**integration testing**), as well as conducting any tests required by the user (**acceptance testing**). The sequence in which these testing activities normally occur is shown in Figure 13.15.

Unit testing is accomplished by developing test data that will force the computer to execute every statement in the program. In addition, each program is tested with abnormal data to determine how it will handle problems. System testing requires the testing of all the programs together. It is not uncommon for

FIGURE 13⊙15

Types of Testing

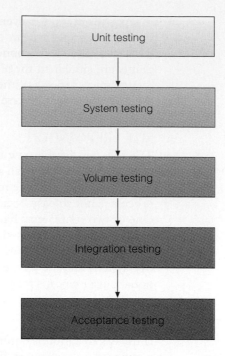

the output from one program to become the input for another. So system testing ensures that the output from one program can be used as input for another program within the system. Volume testing ensures that the entire system can handle a large amount of data under normal operating conditions. Integration testing ensures that the new programs can interact with other major applications. It also ensures that data flows efficiently and without error to other applications. For example, a new inventory control application may require data input from an older order processing application. Integration testing would be done to ensure smooth data flow between the new and existing applications. Integration testing is typically done after unit and system testing. Finally, acceptance testing makes sure that the new or modified system is operating as intended. Run times, the amount of memory required, disk access methods, and more can be tested during this phase. Acceptance testing ensures that all performance objectives defined for the system or application are satisfied. Involving users in acceptance testing may help them understand and effectively interact with the new system. Acceptance testing is the final check of the system before start-up.

START-UP

Start-up begins with the final tested information system. When start-up is finished, the system is fully operational. Various start-up approaches are available (Figure 13.16). **Direct conversion** (also called *plunge* or *direct cutover*) involves stopping the old system and starting the new system on a given date. Direct conversion is usually the least desirable approach because of the potential for problems and errors when the old system is shut off and the new system is turned on at the same instant. The **phase-in approach** is a popular technique preferred by many organizations. In this approach, sometimes called a *piecemeal approach*, components of the new system are slowly phased in while components of the old one are slowly phased out. When everyone is confident that the new system is performing as expected, the old system is completely phased out. This gradual replacement is repeated for each application until the new system is running every application. In some cases, the phase-in approach can take months or years. Sabre, the massive airline and travel reservation system, will phase in a new hardware setup that will take several years.[45] The company is going from large mainframe computers to an array of servers supplied by Compaq.

start-up
the process of making the final tested information system fully operational

direct conversion (also called *plunge* or *direct cutover*)
stopping the old system and starting the new system on a given date

phase-in approach
slowly replacing components of the old system with those of the new one. This process is repeated for each application until the new system is running every application and performing as expected; also called piecemeal approach

pilot start-up
running the new system for one group of users rather than all users

parallel start-up
running both the old and new systems for a period of time and comparing the output of the new system closely with the output of the old system; any differences are reconciled. When users are comfortable that the new system is working correctly, the old system is eliminated

user acceptance document
formal agreement signed by the user that states that a phase of the installation or the complete system is approved

Pilot start-up involves running the new system for one group of users rather than all users. For example, a manufacturing company with a number of retail outlets throughout the country could use the pilot start-up approach and install a new inventory control system at one of the retail outlets. When this pilot retail outlet runs without problems, the new inventory control system can be implemented at other retail outlets. General Motors used pilot start-up to introduce a new wireless handheld device into hospitals and medical facilities.[46] The pilot program involved a group of physicians in Shreveport, Louisiana, and Oklahoma City, Oklahoma. The system included a Palm computer, a printer, and prescription-writing software. **Parallel start-up** involves running both the old and new systems for a period of time. The output of the new system is compared closely with the output of the old system, and any differences are reconciled. When users are comfortable that the new system is working correctly, the old system is eliminated.

USER ACCEPTANCE

Most mainframe computer manufacturers use a formal **user acceptance document**—a formal agreement signed by the user that states that a phase of the installation or the complete system is approved. This is a legal document that usually removes or reduces the information systems vendor from liability for problems that occur after the user acceptance document has been signed. Because this document is so important, many companies get legal assistance before they sign the acceptance document. Stakeholders may also be involved

FIGURE 13.16

Start-Up Approaches

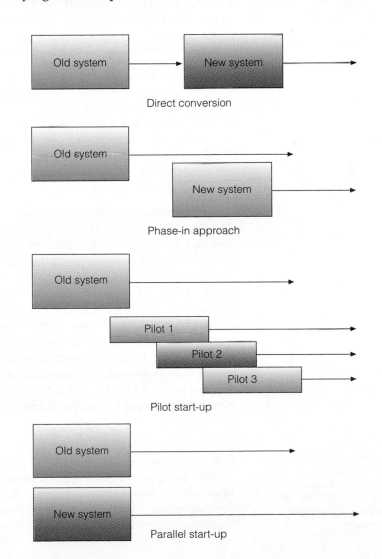

Direct conversion

Phase-in approach

Pilot start-up

Parallel start-up

in acceptance to make sure that the benefits to them are indeed realized. The "IS Principles in Action" box shows how Hon Industries, an office furniture manufacturer, implemented a new supply chain management system, despite problems with user acceptance of the new system.

⊙ SYSTEMS MAINTENANCE

systems maintenance
stage of systems development that involves checking, changing, and enhancing the system to make it more useful in achieving user and organizational goals

Systems maintenance involves checking, changing, and enhancing the system to make it more useful in achieving user and organizational goals. Software maintenance is a major concern for organizations. In some cases, organizations encounter major problems that require recycling the entire systems development process. Mizuho, a Japanese bank formed by a merger of two companies, faced major computer problems that disrupted service to millions of customers and could have seriously damaged the bank's reputation.[47] In other situations, minor modifications are sufficient to remedy problems. Huntington Bancshares decided to perform systems maintenance on its Web site to improve its performance.[48] Now, the average wait for people visiting the company's Web site has been cut from 7 seconds to about 4.7 seconds on average.

REASONS FOR MAINTENANCE

Once a program is written, it is likely to need ongoing maintenance. To some extent, a program is like a car that needs oil changes, tune-ups, and repairs at certain times. Experience shows that frequent, minor maintenance to a program, if properly done, can prevent major system failures later. Some of the reasons for program maintenance are the following:

- Changes in business processes
- New requests from stakeholders, users, and managers
- Bugs or errors in the program
- Technical and hardware problems
- Corporate mergers and acquisitions
- Government regulations
- Change in the operating system or hardware on which the application runs
- Unexpected events, like the terrorist attacks of September 11th[49]

Most companies modify their existing programs instead of developing new ones because existing software performs many important functions, and companies can have millions of dollars invested in their old, legacy systems. So, as new systems needs are identified, the burden of fulfilling the needs most often falls on the existing system. Old programs are repeatedly modified to meet ever-changing needs. Saab, for example, decided to upgrade many of its older programs to enable them to run on the Internet.[50] According to a company representative discussing the older programs, "Throwing them out just didn't make sense." The maintenance allowed the automotive company to more efficiently connect its dealers to the Internet to give them access to the same programs they used to access over slower telecommunications systems. Yet, over time, repeated modifications tend to interfere with the system's overall structure, reducing its efficiency and making further modifications more burdensome.

TYPES OF MAINTENANCE

slipstream upgrade
a minor upgrade—typically a code adjustment or minor bug fix—not worth announcing. It usually requires recompiling all the code and, in so doing, it can create entirely new bugs

Software companies and many other organizations use four generally accepted categories to signify the amount of change involved in maintenance. A **slipstream upgrade** is a minor upgrade—typically a code adjustment or minor bug fix. Many companies don't announce to users that a slipstream upgrade has been made. A slipstream upgrade usually requires recompiling all the code. In so doing, it can

The primary emphasis of systems implementation is to make sure that the right information is delivered to the right person in the right format at the right time.

Supply Chain Management Software Delivers—Despite Resistance

Hon Industries, a $1.8 billion office furniture and hearth products manufacturer in Muscatine, Iowa, was having problems with shipping and warehousing. Its delivery trucks would sometimes show up at facilities with more product than could fit in the warehouse. In other cases, warehouses were left empty, waiting for the delivery of materials. It was shipping and scheduling chaos!

The old legacy systems were unable to accurately measure the capacity of the manufacturer's warehouses. Newly appointed CIO and Vice President Malcolm C. Fields took the bull by the horns, assembled a team, and brainstormed to find a solution. They decided to replace the old distribution system's mainframe code. Rather than start from scratch, Fields and his team decided to use software from SynQuest, a Norcross, Georgia–based maker of supply chain management software. The SynQuest application allows Hon to take a product order, factor in shipping and scheduling variables, then decide which factory could build and ship the product for the least amount of money.

The implementation team achieved its results despite considerable obstacles, including a shake-up in the company's structure, management changes, and resistance from employees who were wedded to traditional processes. Hon Industries completed its advance-planning and scheduling system project at a cost of about $2 million. Though the project ran past its original deadline by six months, it also far exceeded the expectations of the project team, says Fields.

Without offering exact numbers, Fields says the new system has contributed to a drop in freight costs from 6.5 to 5.8 percent of the firm's overall sales revenue. Scheduling accuracy has improved by 20 percent, and there are now 19 inventory turns a year, up from 16, he says.

One major challenge to the project was the constant shifting of business processes at Hon, which meant projects had "to be implemented in short, intensive phases," says Fields. For instance, during the middle of the rollout, the company was split into two separate divisions, and the president of the original operating company was replaced. The business executives who signed off on the project were gone, says Fields. "We had to go out and re-win some hearts and minds," he says. Although work never slowed, for about 30 days the project's fate was uncertain. In the end, project advocates successfully educated the new executive team, and the rollout was a success. Fields says he learned just how tough it is to persuade people to change their way of thinking. "Never underestimate the difficulty of shifting a paradigm," he says.

Discussion Questions

1. Describe how the old system at Hon was unnecessarily costing the company money.
2. Why do you think Hon employees were reluctant to give up the old way of doing things?

Critical Thinking Questions

3. What start-up approach (direct, phase-in, pilot start-up, parallel start-up) do you think would have been wisest to use for Hon's new system? Why?
4. Relate the first two principles of this chapter to this case. Has this project assisted in helping to achieve Hon's corporate goals? Has this project assisted in getting the right information delivered to the right person in the right format at the right time? How so?

Sources: Marc L. Songini, "Computerworld Premier 100, Best in Class: Hon Industries Inc.," *Computerworld*, March 11, 2002, http://www.computer world.com; Derek Slater, "Strategic Planning Don'ts (and Do's)," *CIO Magazine*, June 1, 2002, http://www.cio.com/archive/060102/donts.html; Hon Industries Web site, http://www.honi.com, accessed July 2002.

---

patch
a minor change to correct a problem or make a small enhancement. It is usually an addition to an existing program

release
a significant program change that often requires changes in the documentation of the software

version
a major program change, typically encompassing many new features

request for maintenance form
a form authorizing modification of programs

create entirely new bugs. This maintenance practice can explain why the same computers sometimes work differently with what is supposedly the same software. A **patch** is a minor change to correct a problem or make a small enhancement. It is usually an addition to an existing program. That is, the programming code representing the system enhancement is usually "patched into," or added to, the existing code. Although slipstream upgrades and patches are minor changes, they can cause users and support personnel big problems if the programs do not run as before. A new **release** is a significant program change that often requires changes in the documentation of the software. Finally, a new **version** is a major program change, typically encompassing many new features.

THE REQUEST FOR MAINTENANCE FORM

Because of the amount of effort that can be spent on maintenance, many organizations require a **request for maintenance form** to authorize modification of programs. This form is usually signed by a business manager, who documents the need for the change and identifies the priority of the change relative to other work that has been requested. The IS group reviews the form and identifies the programs to be changed, determines the programmer who will be assigned to the project, estimates the expected completion date, and develops a technical description of the change. A cost/benefit analysis may be required if the change requires substantial resources.

PERFORMING MAINTENANCE

Depending on organizational policies, the people who perform systems maintenance vary. In some cases, the team that designs and builds the system also performs maintenance. This ongoing responsibility gives the designers and programmers an incentive to build systems well from the outset: if there are problems, they will have to fix them. In other cases, organizations have a separate **maintenance team**. This team is responsible for modifying, fixing, and updating existing software.

maintenance team
a special IS team responsible for modifying, fixing, and updating existing software

In the past, companies had to maintain each computer system or server separately. With hundreds or thousands of computers scattered throughout an organization, this task could be very costly and time-consuming. Today, the maintenance function is becoming more automated. Home Depot, for example, is using new maintenance tools and software that will allow the large chain to maintain and upgrade software centrally.[51] Home Depot is expected to have about 90,000 computers using the new approach by the end of 2004.

Regardless of who performs maintenance, the same tools and techniques used for earlier phases of systems development—object-oriented tools, CASE tools, flowcharts, structured programming, and so on—should be used. In addition, all maintenance should be fully documented. Unfortunately, documentation sometimes falls by the wayside. For example, if an order-processing application crashes during peak hours, the company's main objective is to get the application running as soon as possible. Documenting the problem or any changes made to the application may be overlooked in the rush to restart it. But lack of documentation can cause future problems if programmers refer to outdated data-flow diagrams or layout charts when they perform maintenance for the system. Thus, it is essential that the tools used in maintenance allow easy documentation to accurately reflect the changes.

Java and the object-oriented programming languages hold the promise of reducing the program maintenance effort. With the object-oriented approach, existing objects can be modified and used to update an application. This feature of the object-oriented approach can greatly improve programmer productivity, reduce maintenance costs, and dramatically improve the quality of the systems development effort—important advantages.

A number of vendors have developed tools to ease the software maintenance burden. Relativity Technologies recently unveiled RescueWare, a product that converts third-generation code such as COBOL to highly maintainable C++, Java, or Visual Basic object-oriented code. Using RescueWare, maintenance personnel download mainframe code to Windows NT or Windows 2000 workstations. They then use the product's graphical tools to analyze the original system's inner workings. RescueWare lets a programmer see the original system as a set of object views, which visually illustrate module functioning and program structures. IS personnel can choose one of three levels of transformation: revamping the user interface, converting the database access, and transforming procedure logic.

THE FINANCIAL IMPLICATIONS OF MAINTENANCE

The cost of maintenance is staggering. For older programs, the total cost of maintenance can be up to five times greater than the total cost of development.

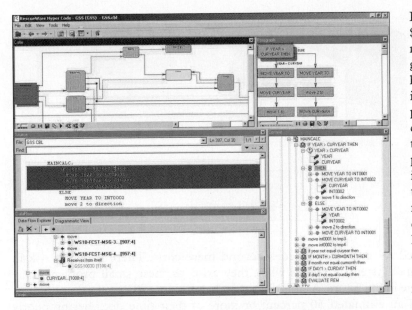

RescueWare provides companies with automated software that accelerates the process of transforming third-generation code, such as COBOL, to Internet or client/server platforms.

(Source: Courtesy of Relativity Technologies, Inc.)

In other words, a program that originally cost $25,000 to develop may cost $125,000 to maintain over its lifetime. The average programmer can spend over 50 percent of his or her time on maintaining existing programs instead of developing new ones. In addition, as programs get older, total maintenance expenditures in time and money increase, as illustrated in Figure 13.17. With the use of newer programming languages and approaches, including object-oriented programming, maintenance costs are expected to decline. Even so, many organizations have literally millions of dollars invested in applications written in older languages (such as COBOL), which are both expensive and time-consuming to maintain.

The financial implications of maintenance make it important to keep track of why systems are maintained, instead of simply keeping cost figures. This is another reason documentation of maintenance tasks is so crucial. A determining factor in the decision to replace a system is the point at which it is costing more to fix it than to replace it.

THE RELATIONSHIP BETWEEN MAINTENANCE AND DESIGN

Programs are expensive to develop, but they are even more expensive to maintain. Programs that are well designed and documented to be efficient, structured, and flexible are less expensive to maintain in later years. Thus, there is a direct relationship between design and maintenance. More time spent on design up front can mean less time spent on maintenance later.

In most cases, it is worth the extra time and expense to design a good system. Consider a system that costs $250,000 to develop. Spending 10 percent more on design would cost an additional $25,000, bringing the total design cost to $275,000. Maintenance costs over the life of the program could be $1,000,000. If this additional design expense can reduce maintenance costs by 10 percent, the savings in maintenance costs would be $100,000. Over the life of the program, the net savings would be $75,000 ($100,000–$25,000). This relationship between investment in design and long-term maintenance savings is graphically displayed in Figure 13.18.

FIGURE 13.17

Maintenance Costs as a Function of Age

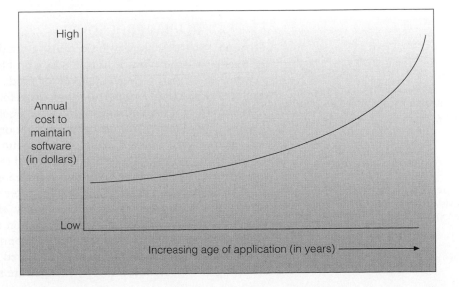

FIGURE 13.18

The Value of Investment in Design

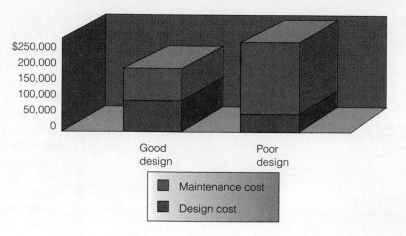

The need for good design goes beyond mere costs. There is a real risk in ignoring small system problems when they arise, as these small problems may become large in the future. As mentioned earlier, because maintenance programmers spend an estimated 50 percent or more of their time deciphering poorly written, undocumented program code, there is little time to spend on developing new, more effective systems. If put to good use, the tools and techniques discussed in this chapter will allow organizations to build longer-lasting, more reliable systems.

SYSTEMS REVIEW

systems review
the final step of systems development, involving the analysis of systems to make sure they are operating as intended

Systems review, the final step of systems development, is the process of analyzing systems to make sure they are operating as intended. This process often compares the performance and benefits of the system as it was designed with the actual performance and benefits of the system in operation. Problems and opportunities uncovered during systems review will trigger systems development and begin the process anew. For example, as the number of users of an interactive system increases, it is not unusual for system response time to increase. If the increase in response time is too great, it may be necessary to redesign some of the system, modify databases, or increase the power of the computer hardware.

Internal employees, external consultants, or both can perform systems review. When the problems or opportunities are industrywide, people from several firms may get together. In some cases, they collaborate at an IS conference or in a private meeting involving several firms.

TYPES OF REVIEW PROCEDURES

event-driven review
review triggered by a problem or opportunity such as an error, a corporate merger, or a new market for products

There are two types of review procedures: event driven and time driven (Table 13.7). An **event-driven review** is triggered by a problem or opportunity such as an error, a corporate merger, or a new market for products. In some cases, companies wait until a large problem or opportunity occurs before a change is made, ignoring minor problems. In contrast, some companies use a continuous improvement approach to systems development. With this approach, an organization makes changes to a system even when small problems or opportunities occur. Although continuous improvement can keep the system current and responsive, doing the repeated design and implementation can be both time-consuming and expensive.

time-driven review
review performed after a specified amount of time

A **time-driven review** is performed after a specified amount of time. Many application programs are reviewed every six months to a year. With this approach, an existing system is monitored on a schedule. If problems or opportunities are uncovered, a new systems development cycle may be initiated. A payroll application, for example, may be reviewed once a year to make sure it is still operating as expected. If it is not, changes are made.

TABLE 13.7

Examples of Review Types

Event Driven	Time Driven
A problem with an existing system	Monthly review
A merger	Yearly review
A new accounting system	Review every few years
An executive decision that an upgraded Internet site is needed to stay competitive	Five-year review

Many companies use both approaches. A billing application, for example, might be reviewed once a year for errors, inefficiencies, and opportunities to reduce operating costs. This is a time-driven approach. In addition, the billing application might be redone if there is a corporate merger, if one or more new managers require different information or reports, or if federal laws on bill collecting and privacy change. This is an event-driven approach.

FACTORS TO CONSIDER DURING SYSTEMS REVIEW

Systems review should investigate a number of important factors, such as the following:

Mission. Is the computer system helping the organization achieve its overall mission? Are stakeholder needs and desires satisfied or exceeded with the new or modified system?

Organizational goals. Does the computer system support the specific goals of the various areas and departments of the organization?

Hardware and software. Are hardware and software up to date and adequate to handle current and future processing needs?

Database. Is the current database up to date and accurate? Is database storage space adequate to handle current and future needs?

Telecommunications. Is the current telecommunications system fast enough, and does it allow managers and workers to send and receive timely messages? Does it allow for fast order processing and effective customer service?

Information systems personnel. Are there sufficient IS personnel to perform current and projected processing tasks?

Control. Are rules and procedures for system use and access acceptable? Are the existing control procedures adequate to protect against errors, invasion of privacy, fraud, and other potential problems?

Training. Are there adequate training programs and provisions for both users and IS personnel?

Costs. Are development and operating costs in line with what is expected? Is there an adequate information systems budget to support the organization?

Complexity. Is the system overly complex and difficult to operate and maintain?

Reliability. Is the system reliable? What is the meantime between failures (MTBF)?

Efficiency. Is the computer system efficient? Are system outputs generated by the right amount of inputs, including personnel, hardware, software, budget, and others?

Response time. How long does it take the system to respond to users during peak processing times?

Documentation. Is the documentation still valid? Are changes in documentation needed to reflect the current situation?

SYSTEM PERFORMANCE MEASUREMENT

Systems review often involves monitoring the system, called **system performance measurement**. The number of errors encountered, the amount of memory

system performance measurement

monitoring the system—the number of errors encountered, the amount of memory required, the amount of processing or CPU time needed, and other problems

required, the amount of processing or CPU time needed, and other problems should be closely observed. If a particular system is not performing as expected, it should be modified, or a new system should be developed or acquired.

system performance products
software that measures all components of the computer-based information system, including hardware, software, database, telecommunications, and network systems

System performance products have been developed to measure all components of the computer-based information system, including hardware, software, database, telecommunications, and network systems. When properly used, system performance products can quickly and efficiently locate actual or potential problems.

A number of products have been developed to assist in assessing system performance. Candle is a leading provider of management tools that monitor mainframe performance and application availability. Its products include Candle Command Center, an advanced systems management tool for optimizing an organization's computing resources and maximizing business application availability, and OMEGAMON II performance monitors, for real-time and historical analysis of performance on a variety of systems.[52] Precise/Pulse is a product from Precise Software Solutions that provides around-the-clock performance monitoring for Oracle database applications. It detects and reports potential problems through systems management consoles. Precise/Pulse monitors the performance of critical database applications and issues alerts about inefficiencies before they turn into application performance problems.

Measuring a system is, in effect, the final task of systems development. The results of this process may bring the development team back to the beginning of the development life cycle, where the process begins again.

SUMMARY

PRINCIPLE *Designing new systems or modifying existing ones should always be aimed at helping an organization achieve its goals.*

The purpose of systems design is to prepare the detailed design needs for a new system or modifications to the existing system. Logical systems design refers to the way the various components of an information system will work together. The logical design includes data specifications for output and input, processing, files and databases, telecommunications, procedures, personnel and job design, and controls and security design. Physical systems design refers to the specification of the actual physical components. The physical design must specify characteristics for hardware and software design, database and telecommunications, and personnel and procedures design.

Logical and physical design can be accomplished using the traditional systems development life cycle or the object-oriented approach. Using the OO approach, we design key objects and classes of objects in the new or updated system. The sequence of events that a new or modified system requires is often called a *scenario*, which can be diagrammed in a sequence diagram.

A number of special design considerations should be taken into account during both logical and physical system design. A sign-on procedure consists of identification numbers, passwords, and other safeguards needed for individuals to gain access to computer resources.

If the system under development is interactive, the design approach must consider using menus, help facilities, table lookup facilities, and restart procedures. A good interactive dialogue will ask for information in a clear manner, respond rapidly, be consistent among applications, and use an attractive format. Also, it will avoid use of computer jargon and treat the user with respect.

Error prevention, detection, and correction should be part of the system design process. Causes of errors include human activities, natural phenomena, and technical problems. Designers should be alert to prevention of fraud and invasion of privacy.

Emergency alternative procedures and disaster recovery are important aspects of systems design. Disaster planning is the process of anticipating and providing for disasters. A disaster can be an act of nature (a flood, fire, or earthquake) or a human act (terrorism, error, labor unrest, or erasure of an important file). The primary tools used in disaster planning and recovery are hardware, software, database, telecommunications, and personnel backup.

Security, fraud, and the invasion of privacy are also important design considerations. Most IS departments establish tight systems controls to maintain data security. Systems controls can help prevent computer misuse, crime, and fraud by employees and others. System controls include input, output, processing, database, telecommunications, and personnel controls.

Whether an individual is purchasing a personal computer or an experienced company is acquiring an expensive mainframe computer, the system could be obtained from one or more vendors. Some of the factors to consider in selecting a vendor are the vendor's reliability and financial stability, the type of service offered after the sale, the goods and services the vendor offers and keeps in stock, the vendor's willingness to demonstrate its products, the vendor's ability to repair hardware, the vendor's ability to modify its software, the availability of vendor-offered training of IS personnel and system users, and evaluations of the vendor by independent organizations.

If new hardware or software will be purchased from a vendor, a formal request for proposal (RFP) is needed. The RFP outlines the company's needs; in response, the vendor provides a written reply. In addition to responding to the company's stated needs, the vendor provides data on its operations. This data might include the vendor's reliability and stability, the type of post sale service offered, the vendor's ability to perform repairs and fix problems, vendor training, and the vendor's reputation. Financial options to consider include purchase, lease, and rent.

RFPs from various vendors are reviewed and narrowed down to the few most likely candidates. In the final evaluation, a variety of techniques—including group consensus, cost/benefit analysis, point evaluation, and benchmark tests—can be used. In group consensus, a decision-making group is appointed and given responsibility for making the final evaluation and selection. With cost/benefit analysis, all costs and benefits of the alternatives are expressed in monetary terms. Benchmarking involves comparing computer systems operating under the same conditions. Point evaluation assigns weights to evaluation factors and each alternative is evaluated in terms of each factor and given a score from 0 to 100. After the vendor is chosen, contract negotiations can begin.

At the end of the systems design step, the final specifications are frozen and no changes are allowed so that implementation can proceed. One of the most important steps in systems design is to develop a good contract if new computer facilities are being acquired. A final design report is developed at the end of the systems design phase.

PRINCIPLE *The primary emphasis of systems implementation is to make sure that the right information is delivered to the right person in the right format at the right time.*

The purpose of systems implementation is to install the system and make everything, including users, ready for its operation. Systems implementation includes hardware acquisition, software acquisition or development, user preparation, hiring and training of personnel, site and data preparation, installation, testing, start-up, and user acceptance. Hardware acquisition requires purchasing, leasing, or renting computer resources from an IS vendor. Hardware is typically obtained from a computer hardware vendor. Increasingly, companies are using service providers to acquire software, Internet access, and other IS resources.

Software can be purchased from vendors or developed in-house—a decision termed the *make-or-buy decision*. A purchased software package usually has a lower cost, less risk regarding the features and performance, and easy installation. The amount of development effort is also less when software is purchased. Developing software can result in a system that more closely meets the business needs and has increased flexibility in terms of customization and changes. Developing software also has greater potential for providing a competitive advantage.

Software development is often performed by a chief programmer team—a group of IS professionals who design, develop, and implement a software program. Programming using traditional programming languages follows a life cycle that includes investigation, analysis, design, language selection, program coding, testing and debugging, documentation, and implementation (conversion). Documentation includes technical and user documentation.

There are many tools and techniques for software development. Structured design is a philosophy of designing and developing application software. Structured programming is not a new programming language; it is a way to standardize computer programming using existing languages. The top-down approach starts with programming a main module and works down to the other modules. Other tools, like cross-platform development and integrated development environments (IDEs), make software development easier and more thorough. CASE tools are often used to automate some of these techniques.

Fourth-generation languages (4GLs) and object-oriented languages offer another alternative to in-house development. Development using these fast and easy-to-use languages requires several steps, much like the programming life cycle. The main difference is that, with object-oriented languages, programmers must identify and select objects and integrate them into an application, instead of using step-by-step coding. Being able to reuse previously developed objects speeds software development and can improve quality when using object-oriented implementation.

Database and telecommunications acquisition involves acquiring the necessary databases, networks, telecommunications, and Internet facilities. Companies have a wide array of choices, including newer object-oriented database systems.

Implementation must address personnel requirements. User preparation involves readying managers, employees, and other users for the new system. New IS personnel may need to be hired, and users must be well trained in the system's functions. Preparation of the physical site of the system must be done, and any existing data to be used in the new system will require conversion to the new format. Hardware installation is done during the implementation step, as is testing. Testing includes program (unit) testing, systems testing, volume testing, integration testing, and acceptance testing.

Start-up begins with the final tested information system. When start-up is finished, the system is fully operational. There are a number of different start-up approaches. Direct conversion (also called *plunge* or *direct cutover*) involves stopping the old system and starting the new system on a given date. With the phase-in approach, sometimes called a *piecemeal approach*, components of the new system are slowly phased in while components of the old one are slowly phased out. When everyone is confident that the new system is performing as expected, the old system is completely phased out. Pilot start-up involves running the new system for one group of users rather than all users. Parallel start-up involves running both the old and new systems for a period of time. The output of the new system is compared closely with the output of the old system, and any differences are reconciled. When users are comfortable that the new system is working correctly, the old system is eliminated. Many IS vendors ask the user to sign a formal user acceptance document that reduces the IS vendor from liability for problems that occur after the document is signed.

PRINCIPLE   *Maintenance and review add to the useful life of a system but can consume large amounts of resources. These activities can benefit from the same rigorous methods and project management techniques applied to systems development.*

Systems maintenance involves checking, changing, and enhancing the system to make it more useful in obtaining user and organizational goals. Maintenance is critical for the continued smooth operation of the system. The costs of performing maintenance can well exceed the original cost of acquiring the system. Some major causes of maintenance are new requests from stakeholders and managers, enhancement requests from users, bugs or errors, technical or hardware problems, newly added equipment, changes in organization structure, and government regulations.

Maintenance can be as simple as a program patch to correct a small problem to the more complex upgrading of software with a new release from a vendor. For older programs, the total cost of maintenance can be greater than the total cost of development. Increased emphasis on design can often reduce maintenance costs. Requests for maintenance should be documented with a request for maintenance form, a document that formally authorizes modification of programs. The development team or a specialized maintenance team may then make approved changes. Maintenance can be greatly simplified with the object-oriented approach.

Systems review is the process of analyzing systems to make sure that they are operating as intended. It involves monitoring systems to be sure they are operating as designed. The two types of review procedures are event-driven review and time-driven review. An event-driven review is triggered by a problem or opportunity. A time-driven review is started after a specified amount of time.

Systems review involves measuring how well the system is supporting the mission and goals of the organization. System performance measurement monitors the system for number of errors, amount of memory and processing time required, and so on.

CHAPTER 13 SELF-ASSESSMENT TEST

Designing new systems or modifying existing ones should always be aimed at helping an organization achieve its goals.

1. _____ details system outputs, inputs, and user interfaces; specifies hardware, software, databases, telecommunications, personnel, and procedures; and shows how these components are related.

2. Determining the needed hardware and software for a new system is an example of
   A. Logical design
   B. Physical design
   C. Interactive design
   D. Object-oriented design

3. Some information systems departments are called deterrence shops, in which only authorized operators can run the computers. True False

4. The _____ often results in a formal bid that is used to determine who gets a contract for designing new or modifying existing systems. It specifies in detail the required resources such as hardware and software.

5. With this approach, a decision-making group is appointed and given the responsibility of making the final evaluation and selection during systems design.
   A. Cost/Benefit
   B. Point evaluation
   C. Group consensus
   D. Nominal evaluation

6. Near the end of the design stage, an organization prohibits further changes in the design of the system. This is called _____.

7. In object-oriented systems design, a sequence of events is called a scenario and can be diagrammed in a sequence diagram. True False

The primary emphasis of systems implementation is to make sure that the right information is delivered to the right person in the right format at the right time.

8. An information systems vendor is a company that offers hardware, software, telecommunications systems, databases, information systems personnel, and/or other computer-related resources. True False

9. Software can be purchased from external developers or developed in-house. This decision is often called the _____ decision.

10. With this type of structure, a program can branch to another part of the program, depending on certain conditions.
    A. Sequence structure
    B. Decision structure
    C. Loop structure
    D. CASE structure

11. _____ testing involves testing the entire system of programs.

12. The phase-in approach to conversion involves running both the old system and the new system for three months or longer. True False

Maintenance and review add to the useful life of a system but can consume large amounts of resources. These activities can benefit from the same rigorous methods and project management techniques applied to systems development.

13. A (An) _____ is a minor change to correct a problem or make a small enhancement to a program or system.

14. Many organizations require a request for maintenance form to authorize modification of programs. True False

15. A systems review that is caused by a problem with an existing system is called
    A. Object review
    B. Structured review
    C. Event-driven review
    D. Critical factors review

16. Java, Visual Basic, and C++ are examples of structured programming languages. True False

17. Monitoring a system after it has been implemented is called _____

Chapter 13 Self-Assessment Test Answers

1. Systems design, 2. B, 3. False, 4. request for proposal (RFP), 5. C, 6. freezing design specifications, 7. True, 8. True, 9. make-or-buy, 10. B, 11. System, 12. False, 13. patch, 14. True, 15. C, 16. False, 17. system performance measurement.

KEY TERMS

REVIEW QUESTIONS

1. What is the purpose of systems design?
2. What is procedures design?
3. What is interactive processing? What design factors should be taken into account for this type of processing?
4. How can the object-oriented approach be used during systems design?
5. What are some of the special design considerations that should be taken into account during both the logical and physical design?
6. What are the different types of software and database backup? Describe the procedure you use to back up your homework files.
7. Identify specific controls that are used to maintain input integrity and security.
8. What is an RFP? What is typically included in one? How is it used?
9. What activities go on during the user preparation phase of system implementation?

10. What are the major steps of systems implementation?
11. What are some tools and techniques for software development?
12. Give three examples of an IS vendor.
13. Explain the three types of structures allowed in structured programming.
14. What are the financial options of acquiring hardware?
15. What are the steps involved in testing the information system?
16. What are some of the reasons for program maintenance? Explain the three types of maintenance.
17. Describe the point evaluation system for selection of the best system alternative.
18. How is systems performance measurement related to the systems review?

DISCUSSION QUESTIONS

1. Describe the participants in the systems design stage. How do these participants compare with the participants of systems investigation?
2. Assume that you are the owner of a company that is about to start marketing and selling bicycles over the Internet. Describe your top three objectives in developing a new Web site for this systems development project.

3. Assume that you are the owner of an on-line stock trading company. Describe how you could design the trading system to recover from a disaster.
4. Identify some of the advantages and disadvantages of purchasing versus leasing hardware.
5. Discuss the relationship between maintenance and system design.

6. Is it equally important for all systems to have a disaster recovery plan? Why or why not?

7. Four approaches were discussed to evaluate a number of systems alternatives. No one approach is always the best. How would you decide which approach to use in a particular instance?

8. What are the advantages and disadvantages of the object-oriented approach to systems implementation?

9. Assume that you are starting an Internet site to sell clothing. Describe how you would design the interactive processing system for this site. Draw a diagram showing the home Web page for the site. Describe the important features of this home page.

10. Identify the various forms of testing used. Why are there so many different types of tests?

11. What is the goal of conducting a systems review? What factors need to be considered during systems review?

12. What features and terms would you insist on in a software package contract?

13. What issues might you expect to arise if you initiate the use of a request for maintenance form where none had been required? How would you deal with these issues?

14. How would you go about evaluating a software vendor?

PROBLEM-SOLVING EXERCISES

1. You have been hired to develop a new computer system for a video rental business using the object-oriented approach. Using the information presented in this chapter and in Chapter 12, describe the approach you would use in a report. Using a graphics program, use the object-oriented approach to document parts of the new computer system.

2. A project team has estimated the costs associated with the development and maintenance of a new system. One approach requires a more complete design and will result in a slightly higher design and implementation cost but a lower maintenance cost over the life of the system. The second approach cuts the design effort, saving some dollars but with a likely increase in maintenance cost.

a. Enter the following data in the spreadsheet. Print the result.

The Benefits of Good Design

	Good Design	Poor Design
Design Costs	$14,000	$10,000
Implementation Cost	$42,000	$35,000
Annual Maintenance Cost	$32,000	$40,000

b. Create a stacked bar graph that shows the total cost, including design, implementation, and maintenance costs. Be sure that the chart has a title and that the costs are labeled on the chart.

c. Use your word processing software to write a paragraph that recommends an approach to take and why.

3. To get a better understanding of the value of a good interactive dialogue, go to the Web site of Amazon.com at http://www.amazon.com. Search for several books based on different subjects and for periodicals. In searching for these materials, use the author as one criterion (e.g., R. M. Stair), the subject (e.g., information systems), and the title (e.g., *Principles of Information Systems*). If you have access to another on-line library catalog, conduct the same searches using that system. Now evaluate the interactive dialogue of the systems, based on the elements discussed in the text. Write up your observations using your word processor. If you can, also send this evaluation to your instructor via e-mail.

1. Assume your project team has been working three months to complete the systems design of a new Web-based customer ordering system. There are two possible options that seem to meet all users' needs. The project team must make a final decision on which option to implement. The table that follows summarizes some of the key facts about each option.

   a. What process would you follow to make this important decision?

   b. Who needs to be involved?
   c. What additional questions need to be answered to make a good decision?
   d. Based on the data, which option would you recommend and why?
   e. How would you account for project risk in your decision making?

Factor	Option #1	Option #2
	(Millions)	(Millions)
Annual gross savings	$1.5	$3.0
Total development cost	$1.5	$2.2
Annual operating cost	$0.5	$1.0
Time required to implement	9 months	15 months
Risk associated with project (expressed in probabilities)		
Benefits will be 50% less than expected	20%	35%
Cost will be 50% greater than expected	25%	30%
Organization will not/cannot make changes necessary for system to operate as expected	20%	25%
Does system meet all mandatory requirements?	Yes	Yes

2. Assume your team works for a medium-sized company that trades treasury bonds in New York City. Your firm has 500 employees in a downtown location. The firm has a local area network that is tied into a global trading network with other firms. What specific recommendations would you and your team members make to the president of the company to allow your firm to recover from a potential disaster or terrorist attack? Make sure to include a detailed description of the backup procedures you would recommend. Prepare your presentation for the president and give your pitch. Document your main points using your word processing program for submission to your instructor.

1. Accenture and IBM are just two of the many systems development outsourcing companies. Find the Web site of one of these companies. Describe the company, its services, and any investor or employment information. You may be asked to develop a report or send an e-mail message to your instructor about what you found.

2. Using the Web, search for information on structured design and programming. Also search the Web for information about the object-oriented approach to systems design and implementation. Write a report on what you found. Under what conditions would you use these approaches to systems development and implementation?

CASES

CASE 1

Lenox Revamps Web Site for Improved Customer Satisfaction

The continuously changing world of Web technology forces companies to update and improve their Web presence frequently. An unattended Web site will be out-of-date within a year and embarrassingly behind the times within two. Not only must companies keep up with current trends in Web technology, but they also must provide Web site visitors with new and interesting content to keep them coming back for more. Companies typically keep their home pages fresh by advertising their latest products and sale items. Every year or two a company may generate excitement by totally revamping its Web site and advertising the change: *come visit our new improved Web site!* With all this activity, how much investment is required for revamping a Web site? Let's examine LenoxCollections.com to find out.

Bob Palmer, vice president of information technology at Lenox Collections, a division of Lenox Inc., set out with a team of developers to revamp LenoxCollections.com. They had four goals:

1. reduce the number of clicks in the checkout process
2. more closely align the site with the company's catalog sales channel
3. provide consumers with more information about their purchases
4. update the site design

After defining their goals, Palmer and his team set out to determine whether it would be more economical to outsource the project or handle it in house. They looked at bigger-budget Web portal software from Redwood City, California–based BroadVision and IBM's WebSphere line. The team ultimately decided on doing the project in house. Palmer's team would use San Francisco–based Macromedia's over-the-counter ColdFusion Web development tools. "If you have experienced, talented Web developers, and we have, then it's not that difficult to do," Palmer says. "You have to ask yourself, 'Do I want to bring in an army of consultants at $150 or $200 an hour when the organization has very talented Web developers who understand our business and cost less?'"

The project budget was set for $105,000, with almost half of that figure slated for design. The team invested 17 months in the project, with the design phase taking 14 months followed by three months of development. By being thorough in the design of the Web site and supportive systems, Palmer and his team could be assured that the development process would progress smoothly and quickly and maintenance of the system would be minimal.

During the development process, Palmer and his team learned that Internet projects expose a lot more than just catalog offerings. "Where you're taking phone orders, there's always been a veil—a human being—between a company's back-end systems and the customer," which hides a lot of system deficiencies, he says. "The Web opens up your business processes to the general public, and if those processes and systems aren't consumer-focused and clear, the deficiencies will be exposed to the world."

The project also ran into some organizational challenges. Lenox had to re-establish who needed to be in the loop when decisions were being made. The firm also needed to better address communication and project goals when members left or joined the development team. "People need to know not only what they're doing, but why they're doing it and how it fits in with what others are doing," Palmer says. "It's basic stuff, but that's what tripped us up."

Palmer says the work has already paid for itself. LenoxCollections.com saw a 115 percent sales increase in the fourth quarter of 2001, compared with the same period in 2000. Much of that increase was driven by a boosted browser-to-buyer conversion rate (the number of visitors to a Web site who purchase something)—up from 4.5 percent to more than 8 percent.

Palmer's team took the experience gained from LenoxCollections.com and applied it to other Lenox Web sites. He says that about 70 percent of the code used on LenoxCollections.com was reusable, allowing his team to put up a site for U.K.-based retailer Brooks & Bentley in less than three months. Palmer says his development team can now quickly implement new features when customers request them. "Now that we've got the basics right, additional features aren't difficult to add," he says.

Bob Palmer was recognized by IDG's *Computerworld*, an information services publication for the information technology community, as one of the business world's Premier 100 IT Leaders. The award honors individuals who have had a positive impact on their organizations through technology. LenoxCollections.com was also recognized by BizRate.com as one of only 24 top e-tailers for outstanding performance over the 2001 holiday season.

Discussion Questions

1. What are the benefits of doing Web development in house? Under what circumstances might a company decide to outsource this work?
2. What management skills were required from Palmer in this project? How do demands on Palmer differ from those placed on his developers?

Critical Thinking Questions

3. Find an example of a poor-quality Web site that embarrasses the organization that it represents. What features make it a bad Web site? What features are likely to motivate customers to return to a Web site and make additional purchases?
4. Palmer and his team spent a considerable amount of time in the design phase of this project. What type of difficulties might have arisen in the development and maintenance of this Web site if the development team were careless in the design stage?

Sources: Michael Meehan, "Lenox Inc.," *Computerworld*, March 11, 2002, http://www.computerworld.com; "EasyAsk Wins Lenox, J. Jill, and SmartBargains; LenoxCollections.com Signs as EasyAsk's First Search Advisor Customer," *Business Wire*, April 15, 2002; LenoxCollections.com Web site, http://www.lenoxcollections.com, accessed July 2002.

CASE 2

Corning Display Technologies Sheds New Light on the Systems Development Process

Corning Display Technologies, a branch of Corning Inc., is the world's leading supplier of the ultra-thin glass substrates used to produce Active Matrix Liquid Crystal Displays (AMLCD)—the displays used for notebook computers, handheld devices, and more recently desktop computers.

In the late 1990s the demand for Corning's displays rose sharply, exposing flaws in the company's manufacturing operations. Corning was unable to keep up with demand. Its manufacturing was divided regionally, for example, the plant in Japan served Japanese customers; the U.S. plant served U.S. customers. "As we looked at the plan, we learned that our existing model just wasn't cutting it," says Corning CIO and Vice President Richard J. Fishburn. In 1999, Corning set about improving its supply chain efficiency.

Most companies would begin such a process by first analyzing existing supply chain management, buying ERP software tools, and adapting their business processes to the technology. The philosophy behind that approach is that the software incorporates best business practices. Jill Jenkins, an analyst at Current Analysis in Sterling, Virginia, describes the thinking behind the old technology-first method as, " 'If I optimize one piece and optimize another piece, when I put it together, it will be optimal'—but it wasn't."

Corning prides itself as being a pioneer in putting business processes first. For Corning, technology was the last aspect of the system it discussed. In fact, when brainstorming better models, Corning first asks managers to "listen to what their operational people are saying," says Richard Fishburn. Only then are opportunities defined, followed by business benefits and, finally, mechanisms to determine whether goals were met.

Corning first created a virtual factory. Display glass is manufactured in two stages: The melting process takes raw silica and produces sheets of glass, and then a finishing line cuts those down to various sizes. Corning used the virtual model of the manufacturing process to experiment with methods of distributing portions of the manufacturing process.

Corning decided that since a melting line costs 10 times as much as a finishing line, it would leave its melting operations unchanged—one in the United States and one in Japan—but break up the global finishing lines into several branch manufacturing units.

Once the new system was fully described, the technology to support the new system was evaluated. Corning decided to add a supply chain module to its existing PeopleSoft ERP software. The module assisted Corning in finding out what displays were needed when and where and applying JIT manufacturing techniques to meet the needs in a timely manner.

The project, three years old and in the final stages of deployment, has stayed on schedule and under budget, and is paying for itself, says Fishburn. "It used to take us five days to do the planning for tomorrow's production. Now we can do it in an hour," he says. Improved planning efficiency meant Corning didn't have to build excess capacity. Instead of using whatever inventory is on hand, creating leftover glass that's expensive to dispose of, Corning can now manufacture only what's needed, in the most optimal sizes.

The entire reorganization of Corning's manufacturing operations and systems development cost $750 million and allowed the company to triple its worldwide market.

Discussion Questions

1. What was the priority in Corning's approach to developing its new system? Why was the solution a success?
2. What factors and metrics should Corning examine to determine whether the new system is being effective?

Critical Thinking Questions

3. How might this story have ended if Corning had not examined the efficiency of its system and simply added a new plant?
4. Name some business processes that are the same for all organizations and can be implemented with off-the-shelf software. Also name processes that are different for each industry and require more specialized software. Under which category does Corning's solution fall?

Sources: Matthew Schwartz, "Corning Inc.," *Computerworld*, March 11, 2002, http://www.computerworld.com; Corning Display Web site, http://www.corning.com/displaytechnologies/, accessed July 2002.

CASE 3

State Street Corp. Speeds Financial Trading with State-of-the-Art Financial Transaction Management System

Most stock traders rely on professional service providers to handle the accounting and settlement and financial transfers required to complete a stock sale. A high percentage of these firms rely on State Street Corp. for this service. With more than $6.3 trillion in assets and in excess of $808 billion under its management, State Street Corp. is a world leader in financial services. Here are just some of its highlights:

- #1 servicer of U.S. mutual funds
- #1 servicer of U.S. pension plans
- #1 investment manager of U.S. pension assets and a leading pension manager globally

- #1 provider of foreign exchange services worldwide
- Third-largest global custodian worldwide
- Seventh-largest investment manager worldwide

Surprisingly, until recently, this financial megacorporation employed low-tech methods for communicating with its clients. The old system of requesting a trade involves faxing the trade request to the financial service to handle. Often faxed trade requests would arrive at State Street Corp. with critical information missing. Using its old system, State Street was often forced to delay a settlement while a staffer confirmed, by phone or fax, the missing information, says John A. Fiore, State Street's executive vice president and CIO. That much manual intervention was tremendously expensive in an operation that makes up as much as two-thirds of State Street's total revenue stream.

Fiore and his team of developers set out to improve the system. A detailed investigation of the current system showed that a large percentage of the follow-up calls were redundant; repeated requests for the same information from the same clients. State Street's solution then was to store the relevant information from its regular clients in its own system so that any missing data could be automatically entered and the request processed without human intervention.

To accomplish this goal, State Street developers had to (1) develop a database to store client information, which would increase in size with each new client, and (2) develop a program that would read incoming fax requests, determine what, if any, information was missing, access the missing information using data from the database, and pass along the request for processing.

The development team chose to develop the system in house using the Java programming language. They set up an Oracle database running on Sun Microsystems servers. The database was tied to underlying processing systems on the mainframe using IBM's MQSeries message-oriented middleware to allow communication among applications.

The resulting Financial Transaction Management (FTM) system now automatically processes more than 80 percent of the requests that flow through it by identifying the source of the fax and automatically filling in missing information from a database of the traders.

Tim Lind, an investment management practice analyst at Needham, Massachusetts–based TowerGroup, says that the FTM system helps put State Street in the forefront of two movements on Wall Street. The first is a push to automate the settlement process to the point where trades can be finalized one day after they're made, also known as T+1. The second is a move to outsource automation to companies like State Street so brokers don't have to pay the cost of automation themselves. Although State Street launched the system to improve its internal efficiency, Lind says FTM puts the bank in a prime position to fulfill the need for outsourcing and automated trade completion.

Of the trades that are now processed through FTM, less than 20 percent must be handled manually. "Even in its early stages, it's showing tremendous business value, even though there are so far no metrics from the operations side of the house," Fiore says. "That it has come off as scheduled and we haven't had problems deploying it has been a tremendous plus for the IT organization."

Discussion Questions

1. How valuable was employee insight in recognizing the need for improvements to this system? How has this system improved life for the employees that monitor incoming trade requests? How has it changed their job descriptions?
2. How might the success of this new system be measured?

Critical Thinking Questions

3. What other methods of requesting trades, besides faxing, might State Street employ to streamline its system?
4. Will increasing reliance on Internet-based communications eventually make the fax machine obsolete? What advantages does fax technology offer over Internet technology?

Sources: Kevin Fogarty, "State Street Corp.," *Computerworld*, March 11, 2002, http://www.computerworld.com; "State Street's Chief Information Officer Honored as 2002 Premier 100 IT Leader by Computerworld," *Business Wire*, January 2, 2002; State Street Web site, http://www.statestreet.com, accessed July 2002.

NOTES

Sources for the opening vignette on p. 561: "Breakthrough on fedex.com Gives Businesses Power to Estimate Costs of Global Shipping," *Business Wire*, June 11, 2002; Linda Rosencrance, "Computerworld Premier 100, Best in Class: FedEx Corp.," *Computerworld*, March 11, 2002, http://www.computerworld.com; FedEx Global Trade Manager Web site, http://www.fedex.com/us/international/, accessed July 2002.

1. Johnson, Maryfran, "Best in Class," *Computerworld*, March 11, 2002, p. 1.
2. Mearian, Lucas, "Merrill Lynch Replacing Global Order Entry System," *Computerworld*, March 11, 2002, p. 9.
3. Mearian, Lucas, "Storage, Server Chargeback," *Computerworld*, February 25, 2002, p. 14.
4. Copeland, Lee, et al., "Companies Urged to Revisit Disaster Recovery Plans," *Computerworld*, October 15, 2001, p. 7.
5. Fox, Pimm, "Making Support Pay," *Computerworld*, March 11, 2002, p. 28.
6. Kuan, Hee, et al., "A Systematic Approach for the Design of Post-Transaction Input Error Handling," *Information and Software Technology*, October 1, 2001, p. 641.
7. Mearian, Lucas, "Quake Rattles IT at Seattle Area-Firms," *Computerworld*, March 5, 2001, p. 6.
8. Smith, Randall, "Morgan Stanley Plans a Backup Trading Floor," *The Wall Street Journal*, October 30, 2001, p. C1.
9. Mearian, Lucas, "Financial Firms Plan Widely Dispersed IT Operations," *Computerworld*, March 4, 2001, p. 7.

10. Latour, Almar, "A Blaze in Albuquerque Sets Off Major Crisis for Cell-Phone Giants," *The Wall Street Journal*, January 29, 2001, p. A1.
11. Tully, Shawn "Rebuilding Wall Street," *Fortune*, October 1, 2001, p. 92.
12. Forbrieve, Janet, "Sun Unveils Monitoring Capability," *The Rocky Mountain News*, June 29, 2001, p. 20B.
13. Disaster Recovery Services at the Web site of Business Recovery Management located at http://www.businessrecords.com, accessed on May 17, 2000.
14. Disaster Recovery at the Guardian Computer Support Web site at http://www.guardiancomputer.com, accessed on May 17, 2000.
15. Millman, Howard, "Virtual Recovery Via Virtual Servers," *Computerworld*, March 4, 2002, p. 46.
16. Kimball, Ralph, "Catastrophic Failure," *Intelligent Enterprise*, November 12, 2001, p. 20.
17. Guidera, Jerry, "GiantLoop Pitches Software to Avert Disaster," *The Wall Street Journal*, January 10, 2002, p. B12.
18. Vijayan, J., "IT Redundancy Helps Bond Trader Rebound from Attacks," *Computerworld*, December 10, 2001, p. 12.
19. Bannan, Karen, "Building Your Safety Net," *PC Magazine*, January 29, 2002, p. iBiz1.
20. Bulkeley, William, "IBM's Disaster-Recovery Business Expands with Software Rollout," *The Wall Street Journal*, October 31, 2001, p. B8.
21. Ulfelder, Steve, "Opting for Outsourcing," *Computerworld*, April 28, 2002, p. 30.

22. Hall, Mark, "Big Blue Is Making a Big Commitment to Linux," *Computerworld*, October 29, 2001, p. 42.

23. Pierce, Lisa, "The ABCs of Preparing Thorough RFPs," *Network World*, July 23, 2001, p. 26.

24. Amato-McCoy, Deena, "Wachovia Discovers Cure for RFP Blues," *Bank Systems & Technology*, February 2002, p. 44.

25. Clermont, Paul, "Cost-Benefit: It's Back in Fashion," *Information Strategy*, Winter 2002, p. 6.

26. Kohli, Rajiv et al., "Managing Customer Relationships," *Decision Support Systems*, December 2001, p. 171.

27. Robertson, Jack, "Rambus Downplays Intel's Growing Interest in Double-Data-Rate SDRAM," *EBN*, March 4, 2002, p. 4.

28. Foster, Ed, "IS IT OK for Microsoft and Others to Forbid Disclosure of Benchmark Tests?" *InfoWorld*, April 16, 2001, p. 103.

29. Auer, Joe, "Manage the Contract," *Computerworld*, September 10, 2001, p. 31.

30. Melymuka, Kathleen, "The Killer Deal Isn't All It's Cracked Up to Be," *Computerworld*, February 25, 2002, p. 32.

31. Meehan, Michael, "IT Managers Turn to eBay to Cut Costs," *Computerworld*, May 6, 2002, p. 14.

32. Murphy, Victoria, "Reinventing Software," *Forbes*, September 17, 2001, p. 134.

33. Buckman, Rebecca, "Microsoft Changes Terms of Pacts under Which Firms Buy Software," *The Wall Street Journal*, May 11, 2001, p. B7.

34. Mearian, Lucas, "Supermarket Dumps $89M SAP Project," *Computerworld*, February 5, 2001, p. 77.

35. Chazan, Guy, "Now Available from Russia: Software Programming," *The Wall Street Journal*, August 6, 2001, p. B1.

36. Einhorn, Bruce, "India 3.0," *Business Week*, February 26, 2001, p. 44.

37. Copeland, Lee, "GM Drives Application Development Offshore," *Computerworld*, January 21, 2001, p. 16.

38. Hailpern, B. et al., "Software Debugging, Testing, and Verification," *IBM Systems Journal*, Volume 41, 2002, p. 4.

39. Ledgard, Henry, "The Emperor with No Clothes," *Communications of the ACM*, October 2001, p. 126.

40. Borck, James, "Set Sights on Services," *InfoWorld*, October 15, 2001, p. 62.

41. Ben-Chorin, Shay, "Speeding Embedded Systems Development," *Electronic News*, April 16, 2001, p. 8.

42. Wong, William, "Open-Source IDE Supports C/C++," *Electronic Design*, March 4, 2002, p. 29.

43. Biggs, Maggie, "Ends Meet the Middle," *InfoWorld*, December 31, 2001, p. 32.

44. Meehan, Michael, "Software Conversion Creates Chaos for Spirit Airlines," *Computerworld*, January 15, 2001, p. 20.

45. Disabatino, Jennifer, "Sabre Sheds Its Mainframe Legacy," *Computerworld*, September 3, 2001, p. 40.

46. Dash, Julekha, "GM Launches Wireless Health Care Pilot," *Computerworld*, June 11, 2001, p. 12.

47. Dvorak, Phred, "Computer Glitches at Japanese Bank Disrupt Millions," *The Wall Street Journal*, April 8, 2002, p. A10.

48. Mearian, Lucas, "Bank Hones Project Management Skills with Redesign," *Computerworld*, April 29, 2002, p.18.

49. Mehta, Stephanie, "Telco on the Frontline," *Fortune*, October 15, 2001, p. 139.

50. Roberts-Witt, Sarah, "Project Renewal," *PC Magazine*, October 30, 2001, p. iBiz1.

51. Meehan, Michael, "Home Depot Seeks Remote Control of Desktops," *Computerworld*, January 7, 2002, p. 12.

52. Products and Services Section of the Candle Corporation Web site at http://www.candle.com, accessed on May 17, 2000.

Effective business information systems are critical to the smooth operation of a large educational institution. When the mission and programs of the institution change, drastic changes can be required in its business information systems. Indeed, whole new systems may be required. As this World Views Case reveals, the University of Nairobi is undergoing such a dramatic change and is facing many difficult challenges in regard to its business systems.

Event-Driven Review of the University of Nairobi's Student Management Information System

Euphraith Muthoni Masinde
Team Leader, Student Management Information System
The University of Nairobi 51; Kenya

The University of Nairobi (UoN) is the oldest, largest, and best-established public university in Kenya and greater East Africa. Its origin can be traced back to 1951 when its precursor, the Royal Technical College of East Africa, was established (www.uonbi.ac.ke). Besides the University of Nairobi, there are 13 other universities in Kenya, both public and private.

The University of Nairobi is located in the heart of the capital city of Nairobi and has an enrollment of about 20,000 students pursuing certificate, diploma, and advanced degree courses in most areas of study. There are about 1,500 academic staff members and 3,000 nonacademic staff who enable the university to fulfill its mission of teaching and conducting research. The university is made up of six colleges that are in turn composed of faculties and institutes (administrative units that fall under colleges, for example, the Faculty of Science and the Institute of Computer Science). Faculties are further divided into departments.

The Institute of Computer Science is the computing unit of the university, with management information system (MIS) support as one of its responsibilities. The Student Management Information System (SMIS) is one of these systems, and it is composed of the following subsystems:

1. **Admissions Subsystem**, which processes admissions for undergraduate students to all public universities in Kenya
2. **Nominal Roll Subsystem**, which keeps track of students' personal records and academic progress
3. **Examinations Subsystem**, which processes undergraduate students' examinations
4. **Fees Subsystem**, which processes students' fees

Until recently, students were enrolled for a degree programme with a fixed duration, fixed admissions fee, and fixed academic periods: one academic year with two semesters. The existence of a standard programme and small student population simplified SMIS subsystems.

Over the years, funding to the university by both the government and other donors has dropped by more than 80 percent, bringing university operations nearly to a halt. To alleviate this funding crisis, self-sponsored students were brought on board in the year 2000. These students met the minimum admission requirements but were not admitted to the public universities because they did not receive government scholarships, which are limited to a small number of top students. To make the programmes at the university attractive, especially to people already employed in the workforce, the university has had to become more flexible, offering variable course workloads, variable completion periods, year-round admissions, and flexible course and examination schedules.

Admissions of the self-sponsored students are carried out for the various faculties at different times using different rules, which the existing Admissions Subsystem could not handle. The academic periods are no longer uniform across faculties; a Quarter System, Semester System, and Term System are all in use. Consequently, students' academic progress is no longer uniform. The existing Nominal Roll Subsystem cannot handle these variations since it assumed a Semester System only. The differing course offerings also resulted in broad ranges of student fees. Some programmes cost as little as US$3,000, while others are as high as US$15,000, and the number of fee components (such as tuition, lab, and examination fees) also vary. The existing Fees Subsystem was not set up to handle this flexibility. It was designed for uniform fees for all programmes. Uniform fees had resulted from all students being government sponsored. They paid a uniform subsidy, and the government met the difference. Self-sponsored students now pay varying fees because they bear the full cost of their various programmes. With the self-sponsored students, examination rules have to be applied at the individual student level as opposed to the earlier system in which they were applied to a whole class (a group of students pursuing the same programme at the same level of study).

The events unfolding from the introduction of the self-sponsored students triggered the need for rigorous review of the entire SMIS. This is an example of *event-driven review.*

With its admissions changes, the university is now generating enough income, and it has sufficient money in its budget for the redevelopment of the SMIS. The university decided to redevelop all SMIS subsystems, except for the Admissions Subsystem, which is jointly owned by all public universities in the country. Three powerful servers running Digital Unix, loaded with Oracle software, were purchased. A high-speed WAN with up to 500 PCs was also put in place. Nine computer science university graduates were hired as system analysts/programmers and trained on Oracle development and administration. Since the existing systems were developed using COBOL and run on VAX/VMS, not much could be gained by revamping them; hence the decision to completely redevelop the systems. Developers chose the traditional structured approach to development. Oracle, Microsoft Office, Developer 2000, Designer 2000, Java, and PROLOG are some of the tools used.

Sadly, not much has been achieved so far. Some streamlining of the separate faculties' operations is urgently needed because no MIS can cope with all of the flexibility in place at the moment. However, convincing everyone to adhere to standards—especially if these standards mean reduction in profits—has been met with resistance.

The SMIS team has a number of options:

- Develop a central SMIS incorporating all the requirements from all the faculties. Coming up with such a system may take several years.
- Focus on decentralized systems with smaller SMISs being customized to cater to individual faculties' needs.
- Convince the University Management Board that purchasing a system off the shelf is the best idea.

Discussion Questions

1. Discuss the following point: The University of Nairobi cannot do without the new Student Management Information System.
2. What are some of the challenges the SMIS development team will face in trying to come up with the new system?

Critical Thinking Questions

3. Assume that you are the head of the University of Nairobi's Institute of Computer Science MIS team. Would you recommend an in-house developed system or an off-the-shelf solution? Why?
4. Assume that in-house development is the method chosen. How would you approach systems development—centralized or decentralized?

Using contractors to develop custom software is a strategy small businesses frequently use to implement effective business information systems. However, the success of such an effort depends on a high level of customer involvement in the investigation, analysis, design, and implementation phases of the project and on the capabilities of the contractor. Both parties must work together effectively, or the results can be quite unsatisfactory, as shown in this World Views Case about a small business in Regina, Saskatchewan.

Song Book Music

M. Gordon Hunter
The University of Lethbridge, Lethbridge, Alberta, Canada

Three years ago two ladies (a music teacher and an accountant) purchased Song Book Music, a sheet music store in Regina, Saskatchewan. Customers are mainly from local high schools and music academies within this western Canadian prairie city with a population of 200,000. Some sales are even made to other centers in the southern part of the province, especially Moose Jaw, renowned for, among other things, its annual Spring Music Festival.

The festival is the major event in the province's celebration of the departure of the long cold winter and the snow that had arrived the previous October. Also, Moose Jaw is the reputed holiday venue of Chicago gangster Al Capone!

During the first year of Song Book Music's operations, the new owners made no changes. However, they soon found that they needed a better inventory control system. They did not know what specific sheet music they had on hand, and it was also difficult to determine what had been ordered. While prairie folk can be quite accommodating, some of the customers were becoming frustrated with the inconsistent service. For many of the rural customers, it is a long drive into Regina, and the trip is usually only made once per week in conjunction with other shopping duties or to deliver cattle to the local market. So, the ladies recognized the potential benefit of a new system and decided to initiate a systems development project.

Because the owners were very busy running the store, they decided to hire a consultant to help them solve their business problem. Buck, the sole proprietor of Buffalo Consulting, convinced the ladies that he could solve their problem. He provided them with an impressive list of former clients from the southern part of Saskatchewan. He explained to them how he could not only provide them with the necessary software to address their business objectives but also custom build the required hardware to operate the software. Buck was so convincing that the owners decided not to investigate his references or to pursue any other alternatives.

In early July of the owners' second year of operation, Buck presented them with a project plan and budget (see Figure 1)—as well as a bill for $250 for his effort in understanding the business problem and developing the specific plan. The owners believed the cost could be easily recovered by the increased revenue from more accurate inventory and the resulting increased customer satisfaction, since those two issues were critically important to the business.

FIGURE 1

PROJECT PLAN	
DATE	**ACTIVITY**
July 15	**Complete Phase 1 ($5,000)**
	Develop software and construct hardware
July 31	**Complete Phase 2 ($3,000)**
	System testing
August 31	**Complete Phase 3 ($2,000)**
	System implementation and data conversion

The plan indicated that the project would be completed by August 31, thus taking a total of two months. During Phase 1 there was very little interaction between the owners and Buck, because he was busy coding the software (using a unique language he had developed as part of his universities studies) and building the hardware from generic components sold locally. When the ladies queried Buck about the project, he would politely indicate that all was well and proceeding as he expected.

During Phase 2 problems started to become evident. The complete system was not available for testing, and what was available did not include all of the data elements required to maintain accurate inventory records. In early August the system testing phase was not complete, but Buck suggested that the owners could start entering data into the system as part of Phase 3. Also, Buck started to ask for payment on work that the ladies felt was not yet completed. Song Book's lawyer recommended withholding payment. Buck started to act unprofessionally, regularly missing meetings and claiming he had other more important clients. System implementation and data conversion were finally completed on April 1 of the following year, a full seven months later than originally planned. At that time Buck was paid in full and informed that his services were no longer required.

Since that time, the owners have had to make many changes to the information system. Newer, more powerful computers have been installed, and more programs have been written to expand the reporting capability of the system. Fortunately, all of this work could be performed by one of the ladies' husbands at no charge. Although the owners consider the system usable, they retain very negative feelings about the entire process.

Discussion Questions

1. What factors contributed directly to the problems experienced in this case?
2. What actions could each of the two businesses have taken to resolve this situation?

Critical Thinking Questions

3. What aspects of small businesses make this situation unique?
4. What action can small businesses take to ensure these types of problems do not arise?

Source: M. Gordon Hunter, "Information Systems Development Outcomes: The Case of Song Book Music," Chapter 3 in S. Burgess (Ed.), *Managing Information Technology in Small Business: Challenges and Solutions* (Hershey, PA: Idea Group Publishing, 2002; ISBN: 1-930708-35-1).

PART 5

INFORMATION SYSTEMS IN BUSINESS AND SOCIETY

Security, Privacy, and Ethical Issues in Information Systems and the Internet

PRINCIPLES	LEARNING OBJECTIVES
Policies and procedures must be established to avoid computer waste and mistakes.	• Describe some examples of waste and mistakes in an IS environment, their causes, and possible solutions. • Identify policies and procedures useful in eliminating waste and mistakes.
Computer crime is a serious and rapidly growing area of concern requiring management attention.	• Explain the types and effects of computer crime. • Identify specific measures to prevent computer crime. • Discuss the principles and limits of an individual's right to privacy.
Jobs, equipment, and working conditions must be designed to avoid negative health effects.	• List the important effects of computers on the work environment. • Identify specific actions that must be taken to ensure the health and safety of employees. • Outline criteria for the ethical use of information systems.

Recommending Strong Medicine to Combat Cybercrime

The U.S. government is recommending a number of strong actions to help combat cyber-crime. Some of these ideas were stimulated by the terrorist acts of September 11, 2001, but many are born out of years of frustration with the dismal track record of software manufacturers for turning out software riddled with security leaks.

Before September 11, Internet service providers were prohibited by federal law from revealing the content of stored e-mail and other electronic communications to the government without proper legal orders based on "probable cause." The USA Patriot Act, which was passed just five weeks after September 11, now allows Internet service providers to disclose such information to law enforcement officials where there is an imminent dangerous situation.

In July 2002, the U.S. House of Representatives voted in favor of the Cyber Security Enhancement Act of 2002, a bill that would impose stiffer penalties on computer hackers. Those who perform cyberattacks that result in bodily injury or death could receive sentences ranging from 20 years to life in prison. The bill has yet to be voted on by the Senate.

In August 2002, President Bush's chief cybersecurity adviser warned IS providers that it's no longer acceptable to sell buggy software with security holes and urged users to stop buying products they know aren't secure. He also accused cable companies, Internet service providers, and telecommunications companies of providing broadband connections to users and failing to warn them of the inherent security vulnerabilities in those constant connections.

In an effort to safeguard valuable government functions and resources, the National Plan for Protecting Cyberspace, to be released in the fall of 2002, may require all federal agencies to adopt a U.S. Department of Defense policy requiring all new IT purchases to be made from a list of independently certified product lines.

The Bush administration also plans to convene a panel of government and private sector labor and legal experts to develop guidelines for conducting background investigations on corporate IS and other employees. Employees in the critical banking, chemical, energy, transportation, telecommunications, shipping, and public health industries would face background checks as a condition of employment.

As you read this chapter, consider the following:
- What motivates an individual to commit a computer crime?
- What can be done to detect and avoid computer crime?

Earlier chapters detailed the amazing benefits of computer-based information systems in business, including increased profits, superior goods and services, and higher quality of work life. Computers have become such valuable tools that today's businesspeople would have difficulty imagining work without them. Yet the information age has also brought some potential problems for workers, companies, and society in general (see Table 14.1).

To a large extent, this book has focused on the solutions—not the thorny issues—presented by information systems. In this chapter we discuss some of the issues as a reminder of the social and ethical considerations underlying the use of computer-based information systems. No business organization, and hence no information system, operates in a vacuum. All IS professionals, managers, and users have a responsibility to see that the potential consequences of IS use are fully considered.

Managers and users at all levels play a major role in helping organizations achieve the positive benefits of IS. These individuals must also take the lead in helping to minimize or eliminate the negative consequences of poorly designed and improperly utilized information systems. For managers and users to have such an influence, they must be properly educated. Many of the issues presented in this chapter, for example, should cause you to think back to some of the systems design and systems control issues we have already discussed. They should also help you look forward to how these issues and your choices might affect your future IS management considerations.

COMPUTER WASTE AND MISTAKES

Computer-related waste and mistakes are major causes of computer problems, contributing as they do to unnecessarily high costs and lost profits. Computer waste involves the inappropriate use of computer technology and resources. Computer-related mistakes refer to errors, failures, and other computer problems that make computer output incorrect or not useful, caused mostly by human error. In this section we explore the damage that can be done as a result of computer waste and mistakes.

COMPUTER WASTE

The U.S. government is the largest single user of information systems in the world. It should come as no surprise then that it is also perhaps the largest misuser. The government is not unique in this regard—the same type of waste and misuse found in the public sector also exists in the private sector. Some companies discard old software and even complete computer systems when they still have value. Others waste corporate resources to build and maintain complex systems never used to their fullest extent. A less dramatic, yet still relevant, example of waste is the amount of company time and money employees may waste playing computer games, sending unimportant e-mail, or accessing the Internet. Junk e-mail, also called *spam*, and junk faxes also cause waste. People receive hundreds of e-mail messages and faxes advertising products and services not wanted or requested. Not only does this waste time, but it also wastes paper and computer resources. When waste is identified, it typically points to one common cause: the improper management of information systems and resources.

TABLE 14.1

Social Issues in Information Systems

• Computer waste and mistakes	• Health concerns
• Computer crime	• Ethical issues
• Privacy	• Patent and copyright violations

COMPUTER-RELATED MISTAKES

Despite many people's distrust, computers themselves rarely make mistakes. Even the most sophisticated hardware cannot produce meaningful output if users do not follow proper procedures. Mistakes can be caused by unclear expectations and a lack of feedback. Or a programmer might develop a program that contains errors. In other cases, a data entry clerk might enter the wrong data. Unless errors are caught early and prevented, the speed of computers can intensify mistakes. As information technology becomes faster, more complex, and more powerful, organizations and individuals face increased risks of experiencing the results of computer-related mistakes. Take, for example, these cases from recent news:

- In late August 2002, iVillage, a Web site devoted to women that ranks among the top 30 Web destinations with 9.5 million users, experienced problems with the free e-mail service it provides subscribers. As a result, subscribers could view other people's e-mail, raising concerns over loss of privacy.[1]

- Over the four-day Easter bank holiday in March 2002, some 20,000 business customers of Barclays Bank PLC did not have full access to their money because their paychecks were not electronically deposited. The problem occurred when the computers that connect Barclays with the United Kingdom's Bankers Automated Clearing System network failed.[2]

- A computer glitch resulted in the Internal Revenue Service sending out incorrect tax refund notices to about 523,000 taxpayers. The notices said the taxpayers would receive a check with the maximum tax refund, when they really wouldn't.[3]

- United Air Lines decided to honor nearly 150 next-to-nothing tickets that were inadvertently sold for international flights on January 31, 2001, on the company's Web site. For example, some travelers were able to book flights from San Francisco to Paris for just $24.98.[4] Apparently not learning from its prior mistake, United Air Lines listed round-trip flights on May 14, 2002, for $5 plus tax for tickets purchased at the United Web site.[5]

- In May 2002, a glitch at Fidelity Investments Canada allowed an Ottawa professor to access the account information of other customers, including investment portfolios and other personal information.[6]

- Customers and businesses were unable to track their packages the week of April 8, 2002, through the U.S. Postal Service due to a problem with the tracking system software.[7]

PREVENTING COMPUTER-RELATED WASTE AND MISTAKES

To remain profitable in a competitive environment, organizations must use all resources wisely. Preventing computer-related waste and mistakes like those just described should therefore be a goal. Today, most organizations use some type of CBIS. To employ IS resources efficiently and effectively, employees and managers alike should strive to minimize waste and mistakes. Preventing waste and mistakes involves (1) establishing, (2) implementing, (3) monitoring, and (4) reviewing effective policies and procedures.

Establishing Policies and Procedures

The first step to prevent computer-related waste is to establish policies and procedures regarding efficient acquisition, use, and disposal of systems and devices. Computers permeate organizations today, and it is critical for organizations to ensure that systems are used to their full potential. As a result, most companies have implemented stringent policies on the acquisition of computer systems and equipment, including requiring a formal justification statement before computer equipment is purchased, a definition of standard computing platforms (operating system, type of computer chip, minimum amount of RAM, etc.), and the use of preferred vendors for all acquisitions.

American Home Products employs 35,000 people in 57 countries around the world. Over a recent three-year period, its central IS purchasing group spent over $400 million on IS acquisitions. This group follows a number of policies and procedures in executing their role. Key to these is a two-fold purchasing strategy. First, they develop a multiyear plan that identifies the key IS components needed to support the business strategy. Second, they perform enterprise asset management so that they know who is using what IS components, where they are, how many they have, and what it is actually costing to run them. Accurate IS asset management is also critical when a form of technology or equipment has become obsolete and it is time to dispose of it.[8]

Prevention of computer-related mistakes begins by identifying the most common types of errors, of which there are surprisingly few (see Table 14.2). To control and prevent potential problems caused by computer-related mistakes, companies have developed preventive policies and procedures that cover the following:

- Acquisition and use of computers, with a goal of avoiding waste and mistakes
- Training programs for individuals and workgroups
- Manuals and documents on how computer systems are to be maintained and used
- Approval of certain systems and applications before they are implemented and used to ensure compatibility and cost-effectiveness
- Requirement that documentation and descriptions of certain applications be filed or submitted to a central office, including all cell formulas for spreadsheets and a description of all data elements and relationships in a database system; such standardization can ease access and use for all personnel

Once companies have planned and developed policies and procedures, they must consider how best to implement them.

Implementing Policies and Procedures

Implementing policies and procedures to minimize waste and mistakes varies according to the business conducted. Most companies develop such policies and procedures with advice from the firm's internal auditing group or its external auditing firm. The policies often focus on the implementation of source data automation and the use of data editing to ensure data accuracy and completeness, and assigning clear responsibility for data accuracy within each information system. Table 14.3 lists some useful policies to minimize waste and mistakes.

TABLE 14.2

Types of Computer-Related Mistakes

- Data entry or capture errors
- Errors in computer programs
- Errors in handling files, including formatting a disk by mistake, copying an old file over a newer one, and deleting a file by mistake
- Mishandling of computer output
- Inadequate planning for and control of equipment malfunctions
- Inadequate planning for and control of environmental difficulties (electrical problems, humidity problems, etc.)
- Installing computing capacity inadequate for the level of activity on corporate Web sites
- Failure to provide access to the most current information by not adding new and deleting old URL links

TABLE 14.3

Useful Policies to Eliminate Waste and Mistakes

- Changes to critical tables, HTML, and URLs should be tightly controlled, with all changes authorized by responsible owners and documented.

- A user manual should be available that covers operating procedures and that documents the management and control of the application.

- Each system report should indicate its general content in its title and specify the time period it covers.

- The system should have controls to prevent invalid and unreasonable data entry.

- Controls should exist to ensure that data input, HTML, and URLs are valid, applicable, and posted in the right time frame.

- Users should implement proper procedures to ensure correct input data.

Training is another key aspect of implementation. Many users are not properly trained in developing and implementing applications, and their mistakes can be very costly. Since more and more people use computers in their daily work, it is important that they understand how to use them. Training is often the key to acceptance and implementation of policies and procedures. Because of the importance of maintaining accurate data and of people understanding their responsibilities, companies converting to ERP and e-commerce systems invest weeks of training for key users of the system's various modules.

Monitoring Policies and Procedures

To ensure that users throughout an organization are following established procedures, the next step is to monitor routine practices and take corrective action if necessary. By understanding what is happening in day-to-day activities, organizations can make adjustments or develop new procedures. Many organizations implement internal audits to measure actual results against established goals for things such as percentage of end-user reports produced on time, percentage of data input errors rejected, number of input transactions entered per eight-hour shift, and so on.

The passing of the Sarbanes-Oxley Act has caused many companies to monitor their policies and procedures and to plan changes in information systems that could have a profound effect on many business activities. As mentioned in Chapter 9, this act set deadlines for public companies to implement procedures to ensure that their audit committees can document underlying financial data to validate earnings reports. In August 2002, AOL Time Warner disclosed that it was investigating three transactions in which it may have improperly accounted for revenue. The total amount was $49 million, a relatively insignificant amount compared to the company's total revenues, and was spread over six financial quarters. Embarrassingly, the problem was found just days after its chief executive officer and chief financial officer certified the company's financial reports in compliance with an SEC order and the recently passed Sarbanes-Oxley Act. AOL Time Warner planned to continue investigating these and other AOL transactions involving advertising and revenue. In addition, AOL Time Warner business managers are demanding that their financial systems provide real-time data feeds and updates of expenditures and sales so they can monitor the accuracy of their numbers.[9]

Reviewing Policies and Procedures

The final step is to review existing policies and procedures and determine whether they are adequate. During review, people should ask the following questions:

- Do current policies cover existing practices adequately? Were any problems or opportunities uncovered during monitoring?

- Does the organization plan any new activities in the future? If so, does it need new policies or procedures on who will handle them and what must be done?
- Are contingencies and disasters covered?

This review and planning allows companies to take a proactive approach to problem solving, which can increase productivity and improve customer service. During such a review, companies are alerted to upcoming changes in information systems that could have a profound effect on many business activities. An example is the need for health care organizations to meet the requirements of the Health Insurance Portability and Accountability Act of 1996 (HIPAA). The goal of this act is to require health care organizations to implement cost-effective procedures for exchanging medical data. Health care organizations must employ standard electronic transactions, codes, and identifiers designed to enable them to fully "digitize" medical records and make it possible to use the Internet rather than expensive private networks for electronic data interchange. The regulations affect 1.5 million health care providers, 7,000 hospitals, and 2,000 health care plans. Companies have until April 2003 to comply.[10] Now that the full details of HIPAA are becoming clear, many experts are concerned. Some fear that the HIPAA provisions are too complicated and will not meet the original objective of reducing medical industry costs and will instead increase costs and paperwork for doctors without improving medical care. On the other hand, the Agency for Healthcare Research and Quality (the research arm of the Department of Health and Human Services) states that HIPAA will require computer systems that, if implemented correctly, can greatly reduce the adverse reactions caused by medication errors. The agency estimates that hospitals will save $500,000 in direct costs annually.[11]

Information systems professionals and users still need to be aware of the misuse of resources throughout an organization. Preventing errors and mistakes is one way to do so. Another is implementing in-house security measures and legal protections to detect and prevent a dangerous type of misuse: computer crime.

COMPUTER CRIME

Even good IS policies may not be able to predict or prevent computer crime. A computer's ability to process millions of pieces of data in less than a second can help a thief steal data worth thousands or millions of dollars. Compared with the physical dangers of robbing a bank or retail store with a gun, a computer criminal with the right equipment and know-how can steal large amounts of money from the privacy of a home. Computer crime often defies detection, the amount stolen or diverted can be substantial, and the crime is "clean" and nonviolent.

Here is a sample of computer crimes from 2002. In January, a man admitted that he committed computer intrusion and sent electronic mail to approximately 30,000 employees and associates of Catholic Healthcare West (CHW). The e-mail purported to be from a named employee of CHW and contained insulting statements about that named employee and other CHW employees. The intruder caused damage to CHW of more than $25,000.[12] In February, a former computer network administrator was sentenced to 41 months in prison for causing over $10 million in damage when he deleted all the production programs of a New Jersey–based high-tech measurement and control instruments manufacturer.[13] In March, the former controller of a Manhattan apparel manufacturer and designer was arrested and charged with unauthorized intrusion into the computer network of his former employer. About two months after the employee was let go, he accessed the firm's database containing customer orders and deleted them all.[14] In April, the members of the admission staff at Princeton University improperly and repeatedly accessed a Web site set up to let Yale applicants know if they had been accepted as students. The Web site included information about students' academic and extracurricular activities.[15] In May, two Kazakhstan citizens were extradited to the United States on charges of breaking into a Bloomberg (provider of financial database services) computer system in an attempt to extort money from Bloomberg. They sent a number of e-mails to Michael Bloomberg, the company's founder, demanding that he pay them $200,000 in exchange for details on how they were able to infiltrate Bloomberg's computer system.[16] In June 2002, two men were sentenced to federal prison after pleading guilty to conspiring to traffic over $1.5 million of counterfeit Microsoft software. The bogus products included Windows 98, Windows Office 2000, and Windows Millennium Edition.[17] In July, hackers broke into *USA Today*'s Web site and replaced legitimate news stories with phony articles. The bogus pages were available to USAToday.com readers for about 15 minutes before being discovered and taken off-line.[18]

Although no one really knows how pervasive cybercrime is, the number of IT-related security incidents is increasing dramatically. The Computer Emergency Response Team Coordination Center (CERT/CC) is located at the Software Engineering Institute (SEI), a federally funded research and development center at Carnegie Mellon University in Pittsburgh, Pennsylvania. It is charged with coordinating communication among experts during computer security emergencies and helping to prevent future incidents. CERT employees study Internet security vulnerabilities, handle computer security incidents, publish security alerts, research long-term changes in networked systems, develop information and training to help organizations improve security at their sites, and conduct an ongoing public awareness campaign. The number of security problems reported to CERT increased 25-fold between 1997 and 2001, as shown in Figure 14.1. Many attacks go undetected—as many as 60 percent, according to security experts. What's more, of the attacks that are exposed, only an estimated 15 percent are reported to law enforcement agencies. Why? Companies don't want the bad press. Such publicity makes the job even tougher for law enforcement. Most companies that have been electronically attacked won't talk to the press. A big concern is loss of public trust and image—not to mention the fear of encouraging copycat hackers.

The increase in computer security breaches is raising concern that the American workforce may not have enough troops to battle cybersnoops. Too few colleges offer security courses. Only about a half-dozen U.S. academic institutions have graduate programs in computer security—and that number hasn't changed much in ten years. Even more unsettling is that no quick solution is in sight. Computer assurance still isn't a recognized discipline at most colleges.

FIGURE 14.1

Number of Incidents Reported to CERT

Source: Data from CERT Web site at http://www.CERT.org/stats, accessed on May 29, 2002.

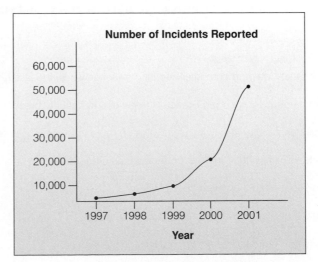

Number of Incidents Reported

Highlights of the annual Computer Crime and Security Survey are shown in Table 14.4. The 2002 survey results are based on responses from 503 companies and government agencies. The Computer Security Institute, with the participation of the San Francisco Federal Bureau of Investigation (FBI) Computer Intrusion Squad, conducts this survey. The aim of the survey is to raise awareness of security, as well as to determine the scope of computer crime in the United States.

Today, computer criminals are a new breed—bolder and more creative than ever. With the increased use of the Internet, computer crime is now global. It's not just on U.S. shores that law enforcement has to battle cybercriminals. Regardless of its nonviolent image, computer crime is different only because a computer is used. It is still a crime. Part of what makes computer crime unique and so hard to combat is its dual nature—the computer can be both the tool used to commit a crime and the object of that crime.

THE COMPUTER AS A TOOL TO COMMIT CRIME

A computer can be used as a tool to gain access to valuable information and as the means to steal thousands or millions of dollars. It is, perhaps, a question of motivation—many individuals who commit computer-related crime claim they do it for the challenge, not for the money. Credit card fraud—whereby a criminal illegally gains access to another's line of credit with stolen credit card numbers—is a major concern for today's banks and financial institutions. In general, criminals need two capabilities to commit most computer crimes. First, the criminal needs to know how to gain access to the computer system. Sometimes obtaining access requires knowledge of an identification number and a password. Second, the criminal must know how to manipulate the system to produce the desired result. Frequently, a critical computer password has been talked out of an individual, a practice called **social engineering**. Or, the attackers simply go through the garbage—**dumpster diving**—for important pieces of information that can help crack the computers or convince someone at the company to give them more access. In addition, there are over 2,000 Web sites that offer the digital tools—for free—that will let people snoop, crash computers, hijack control of a machine, or retrieve a copy of every keystroke.

Also, with today's sophisticated desktop publishing programs and high-quality printers, crimes involving counterfeit money, bank checks, traveler's checks, and stock and bond certificates are on the rise. As a result, the U.S. Treasury Department redesigned and printed new currency that is much more difficult to counterfeit.

Cyberterrorism

Government officials and IS security specialists have documented a significant increase in Internet probes and server scans since early 2001. There is a growing

social engineering
the practice of talking a critical computer password out of an individual

dumpster diving
searching through the garbage for important pieces of information that can help crack an organization's computers or be used to convince someone at the company to give them access to the computers

TABLE 14.4

Summary of Key Data from 2002 Computer Crime and Security Survey

Source: Data from Richard Power, "2002 CSI/FBI Computer Crime and Security Survey," *Computer Security Issues & Trends*, Vol. VIII, No. 1, Spring 2002, p. 4.

Incident	2002 Results
Respondents that detected computer security breaches within the last 12 months	90%
Respondents that acknowledged financial losses due to security breaches	80%
Average dollar loss of the 44% who were willing or able to quantify their financial losses	$2.0 million
Respondents that cited their Internet connection as a frequent point of attack	74%
Respondents that cited their internal systems as a frequent point of attack	33%
Respondents that reported intrusions to law enforcement	34%
Respondents that detected computer viruses	85%

concern among federal officials that such intrusions are part of an organized effort by cyberterrorists, foreign intelligence services, or other groups to map potential security holes in critical systems. For example, hackers broke into computer systems owned by California Independent System Operator (Cal-ISO), California's primary electric power grid operator, and remained undetected for 17 days in April 2001. The intent of the network break-in wasn't clear; however, the lack of apparent damage indicates that it may have been conducted by hackers whose intent was to collect information about how the systems work and to document their vulnerabilities. Cal-ISO officials said they managed to trace the attack to a system in China.[19]

Even before the September 11, 2001, terrorist attacks, the U.S. government considered the potential threat of cyberterrorism serious enough that it established the National Infrastructure Protection Center in February 1998. This branch of the FBI serves as a focal point for threat assessment, warning, investigation, and response for threats or attacks against our country's critical infrastructure that provides telecommunications, energy, banking and finance, water systems, government operations, and emergency services. Successful cyberattacks against the facilities that provide these services could cause widespread and massive disruptions to the normal function of our society.[20]

Identity Theft

identity theft
a crime in which an imposter obtains key pieces of personal identification information, such as social security or driver's license numbers, in order to impersonate someone else. The information is then used to obtain credit, merchandise, and services in the name of the victim, or to provide the thief with false credentials

Identity theft is a crime in which an imposter obtains key pieces of personal identification information, such as social security or driver's license numbers, in order to impersonate someone else. The information is then used to obtain credit, merchandise, and services in the name of the victim, or to provide the thief with false credentials.

In some cases, the identity thief uses personal information to open new credit accounts, establish cellular phone service, or open a new checking account to obtain blank checks. In other cases, the identity thief uses personal information to gain access to the person's existing accounts. Typically, the thief will change the mailing address on an account and run up a huge bill before the person whose identity has been stolen realizes there is a problem. The Internet has made it easier for an identity thief to use the information he or she has stolen because transactions can be made without any personal interaction.

According to the Identity Theft Resource Center in San Diego, identity theft, facilitated by the Internet, is rising rapidly. Some 700,000 consumers became victims of identity theft during 2001.[21] In an example of corporate identity theft, a securities dealer stole the password and registration number of the Hyundai Investment Trust Company to buy 5 million shares of Delta Information and Communications Company (worth $22 million) on the Korean Kosdaq stock exchange. Evidently, the purpose was to pump up the value of Delta stock, which rose by over 7 percent soon after the order was placed. The stock dropped by 12 percent the next day when the deal was discovered to be fraudulent. By that time, the dealer had fled the country leaving his firm, Daewoo Securities, responsible for sorting out the whole affair and repaying the cost of the shares.[22]

"Dumpster diving"—criminals rummaging through trash seeking social security numbers to steal—is a common way for identity thieves to gain the information they need. Another popular method to get information is "shoulder surfing"—the identity thief simply stands next to someone at a public office, such as the Bureau of Motor Vehicles, and watches as the person fills out personal information on a form. Consumers can help protect themselves by regularly checking their credit reports with major credit bureaus, following up with creditors if their bills do not arrive on time, not revealing any personal information in response to unsolicited e-mail or phone calls (especially social security numbers and credit card account numbers), and shredding bills and other documents that contain sensitive information.[23]

For its part, Congress passed the Identity Theft and Assumption Deterrence Act of 1998 to fight identity theft. Under this act, the Federal Trade Commission (FTC) is assigned responsibility to help victims restore their credit and erase the impact of the imposter. It also makes identity theft a federal felony punishable by a prison term ranging from 3 to 25 years.[24]

THE COMPUTER AS THE OBJECT OF CRIME

A computer can also be the object of the crime, rather than the tool for committing it. Tens of millions of dollars of computer time and resources are stolen every year. Each time system access is illegally obtained, data or computer equipment is stolen or destroyed, or software is illegally copied, the computer becomes the object of crime. These crimes fall into several categories: illegal access and use, data alteration and destruction, information and equipment theft, software and Internet piracy, computer-related scams, and international computer crime.

Illegal Access and Use

Crimes involving illegal system access and use of computer services are a concern to both government and business. Since the outset of information technology, computers have been plagued by criminal hackers. A **hacker** is a person who enjoys computer technology and spends time learning and using computer systems. In many cases, criminal hackers are people who are looking for fun and excitement—the challenge of beating the system. A **criminal hacker**, also called a **cracker**, is a computer-savvy person who attempts to gain unauthorized or illegal access to computer systems to steal passwords, corrupt files and programs, or even transfer money. In the summer of 2002, a hacker known as RaFa broke into NASA computers and stole detailed design information about the space shuttle program and the ground control system for the next generation of space shuttle. This information was then posted on a Web site for all to see. Defense and intelligence experts said that the incident could have both national security and political ramifications.[25] **Script bunnies** are wannabe crackers with little technical savvy—crackers who download programs, or scripts, that automate the job of breaking into computers. **Insiders** are employees, disgruntled or otherwise, working solo or in concert with outsiders to compromise corporate systems.

Catching and convicting criminal hackers remains a difficult task. The method behind these crimes is often hard to determine. Even if the method behind the crime is known, tracking down the criminals can take a lot of time. It took years for the FBI to arrest one criminal hacker for the alleged "theft" of almost 20,000 credit card numbers that had been sent over the Internet. Table 14.5 provides some guidelines to follow in the event of a computer security incident.

Data Alteration and Destruction

Data and information are valuable corporate assets. The intentional use of illegal and destructive programs to alter or destroy data is as much a crime as destroying tangible goods. Most common of these types of programs are viruses and worms, which are software programs that, when loaded into a computer system, will destroy, interrupt, or cause errors in processing. There are over 60,000 known computer viruses today, with over 5,000 new viruses and worms being discovered each year. A **virus** is a program that attaches itself to other programs. An infected user must take some sort of action to spread a virus to others. A **worm** functions as an independent program, replicating its own program files until it interrupts the operation of networks and computer systems. A worm has the ability to self-propagate from an infected user's computers to other computers, for example, by e-mailing itself to everyone in a user's address book. In some cases, a virus or a worm can completely halt the operation of a computer system or network for days or longer until the problem is found and repaired. In other cases, a virus or a worm can destroy important data and programs. If backups are

---

hacker
a person who enjoys computer technology and spends time learning and using computer systems

criminal hacker (cracker)
a computer-savvy person who attempts to gain unauthorized or illegal access to computer systems

script bunnies
wannabe crackers with little technical savvy who download programs—scripts—that automate the job of breaking into computers

insiders
employees, disgruntled or otherwise, working solo or in concert with outsiders to compromise corporate systems

virus
a program that attaches itself to other programs

worm
an independent program that replicates its own program files until it interrupts the operation of networks and computer systems

TABLE 14.5

How to Respond to a Security Incident

- Follow your site's policies and procedures for a computer security incident. (They are documented, aren't they?)

- Contact the incident response group responsible for your site as soon as possible.

- Inform others, following the appropriate chain of command.

- Further communications about the incident should be guarded to ensure intruders do not intercept information.

- Document all follow-up actions (phone calls made, files modified, system jobs that were stopped, etc.).

- Make backups of damaged or altered files.

- Designate one person to secure potential evidence.

- Make copies of possible intruder files (malicious code, log files, etc.) and store them off-line.

- Evidence, such as tape backups and printouts, should be secured in a locked cabinet, with access limited to one person.

- Get the National Computer Emergency Response Team involved if necessary.

- If you are unsure of what actions to take, seek additional help and guidance before removing files or halting system processes.

inadequate, the data and programs may never be fully functional again. Table 14.6 shows the worldwide economic impact of six infamous computer incidents. The costs include the value of the effort required to identify and neutralize the virus or worm, and to restore computer files and data, as well as the value of business lost because of unscheduled computer downtime.

In 2000, the infamous ILOVEYOU worm automatically mailed itself to all contacts in a user's Microsoft Outlook address book after the person opened an infected e-mail attachment. In the two days it took to develop and distribute a patch to overcome it, the worm had copied and distributed itself to some 10 million computers worldwide, including those at the British Parliament, the United States Senate, and the Pentagon. The source of the bug was an embittered student in the Philippines; the program began its life as a project that was rejected by a teacher.

The Code Red worm targeted computers running Windows 2000 or Windows NT 4.0 operating systems, along with Microsoft Corp.'s widely used Web server software, Internet Information Server (IIS). As a result, not as many

TABLE 14.6

The Six Computer Incidents with the Greatest Worldwide Economic Impact

Source: *Press Releases*, "Malicious Code Attacks Had $13.2 Billion Economic Impact in 2001," January 4, 2002, Computer Economics Web site, http://www.computereconomics.com.

Year	Code name	Worldwide Economic Impact
2001	Nimda	$.635 billion
2001	Code Red	$2.62 billion
2001	SirCam	$1.15 billion
2000	ILoveYou	$8.75 billion
1999	Melissa	$1.10 billion
1999	Explorer	$1.02 billion

computers were affected as with the ILOVEYOU worm. Code Red exploited a known software problem in Versions 4.0 and 5.0 of IIS. Although a software patch existed and could correct this problem, many companies had failed to install the patch on their IIS servers. As a result, Code Red claimed many victims in various locales worldwide—including Microsoft itself. In some countries, employees of consulting company Cap Gemini Ernst & Young LLP lost access to their corporate intranet for several hours because of the worm. Also, some 30,000 digital subscriber line users in Taiwan were attacked by the worm, slowing down their Internet connections.[26]

Some viruses and worms attack personal computers, while others attack network and client/server systems. A personal computer can get a virus from an infected disk, an application, or e-mail attachment received from the Internet. A virus or worm that attacks a network or client/server system is usually more severe because it can affect hundreds or thousands of personal computers and other devices attached to the network. Workplace computer virus infections are increasing rapidly because of the increased spread of viruses in e-mail attachments. The primary way to avoid viruses and worms is to install antivirus software on all systems, update it routinely, and abstain from using disks or files from unknown or unreliable sources. You should also avoid opening files even from people you know unless you are expecting them. Many worms are sent as e-mails to people in the initial victim's address book so that it appears as a file received from someone you know.

The two most common kinds of viruses are application viruses and system viruses. **Application viruses** infect executable application files such as word processing programs. When the application is executed, the virus infects the computer system. Because these types of viruses normally attach themselves to application files, they can often be detected by checking the length or size of the file. If the file is larger than it should be, a virus may be attached. A **system virus** typically infects operating system programs or other system files. These types of viruses usually infect the system as soon as the computer is started.

Another type of program that can destroy a system is a **logic bomb**, an application or system virus designed to "explode" or execute at a specified time and date. Logic bombs are often disguised as a **Trojan horse**, a program that appears to be useful but actually masks the destructive program. Some of these programs execute randomly; others are designed to remain inert in software until a certain cue is given. When it detects the cue, the bomb will explode months, or even years, after being "planted."

A **macro virus** is a virus that uses an application's own macro programming language to distribute itself. Unlike the viruses discussed so far, macro viruses do not infect programs; they infect documents. The document could be a letter created using a word processing application, a graphics file developed for a presentation, or a database file. Worm/Klez.E, for example, is a macro virus that attaches itself to e-mail messages and shuts down any antivirus program running in your computer so it can continue to function. Macro viruses that are hidden in a document file can be difficult to detect. As with other viruses, however, virus detection and correction programs can be used to find and remove macro viruses.

Most macro viruses are written for Microsoft's Word for Windows and Excel for Windows. However, there are also macro viruses for Lotus AmiPro (APM/Greenstripe). If you count every single-bit difference as a virus variant, the total number is well above 2,000 and growing at the rate of a handful of new macro viruses every day.

Hoax, or false, viruses are another problem. Criminal hackers sometimes warn the public of a new and devastating virus that doesn't exist. Companies can spend hundreds of hours warning employees and taking preventive action against

application virus
a virus that infects executable application files such as word processing programs

system virus
a virus that typically infects operating system programs or other system files

logic bomb
an application or system virus designed to "explode" or execute at a specified time and date

Trojan horse
a program that appears to be useful but actually masks a destructive program

macro virus
a virus that infects documents by using an application's own macro programming language to distribute itself

a nonexistent virus. Security specialists recommend that IS personnel establish a formal paranoia policy to thwart virus panic among gullible end users. Such policies should stress that before users forward an e-mail alert to colleagues and higher-ups, they should send it to the help desk or the security team. The corporate intranet can be used to explain the difference between real viruses and fakes, and it can provide links to Web sites to set the record straight. Table 14.7 lists the seven most active viruses during the month of July 2002, according to Central Command, a provider of antivirus software.

Information and Equipment Theft

Data and information represent assets or goods that can also be stolen. Individuals who illegally access systems often do so to steal data and information. To obtain illegal access, criminal hackers require identification numbers and passwords. Some criminals try different identification numbers and passwords until they find ones that work. Using password sniffers is another approach. A **password sniffer**

password sniffer
a small program hidden in a network or a computer system that records identification numbers and passwords

TABLE 14.7

Top Viruses—July 2002

(Source: Data from Central Command Web site, http://www.centralcommand. com, accessed August 25, 2002.)

Rank	Virus	Partial Description	Percentage of Virus Occurrences Confirmed
1	Worm/Klez.E	If the system date is an odd-numbered month (January, March, etc.) and the day is the 13th, the virus starts scanning local disks (or drives on the network) and fills the files it finds with random data, permanently destroying them.	57.3%
2	W32Elkem.C	The virus monitors all running applications, and if there are any applications belonging to an antivirus program, it closes them.	16.8%
3	Worm/W32.SirCam	Displays a screensaver with a multicolor message that shakes the screen after it is complete. The display messages are: • True Love never Ends • U r My Best Friend • U r so cute today #!#!	4.4%
4	W32/Yaha.E	Arrives as an e-mail with an attachment that begins with one of the following names: loveletter, resume, love, weeklyreport, goldfish, report, mountan, biodata, dailyreport, lovegreetings, or shakingfriendship.	4.2%
5	W32/Nimda	Arrives through e-mail as an attached file with the name README.EXE. The body of the message appears empty but actually contains code to execute the virus when the user views the message.	2.6%
6	Worm/Frethem.L	Arrives as an e-mail attachment that when the attachment is opened, collects e-mail addresses from the Windows Address Book and files with .DBX, .MBX, .EML, .WAB, and .MDB extensions. It then sends infected messages.	2.2%
7	W32.Magistar.B	Checks for existence of the ZoneAlarm firewall software and if it exists, terminates it.	2.0%
8	Others		10.5%

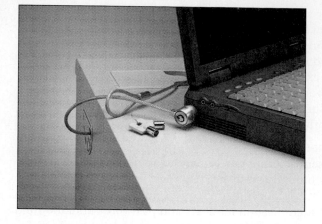

To fight computer crime, many companies use devices that disable the disk drive and lock the computer to the desk.

(Source: Courtesy of Kensington Technology Group.)

is a small program hidden in a network or a computer system that records identification numbers and passwords. In a few days, a password sniffer can record hundreds or thousands of identification numbers and passwords. Using a password sniffer, a criminal hacker can gain access to computers and networks to steal data and information, invade privacy, plant viruses, and disrupt computer operations. In April 2001, a 15-year-old youth was charged with hacking into a secure connection between the Air Mobility Command system at Scott Air Force Base in Belleville, Illinois, and a U.S. Department of Transportation computer system at the John A. Volpe National Transportation Systems Center in Cambridge, Massachusetts. The intruder was able to gain access to the Volpe Center's system that tracks the positions of U.S. Air Force planes worldwide and used a "sniffer" program to secretly intercept all wire communications. The hacker also ran a program to destroy all the electronic data files that recorded his presence on the system.[27]

In addition to theft of data and software, all types of computer systems and equipment have been stolen from offices. Computer theft is now second only to automobile theft, according to recent U.S. crime statistics. In the United Kingdom more than 30 percent of all reported thefts are computer related. Printers, desktop computers, and scanners are often targets. Portable computers such as laptops (and the data and information stored in them) are especially easy for thieves to take. In some cases, the data and information stored in these systems are more valuable than the equipment. Without adequate protection and security measures, equipment can easily be stolen.

Software and Internet Piracy

Each time you use a word processing program or access software on a network, you are taking advantage of someone else's intellectual property. Like books and movies—other intellectual properties—software is protected by copyright laws. Often, people who would never think of plagiarizing another author's written work have no qualms about using and copying software programs they have not paid for. Such illegal duplicators are called *pirates*; the act of illegally duplicating software is called **software piracy**. Technically, software purchasers are granted the right only to use the software under certain conditions; they don't really own the software. Read the "IS Principles in Action" special feature to learn more about how software manufacturers are combating software piracy.

software piracy
the act of illegally duplicating software

Internet piracy
illegally gaining access to and using the Internet

Internet piracy involves illegally gaining access to and using the Internet. Although not yet as prevalent as software piracy, the amount of Internet piracy is growing rapidly. Many companies on the Internet receive fees from customers for their information, services, and even products. Some investment firms, for example, offer market analysis and investment information for a monthly or annual fee. Other companies offer information on sports or provide research information on a variety of topics for a fee. For some services, the fees can be thousands of dollars annually. Typically, Internet companies give customers identification numbers or passwords. Like customers illegally copying software, some customers illegally share their identification numbers and passwords with others. In other cases, criminal hackers obtain these numbers illegally on the Internet or from other sources. When unauthorized people use these services, Internet firms lose valuable revenues. Internet piracy can also be directed against individuals. While users are surfing the Web, outsiders can download an applet to the browser's machine, use its processor to perform computations, and send the results back to a host. This technique is called *MIPs-sucking*.

Computer crime is a serious and rapidly growing area of concern requiring management attention.

Building Alliances to Combat Software Piracy

The U.S. Copyright Office began to register software as a form of literary expression in 1964. According to Title 17 of the United States Code, it is illegal to make or distribute copies of copyrighted material without authorization, except to make a single backup copy. Penalties for software violations include fines of up to $150,000 per copyright work infringed. In spite of similar copyright protection in many other countries, nearly one in four copies of commercial software worldwide is illegal. Software piracy costs the software industry almost $11 billion a year. It also costs local, state, and federal governments hundreds of millions of dollars in lost tax revenue. Unsuspecting customers can also suffer losses by getting defective pirated products that are ineligible for customer support.

The Business Software Alliance (BSA) is a watchdog group established in 1988, which represents many leading U.S. software manufacturers including Adobe, Apple, Autodesk, Bentley Systems, Borland, CNC Software/Mastercam, Macromedia, Microsoft, Symantec, and Unigraphic Solutions. Additional members of BSA's Policy Council include Dell, Entrust, Hewlett-Packard, IBM, Intel, Intuit, Network Associates, Novell, and Sybase. With operations in 65 countries, BSA's goals are to educate computer users on software copyrights and cybersecurity, advocate public policy that fosters innovation and expands trade opportunities, and fight software piracy.

Since 1990, BSA has collected more than $78 million in penalties from companies using unlicensed software. In addition to making payments to avoid legal action, the companies typically agree to delete any unlicensed copies, purchase any needed replacement software, and strengthen their software management practices.

From time to time, the Business Software Alliance will grant a 30-day grace period for businesses in specific areas of the country to rid their systems of illegally installed software. For example, in January 2002, BSA offered the grace period to about 800,000 businesses in Billings (Montana), Houston, Indianapolis, Nashville, Norfolk/Richmond, Oakland, Orlando, and San Francisco. During this period, companies that become compliant with software copyright laws are not penalized for violations. BSA sends postcards explaining the program to the affected companies and later sends out a second "reminder" mailing. It also runs radio spot ads in the targeted markets. Businesses trying to determine whether their organization is using unlicensed software can download the free BSA Software Audit tool at http://www.bsagrace.com or call the special Grace Period hotline at 1-877-536-4BSA for additional information.

BSA averages 6 to 12 calls per day reporting the use of illegal copies of software, mostly from disgruntled former employees of the offending company. Most companies are not intentionally trying to cheat vendors by avoiding software licensing; they just have sloppy procurement practices and poor management of IT assets. Every once in while, however, BSA finds companies that are severely underlicensed—a clear indication the company has decided to operate illegally or it has a reckless disregard for how its software is being managed.

Discussion Questions

1. Imagine that you work for BSA and answer calls from individuals wishing to report a company that is using pirated software. What questions would you need to ask a caller to evaluate the seriousness of the situation and to assess whether the report is legitimate?

2. What actions would you expect an offending company to take to strengthen its software purchasing and management practices after being fined by the BSA for use of pirated software?

Critical Thinking Questions

3. Would you report your company to the BSA if it were guilty of having many unlicensed copies of software? What if a reward was offered? What if you had just been fired?

4. Describe two real-world scenarios that you think frequently lead to an individual installing a pirated software license.

Sources: Adapted from "About Us," Business Software Alliance Web site, http://www.bsa.org, accessed September 1, 2002; George V. Hulme, "Four Arrested in Software Piracy Raids," *InformationWeek*, December 12, 2001, http://www.informationweek.com; George V. Hulme, "Software Piracy Ring Cornered," *InformationWeek*, December 17, 2001, http://www.information week.com; Linda Rosencrance, "BSA Offers One-Month Grace Period to Software Pirates," *Computerworld*, January 2, 2002, http://www.computer world.com.

The vast majority of wireless networks in operation today are based on the 802.11b (Wi-Fi) standard. It is relatively easy for someone to eavesdrop on an organization's wireless network from outside its actual premises without establishing a physical network connection if the following safeguards are not taken:

- The Service Set Identifier (SSID) is an ID of up to 32 characters that is continuously transmitted by default by a Wi-Fi LAN access point. Wi-Fi LAN security begins by turning off the default transmission of a LAN access point SSID.

- Each piece of hardware on a Wi-Fi network has a unique Media Access Control (MAC) address. It is imperative that these addresses be filtered to make it more difficult for a hacker to penetrate the network.
- Wired Equivalent Privacy (WEP) is a security mechanism that enables encryption of wireless traffic. However, this encryption is turned off by default in wireless devices and software. WEP needs to be enabled, and organizations must avoid using a single key that's shared by all mobile stations and access points.[28]

Some so-called "hobbyists" conducted the Worldwide Wardrive the week starting August 31, 2002. Their goal was to detect wireless LANs using NetStumbler freeware available on the Web. Results of the exercise were posted to the SecurityTribe Web site (http://www.securitytribe.com) and included latitude and longitude data of the wireless LAN access points they detected. Unfortunately, the results show that many wireless LAN users fail to use even the most elementary form of security to protect their systems and that these networks are wide open to attack by cyberterrorists, hackers, and industrial spies.[29]

Computer-Related Scams

People have lost hundreds of thousands of dollars on real estate, travel, stock, and other business scams. Today, many of these same types of scams are being performed using computers. Using the Internet, scam artists offer get-rich-quick schemes involving real estate deals, "free" vacations, bank fraud, telephone lotteries, penny stocks, and tax avoidance. In one Internet scam, a supposed member of a Nigerian government committee asks the receivers of his e-mail to "participate in a strictly confidential business proposal" to help transfer $21.5 million of illegally gotten money out of the country promising a commission of 10% for their effort.

Here are some general tips to help you avoid becoming a victim:

- Don't agree to anything in a high-pressure meeting or seminar. Insist on having time to think it over and to discuss things with a spouse, partner, or even your lawyer. If a company won't give you the time you need to check it out and think things over, you don't want to do business with it. A good deal now will be a good deal tomorrow; the only reason for rushing you is if the company has something to hide.
- Don't judge a company based on appearances. Flashy Web sites can be created and put up on the Net in a matter of days. After a few weeks of taking money, a site can vanish without a trace in just a few minutes. You may find that the perfect money-making opportunity offered on a Web site was a money maker for the crook and a money loser for you.
- Avoid any plan that pays commissions simply for recruiting additional distributors. Your primary source of income should be your own product sales. If the earnings are not made primarily by sales of goods or services to consumers or sales by distributors under you, you may be dealing with an illegal pyramid.
- Beware of shills, people paid by a company to lie about how much they've earned and how easy the plan was to operate. Check with an independent source to make sure that you aren't having the wool pulled over your eyes.
- Beware of a company's claim that it can set you up in a profitable home-based business but that you must first pay up front to attend a seminar and buy expensive materials. Frequently, seminars are high-pressure sales pitches, and the material is so general that it is worthless.
- If you are interested in starting a home-based business, get a complete description of the work involved before you send any money. You may find that what you are asked to do after you pay is far different from what was stated in the ad. You should never have to pay for a job description or for needed materials.
- Get in writing the refund, buy-back, and cancellation policies of any company you deal with. Do not depend on oral promises.

- Do your homework. Check with your state attorney general and the National Fraud Information Center before getting involved, especially when the claims about a product or potential earnings seem too good to be true.

If you need advice about an Internet or on-line solicitation, or you want to report a possible scam, use the Online Reporting Form or Online Question & Suggestion Form features on the Web site for the National Fraud Information Center at http://fraud.org, or call the NFIC hotline at 1-800-876-7060.

International Computer Crime

Computer crime is also an international issue, and it becomes more complex when it crosses borders. Estimates of software piracy in the global marketplace indicate that over one-third of software is pirated, adding up to more than $11 billion in lost revenue worldwide. In Vietnam and China over 90% of software is pirated, with lost revenue totaling well over $1 billion.[30]

With cash and funds being transferred electronically, some are concerned that terrorists, international drug dealers, and other criminals are using information systems to launder illegally obtained funds. The Bank Secrecy Act and Anti-Money Laundering Compliance System was developed by the United States Postal Service (USPS) and Information Builders Inc. to detect patterns that indicate potential money-laundering activity. The system can identify suspicious money orders at the point of sale, track those orders through the banking system, and identify accounts through which they've passed, even after they've been deposited. The system provides drill-down, querying, and reporting functions to enable law enforcement officials to identify terrorists, corrupt investment bankers, drug dealers, and others who are engaged in money-laundering schemes.[31]

PREVENTING COMPUTER-RELATED CRIME

Because of increased computer use today, greater emphasis is placed on the prevention and detection of computer crime. Although more than 45 states have passed computer crime bills, some believe that they are not effective because companies do not always actively detect and pursue computer crime, security is inadequate, and convicted criminals are not severely punished. However, all over the United States, private users, companies, employees, and public officials are making individual and group efforts to curb computer crime, and recent efforts have met with some success.

Crime Prevention by State and Federal Agencies

State and federal agencies have begun aggressive attacks on computer criminals, including criminal hackers of all ages. In 1986, Congress enacted the Computer Fraud and Abuse Act, which mandates punishment based on the victim's dollar loss. The Department of Defense also supports the Computer Emergency Response Team (CERT), which responds to network security breaches and monitors systems for emerging threats. Law enforcement agencies are also increasing their efforts to stop criminal hackers, and many states are now passing new, comprehensive computer crime bills to help eliminate computer crimes. Recent court cases and police reports involving computer crime show that lawmakers are ready to introduce newer and tougher computer crime legislation. Several states have passed laws in an attempt to outlaw spamming, the practice of sending large amounts of unsolicited e-mail to overwhelm users' e-mail boxes or the e-mail servers on a network. In one notable case, a group of California activists filed a mind-boggling $2.2 trillion set of lawsuits against Fax.com, a facsimile marketer, claiming that "millions of 'junk faxes' are clogging the nation's fax machines, jamming communications, and possibly endangering lives."[32]

Crime Prevention by Corporations

Companies are also taking crime-fighting efforts seriously. Many businesses have designed procedures and specialized hardware and software to protect their corporate data and systems. Specialized hardware and software, such as encryption devices, can be used to encode data and information to help prevent unauthorized use. As discussed in Chapter 7, encryption is the process of converting an original electronic message into a form that can be understood only by the intended recipients. A key is a variable value that is applied using an algorithm to a string or block of unencrypted text to produce encrypted text, or to decrypt encrypted text. Encryption methods rely for their security on the limitations of computing power—if breaking a code requires too much computing power, even the most determined code crackers will not be successful. The length of the key used to encode and decode messages determines the strength of the encryption algorithm.

Public Key Infrastructure (PKI)

A means to enable users of an unsecure public network such as the Internet to securely and privately exchange data through the use of a public and a private cryptographic key pair that is obtained and shared through a trusted authority

Public Key Infrastructure (PKI) enables users of an unsecure public network such as the Internet to securely and privately exchange data through the use of a public and a private cryptographic key pair that is obtained and shared through a trusted authority. PKI is the most common method on the Internet for authenticating a message sender or encrypting a message. PKI uses two keys to encode and decode messages. One key of the pair, the message receiver's public key, is readily available to the public and is used by anyone to send that individual encrypted messages. The second key, the message receiver's private key, is kept secret and is known only by the message receiver. Its owner uses the private key to *decrypt* messages—convert encoded messages back into the original message. Knowing an individual's public key does not enable you to decrypt an encoded message to that individual.

biometrics

the measurement of a living trait, whether physical or behavioral

Using biometrics is another a way to protect important data and information systems. **Biometrics** involves the measurement of a living trait, whether physical or behavioral. Biometric techniques compare a person's unique characteristics against a stored set for the purpose of detecting differences between them. Biometric systems can scan fingerprints, faces, handprints, and retinal images to prevent unauthorized access to important data and computer resources. Most of the interest among corporate users is in fingerprint technology, followed by face recognition. Fingerprints hit the middle ground between price and usability. Iris and retina scans are more accurate, but they are more expensive, and they involve more equipment.

Many of the early adopters of biometric technologies have been health care organizations, because new federal legislation requires such organizations to protect the privacy of patients. Instead of typing in a password, a doctor or hospital employee puts his or her finger on a scanner connected to a personal computer and the image is matched against a set of authorized fingerprints. With some systems, instead of an image of a fingerprint, software uses an algorithm to create a personal identification number that equates to a person's fingerprint. This saves data storage and avoids having actual fingerprints stored on file.

This fingerprint authentication device provides security in the PC environment by using fingerprint information instead of passwords.

(Source: Courtesy of Bioscrypt, Inc.)

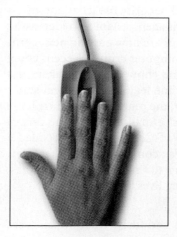

As employees move from one position to another at the same company, they tend to build up access to multiple systems because security procedures fail to revoke access privileges. Clearly it is not appropriate for people who have changed positions and responsibilities to still have access to systems they no longer use. To avoid this problem, many organizations are creating role-based system access lists so that a person filling a particular job function has access to only specific systems. Huntington Bancshares in Columbus, Ohio, which has 10,000 employees and $26 billion in assets, has implemented such role-based access lists. Huntington is also evaluating other forms of access identification including card-stripe readers, smart cards, and credit-card-type tokens.[33]

Crime-fighting procedures usually require additional controls on the information system. Before designing and implementing controls, organizations must consider the types of computer-related crime that might occur, the consequences of these crimes, and the cost and complexity of needed controls. In most cases, organizations conclude that the trade-off between crime and the additional cost and complexity weighs in favor of better system controls. Having knowledge of some of the methods used to commit crime is also helpful in preventing, detecting, and developing systems resistant to computer crime (see Table 14.8). Some companies actually hire former criminals to thwart other criminals. Table 14.9 provides a set of useful guidelines to protect your computer from hackers.

Companies are also joining together to fight crime. The Software Publishers Association (SPA), which was formed by a number of leading software companies, audits companies and checks for software licenses. Organizations that are found to be illegally using software can be fined or sued. Depending on the violation, the fine can be hundreds of thousands of dollars. While the SPA is an effective deterrent in the United States, efforts to curb software abuse in other countries is much more difficult.

Using Antivirus Programs

antivirus programs
programs or utilities that prevent viruses and recover from them if they infect a computer

As a result of the increasing threat of viruses and worms, most computer users and organizations have installed **antivirus programs** on their computers. Such software runs in the background to protect your computer from dangers lurking on the Internet and other possible sources of infected files. Some antivirus software is even capable of repairing common virus infections automatically, without interrupting your work. The latest virus definitions are downloaded automatically when you connect to the Internet, ensuring that your PC's protection is current. To safeguard your PC and prevent it from spreading viruses to your friends and co-workers, some antivirus software scans and cleans both incoming and outgoing e-mail messages. Table 14.10 lists some of the most popular antivirus software.

TABLE 14.8

Common Methods Used to Commit Computer Crimes

Even though the number of potential computer crimes appears to be limitless, the actual methods used to commit crimes are limited.

Methods	Examples
Add, delete, or change inputs to the computer system.	Delete records of absences from class in a student's school records.
Modify or develop computer programs that commit the crime.	Change a bank's program for calculating interest to make it deposit rounded amounts in the criminal's account.
Alter or modify the data files used by the computer system.	Change a student's grade from C to A.
Operate the computer system in such a way as to commit computer crime.	Access a restricted government computer system.
Divert or misuse valid output from the computer system.	Steal discarded printouts of customer records from a company trash bin.
Steal computer resources, including hardware, software, and time on computer equipment.	Make illegal copies of a software program without paying for its use.
Offer worthless products for sale over the Internet.	Send e-mail requesting money for worthless hair growth product.
Blackmail executives to prevent release of harmful information.	Eavesdrop on organization's wireless network to capture competitive data or scandalous information.
Blackmail company to prevent loss of computer-based information.	Plant logic bomb and send letter threatening to set it off unless paid considerable sum.

T A B L E   14 ⊙ 9

*How to Protect Your Corporate
Data From Hackers*

• Install strong user authentication and encryption capabilities on your firewall.
• Install the latest security patches, which are often available at the vendor's Internet site.
• Disable guest accounts and null user accounts that let intruders access the network without a password.
• Do not provide overfriendly log-in procedures for remote users (e.g., an organization that used the word *welcome* on their initial log-on screen found they had difficulty prosecuting a hacker).
• Give an application (e-mail, file transfer protocol, and domain name server) its own dedicated server.
• Restrict physical access to the server and configure it so that breaking into one server won't compromise the whole network.
• Turn audit trails on.
• Consider installing caller ID.
• Install a corporate firewall between your corporate network and the Internet.
• Install antivirus software on all computers and regularly download vendor updates.
• Conduct regular IS security audits.
• Verify and exercise frequent data backups for critical data.

Proper use of antivirus software requires the following steps:

1. *Install a virus scanner and run it often.* Many of these programs automatically check for viruses each time you boot up your computer or insert a diskette or CD, and some even monitor all transmissions and copying operations.
2. *Update the virus scanner often.* Old programs may fail to detect new viruses.
3. *Scan all diskettes and CDs before copying or running programs from them.* Hiding on diskettes or CDs, viruses often move between systems. If you carry document or program files on diskettes or CDs between computers at school or work and your home system, always scan them.

T A B L E   14 ⊙ 10

Antivirus Software

Antivirus Software	Software Manufacturer	Approximate Cost for Single License (Fall 2002)	Web Site
Symantec Norton AntiVirus 2002	Symantec	$40	http://www.symantec.com
McAfee VirusScan	McAfee	$50	http://www.mcafee.com
Panda Antivirus Platinum	Panda Software	$30	http://www.computervirusprotection.com
Vexira Antivirus	Central Command	$50	http://www.centralcommand.com
Sophos Antivirus	Sophos	$45	http://www.sophos.com
PC-cillen	Trend Micro	$40	http://www.trendmicro.com

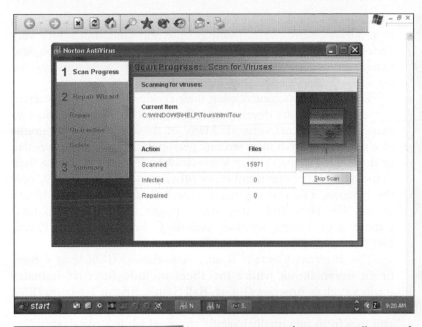

Antivirus software should be used and updated often.

intrusion detection system (IDS)
Security system that monitors system and network resources and notifies network security personnel when it senses a possible intrusion

managed security service provider (MSSP)
An organization that monitors and maintains network security hardware and software for its client companies

4. *Install software only from a sealed package or secure Web site of a known software company.* Even software publishers can unknowingly distribute viruses on their program disks or software downloads. Most scan their own systems, but viruses may still remain.

5. *Follow careful downloading practices.* If you download software from the Internet or a bulletin board, check your computer for viruses immediately after completing the transmission.

6. *If you detect a virus, take immediate action.* Early detection often allows you to remove a virus before it does any serious damage.

Despite careful precautions, viruses can still cause problems. They can elude virus-scanning software by lurking almost anywhere in a system. Future antivirus programs may incorporate "nature-based models" that check for unusual or unfamiliar computer code. The advantage of this type of virus program is the ability to detect new viruses that are not part of an antivirus database.

Using Intrusion Detection Software

An **intrusion detection system (IDS)** monitors system and network resources and notifies network security personnel when it senses a possible intrusion. Examples of suspicious activities include repeated failed log-in attempts, attempts to download a program to a server, and access to a system at unusual hours. Such activities generate alarms that are captured on log files and may send e-mails to security personnel. Use of an IDS provides another layer of protection in the event that an intruder gets past the outer security layers—passwords, security procedures, and the corporate firewall. Mount Sinai School of Medicine in New York implemented the OpenSnort Sensor IDS appliance to protect sensitive medical information from hackers. The IDS has identified a surprisingly large number of probes and viruses that have tried and failed to get into the school's systems.[34]

Use of Managed Security Service Providers (MSSPs)

For most small and mid-size organizations, the cost of acquiring and maintaining in-house network security expertise to protect their business operations has forced them to consider other options. As a result, many are outsourcing their network security operations to **managed security service providers (MSSPs)** such as Counterpane, Guardent, Internet Security Services, Riptech, and Symantec. MSSPs monitor and maintain network security hardware and software. They provide a valuable service for IS departments drowning in reams of alerts and false alarms coming from virtual private networks (VPNs) and antivirus, firewall, intrusion detection systems, and other security monitoring systems. In addition, some provide vulnerability scanning and Web blocking/filtering capabilities. Denver-based Newmont Mining, a gold-mining company, contracted with MSSP Guardent to help manage its firewalls and provide intrusion-detection system (IDS) services at its seven operations sites around the globe including South America, Indonesia, and Asia. While Newmont Mining initially considered building its own security operations center, cost considerations and a dearth of in-house security skills made outsourcing the way to go.[35]

Internet Laws for Libel and Protection of Decency

The Telecommunications Act of 1996 included an act called the Communications Decency Act. One of the original provisions of this act was the government's ability

Net Nanny is a filtering software program that helps block unwanted Internet content from children and young adults.

(Source: Courtesy of Net Nanny Software International.)

ICRA rating system
system to protect individuals from harmful or objectionable Internet content while safeguarding the free speech rights of others

to jail or fine anyone up to $100,000 for sending indecent materials to a minor electronically. Many people and companies were very concerned about the free speech implications of this provision, and many turned their Web pages to black in protest. Subsequently, the Supreme Court ruled the law unconstitutional.

To help parents control what their children see on the Internet, some companies are developing software called *filtering software* to help screen Internet content. Many of these screening programs also prevent children from sending personal information over e-mail or through chat groups. This stops children from broadcasting their name, address, phone number, or other personal information over the Internet. The two approaches used are filtering, which blocks certain Web sites, and rating, which places a rating on Web sites. Examples of filtering software include Cybersitter, Cyber Patrol, Net Nanny, SurfGuard, and SurfWatch.

The Internet Content Rating Association (ICRA) is a non-profit organization whose members include Internet industry leaders such as America Online, Bell South, British Telecom, IBM, Microsoft, UUNet, and Verizon. Its specific goals are to protect children from potentially harmful material while also safeguarding free speech on the Internet. Using the **ICRA rating system,** Web authors fill out an on-line questionnaire describing the content of their site—what is and isn't present. The broad topics covered are the following: chat capabilities, the language used on the site, the nudity and sexual content of a site, the violence depicted on the site, and other topics such as alcohol, drugs, gambling, and suicide. Based on the authors' responses, ICRA then generates a content label (a short piece of computer code) that the authors add to their site. Internet users (and parents) can then set their browser to allow or disallow access to Web sites based on the objective rating information declared in the content label and their own subjective preferences. Reliance on Web site authors to do their own rating has its weaknesses, though. Web site authors can lie when completing the ICRA questionnaire so that their site receives a content label that doesn't accurately reflect the site's content. In addition, many hate group and sexually explicit sites don't have an ICRA rating, so they would not be blocked unless a browser is set to block all unrated sites. Yet this option would block out so many acceptable sites that it could make Web surfing useless. For these reasons, at this time, site labeling is at best a complement to other filtering techniques.

The Children's Internet Protection Act (CIPA) is a federal law passed in December 2000 that required federally funded schools and libraries to use some form of technology prevention measure (such as Internet filters) to block access to obscene material or other material considered harmful to minors. Opponents of the law feared that it transferred power over the education process and which sites were to be blocked to private software companies that develop the Internet filters. In May 2002, a three-judge panel declared the act unconstitutional because the blocking programs cannot effectively screen out only material deemed harmful to minors. The current state of the art is that such filters block out access to legitimate sites while allowing users to access objectionable sites. The erroneous filtering caused by the law resulted in violation of the First Amendment.[36]

With the increased popularity of networks and the Internet, libel and decency become important legal issues. A publisher, such as a newspaper, can be sued for libel, which involves publishing a written statement that is damaging to a person's reputation. Generally, a bookstore cannot be held liable for statements made in a newspaper or other publications it sells. On-line services, such as CompuServe and America Online, may exercise some control over who puts information on their service but may not have direct control over the content of what is published by others on their service. Can on-line services be sued for libel for content that someone else publishes on their service? Are on-line services more like a newspaper or a bookstore? This legal issue has not been completely resolved, but some court cases have been decided. The *Cubby, Inc. v. CompuServe* case ruled that CompuServe was more like a bookstore and not liable for content put on its service by others. In this case, the judge stated, "While CompuServe may decline to carry a given publication altogether, in reality, once it does decide to carry a given publication, it will have little or no editorial control over that publication's content." This case set a legal precedent that has been applied in similar, subsequent cases.

Individuals must be careful what they post on the Internet to avoid libel charges. There have been many cases of disgruntled former employees being sued by their former employers for material posted on the Internet. For example, in December 2001, a jury ordered two former employees to pay $775,000 in damages for defaming and harassing Varian Medical Systems managers and their families. In addition, the judge issued an injunction that prohibits the former employees from continuing to post defamatory and harassing statements on the Internet. Following their terminations, the two employees posted thousands of messages on Internet message boards and their own Web site falsely accusing various members of Varian management of being homophobic, discriminating against pregnant women, having sexual affairs, and secretly videotaping employees while they were in office restrooms.[37]

Preventing Crime on the Internet

As mentioned in Chapter 7, Internet security can include firewalls and a number of methods to secure financial transactions. A firewall can include both hardware and software that act as a barrier between an organization's information system and the outside world. A number of systems have been developed to safeguard financial transactions on the Internet.

To help prevent crime on the Internet, the following steps can be taken:

1. Develop effective Internet and security policies for all employees.
2. Use a stand-alone firewall (hardware and software) with network monitoring capabilities.
3. Monitor managers and employees to make sure they are using the Internet for business purposes only.
4. Use Internet security specialists to perform audits of all Internet and network activities.

Even with these precautions, computers and networks can never be completely protected against crime. One of the biggest threats is from employees. Although firewalls provide good perimeter control to prevent crime from the outside, procedures and protection measures are needed to protect against computer crime by employees. Passwords, identification numbers, and tighter control of employees and managers also help prevent Internet-related crime.

PRIVACY

Another important social issue in information systems involves privacy. In 1890, U.S. Supreme Court Justice Louis Brandeis stated that the "right to be left alone" is one of the most "comprehensive of rights and the most valued by civilized man." Basically, the issue of privacy deals with this right to be left alone or to be withdrawn from public view. With information systems, privacy deals with the collection and use or misuse of data. Data is constantly being collected and stored on each of us. This data is often distributed over easily accessed networks and without our knowledge or consent. Concerns of privacy regarding this data must be addressed.

With today's computers, the right to privacy is an especially challenging problem. More data and information are produced and used today than ever before. A difficult question to be answered is, "Who owns this information and knowledge?" If a public or private organization spends time and resources to obtain data on you, does the organization own the data, and can it use the data in any way it desires? Government legislation answers these questions to some extent for federal agencies, but the questions remain unanswered for private organizations.

PRIVACY ISSUES

The issue of privacy is important because data on an individual can be collected, stored, and used without that person's knowledge or consent. When someone is born, takes certain high school exams, starts working, enrolls in a college course, applies for a driver's license, purchases a car, serves in the military, gets married, buys insurance, gets a library card, applies for a charge card or loan, buys a house, or merely purchases certain products, data is collected and stored somewhere in computer databases.

Privacy and the Federal Government

The federal government is perhaps the largest collector of data. Close to 4 billion records exist on individuals, collected by about 100 federal agencies, ranging from the Bureau of Alcohol, Tobacco, and Firearms to the Veterans Administration. Other data collectors include state and local governments and profit and nonprofit organizations of all types and sizes. Read the "Ethical and Societal Issues" special-interest feature to learn more about the selling of personal data to law enforcement agencies.

In recent years, a number of actions by the federal government have caused concern for the privacy of individuals' data. More than 50 laws are before Congress regarding various aspects of Internet privacy. Here are a few:

- The terrorist attacks of September 11, 2001, revived proposals for a national identity card system as a means to verify the identity of airline passengers and prevent terrorists from entering the country. This is the latest in a number of proposals advanced over the past decade for a national identity card for various uses, among them immigration policy, gun control, and health care reform. Civil libertarians argue that the creation of a national identity card offers only a false sense of security and poses serious threats to our civil liberties and civil rights.

- Carnivore is a highly controversial system used by the FBI to monitor selected e-mail messages going to or from specific individuals. Although a court order is required to monitor the specified messages, privacy and civil rights advocates argue that the use of Carnivore violates U.S. citizens' right to privacy. Furthermore, the FBI has been unable to prove that e-mails not covered by a court order are left alone.[38]

- Echelon, a global surveillance network run by the National Security Agency and allied intelligence bureaus, listens to every electronic communication in the world—cellphone calls, satellite transmissions, e-mail messages. The system operates with minimal oversight and the various agencies involved provide few details of the procedures taken to safeguard individuals' right to privacy. In fact, the governments of the countries thought to be involved will not even acknowledge the existence of Echelon. Although the system may have been operating since the early 1970s, it is only recently that journalists using Freedom of Information Act requests were able to confirm its existence.

The European Union has already passed a data-protection directive that requires firms transporting data across national boundaries to have certain privacy procedures in place. This directive affects virtually any company doing business in Europe, and it is driving much of the attention being given to privacy in the United States.

Personal Data Sold to Law Enforcement Agencies

Anybody can view public records, but the legal definition of what records are public or private depends on both state and federal law. Public records include a wealth of personal information about individuals, ranging from their birth records, arrest records, court files, driver's license information, marriage and divorce records, occupational licenses, property ownership and tax information, and death records. People must reveal much personal information when they interact with state or federal governments for voting or obtaining public benefits. Once a record becomes public, there is generally no restriction on the purpose for which the personal information can be used. The availability of this information, combined with the lack of restriction on its use, exposes individuals to potential abuses.

Private companies are building thick electronic dossiers of personal information on citizens for the government by mining public records and purchasing credit-reporting data. Also, government access to personal data has become more widespread and controversial since September 11, as antiterrorism investigative powers have been expanded. Serious data privacy questions have arisen regarding access to people's dossiers, their accuracy, and their potential for misuse.

In February 2002, the Electronic Privacy Information Center (EPIC) filed a lawsuit seeking information concerning the type of data U.S. government agencies have purchased from companies that sell personal data. The EPIC lawsuit singled out ChoicePoint and Experian Information Systems, claiming that both are selling profiles on U.S. citizens to federal law enforcement agencies with little public awareness or oversight. ChoicePoint and Experian sell the IRS credit data, property records, state motor vehicle records, marriage and divorce data, and international asset location data. IRS employees have access to this personal data from their desktop computers.

The Washington, D.C.–based privacy advocacy group filed for information under the Freedom of Information Act (FOIA) after several agencies in the U.S. Treasury and Justice Departments failed to provide EPIC with information about contracts with data-collection companies. The FOIA requests sought records relating to "transactions, communications, and contracts" between law enforcement agencies and private firms that are engaged in the sale of personal information. EPIC's ultimate goal is to regulate such companies so that citizens have the right and ability to monitor and correct their profiles in these databases.

Two existing federal laws cover individual privacy. The Fair Credit Reporting Act of 1970 regulates the operations of credit-reporting bureaus, including how they collect, store, and use credit information. It also provides processes that allow individuals to review their credit reports and request changes. The Privacy Act of 1974 limits how the U.S. government collects, maintains, uses, and disseminates personal information. Its purpose is to provide some safeguards for individuals against federal agencies' invasion of personal privacy. The Central Intelligence Agency and law enforcement agencies are excluded from this act, however. This act also excludes the actions of private industry.

ChoicePoint maintains many databases of public records, including court documents and property records that are used by some 7,500 U.S. law enforcement agencies to help in their investigations. ChoicePoint has even created a federal government Web portal at http://www.cpgov.com to facilitate IRS and law enforcement access to its 10 billion records on individuals and businesses. In addition, private businesses, such as insurance companies, use ChoicePoint databases to reduce fraud and conduct background checks. Because ChoicePoint is not a credit-reporting bureau, it is not regulated by the Fair Credit Reporting Act.

Experian is a global, $1.5 billion subsidiary of Britain's Great Universal Stores. In addition to being a major credit report provider, Experian maintains the largest database of U.S. demographic marketing data. Called INSOURCE, the database contains information about 215 million U.S. consumers, including names, addresses, phone numbers, real estate ownership, mortgage balances, motor vehicle records, legal relationships, and other information. Among the agencies purchasing Experian's data are the FBI, U.S. Drug Enforcement Agency, U.S. Marshals Service, Internal Revenue Service, U.S. Immigration and Naturalization Service, and the Bureau of Alcohol, Tobacco, and Firearms. Although Experian's business is regulated under the Fair Credit Reporting Act, citizens may not even be aware that the government is using that data. Furthermore, its interactions with the CIA and federal law enforcement agencies are explicitly excluded from coverage under the Privacy Act.

Discussion Questions

1. What is EPIC's ultimate goal in requesting information about agreements and contracts between law enforcement agencies and private firms that are engaged in the sale of personal information?
2. Visit the EPIC Web site at http://www.epic.org to learn the current status of its effort in these matters. Write a paragraph summarizing your findings.

Critical Thinking Questions

3. Do you think that the Privacy Act exemption for the CIA and law enforcement agencies is reasonable? Why or why not?
4. Use your imagination to create a scenario where the use of public data and credit information about private citizens leads to an abuse of individual privacy.

Sources: Adapted from Jennifer DiSabatino, "Privacy Group Files Suit over Profiling Data Sold to Government," *Computerworld*, January 16, 2002, accessed at http://www.computerworld.com; Jennifer DiSabatino, "Unregulated Databases Hold Personal Data," *Computerworld*, January 21, 2002, accessed at http://www.computerworld.com; "EPIC Files FOIA Suit for Profiling Records," accessed at EPIC Web site at http://www.epic.org on February 22, 2002.

Most companies and computer vendors are wary of having the federal government dictate Internet privacy standards. A group called the Online Privacy

E-mail has changed how workers and managers communicate. With e-mail, people can communicate in the same building or around the world. E-mail, however, can be monitored and intercepted. As with other services—like cellular phones—the convenience of e-mail must be balanced with the potential of privacy invasion.

(Source: © 2002 PhotoDisc.)

Alliance is developing a voluntary code of conduct. It is backed by companies such as AOL Time Warner, AT&T, Boeing, DoubleClick, eBay, Equifax, IBM, Microsoft, Verizon Communications, and Yahoo. The alliance's guidelines call on companies to notify users when they are collecting data at Web sites to gain consent for all uses of that data, to provide for the enforcement of privacy policies, and to have a clear process in place for receiving and addressing user complaints. The alliance's policy can be found at http://www.privacyalliance.org.

Privacy at Work

The right to privacy at work is also an important issue. Currently, the rights of workers who want their privacy and the interests of companies that demand to know more about their employees are in conflict. Recently, companies that have been monitoring their employees have raised concerns. For example, workers may find that they are being closely monitored via computer technology. These computer-monitoring systems tie directly into workstations; specialized computer programs can track every keystroke made by a user. This type of system can determine what workers are doing while at the keyboard. The system also knows when the worker is not using the keyboard or computer system. These systems can estimate what a person is doing and how many breaks he or she is taking. Needless to say, many workers consider this close supervision very dehumanizing.

E-Mail Privacy

E-mail also raises some interesting issues about work privacy. Federal law permits employers to monitor e-mail sent and received by employees. Furthermore, e-mail messages that have been erased from hard disks may be retrieved and used in lawsuits because the laws of discovery demand that companies produce all relevant business documents. On the other hand, the use of e-mail among public officials may violate "open meeting" laws. These laws, which apply to many local, state, and federal agencies, prevent public officials from meeting in private about matters that affect the state or local area.

Privacy and the Internet

Some people assume that there is no privacy on the Internet and that you use it at your own risk. Others believe that companies with Web sites should have strict privacy procedures and be accountable for privacy invasion. However, the courts are not clear on this issue. In May 2002, the Supreme Court left in place an injunction barring enforcement of The Children's Online Privacy Protection Act. This act was directed at Web sites catering to children, requiring them to post comprehensive privacy policies on their sites and to obtain parental consent before they collect any personal information from children under 13 years of age. Thus, at this time, within the United States, the communication and implementation of Web site privacy policies is still voluntary. Regardless of your view, the potential for privacy invasion on the Internet is huge. People wanting to invade your privacy could be anyone from criminal hackers to marketing companies to corporate bosses. Your personal and professional information can be seized on the Internet without your knowledge or consent. E-mail is a prime target, as discussed previously. Sending an e-mail message is like having an open conversation in a large room—people can listen to your messages. When you visit a Web site on the Internet, information about you and your computer can be captured. When this information is combined with other information, companies can know what you read, what products you buy, and what your interests are. According to an executive of an Internet software monitoring company, "It's a marketing person's dream."

Most people who buy products on the Web say it's very important for a site to have a policy explaining how personal information is used, and the policy statement must make people feel comfortable and be extremely clear about what information is collected and what will and will not be done with it. However, many Web sites still do not prominently display their privacy policy or implement practices completely consistent with that policy. The real issue that Internet users need to be concerned with is—what do content providers want with their personal information? If a site requests that you provide your name and address, you have every right to know why and what will be done with it. If you buy something and provide a shipping address, will it be sold to other retailers? Will your e-mail address be sold on a list of active Internet shoppers? And if so, you should realize that it's no different than the lists compiled from the orders you place with catalog retailers. You have the right to be taken off any mailing list.

A potential solution to some consumer privacy concerns is the screening technology called the **Platform for Privacy Preferences (P3P)** being proposed to shield users from sites that don't provide the level of privacy protection they desire. Instead of forcing users to find and read through the privacy policy for each site they visit, P3P software in a computer's browser will download the privacy policy from each site, scan it, and notify the user if the policy does not match his or her preferences. (Of course, unethical marketers can post a privacy policy that does not accurately reflect the manner in which the data is treated.) The World Wide Web Consortium, an international industry group whose members include Apple, Commerce One, Ericsson, and Microsoft, is supporting the development of P3P. Version 1.0 of the P3P was released in April 2002 and can be found at http://www.w3.org/TR/P3P/.

Platform for Privacy Preferences (P3P)
a screening technology that shields users from Web sites that don't provide the level of privacy protection they desire

FAIRNESS IN INFORMATION USE

Selling information to other companies can be so lucrative that many companies will continue to store and sell the data they collect on customers, employees, and others. When is this information storage and use fair and reasonable to the individuals whose data is stored and sold? Do individuals have a right to know about data stored about them and to decide what data is stored and used? As shown in Table 14.11, these questions can be broken down into four issues that should be addressed: knowledge, control, notice, and consent.

FEDERAL PRIVACY LAWS AND REGULATIONS

In the past few decades, significant laws have been passed regarding an individual's right to privacy. Others relate to business privacy rights and the fair use of data and information.

The Privacy Act of 1974

The major piece of legislation on privacy is the Privacy Act of 1974 (PA74), enacted by Congress during Gerald Ford's presidency. PA74 applies only to certain federal agencies. The act, which is about 15 pages long, is straightforward and easy to understand. The purpose of this act is to provide certain safeguards for individuals against an invasion of personal privacy by requiring federal agencies (except as otherwise provided by law) to do the following:

- Permit individuals to determine what records pertaining to them are collected, maintained, used, or disseminated by such agencies
- Permit individuals to prevent records pertaining to them from being used or made available for another purpose without their consent

Fairness Issues	Database Storage	Database Usage
The right to know	Knowledge	Notice
The ability to decide	Control	Consent

Knowledge. Should individuals have knowledge of what data is stored on them? In some cases, individuals are informed that information on them is stored in a corporate database. In others, individuals do not know that their personal information is stored in corporate databases.

Control. Should individuals have the ability to correct errors in corporate database systems? This is possible with most organizations, although it can be difficult in some cases.

Notice. Should an organization that uses personal data for a purpose other than the original purpose notify individuals in advance? Most companies don't do this.

Consent. If information on individuals is to be used for other purposes, should these individuals be asked to give their consent before data on them is used? Many companies do not give individuals the ability to decide if information on them will be sold or used for other purposes.

TABLE 14○11

The Right to Know and the Ability to Decide

- Permit individuals to gain access to information pertaining to them in federal agency records, to have a copy of all or any portion thereof, and to correct or amend such records
- Ensure that they collect, maintain, use, or disseminate any record of identifiable personal information in a manner that ensures that such action is for a necessary and lawful purpose, that the information is current and accurate for its intended use, and that adequate safeguards are provided to prevent misuse of such information
- Permit exemptions from this act only in cases where there is an important public need for such exemption, as determined by specific law-making authority
- Be subject to civil suit for any damages that occur as a result of willful or intentional action that violates any individual's rights under this act

PA74, which applies to all federal agencies except the CIA and law enforcement agencies, also established a Privacy Study Commission to study existing databases and to recommend rules and legislation for consideration by Congress. PA74 also requires training for all federal employees who interact with a "system of records" under the act. Most of the training is conducted by the Civil Service Commission and the Department of Defense. Another interesting aspect of PA74 concerns the use of social security numbers—federal, state, and local governments and agencies cannot discriminate against any individual for not disclosing or reporting his or her social security number.

Gramm-Leach-Bliley Act

This act was passed in 1999 and requires all financial institutions to protect and secure customers' nonpublic data from unauthorized access or use. Under terms of this act, it was assumed that all customers approve of the financial institutions' collecting and storing their personal information. The institutions were required to contact their customers and inform them of this fact. Customers were required to write separate letters to each of their financial institutions if they wanted to opt out of the data collection and storage process. Most people were overwhelmed with the mass mailings they received from their financial institutions and simply discarded them without ever understanding their importance.

USA Patriot Act

As discussed in the opening vignette, the 2001 Uniting and Strengthening America by Providing Appropriate Tools Required to Intercept and Obstruct Terrorism (USA Patriot Act) was passed in response to the September 11 terrorism acts. Proponents argue that it gives necessary new powers to both domestic law enforcement and international intelligence agencies. Critics argue that the law removes many of the checks and balances that previously allowed the courts to ensure law enforcement agencies did not abuse their powers. For example, under this act, Internet service providers and telephone companies must turn over customer information, including numbers called, without a court order, if the FBI claims that the records are relevant to a terrorism investigation. Also, the company is forbidden to disclose that the FBI is conducting an investigation. Only time will tell how this act will be applied in the future.

Other Federal Privacy Laws

In addition to PA74, other pieces of federal legislation relate to privacy. A federal law that was passed in 1992 bans unsolicited fax advertisements. This law was upheld in a 1995 ruling by the Ninth U.S. Circuit Court of Appeals, which concluded that the law is a reasonable way to prevent the shifting of advertising costs to customers. Table 14.12 lists additional laws related to privacy.

TABLE 14.12

Federal Privacy Laws and Their Provisions

Law	Provisions
Fair Credit Reporting Act of 1970 (FCRA)	Regulates operations of credit-reporting bureaus, including how they collect, store, and use credit information
Tax Reform Act of 1976	Restricts collection and use of certain information by the Internal Revenue Service
Electronic Funds Transfer Act of 1979	Outlines the responsibilities of companies that use electronic funds transfer systems, including consumer rights and liability for bank debit cards
Right to Financial Privacy Act of 1978	Restricts government access to certain records held by financial institutions
Freedom of Information Act of 1970	Guarantees access for individuals to personal data collected about them and about government activities in federal agency files
Education Privacy Act	Restricts collection and use of data by federally funded educational institutions, including specifications for the type of data collected, access by parents and students to the data, and limitations on disclosure
Computer Matching and Privacy Act of 1988	Regulates cross-references between federal agencies' computer files (e.g., to verify eligibility for federal programs)
Video Privacy Act of 1988	Prevents retail stores from disclosing video rental records without a court order
Telephone Consumer Protection Act of 1991	Limits telemarketers' practices
Cable Act of 1992	Regulates companies and organizations that provide wireless communications services, including cellular phones
Computer Abuse Amendments Act of 1994	Prohibits transmissions of harmful computer programs and code, including viruses
Gramm-Leach-Bliley Act of 1999	Requires all financial institutions to protect and secure customers' nonpublic data from unauthorized access or use
USA Patriot Act of 2001	Internet service providers and telephone companies must turn over customer information, including numbers called, without a court order, if the FBI claims that the records are relevant to a terrorism investigation

STATE PRIVACY LAWS AND REGULATIONS

State legislatures have been considering and passing privacy legislation that is far-reaching and potentially more burdensome to business than existing federal legislation. The use of social security numbers, access to medical records, the disclosure of unlisted telephone numbers, the sharing of credit reports by credit bureaus, the disclosure of bank and personal financial information, and the use of criminal files are some of the issues being considered by state legislators. These state proposals could have an enormous effect on companies that do business within their borders. For example, the federal Gramm-Leach-Bliley law allows financial services groups to share customer data unless the customer says no. But the law allowed states to set a tougher standard, by first requiring customer consent or "opt in." In the summer of 2002, California's San Mateo County and the states of North Dakota and Vermont approved laws requiring banks to get customer permission before sharing their data with third parties.[39] Such state-by-state and even county-by-county exceptions to the federal law greatly complicate financial record keeping and data sharing.

CORPORATE PRIVACY POLICIES

Even though privacy laws for private organizations are not very restrictive, most organizations are very sensitive to privacy issues and fairness. They realize that invasions of privacy can hurt their business, turn away customers, and dramatically reduce revenues and profits. Consider a major international credit card company. If the company sold confidential financial information on millions of customers to other companies, the results could be disastrous. In a matter of days, the firm's business and revenues could be reduced dramatically. Thus, most organizations maintain privacy policies, even though they are not required by law. Some companies even have a privacy bill of rights that specifies how the privacy of employees, clients, and customers will be protected. Corporate privacy policies should address a customer's knowledge, control, notice, and consent over the storage and use of information. They may also cover who has access to private data and when it may be used.

Multinational companies face an extremely difficult challenge in implementing data collection and dissemination processes and policies because of the multitude of differing country or regional statutes. For example, Australia requires companies to destroy customer data (including backup files) or make it anonymous once it's no longer needed. Firms that transfer customer and personnel data out of Europe must comply with European privacy laws that allow customers and employees to access data about them and let them determine how that information can be used.[40]

A good database design practice is to assign a single unique identifier to each customer—so that each has a single record describing all relationships with the company across all its business units. That way, the organization can apply customer privacy preferences consistently throughout all databases. Failure to do so can expose the organization to legal risks—aside from upsetting customers who opted out of some collection practices. Again, the 1999 Gramm-Leach-Bliley Financial Services Modernization Act required all financial-service institutions to communicate their data privacy rules and honor customer preferences. Key Bank redesigned its massive customer databases to conform to this act. Some 50 million customer records held by various business units were summarized into a single database of 11 million records, with each customer having a single record describing all relationships with the bank (e.g., savings accounts, checking accounts, home mortgage, auto loans, etc.).[41]

INDIVIDUAL EFFORTS TO PROTECT PRIVACY

Although numerous state and federal laws deal with privacy, privacy laws do not completely protect individual privacy. In addition, not all companies have privacy policies. As a result, many people are taking steps to increase their own privacy protection. Some of the steps that individuals can take to protect personal privacy include the following:

- *Find out what is stored about you in existing databases.* Call the major credit bureaus to get a copy of your credit report (you can obtain one free if you have been denied credit in the last 60 days). The major companies are Equifax (800-567-8688, http://www.equifax.com), TransUnion (800-888-4213, http://www.transunion.com), and Experian (888-397-3742, http://www.experian.com). You can also submit a Freedom of Information Act request to a federal agency that you suspect may have information stored on you.
- *Be careful when you share information about yourself.* Don't share information unless it is absolutely necessary. Every time you give information about yourself through an 800, 888, or 900 call, your privacy is at risk. You can ask your doctor, bank, or financial institution not to share information about you with others without your written consent.
- *Be proactive to protect your privacy.* You can get an unlisted phone number and ask the phone company to block caller ID systems from reading your phone number. If you change your address, don't fill out a change-of-address form with the U.S. Postal Service; you can notify the people and companies that you want to have your new address. Destroy carbon copies of your charge card bills and shred monthly statements before disposing of them in the garbage. Be careful about sending personal e-mail messages over a corporate e-mail system. You can also get help in avoiding junk mail and telemarketing calls by visiting the Direct Marketing Association Web site at http://www.the-dma.org/. Go to the Web site and look under Consumer Help—Remove Name for Lists.
- *When purchasing anything from a Web site, make sure you safeguard your credit card numbers, passwords, and personal information.* Do not do business with a site unless you know that it handles credit card information securely (with Netscape Navigator, look for a solid blue key in a small blue rectangle; with Microsoft Explorer, look for the words "Secure Web Site"). Do not provide personal information without reviewing the site's data privacy policy.

THE WORK ENVIRONMENT

The use of computer-based information systems has changed the makeup of the workforce. Jobs that require IS literacy have increased, and many less-skilled positions have been eliminated. Corporate programs, such as reengineering and continuous improvement, bring with them the concern that, as business processes are restructured and ISs are integrated within them, the people involved in these processes will be removed.

However, the growing field of computer technology and IS has opened up numerous avenues to professionals and nonprofessionals of all backgrounds. Enhanced telecommunications has been the impetus for new types of business and has created global markets in industries once limited to domestic markets. Even the simplest tasks have been aided by computers, making cash registers faster, smoothing order processing, and allowing people with disabilities to participate more actively in the workforce. As computers and other IS components drop in cost and become easier to use, more workers will benefit from the increased productivity and efficiency provided by computers. Yet, despite these increases in productivity and efficiency, information systems can raise other concerns.

HEALTH CONCERNS

Organizations can increase employee effectiveness by paying attention to the health concerns in today's work environment. For some people, working with computers can cause occupational stress. Workers' anxieties about job insecurity, loss of control, incompetence, and demotion are just a few of the fears they might experience. In some cases, the stress may become so severe that workers may sabotage computer systems and equipment. Monitoring of employee stress may alert companies to potential problems. Training and counseling can often help the employee and deter problems.

Computer use may affect physical health as well. Strains, sprains, tendonitis, and other problems account for more than 60 percent of all occupational illnesses and about a third of workers' compensation claims, according to the Joyce Institute in Seattle. The cost to U.S. corporations for these types of health problems is as high as $27 billion annually. Claims relating to **repetitive motion disorder**, which can be caused by working with computer keyboards and other equipment, have increased greatly. Also called **repetitive stress injury (RSI)**, the problems can include tendonitis, tennis elbow, the inability to hold objects, and sharp pain in the fingers. Also common is **carpal tunnel syndrome (CTS)**, which is the aggravation of the pathway for nerves that travel through the wrist (the carpal tunnel). CTS involves wrist pain, a feeling of tingling and numbness, and difficulty grasping and holding objects. It may be caused by a number of factors, such as stress, lack of exercise, and the repetitive motion of typing on a computer keyboard. Decisions on workers' compensation related to repetitive stress injuries have been decided both for and against employees.

Other work-related health hazards involve emissions from improperly maintained and used equipment. Some studies show that poorly maintained laser printers may release ozone into the air; others dispute the claim. Numerous studies on the impact of emissions from display screens have also resulted in conflicting theories. Although some medical authorities believe that long-term exposure can cause cancer, studies are not conclusive at this time. In any case, many organizations are developing conservative and cautious policies.

Most computer manufacturers publish technical information on radiative emissions from their screens, and many companies pay close attention to this information. San Francisco was one of the first cities to propose a video display terminal (VDT) bill. The bill requires companies with 15 or more employees who spend at least four hours a day working with computer screens to give 15-minute breaks every two hours. In addition, adjustable chairs and workstations are required if employees request them.

A story initially distributed by the *Washington Post* in May 1999, which was published in the *Boston Globe* under the headline "Study suggests cellphones tied to cancer," cited "possible connections" between cell-phone use and brain cancer. It used a statistical study examining rates of a rare brain cancer called neurocytoma, as well as a laboratory study. The work was overseen by a Washington consulting firm, Wireless Technology Research LLC, under a six-year, $27 million contract that was funded by a blind trust established by cellular industry companies. The trust arrangement was aimed at enhancing the study's credibility. The scientists involved in the study were furious that preliminary data were leaked from their research in a way that, they say, falsely suggested that the phones may be linked to brain tumors. In fact, investigators say, the data show no clear link.

The World Health Organization (WHO), U.S. Food and Drug Administration (FDA), and the U.S. General Accounting Office (GAO) have also analyzed health data on cell-phone use, but they cannot definitively say whether cell phones pose any health risk. WHO states that gaps in knowledge need further research to better assess health risks. It expects that it will take until 2006 for the required research to be completed, evaluated, and the final results published. At this point, WHO states that radio frequency–absorbing covers or other absorbing devices on mobile phones cannot be justified on health grounds. The FDA states that while high levels of radio frequencies can produce biological damage, it is not known

repetitive motion disorder (repetitive stress injury; RSI) an injury that can be caused by working with computer keyboards and other equipment

carpal tunnel syndrome (CTS) the aggravation of the pathway for nerves that travel through the wrist (the carpal tunnel)

Research has shown that developing certain ergonomically correct habits can reduce the risk of RSI when using a computer.

(Source: Courtesy of Balt, Inc.)

ergonomics
the study of designing and positioning computer equipment for employee health and safety

whether lower levels such as those associated with cell phones can cause adverse health effects.[42] A GAO report on the potential health hazards of mobile phones has concluded that research conducted by the United States and international organizations shows that radio-frequency energy emitted by cell phones doesn't produce adverse health effects. But, the report noted, "There is not enough information to conclude they pose no risk."[43] Experts recommend that if consumers are concerned, they should use a "hands-free" set-up such as a headset or ear bud.

In addition to the possible health risks from radio frequencies, cell-phone use has raised a safety issue—an increased risk of traffic accidents as vehicle operators become distracted by talking on their cellphones (or operating their laptop computers, car navigation systems, or other computer devices) while driving. As a result, some states have made it illegal to operate a cell phone while driving.

AVOIDING HEALTH AND ENVIRONMENTAL PROBLEMS

Many computer-related health problems are minor and caused by a poorly designed work environment. The computer screen may be hard to read, with glare and poor contrast. Desks and chairs may also be uncomfortable. Keyboards and computer screens may be fixed in place or difficult to move. The hazardous activities associated with these unfavorable conditions are collectively referred to as *work stressors*. Although these problems may not be of major concern to casual users of computer systems, continued stressors such as repetitive motion, awkward posture, and eyestrain may cause more serious and long-term injuries. If nothing else, these problems can severely limit productivity and performance.

The study of designing and positioning computer equipment to improve worker productivity and minimize worker injuries, called **ergonomics**, has suggested a number of approaches to reducing these health problems. The objective is to have "no pain" computing. The slope of the keyboard, the positioning and design of display screens, and the placement and design of computer tables and chairs have been carefully studied. Flexibility is a major component of ergonomics and an important feature of computer devices. People come in many sizes, have differing preferences, and require different positioning of equipment for best results. Some people, for example, want to place the keyboard in their laps; others prefer it on a solid table. Because of these individual differences, computer designers are attempting to develop systems that provide a great deal of flexibility. In fact, the revolutionary design of Apple's iMac computer came about by concerns for users' comfort.

In addition to steps taken by companies, individuals can also reduce RSI and develop a better work environment. A number of excellent ideas can be found at the U.S. Department of Labor's Occupational Safety & Health Administration Web site using the Ergonomics link at http://www.OSHA-slc.gov or at the Carpal Tunnel Syndrome Web page at http://www.ctsplace.com. Here is a brief set of recommendations.

- Maintain good posture and positioning. In addition to good equipment, good posture and work habits can eliminate or reduce the potential of RSI.
- Don't ignore pain or discomfort. Many workers ignore early signs of RSI, and as a result, the problem becomes much worse and more difficult to treat.
- Use stretching and strengthening exercises. Often, such exercises can prevent RSI.
- Find a good physician who is familiar with RSI and how to treat it.
- After treatment, start back slowly and pace yourself. Many people who are treated for RSI start back to work too soon and injure themselves again.

We have investigated how computers may be harmful to your health, but the computer can also be used to help prevent and treat general health problems. As discussed in Part III on business information systems, we have seen how computers can be used to assist doctors and other medical professionals by diagnosing medical problems and suggesting potential treatments. People can also use computers to get medical information. Special medical software for personal computers can help people get medical information and determine whether they need to see a doctor. A wealth of information is also available on the Internet on a variety of medical topics. See Table 14.13 for a few examples.

ETHICAL ISSUES IN INFORMATION SYSTEMS

As you've seen throughout the book in our "Ethical and Societal Issues" boxes, ethical issues deal with what is generally considered right or wrong. Some IS professionals believe that their field offers many opportunities for unethical behavior. They also believe that unethical behavior can be reduced by top-level managers developing, discussing, and enforcing codes of ethics. Information systems professionals are usually more satisfied with their jobs when top management stresses ethical behavior.

According to one view of business ethics, the "old contract" of business, the only responsibility of business is to its stockholders and owners. According to another view, the "social contract" of business, businesses are responsible to society. At one point or another in their operations, businesses may have employed one or both philosophies.

Various organizations and associations promote ethically responsible use of information systems and have developed codes of ethics. These organizations include the following:

- The Association of Information Technology Professionals (AITP), formerly the Data Processing Management Association (DPMA)
- The Association for Computing Machinery (ACM)
- The Institute of Electrical and Electronics Engineers (IEEE)
- Computer Professionals for Social Responsibility (CPSR)

TABLE 14.13

Medical Topics on the Internet

Internet Address	Description
http://www.neoforma.com	Enables users to conduct electronic commerce with healthcare vendors, automatically send out requests-for-proposals via broadcast e-mail, establish free e-mail accounts, post classified advertisements, obtain career information and job postings, and participate in topical discussion groups.
http://www.nlm.nih.gov/research/ visible/visible_human.html	Anatomy and Medical Graphics Visible Human Project is a complete, anatomically detailed, three-dimensional representation of the male and female human body. The current phase of the project is collecting transverse CAT, MRI, and frozen section images of representative male and female cadavers at one-millimeter intervals.
http://www.WebMD.com	Provides access to reference material and on-line professional publications from Thomas Healthcare Information Group, Stamford, Connecticut.
http://www.cancer.org	Web site of the American Cancer Society.
http://www.mayo.edu	A tour of the Mayo Clinic.
http://oncolink.upenn.edu	A University of Pennsylvania site that deals with cancer information.

The AITP Code of Ethics

The AITP has developed a code of ethics, standards of conduct, and enforcement procedures that give broad responsibilities to AITP members (Figure 14.2). In general, the code of ethics is an obligation of every AITP member in the following areas:

- Obligation to management
- Obligation to fellow AITP members
- Obligation to society
- Obligation to college or university
- Obligation to the employer
- Obligation to country

For each area of obligation, standards of conduct describe the specific duties and responsibilities of AITP members. In addition, enforcement procedures stipulate that any complaint against an AITP member must be in writing, signed by the individual making the complaint, properly notarized, and submitted by certified or registered mail. Charges and complaints may be initiated by any AITP member in good standing.

The ACM Code of Professional Conduct

The ACM has developed a number of specific professional responsibilities. These responsibilities include the following:

- Strive to achieve the highest quality, effectiveness, and dignity in both the process and products of professional work
- Acquire and maintain professional competence

FIGURE 14.2

AITP Code of Ethics

(Source: Courtesy of AITP—
www.aitp.org.)

Code of Ethics

I acknowledge:

That I have an obligation to management, therefore, I shall promote the understanding of information processing methods and procedures to management using every resource at my command.

That I have an obligation to my fellow members, therefore, I shall uphold the high ideals of AITP as outlined in the Association Bylaws. Further, I shall cooperate with my fellow members and shall treat them with honesty and respect at all times.

That I have an obligation to society and will participate to the best of my ability in the dissemination of knowledge pertaining to the general development and understanding of information processing. Further, I shall not use knowledge of a confidential nature to further my personal interest, nor shall I violate the privacy and confidentiality of information entrusted to me or to which I may gain access.

That I have an obligation to my College or University, therefore, I shall uphold its ethical and moral principles.

That I have an obligation to my employer whose trust I hold, therefore, I shall endeavor to discharge this obligation to the best of my ability, to guard my employer's interests, and to advise him or her wisely and honestly.

That I have an obligation to my country, therefore, in my personal, business, and social contacts, I shall uphold my nation and shall honor the chosen way of life of my fellow citizens.

I accept these obligations as a personal responsibility and as a member of this Association. I shall actively discharge these obligations and I dedicate myself to that end.

- Know and respect existing laws pertaining to professional work
- Accept and provide appropriate professional review
- Give comprehensive and thorough evaluations of computer systems and their impacts, including analysis of possible risks
- Honor contracts, agreements, and assigned responsibilities
- Improve public understanding of computing and its consequences
- Access computing and communication resources only when authorized to do so

The mishandling of the social issues discussed in this chapter—including waste and mistakes, crime, privacy, health, and ethics—can devastate an organization. The prevention of these problems and recovery from them are important aspects of managing information and information systems as critical corporate assets. Increasingly, organizations are recognizing that people are the most important component of a computer-based information system and that long-term competitive advantage can be found in a well-trained, motivated, and knowledgeable workforce.

SUMMARY

PRINCIPLE *Policies and procedures must be established to avoid computer waste and mistakes.*

Computer waste is the inappropriate use of computer technology and resources in both the public and private sectors. Computer mistakes relate to errors, failures, and other problems that result in output that is incorrect and without value. Waste and mistakes occur in government agencies as well as corporations. At the corporate level, computer waste and mistakes impose unnecessarily high costs for an information system and drag down profits. Waste often results from poor integration of IS components, leading to duplication of efforts and overcapacity. Inefficient procedures also waste IS resources, as do thoughtless disposal of useful resources and misuse of computer time for games and personal processing jobs. Inappropriate processing instructions, inaccurate data entry, mishandling of IS output, and poor systems design all cause computer mistakes.

Careful programming practices, thorough testing, flexible network interconnections, and rigorous backup procedures can help an information system prevent and recover from many kinds of mistakes. Companies should develop manuals and training programs to avoid waste and mistakes. Company policies should specify criteria for new resource purchases and user-developed processing tools to help guard against waste and mistakes.

PRINCIPLE *Computer crime is a serious and rapidly growing area of concern requiring management attention.*

Some crimes use computers as tools (e.g., to manipulate records, counterfeit money and documents, commit fraud via telecommunications links, and make unauthorized electronic transfers of money). Identity theft is a crime in which an imposter obtains key pieces of personal identification information in order to impersonate someone else. The information is then used to obtain credit, merchandise, and services in the name of the victim, or to provide the thief with false credentials.

Other crimes target computer systems, including illegal access to computer systems by criminal hackers, alteration and destruction of data and programs by viruses (system, application, and document), and simple theft of computer resources. A virus is a program that attaches itself to other programs. A worm functions as an independent program, replicating its own program files until it destroys other systems and programs or interrupts the operation of computer systems and networks. Application viruses infect executable application files, and a system virus infects operating system programs. A macro virus uses an application's own macro programming language to distribute itself. Unlike other viruses, macro viruses do not infect programs; they infect documents. A logic bomb is designed to "explode" or execute at a specified time and date.

Because of increased computer use, greater emphasis is placed on the prevention and detection of computer crime. Antivirus software is used to detect the presence of viruses, worms, and logic bombs. Use of an intrusion detection system (IDS) provides another layer of protection in the event that an intruder gets past the outer security layers—passwords, security procedures, and corporate firewall. It monitors system and network resources and notifies network security personnel when it senses a possible intrusion. Many small and mid-size organizations are outsourcing their network security operations to managed security service providers (MSSPs) that monitor and maintain network security hardware and software.

Software and Internet piracy may represent the most common computer crime. Computer scams have cost individuals and companies thousands of dollars. Computer crime is also an international issue.

Many organizations and people help prevent computer crime, among them state and federal agencies, corporations, and individuals. Security measures, such as using passwords, identification numbers, and data encryption, help to guard against illegal computer access, especially when supported by effective control procedures. Public Key Infrastructure (PKI) enables users of an unsecure public network such as the Internet to securely and privately exchange data through the use of a public and a private cryptographic key pair that is obtained and shared through a trusted authority. Virus scanning software identifies and removes damaging computer programs. Law enforcement agencies armed with new legal tools enacted by Congress now actively pursue computer criminals.

Although most companies use data files for legitimate, justifiable purposes, opportunities for invasion of privacy abound. Privacy issues are a concern with government agencies, e-mail use, corporations, and the Internet. The Privacy Act of 1974, with the support of other federal laws, establishes straightforward and easily understandable requirements for data collection, use, and distribution by federal agencies; federal law also serves as a nationwide moral guideline for privacy rights and activities by private organizations. The USA Patriot Act, passed just five weeks after September 11, requires Internet service providers and telephone companies to turn over customer information, including numbers called, without a court order, if the FBI claims that the records are relevant to a terrorism investigation. Also, the company is forbidden to disclose that the FBI is conducting an investigation. Only time will tell how this act will be applied in the future. The Gramm-Leach-Bliley Act requires all financial institutions to protect and secure customers' nonpublic data from unauthorized access or use. Under terms of this act, it was assumed that all customers approve of the financial institutions, collecting and storing their personal information and need to opt out if not.

Some states supplement federal protections and limit private organizations' activities within their jurisdictions. A business should develop a clear and thorough policy about privacy rights for customers, including database access. That policy should also address the rights of employees, including electronic monitoring systems and e-mail. Fairness in information use for privacy rights emphasizes knowledge, control, notice, and consent for people profiled in databases. Individuals should have knowledge of the data that is stored about them and have the ability to correct errors in corporate database systems. If information on individuals is to be used for other purposes, these individuals should be asked to give their consent beforehand. Each individual has the right to know and the ability to decide. Platform for Privacy Preferences (P3P) is a screening technology that shields users from Web sites that don't provide the level of privacy protection they desire.

PRINCIPLE *Jobs, equipment, and working conditions must be designed to avoid negative health effects.*

Computers have changed the makeup of the workforce and even eliminated some jobs, but they have also expanded and enriched employment opportunities in many ways. Computers and related devices affect employees' emotional and physical health, especially by causing repetitive stress injury (RSI). Some critics blame computer systems for emissions of ozone and electromagnetic radiation. There is no conclusive data connecting cell-phone use and cancer; however, heavy cell phone users may wish to use "hands free" phone sets. "Hands free" usage can also lower the risk of accidents while driving.

The study of designing and positioning computer equipment, called *ergonomics*, has suggested a number of approaches to reducing these health problems. Ergonomic design principles help to reduce harmful effects and increase the efficiency of an information system. The slope of the keyboard, the positioning and design of display screens, and the placement and design of computer tables and chairs are essential for good health. RSI prevention includes keeping good posture, not ignoring pain or problems, performing stretching and strengthening exercises, and seeking proper treatment. In addition to these negative health consequences, information systems can be used to provide a wealth of information on health topics through the Internet and other sources.

Ethics determine generally accepted and discouraged activities within a company and the larger society. Ethical computer users define acceptable practices more strictly than just refraining from committing crimes; they also consider the effects of their IS activities, including Internet usage, on other people and organizations. The Association for Computing Machinery and the Association of Information Technology Professionals have developed guidelines and a code of ethics. Many IS professionals join computer-related associations and agree to abide by detailed ethical codes.

CHAPTER 14 SELF-ASSESSMENT TEST

Policies and procedures must be established to avoid computer waste and mistakes.

1. It is only managers and users at the senior level who play a major role in helping organizations achieve the positive benefits of IS. True False

2. As information technology becomes faster, more complex, and more powerful, organizations and individuals face a (an) _____ risk of experiencing the results of computer-related mistakes.

3. The first step to prevent computer-related waste is to:
   A. establish policies and procedures regarding efficient acquisition, use, and disposal of systems and devices
   B. implement policies and procedures to minimize waste and mistakes according to the business conducted
   C. monitor routine practices and take corrective action if necessary
   D. review existing policies and procedures and determine whether they are adequate

Computer crime is a serious and rapidly growing area of concern requiring management attention.

4. The number of security problems reported to CERT between 1997 and 2001:
   A. increased twenty-five-fold
   B. decreased fifty percent
   C. doubled
   D. stayed about the same

5. The National Infrastructure Protection Center is a branch of the _____ that serves as a focal point for threat assessment, warning, investigation, and response for threats or attacks against our country's critical infrastructure that provides telecommunications, energy, banking and finance, water systems, government operations, and emergency services.

6. Frequently, a critical computer password has been talked out of an individual, a practice called dumpster diving. True False

7. _____ is a crime in which an imposter obtains key pieces of personal identification information, such as social security or driver's license numbers, in order to impersonate someone else.

8. A computer-savvy person who attempts to gain unauthorized or illegal access to computer systems to steal passwords, corrupt files and programs, or even transfer money.
   A. script bunny
   B. hacker
   C. criminal hacker or cracker
   D. social engineer

9. Essentially, there is no significant difference between a virus and a worm. True False

10. A program that appears to be useful but actually masks the destructive program is called a:
    A. logic bomb
    B. Trojan horse
    C. macro virus
    D. worm

11. A (An) _____ is a small program hidden in a network or a computer system that records identification numbers and passwords.

12. Countries other than the United States rarely have problems with computer crime. True False

13. Federal law permits employers to monitor e-mail sent and received by employees. True False

Jobs, equipment, and working conditions must be designed to avoid negative health effects.

14. RSI or _____ problems can include tendonitis, tennis elbow, the inability to hold objects, and sharp pain in the fingers.

15. There is positive evidence that excessive use of cell phones increases the risk of brain cancer. True False

16. The study of designing and positioning computer equipment to improve worker productivity and minimize worker injuries is called _____.

Chapter 14 Self-Assessment Test Answers:

1. False, 2. increased (or greater), 3. A, 4. A, 5. FBI, 6. False, 7. Identity theft, 8. C, 9. F, 10. B, 11. password sniffer, 12. False, 13. True, 14. repetitive stress injury, 15. False, 16. ergonomics.

KEY TERMS

antivirus programs, 633
application virus, 626
biometrics, 632
carpal tunnel syndrome (CTS), 646
criminal hacker (cracker), 624
dumpster diving, 622
ergonomics, 647
hacker, 624
ICRA rating system, 636
identity theft, 623

insiders, 624
Internet piracy, 628
intrusion detection system (IDS), 635
logic bomb, 626
macro virus, 626
managed security service provider (MSSP), 635
password sniffer, 627
Platform for Privacy Preferences (P3P), 641

Public Key Infrastructure (PKI), 632
repetitive motion disorder (repetitive stress injury; RSI), 646
script bunnies, 624
social engineering, 622
software piracy, 628
system virus, 626
Trojan horse, 626
virus, 624
worm, 624

REVIEW QUESTIONS

1. What is the USA Patriot Act, and why was it implemented?
2. What can organizations do to prevent computer-related waste and mistakes?
3. Identify four specific actions that can be taken to reduce crime on the Internet.
4. What are four key tenets of the concept of fairness in information use for privacy rights for people profiled in computer databases?
5. What is the difference between a cracker and a script bunny? What are the major problems caused by criminal hackers?
6. What is the difference between a worm and a virus?
7. What are application viruses, system viruses, and macro viruses?

8. What is software piracy, and why is it so common?
9. What is ergonomics? How can it be applied to office workers?
10. What is identity theft and what actions can a person take to avoid it?
11. What are the provisions of the Privacy Act of 1974?
12. What specific actions can you take to avoid RSI?
13. What is the difference between CTS and RSI?
14. Under what conditions is the monitoring of e-mail not considered an invasion of privacy?
15. Describe the traditional views of ethics in business.
16. What is a code of ethics? Give an example.

DISCUSSION QUESTIONS

1. How can a criminal use dumpster diving to enhance his or her opportunity of success at social engineering?
2. You are surprised when you receive a check from the IRS for a tax refund for $10,000 more than you are owed. How could this have happened? What would you do?
3. What is HIPAA? What is its intended goal? What impact might it have on healthcare organizations?
4. Your marketing department has just opened a Web site and is requesting visitors to register at

the site to enter a promotional contest where the chances of winning a prize are better than one in three. Visitors must provide the information necessary to contact them plus fill out a brief survey about the use of your company's products. What data privacy issues may arise?
5. Briefly discuss the potential for cyberterrorism to cause a major disruption in our daily life. What are some likely targets of a cyberterrorist? What sort of action could a cyberterrorist take against these targets?

6. Give three reasons why a national identification card is a good idea. Now argue the other side and give three reasons against the creation of a national identification card.

7. How could you use the Internet to help improve your health?

8. During 2002, a number of corporations were forced to restate earnings because they had used unethical accounting practices. Briefly discuss the extent to which you feel that these problems were caused by a failure in the firm's accounting information systems.

9. Using information presented in this chapter on federal privacy legislation, identify which federal law regulates the following areas and situations: cross-checking IRS and social security files to verify the accuracy of information; credit bureaus processing home loans; customer liability for debit cards; individuals' right to access data contained in federal agency files; the IRS obtaining personal information; the government obtaining financial records; and employers' access to university transcripts.

10. Briefly discuss the difference between acting morally and acting legally. Give an example of acting legally and yet immorally.

PROBLEM-SOLVING EXERCISES

1. Access the CSI-FBI Survey Results for the past four years (start at http://www.gocsi.com and select the link to the CSI/FBI Computer Crime and Security Survey) to get statistics to quantify the number of incidents and their dollar impact. Choose one of the parameters tracked by this survey and use the graphics routine in your spreadsheet software to graph the variation over time.

2. Using your word processing software, write a few brief paragraphs summarizing the trends you see from reviewing the CSI-FBI Survey Results for the past four years. Then cut and paste the graph from the previous exercise into your report.

TEAM ACTIVITIES

1. Interview members of your school's IS security organization to learn more about their role and the types and frequency of computer incidents with which they must deal. What was the most significant incident with which they had to deal in the past year? Do they think that schools have unique security issues not faced by private industry? Write a brief paper summarizing the interview.

2. Your team has been hired by a large telemarketing firm to help identify actions to reduce the number of worker's compensation claims related to eyestrain, RSI, and CTS. The hundreds of employees are divided into three shifts that work around the clock to support the telemarketing campaigns of various companies nationwide. Each telemarketer works a minimum of 40 hours per week, and 8 to 10 hours of overtime is not uncommon. The telemarketers read from prepared scripts displayed on their computer monitors and then enter data into predefined screens in response to what the customer says. How would you and your team determine what needs to be done? Develop a list of half a dozen recommendations you think might help.

3. Have each member of your team access ten different Web sites and summarize their findings in terms of the existence of data privacy policy statements—did the site have such a policy, was it easy to find, was it complete and easy to understand? Did you find any sites using the P3P standard?

WEB EXERCISES

1. Do research on the Web to find a number of opinions, both pro and con, on the USA Patriot Act. Write a paragraph for each side summarizing their key points. Write a paragraph that captures your thoughts on this act.
2. Visit the Web site of the Internet Content Rating Association (ICRA) and obtain a copy of the on-line questionnaire Web site authors are asked to complete describing the content of their site. Next visit another Web site of interest to you. Complete the questionnaire for this Web site. Write a paragraph stating your opinion of the effectiveness of this means of Web site screening.

3. Echelon is a top-secret electronic eavesdropping system managed by the U.S. National Security Agency that is capable of intercepting and decrypting almost any electronic message sent anywhere in the world. It may have been in operation as early as the 1970s, but it wasn't until the 1990s that journalists were able to confirm its existence and gain insight into its capabilities. Do Web research to find out more about this system and its capabilities. Write a paragraph or two summarizing your findings.

CASES

CASE 1

The Role of the CERT/CC

The Computer Emergency Response Team Coordination Center (CERT/CC) is a government-funded group at Carnegie Mellon University in Pittsburgh. It has been monitoring Internet security since the late 1980s. Its primary sponsor, the U.S. Department of Defense Office of the Secretary of Research and Development, provides about $3.5 million per year in funding.

CERT/CC's responsibilities include analyzing flaws in Internet systems, assisting in their remediation, and studying intruder-developed code. It serves as a central clearinghouse for reports related to Internet security breaches including attack attempts, probes, scans, and successful attacks. As new reports come in, they are analyzed to determine the attack method and then correlated with other reports to determine the scope and magnitude of the attack. CERT/CC then determines whether any attack represents a new type that needs to be investigated, whether a change in frequency of attack method has occurred, and whether new defenses or countermeasures are needed. CERT/CC pledges not to identify any specific victim but does share information anonymously and describes activity without attribution.

CERT/CC focuses specifically on technical issues related to Internet security—not enforcement. It does not attempt to identify intruders, their locations, or their motivations. But it does inform the Internet community about current activity and new types of attacks and provides defense tips against them. This information is collected by analyzing code written by intruders to determine what it does and what vulnerabilities are exploited and then determining how to defend against it and assessing who its victims or targets might be. CERT/CC also attempts to predict trends in malicious code development and functionality.

In 2001, more than 52,000 incidents were reported to the Computer Emergency Response Team Coordination Center (CERT/CC), more than double the number reported in 2000. Most reports are filed by individuals in private industry; only 10 percent of the reports come from the public sector. Because businesses are interconnected via the Internet and use common hardware and software platforms and applications to conduct business, the risk of a single computer incident affecting many organizations is greatly increased. Indeed, a corporate network's defense to an Internet attack is only as strong as its weakest business partner. The result is that it is highly likely that almost every company will experience a computer security breach in the next year.

Discussion Questions
1. Why do you think CERT/CC pledges not to identify the victims of a computer incident?
2. What factors can you identify that are leading to an increase in the number of Internet security incidents?

Critical Thinking Questions
3. Why do you think that CERT/CC focuses specifically on technical issues related to Internet security and does not attempt to identify intruders, their locations, or their motivations? Do you think that it should? Why or why not?
4. Imagine that you are the Director of the CERT/CC operations. Prepare an argument justifying at least a 50 percent increase in your organization's budget from 2001 to 2000.

Sources: Adapted from Sam Costello, "CERT to Sell Security Threat Information," *Computerworld*, April 19, 2001, accessed at http://www.computerworld.com; Sam Costello, "CERT: Flaws in CDE Could Lead to Denial of Service," *Computerworld*, July 11, 2002, http://www.computerworld.com; Dan Verton, "As Another Security Coalition Launches, Some Ask Why," *Computerworld*, April 23, 2001, accessed at http://www.computerworld.com; "Collaboration Between CERT Coordination Center and Internet Security Alliance," CERT/CC Web site at http://www.cert.org on September 1, 2002.

Australia Struggles to Implement Internet Laws

Australia, just like the United States and many other countries, is struggling to find the right balance between individuals' right to privacy and legitimate law enforcement agency needs in the policing of Internet content and freedom of speech. Here is a sample of recent legislative actions.

The Privacy and Personal Information Protection (PPIP) Act of 1998 introduced a set of binding privacy standards known as information protection principles that regulate the way public-sector agencies collect, use, store, and disclose personal information. Under the PPIP Act Australians can apply for access to any personal information a government department holds about them. They can make a complaint if they believe a department has breached a provision of the act or its own Code of Practice.

The Broadcasting Services Amendment (Online Services) Act of 1999 prohibits the publication of X-rated material on-line. Under this law, content hosts are required to delete Australian hosted content that is deemed "objectionable" or "unsuitable for minors" from their server (Web, Usenet, FTP, etc.) when they receive a removal notice from the government regulator, the Australian Broadcasting Authority (ABA). In the first 18 months after the laws were introduced, 706 complaints were received from the public, and 372 were referred to police.

State and territory criminal laws enable prosecution of X-rated content providers and creators (as opposed to host sites) for making available material that is deemed "objectionable" or "unsuitable for minors." Not all jurisdictions have such laws in force; several are still pending in some jurisdictions.

A proposed Telecommunications Interception Legislation Amendment Bill of 2002 would have allowed government agencies to intercept and read the contents of communications passing over a telecommunications system without a warrant of any type. E-mail and voice-mail messages stored on a service provider's equipment pending delivery to the intended recipient could be read by a government agency before the intended recipient even knew a message had been sent. On June 27, 2002, the Australian Senate rejected this bill.

Similiar to the Electronic Frontier Foundation in the United States, Australia has its own Electronic Frontiers Australia (EFA)—a nonprofit national organization representing Internet users concerned with on-line freedoms and rights. The EFA was formed in 1994, and it is funded by membership subscriptions and donations from individuals and organizations interested in promoting civil liberties. Its goals are to protect and promote the civil liberties of users and operators of computer based communications systems, to advocate the amendment of laws and regulations which restrict free speech or limit access to information, and to educate the community at large about social, political, and civil liberties.

Discussion Questions

1. Identify the equivalent U.S. laws for the Australian laws mentioned in this case.
2. Given today's social climate and the threat of terrorism, do you think that the Telecommunications Interception Legislation Amendment Bill of 2002 should have been passed? Why or why not?

Critical Thinking Questions

3. What fundamental principles do you think should apply to laws concerning access to private information about individuals and X-rated material? Why is it so difficult for countries to enact legislation in these areas?
4. Visit the Web site for Electronic Frontiers Australia (http://www.efa.org.au) and identify the current "hot topics." Write a paragraph or two summarizing your findings.

Sources: Adapted from Adrian Lynch, "Working Safely among Security Dangers," *Australian IT*, August 20, 2002, accessed at http://www.australianIT.news.com.au; Chantel Rumble, "Guideline Heralds Crackdown," *Australian IT*, July 30, 2002, accessed at http://www.australianIT.news.com.au; Caitlin Fitzsimmons, "Net Censorship Laws in Limbo," *Australian IT*, July 15, 2002, accessed at http://www.australianIT.news.com.au; Caitlin Fitzsimmons, "Email Snooping Bill Knocked Down," *Australian IT*, June 28, 2002, accessed at http://www.australianIT.news.com.au; Kate Mackenzie, "Censor Law Not So Bad: Survey," *Australian IT*, June 4, 2002, accessed at http://www.australianIT.news.com.au; "Privacy," The New South Wales Department of Fair Trading Web site accessed at http://www.dft.nsw.gov.au on September 3, 2002.

E-Mail Monitoring—A Necessary Evil?

Nearly half of U.S. companies monitor e-mail, according to a 2001 American Management Association survey. Concern over lawsuits is a key factor in their decision to monitor. Recent U.S. Supreme Court decisions have found that once a case of harassment comes to an employer's attention, the company must stop the abuse. Otherwise, it can be held liable. As a result, Dow Chemical fired 50 workers and disciplined another 200 for sending explicit pornographic images through the company's e-mail system. Pharmaceutical giant Merck fired or disciplined an undisclosed number of employees for inappropriate use of e-mail. The *New York Times* fired 23 workers because they had allegedly distributed offensive jokes on the company's e-mail system.

Employers also monitor employee e-mail because they are becoming increasingly concerned about losses in worker productivity. Too much time spent on nonbusiness e-mail can seriously detract from an individual's productivity. Yet from the employees' perspective, workers often argue that e-mail provides a welcome relief from work and can improve working relationships with coworkers.

Another concern about excessive e-mail use is the wasting of network capacity. After all, many companies conduct much of their business over the Internet or a corporate network, and when casual use takes up much of the network capacity, that's a problem.

While employers have the legal right to monitor employee e-mail, concerns about the negative effects of unannounced employee monitoring have raised a firestorm of debate. The federal Notice of Electronic Monitoring Act was discussed in

Congress in 2000 and 2001, but it never passed. Currently, Connecticut is the only state that requires employers to tell employees if they're being monitored.

To monitor e-mail, companies install basic software to scan incoming and outgoing mail for words and key phrases that managers have compiled in lists. The software can also identify e-mail viruses. Worldwide sales of employee-monitoring software are estimated at $145 million per year. Two interesting software packages include Cameo and PornSweeper. Cameo from MicroData Software enables e-mail administrators to identify up to 200 keywords or phrases to search e-mail for. When a message is found to contain these words, the e-mail may be automatically deleted or sent to its destination, with a copy sent to a designated address or distribution list. PornSweeper analyzes e-mail attachments for nude images or pornographic content. When an inappropriate message is detected, both the sender and recipient get e-mail letting them know that PornSweeper blocked the message. Occasionally, however, PornSweeper picks up false alarms, such as photos of newborn babies.

Successful implementation of an e-mail monitoring policy requires advance written notice and a solid statement of the business reasons for doing it. And employees need to understand that the company owns any e-mail on its system—it is not theirs. Failure to do so can negatively affect worker morale and create a workplace full of paranoia and suspicion.

Discussion Questions

1. Obtain and read a copy of your company's or school's e-mail policy. Does it outline the steps you should take if you receive "hate mail" or pornographic pictures?
2. Visit the Web site of a firm that makes e-mail monitoring software. Make a list of the features and capabilities of its software.

Critical Thinking Questions

3. Draft a paragraph outlining an employer's rationale for monitoring employees' e-mail.
4. Now present an argument against the monitoring of employee e-mail.

Sources: Adapted from Sam Costello, "New Software Lets Managers Search E-Mail," *Computerworld*, August 30, 2001, accessed at http://www. computerworld.com; Diane Rezendes Khirallah, "Employee Monitoring IS Growing Trend, Study Shows," *InformationWeek*, July 10, 2001, accessed at http://www.informationweek.com; Linda Rosencrance, "Study: Monitoring of Employee E-Mail, Web Use Escalates," *Computerworld*, July 9, 2001, accessed at http://www.computerworld.com; Sandar Swanson, "Beware: Employee Monitoring Is on the Rise," *InformationWeek*, August 20, 2001, accessed at http://www.informationweek.com; Jude Thaddeus, "Reading Employees Their E-Mail Rights," *Computerworld*, January 15, 2001, accessed at http://www.computerworld.com.

NOTES

Sources for the opening vignette on p. 615: Adapted from Dan Verton, "IT Pros May Face Background Check," *Computerworld*, July 29, 2002, http://www.computerworld.com; Jaikumar Vijayan, "Bill with Tougher Penalties Passes House," *Computerworld*, July 16, 2002, http://www.computerworld.com; Dan Verton, "Cybersecurity Czar Takes Stand on Software Quality," *Computerworld*, August 5, 2002, http://www.computerworld.com; Tish Keefe, "Software Insecurity," *Computerworld*, August 5, 2002, http://www.computerworld.com.

1. D. Ian Hopper, The Associated Press, "E-Mail Glitch Has Site Users Worried," *Cincinnati Enquirer*, August 30, 2002, p. A-6.
2. Brian Sullivan, "Computer Glitch Disrupts Deposits at Barclays," *Computerworld*, March 28, 2002, accessed at http://www.computerworld.com.
3. Linda Rosencrance, "IRS Sends Out 523,000 Incorrect Refund Check Notices," *Computerworld*, July 17, 2001, accessed at http://www.computerworld.com.
4. Linda Rosencrance, "United to Honor Dirt Cheap Online Ticket Fares," *Computerworld*, February 20, 2001, accessed at http://www.computerworld.com.
5. Jennifer Disabatino, "Brief: Glitch Offers Domestic Travel for $5 at United.com," *Computerworld*, May 15, 2002, accessed at http://www.computerworld.com.
6. Brian Sullivan, "Glitch at Fidelity Canada Exposes Customer Info," *Computerworld*, May 30, 2002, accessed at http://www.computerworld.com.
7. Jennifer Disabatino, "Glitch Disrupts Tracking Systems at Post Office," *Computerworld*, April 19, 2002, accessed at http://www.computerworld.com.
8. Kevin Fogarty, "Technical Agility," *Computerworld*, May 21, 2001, accessed at http://www.computerworld.com.
9. Stephen Lawson, "AOL Time Warner Raises Red Flag over Three Past Deals," *Computerworld*, August 15, 2002, accessed at http://www.computerworld.com.

10. Tracy Mayor, "The Privacy Problem," *CIO*, January 15, 2001, pp. 75–84.
11. Robert Pear, "Medical Industry Lobbies to Rein in New Patients Privacy Rules," *The New York Times on the Web*, February 12, 2001, accessed at http://www.nytimes.com.
12. "San Francisco Man Pleads Guilty to Unauthorized Access to Catholic Healthcare West Computer Causing Damage," Department of Justice Press Release, January 18, 2002, accessed at http://www.usdoj.gov.
13. "Former Computer Network Administrator at New Jersey High-Tech Firm Sentenced to 41 Months for Unleashing $10 Million Computer 'Time Bomb,'" Department of Justice Press Release, February 26, 2002, accessed at http://www.usdoj.gov.
14. "U.S Charges Engineer with Computer Intrusion, Destruction of Database at Manhattan Apparel Company," Department of Justice Press Release, April 26, 2002, accessed at http://www.usdoj.gov.
15. Karen W. Arenson, "Princeton Pries into Web Site for Yale Applicants," *Computerworld*, July 26, 2002, accessed at http://www.computerworld.com.
16. "Two Kazakhstan Citizens Accused of Breaking into Bloomberg L.P.'s Computer and Extortion Are Extradited," Department of Justice Press Release, May 21, 2002, accessed at http://www.usdoj.gov.
17. "San Gabriel Valley Men Sentenced for Conspiring to Traffic in Counterfeit Microsoft Software," Department of Justice Press Release, June 25, 2002, accessed at http://www.usdoj.gov.
18. "Hackers Hit USA Today Web Site," *Computerworld*, July 12, 2002, accessed at http://www.computerworld.com.
19. Dan Verton, "California Hack Points to Potential IT Surveillance Threat," *Computerworld*, June 12, 2001, accessed at http://www.computerworld.com.
20. John Schwartz, "Cyberspace Seen as Potential Battleground," *The New York Times on the Web*, November 23, 2001, accessed at http://www.nytimes.com.

21. Stephan Chiger, "ID Theft Increase Doesn't Deter E-Shoppers," *Computerworld*, August 8, 2002, accessed at http://www.computer world.com.

22. Don Kirk, "Impostor's Stock Trade Roils Korea Market," *The New York Times on the Web*, August 27, 2002, accessed at http://www.nytimes.com.

23. Brian Sullivan, "FTC Pushing for Stiffer Penalties for ID Theft," *Computerworld*, July 10, 2002, accessed at http://www.computerworld.com.

24. "What's The Department of Justice Doing about Identity Theft and Fraud?," accessed at the Department of Justice Web site, http://www.usdoj.gov, on May 30, 2002.

25. Dan Verton, "Update: NASA Investigating Hacker Theft of Sensitive Documents," *Computerworld*, August 8, 2002, accessed at http://www.computerworld.com.

26. Joris Evers, "Having Failed to Patch Servers, Microsoft Hit by Code Red," *Computerworld*, August 9, 2001, accessed at http://www.computer world.com.

27. Linda Rosencrance, "Connecticut Teen Charged with Hacking into Air Force Computer System," *Computerworld*, April 24, 2001.

28. Russell Kay, "Wireless Security," *Computerworld*, June 24, 2002, accessed at http://www.computerworld.com.

29. Bob Brewin, "Worldwide 'War Drive' Exposes Insecure Wireless LANs," *Computerworld*, September 9, 2002, accessed at http://www.computer world.com.

30. David Legard, "Software Piracy Losses Fell to $11 B in 2001, Says BSA," *Computerworld*, June 10, 2002, accessed at http://www.computer world.com.

31. Dan Verton, "Antilaundering System Offers USPS Real-Time Intelligence," *Computerworld*, October 29, 2001, accessed at http://www.computer world.com.

32. "Lawsuits Seek $2.2 Trillion over 'Junk Faxes,'" *Computerworld*, August 22, 2002, accessed at http://www.computerworld.com.

33. Lucas Mearian, "Security: An Internal Affair," *Computerworld*, August 5, 2002, accessed at http://www.computerworld.com.

34. Mitch Betts, "Reporter's Notebook: IT Security," *Computerworld*, July 15, 2002, accessed at http://www.computerworld.com.

35. Brian Fonseca and Wayne Rash, "Security at Your Service," *Infoworld*, August 26, 2002, accessed at http://www.infoworld.com.

36. "Federal Court Rejects Government Censorship in Libraries, Citing Free Speech Rights of Patrons," May 31, 2002, ACLU Web site at http://www.aclu.org.

37. "Varian Wins $775,000 Jury Verdict in Internet Libel Case," *BusinessWire*, December 18, 2001, accessed at http://www.findarticles.com.

38. Larry Kahaner, "Hungry for Your E-Mail," *InformationWeek*, April 23, 2001, pp. 59–64.

39. Patrick Thibodeau, "California County Opts-In for Tougher Privacy Law," *Computerworld*, August 16, 2002, accessed at http://www.computer world.com.

40. Patrick Thibodeau, "Profitable Privacy," *Computerworld*, February 18, 2002, accessed at http://www.computerworld.com.

41. Patrick Thibodeau, "Profitable Privacy," *Computerworld*, February 18, 2002, accessed at http://www.computerworld.com.

42. Elisa Batista, "Radiation Still Hard to Prove," *Wired News*, February 22, 2002, accessed at http://www.wired.com.

43. Bob Brewin, "Report on Health Risks From Cell Phones Inconclusive," *Computerworld*, May 28, 2001, accessed at http://www.computerworld.com.

acceptance testing conducting any tests required by the user

accounting MIS information system that provides aggregate information on accounts payable, accounts receivable, payroll, and many other applications

accounting systems systems that include budget, accounts receivable, payroll, asset management, and general ledger

accounts payable system system that increases an organization's control over purchasing, improves cash flow, increases profitability, and provides more effective management of current liabilities

accounts receivable system system that manages the cash flow of the company by keeping track of the money owed the company on charges for goods sold and services performed

ad hoc DSS a DSS concerned with situations or decisions that come up only a few times during the life of the organization

analog signal a continuous, curving signal

antivirus programs programs or utilities that prevent viruses and recover from them if they infect a computer

applet small program embedded in Web pages

application flowcharts diagrams that show relationships among applications or systems

application program interface (API) interface that allows applications to make use of the operating system

application servers Software packages, often written in the Java programming language for use on computers running the Windows NT operating system, that connect end users to the databases holding the information they need to access

application service provider a company that provides both end-user support and the computers on which to run the software from the user's facilities

application software programs that help users solve particular computing problems

application virus a virus that infects executable application files such as word processing programs

arithmetic/logic unit (ALU) portion of the CPU that performs mathematical calculations and makes logical comparisons

ARPANET project started by the U.S. Department of Defense (DOD) in 1969 as both an experiment in reliable networking and a means to link DOD and military research contractors, including a large number of universities doing military-funded research

artificial intelligence (AI) a field in which the computer system takes on the characteristics of human intelligence

artificial intelligence systems people, procedures, hardware, software, data, and knowledge needed to develop computer systems and machines that demonstrate the characteristics of intelligence

asking directly an approach to gather data that asks users, stakeholders, and other managers about what they want and expect from the new or modified system

asset management transaction processing system system that controls investments in capital equipment and manages depreciation for maximum tax benefits

asynchronous communications communication in which the receiver gets the message minutes, hours, or days after it is sent

attribute a characteristic of an entity

audit trail documentation that allows the auditor to trace any output from the computer system back to the source documents

auditing analyzing the financial condition of an organization and determining whether financial statements and reports produced by the financial MIS are accurate

backbone one of the Internet's high-speed, long-distance communications links

backward chaining The process of starting with conclusions and working backward to the supporting facts

bandwidth the width of the range of frequencies that an electronic signal occupies on a given transmission medium

batch processing system method of computerized processing in which business transactions are accumulated over a period of time and prepared for processing as a single unit or batch

benchmark test an examination that compares computer systems operating under the same conditions

best practices the most efficient and effective ways to complete a business process

biometrics the measurement of a living trait, whether physical or behavioral

bit BInary digiT—0 or 1

bot a software tool that searches the Web for information, products, prices, etc.

brainstorming decision-making approach which often consists of members offering ideas "off the top of their heads"

bridge connection between two or more networks at the media access control portion of the data link layer; the two networks must use the same communications protocol

broadband telecommunications in which a wide band of frequencies is available to transmit information, allowing more information to be transmitted in a given amount of time

budget transaction processing system system that automates many of the tasks required to amass budget data, distribute it to users, and consolidate the prepared budgets

bus line the physical wiring that connects the computer system components

bus network a type of topology that contains computers and computer devices on a single line; each device is connected directly to the bus and can communicate directly with all other devices on the network; one of the most popular types of personal computer networks

business intelligence the process of gathering enough of the right information in a timely manner and usable form and analyzing it to have a positive impact on business strategy, tactics, or operations

business resumption planning the process of anticipating and providing for disasters

business-to-business (B2B) e-commerce a form of e-commerce in which the participants are organizations

business-to-consumer (B2C) e-commerce a form of e-commerce in which customers deal directly with the organization, avoiding any intermediaries

byte (B) eight bits together that represent a single character of data

cache memory a type of high-speed memory that a processor can access more rapidly than main memory

carpal tunnel syndrome (CTS) the aggravation of the pathway for nerves that travel through the wrist (the carpal tunnel)

CASE repository a database of system descriptions, parameters, and objectives

catalog management software software that automates the process of creating a real-time interactive catalog and delivering customized content to a user's screen

CD-rewritable (CD-RW) disk an optical disk that allows personal computer users to replace their diskettes with high-capacity CDs that can be written upon and edited over

CD-writable (CD-W) disk an optical disk that can be written upon but only once

central processing unit (CPU) the part of the computer that consists of three associated elements: the

arithmetic/logic unit, the control unit, and the register areas

centralized processing processing alternative in which all processing occurs in a single location or facility

certificate authority (CA) a trusted third party that issues digital certificates

certification process for testing skills and knowledge that results in a statement by the certifying authority that says an individual is capable of performing a particular kind of job

change model representation of change theories that identifies the phases of change and the best way to implement them

character basic building block of information, consisting of uppercase letters, lowercase letters, numeric digits, or special symbols

chat room a facility that enables two or more people to engage in interactive "conversations" over the Internet

chief programmer team a group of skilled IS professionals with the task of designing and implementing a set of programs. This team has total responsibility for building the best software possible

choice stage the third stage of decision making, which requires selecting a course of action

clickstream data data gathered based on the Web sites you visit and what items you click on

client/server an architecture in which multiple computer platforms are dedicated to special functions such as database management, printing, communications, and program execution

clock speed a series of electronic pulses produced at a predetermined rate that affect machine cycle time

closed shops IS department in which only authorized operators can run the computers

cold site a computer environment that includes rooms, electrical service, telecommunications links, data storage devices, and the like; also called a shell

collaborative computing software software that helps teams of people to work together toward a common goal

command-based user interface a user interface that requires that text commands be given to the computer to perform basic activities

common carriers long-distance telephone companies

communications software software that provides a number of important functions in a network, such as error checking and data security

compact disk read-only memory (CD-ROM) a common form of optical disk on which data, once it has been recorded, cannot be modified

competitive advantage a significant and (ideally) long-term benefit to a company over its competition

competitive intelligence a continuous process involving the legal and ethical collection of information, analysis, and controlled dissemination of information to decision makers

compiler a special software program that converts the programmer's source code into the machine language instructions consisting of binary digits

complex instruction set computing (CISC) a computer chip design that places as many microcode instructions into the central processor as possible

computer literacy knowledge of computer systems and equipment and the ways they function; it stresses equipment and devices (hardware), programs and instructions (software), databases, and telecommunications

computer network the communications media, devices, and software needed to connect two or more computer systems and/or devices

computer programs sequences of instructions for the computer

computer server a computer designed for a specific task, such as network or Internet applications

computer system architecture the structure, or configuration, of the hardware components of a computer system

computer system platform the combination of a particular hardware configuration and systems software package

computer-aided software engineering (CASE) tools that automate many of the tasks required in a systems development effort and enforce adherence to the SDLC

computer-assisted manufacturing (CAM) a system that directly controls manufacturing equipment

computer-based information system (CBIS) consists of hardware, software, databases, telecommunications, people, and procedures that are configured to collect, manipulate, store, and process data into information

computer-integrated manufacturing (CIM) using computers to link the components of the production process into an effective system

concurrency control a method of dealing with a situation in which two or more people need to access the same record in a database at the same time

consumer-to-consumer (C2C) e-commerce a form of e-commerce in which the participants are individuals, with one serving as the buyer and the other as the seller

content streaming a method for transferring multimedia files over the Internet so that the data stream of voice and pictures plays more or less continuously without a break, or very few of them; enables users to browse large files in real time

continuous improvement constantly seeking ways to improve the business processes to add value to products and services

contract software software developed for a particular company

control unit part of the CPU that sequentially accesses program instructions, decodes them, and coordinates the flow of data in and out of the ALU, the registers, primary storage, and even secondary storage and various output devices

cookie a text file that an Internet company can place on the hard disk of a computer system

coprocessor part of the computer that speeds processing by executing specific types of instructions while the CPU works on another processing activity

cost centers divisions within a company that do not directly generate revenue

cost/benefit analysis an approach that lists the costs and benefits of each proposed system. Once expressed in monetary terms, all the costs are compared with all the benefits

counterintelligence the steps an organization takes to protect information sought by "hostile" intelligence gatherers

creative analysis the investigation of new approaches to existing problems

criminal hacker (cracker) a computer-savvy person who attempts to gain unauthorized or illegal access to computer systems

critical analysis the unbiased and careful questioning of whether system elements are related in the most effective or efficient ways

critical path activities that, if delayed, would delay the entire project

critical success factors (CSFs) factors that are essential to the success of a functional area of an organization

cross-platform development development technique that allows programmers to develop programs that can run on computer systems having different hardware and operating systems, or platforms

cryptography the process of converting a message into a secret code and changing the encoded message back to regular text

culture set of major understandings and assumptions shared by a group

customer relationship management (CRM) system a collection of people, processes, software, and Internet capabilities that help an enterprise manage customer relationships effectively and systematically

cybermall a single Web site that offers many products and services at one Internet location

data raw facts, such as an employee's name and number of hours worked in a week, inventory part numbers, or sales orders

data administrator a non-technical, but important role that ensures that data is managed as an important organizational resource

data analysis manipulation of the collected data so that it is usable for the development team members who are participating in systems analysis

data cleanup the process of looking for and fixing inconsistencies to ensure that data is accurate and complete

data collection the process of capturing and gathering all data necessary to complete transactions

data communications a specialized subset of telecommunications that refers to the electronic collection, processing, and distribution of data—typically between computer system hardware devices

data correction the process of reentering miskeyed or misscanned data that was found during data editing

data definition language (DDL) a collection of instructions and commands used to define and describe data and data relationships in a specific database

data dictionary a detailed description of all the data used in the database

data editing the process of checking data for validity and completeness

data entry process by which human-readable data is converted into a machine-readable form

data input process that involves transferring machine-readable data into the system

data integrity the degree to which the data in any one file is accurate

data item the specific value of an attribute

data manipulation the process of performing calculations and other data transformations related to business transactions

data manipulation language (DML) the commands that are used to manipulate the data in a database

data mart a subset of a data warehouse

data mining an information analysis tool that involves the automated discovery of patterns and relationships in a data warehouse

data model a diagram of data entities and their relationships

data normalization the process of taking a complex set of data and converting it into a set of simple two-dimensional tables

data preparation, or data conversion conversion of manual files into computer files

data redundancy duplication of data in separate files

data storage the process of updating one or more databases with new transactions

data store representation of a storage location for data

data warehouse a database that collects business information from many sources in the enterprise, covering all aspects of the company's processes, products, and customers

data-flow diagram (DFD) a model of objects, associations, and activities by describing how data can flow between and around various objects

data-flow line arrows that show the direction of data element movement

database an organized collection of facts and information

database administrator (DBA) a highly skilled and trained systems professional who directs or performs all activities related to maintaining a successful database environment

database approach to data management an approach whereby a pool of related data is shared by multiple application programs

database management system (DBMS) a group of programs that manipulate the database and provide an interface between the database and the user of the database and other application programs

decentralized processing processing alternative in which processing devices are placed at various remote locations

decision room a room that supports decision making, with the decision makers in the same building, combining face-to-face verbal interaction with technology to make the meeting more effective and efficient

decision structure a programming structure that allows the computer to branch, depending on certain conditions. Normally, there are only two possible branches

decision support system (DSS) an organized collection of people, procedures, software, databases, and devices used to support problem-specific decision making

decision-making phase the first part of problem solving, including three stages: intelligence, design, and choice

dedicated line a communications line that provides a constant connection between two points; no switching or dialing is needed, and the two devices are always connected

delphi approach a decision-making approach in which group decision makers are geographically dispersed; this approach encourages diversity among group members and fosters creativity and original thinking in decision making

demand reports reports developed to give certain information at a manager's request

denial-of-service attack an on-line attack of a Web site in which the attacker takes command of many computers on the Internet and causes them to flood the target site with requests for data and other tasks, keeping it too busy to serve legitimate users

design report the primary result of systems design, reflecting the decisions made for system design and preparing the way for systems implementation

design stage the second stage of decision making, in which alternative solutions to the problem are developed

deterrence controls rules and procedures to prevent problems before they occur

dialogue manager user interface that allows decision makers to easily access and manipulate the DSS and to use common business terms and phrases

digital certificate an attachment to an e-mail message or data embedded in a Web page that verifies the identity of a sender or a Web site

digital computer camera input device used with a PC to record and store images and video in digital form

digital signal a signal represented by bits

digital signature encryption technique used to verify the identity of a message sender for processing on-line financial transactions

digital subscriber line (DSL) a communications line that uses existing phone wires going into today's homes and businesses to provide transmission speeds exceeding 500 Kbps at a cost of $20 or more per month

digital versatile disk (DVD) storage medium used to store digital video or computer data

direct access retrieval method in which data can be retrieved without the need to read and discard other data

direct access storage device (DASD) device used for direct access of secondary storage data

direct conversion (also called *plunge* **or** *direct cutover*) stopping the old system and starting the new system on a given date

direct observation watching the existing system in action by one or more members of the analysis team

disaster recovery the implementation of the business resumption plan

disintermediation the elimination of intermediate organizations between the producer and the consumer

disk mirroring a process of storing data that provides an exact copy that protects users fully in the event of data loss

distance learning the use of telecommunications to extend the classroom

distributed database a database in which the data may be spread across several smaller databases connected via telecommunications systems

distributed processing processing alternative in which computers are placed at remote locations but connected to each other via telecommunications devices

document production the process of generating output records and reports

documentation text that describes the program functions to help the user operate the computer system

domain the allowable values for data attributes

domain expert The individual or group who has the expertise or knowledge one is trying to capture in the expert system

downsizing reducing the number of employees to cut costs

drill down reports reports providing increasingly detailed data about a situation

dumpster diving searching through the garbage for important pieces of information that can help crack an organization's computers or be used to convince someone at the company to give them access to the computers

dynamic Web pages Web pages containing variable information that are built in response to a specific Web visitor's request

e-commerce any business transaction executed electronically between parties such as companies (business-to-business), companies and consumers (business-to-consumer), business and the public sector, and consumers and the public sector

e-commerce software software that supports catalog management, product configuration, shopping cart facilities, e-commerce transaction processing, and Web traffic data analysis

e-commerce transaction processing software software that provides the basic connection between participants in the e-commerce economy, enabling communications between trading partners, regardless of their technical infrastructure

economic feasibility determination of whether the project makes financial sense and whether predicted benefits offset the cost and time needed to obtain them

economic order quantity (EOQ) the quantity that should be reordered to minimize total inventory costs

effectiveness a measure of the extent to which a system achieves its goals; it can be computed by dividing the goals actually achieved by the total of the stated goals

efficiency a measure of what is produced divided by what is consumed

electronic bill presentment a method of billing whereby the biller posts an image of your statement on the Internet and alerts you by e-mail that your bill has arrived

electronic cash an amount of money that is computerized, stored, and used as cash for e-commerce transactions

electronic data interchange (EDI) an intercompany, application-to-application communication of data in standard format, permitting the recipient to perform a standard business transaction

electronic document distribution process that involves transporting documents—such as sales reports, policy manuals, and advertising brochures—over communications lines and networks

electronic exchange an electronic forum where manufacturers, suppliers, and competitors buy and sell goods, trade market information, and run back-office operations

electronic retailing (e-tailing) the direct sale from business to consumer through electronic storefronts, typically designed around an electronic catalog and shopping cart model

electronic shopping cart a model commonly used by many e-commerce sites to track the items selected for purchase, allowing shoppers to view what is in their cart, add new items to it, and remove items from it

electronic software distribution process that involves installing software on a file server for users to share by signing onto the network and requesting that the software be downloaded onto their computers over a network

electronic wallet a computerized stored value that holds credit card information, electronic cash, owner identification, and address information

empowerment giving employees and their managers more responsibility and authority to make decisions, take certain actions, and have more control over their jobs

encryption the conversion of a message into a secret code

end-user systems development any systems development project in which the primary effort is undertaken by a combination of business managers and users

enterprise data modeling data modeling done at the level of the entire enterprise

enterprise resource planning (ERP) system a set of integrated programs capable of managing a company's vital business operations for an entire multisite, global organization

enterprise sphere of influence sphere of influence that serves the needs of the firm in its interaction with its environment

entity generalized class of people, places, or things for which data is collected, stored, and maintained

entity symbol representation of either a source or destination of a data element

entity-relationship (ER) diagrams data models that use basic graphical symbols to show the organization of and relationships between data

ergonomics the study of designing and positioning computer equipment for employee health and safety

event-driven review review triggered by a problem or opportunity such as an error, a corporate merger, or a new market for products

exception reports reports automatically produced when a situation is unusual or requires management action

execution time (E-time) the time it takes to execute an instruction and store the results

executive support system (ESS), or executive information system (EIS) specialized DSS that includes all hardware, software, data, procedures, and people used to assist senior-level executives within the organization

expandable storage devices storage that uses removable disk cartridges to provide additional storage capacity

expert system a system that gives a computer the ability to make suggestions and act like an expert in a particular field

expert system shells A collection of software packages and tools used to develop expert systems

explanation facility Component of an expert system that allows a user or decision maker to understand how the expert system arrived at certain conclusions or results

Extensible Markup Language (XML) markup language for Web documents containing structured information, including words, pictures, and other elements

external auditing auditing performed by an outside group

extranet a network based on Web technologies that allows selected outsiders, such as business partners and customers, to access authorized resources of the intranet of a company

feasibility analysis assessment of the technical, operational, schedule, economic, and legal feasibility of a project

feedback output that is used to make changes to input or processing activities

field typically a name, number, or combination of characters that describes an aspect of a business object or activity

file a collection of related records

file server an architecture in which the application and database reside on the one host computer, called the file server

file transfer protocol (FTP) a protocol that describes a file transfer process between a host and a remote computer and allows users to copy files from one computer to another

final evaluation a detailed investigation of the proposals offered by the vendors remaining after the preliminary evaluation

financial management information system an information system that provides financial information to all financial managers within an organization and a broader set of people who need to make better decisions

firewall a device that sits between an internal network and the Internet, limiting access into and out of a network based on access policies

five-force model a widely accepted model that identifies five key factors that can lead to attainment of competitive advantage including (1) rivalry among existing competitors, (2) the threat of new entrants, (3) the threat of substitute products and services, (4) the bargaining power of buyers, and (5) the bargaining power of suppliers

flash memory a silicon computer chip that, unlike RAM, is nonvolatile and keeps its memory when the power is shut off

flat organizational structure organizational structure with a reduced number of management layers

flexible manufacturing system (FMS) an approach that allows manufacturing facilities to rapidly and efficiently change from making one product to another

forecasting predicting future events to avoid problems

forward chaining The process of starting with the facts and working forward to the conclusions

front-end processor a special-purpose computer that manages communications to and from a computer system

full-duplex channel a communications channel that permits data transmission in both directions at the same time, thus the full-duplex channel is like two simplex lines

fuzzy logic A special research area in computer science that allows shades of gray and does not require everything to be simple black or white, yes/no, or true/false

Gantt chart a graphical tool used for planning, monitoring, and coordinating projects

gateway connection that operates at or above the OSI transport layer and links LANs or networks that employ different, higher-level protocols and allows networks with very different architectures and using dissimilar protocols to communicate

general ledger system system designed to automate financial reporting and data entry

general-purpose computers computers used for a wide variety of applications

geographic information system (GIS) a computer system capable of assembling, storing, manipulating, and displaying geographic information, i.e., data identified according to their locations

gigahertz (GHz) billions of cycles per second

goal-seeking analysis the process of determining the problem data required for a given result

graphical user interface (GUI) an interface that uses icons and menus displayed on screen to send commands to the computer system

grid chart table that shows relationships among the various aspects of a systems development effort

group consensus decision making by a group that is appointed and given the responsibility of making the final evaluation and selection

group consensus approach decision-making approach that forces members in the group to reach a unanimous decision

group decision support system (GDSS) software application that consists of most elements in a DSS, plus software needed to provide effective support in group decision making; also called *group support system* or *computerized collaborative work system*

groupware software that helps groups of people work together more efficiently and effectively

hacker a person who enjoys computer technology and spends time learning and using computer systems

half-duplex channel a communications channel that can transmit data in either direction, but not simultaneously

hardware computer equipment used to perform input, processing, and output activities

help facility a program that provides assistance when a user is having difficulty understanding what is happening or what type of response is expected

hertz one cycle or pulse per second

heuristics commonly accepted guidelines or procedures that usually find a good solution

hierarchical database model a data model in which data is organized in a top-down, or inverted tree, structure

hierarchical network a type of topology that uses a treelike structure with messages passed along the branches of the hierarchy until they reach their destination

hierarchy of data bits, characters, fields, records, files, and databases

highly structured problems problems that are straightforward and require known facts and relationships

home page a cover page for a Web site that has graphics, titles, and text

hot site a duplicate, operational hardware system or immediate access to one through a specialized vendor

HTML tags codes that let the Web browser know how to format text—as a heading, as a list, or as body text—and whether images, sound, and other elements should be inserted

human resource MIS an information system that is concerned with activities related to employees and potential employees of an organization, also called a personnel MIS

hybrid network a network topology that is a combination of other network types

hypermedia tools that connect the data on Web pages, allowing users to access topics in whatever order they wish

hypertext markup language (HTML) the standard page description language for Web pages

icon picture

ICRA rating system system to protect individuals from harmful or objectionable Internet content while safeguarding the free speech rights of others

identity theft a crime in which an imposter obtains key pieces of personal identification information, such as social security or driver's license numbers, in order to impersonate someone else. The information is then used to obtain credit, merchandise, and services in the name of the victim, or to provide the thief with false credentials

if-then statements Rules that suggest certain conclusions

image log a separate file that contains only changes to applications

implementation stage a stage of problem solving in which a solution is put into effect

in-house development development of application software using the company's resources

incremental backup making a backup copy of all files changed during the last few days or the last week

inference engine Part of the expert system that seeks information and relationships from the knowledge base and provides answers, predictions, and suggestions the way a human expert would

information a collection of facts organized in such a way that they have additional value beyond the value of the facts themselves

information center a support function that provides users with assistance, training, application development, documentation, equipment selection and setup, standards, technical assistance, and troubleshooting

information service unit a miniature IS department

information system (IS) a set of interrelated components that collect, manipulate, and disseminate data and information and provide a feedback mechanism to meet an objective

information systems literacy knowledge of how data and information are used by individuals, groups, and organizations

information systems planning the translation of strategic and organizational goals into systems development initiatives

input the activity of gathering and capturing raw data

insiders employees, disgruntled or otherwise, working solo or in concert with outsiders to compromise corporate systems

installation the process of physically placing the computer equipment on the site and making it operational

instant messaging a method that allows two or more individuals to communicate on-line using the Internet

institutional DSS a DSS that handles situations or decisions that occur more than once, usually several times a year or more. An institutional DSS is used repeatedly and refined over the years

instruction time (I-time) the time it takes to perform the fetch-instruction and decode-instruction steps of the instruction phase

integrated development environments (IDEs) development approach that combines the tools needed for programming with a programming language into one integrated package

integrated services digital network (ISDN) a technology that uses existing common-carrier lines to simultaneously transmit voice, video, and image data in digital form

integrated-CASE tools (I-CASE) tools that provide links between upper- and lower-CASE packages, thus allowing lower-CASE packages to generate program code from upper-CASE package designs

integration testing testing all related systems together

intellectual property music, books, inventions, paintings, and other special items protected by patents, copyrights, or trademarks

intelligence stage the first stage of decision making, in which potential problems or opportunities are identified and defined

intelligent behavior the ability to learn from experiences and apply knowledge acquired from experience, handle complex situations, solve problems when important information is missing, determine what is important, react quickly and correctly to a new situation, understand visual images, process and manipulate symbols, be creative and imaginative, and use heuristics

internal auditing auditing performed by individuals within the organization

international network a network that links systems between countries

Internet the world's largest computer network, actually consisting of thousands of interconnected networks, all freely exchanging information

Internet piracy illegally gaining access to and using the Internet

Internet protocol (IP) communication standard that enables traffic to be routed from one network to another as needed

Internet service provider (ISP) any company that provides individuals or organizations with access to the Internet

intranet an internal network based on Web technologies that allows people within an organization to exchange information and work on projects

intrusion detection system (IDS) Security system that monitors system and network resources and notifies network security personnel when it senses a possible intrusion

inventory control system system that updates the computerized inventory records to reflect the exact quantity on hand of each stock-keeping unit

Java an object-oriented programming language from Sun Microsystems based on C++ that allows small programs (applets) to be embedded within an HTML document

joining data manipulation that combines two or more tables

joint application development (JAD) process for data collection and requirements analysis

just-in-time (JIT) inventory approach a philosophy of inventory management in which inventory and materials are delivered just before they are used in manufacturing a product

kernel the heart of the operating system which controls the most critical processes

key a field or set of fields in a record that is used to identify the record

key-indicator report summary of the previous day's critical activities; typically available at the beginning of each workday

knowledge an awareness and understanding of a set of information and ways that information can be made useful to support a specific task or reach a decision

knowledge acquisition facility Part of the expert system that provides convenient and efficient means of capturing and storing all the components of the knowledge base

knowledge base the collection of data, rules, procedures, and relationships that must be followed to achieve value or the proper outcome

knowledge engineer An individual who has training or experience in the design, development, implementation, and maintenance of an expert system

knowledge management the process of capturing a company's collective expertise wherever it resides—in computers, on paper, in people's heads—and distributing it wherever it can help produce the biggest payoff

knowledge user The individual or group who uses and benefits from the expert system

learning systems A combination of software and hardware that allows the computer to change how it functions or reacts to situations based on feedback it receives

legal feasibility determination of whether laws or regulations may prevent or limit a systems development project

linking data manipulation that combines two or more tables using common data attributes to form a new table with only the unique data attributes

local area network (LAN) a network that connects computer systems and devices within the same geographic area

logic bomb an application or system virus designed to "explode" or execute at a specified time and date

logical design description of the functional requirements of a system

lookup tables tables containing data that are developed and used by computer programs to simplify and shorten data entry

loop structure a programming structure with two commonly used structures for loops: do-until and do-while. In the do-until structure, the loop is done until a certain condition is met. For the do-while structure, the loop is done while a certain condition exists

lower-CASE tools tools that focus on the later implementation stage of systems development

machine cycle the instruction phase followed by the execution phase

macro virus a virus that infects documents by using an application's own macro programming language to distribute itself

magnetic disk common secondary storage medium, with bits represented by magnetized areas

magnetic tape common secondary storage medium, Mylar film coated with iron oxide with portions of the tape magnetized to represent bits

magneto-optical disk a hybrid between a magnetic disk and an optical disk

mainframe computer large, powerful computer often shared by hundreds of concurrent users connected to the machine via terminals

maintenance team a special IS team responsible for modifying, fixing, and updating existing software

make-or-buy decision the decision regarding whether to obtain the necessary software from internal or external sources

managed security service provider (MSSP) An organization that monitors and maintains network security hardware and software for its client companies

management information system (MIS) an organized collection of people, procedures, software, databases, and devices used to provide routine information to managers and decision makers

manufacturing resource planning (MRPII) an integrated, companywide system based on network scheduling that enables people to run their business with a high level of customer service and productivity

market segmentation the identification of specific markets to target them with advertising messages

marketing MIS information system that supports managerial activities in product development, distribution, pricing decisions, and promotional effectiveness

massively parallel processing a form of multiprocessing that speeds processing by linking hundreds or thousands of processors to operate at the same time, or in parallel, with each processor having its own bus, memory, disks, copy of the operating system, and applications

material requirements planning (MRP) a set of inventory control techniques that help coordinate thousands of inventory items when the demand of one item is dependent on the demand for another that determines when to order more inventory

megahertz (MHz) millions of cycles per second

menu-driven system system in which users simply pick what they want to do from a list of alternatives

meta tag a special HTML tag, not visible on the displayed Web page, that contains keywords representing your site's content, which search engines use to build indexes pointing to your Web site

meta-search engine a tool that submits keywords to several individual search engines and returns the results from all search engines queried

microcode predefined, elementary circuits and logical operations that the processor performs when it executes an instruction

midrange computer formerly called minicomputer, a system about the size of a small three-drawer file cabinet that can accommodate several users at one time

MIPS millions of instructions per second

mission-critical systems systems that play a pivotal role in an organization's continued operations and goal attainment

model an abstraction or an approximation that is used to represent reality

model base part of a DSS that provides decision makers access to a variety of models and assists them in decision making

model management software (MMS) software that coordinates the use of models in a DSS

modem a device that translates data from digital to analog and analog to digital

monitoring stage final stage of the problem-solving process, in which decision makers evaluate the implementation

Moore's Law a hypothesis that states that transistor densities on a single chip will double every 18 months

multidimensional organizational structure structure that may incorporate several structures at the same time

multifunction device a device that can combine a printer, fax machine, scanner, and copy machine into one device

multiplexer a device that allows several telecommunications signals to be transmitted over a single communications medium at the same time

multiprocessing simultaneous execution of two or more instructions at the same time

multitasking capability that allows a user to run more than one application at the same time

music device a device that can be used to download music from the Internet and play the music

natural language processing Processing that allows the computer to understand and react to statements and commands made in a "natural" language, such as English

net present value the preferred approach for ranking competing projects and determining economic feasibility

network computer a cheaper-to-buy and cheaper-to-run version of the personal computer that is used primarily for accessing networks and the Internet

network management software software that enables a manager on a networked desktop to monitor the use of individual computers and shared hardware (like printers), scan for viruses, and ensure compliance with software licenses

network model an expansion of the hierarchical database model with an owner-member relationship in which a member may have many owners

network operating system (NOS) systems software that controls the computer systems and devices on a network and allows them to communicate with each other

network topology logical model that describes how networks are structured or configured

networks connected computers and computer equipment in a building, around the country, or around the world to enable electronic communications

neural network A computer system that can simulate the functioning of a human brain

newsgroups on-line discussion groups that focus on specific topics

nominal group technique decision-making approach that encourages feedback from individual group members, and the final decision is made by voting, similar to the way public officials are elected

nonoperational prototype a mock-up, or model, that includes output and input specifications and formats

nonprogrammed decisions decisions that deal with unusual or exceptional situations

object a collection of data and programs

object-oriented systems development (OOSD) approach to systems development that combines the logic of the systems development life cycle with the power of object-oriented modeling and programming

object-relational database management system (ORDBMS) a DBMS capable of manipulating audio, video, and graphical data

off-the-shelf software existing software program

on-line analytical processing (OLAP) software that allows users to explore data from a number of different perspectives

on-line transaction processing (OLTP) computerized processing in which each transaction is processed immediately, without the delay of accumulating transactions into a batch

open database connectivity (ODBC) standards that ensure that software can be used with any ODBC-compliant database

open shops IS department in which other people, such as programmers and systems analysts, are also authorized to run the computers

open source software software that is freely available to anyone in a form that can be easily modified

Open Systems Interconnection (OSI) model a standard model for network architectures that divides data communications functions into seven distinct layers to promote the development of modular networks that simplify the development, operation, and maintenance of complex telecommunications networks

opensourcing extending software development beyond a single organization by finding others who share the same problem and involving them in a common development effort

operating system (OS) a set of computer programs that controls the computer hardware and acts as an interface with application programs

operational feasibility measure of whether the project can be put into action or operation

operational prototype a functioning prototype that accesses real data files, edits input data, makes necessary computations and comparisons, and produces real output

optical disk a rigid disk of plastic onto which data is recorded by special lasers that physically burn pits in the disk

optical processors computer chips that use light waves instead of electrical current to represent bits

optimization model a process to find the best solution, usually the one that will best help the organization meet its goals

order entry system process that captures the basic data needed to process a customer order

order processing systems systems that process order entry, sales configuration, shipment planning, shipment execution, inventory control, invoicing, customer relationship management, and routing and scheduling

organization a formal collection of people and other resources established to accomplish a set of goals

organizational change the responses that are necessary for profit and non-profit organizations to plan for, implement, and handle change

organizational culture the major understandings and assumptions for a business, a corporation, or an organization

organizational learning adaptations to new conditions or alterations of organizational practices over time

organizational structure organizational subunits and the way they are related to the overall organization

output production of useful information, usually in the form of documents and reports

outsourcing contracting with outside professional services to meet specific business needs

paging process of swapping programs or parts of programs between memory and one or more disk devices

parallel start-up running both the old and new systems for a period of time and comparing the output of the new system closely with the output of the old system; any differences are reconciled. When users are comfortable that the new system is working correctly, the old system is eliminated

password sniffer a small program hidden in a network or a computer system that records identification numbers and passwords

patch a minor change to correct a problem or make a small enhancement. It is usually an addition to an existing program

payroll journal a report that contains employees' names, the area where employees worked during the week, hours worked, the pay rate, a premium factor for overtime pay, earnings, earnings type, various deductions, and net pay calculations

perceptive system A system that approximates the way a human sees, hears, and feels objects

personal computer (PC) relatively small, inexpensive computer system, sometimes called a microcomputer

personal productivity software software that enables users to improve their personal effectiveness, increasing the amount of work they can do and its quality

personal sphere of influence sphere of influence that serves the needs of an individual user

phase-in approach slowly replacing components of the old system with those of the new one. This process is repeated for each application until the new system is running every application and performing as expected; also called piecemeal approach

physical design specification of the characteristics of the system components necessary to put the logical design into action

pilot start-up running the new system for one group of users rather than all users

pipelining a form of CPU operation in which there are multiple execution phases in a single machine cycle

pixel a dot of color on a photo image or a point of light on a display screen

planned data redundancy a way of organizing data in which the logical database design is altered so that certain data entities are combined, summary totals are carried in the data records rather than calculated from elemental data, and some data attributes are repeated in more than one data entity to improve database performance

Platform for Privacy Preferences (P3P) a screening technology that shields users from Web sites that don't provide the level of privacy protection they desire

plotter a type of hard-copy output device used for general design work

point evaluation system an evaluation process in which each evaluation factor is assigned a weight, in percentage points, based on importance. Then each proposed information system is evaluated in terms of this factor and given a score ranging from 0 to 100. The scores are totaled, and the system with the greatest total score is selected

point-of-sale (POS) device terminal used in retail operations to enter sales information into the computer system

point-to-point protocol (PPP) a communications protocol that transmits packets over telephone lines

predictive analysis a form of data mining that combines historical data with assumptions about future conditions to predict outcomes of events such as future product sales or the probability that a customer will default on a loan

preliminary evaluation an initial assessment whose purpose is to dismiss the unwanted proposals; begins after all proposals have been submitted

primary key a field or set of fields that uniquely identifies the record

primary storage (main memory; memory) part of the computer that holds program instructions and data

private branch exchange (PBX) a communications system that can manage both voice and data transfer within a building and to outside lines

problem solving a process that goes beyond decision making to include the implementation stage

procedures the strategies, policies, methods, and rules for using a CBIS

process a set of logically related tasks performed to achieve a defined outcome

process symbol representation of a function that is performed

processing converting or transforming data into useful outputs

product configuration software software used by buyers to build the product they need on-line

productivity a measure of the output achieved divided by the input required

programmer specialist responsible for modifying or developing programs to satisfy user requirements

profit centers departments within an organization that track total expenses and net profits

Program Evaluation and Review Technique (PERT) a formalized approach for developing a project schedule

program-data dependence concept according to which programs and data developed and organized for one application are incompatible with programs and data organized differently for another application

programmed decisions decisions made using a rule, procedure, or quantitative method

programming languages sets of keywords, symbols, and a system of rules for constructing statements by which humans can communicate instructions to be executed by a computer

programming life cycle a series of steps and planned activities developed to maximize the likelihood of developing good software

project deadline date the entire project is to be completed and operational

project milestone a critical date for the completion of a major part of the project

project organizational structure structure centered on work teams or groups

project schedule detailed description of what is to be done

projecting data manipulation that eliminates columns in a table

proprietary software a one-of-a-kind program for a specific application

protocols rules that ensure communications among computers of different types and from different manufacturers

prototyping an iterative approach to the systems development process

Public Key Infrastructure (PKI) A means to enable users of an unsecure public network such as the Internet to securely and privately exchange data through the use of a public and a private cryptographic key pair that is obtained and shared through a trusted authority

public network services systems that give personal computer users access to vast databases and other services, usually for an initial fee plus usage fees

purchase order processing system system that helps purchasing departments complete their transactions quickly and efficiently

purchasing transaction processing systems systems that include inventory control, purchase order processing, receiving, and accounts payable

push technology automatic transmission of information over the Internet rather than making users search for it with their browsers

quality the ability of a product (including services) to meet or exceed customer expectations

quality control a process that ensures that the finished product meets the customers' need

questionnaires a method of gathering data when the data sources are spread over a wide geographic area

random access memory (RAM) a form of memory in which instructions or data can be temporarily stored

rapid application development (RAD) a systems development approach that employs tools, techniques, and methodologies designed to speed application development

read-only memory (ROM) a nonvolatile form of memory

receiving system system that creates a record of expected receipts

record a collection of related data fields

reduced instruction set computing (RISC) a computer chip design based on reducing the number of microcode instructions built into a chip to an essential set of common microcode instructions

redundant array of independent/inexpensive disks (RAID) method of storing data that generates extra bits of data from existing data, allowing the system to create a "reconstruction map" so that if a hard drive fails, it can rebuild lost data

reengineering (process redesign) the radical redesign of business processes, organizational structures, information systems, and values of the organization to achieve a breakthrough in business results

register high-speed storage area in the CPU used to temporarily hold small units of program instructions and data immediately before, during, and after execution by the CPU

relational model a database model that describes data in which all data elements are placed in two-dimensional tables, called *relations*, that are the logical equivalent of files

release a significant program change that often requires changes in the documentation of the software

reorder point (ROP) a critical inventory quantity level

repetitive motion disorder (repetitive stress injury; RSI) an injury that can be caused by working with computer keyboards and other equipment

replicated database a database that holds a duplicate set of frequently used data

report layout technique that allows designers to diagram and format printed reports

request for maintenance form a form authorizing modification of programs

request for proposal (RFP) a document that specifies in detail required resources such as hardware and software

requirements analysis determination of user, stakeholder, and organizational needs

restart procedures simplified process to access an application from where it left off

return on investment (ROI) one measure of IS value that investigates the additional profits or benefits that are generated as a percentage of the investment in information systems technology

revenue centers divisions within a company that track sales or revenues

ring network a type of topology that contains computers and computer devices placed in a ring, or circle; there is no central coordinating computer; messages are routed around the ring from one device or computer to another

robotics Mechanical or computer devices that perform tasks requiring a high degree of precision or that are tedious or hazardous for humans

router connection that operates at the network level of the OSI model and features more sophisticated addressing software than bridges; whereas bridges simply pass along everything that comes to them, routers can determine preferred paths to a final destination

routing system system that determines the best way to get products from one location to another

rule A conditional statement that links given conditions to actions or outcomes

safe harbor principles a set of principles that address the e-commerce data privacy issues of notice, choice, and access

sales configuration system process that ensures that the products and services ordered are sufficient to accomplish the customer's objectives and will work well together

satisficing model a model that will find a good—but not necessarily the best—problem solution

scalability the ability to increase the capability of a computer system to process more transactions in a given period by adding more, or more powerful, processors

schedule feasibility determination of whether the project can be completed in a reasonable amount of time

scheduled reports reports produced periodically, or on a schedule, such as daily, weekly, or monthly

scheduling system system that determines the best time to deliver goods and services

schema a description of the entire database

screen layout a technique that allows a designer to quickly and efficiently design the features, layout, and format of a display screen

script bunnies wannabe crackers with little technical savvy who download programs—scripts—that automate the job of breaking into computers

search engine a Web search tool

secondary storage (permanent storage) devices that store larger amounts of data, instructions, and information more permanently than allowed with main memory

secure sockets layer (SSL) a communications protocol used to secure sensitive data

selecting data manipulation that eliminates rows according to certain criteria

selective backup creating backup copies of only certain files

semistructured or unstructured problems more complex problems in which the relationships among the data are not always clear, the data may be in a variety of formats, and the data is often difficult to manipulate or obtain

sequence structure a programming structure in which, after starting the sequence, programming statements are executed one after another until all the statements in the sequence have been executed. Then the program either ends or continues on to another sequence

sequential access retrieval method in which data must be accessed in the order in which it is stored

sequential access storage device (SASD) device used to sequentially access secondary storage data

serial line Internet protocol (SLIP) a communications protocol that transmits packets over telephone lines

Shannon's fundamental law of information theory the law of telecommunications that states that the information-carrying capacity of a channel is directly proportional to its bandwidth—the broader the bandwidth, the more information can be carried

shipment execution system system that coordinates the outflow of all products from the organization, with the objective of delivering quality products on time to customers

shipment planning system system that determines which open orders will be filled and from which location they will be shipped

sign-on procedure identification numbers, passwords, and other safeguards needed for an individual to gain access to computer resources

simplex channel a communications channel that can transmit data in only one direction

simulation the ability of the DSS to duplicate the features of a real system

site preparation preparation of the location of the new system

slipstream upgrade a minor upgrade—typically a code adjustment or minor bug fix—not worth announcing. It usually requires recompiling all the code and, in so doing, it can create entirely new bugs

smart card a credit card–sized device with an embedded microchip to provide electronic memory and processing capability

social engineering the practice of talking a critical computer password out of an individual

software the computer programs that govern the operation of the computer

software bug a defect in a computer program that keeps it from performing in the manner intended

software interface programs or program modifications that allow proprietary software to work with other software used in the organization

software piracy the act of illegally duplicating software

software suite a collection of single-application software packages in a bundle

source data automation capturing and editing data where the data is originally created and in a form that can be directly input to a computer, thus ensuring accuracy and timeliness

spam e-mail sent to a wide range of people and Usenet groups in discriminately

special-purpose computers computers used for limited applications by military and scientific research groups

sphere of influence the scope of problems and opportunities addressed by a particular organization

split-case distribution a distribution system that requires cases of goods to be opened on the receiving dock and the individual items from the cases are stored in the manufacturer's warehouse

stakeholders individuals who, either themselves or through the organization they represent, ultimately benefit from the system development project

star network a type of topology that has a central hub or computer system, and other computers or computer devices are located at the end of communications lines that originate from the central hub or computer

start-up the process of making the final tested information system fully operational

static Web pages Web pages that always contain the same information

statistical sampling selection of a random sample of data and applying the characteristics of the sample to the whole group

steering committee an advisory group consisting of senior management and users from the IS department and other functional areas

storage area network (SAN) technology that provides high-speed connections between data storage devices and computers over a network using the Fibre Channel communications protocol

storefront broker companies that act as middlemen between your Web site and on-line merchants that have the products and retail expertise

strategic alliance (strategic partnership) an agreement between two or more companies that involves the joint production and distribution of goods and services

strategic planning determining long-term objectives by analyzing the strengths and weaknesses of the organization, predicting future trends, and projecting the development of new product lines

structured interview an interview where the questions are written in advance

structured walkthrough a planned and preannounced review of the progress of a program module, a structure chart, or a human procedure

subschema a file that contains a description of a subset of the database and identifies which users can view and modify the data items in the subset

supercomputers the most powerful computer systems, with the fastest processing speeds

superconductivity a property of certain metals that allows current to flow with minimal electrical resistance

supply chain management a key value chain composed of demand planning, supply planning, and demand fulfillment

switch a device that routes or switches data to its destination

switched line a communications line that uses switching equipment to allow one transmission device to be connected to other transmission devices

symmetrical multiprocessing (SMP) another form of parallel processing in which multiple processors run a single copy of the operating system and share the memory and other resources of one computer

synchronous communications communication in which the receiver gets the message instantaneously

syntax a set of rules associated with a programming language

system a set of elements or components that interact to accomplish goals

system boundary the limits of the system; it defines the system and distinguishes it from everything else (the environment)

system parameter a value or quantity that cannot be controlled, such as the cost of a raw material

system performance measurement monitoring the system—the number of errors encountered, the amount of memory required, the amount of processing or CPU time needed, and other problems

system performance products software that measures all components of the computer-based information system, including hardware, software, database, telecommunications, and network systems

system performance standard a specific objective of the system

system testing testing the entire system of programs

system variable a quantity or item that can be controlled by the decision maker

system virus a virus that typically infects operating system programs or other system files

systems analyst professional who specializes in analyzing and designing business systems

systems analysis a stage of systems development during which the problems and opportunities of the existing system are defined

systems controls rules and procedures to maintain data security

systems design a stage of systems development that determines how the new system will work to meet the business needs defined during systems analysis

systems development the activity of creating or modifying existing business systems

systems implementation a stage of systems development during which the various system components (hardware, software, databases, etc.) defined in the design step are created or acquired and then assembled and the new system is put into operation

systems investigation a stage of systems development that has as its goal to gain a clear understanding of the problem to be solved or opportunity to be addressed

systems investigation report summary of the results of the systems investigation and the process of feasibility analysis and recommendation of a course of action

systems maintenance stage of systems development that involves checking, changing, and enhancing the system to make it more useful in achieving user and organizational goals

systems maintenance and review a stage of the systems development process that has as its goal to check and modify the system so that it continues to meet changing business needs

systems request form document filled out by someone who wants the IS department to initiate systems investigation

systems review the final step of systems development, involving the analysis of systems to make sure they are operating as intended

systems software the set of programs designed to coordinate the activities and functions of the hardware and various programs throughout the computer system

T1 carrier a line or channel developed by AT&T and used in North America to increase the number of voice calls that can be handled through existing cables

team organizational structure structure centered on work teams or groups

technical documentation written details used by computer operators to execute the program and by analysts and programmers in case there are problems with the program or the program needs modification

technical feasibility assessment of whether the hardware, software, and other system components can be acquired or developed to solve the problem

technology acceptance model (TAM) a model that describes the factors that lead to higher levels of acceptance and usage of technology

technology diffusion a measure of how widely technology is spread throughout the organization

technology infrastructure all the hardware, software, databases, telecommunications, people, and procedures that are configured to collect, manipulate, store, and process data into information

technology infusion the extent to which technology is deeply integrated into an area or department

technology-enabled relationship management the use of detailed information about a customer's behavior, preferences, needs, and buying patterns to set prices, negotiate terms, tailor promotions, add product features, and otherwise customize the entire relationship with that customer

telecommunications the electronic transmission of signals for communications; enables organizations to carry out their processes and tasks through effective computer networks

telecommunications medium anything that carries an electronic signal and interfaces between a sending device and a receiving device

telecommuting a work arrangement whereby employees work away from the office using personal computers and networks to communicate via e-mail with other workers and to pick up and deliver results

telnet a terminal emulation protocol that enables users to log on to other computers on the Internet to gain access to public files

terminal-to-host an architecture in which the application and database reside on one host computer, and the user interacts with the application and data using a "dumb" terminal

time-driven review review performed after a specified amount of time

time-sharing capability that allows more than one person to use a computer system at the same time

top-down approach a good general approach to writing a large program, starting with the main module and working down to the other modules

total cost of ownership (TCO) measurement of the total cost of owning computer equipment, including desktop computers, networks, and large computers

total quality management (TQM) a collection of approaches, tools, and techniques that offers a commitment to quality throughout the organization

traditional approach to data management an approach whereby separate data files are created and stored for each application program

traditional organizational structure organizational structure in which major department heads report to a president or top-level manager

transaction any business-related exchange such as payments to employees, sales to customers, or payments to suppliers

transaction processing cycle the process of data collection, data editing, data correction, data manipulation, data storage, and document production

transaction processing system (TPS) an organized collection of people, procedures, software, databases, and devices used to record completed business transactions

transaction processing system audit an examination of the TPS to answer whether the system meets the business need for which it was implemented, what procedures and controls have been established, whether these procedures and controls are being used properly, and whether the information systems and procedures are producing accurate and honest reports

Transmission Control Protocol/Internet Protocol (TCP/IP) the primary communications protocol of the Internet, originally developed to link defense research agencies

transport control protocol (TCP) widely used transport layer protocol that is used in combination with IP by most Internet applications

Trojan horse a program that appears to be useful but actually masks a destructive program

tunneling the process by which VPNs transfer information by encapsulating traffic in IP packets over the Internet

uniform resource locator (URL) an assigned address on the Internet for each computer

unit testing testing of individual programs

unstructured interview an interview where the questions are not written in advance

upper-CASE tools tools that focus on activities associated with the early stages of systems development

usenet a system closely allied with the Internet that uses e-mail to provide a centralized news service; a protocol that describes how groups of messages can be stored on and sent between computers

user acceptance document formal agreement signed by the user that states that a phase of the installation or the complete system is approved

user documentation written description developed for individuals who use a program, showing users, in easy-to-understand terms, how the program can and should be used

user interface element of the operating system that allows individuals to access and command the computer system

user preparation the process of readying managers, decision makers, employees, other users, and stakeholders for the new systems

users individuals who will interact with the system regularly

utility programs programs used to merge and sort sets of data, keep track of computer jobs being run, compress data files before they are stored or transmitted over a network, and perform other important tasks

value chain a series (chain) of activities that includes inbound logistics, warehouse and storage, production, finished product storage, outbound logistics, marketing and sales, and customer service

value-added carriers companies that have developed private telecommunications systems and offer their services for a fee

version a major program change, typically encompassing many new features

videoconferencing a telecommunication system that combines video and phone call capabilities with data or document conferencing

virtual memory memory that allocates space on the hard disk to supplement the immediate, functional memory capacity of RAM

virtual organizational structure structure that employs individuals, groups, or complete business units in geographically dispersed areas

virtual private network (VPN) a secure connection between two points across the Internet

virtual reality originally, the term referred to immersive virtual reality, which means the user becomes fully immersed in an artificial, three-dimensional world that is completely generated by a computer

virtual reality system A system that enables one or more users to move and react in a computer-simulated environment

virtual workgroups teams of people located around the world working on common problems

virus a program that attaches itself to other programs

vision systems The hardware and software that permit computers to capture, store, and manipulate visual images and pictures

voice mail technology that enables users to leave, receive, and store verbal messages for and from other people around the world

voice-over-IP (VOIP) technology that enables network managers to route phone calls and fax transmissions over the same network they use for data

voice-recognition device an input device that recognizes human speech

volume testing testing the application with a large amount of data

Web appliance a device that can connect to the Internet, typically through a phone line

Web auction an Internet site that matches people who want to sell products and services with people who want to purchase these products and services

Web browser software that creates a unique, hypermedia-based menu on a computer screen, providing a graphical interface to the Web

Web log file a file that contains information about visitors to a Web site

Web page construction software software that uses Web editors and extensions to produce both static and dynamic Web pages

Web site development tools tools used to develop a Web site, including HTML or visual Web page editor, software development kits, and Web page upload support

Web site hosting companies companies that provide the tools and services required to set up a Web page and conduct e-commerce within a matter of days and with little up-front cost

Web site traffic data analysis software software that processes and analyzes data from the Web log file to provide useful information to improve Web site performance

what-if analysis the process of making hypothetical changes to problem data and observing the impact on the results

wide area network (WAN) a network that ties together large geographic regions using microwave and satellite transmission or telephone lines

wordlength the number of bits the CPU can process at any one time

workflow system rule-based management software that directs, coordinates, and monitors execution of an interrelated set of tasks arranged to form a business process

workgroup two or more people who work together to achieve a common goal

workgroup sphere of influence sphere of influence that serves the needs of a workgroup

workstation computer that fits between high-end personal computers and low-end midrange computers in terms of cost and processing power

World Wide Web (WWW, or W3) a collection of tens of thousands of independently owned computers that work together as one in an Internet service

worm an independent program that replicates its own program files until it interrupts the operation of networks and computer systems

SUBJECT

A boldface page number indicates a key term and the location where its definition can be found.

NAME & COMPANY

A

Accenture Technology, 522
Accenture Technology Labs, 488, 489
Ace Hardware, 370
ACNielsen, 221
Acohido, Byron, 176n
Actuality Systems, 485
Adobe, 257
Advanced Micro Devices, 91
Aerosmith, 487–488
Aerospazio, 233
Aetna, 174, 346–347, 510
Agency for Healthcare Research and
    Quality, 620
Agile Software, 270
Airborne, 340
Airbus, 333
Aircuity, Inc., 479
AI Trilogy, 469–470
Alaska Airlines, 429
Alberston's, 319
Alcatel, 233
Alenia, 233
Aleri Inc., 367
AllEventsTickets.com, 369–370
Allied Irish Bank, 367
AltaVista, 295, 334
Altra Energy Technologies, 18
Alventive, 270
Amazon.com, 21, 54, 55, 292, 320, 331,
    338, 352
Amdocs Limited, 384
American Airlines, 17, 60
American Blind and Wallpaper Factory, 45
American Cancer Society, The, 249
American Eagle Outfitters Inc., 379
American Express, 344, 522
American Home Products Corporation,
    168, 618
American National Standards Institute
    (ANSI), 197
American Outdoor Products, 321
American Telephone & Telegraph (AT&T). *See*
    AT&T
America Online, 60, 144, 260, 261, 280, 281,
    284, 636
Ameritrade, 60, 70
AMR Research, 426
Anderson, Bruce, 444
Anderson, Robert, 415
Anthes, Gary H., 466n
AOL Time Warner, 166, 236, 288, 294, 346,
    619. *See also* America Online
Apparent Technologies, 488
Apple Computer, 60, 93, 111, 113–114, 140,
    142, 261
Applied Industrial Technologies, 322
Applix, Inc., 384
Ariba, 333
Art.com, 431
Artificial Life, Inc., 461
Ascend Software, 420
Ask Jeeves, 295
Aspect Development, 515–516
Associated Food Stores, 488
Association for Computing Machinery (ACM),
    648, 649–650
Association of Information Technology
    Professionals (AITP), 648, 649
AT&T, 57, 141, 237, 240, 281, 483, 522

AT&T Broadband, 345
@Home, 236
Atlanta Journal-Constitution, 114–115
Austin-Hayne, 483
Australian Broadcasting Authority (ABA), 656
Automated Data Processing, Inc., 19
AutoNation.com, 331
Avaya Inc., 26
AXA Group, 207

B

Baan, 159, 395
BankAmerica, 581
Bank One Corp., 17
Barclays Bank PLC, 617
Barlas, Pete, 78n
Barnes and Noble, 112, 320
Barrett, Rick, 76n
BASF, 8
Batista, Elisa, 269n
Bears, Stearns & Co., 269
BEA Systems, 162
Bechtel, 50
Bednarz, Ann, 222n
Bell Atlantic, 408
Bell Canada, 164
Bell Harbor Marina, 359
Bell Labs, 231
BellSouth, 86, 281
Benioff, Marc, 578–579
Berg, Jeff, 43
Berger, Matt, 208n
Berners-Lee, Tim, 291
Best Buy, 26, 184, 185–186
Better Business Bureau, 346, 347
Bigstep, 350
Blair Corp., 161
Blanchet, Marcel, 55
Blockbuster Inc., 60, 210–211
Bloomberg, 621
Bloomberg, Michael, 621
Bloomberg.com, 334
Blue Cross/Blue Shield, 320
Blue Cross Blue Shield of Massachusetts, 382
Blyskal, Bob, 3
Bob Evans Farms, 17, 232
Boeing, 14, 275, 379, 541–542, 561
Boothe, Lisa, 17
Borland, 161
Bose Corporation, 257
Boulton, Clint, 118n
BP, 438
Brady, Mike, 17
Brandeis, Louis, 637
Brewin, Bob, 226n, 233n
Britannica, 292
British Telecommunications PLC, 510, 511
BroadVision, 300, 330
Broughton, Donald, 557
Browse3DCorp, 312–313
Buckley, John, 294
Buckman Laboratories, 18
Burdean, Grigore, 486
Bureau of Motor Vehicles, 352
Bureau of Vital Statistics, 352
Burgess, S., 612n
Burlington Coat Factory, 558
Burlington Northern Santa Fe Corp.
    (BNSF), 262
Burst.com, Inc., 166

Bush, George W., 421, 615
Business Software Alliance (BSA), 629
Business Winstone, 91
Buter, Ed, 26
Buy.com, 435
Buzek, Greg, 26

C

Cable & Wireless USA, 345
Calico Commerce, 553
California Independent System Operator (Cal-
    ISO), 623
CampusShip (UPS), 457–458
Candle, 598
Cannon, 577
Cantor Fitzgerald, 568
Carey, John, 580
Carlson Wagonlit Travel, 297
Carnegie Mellon University, 527, 621
    Software Engineering Institute, 163
Carrefour, 499
Carter, Robert B., 557–558a
Casio, 145
Caterpillar, 333, 443–444
Caterpillar Financial Services Corporation, 295
Catholic Healthcare West (CHW), 620
CBS, 411
Celanese Chemicals, 228
Celera Genomics, 85, 96
Celestica, 131
Center for Advanced Information Processing,
    Rutgers University, 486
Center for Digital Government, 36–37
Centers for Disease Control (CDC), 292, 509
CERN, 201, 291
CertifiedMail.com, 299
Cervalis, 338
Chamberlain, D.D., 197
Charette, Brian, 442
Charles Schwab & Co. *See* Schwab,
    Charles, & Co.
CheckFree, 323, 336
Chevron, 25
Chevron Phillips Chemical Company, 469
Chick-fil-A, 396
China Merchants Bank, 145
China Telecom, 233
ChoicePoint, 639
Christopher, T. Heun, 319n
Chun, Andy, 500
Chunghwa, 415
CIA, 509
Cinergy Corporation, 439
Circuit City Stores, Inc., 21, 26, 331
Cisco Systems, 131, 256, 320–321
Citibank, 43–44, 344
Citigroup, 43
City University of Hong Kong, 500
Clark, Richard T., 3
Claus, Matt, 568
Clear Channel, 288
Cline, Davis & Mann Inc., 140
CMS, 300
CNN, 338
Coca-Cola Co., 377
Coca-Cola de Mexico, 378
Cocke, John, 92
Cognos, 210
Colgate-Palmolive, 430
Columbia Business School, 287